Handbook of Adhesive Technology
Third Edition

Handbook of Adhesive Technology
Third Edition

Edited By
A. Pizzi and K. L. Mittal

CRC Press
Taylor & Francis Group
Boca Raton   London   New York

CRC Press is an imprint of the
Taylor & Francis Group, an **informa** business

CRC Press
Taylor & Francis Group
6000 Broken Sound Parkway NW, Suite 300
Boca Raton, FL 33487-2742

First issued in paperback 2020

© 2018 by Taylor & Francis Group, LLC
CRC Press is an imprint of Taylor & Francis Group, an Informa business

No claim to original U.S. Government works

ISBN-13: 978-0-367-57239-6 (pbk)
ISBN-13: 978-1-4987-3644-2 (hbk)

Library of Congress Cataloging-in-Publication Data

Names: Pizzi, A. (Antonio), 1946- editor. | Mittal, K. L., 1945- editor.
Title: Handbook of adhesive technology / editors, A. Pizzi & K.L. Mittal.
Description: Third edition. | Boca Raton : CRC Press, 2018. | Includes
bibliographical references.
Identifiers: LCCN 2017026609| ISBN 9781498736442 (hardback : alk. paper) |
ISBN 9781351647267 (epub) | ISBN 9781498736473 (web pdf) | ISBN
9781351637763 (mobi/kindle)
Subjects: LCSH: Adhesives. | Sealing (Technology)
Classification: LCC TP968 .H347 2018 | DDC 668/.3--dc23
LC record available at https://lccn.loc.gov/2017026609

**Visit the Taylor & Francis Web site at
http://www.taylorandfrancis.com**

**and the CRC Press Web site at
http://www.crcpress.com**

Contents

PART 1   Fundamental Aspects

PART 2   Adhesive Classes

PART 3   *Applications of Adhesives*

Preface

This volume constitutes the Third Edition of the popular book *Handbook of Adhesive Technology* which made its debut in 1994 and the second edition appeared in 2003.

Adhesives are used in a myriad of applications to bond similar or dissimilar materials. The applications of adhesives range from mundane DIY (Do-It-Yourself) projects to bonding composite materials to fabricate structural elements. The use of adhesives is ubiquitous and adhesives find their utility in a legion of varied and diverse industries, e.g., construction, aerospace, automotive, packaging, microelectronics, dentistry, and medical (surgical procedures).

As a matter of fact, adhesive bonding is a branch of the general discipline called "zygology" (the science of joining). The popularity of the use of adhesives for bonding is owing to the fact that adhesive bonding offers many advantages *vis-à-vis* other methods of joining, e.g., riveting, welding, soldering, nailing which, *inter alia*, can be enumerated as: uniform distribution of stress, lightweight assemblies, cost effectiveness, and no danger from corrosion as is the case with nails.

Recently, there has been a flurry of R&D activity in ameliorating the existing adhesive products and to come up with new and improved adhesives with desirable new functionalities and performance characteristics. The holy grail in the field of adhesive bonding is the durability of adhesively bonded joints against the deleterious effect of water. Water has been called the "God's cruelest liquid" or the worst enemy of adhesion. Also, there is presently an accelerated pace of research to understand and control the factors affecting the performance of adhesively bonded joints.

This book (Third Edition) containing 24 chapters written by internationally renowned authors is profusely illustrated and copiously referenced. It differs from its predecessor volumes in many respects and its particular and important characteristics are: some completely new chapters are included; some topics are the same but written by other subject matter experts with different perspective; and some topics are the same but these have been revised and updated. These days the mantras are: nano, green, renewable, sustainable, biobased and biomimetic, and some new as well as certain chapters on established adhesives address these catchwords.

The book is divided into three parts as follows: Part 1: Fundamental Aspects; Part 2: Adhesive Classes; and Part 3: Applications of Adhesives. The topics covered include: theories and mechanisms of adhesion; surface mechanical (physical) treatments prior to bonding; plasma surface treatment to enhance adhesive bonding; applications of nanoparticles in adhesives; protein adhesives, rubber-based adhesives; elastic adhesives; phenolic resin adhesives; natural phenolic adhesives derived from tannins and lignin; urea and melamine aminoresin adhesives; polyurethane adhesives; reactive acrylic adhesives; anaerobic adhesives; aerobic acrylic adhesives; biobased acrylic adhesives; silicone adhesives; epoxy adhesives; biosourced epoxy resins; pressure-sensitive adhesives and products; adhesives in the wood industry; bioadhesives in drug delivery; adhesives in dentistry; new adhesive technologies in the footwear industry; and adhesives in the automotive industry.

This compendium contains a wealth of information in the arena of adhesives and adhesive bonding and represents a commentary on the current state of the knowledge and R&D activity in the domain of adhesives—a fascinating class of materials. Anyone interested centrally or tangentially in adhesives should find this volume of much interest and value. The field of adhesive bonding is veritably inter-, multi- and transdisciplinary, so researchers in seemingly different disciplines should find the information compiled here useful, particularly it should appeal to polymer scientists, surface chemists, materials scientists, adhesionists and those who need to use adhesives.

The editors sincerely hope that this Third Edition will receive the same warm welcome from the scientific and technological community as its predecessors. Here we have attempted to take due cognizance of the many important aspects of adhesives and adhesive bonding.

Now comes the important but fun part of writing the Preface. First and foremost, we would like to profusely thank all the authors for their sustained interest, enthusiasm, unwavering cooperation and contribution which were *sine qua non* for materializing this book. Also we very much appreciate the steadfast interest and support of Barbara Knott and other staff members of Taylor & Francis to give this book a body form.

A. Pizzi and K. L. Mittal

Editors

Antonio Pizzi is full professor of industrial chemistry at the ENSTIB, University of Lorraine. Prof. Pizzi, who holds a Dr. Chem. (Polymers, Rome, Italy), a Ph.D. (Organic Chemistry, South Africa) and a D.Sc. (Wood Chemistry, South Africa), is the author of more than 700 research and technical articles, patents, contract reports and international conference papers as well as 7 books on adhesion and adhesives published in New York. He is the recipient of numerous prestigious international prizes for new industrial developments in his fields of specialization, such as, among others, the Descartes Prize of the European Commission and the Schweighofer Prize for Wood Research Innovation. His best-known area of specialization is on wood and fiber glueing and wood adhesives chemistry, formulation and application, in particular in bioadhesives and their application to composite products based on natural materials.

Kashmiri Lal (Kash) Mittal received his Ph.D. from the University of Southern California in 1970 and was associated with the IBM Corp. from 1972 to 1994. He is currently teaching and consulting worldwide in the areas of adhesion and surface cleaning. He is the editor of more than 125 published books, as well as others that are in the process of publication within the realms of surface and colloid science, and adhesion. He has received many awards and honors and is listed in many biographical reference works. Dr. Mittal was a founding editor of the *Journal of Adhesion Science and Technology* in 1987 and was its editor-in-chief until April 2012. He has served on the editorial boards of a number of scientific and technical journals. He was recognized for his contributions and accomplishments by the international adhesion community with the First International Congress on Adhesion Science and Technology in Amsterdam organized in his honor on the occasion of his 50th birthday in 1995 (235 papers from 38 countries were presented). In 2002, he was honored by the global surfactant community, which instituted the Kash Mittal Award in the surfactant field in his honor. In 2003, he was honored by the Maria Curie-Sklodowska University, Lublin, Poland, which awarded him the title of doctor *honoris causa*. In 2010, he was honored by both adhesion and surfactant communities on the occasion of publication of his 100th edited book. In 2013, he initiated a new journal titled *Reviews of Adhesion and Adhesives*. In 2014, 2 books entitled *Recent Advances in Adhesion Science and Technology, and Surfactant Science and Technology: Retrospects and Prospects were* published in his honor.

List of Contributors

Francisca Arán-Aís
INESCOP
Spanish Footwear Technology Institute
Alicante, Spain

István Benedek
Pressure-Sensitive Consulting
Wuppertal, Germany

Nitu Bhatnagar
Department of Chemistry
Manipal University Jaipur,
Jaipur, Rajasthan, India

David Birkett
Henkel Adhesive Technologies
Technology Centre Europe
Dublin, Ireland

Bernard Boutevin
Institut Charles Gerhardt
Université de Montpellier
Montpellier, France

Janette Brezinová
Faculty of Mechanical Engineering
Technical University of Košice
Košice, Slovakia

Sylvain Caillol
Institut Charles Gerhardt
Université de Montpellier
Montpellier, France

David Condron
Henkel Adhesive Technologies
Technology Centre Europe
Dublin, Ireland

Paul Cranley
The Dow Chemical Company
Freeport, Texas

Christina Despotopoulou
Henkel AG & Co
Düsseldorf, Germany

Klaus Dilger
Institute of Welding and Joining
Technical University of Brunswick
Braunschweig, Germany

Dagmar Draganovská
Faculty of Mechanical Engineering
Technical University of Košice
Košice, Slovakia

Manfred Dunky
Institute for Wood Technology and Renewable
    Resources
University of Natural Resources and Life
    Sciences
Vienna, Austria

Michael Frauenhofer
Audi AG
Ingolstadt, Germany

Charles R. Frihart
USDA Forest Service, Forest Products Laboratory
Madison, Wisconsin

Douglas J. Gardner
University of Maine
Advanced Structures and Composites Center
Orono, Maine

Anna Guzanová
Faculty of Mechanical Engineering
Technical University of Košice
Košice, Slovakia

John Hill
Lord Corporation
Cary, North Carolina

Johann Klein
Henkel AG & Co
Düsseldorf, Germany

Jerome M. Klosowski
Klosowski Scientific Inc
Bay City, Michigan

Dennis G. Lay
The Dow Chemical Company
Freeport, Texas

Linda F. Lorenz
USDA Forest Service, Forest Products
    Laboratory
Madison, Wisconsin

Hanna J. Maria
International and Inter University Centre for
    Nanoscience and Technology
Mahatma Gandhi University
Kottayam, Kerala, India

César Orgilés-Barceló
INESCOP
Spanish Footwear Technology Institute
Alicante, Spain

Elena Orgilés-Calpena
INESCOP
Spanish Footwear Technology Institute
Alicante, Spain

Erdem Özdemir
Metin Kasapoğlu Caddesi
Antalya, Turkey

Paramjot
Department of Pharmaceutics
Chitkara College of Pharmacy
Chitkara University
Raypura (Patiala), Punjab, India

Jean-Pierre Pascault
INSA-Lyon
Villeurbanne, France
and
Université de Lyon
Lyon, France

Emmanuel Pitia
Lord Corporation
Erie, Pennsylvania

Antonio Pizzi
LERMAB
University of Lorraine
Epinal, France

Anna Rudawska
Lublin University of Technology
Mechanical Engineering
Department of Production Engineering
Lublin, Poland

A. A. Shybi
Rubber Research Institute of India
Kottayam, Kerala, India

Inderbir Singh
Department of Pharmaceutics
Chitkara College of Pharmacy
Chitkara University
Raypura (Patiala), Punjab, India

Nigel Sweeney
Henkel Ireland Operations and Research Ltd
Dublin, Ireland

Farid Taheri
Department of Mechanical Engineering
Dalhousie University
Halifax, Canada

Sabu Thomas
International and Inter University Centre for
    Nanoscience and Technology
Mahatma Gandhi University
Kottayam, Kerala, India
and
School of Chemical Sciences
Mahatma Gandhi University
Kottayam, Kerala, India

Ana M. Torró-Palau
INESCOP
Spanish Footwear Technology Institute
Alicante, Spain

Siby Varghese
Rubber Research Institute of India
Kottayam, Kerala, India

Part 1

Fundamental Aspects

1 Theories and Mechanisms of Adhesion

Douglas J. Gardner

CONTENTS

1.1 INTRODUCTION

The concept of joining things together through the use of sticky or glue-like substances has been around for thousands of years [1]. Early humans were quite adept at utilizing products found in nature that are sticky, such as pitches and bitumen, and that could contribute to the manufacture of useful bonded articles such as tools and building materials as well as artisanal objects. It is only more recently, within the past century or so, that man has tried to classify adhesion based on the fundamental behavior of materials. As such, the study of adhesion has gained importance in the fields of materials science, engineering, and biomedical science. It is the goal of this chapter to provide an overview of the current theories and mechanisms of adhesion.

1.1.1 ADHEREND MATERIAL PROPERTIES RELEVANT TO ADHESION

In the adhesion science and technology community, most materials to be adhesively bonded or glued are referred to as *adherends*. Adherends being bonded are usually in a solid form, while adhesives can be in either solid or liquid form (Table 1.1). There are a wide variety of adherend and adhesive types, as well as different processes to bond materials, such that many adhesion scientists will specialize in a specific area of adhesion/adhesives. A list of common adherend materials is found in Table 1.2. Examples of adherend materials include plastics, textiles, wood, tapes, coated abrasives, building materials, and materials in the automotive and aerospace industries.

The processes of joining materials through adhesive bonding to form a bonded assembly are quite variable in terms of adherend materials and bonding processes, as well as the strength and

TABLE 1.1
Examples of Adherend and Adhesive Types

Adherend Type	Examples	Adhesive Type	Examples
Dense solid	Metals, polymers	Highly viscous elastomeric liquid	Sealants, caulks
Brittle solid	Glass	Medium-viscosity liquid	Thermosetting or cold setting catalyzed polymer solutions
Porous solid	Wood, foams	Low-viscosity liquid	Adhesion promoters or highly reactive low molecular weight polymer adhesives such as Superglue
Soft solid	Elastomers	Solids	Hot-melt adhesives, powdered adhesives typically require heat to achieve liquid state to facilitate adhesion and curing
Biological solid	Teeth	Low- and medium-viscosity liquid	Acrylate adhesives

TABLE 1.2
List of Common Adherend Materials with Product Examples

Adherend Materials	Product Examples
Plastics	Consumer goods, composites
Textiles	Waterproof clothing
Wood	Furniture
Laminates	Tapes, labels
Coated abrasives	Sandpaper
Building materials	Tiles, flooring
Automotive composites	Vehicle bumpers
Aerospace composites	Fuselage assembly

durability requirements of the resulting adhesive bond. Because of the variability in adhesive bonding processes, there is no single adhesive bonding mechanism that describes all adhesive bond types. To better understand adhesive bonding processes, adhesion scientists have categorized adhesion mechanisms or theories based on the fundamental behavior of materials being bonded (adherends) as well as the adhesives used to bond the materials. Understanding adhesion requires an intimate knowledge of the bulk and surface material properties of the adherend, as well as the material property behavior of the adhesive. A list of general material property characteristics to be considered in studying or assessing adhesion is shown in Table 1.3. Surface properties of interest related to adhesion include topography, surface thermodynamics, chemical functionality, and hardness. Adhesive characteristics to be considered include molecular weight, rheology, curing characteristics, thermal transition of polymers, and viscoelasticity. For the bonded assembly, the ultimate mechanical properties and durability characteristics are of prime importance.

1.1.2 Length Scale of Adherend–Adhesive Interactions

The prevailing adhesion theories can be grouped into two types of interactions: (1) those that rely on interlocking or entanglement, and (2) those that rely on charge interactions. Furthermore, it is useful to know the length scale(s) over which the adhesion interactions occur. The comparison of adhesion interactions relative to length scale is listed in Table 1.4.

TABLE 1.3

General Materials Related to Adhesion and Their Assessment Methods

Material	Assessment Methods
Adherend	Topography, wettability, chemical functionality, hardness
Adhesive	Molecular weight, rheology, curing characteristics, thermal transitions, viscoelasticity
Bonded assembly	Mechanical properties, durability, creep behavior

TABLE 1.4

Comparison of Adhesion Interactions Relative to Length Scale

Category of Adhesion Mechanism	Type of Interaction	Length Scale
Mechanical	Interlocking or entanglement	0.01–1000 μm
Diffusion	Interlocking or entanglement	10 nm–2 μm
Electrostatic	Charge	0.1–1 μm
Covalent bonding	Charge	0.1–0.2 nm
Acid–base interaction	Charge	0.1–0.4 nm
Hydrogen bonding	Charge	0.235–0.27 nm
Lifshitz–van der Waals	Charge	0.5–1 nm

TABLE 1.5

Orders of Scale for Adherend–Adhesive Interactions

Scale	Test Specimen or Material Characteristics for Determining Adherend–Adhesive Interactions
1 m, 100 cm	Glulam beam laminates
10^{-1} m, 10 cm	Furniture bondlines
10^{-3} m, 1 mm	Polymer microdroplet on a glass fiber
10^{-4} m, 100 μm	Microscopic evaluation of adherend–adhesive bondline
10^{-6} m, 1 μm	Small paint droplets on automobile panels
10^{-7} m, 100 nm	Scale of cellulose nanofibrils
10^{-8}–10^{-9} m, 1–100 nm	Scale of adhesive polymer chains

Source: Adapted from Gardner, D. J. et al., *Rev. Adhesion Adhesives*, 2, 127–172, 2014.

It is apparent that the adhesion interactions relying on interlocking or entanglement (mechanical and diffusion) can occur over greater length scales than the adhesion interactions relying on charge interactions. Most charge interactions involve interactions on the molecular level or nanolength scale.

The length scale of adherend–adhesive interactions is also of importance in understanding adhesion mechanisms, because although many practical aspects of adhesion occur on the macroscopic length scale (millimeter to meter), many of the basic adhesion interactions occur on a much smaller length scale (nanometer to micrometer) (Table 1.5). Evaluations of laminate adhesion failure in wood glulam beams are determined on the meter length scale, whereas many gluelines in furniture occur on the centimeter length scale. Interactions between polymer droplets on individual glass fibers occur on the millimeter length scale, and microscopic evaluation of the adherend–adhesive bondline is carried out on the 100 μm length scale. The smallest paint droplets on automobile panels are of the order of 1–10 μm in diameter. Cellulose nanofibrils are on the scale of 100 nm in length and 10–20 nm in diameter. The smaller molecular weight fraction(s) of many thermosetting adhesive polymers range from 1 to 100 nm in length.

1.2   THEORIES OF ADHESION

There are seven accepted theories of adhesion [3–5]. These are:

1. Mechanical interlocking or hooking
2. Electronic, electrostatic, or electrical double layer
3. Adsorption (thermodynamic) or wetting
4. Diffusion
5. Chemical (covalent) bonding
6. Acid–base
7. Weak boundary layers

It should be noted that these mechanisms are not self-excluding, and several can occur simultaneously in an adhesive bond depending on the specific bonding situation. An additional adhesive mechanism for pressure-sensitive or elastomeric adhesives should be included in this list given the nature of that particular bonding mechanism, although some adhesion scientists have attempted to explain the bonding behavior of pressure-sensitive adhesives using surface energetics and the concept of tack [6]. We will discuss the issue regarding elastomeric adhesives in greater detail later.

1.2.1   MECHANICAL INTERLOCKING THEORY

Conceptually, the ubiquity of mechanical interlocking has long been a topic of interest in nature, art, and society [7]. In the field of adhesion, mechanical interlocking was first proposed in the early part of the last century [8,9]. There have been changing perceptions on the importance of mechanical interlocking in adhesion as analytical methods to study adhesion and our fundamental understanding have improved [10]. Essentially, mechanical interlocking can be divided into two groups: locking by friction and locking by dovetailing (Figure 1.1). For mechanically interlocked adherends, there are irregularities, pores, or crevices where adhesives penetrate or absorb into, and thus the mechanical properties of the adherends are involved [11]. In addition to geometry factors, surface roughness has a considerable influence on adhesion. Rougher adherend surfaces produce better adhesion than smooth surfaces. High-level adhesion can be attained by

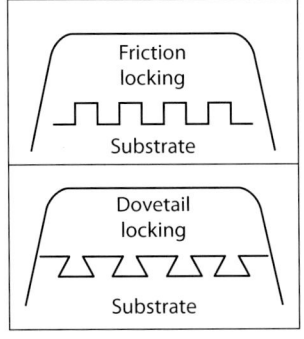

FIGURE 1.1   Schematic diagram of mechanical interlocking mechanisms.

improving the adherend surface properties, and mechanical keying can be enhanced by increasing the surface area [12].

Absorption is an important factor in mechanical interlocking, because it affects penetration of adhesives into pores or irregularities on adherend surfaces. Greater absorption produces better adhesion in mechanical interlocking systems [13]. The length scale, which changes according to the type of interaction, is another factor that affects adhesion. Mechanical interlocking is strongly dependent on the surface properties. When studying mechanical interlocking, the adherend surface properties, including the presence of crevices, pores, roughness, and irregularities, should be well characterized. Optimizing the surface properties—for instance, increasing the roughness of the surface—will produce stronger or enhanced mechanical interlocking. A primary limitation of the mechanical interlocking theory is that it does not inherently take into account charge interactions that may also occur in the creation of an adhesive bond.

Over the past several decades, the focus of mechanical interlocking in the adhesion field has been in the area of micro- and nanomaterials [14,15]. There are two popular research areas in polymer materials that address the mechanical interlocking theory: mechanically interlocked molecules (MIMs) [16] including dendrimers [17], and surface microstructuring to enhance adhesion in polymer composites [14]. In evaluating the effect of mechanical interlocking on adhesion strength of polymer–metal interfaces, micropatterned topographies were introduced on metal surfaces via a machining process. It was found that the molecular dissipation of the polymer in the vicinity of the interface is the major cause of the practical energy of separation during mechanical testing [13]. Mechanical interlocking also provides a simple and effective means of enhancing adhesion between dissimilar materials in microelectromechanical systems (MEMS) [18,19].

The morphological properties of nanoparticles are also germane to the understanding of mechanical interlocking on the nanoscale. Nanoporous gold particles added to silicone in film applications show excellent adhesion to the silicone attributable to mechanical interlocking with the elastomer substrate [20]. In polymer nanocomposites with low nanofiller content, graphene platelets perform better than carbon nanotubes in terms of enhancing mechanical properties, and this is partially attributed to improved mechanical interlocking/adhesion at the nanofiller–matrix interface [21]. Indeed, the role of wrinkles in thermally exfoliated chemically modified graphene may possibly contribute to nanoscale surface roughness that could enhance mechanical interlocking in polymer nanocomposite applications [22]. Nanomechanical interlocking has been observed at the nanotube–polymer interface, and this contributes to improved mechanical properties in polymer nanocomposites [15].

1.2.2 ELECTROSTATIC THEORY

The electrostatic mechanism of adhesion was proposed in 1948 [23]. The primary tenet of the electrostatic mechanism is that the two adhering materials are viewed as akin to the plates of an electrical condenser across which charge transfer takes place and adhesion strength is attributed to electrostatic forces (Figure 1.2) [4]. The concepts and quantities important in electrostatic adhesion are listed in Table 1.6.

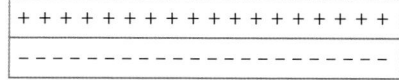

FIGURE 1.2 Schematic of the formation of an adhesion bond attributed to transfer of charge from an electropositive material to an electronegative material.

TABLE 1.6
Concepts and Quantities Important in Electrostatic Adhesion

Concept	Definition
Electric field	Generated by electrically charged particles.
Coulomb's law	Electrostatic interaction between electrically charged particles.
Capacitor	Consists of two conductors separated by a nonconductive area.
Charge density	Measure of electric charge per unit volume of space, in one, two, or three dimensions.
Van der Waals force	Close-range force between two molecules attributed to their dipole moments.
Hamaker constant	Augmentation factor for van der Waals force when many molecules are involved, as in the case of nanoparticles.
DLVO theory	Named after Derjaguin, Landau, Verwey, and Overbeek. Theory explains the aggregation of particles in aqueous dispersions quantitatively and describes the force between charged surfaces interacting through a liquid medium. It combines the effects of the van der Waals attraction and the electrostatic repulsion due to the so-called double layer of counterions.
Zeta potential	The potential difference between the dispersion medium and the stationary layer of liquid attached to the dispersed particle.
Smoluchowsky approximation	Used to calculate the zeta potentials of dispersed spherical nanoparticles.

Source: Adapted and augmented from Horenstein, M.N., *J. Electrostatics*, 67, 384–393, 2009.

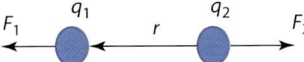

FIGURE 1.3   Interaction between electrically charged particles. F_1 and F_2 are the forces of interaction between two point charges (q_1 and q_2) and the distance (r) between them.

Coulomb's law describes the electrostatic interaction between electrically charged particles (Figure 1.3) as:

$$|F| = k_e \frac{|q_1 q_2|}{r^2} \tag{1.1}$$

where:

F is the force
k_e is Coulomb's constant
q_1 and q_2 are the charges
r is the distance between the charges

Capacitance C is defined as the ratio of charge Q on each conductor to the voltage V between them:

$$C = \frac{Q}{V} \tag{1.2}$$

Derjaguin conveyed the force $F(h)$ acting between two charges separated from one another to the strength of an adhesion bond where:

$$F(h) = 2\pi R_{\text{eff}} W(h) \tag{1.3}$$

where $W(h)$ is the interaction energy per unit area between the two planar walls and R_{eff} the effective radius.

In considering electrostatic interactions in liquids, the Derjaguin, Landau, Verwey, and Overbeek (DLVO) theory defines the interactions between charged surfaces where the total adhesion force F_A is equal to the sum of the van der Waals force F_{vdW} and the electric double layer force F_{EDL}:

$$F_A = F_{vdW} + F_{EDL} \tag{1.4}$$

The van der Waals force is a function of the system Hamaker constant, particle diameter, contact radius, and particle–surface separation distance. The electric double layer force is a function of the liquid medium dielectric constant, zeta potential, reciprocal double layer thickness, particle diameter, and particle–surface separation distance.

The electrostatic theory is often used to describe adhesion behavior of powders to solid surfaces [23–26]. Electrostatic adhesion that occurs in the liquid phase through colloidal interactions has received much greater emphasis in the scientific literature, and practical applications are plentiful in various fields. Electrostatic self-assembly in liquids is an important area in nanoscience applications [24,27]. A primary limitation of the electrostatic theory is that charge neutralization through grounding or a similar mechanism can potentially disrupt bonding.

Recent research in electrostatic adhesion has focused on the biomimetic aspects of gecko lizard toe adhesion using synthetic materials. Dry adhesives inspired by biomimetic gecko lizard toe pad adhesion using Teflon amorphous fluoropolymer nanopillar sheets was attributed to electrostatic adhesion [28]. Improved controllable adhesion on both rough and smooth surfaces can be achieved with a hybrid/gecko-like adhesive [29]. Other research has addressed comparisons of adhesion forces between electrostatic and Coulombic attraction [30], electrostatic adhesion of nanosized particles, and the cohesive role of water [31]. Electrostatic forces greatly impact adhesion interactions from the micro- to the nanoscale [32] including micromanipulation of micrometer-scale objects [33].

1.2.3 Wettability, Surface Free Energy, Thermodynamic Adhesion Theory

Thermodynamic adhesion or wetting refers to the atomic and molecular interactions between adhesives and adherends. Surface tension or surface free energy are manifestations of these forces and are regarded as fundamental material properties to understand adhesion, because they are associated with adhesive bond formation [3]. Bond formation arises from the highly localized intermolecular interaction forces between materials. Therefore, good wetting is beneficial to strong adhesive bonding. It is well known that the dominant surface chemical and energetic factor influencing joint strength is interfacial tension between the adhesive and the adherend (γ_{sl}): the joint strength increases as γ_{sl} decreases [34]. The atomic and molecular forces involved in wetting include: (1) acid–base interactions, (2) weak hydrogen bonding, or (3) van der Waals forces (dipole–dipole and dispersion forces) [3]. The condition necessary for spontaneous wetting is given as:

$$\gamma_{sg} \geq \gamma_{sl} + \gamma_{lg} \tag{1.5}$$

where γ_{sg}, γ_{sl}, and γ_{lg} are, respectively, the interfacial free energies for solid–gas, solid–liquid, and liquid–gas interfaces.

If γ_{sl} is insignificant, the criterion can be simplified to:

$$\gamma_{sg} \geq \gamma_{lg} \text{ or } \gamma_{substrate} \geq \gamma_{adhesive} \tag{1.6}$$

which means that the adhesive will wet the surface of the adherend when the surface free energy of the substrate is greater.

The surface free energies of solids can be determined by measuring the contact angles of appropriate probe liquids on a solid surface. Different contact angle analysis techniques are applied in the measurements of various forms of substrates. One is the sessile drop method, which is also referred to as the static contact angle technique. Another method is the Wilhelmy plate technique, which is suitable for making contact angle measurements on thin plates and single fibers. The contact angle can be calculated using the Wilhelmy equation (Equation 1.7) [35].

$$F = \gamma_L P \cos\theta + mg - \rho_L Ahg \qquad (1.7)$$

where:

F	is the advancing or receding force on the sample in liquid
γ_L	is the surface tension of the liquid
P	is the perimeter of the wetted cross-section
M	is the mass of the specimen
g	is the acceleration due to gravity
ρ_L	is the liquid density
A	is the cross-sectional area of the specimen
h	is the depth of immersion

For particles (also fibers), by recording the process of liquid going through a column attributed to capillary forces where particles of interest are packed inside, the contact angle can be calculated from the Washburn equation (Equation 1.8) [36], which governs the capillary wicking process:

$$h^2 = \frac{tR\gamma_L \cos\theta}{2\eta} \qquad (1.8)$$

where:

h	is the height to which liquid has risen as a function of time t
R	is the effective interstitial pore radius between the packed particles
γ_L	is the surface tension of the liquid
η	is the viscosity of the liquid

The methods of determining surface free energy of solids based on contact angles are various; for example, the Zisman approach [37], the equation of state [38], the Chibowski approach, the harmonic mean approach, the Owens and Wendt approach (the geometric mean), and the acid–base approach, which are described in a recent review [39]. Although satisfactory wetting or intrinsic adhesion is desirable in the creation of an adhesive bond, it does not necessarily ensure that the final mechanical bond strength will be optimal for a given bonding situation.

1.2.4 Diffusion Theory

The diffusion theory is based on the concept that two materials are soluble in one another, that is, compatible, and if they are brought into close contact, they dissolve in one another and form an interphase, which is a solution of both materials in one another and therefore does not form a discontinuity of physical properties between the two materials (Figure 1.4) [6]. The diffusion theory was first mentioned by Voyutskii and Vakula, and considered the role of polymer–polymer interactions in the creation of an adhesive bond based on the diffusion phenomenon [40].

For the diffusion mechanism of adhesion to occur, there must be similar solubility parameters for the adhesive and adherend [41]. This phenomenon is well illustrated by solvent welding in thermoplastic systems. The adhesive is typically a low molecular weight polymer solution in a compatible solvent that is applied to the adherend, and the solvent–polymer solution will diffuse into the

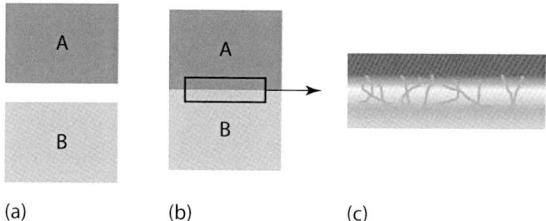

(a)   (b)   (c)

FIGURE 1.4   Schematic of diffusion theory of adhesion: (a) two compatible materials are brought into close contact (b) and an interphase (c) is formed where both materials mix and/or entangled with one another.

adherend to create molecular entanglement characterizing a diffusion bond. Thermal welding of thermoplastic polymers by various heating techniques is an adhesion bonding subject area in itself [42]. Thermal welding offers a way to create an adhesive bond between two adherends without the addition of a separate adhesive, because the adherends themselves essentially contribute to the adhesive bond. Polymer–polymer adhesion of plastic parts made by the additive manufacturing process of fused layer or fused deposition modeling is also dependent on diffusion bonding (welding) interactions [43]. Diffusion bonding is not applicable in situations where an adherend is not capable of absorbing a polymer adhesive, as in the case of bonding glass.

1.2.4.1   Interpenetrating Polymer Network (IPN)

There is a class of polymer interactions where two different polymer types will overlap in the same three-dimensional space on a molecular length scale. These overlapping polymers comprise a class of materials known as interpenetrating polymer networks (IPNs). The International Union of Pure and Applied Chemistry (IUPAC) defines an IPN as "A polymer comprising two or more networks which are at least partially interlaced on a molecular scale but not covalently bonded to each other and cannot be separated unless chemical bonds are broken." A more detailed description of IPNs can be found in Lipatov [44] and Sperling [45]. In many instances, the formation of IPNs requires interdiffusion among polymer types, so it is worth mentioning them here. In addition, the concept of IPNs has been explored extensively in adhesive bonding of different materials such as wood bonding [46], fiber-reinforced polymer (FRP) composites in dental applications [47], and combined thermoset/thermoplastic FRP composites [48].

1.2.5   Chemical (Covalent) Bonding Theory

A covalent bond is a bond where two atoms share an electron pair, and is believed to improve the bond durability between the adherend and an adhesive. The bond strength of covalent bonds is tantamount to its importance in adhesion and adhesive bond strength. In a given material, the bond energy of a covalent bond (cohesive bond strength) is approximately 1000 times greater than the surface free energy of the same material. Therefore, creating a covalent bond between adhesive and adherend should provide a high-strength adhesive bond.

In composite material systems where two dissimilar materials are being joined, the use of coupling agents that bridge the chemical interaction between two substances has been an important area of adhesion technology development [49–51]. An example of a silane coupling agent undergoing (1) hydrolysis and (2) reaction with a hydroxyl functional substrate (glass) is depicted in Figure 1.5.

Coupling agents enable the creation of strong adhesive bonds between materials that are chemically dissimilar, such as glass fibers and polyester, epoxy and aluminum, and polypropylene and talc.

1.2.5.1   Hydrogen Bonding

The role of hydrogen bonding in adhesion is well recognized, but the historical interpretation of hydrogen bond strength typically placed it in the range of Lifshitz–van der Waals or acid–base

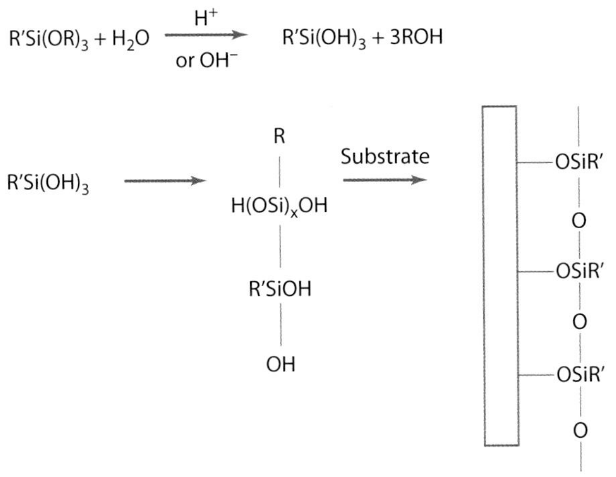

$$R'Si(OR)_3 + H_2O \xrightarrow[\text{or } OH^-]{H^+} R'Si(OH)_3 + 3ROH$$

$$R'Si(OH)_3 \longrightarrow \begin{array}{c} R \\ | \\ H(OSi)_xOH \\ | \\ R'SiOH \\ | \\ OH \end{array} \xrightarrow{\text{Substrate}}$$

FIGURE 1.5 Hydrolysis of organofunctional silane and reaction of hydrolyzed organosilane with hydroxyl functional substrate.

TABLE 1.7

Bond Strength of Various Types of Chemical Bonds and Intermolecular Forces

Chemical Bond or Intermolecular Force	Bond Strength (kJ/mol)	Bond Length
Electrostatic (ionic)	418	0.1–1 μm
Lifshitz–van der Waals	8.4–21	0.5–1.0 nm
Covalent bonding	147–628	0.1–0.2 nm
Hydrogen bonding (new)[a]	4.2–188	0.15–0.45 nm
Hydrogen bonding (old)	12.6–25.1	0.1–0.3 nm

Source: Adapted from Gardner, D. J. et al., *Rev. Adhesion Adhesives*, 2, 127–172, 2014.

[a] Gilli, G. and Gilli, P. *The Nature of the Hydrogen Bond. Outline of a Comprehensive Hydrogen Bond Theory*, Oxford University Press, New York, 2009.

---

interaction bond strengths (8–25 kJ/mol) (Table 1.7). Recent evidence suggests that hydrogen bond strengths (4–188 kJ/mol) approach the range of covalent bond strength (147–628 kJ/mol) [52]. Many common synthetic and biobased adhesives such as epoxies, polyurethanes, proteins, and formaldehyde-based resins have strong hydrogen bonding functionalities. The new bond strength data elevate the importance of hydrogen bonding in regard to the chemical bonding theory of adhesion.

1.2.6 ACID–BASE THEORY

Based on the correlation of acid–base interactions by Drago et al. [53], Fowkes and Mostafa [54] proposed a new method to interpret the interactions during polymer adsorption where the polar interaction is referred to as an acid–base interaction. In this interaction, an acid (electron-acceptor) is bonded to a base (electron-donor) by sharing the electron pair offered by the latter, which forms a coordinate bond.

The following briefly summarizes the Lewis acid–base concept in wetting-related phenomena. According to Fowkes [55] and van Oss et al. [56], the total work of adhesion in interfacial interaction between solids and liquids can be expressed as the sum of the Lifshitz–van der Waals (LW) and the Lewis acid–base (AB) interactions, namely,

$$W_a = W_a^{LW} + W_a^{AB} \tag{1.9}$$

The separation of the work of adhesion into LW and AB components is also applicable to the surface free energies according to:

$$\gamma_i = \gamma_i^{LW} + \gamma_i^{AB} \tag{1.10}$$

An advance in the understanding of wetting phenomena was the Good–Girifalco–Fowkes "geometric mean" combination rule for the LW interactions between two compounds i and j, which can be expressed as [57, 58]:

$$W_a^{LW} = 2\sqrt{\gamma_i^{LW}\gamma_j^{LW}} \tag{1.11}$$

Hence, if the contact angle (θ) is determined for both a nonpolar and a polar liquid, with known γ^{LW} parameters on the same surface, then W_a^{LW} and W_a^{AB} can be determined using Equations 1.9 through 1.11.

The acid–base theory plays a critical role in surface chemistry and adhesion, and it has been exploited broadly on different materials [59–61]. Several models of calculating the surface free energy of solids were proposed where acid–base theory was applied, including the Fowkes method, the Good method, the van Oss method, and the Chang-Chen method [39]. On the nanoscale, LW forces are important in the bonding of silver nanoparticles to polyimide in printed electronic applications [62]. The importance of acid–base interactions in the adhesion field can be assessed by comparing the adhesive bond strength of nonpolar versus polar polymer substrates.

1.2.7 WEAK BOUNDARY LAYERS CONCEPT

Bikerman [63] first introduced the concept of a weak boundary layer (WBL) in adhesion science. Three different classes of WBLs were specified: air bubbles, impurities at the interface, and reactions between components and the medium. Good [64] further implied a WBL on the surface of adherends to be responsible for lower mechanical strength. The interface is the location of adhesion failure of a bonded assembly when a WBL is present. If the tenets of proper adherend preparation are followed in the creation of an adhesive bond, especially the bonding of a freshly prepared surface, then the concept of WBLs is not an issue. However, in many bonding situations, a freshly prepared, clean adherend surface may not be possible. It simplifies our understanding of WBLs to categorize them as being mechanical or chemical in nature (Figure 1.6).

Mechanical WBLs can arise from improper machining and lack of cleaning of an adherend surface prior to bonding, while chemical WBLs can be attributed to processing aids or lubricants used to prepare a surface. Examples of mechanical WBLs are common in wood adhesion [2,65], while chemical WBLs are common in preparing metal surfaces (oils) and extruded plastic surfaces (lubricants) for bonding. In addition, "aged" surfaces are often chemically altered because of environmental influences such as exposure to moisture, ultraviolet light, oxygen, or heat. Aged surfaces tend to have lower surface free energies and are thus more difficult to be wetted by adhesives.

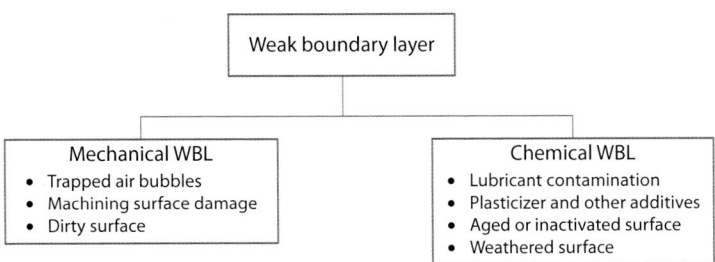

FIGURE 1.6   Characteristics of mechanical and chemical WBLs.

Adhesives can be formulated to accommodate WBLs in certain bonding situations, but it is recommended to try to remove WBLs prior to bonding if at all possible. A great example of an adhesive group that can tolerate moisture in a "wet" WBL is based on isocyanate functionality. Isocyanates can chemically react with water (hydroxyl groups) to form urea linkages that contribute to the adhesive bond. Adhesives that are catalyzed by strong acids or bases for the curing process can also impact the adherend surface and help "activate" an aged surface.

1.2.8   SPECIAL MECHANISM OF ELASTOMERIC-BASED ADHESIVES

An important class of adhesives that exhibit characteristics of both a solid and liquid are the elastomeric-based adhesives, which include pressure-sensitive and contact bond adhesives. Many elastomeric-based adhesives are in the form of highly viscous liquids that are combined with flexible substrates in the form of tapes that can be bonded to a variety of material substrates in an instantaneous manner using low bonding pressure (pressure-sensitive adhesives). Contact bond adhesives are represented by the extrudable construction-based adhesives and caulks and sealants that are highly viscous and also form relatively instantaneous semistructural bonds. The major differences between the pressure-sensitive and contact bond adhesives are the bond strength of the adhesive application and the length of time required to hold a bond [4].

The elastomeric-based adhesives have a characteristic adhesion behavior described as tackiness or stickiness that aids in the creation of an almost instantaneous adhesive bond. Tackiness is generated by adding low molecular weight, resinous tackifiers to elastomeric polymers used in the formulation of elastomeric-based adhesives [4,6]. The glass transition and softening temperatures of tackifiers are often much above room temperature. There are several definitions for tack, including one promulgated by the Pressure-Sensitive Tape Council, "the condition of the adhesive when it feels sticky or highly adhesive" and the ASTM definition, "the property of an adhesive that enables it to form a bond of measurable strength immediately after the adherend and the adhesive are brought into contact under low pressure." A visual example of tackiness is shown in Figure 1.7.

An interesting characteristic of elastomeric-based adhesives is that the magnitude of stickiness or tackiness that is formulated to occur in a particular adhesive is greatest at the application or use temperature and that tackiness will decrease both below and above the formulated application temperature.

Elastomeric-based adhesives—and any adhesive that exhibits tackiness, for that matter—will also need to consider other adhesion characteristics, including surface tension, wettability, mechanical interlocking, and so forth in creating proper adhesion with a substrate. However, in this author's opinion, the concept of stickiness or tackiness deserves to be considered among adhesion mechanisms.

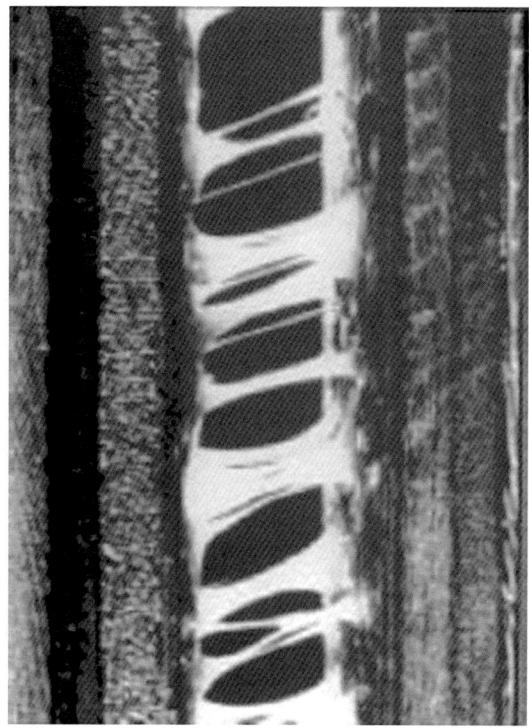

FIGURE 1.7 Behavior of a tacky (sticky) adhesive in bonding two pieces of wood adherends.

1.3 SUMMARY

At present, no practical unifying theory describing all adhesive bonds exists, although a unified adhesion theory was proposed [66]. However, adhesion phenomena are too complex in terms of the materials to be bonded and the diversity of bonding conditions encountered to be simplified into a single theory [5]. Understanding adhesion requires an intimate knowledge of the bulk and surface material properties of the particular adherend to be bonded, as well as the material property behavior of the particular adhesive used in the bonding process. The length scale over which practical adhesion occurs also impacts the evaluation and study of adhesive bonding. Adhesion mechanisms relying on entanglement occur over a larger length scale than those relying only on charge interactions. More recently, the focus of adhesion research has shifted to the challenges and opportunities associated with measurement and evaluation of adhesion on the nanolength scale. It is envisioned that much of the new knowledge being generated regarding adhesion theories will be incremental in nature, unless researchers adopt nonconventional approaches and experimental methodologies to address this subject area. The study of adhesion theories has been and will continue to be an important topic for researchers, as well as practitioners of adhesive bonding.

REFERENCES

1. P. A. Fay. History of adhesive bonding. In: *Adhesive Bonding Science, Technology and Applications*, R. D. Adams (Ed.) Woodhead Publishing, Cambridge, UK (2005).
2. D. J. Gardner, M. Blumentritt, L. Wang and N. Yildirim. Adhesion theories in wood adhesive bonding: A critical review. *Rev. Adhesion Adhesives* 2, 127–172 (2014).
3. A. Baldan. Adhesion phenomena in bonded joints. *Int. J. Adhesion Adhesives* 38, 95–116 (2012).

4. A. V. Pocius. *Adhesion and Adhesives Technology: An Introduction*, 3rd edn, p. 370, Carl Hanser Verlag, Munich (2012).

5. J. Schultz and M. Nardin. Theories and mechanisms of adhesion. In: *Handbook of Adhesive Technology*, A. Pizzi and K. L. Mittal (Eds.), pp. 19–33, Marcel Dekker, New York (1994).

6. S. C. Temin. Pressure-sensitive adhesives for tapes and labels. In: *Handbook of Adhesives*, I. Skeist (Ed.), 3rd edn, p. 641, Van Nostrand Reinhold, New York (1990).

7. C. J. Bruns and J. F. Stoddart. The mechanical bond: A work of art. *Topics Current Chem.* 323, 19–72 (2012).

8. J. W. McBain and D. J. Hopkins. On adhesives and adhesive action. *J. Phys. Chem.* 29, 188–204 (1925).

9. D. E. Packham. The mechanical theory of adhesion. In: *Handbook of Adhesive Technology*, A. Pizzi and K. L. Mittal (Eds.), 2nd edn, pp. 69–93, Marcel Dekker, New York (2003).

10. D. E. Packham. The mechanical theory of adhesion: Changing perceptions 1925–1991. *J. Adhesion* 39, 137–144 (1992).

11. H. Weiss. Adhesion of advanced overlay coatings: Mechanisms and quantitative assessment. *Surface Coatings Technol.* 71, 201–207 (1995).

12. W. C. Wake. *Adhesion and the Formulation of Adhesives*, p. 332, Applied Science Publishers, London (1982).

13. W. Kim, I. Yun, L. Jung and H. Jung. Evaluation of mechanical interlock effect on adhesion strength of polymer-metal interfaces using micro-patterned topography. *Int. J. Adhesion Adhesives* 30, 408–417 (2010).

14. J. Byskov-Nielsen, J. V. Boll, A. H. Holm, R. Hojsholt, and P. Balling. Ultra-high-strength micro-mechanical interlocking by injection molding into laser-structured surfaces. *Int. J. Adhesion Adhesives* 30, 485–488 (2010).

15. K.-T. Lau, C. Gu and D. Hui. A critical review on nanotube/nanoclay related polymer composite materials. *Composites Part B* 37, 425–436 (2006).

16. B. M. Rambo, H.-Y. Gong, M. Oh and J. L. Sessler. The "Texas-sized" molecular box: A versatile building block for the construction of anion-directed mechanically interlocked structures. *Acc. Chem. Res.* 45, 1390–1401 (2012).

17. K. C.-F. Leung, F. Arico, S. J. Cantrill and J. F. Stoddart. Template-directed dynamic synthesis of mechanically interlocked dendrimers. *J. Amer. Chem. Soc.* 127, 5808–5810 (2005).

18. M. P. Larsson and M. M. Ahmad. Improved polymer-glass adhesion through micro-mechanical interlocking. *J. Micromech. Microeng.* 16, S161–S168 (2006).

19. S. H. Kim, M. T. Dugger and K. L. Mittal (Eds.) *Adhesion Aspects in MEMS/NEMS*, CRC Press, Boca Raton, FL (2011).

20. E. Seker, M. Reed, M. Utz and M. R. Begley. Flexible and conductive bilayer membranes of nanoporous gold and silicon: Synthesis and characterization. *Appl. Phys. Letters* 92, 154101, 1–3 (2008).

21. M. A. Rafiee, J. Rafiee, Z. Wang, H. Song, Z.-Z. Yu and N. Koratkar. Enhanced mechanical properties of nanocomposites at low graphene content. *ACS Nano* 3, 3884–3890 (2009).

22. R. Verdejo, M. M. Bernal, L. J. Romasanta and M. A. Lopez-Manchado. Graphene filled polymer nanocomposites. *J. Mater. Chem.* 21, 3301–3310 (2011).

23. B. V. Derjaguin, I. N. Aleinikova and Y. P. Toporov. On the role of electrostatic forces in the adhesion of polymer particles to solid surfaces. *Powder Technol.* 2, 154–158 (1969).

24. M. N. Horenstein. Electrostatics and nanoparticles: What's the same, what's different? *J. Electrostatics* 67, 384–393 (2009).

25. A. G. Bailey. The science and technology of electrostatic powder spraying, transport and coating. *J. Electrostatics* 45, 85–120 (1998).

26. I. I. Inculet. Electrostatics in industry. *J. Electrostatics* 4, 175–192 (1978).

27. K. Ariga, J. P. Hill, M. V. Lee, A. Vinu, R. Charvet and S. Acharya. Challenges and breakthroughs in recent research on self-assembly. *Sci. Technol. Adv. Mater.* 9, 014109 (2008).

28. H. Izadi, M. Golmakani and A. Penlidis. Enhanced adhesion and friction by electrostatic interactions of double-level Teflon nanopillars. *Soft Matter* 9, 1985–1996 (2013).

29. D. Ruffatto III, A. Parness and M. Spenko. Improving controllable adhesion on both rough and smooth surfaces with a hybrid electrostatic/gecko-like adhesive. *J. R. Soc. Interface* 11, 20131089 (2014).

30. B. A. Kemp, and J. G. Whitney. Electrostatic adhesion of multiple non-uniformly charged dielectric particles. *J. Appl. Phys.* 113, 044903, 1–6 (2013).

31. L. F. Valadares, E. M. Linares, F. C. Braganca and F. Galembeck. Electrostatic adhesion of nanosized particles: The cohesive role of water. *J. Phys. Chem.* 112, 8534–8544 (2008).

32. G. Kumar, S. Smith, R. Jaiswal and S. Beaudoin. Scaling of van der Waals and electrostatic adhesion interactions from the micro- to the nano-scale. *J. Adhesion Sci. Technol.* 22, 407–428 (2008).

33. M. S. Lhernould, P. Berke, T. J. Massart, S. Regnier and P. Lambert. Variation of the electrostatic adhesion force on a rough surface due to the deformation of roughness asperities during micromanipulation of a spherical rigid body. *J. Adhesion Sci. Technol.* 23, 1303–1325 (2009).

34. K. L. Mittal. The role of the interface in adhesion phenomena. *Polym. Eng. Sci.* 17, 467–473 (1977).

35. L. Wilhelmy. Über die Abhängigkeit der Capillaritätskonstanten des Alkohols von Substanz und Gestalt des benetzten festen Körpers. *Annalen der Physik* 195, 177–217 (1863).

36. E. W. Washburn. The dynamics of capillary flow. *Phys. Rev.* 17, 273–282 (1921).

37. W. A. Zisman. Influence of constitution on adhesion. *Ind. Eng. Chem. Res.* 55, 18–38 (1963).

38. A. W. Neumann, R. Good, C. Hope and M. Sejpal. An equation-of-state approach to determine surface tensions of low-energy solids from contact angles. *J. Colloid Interface Sci.* 49, 291–304 (1974).

39. F. M. Etzler. Determination of the surface free energy of solids: A critical review. *Rev. Adhesion Adhesives* 1, 3–45 (2013).

40. S. S. Voyutskii and V. L. Vakula. The role of diffusion in polymer-to-polymer adhesion. *J. Appl. Polym. Sci.* 7, 475–491 (1963).

41. B. Lin, S. Lee and K. S. Liu. The microstructure of solvent-welding of PMMA. *J. Adhesion* 43, 221–240 (1991).

42. R. J. Wise. *Thermal Welding of Polymers*. Woodhead Publishing, Cambridge, UK(1999).

43. Q. Sun, G. M. Rizvi, C. T. Bellehumeur and P. Gu. Effect of processing conditions on the bonding quality of FDM polymer filaments. *Rapid Prototyping J.* 14, 72–80 (2008).

44. Y. S. Lipatov. Polymer blends and interpenetrating polymer networks at the interface with solids. *Prog. Polym. Sci.* 27, 1721–1801 (2002).

45. L. H. Sperling. *Interpenetrating Polymer Networks and Related Materials*. Softcover reprint of the hardcover 1st edition 1981, Plenum Press, New York (2012).

46. D. J. Gardner. Adhesion mechanisms of durable wood adhesive bonds. In: *Characterization of the Cellulosic Cell Wall*, D. D. Stokke and L. H. Groom (Eds.), pp. 254–265, Blackwell Publishing, Ames, IA (2006).

47. T. M. Lastumaki, L. V. J. Lassila and P. K. Vallittu. The semi-interpenetrating polymer network matrix of fiber-reinforced composite and its effect on the surface adhesive properties. *J. Mater. Sci.: Mater. Medicine* 14, 803–809 (2003).

48. S. P. Lin, J. L. Han, J. T. Yeh, F. C. Chang and K. H. Hsieh. Composites of UHMWPE fiber reinforced PU/epoxy grafted interpenetrating polymer networks. *European Polym. J.* 43, 996–1008 (2007).

49. H. Ishida. A review of recent progress in the studies of molecular and microstructure of coupling agents and their functions in composites, coatings and adhesive joints. *Polymer Composites* 5, 101–113 (1984).

50. J. G. Marsden. Organofunctional silane coupling agents. In: *Handbook of Adhesives*, I. Skeist (Ed.) 3rd edn, pp. 536–548, Van Nostrand Reinhold, New York (1990).

51. H. S. Katz. Non-silane coupling agents. In: *Handbook of Adhesives*, I. Skeist (Ed.) 3rd edn, pp. 549–555, Van Nostrand Reinhold, New York (1990).

52. G. Gilli and P. Gilli. *The Nature of the Hydrogen Bond. Outline of a Comprehensive Hydrogen Bond Theory*. IUCr Monographs on Crystallography 23, Oxford University Press, New York. p. 317 (2009).

53. R. S. Drago, G. C. Vogel and T. E. Needham. Four-parameter equation for predicting enthalpies of adduct formation. *J. Am. Chem. Soc.* 93, 6014–6026 (1971).

54. F. M. Fowkes and M. A. Mostafa. Acid-base interactions in polymer adsorption. *Ind. Eng. Chem. Prod. R+D* 17, 3–7 (1978).

55. F. M. Fowkes. Acid-base interactions in polymer adhesion. In: *Physicochemical Aspects of Polymer Surfaces*, Vol. 2, K. L. Mittal (Ed.), pp. 583–603, Plenum Press, New York (1983).

56. C. J. van Oss, M.K. Chaudhury and R. J. Good. Monopolar surfaces. *Adv. Colloid Interface Sci.* 28, 35–64 (1987).

57. R. J. Good and L. A. Girifalco. A theory for estimation of surface and interfacial energies, III. Estimation of surface energies of solids from contact angle data. *J. Phys. Chem.* 64, 561–565 (1960).

58. F. M. Fowkes. Additivity of intermolecular forces at interfaces. I. Determination of the contribution to surface and interfacial tension of dispersion forces in various liquids. *J. Phys. Chem.* 67, 2538–2541 (1963).

59. K. L. Mittal (Ed.). *Acid-Base Interactions: Relevance to Adhesion Science and Technology*, Vol. 2, CRC Press, Boca Raton, FL (2000).

60. K. L. Mittal and H. R. Anderson, Jr. (Eds.). *Acid-Base Interactions: Relevance to Adhesion Science and Technology*, CRC Press, Boca Raton, FL (1991).

61. M. M. Chehimi, A. Azioune and E. Cabet-Deliry. Acid-base interactions: Relevance to adhesion and adhesive bonding. In: *Handbook of Adhesive Technology*, A. Pizzi and K. L. Mittal (Eds.), 2nd edn, pp. 95–144, Marcel Dekker, New York (2003).

62. S. Joo and D. F. Baldwin. Adhesion mechanisms of nanoparticle silver to substrate materials: Identification. *Nanotechnology* 21, 055204, (12 pp.) (2010).

63. J. J. Bikerman. *The Science of Adhesive Joints*, p. 258. Academic Press, New York (1961).

64. R. J. Good. Theory of "cohesive" vs "adhesive" separation in an adhering system. *J. Adhesion* 4, 133–154 (1972).

65. M. Stehr and I. Johansson. Weak boundary layers on wood surfaces. *J. Adhesion Sci. Technol.* 14, 1211–1224 (2000).

66. F. H. Chung. Unified theory and guidelines on adhesion. *J. Appl. Polym. Sci.* 42, 1319–1331 (1991).

2 Surface Mechanical (Physical) Treatments Prior to Bonding

Janette Brezinová, Anna Guzanová, and Dagmar Draganovská*

CONTENTS

2.1 INTRODUCTION

Surface pretreatment is an essential and very important part of adhesive bonding technology [1–11]. Its aim is to prepare the surface so the prepared bonded joint achieves the required properties and is able to perform its function. The resulting strength of an adhesive joint is affected mainly by:

- Pretreatment of the substrate
- Choice of an appropriate adhesive
- Compliance with adhesive bonding process specifications

There are different types of contaminants on the surface (Figure 2.1):

- *Intrinsic*: Contaminants chemically bonded to the surface (chemisorption)
- *Extrinsic*: Contaminants bonded to the surface by adhesion and adsorption forces

Selecting the most appropriate surface pretreatment, and thus the way to remove intrinsic and extrinsic contaminants, depends on the condition of the substrate (Table 2.1), the extent of corrosion attack, and the corrosion aggressiveness of the environment (Table 2.2) in which the treated part will operate.

Methods of surface pretreatment prior to bonding are as follows.

---

* Indicates corresponding author

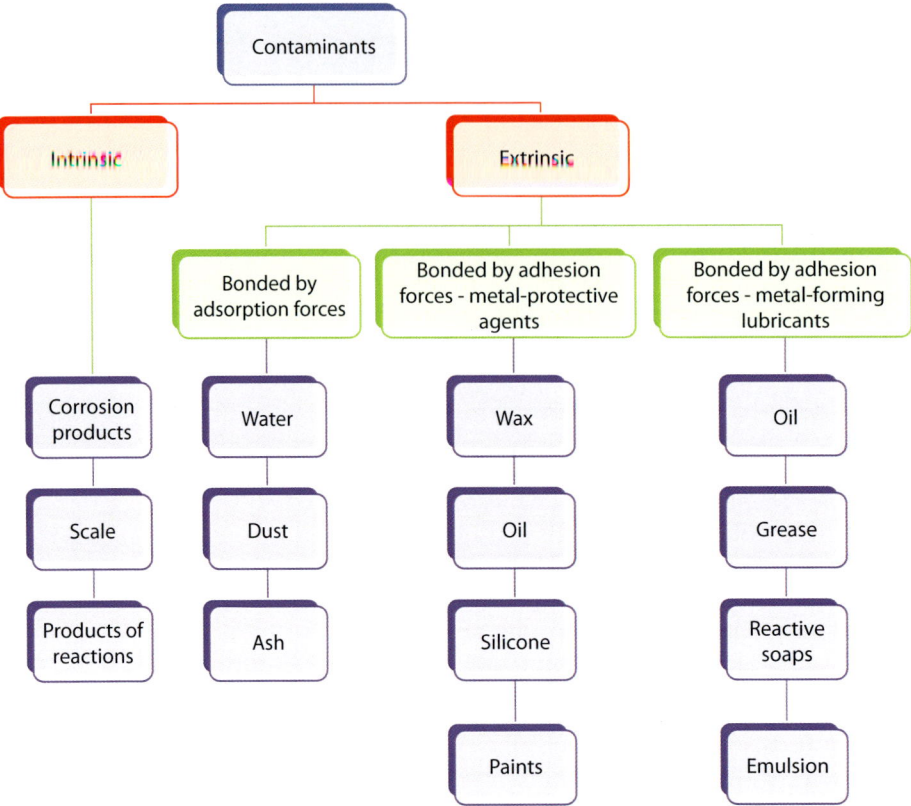

FIGURE 2.1 Contaminants on the surface of a real metal part. (From D. Jankura, J. Brezinova, J. Ševčikova, D. Draganovska and A. Guzanova. *Materials in Mechanical Engineering and Technology of their Finalization,* 1st edn, Technical University of Košice, Košice, Slovakia, 2011.)

2.1.1 CHEMICAL–PHYSICAL TREATMENT

The aim of a chemical–physical surface treatment is to:

- Remove intrinsic and extrinsic contaminants by chemical or chemical–physical disruption of their bond to the substrate using appropriate agents
- Increase the effective surface area
- Activate the surface

This includes degreasing and acid pickling.

2.1.2 MECHANICAL PRETREATMENT

The aim of mechanical surface treatment is to:

- Remove intrinsic and extrinsic contaminants by mechanical disruption of their bond to the substrate
- Remove any surface defects and structural surface defects
- Increase the effective surface area
- Activate the surface

This includes methods such as abrasive blast cleaning, grinding, and brushing.

TABLE 2.1

Methods of Removing Weakly Bonded Layers and Extrinsic Contaminants in Accordance with ISO 12944

Contaminant	Cleaning Method	Comments
Grease and oil	Cleaning with a high-pressure water jet	Pure water with detergent, pressure <70 MPa, followed by rinsing with pure water
	Steam cleaning	Pure water. If a detergent is added, then rinsing with pure water follows
	Emulsion cleaning	Rinsing with pure water
	Cleaning with alkaline solutions	Aluminum, zinc, and other metal surfaces are prone to corrosion when alkaline cleaning agents are used. Rinsing with pure water
	Cleaning with organic solvents	Organic solvents are a health hazard. When using rags for cleaning, they must always be clean or frequently replaced, otherwise grease spreads evenly on the surface and, after solvent evaporation, creates an even layer
Water-soluble contaminants (salts)	Cleaning with a high-pressure water jet	Pure water, pressure <70 MPa
	Steam cleaning	Rinsing with pure water
	Cleaning with alkaline solutions	Aluminum, zinc, and other metal surfaces are prone to corrosion when alkaline cleaning agents are used. Rinsing with pure water
Scale	Acid pickling	Rinsing with pure water
	Dry abrasive blasting	Abrasive: steel shot or grit. Dust and dirt residues removed by dry compressed air or vacuuming
	Wet abrasive blasting	Rinsing with pure water
	Flame cleaning	Mechanical cleaning is required for removing the products formed in the process of flaming
	Manual descaling	Mechanical cleaning can be used for weakly bonded corrosion products; grinding for strongly bonded corrosion products
	Water blasting	For removing nonbonded corrosion products. It does not produce the desired surface roughness on steel surfaces
	Local blasting	For local removal of corrosion products
Old paint	Paint strippers/removers	Solvent-based removers for solvent-based coatings. Rinse residues with a solvent. Paint remover is recommended for use in small areas
	Dry abrasive blasting	Abrasive: steel shot or grit. Dust and dirt residues are removed by dry compressed air or vacuuming
	Wet abrasive blasting	Rinsing with pure water
	Water abrasive blasting	For removing weakly bonded coatings. The ultrahigh pressure (>170 MPa) can be used for removing strongly bonded coatings
	Sweep blasting	For removing nonbonded coatings
	Local blasting	For local coating removal
Zinc corrosion products	Sweep blasting	Light blasting by corundum
	Alkaline cleaning	5% ammonia solution with abrasive additives can be used for removing local zinc corrosion products. Alkaline cleaning agents are used for larger areas. Increasing the pH leads to zinc corrosion

TABLE 2.2
Classification of Corrosive Environments According to ISO 12 944

| Corrosive Category | Mass Loss/Thickness Loss After the First Year of Exposure | | | | Typical Environments | |
| | Low-Carbon Steel | | Zinc | | | |
	Mass Loss (g/m²)	Thickness Loss (μm)	Mass Loss (g/m²)	Thickness Loss (μm)	Exterior Environment	Interior Environment
C1 Very low	<10	<1.3	<0.7	<0.1	—	Heated buildings, offices, shops, schools, hotels, etc.
C2 Low	10–200	1.3–25	0.7–5	0.1–0.7	Areas with low levels of pollution, mostly in rural areas	Unheated buildings where condensation may occur, e.g., depots, sports halls, etc.
C3 Medium	200–400	25–50	5–15	0.7–2.1	Urban and industrial areas, moderate sulfur dioxide levels. Coastal areas with low salinity	Production rooms with high humidity and some air pollution, e.g., food processing plants, laundries, breweries, etc.
C4 High	400–650	50–80	15–30	2.1–4.2	Industrial and coastal areas with moderate salinity	Chemical plants, swimming pools, coastal ship- and boatyards
C5-I Very high (industrial)	650–1500	80–200	30–60	4.2–8.4	Industrial areas with high humidity and aggressive environments	Buildings or areas with almost permanent condensation and with high pollution
C5-M Very high (marine)	650–1500	80–200	30–60	4.2–8.4	Coastal and offshore areas with high salinity	Buildings or areas with almost permanent condensation and with high pollution

2.2 VARIOUS SURFACE PRETREATMENTS

2.2.1 ABRASIVE BLAST CLEANING

The cleaning process using abrasives is usually denoted by two terms: *shot peening* and *blasting*. The term *shot peening* is used to denote a process whose main objective is to alter the mechanical properties or to create a compressive residual stress layer on the surface of metal parts. In the process of shot peening, the surface is exposed to a stream of accelerated steel shot. Each particle of the shot acts as a ball-peen hammer that induces surface strengthening at room temperature (strain hardening), and thus prevents the formation of stress cracks that may lead to fatigue damage [12]. The surface layer, with a compressive stress, increases the fatigue strength of the component. *Blasting* is a term used to describe processes involving surface finishing procedures, cleaning, deburring, or other specific surface treatments of metal parts. In the process of blasting, the surface is exposed to a stream of accelerated irregular sharp-edged grains of an abrasive, and removal of the material surface layer occurs. The blasting process also results in the creation of a compressive residual stress layer on the surface; however, the effect is only secondary. To achieve the desired changes on the surface of the substrate, one must choose an appropriate type of blasting abrasive, because the resulting surface morphology depends on the abrasive material and shape and grain composition.

Blasting is currently the technology with significant application in practice, with a variety of materials used as blasting abrasives. An overview of blasting abrasives used for various application areas is given in Table 2.3.

Abrasive blasting is a mechanical treatment of the surface in which the kinetic energy of the abrasive is used for:

- Surface cleaning by mechanical disruption of contaminants (corrosion products, scale, old paint, etc.) bonded to the substrate
- Roughening, achieving the same morphology on different surfaces, elimination of any defects and structural imperfections on the surface
- Surface strengthening
- Surface area increase
- Activation of the surface

2.2.1.1 Blasting Abrasives

ISO 11124-1 defines a blasting abrasive as a material composed of a set of solid particles. It is a free-flowing polydispersed material.

Desired properties of blasting abrasives can be summarized as follows:

- Grain material
- Grain shape
- Grain size
- Grain resistance to wear
- Hardness
- Tensile strength
- Ability to remove contaminants
- Roughening effect

2.2.1.1.1 Materials of Blasting Abrasives

EN ISO 11124-1 defines types of blasting abrasives (see Table 2.4). Some properties of selected types of blasting abrasives are given in Table 2.5.

Blasting abrasives should be made of the same or similar material as the surface to be cleaned. If this principle is not obeyed, particles of abrasive will remain on the cleaned surface after blasting, and due to the potential difference between them, electrical cells would be formed, which could lead to the initiation of corrosion processes [14–16].

2.2.1.1.2 Shape of Blasting Abrasives

The shape of the particles is determined by their production technique. Basic shapes of blasting abrasives are given in Table 2.6 and characteristics of the most commonly used blasting abrasives are given in Table 2.7.

Shot is produced by gas or water atomization.

Grit is produced by crushing the shot in special mills. Great emphasis is placed on particle size sorting and eliminating uncrushed or hemispherical particles.

Cut wire is a blasting abrasive of cylindrical shape, produced by cutting steel wire with a length:diameter ratio of about 1.

TABLE 2.3

Overview of Blasting Abrasives Used for Various Applications

Applications	Blasting Abrasive						
	Steel Grit	Corundum	Steel Shot	Glass Beads	Plastics	Ceramics	Organic Blasting Media
Adhesive removal					X		
Preparation prior to anodizing		X		X		X	
Matting of surface				X		X	
Artifact renovation		X			X		X
Preparation prior to bonding	X	X					
Cleaning of castings	X		X				
Cleaning of composites					X		
Cleaning of concrete	X						
Removal of corrosion products	X	X	X	X		X	
Removal of flash		X		X		X	
Deburring					X		X
Routine cleaning	X	X	X				
Glass decoration		X				X	
Cleaning molds for pressure casting		X		X	X	X	X
Removal of old paint	X	X	X		X		X
Preparation prior to coating	X	X	X		X		
Steel sheet flattening and shot peen forming [13]			X	X			
Machinery maintenance	X	X		X		X	
Preparation prior to metallization	X	X					
Preparation prior to galvanizing				X		X	
Cleaning before and after welding	X	X		X		X	
Cleaning of runway lights							X
Descaling	X	X	X	X		X	
Surface shaping	X	X	X	X		X	
Cleaning of turbines							X
Cleaning of wood surfaces		X					

TABLE 2.4
Materials of Blasting Abrasives

Metallic	Nonmetallic Inorganic	Nonmetallic Organic
Chilled iron grit	Silica sand	Plastics
High-carbon cast steel	Copper refinery slag	Wheat starch
Low-carbon cast steel	Coal furnace slag	Corn grit
Cut steel wire	Nickel refinery slag	Milled corn cob
Cut aluminum wire	Iron furnace slag	Crushed walnut shells
Cut zinc or galvanized wire	Fused aluminum oxide	Wood pellets
	Olivine sand	Dry ice
	Staurolite	Baking soda
	Almandite garnet	
	Ceramic shot	
	Glass beads	

The shape of the dimple after abrasive impact corresponds to the shape of the abrasive (Figures 2.2 and 2.3).

Sharp-edged particles remove surface material on impact, while round particles without cutting edges are used to pound or *peen* a surface. Angular abrasives such as grit and cut wire may be recommended for surface preparation prior to bonding.

2.2.1.1.3 *Granularity of Blasting Abrasives*

The grain size of the abrasive used significantly influences the roughness and, in the case of thin substrates, also the deformation of the prepared surface. Larger particles generate higher impact force, removing the material surface more quickly and producing a deeper texture. The granularity of an abrasive may be characterized by *grit number* or *grain size*. The grit number is the designation for the grain size of the abrasive, reflecting the smallest number of openings per linear inch in the screen through which the grains will pass. The greater the grit number, the smaller the holes in the sieve and the grains themselves. Grain size is expressed in micrometers or millimeters, and expresses the greatest dimension of orthogonal projection of the grain.

The granularity range of blasting abrasives is determined by *sieve analysis*—sieving of a representative test portion of blasting abrasive on the set of control sieves arranged from largest to smallest mesh size and registering the mass fractions collected on individual sieves after the specified sieving time.

A graphical representation of sieve analysis results in a *particle size distribution curve*, as shown in Figure 2.4. Using the particle size distribution curve of granularity, the average grain size of the abrasive can be determined.

The characteristics of the blasted surface profile (type of irregularities, roughness) are influenced mainly by the shape and grain size of the abrasive used.

2.2.1.1.4 *Wear Resistance: Durability*

Repeated use of blasting abrasive causes shape transformations due to the processes of plastic deformation, fracturing, and progressive degradation of the grain material.

- *Metallic shot* disintegrates into smaller sharp angular grains due to repeated use, which can round off in the next use.
- *Metallic grits* round off after reuse and can then disintegrate into smaller sharp angular grains (Figure 2.5).
- *Nonmetallic grits* also disintegrate into smaller sharp angular grains, mostly by brittle fracture, due to the material characteristics of the abrasive (Figure 2.6).

TABLE 2.5
Basic Properties of Blasting Abrasives

Material	Mesh Size	Shape	Density (kg/dm³)	Hardness (Mohs)	Brittleness	Number of Cycles Durability	Cost	Origin
Silica sand	6–270	✴	1.66	5.0–6.0	High	1	Medium	Natural
Mineral slag	8–80	✴	1.4–1.86	7.0–7.5	High	1–2	Medium	By-product
Steel grit	10–325	✴	3.83	8.0	Low	200+	Medium	Manufactured
Steel shot	8–200	●	4.6	8.0	Low	200+	Medium	Manufactured
Corundum	12–325	✴	2.08	9.0	Medium	6–8	Medium	Manufactured
Silicon carbide	12–325	✴	1.83	9.5	Medium	5–6	Medium	Manufactured
Glass beads	10–400	●	1.4–15	5.5–6.0	Medium	8–10	Low	Manufactured
Plastics	12–80	✴	0.75–1	3.0–4.0	Low/medium	8–10	Medium	Manufactured
Wheat starch	12–80	✴	0.75	3.0	Medium	12–15	High	Manufactured
Corn grit	16–60	✴	0.75	3.0	Low	14–17	Medium	Manufactured
Milled corn cob	8–40	✴	0.58–0.75	2.0–4.5	Medium	4–5	Low	By-product

Source: https://www.theshotpeenermagazine.com/wp-content/uploads/tsp23no2.pdf H. Tobben. Back to Basics: Choosing the Right Media. *The Shot Peener* 23, 18–20 (2009). Printed with permission from Clemco Industries Corp.

✴ - angular, ● - round

TABLE 2.6
Basic Shapes of Blasting Abrasives

Initial Particle Shape	Designation
Shot: round	S (shot)
Grit: angular, irregular	G (grit)
Cylindrical: sharp-edged (cut wire)	C (cut wire)

Any change in the shape of the blasting abrasive is reflected in a change in quality and geometry of the surface, so it is important to know the lifetime of particular types of blasting abrasives and regularly exchange them before such changes occur.

The appearance of some metallic blasting abrasives in different stages of abrasion, and also that of the mild steel substrate blasted by them, are shown in Figure 2.7.

Small particles, as a product of grain disintegration, can embed in cleaned surfaces and deteriorate joint strength (secondary contamination: see Section 2.2.1.6.4).

All abrasives, in particular nonmetallic blasting abrasives, are characterized by progressive fragmentation. Dust particles are fragments of blasting abrasive of a particular minimum size (the critical particle size of metallic abrasives is <0.2 mm and of nonmetallic abrasives <0.09 mm).

Figure 2.8 shows the increase of dust particles (<0.09 mm) of almandite garnet (AG) and brown corundum (BC) during 20 impact cycles.

Whereas wear of an abrasive causes change in grain size due to its disintegration, the durability of an abrasive can be expressed in the number of cycles at which 50% of grains disintegrate (Figure 2.9). It is interesting that the durability of an abrasive with lower grain size is greater than that of the same abrasive with higher grain size.

Figure 2.10 shows a comparison of the durability of cut wire and steel shot with the same grain size obtained using sieve analysis.

The roughness of the surface after blasting depends on the state of wear of the abrasive used. The durability of the abrasive can also be expressed through the achieved surface roughness as the number of cycles in which the achieved roughness of the substrate decreases by 15% compared with the roughness achieved by fresh abrasive (Figure 2.11).

The dependence of the R_a (arithmetical mean deviation of the assessed profile) and R_z (maximum height of profile) parameters of mild steel substrate blasted by BC and demetallized steelmaking slag (DSS) (Figure 2.12) [1,17] on the number of cycles shows the decrease in roughness after repeated use of blasting abrasive, which is caused by the fragmentation of grains, thus reducing their lifetime.

By comparing the achieved values of roughness, it can be concluded that with increasing grain size of the same blasting abrasive (DSS in Figure 2.12), higher roughness values (R_a, R_z) are achieved. This is due to deeper indentations made by grains with greater weight, which at the time of impact have higher kinetic energy.

2.2.1.1.5 Hardness of Blasting Abrasive

The hardness of an abrasive affects its disintegration as well as its effect on the substrate. The hardness of nonmetallic abrasives can be expressed through the Mohs hardness scale. Plastic blasting abrasives have a hardness of 3–4, glass beads 5–6, steel grit 8, corundum 8.9–9.2, and silicon carbide 9.5 on this scale.

2.2.1.1.6 Tensile Strength of Blasting Abrasives

The tensile strength of cut wire can only be determined before it is cut. It may be influenced by the manufacturing technology of the wire, thermal treatment, chemical composition, and so forth. The tensile strength of metallic abrasives influences their durability [19].

TABLE 2.7
Characteristics of the Most Commonly Used Blasting Abrasives

Appearance	Characteristics	Chemical Composition
Steel shot	Steel shot of type S is made from hypereutectoid specially heat-treated steel. It has fine homogeneous structure of tempered martensite that exhibits optimal elasticity and fatigue resistance	C:0.75%–1.20% Mn: 0.60%–1.10 % Si: 0.60%–1.10% P: max. 0.04% S: max. 0.04%
Steel grit	Angular SG is produced by crushing specially heat-treated granules of a larger diameter	C: 0.75%–1.20% Mn: 0.60%–1.10% Si: 0.60%–1.10% P: max. 0.04% S: max. 0.04%
Aluminum cut wire	Especially suitable for matting products and cleaning castings of aluminum	Fe: 0.16%–0.40% Al: 99.7% Zn: 0–0.1% Mg: 0–5.6% Cu: 0.01–0.1% Cr: 0–0.2%
Zinc cut wire	Especially suitable for matting products and cleaning castings of zinc. It can also be used for blasting steel surfaces. In this process deposition of zinc on the steel surface occurs ensuring temporary protection of steel against corrosion	Fe: 0.004% Zn: 99.9% Al: 0.003% Cu: 0.001% Cd: 0.002%
Brown corundum	Very aggressive and hardest material used for blasting, made by fusing bauxite in the induction furnace at a temperature of 1600°C. Often used for cleaning steel or gray cast iron, deburring of hardened steel, processing of wood and plastics, removal of corrosion products, roughening, matting. Presents a health risk if inhaled	Al_2O_3: 95.50% min. SiO_2: 1.40% max. Fe_2O_3: 0.60% max. CaO: 0.40% max. TiO_2: 1.80%–2.80%
Glass beads	Mainly used for fine blasting, polishing, blasting stainless materials, final treatment of surface, etc. A glass bead is an inert material that does not react with the substrate. It is toxicologically and environmentally friendly	SiO_2: min. 65.0% Na_2O: min. 14.0% CaO: min. 8.0% MgO: min. 2.5% Al_2O_3: min. 0.5–2.0% Fe_2O_3: max. 0.15% Other: max. 2.0%
Baking soda (sodium bicarbonate)	White crystalline water-soluble, nonflammable, nonexplosive, odorless, nontoxic material that does not damage the cleaned surface. It is environmentally safe and is used for cleaning all kinds of materials: aluminum, metal alloys, plated materials including chrome-plated, glass, rubber, plastics, ceramics, concrete, tiles, etc. It easily removes grease, oil, scale, rust, carbon, graffiti, old paint and different layers from the surface [1]	$NaHCO_3$
Plastic blasting abrasive (e.g., Duroplast)	Used for removing scale from electronic components and removing synthetic residues from connectors so the metal surface is not damaged. It does not release silicates and is nontoxic	Unsaturated polyester

(*Continued*)

TABLE 2.7 (CONTINUED)

Characteristics of the Most Commonly Used Blasting Abrasives

Appearance	Characteristics	Chemical Composition
 Dry ice	Cylindrical dry ice pellets with a diameter of 1–6 mm, a length of 5–15 mm and Mohs scale hardness 2–3. After blasting they leave no secondary residues, because straight after the impact on the surface they sublimate into gas phase and have few abrasive and corrosive effects	Solid form of CO_2

FIGURE 2.2 Dimple caused by shot impact. (From Brezinová, J. et al., *Abrasive Blast Cleaning and its Application*, Trans Tech Publications, Pfäffikon, Switzerland, 2015.)

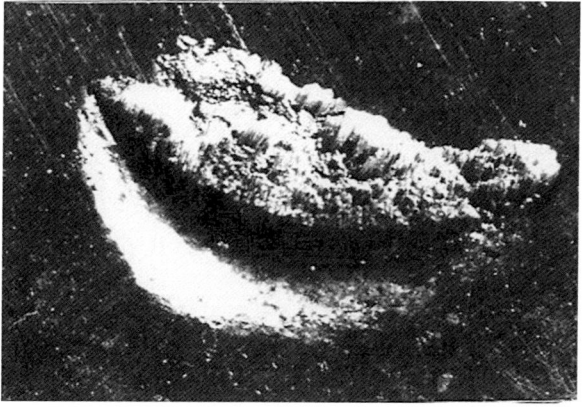

FIGURE 2.3 Dimple caused by cut wire impact. (From Brezinová, J. et al., *Abrasive Blast Cleaning and its Application*, Trans Tech Publications, Pfäffikon, Switzerland, 2015.)

2.2.1.1.7 Ability of Blasting Abrasive to Remove Surface Layer

Pretreatment by blasting is a mechanical treatment of metal surfaces that does not have significant influence on the shape or dimension of treated products. The blasting process is primarily characterized by release of dirt, corrosion products, mill scale, old paint, and so forth from the surface, mostly due to the impact effect of the grain.

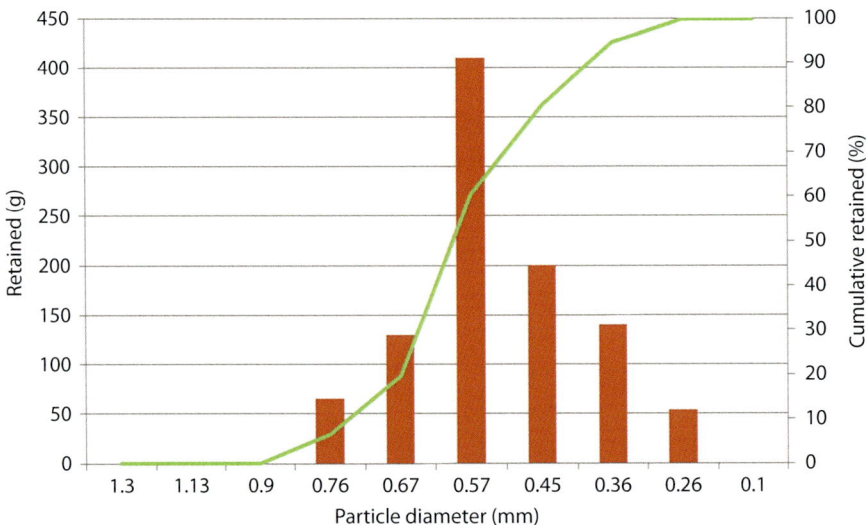

FIGURE 2.4   Particle size distribution curve; mean grain size $d_z = 0.60$ mm.

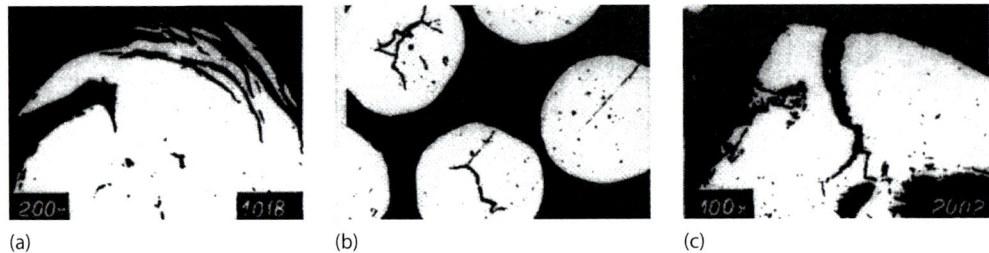

FIGURE 2.5   Wear of metallic abrasives: (a) cut wire, (b) steel shot, (c) cast iron grit. (From Brezinová, J. et al., *Abrasive Blast Cleaning and its Application*, Trans Tech Publications, Pfäffikon, Switzerland, 2015.)

FIGURE 2.6   Sodium bicarbonate: (a) before blasting, (b) after one impact cycle. (From Jankura, D. et al., *Materials in Mechanical Engineering and Technology of their Finalization*, Technical University of Košice, Košice, Slovakia, 2011.)

After the impact of an abrasive on a tough surface coated with scale, slashing of the scale layer, local plastic deformation of the substrate, and peeling of the scale layer close to the edge of the notch occur and the material of the substrate protrudes over the edge of the notch above the original surface level. Individual grains indent the surface and cause surface impurities to break loose (Figures 2.13 and 2.14).

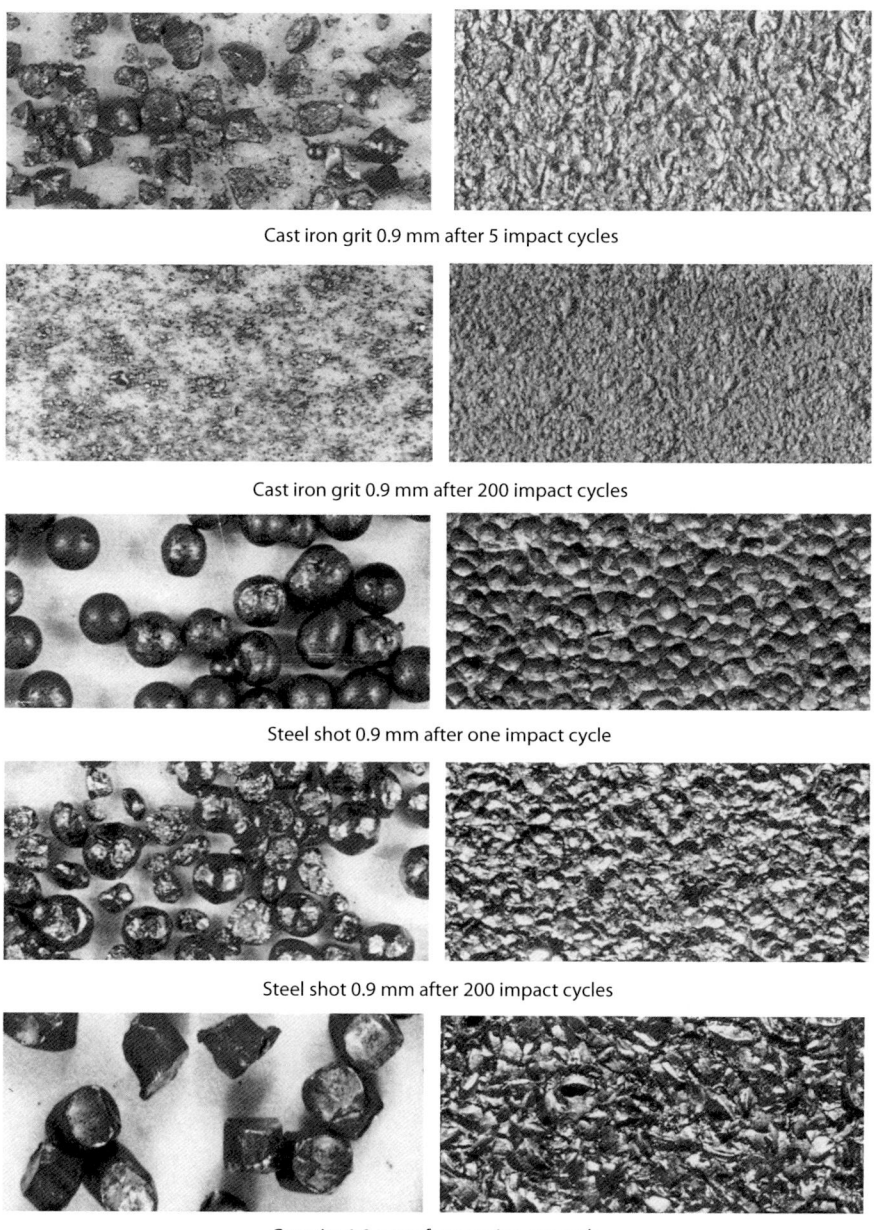

Cast iron grit 0.9 mm after 5 impact cycles

Cast iron grit 0.9 mm after 200 impact cycles

Steel shot 0.9 mm after one impact cycle

Steel shot 0.9 mm after 200 impact cycles

Cut wire 1.0 mm after one impact cycle

FIGURE 2.7 Appearance of selected blasting abrasives after a particular number of cycles (a) and appearance of mild steel substrate blasted by them (b). (From Jankura, D. et al., *Materials in Mechanical Engineering and Technology of their Finalization*, Technical University of Košice, Košice, Slovakia, 2011.)

The mechanisms of scale layer removal by round and sharp angular blasting abrasives have some differences with regard to their shape:

- *Round grains*: On the surface of tough materials, indentations are created in the form of dimples, whose sides rise above the level of the original surface of the substrate. Scales are removed at the point of impact, and also in neighboring areas of scale where compressive stress is induced.

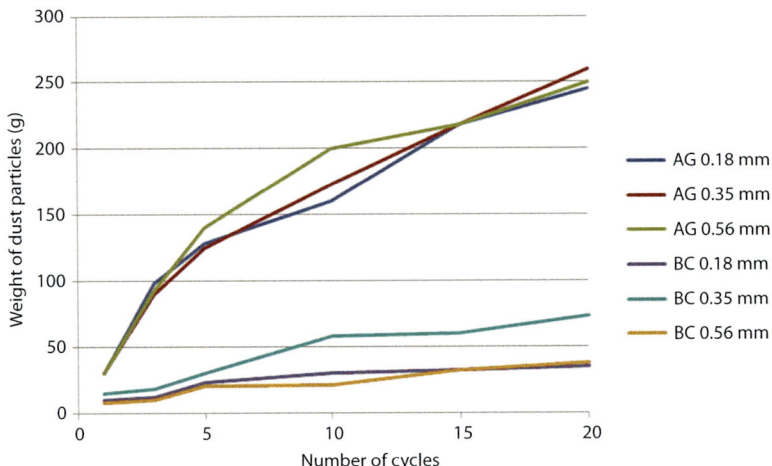

FIGURE 2.8   Increase in dust particles of almandite garnet (AG) and brown corundum (BC) depending on the number of cycles. (From Jankura, D. et al., *Materials in Mechanical Engineering and Technology of their Finalization*, Technical University of Košice, Košice, Slovakia, 2011.)

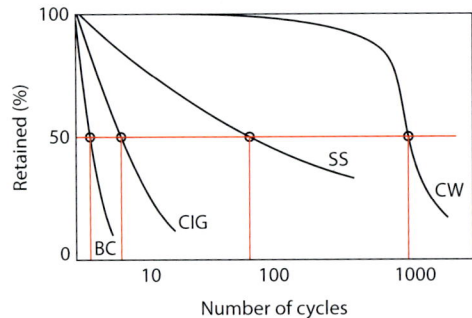

FIGURE 2.9   Durability curves for monodisperse. BC: brown corundum; CIG: chilled iron grit; SS: steel shot; CW: cut wire. (From Jankura, D. et al., *Materials in Mechanical Engineering and Technology of their Finalization*, Technical University of Košice, Košice, Slovakia, 2011.)

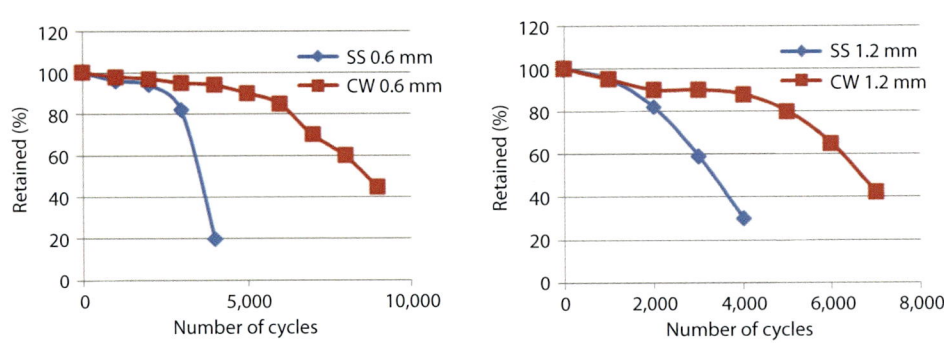

FIGURE 2.10   Durability curves obtained for steel shot and cut wire of sizes 0.6 and 1.2 mm, respectively. (From Jankura, D. et al., *Materials in Mechanical Engineering and Technology of their Finalization*, Technical University of Košice, Košice, Slovakia, 2011.)

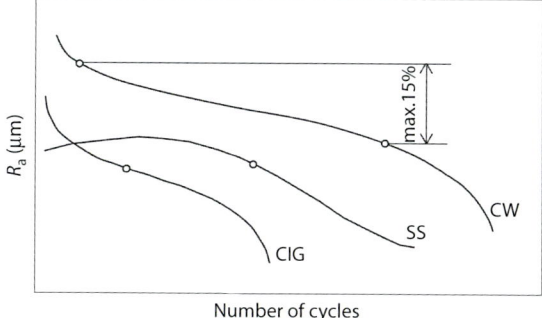

FIGURE 2.11 Roughness curves R_a depending on the number of cycles. CIG: chilled iron grit; SS: steel shot; CW: cut wire. (From Jankura, D. et al., *Materials in Mechanical Engineering and Technology of their Finalization*, Technical University of Košice, Košice, Slovakia, 2011.)

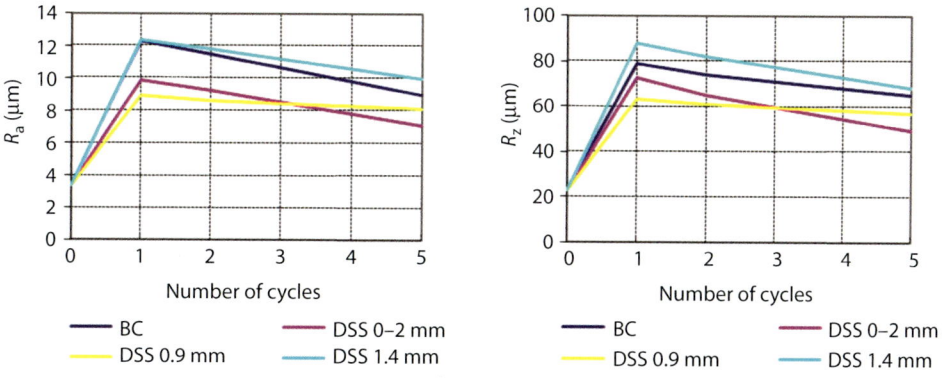

FIGURE 2.12 Dependence of R_a and R_z on the number of cycles. DSS: demetallized steelmaking slag compared with brown corundum (BC) [18].

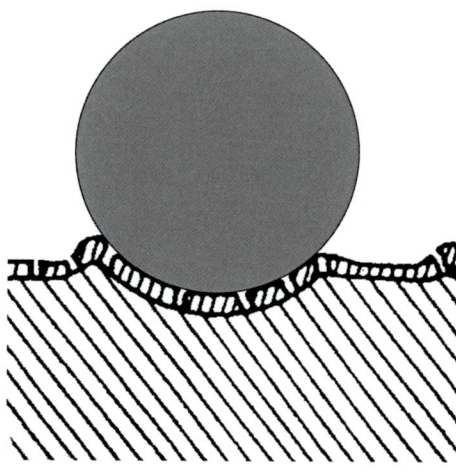

FIGURE 2.13 Mechanism of scale removal using a round abrasive. (From Brezinová, J. et al., *Abrasive Blast Cleaning and its Application*, Trans Tech Publications, Pfäffikon, Switzerland, 2015.)

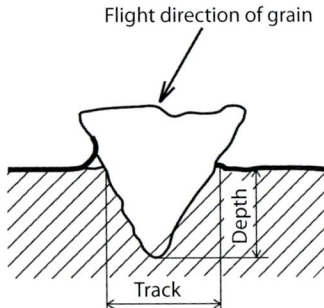

FIGURE 2.14 Mechanism of scale removal using a sharp angular abrasive. (From Brezinová, J. et al., *Abrasive Blast Cleaning and its Application*, Trans Tech Publications, Pfäffikon, Switzerland, 2015.)

- *Sharp angular*: Grains break the scale layer at the point of impact, while the scale near the edge of the notch partially peels off, but may also be pushed into the notches formed. In the case of nonmetallic abrasives, the grains themselves may become jammed, and are broken due to the impact of other grains and remain in the substrate. This causes secondary residues on the surface due to the blasting abrasive.

To demonstrate the removal characteristics of individual abrasives, it is possible to construct a *removal curve* for each type of abrasive and type of cleaned surface (Figure 2.15).

This shows the functional dependence of the amount of scale (or other contaminants) removed from the surface on the amount of abrasive applied per unit of blasted surface. On the curve, there are two visible areas: An initial steep increase in the amount of removed material represents descaling or removal of other contaminants from the surface. In the second area, the rise of the curve is substantially mitigated, because contaminants from the surface are removed and the removal of the substrate occurs. In terms of blasting efficiency, it is necessary to apply to the cleaned surface only such amount of abrasive that delivers removal of the contaminants but does not lead to removal of the substrate. This amount of abrasive is called the *necessary amount of blasting abrasive to descale the surface* (q_n).

When cleaning surfaces, it is also necessary to choose an appropriate abrasive grain size because of the effects of large and small grains on the substrate are different:

- Large grain size:
  - Considerable energy concentrated at the point of impact
  - Significant deformation of scales and substrate
  - Local work hardening of the substrate and scales

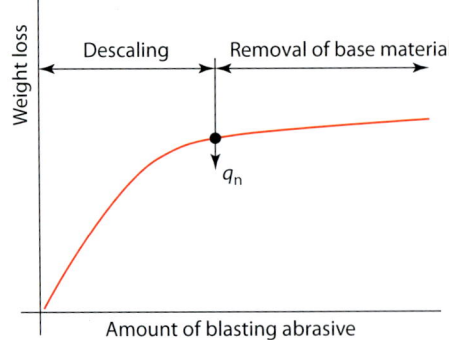

FIGURE 2.15 Schematic representation of removal curve. (From Jankura, D. et al., *Materials in Mechanical Engineering and Technology of their Finalization*, Technical University of Košice, Košice, Slovakia, 2011.)

- Small grain size:
  - Lower kinetic energy of the grain
  - Smaller deformation of the substrate
  - Higher even distribution of compressive stress in the scale layer

2.2.1.1.8 Roughening Effect of Blasting Abrasives

The roughening ability of a blasting abrasive can be expressed by a *roughness curve*, which shows the functional dependence of the arithmetical mean deviation of the substrate R_a on the amount of abrasive applied on a unit area (Figure 2.16).

The curve does not start from the beginning of the coordinate system, but from the value that corresponds to the initial roughness R_a of the surface before blasting. Three phases can be seen on the roughness curve:

1. An increase of the amount of blasting abrasive leads to an increase in surface roughness also.
2. The surface roughness reaches a maximum value. This value can be explained by the plastic properties of the substrate. This state remains until the deformation capability of the material is exhausted, resulting in work hardening of the substrate and partial reduction in roughness.
3. A further increase in the amount of blasting abrasive applied causes fatigue disruption of the surface layers (delamination): see Figure 2.17.

The boundary between Phases I and II (see Figure 2.16a) for these curves is called the *necessary amount of abrasive needed for full coverage of the surface by notches after abrasive impact* (q_{nR})

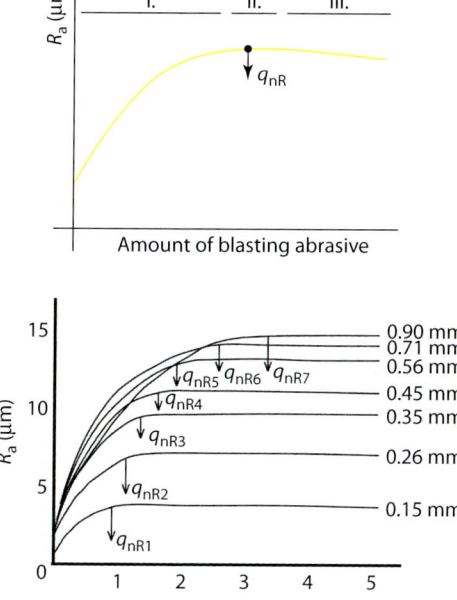

(b)

FIGURE 2.16 (a) Schematic representation of surface roughness curve; (b) actual roughness curves for brown corundum of different grain sizes. q_{nR}: necessary amount of abrasive needed for full coverage of the surface by dimples after abrasive impact; q: amount of blasting abrasive applied on a unit area. (From Jankura, D. et al., *Materials in Mechanical Engineering and Technology of their Finalization*, Technical University of Košice, Košice, Slovakia, 2011.)

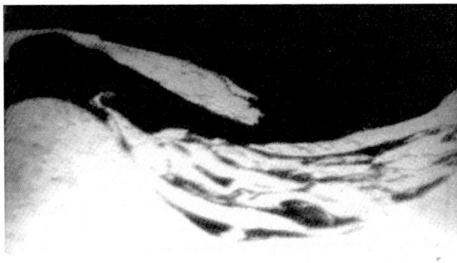

FIGURE 2.17 Delamination of mild steel surface due to overblasting; metallographic cross-section. (From Jankura, D. et al., *Materials in Mechanical Engineering and Technology of their Finalization*, Technical University of Košice, Košice, Slovakia, 2011.)

when the maximum roughness of substrate is reached. The decrease in roughness at the end of the curve is caused by the so-called *overblasting* of the surface, when a further increase of the amount of abrasive leads to fatigue disruption of surface layers and their peeling. This causes a decrease in bonded joint strength. Therefore, it is important to ensure that the coverage level is equal to 1, which corresponds to blasting by the necessary amount of abrasive q_{nR}. The given curve is valid only for a specific combination of blasting abrasive and substrate. For each abrasive granularity, impact speed, angle of impact, and for each cleaned material, it is necessary to determine q_{nR} experimentally.

2.2.1.2 Influence of Blasting Abrasive on Quality of Blasted Surface

2.2.1.2.1 *Impact of Abrasive Shape*

The nature of the blasted surface depends on the shape of the abrasive used. When using *round grains* of blasting abrasive, a relatively uniform deformation of the surface is achieved. The surface consists of intersecting spherical dimples (Figure 2.18a). Sharp *angular grains* of abrasive cause the formation of notches in the substrate whose orientation on the surface is stochastic. In their random movement within the stream of abrasive, grains can impact the surface with their edge, straight surface, or tip (Figure 2.18b).

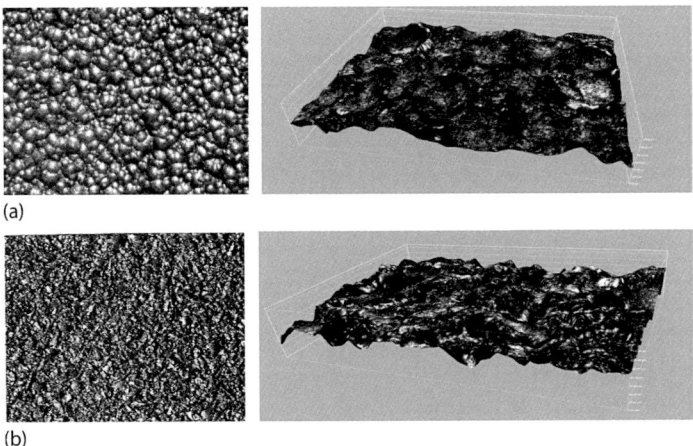

(a)

(b)

FIGURE 2.18 Surface of mild steel after blasting (left) and its 3-D reconstruction (right) after the impact of (a) round (b) sharp angular blasting abrasive. (From Guzanová, A. et al., *J. Adhes. Sci. Technol.*, 28, 1–18, 2014.)

2.2.1.2.2 Impact of Abrasive Grain Size

2.2.1.2.2.1 *Large Grains* Large grains concentrate significant impact energy at the point of impact, while their effect depends on the use of this impact energy. An impacting grain not only causes local disintegration of the oxide layer, but also significantly hammers the substrate. The effect on the substrate is reflected mainly in its plastic deformation and corresponding work hardening; or, depending on the impact angle, the removal of the substrate may also occur. With larger grains of abrasive, a higher surface roughness is achieved, but also a greater amount of blasting abrasive is required for complete surface coverage.

2.2.1.2.2.2 *Small Grains* On the other hand, small grains have a lower kinetic energy, which is consumed mainly by the removal of the scale layer from the surface and has a smaller effect on the substrate. Smaller grains require lower necessary quantities (q_{nR}) and the surface is smoother and evenly covered. Surfaces blasted by different grain sizes of blasting abrasives are shown in Figure 2.19.

Even blasting with abrasives of the same grain size but different materials give different values of surface roughness, due to their different densities (Figure 2.20).

The effect of abrasive grain size on the strength of single-lap joints is shown in Figure 2.21.

2.2.1.2.3 Impact of Abrasive Hardness

The impact of abrasive hardness on the substrate is reflected in the characteristics of the blasted surface. The impact of substrate hardness is suppressed when blasted by very hard abrasives. When selecting the blasting abrasive for pretreatment of the substrate, the following principle applies: The selected abrasive must be softer than the material to be retained after blasting, but harder than the material that is to be removed by blasting.

2.2.1.2.4 Impact of Abrasive Grain Size Distribution

The homogeneity of grain size distribution affects the appearance and roughness of the blasted surface. The narrower the grain size distribution, the more marked the surface relief. In wide grain size distribution, the characteristic surface relief is not so marked (Figure 2.22).

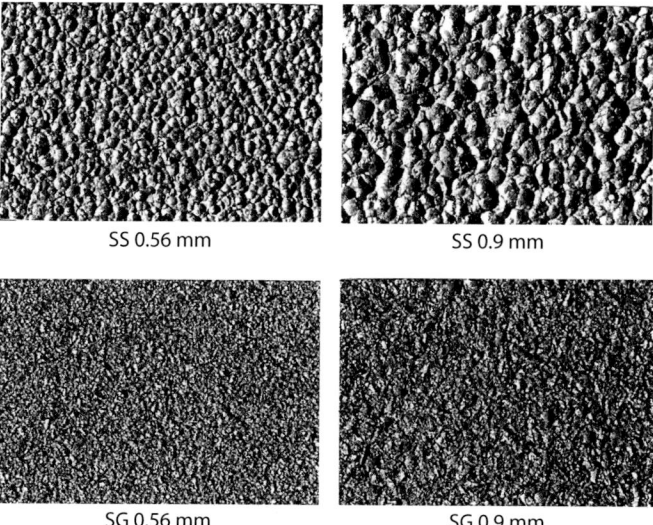

FIGURE 2.19 Surfaces of mild steel blasted by different grain sizes of blasting abrasives. SS: steel shot; SG: steel grit.

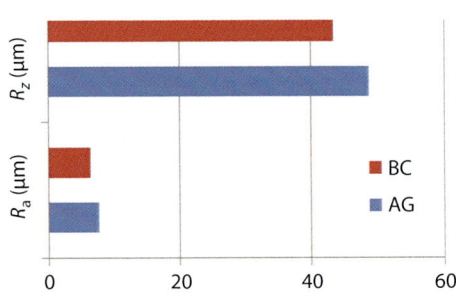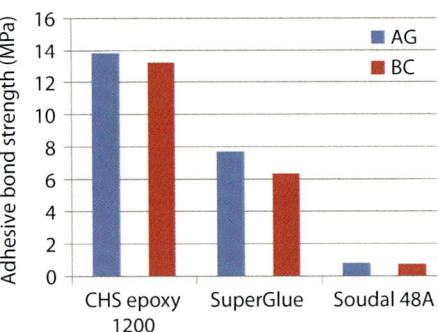

FIGURE 2.20 Roughness of mild steel surfaces blasted with brown corundum (BC) and almandite garnet (AG) with grain size 0.56 mm (left). Strength of single-lap adhesive joints using three adhesives: CHS Epoxy 1200, SuperGlue and Soudal 48A (right).

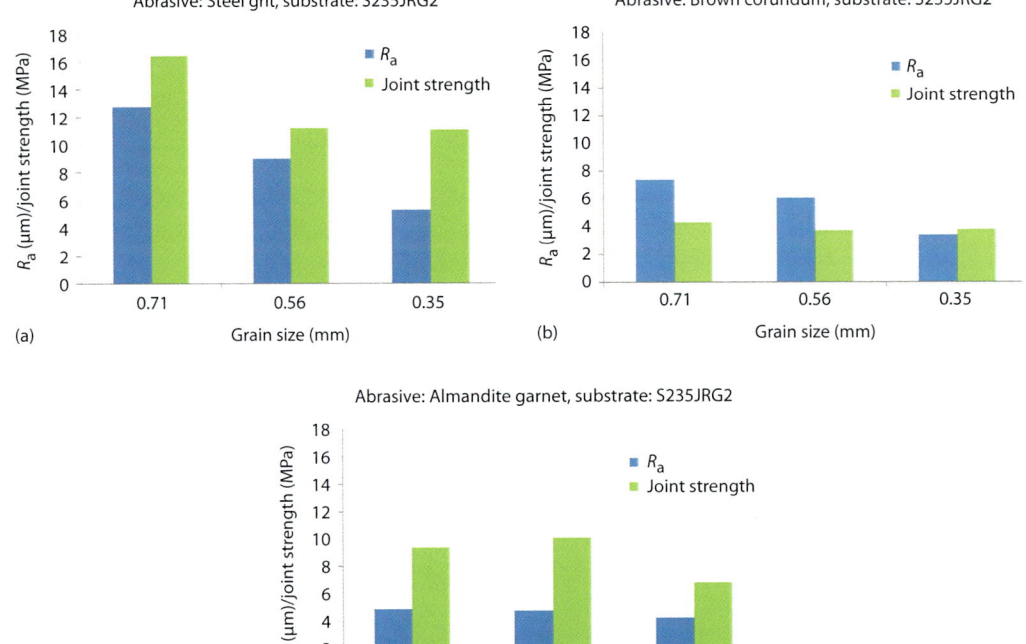

FIGURE 2.21 Effect of abrasive grain size on the strength of single-lap joints. Adhesive: epoxy resin; substrate: mild steel S235JRG2.

The appearance of surfaces blasted by polydispersed mixtures of blasting abrasives (steel shot and chilled iron grit) is shown in Figure 2.23.

2.2.1.3 Influence of Blasting Parameters on Quality of Blasted Surface

2.2.1.3.1 Impact of Abrasive Velocity

From Figures 2.24 and 2.25, it is clear that the higher the speed of the grain, the greater the roughness of the substrate and the smaller the amount of blasting abrasive necessary to cover a certain

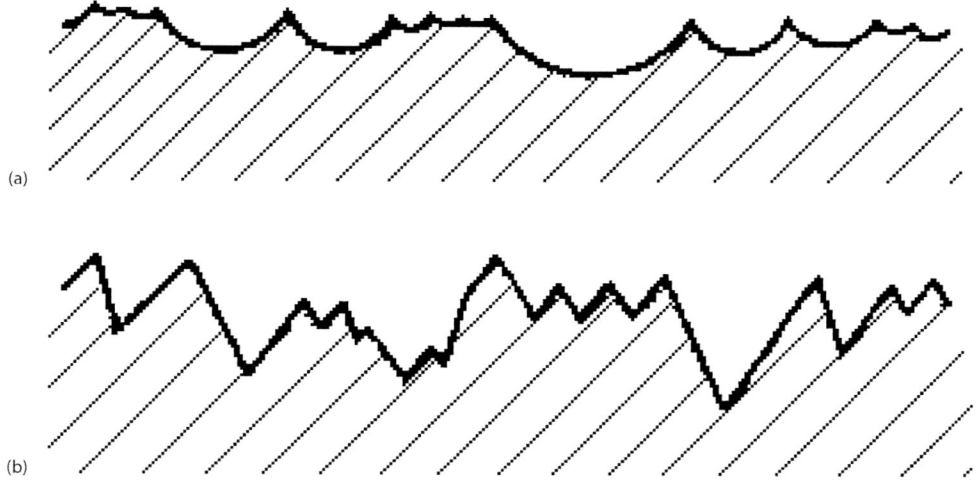

FIGURE 2.22 Surface of metallic substrate blasted by polydispersed (a) shot; (b) grit. (From Jankura, D. et al., *Materials in Mechanical Engineering and Technology of their Finalization*, Technical University of Košice, Košice, Slovakia, 2011.)

substrate area. To achieve a high quality of surface, one chooses a higher blasting speed for a larger abrasive grain size and a lower blasting speed for a smaller grain size.

In mechanical blasting systems, the speed of abrasive grains is controlled by the speed of the blasting wheel, and in air blasting systems by the air pressure. The influence of air pressure on the surface roughness is shown in Figure 2.26.

2.2.1.3.2 Impact of Blasting Distance
The effect of nozzle–substrate distance on surface roughness is not clearly determined. While some tend to believe that the distance of the nozzle from the surface does not affect the resulting surface roughness, others have found a clear, almost linear, increase in roughness with an increasing nozzle–substrate distance. It has been found that nozzle–substrate distance affects the final roughness only when a specific distance is exceeded. The value of the specific distance depends on air pressure; for instance, for a pressure of 0.4 MPa, the critical distance is 250 mm. It is possible to determine an optimal nozzle–substrate distance to achieve the maximum R_a. This optimal distance depends on the hardness of the substrate (Figure 2.27).

For materials with greater hardness, the optimal range of nozzle–substrate distances is wider. At a distance greater than the optimal, the lower kinetic energy of the impacting particles causes a decrease in roughness.

2.2.1.3.3 Impact of Hardness of Substrate
Figure 2.28 shows that with increasing hardness of the substrate at constant blasting conditions, a decrease in the blasted surface roughness occurs.

The appearance of two substrates of different hardnesses blasted by the same blasting regime is shown in Figure 2.29.

Depending on the combination of substrate and abrasive hardness, four different states can occur:

1. Abrasive and substrate are relatively soft; deformation of both occurs.
2. Abrasive is relatively soft, substrate is relatively hard; abrasive deformation is significant, substrate will be polished.
3. Abrasive is relatively hard, substrate is relatively soft; substrate is fully covered by notches or dimples.

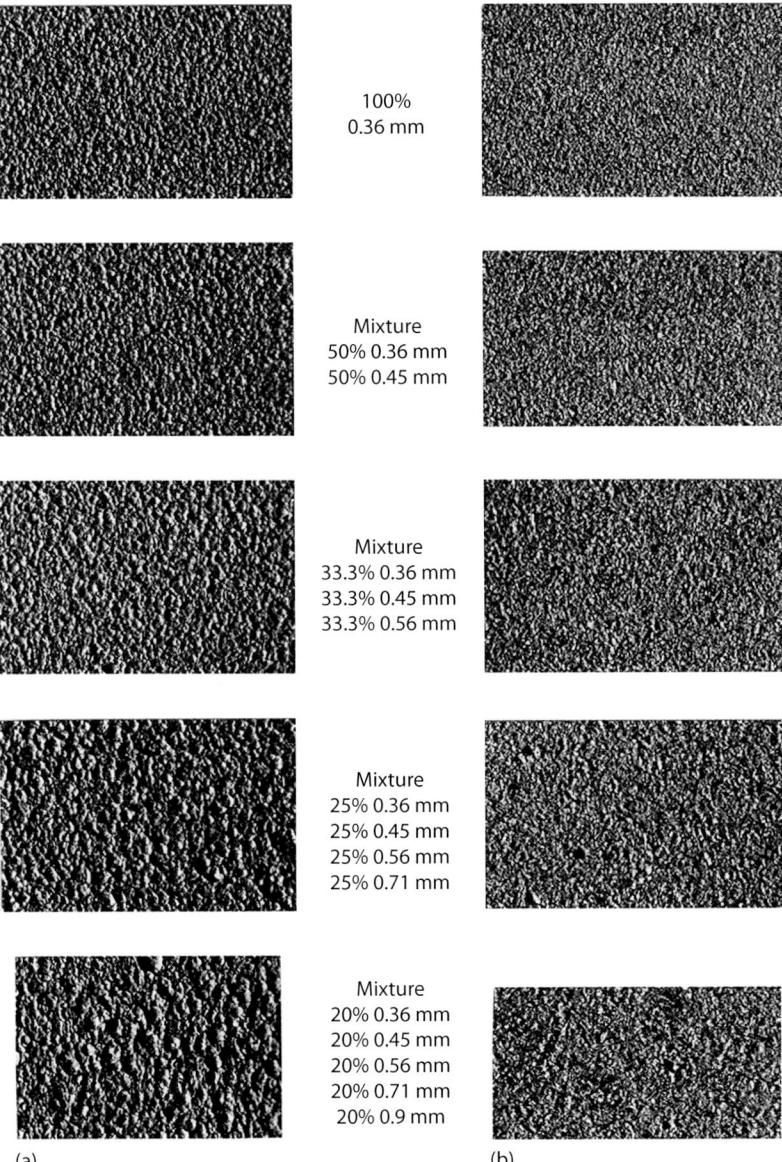

100%
0.36 mm

Mixture
50% 0.36 mm
50% 0.45 mm

Mixture
33.3% 0.36 mm
33.3% 0.45 mm
33.3% 0.56 mm

Mixture
25% 0.36 mm
25% 0.45 mm
25% 0.56 mm
25% 0.71 mm

Mixture
20% 0.36 mm
20% 0.45 mm
20% 0.56 mm
20% 0.71 mm
20% 0.9 mm

(a)   (b)

FIGURE 2.23 Appearance of mild steel surfaces blasted by polydispersed mixtures of (a) steel shot; (b) chilled iron grit. (From Brezinová, J. et al., *Abrasive Blast Cleaning and its Application*, Trans Tech Publications, Pfäffikon, Switzerland, 2015.)

4. Both abrasive and substrate are relatively hard; intense fragmentation of the blasting grain material occurs with only little roughening of substrate.

The resulting surface morphology is determined by the type of substrate, material of blasting abrasive, and blasting conditions.

2.2.1.3.4 Impact of Blasting Angle

The blasting angle α for air blasting is characterized as the angle between the surface of the substrate and the blasting stream (Figure 2.30a); or, for mechanical blasting, the angle between the trajectory of the grains flying out of the blasting wheel and the substrate (Figure 2.30b).

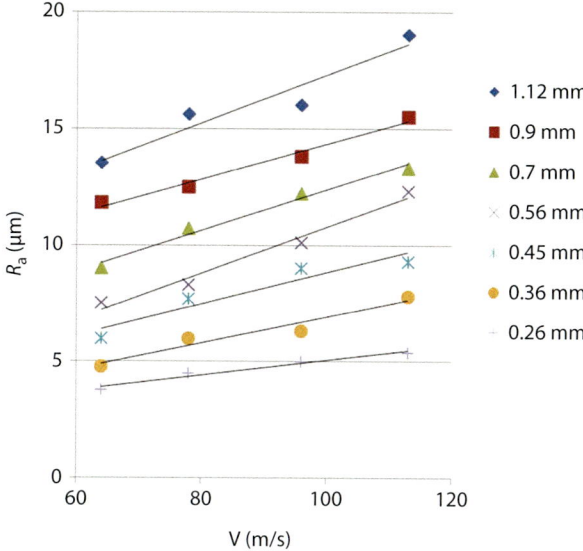

FIGURE 2.24 Dependence of R_a on grain speed v at various grain sizes. (From Brezinová, J. et al., *Abrasive Blast Cleaning and its Application*, Trans Tech Publications, Pfäffikon, Switzerland, 2015.)

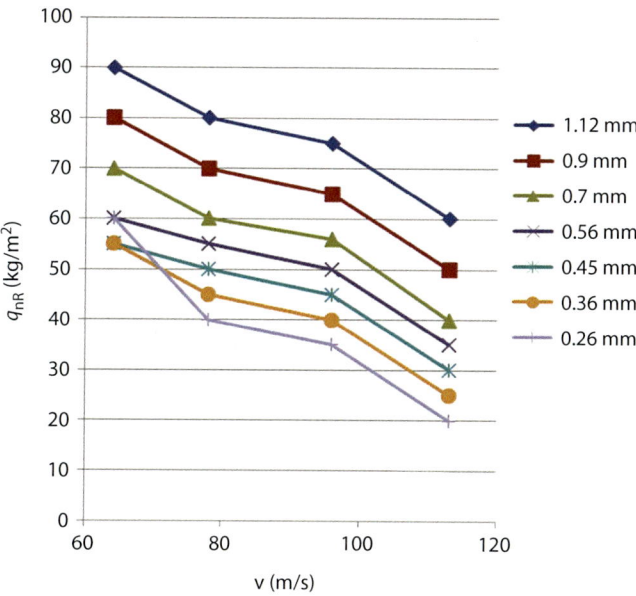

FIGURE 2.25 Dependence of the necessary amount of abrasive q_{nR} on grain speed v for various grain sizes. (From Brezinová, J. et al., *Abrasive Blast Cleaning and its Application*, Trans Tech Publications, Pfäffikon, Switzerland, 2015.)

The blasting angle affects changes caused by the impact of blasting abrasive. During blasting, either the process of creating indentations or that of the removal of grooving prevails. If the blasting angle is less than 45°, the grooving effect of the abrasive prevails, and the smaller the blasting angle, the greater the length of the grooves. The surface of the substrate with grooves does not lead to good strength of bonded joints. At a blasting angle of 75°, the removal process prevails and the resulting surface is roughened, containing many sites suitable for mechanical anchoring of the adhesive [20–24].

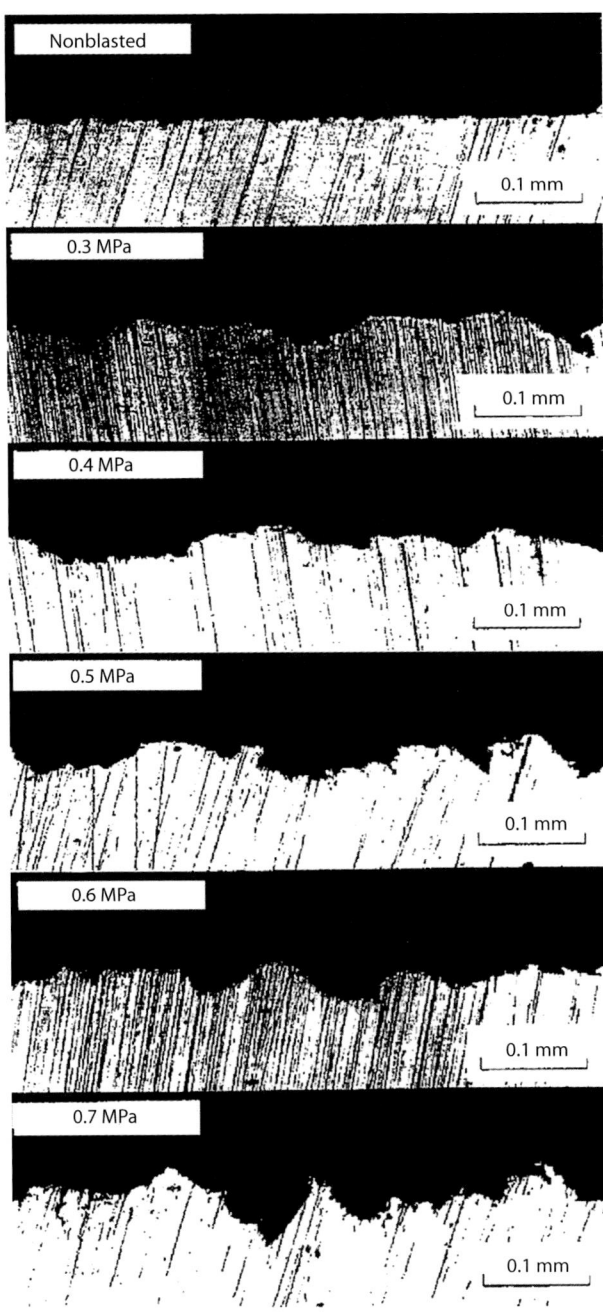

FIGURE 2.26 Surface profiles of mild steel obtained for blasting at various air pressures. (From Amada, S. and Yamada, H., *Surf. Coat. Technol.*, 78, 50–55, 1996.)

At a blasting angle of 90°, the indentation process prevails, and the resulting surface is not suitable for mechanical anchoring of the adhesive.

The different impact angles of blasting abrasive lead to changes in the blasting track on the cleaned surface (Figure 2.31).

The distribution of grain impact density at different impact angles is shown in Figure 2.32.

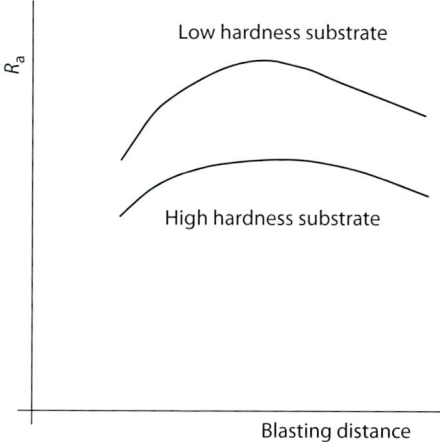

FIGURE 2.27 Influence of nozzle–substrate distance on surface roughness for substrates with low and high hardness.

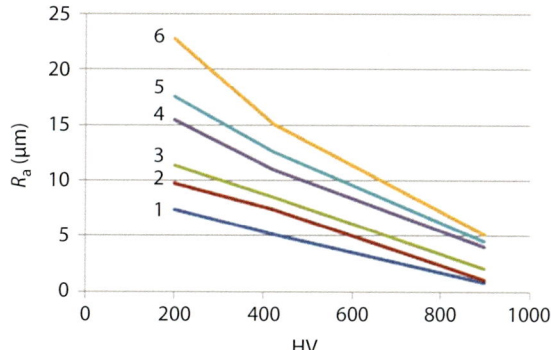

FIGURE 2.28 Dependence of blasted surface roughness R_a on substrate Vickers hardness (HV) for different abrasive grain sizes (1: low grain size; 6: high grain size). (From Brezinová, J. et al., *Abrasive Blast Cleaning and its Application*, Trans Tech Publications, Pfäffikon, Switzerland, 2015.)

(a) (b)

FIGURE 2.29 Surface of (a) aluminum sheet (EW AW 1050 A, Al 99.5) and (b) Cr–Ni steel sheet (X5CrNi 18-10) blasted by glass beads (<0.4 mm).

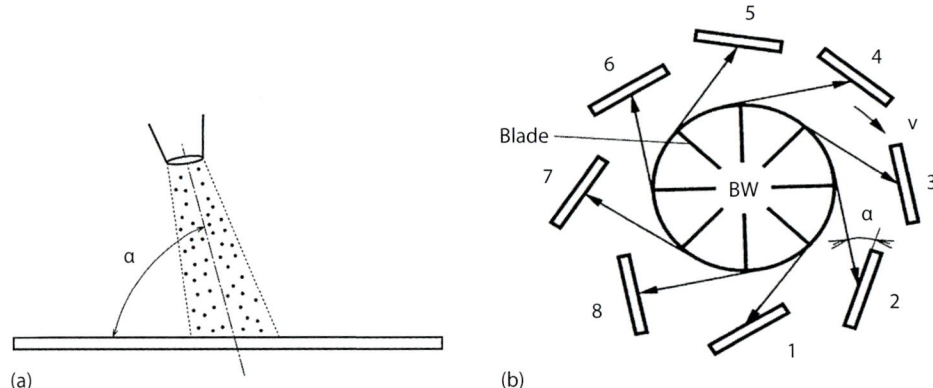

FIGURE 2.30 Depiction of blasting angle for air blasting (a) and mechanical blasting (b). BW: blasting wheel; *v*: abrasive speed; 1–8: blasted samples positioned around blasting wheel.

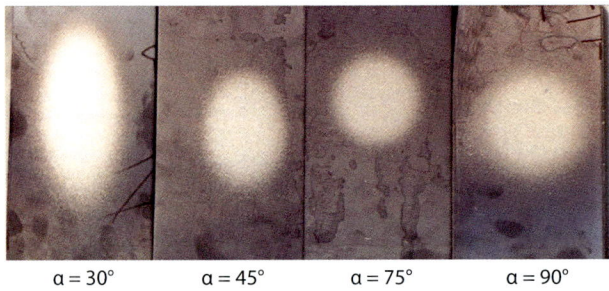

$\alpha = 30°$ $\qquad$ $\alpha = 45°$ $\qquad$ $\alpha = 75°$ $\qquad$ $\alpha = 90°$

FIGURE 2.31 Blasting tracks at different blasting angles α. Abrasive: steel shot; grain size: 0.9 mm; blasting distance: 300 mm.

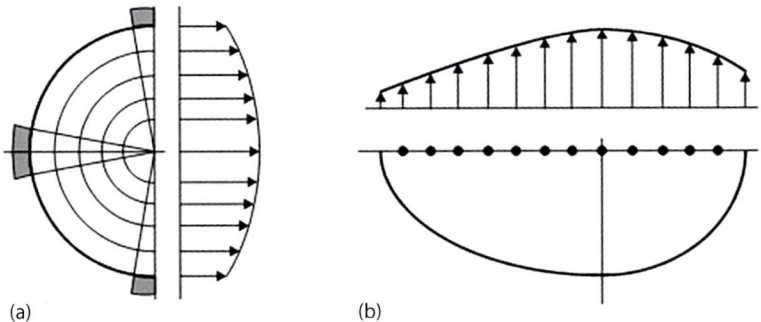

FIGURE 2.32 Schematic representation of the distribution of grain impact density at an impact angle of 90° (a) and 30° (b). (From Jankura, D. et al., *Materials in Mechanical Engineering and Technology of their Finalization*, Technical University of Košice, Košice, Slovakia, 2011.)

Values of roughness along the blasted area are not constant. In the middle of the track, roughness reaches its highest value, because the stream of accelerated grains is stable there. At the periphery of the track, the grains are deflected from their trajectory by other grains bouncing back from the substrate and, moreover, the roughness values are smaller (Figure 2.33).

In Figure 2.34, metallographic sections of the substrate blasted by steel grit at different impact angles are shown.

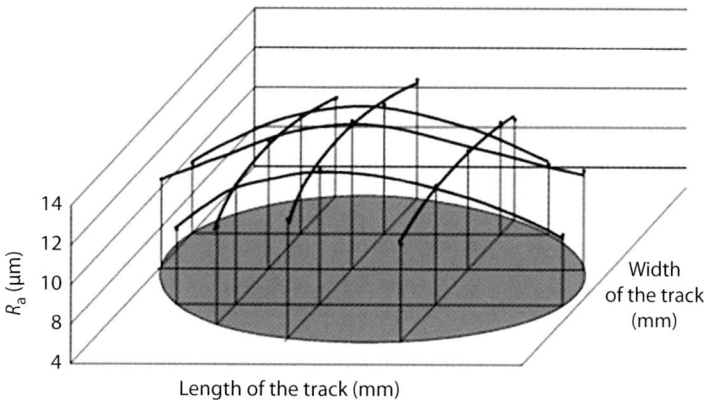

FIGURE 2.33 Distribution of surface roughness in affected area (steel grit; blasting angle: 30°; grain size: 0.9 mm). (From Jankura, D. et al., *Materials in Mechanical Engineering and Technology of their Finalization*, Technical University of Košice, Košice, Slovakia, 2011.)

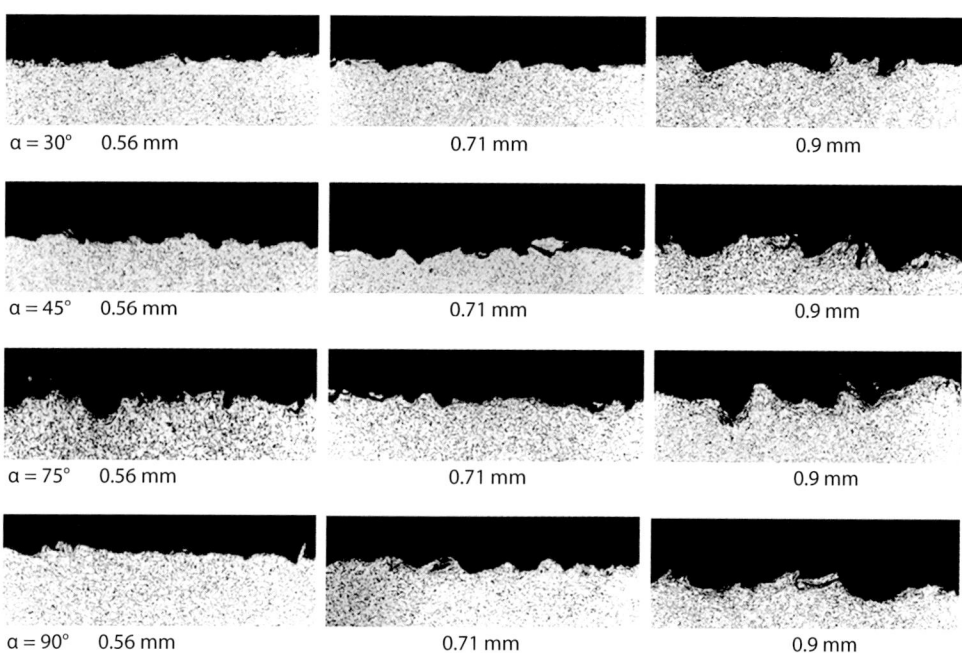

FIGURE 2.34 Metallographic sections of mild steel substrate blasted by steel grit with different grain sizes (indicated in millimeters) and different impact angles α. (From Jankura, D. et al., *Materials in Mechanical Engineering and Technology of their Finalization*, Technical University of Košice, Košice, Slovakia, 2011.)

This figure shows that at an impact angle of 75°, the morphology of the surface is most indented. The rougher the surface, the better the conditions for mechanical anchoring of the adhesive and the better the joint strength.

Similarly, studies with mechanical blasting using a blasting wheel show that an impact angle of 75° is an optimum angle at which the maximum amount of material is removed, and cleaning and roughening occurs with minimum consumption of abrasive.

An impact angle of 75° is used for surface preparation prior to adhesive bonding. The change in surface roughness at different impact angles is shown in Figure 2.35.

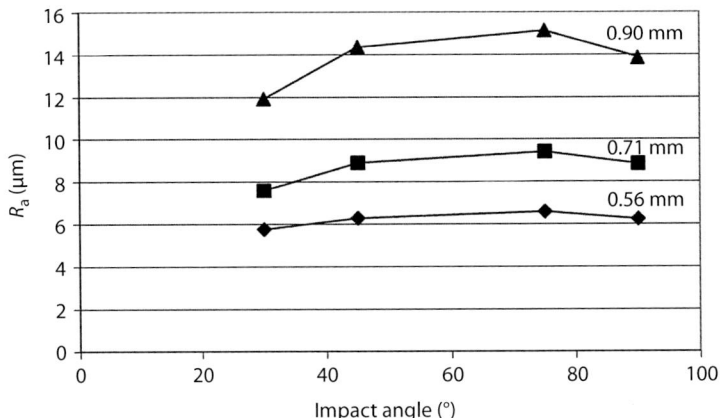

FIGURE 2.35 Dependence of R_a on impact angle of steel grit with different grain sizes. (From Jankura, D. et al., *Materials in Mechanical Engineering and Technology of their Finalization*, Technical University of Košice, Košice, Slovakia, 2011.

2.2.1.4 Choosing a Blasting Abrasive

The selection of a proper blasting abrasive depends on the purpose of the pretreatment. It is necessary to take into account the state of the surface, the type of contaminants, the objective of pretreatment (cleaning, roughening, and strengthening of the surface), and also the shape and size of the area to be treated.

Blasting abrasives can be classified into the following categories, according to their impact intensity on a substrate:

- Noninvasive blasting (without removal):
  - Plastic shot
  - Crushed shells
  - Crushed fruit stones
- Decorative blasting:
  - Glass beads
  - Ceramic shot
  - Stainless steel shot
  - Brass shot
- Cleaning, matting, and material removal:
  - Corundum
  - Garnet
  - Steel grit
  - Chilled iron grit
- Surface strengthening:
  - Steel shot
  - Ceramic shot
  - Cut wire

2.2.1.5 Determination of Surface Coverage

Surface coverage in blasting is the rate of original surface relief removal and its replacement by a new relief specific to blasting (Figure 2.36). The blasting time (under air blasting) to achieve 100% coverage of the material surface is different for materials with different hardnesses.

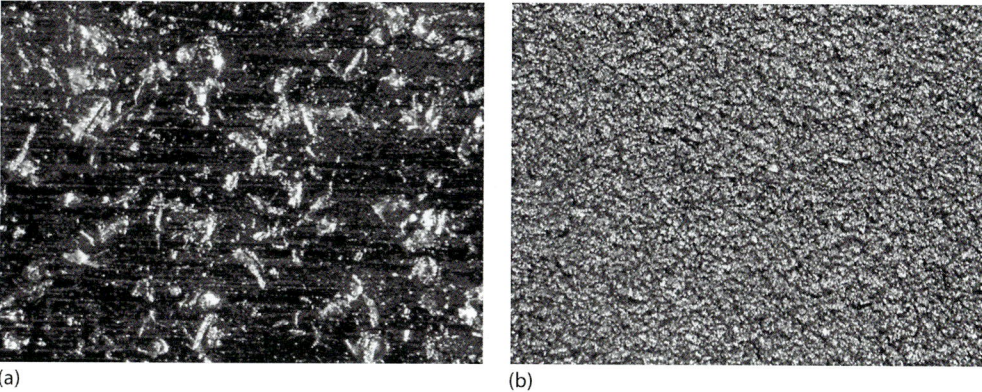

(a)   (b)

FIGURE 2.36 Surface coverage in blasting (BC, mild steel): (a) 35%, (b) 100%.

2.2.1.5.1 Experimental–Analytical Determination of Surface Coverage

In experimental–analytical determination of the necessary blasting abrasive amounts q_{nR} (100% coverage), it is important to recognize the purpose for which the blasting will be carried out. If blasting is used to descale the surface or remove a foreign layer from the surface, the necessary amounts are determined using removal curves; then, we refer to the necessary amount to descale the surface q_n. For pretreatment by blasting before other treatments (including adhesive bonding), in terms of achieving the required morphology of the substrate, it is necessary to completely cover the surface by dimples after abrasive grain impacts. The necessary amounts q_{nR} to achieve complete coverage are, in this case, determined from the so-called *roughness curves* (Figure 2.37).

Removal curves are constructed by the sequential blasting of a surface covered with scale by smaller doses of an abrasive and, after each partial blasting, the weight of the scale removed is determined.

Roughness curves are obtained by sequential blasting of a surface by smaller doses of an abrasive and, after each partial blasting, the roughness of the substrate is determined. The necessary amount to completely cover the surface is then the minimum amount of blasting abrasive that leads to complete coverage of the surface. Such complete coverage is reflected in the maximum value of the arithmetical mean deviation of the blasted surface R_a, and is verified by visual inspection of the surface at a low magnification (Figure 2.38) [40].

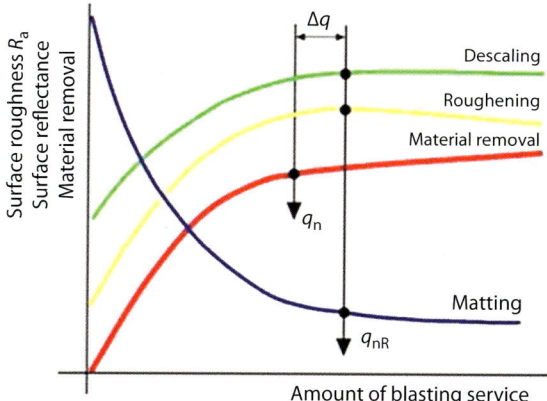

FIGURE 2.37 Removal curve (red), roughness curve (yellow), light reflectance curve when descaling (green), light reflectance curve when matting (blue). (From Brezinová, J. et al., *Abrasive Blast Cleaning and its Application*, Trans Tech Publications, Pfäffikon, Switzerland, 2015.)

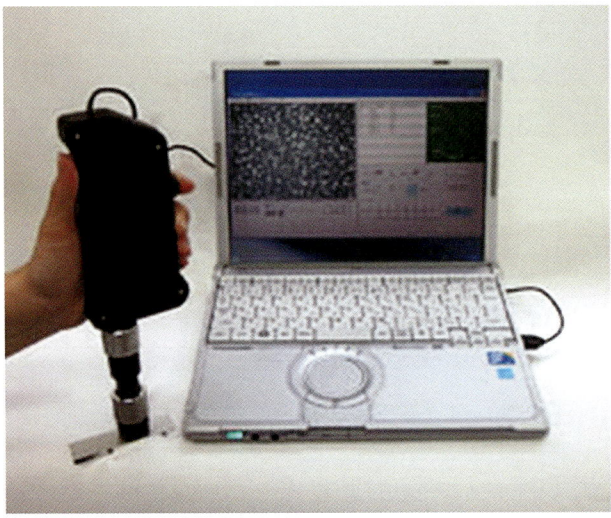

FIGURE 2.38 Equipment for checking the coverage of the surface after blasting. (From Toyo Seiko Co., Ltd., http://www.toyoseiko.co.jp/english/product/product08.html, 2017).

Figure 2.37 shows that the necessary amount of abrasive to descale the surface q_n is smaller than that necessary to completely cover the surface q_{nR}. The difference Δq can be explained because in blasting a surface with scales, the layer of scales cracks and disintegrates not only at the point of impact of the grain, but also in the surrounding area. Therefore, when descaling, 100% coverage of the surface is not necessary. To roughen the surface, 100% coverage of the surface is necessary; hence, a greater amount is applied per unit of area. The necessary amount in mechanical abrasive blasting using a blasting wheel is expressed in kilograms per square meter. In air blasting, it is expressed by the blasting time in seconds.

A *light reflectance curve* shows the functional dependence of reflected testing light from the surface on the amount of the abrasive applied. During cleaning of the surface, reflectance initially increases; then, after the removal of corrosion products or old paints from the metal substrate, the reflectance gradually settles. Further blasting of the surface increases the reflectance, but only very slightly. The values of maximum reflectance are different for different substrates and blasting abrasives used, but the reflectance curve is always characterized by a rapid increase of reflectance and a subsequent slight increase of the curve toward the asymptotic value.

In cases where blasting is applied for other reasons such as cleaning, for example for the purpose of matting—roughening of smooth surfaces such as glass or polished surfaces— the reflectance of the substrate decreases (Figure 2.39).

2.2.1.5.2 Experimental Determination of Surface Coverage

To check surface coverage in practice, fluorescent liquids can be used. The liquid is applied on the surface of the substrate that is subsequently blasted. The extent of coverage is then checked using ultraviolet light. The liquid contains a phosphorescent pigment that remains on the insufficiently blasted surface and can be indicated by ultraviolet light. Surface coverage is thus given by the degree of indicator removal. For smaller areas, the liquid can be applied in the form of a felt pen; for larger areas, the indicator is available in a liquid form (with methyl ethyl ketone) or powder form (the customer must add methyl ethyl ketone) and is applied to the controlled surface using a paintbrush.

In air blasting, complete surface coverage is closely related to blasting intensity. The blasting intensity is a measure of energy of the blasting beam, which is closely related to compressive stresses that are introduced into the surface.

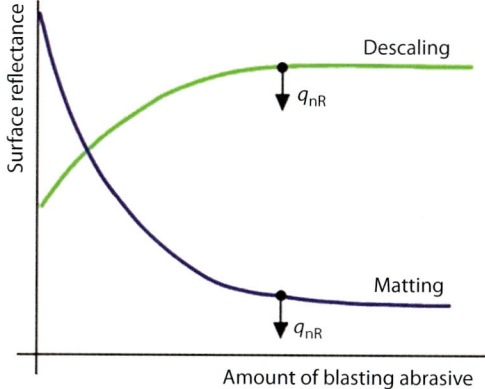

FIGURE 2.39 Reflectance curve for descaling and matting. (From Brezinová, J. et al., *Abrasive Blast Cleaning and its Application*, Trans Tech Publications, Pfäffikon, Switzerland, 2015.)

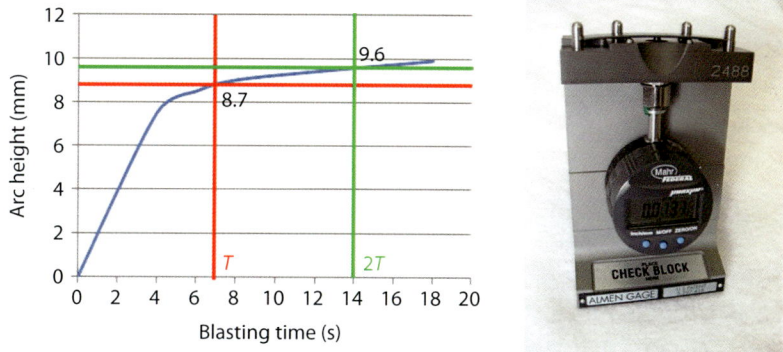

FIGURE 2.40 (a) Dependence of Almen strip arc height on blasting time. T: time needed for surface coverage; $2T$: double blasting time, (b) scheme of Almen gauge.

Blasting intensity is determined using Almen strips that are blasted on one side. The blasting abrasive roughens the metal on the surface of the Almen strip. This causes its deformation by bending. The arc height of the bend is proportional to the intensity of the blasting beam. The bending of the Almen strip can be measured by an Almen gauge. When the surface is blasted for a long time, the intensity of the arc height increases until it stabilizes. Saturation of the surface or full coverage (T) is defined as a point when doubling of the blasting time ($2T$) results in an increase in the arc height of the Almen strip by up to 10% (Figure 2.40).

2.2.1.6 Characteristics of Blasted Surface Quality

The surface created by the blasting process during repeated abrasive impact undergoes a plastic deformation. Under the effect of repeated impact, surface layers change their sizes, geometrical characteristics, structural properties, state of stress; an increase of substrate temperature occurs, and so forth. By blasting, modification of certain mechanical and technological properties is achieved (e.g., toughness, resistance to fatigue and abrasion, and others) [25–30]. From the point of view of the blasting effect on the substrate it is necessary to understand the blasting process as one of surface deformation or elastic–plastic deformation of the surface.

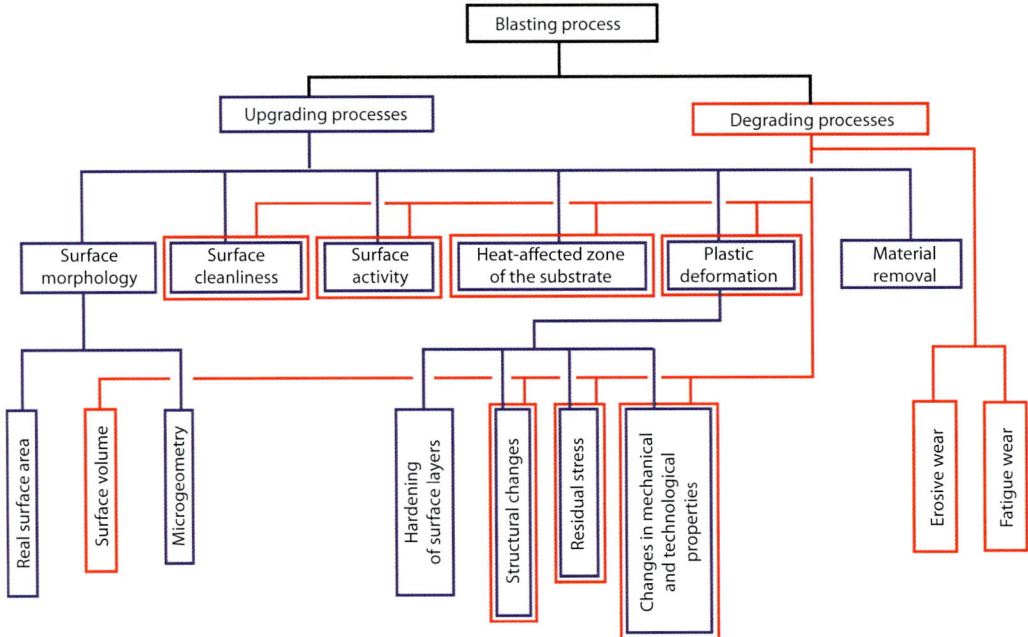

FIGURE 2.41 3. Schematic classification of blasting effects on the substrate. (From Brezinová, J. et al., *Abrasive Blast Cleaning and its Application*, Trans Tech Publications, Pfäffikon, Switzerland, 2015.)

The quality of the blasted surface is characterized by the following features:

1. *Geometrical shape of the blasted surface*: Involves the microgeometry, the surface volume, and the real size of the surface.
2. *Condition of the blasted surface*: Involves plastic deformation, strengthening, thermal effect, structural changes, changes of mechanical and technological properties and residual stresses.
3. *Cleanliness of the blasted surface*: The blasting process involves several simultaneously occurring processes. Some of them shift a blasted surface to a qualitatively higher level (upgrading processes); others, on the other hand, degrade the substrate (degrading processes) (Figure 2.41). However, it is not possible to clearly classify these phenomena as upgrading or degrading. Their classification depends on the purpose of blasting. For instance, the removal of substrate during blasting is a degrading phenomenon, but when considering the need to clean the surface, it is an upgrading phenomenon.

2.2.1.6.1 Microgeometry of Blasted Surface

From the microgeometry point of view, surfaces can be divided into:

- *Oriented surface*: This has noticeably different values of roughness in two perpendicular directions (anisotropic surface).
- *Unoriented surface*: The roughness and spacing of peaks in two perpendicular directions are not noticeably different (isotropic surface).

On the basis of the stated facts, a blasted surface is an unoriented (isotropic) surface, while its creation is mainly conditioned by the grain shape of the blasting material used. The selection of the type, size, and shape of blasting material depends on the purpose of the newly created surface. This selection is the basic issue for the process of blasting.

2.2.1.6.1.1 2-D Evaluation of Microgeometry of Blasted Surfaces The ISO 4287 standard—the accepted international norm nowadays—defines the terms and parameters of a surface. The computing system for profile parameter evaluation is part of the standard, based on a system of the mean line for the roughness profile, the waviness profile, and the mean line for the primary profile.

The new surface, obtained by the blasting process, has a specific character. The microgeometry of a blasted surface depends on:

- *Blasted material properties (mainly hardness)*: A material with a greater hardness will not be affected during blasting to such an amount as a soft material from the point of view of its surface microgeometry.
- *Type of blasting material used:* Blasting materials with a greater hardness and with a larger grain size roughen a surface more intensively than fine-grained blasting materials. It is evident that during the blasting process, with a demand to obtain a specified roughness, it is necessary to select a proper combination of substrate hardness with appropriate hardness and grain size of blasting material. The ratio of hardness of the blasting material to hardness of the substrate is important from the point of view of the surface quality as well as in terms of removal of contaminants and other deposits from the substrate.
- *Blasting parameters*: In the selection of blasting material type, blasting parameters are also important. The same blasting material creates different surface roughnesses at different blasting parameters (abrasive speed, impact angle, etc.).

To determine the roughness parameters, various criteria have been set. The most elaborated 2-D evaluation of surfaces is the profile method using contact profilometers. In mechanical engineering, the mean line system is generally accepted and is widespread.

In terms of worldwide technical practice, the most widely used parameter of roughness is the arithmetical mean deviation of the assessed profile R_a, which is often confused with the general term of *roughness*. It is necessary to note that this parameter has only an informative character and has a low predictive value about the nature of the surface. It does not allow distinction between tapering or rounding of profile peaks (a very important factor during wear), is notably confused with cracks and other surface defects, and does not give a picture of surface rugosity.

In Table 2.8, average values of selected roughness parameters of blasted surfaces using various blasting materials are listed as steel grit (SG) with grain size $d_z = 0.71$ mm, BC: $d_z = 0.9$ mm, steel shot (SS): $d_z = 0.9$ mm, and DSS: $d_z = 0.9$ mm [21]. For all evaluated surfaces, a comparable R_a parameter was achieved; however, values of other roughness parameters (R_z: maximum height of profile; R_{Sm}: mean width of profile elements; R_{Pc}: density of peaks per unit of length) are different, and these depend on the type of blasting material used. Individual profiles with bearing area curves of blasted profiles also show differences between the surfaces (Figure 2.42).

TABLE 2.8
Arithmetic Average of Roughness Parameters of Blasted Surfaces

Type and Granularity of Blasting Abrasive	R_a (µm)	R_z (µm)	R_{Sm} (µm)	R_{Pc} (peaks/cm)
Steel shot 0.9	11.22	60.79	642.7	15.88
Steel grit 0.71	10.91	68.7	355.05	28.6
Brown corundum 0.9	11.43	70.29	339.45	29.21
Demetallized steelmaking slag 0.9	10.59	64.31	401.4	25.31

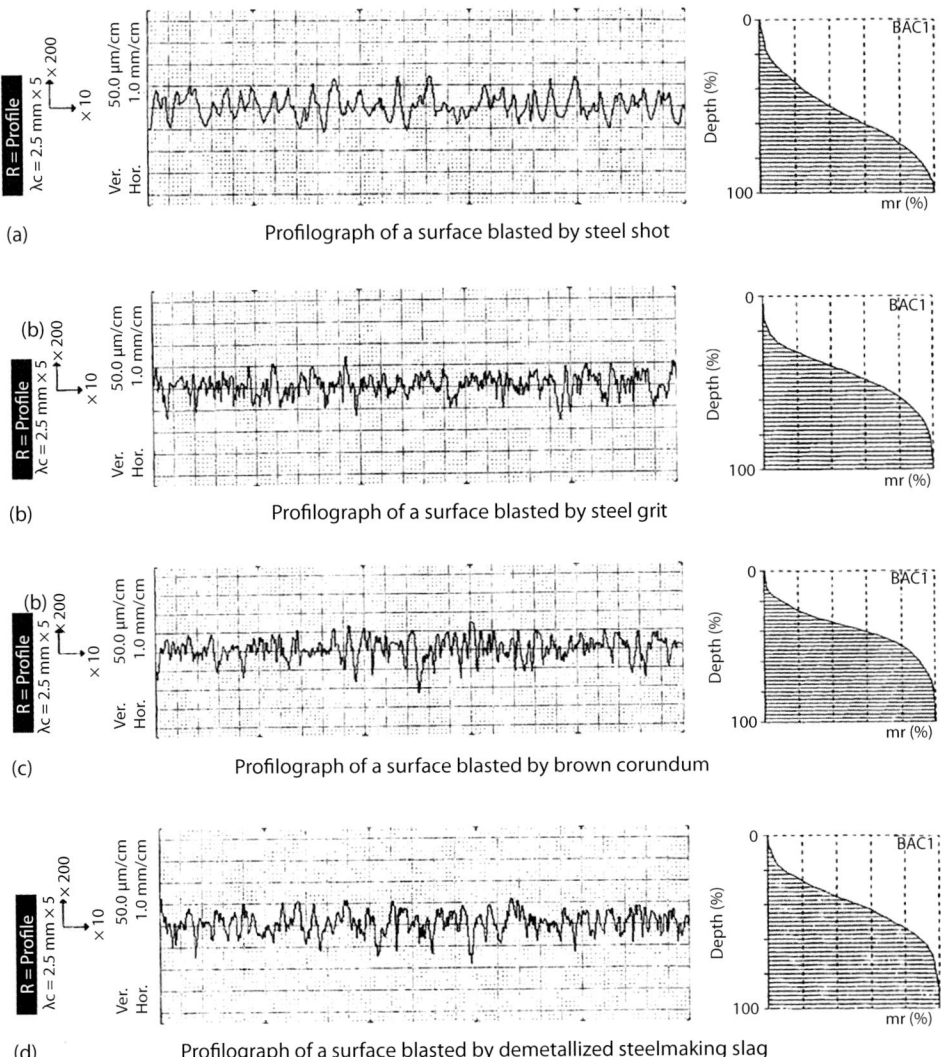

(a)   Profilograph of a surface blasted by steel shot

(b)   Profilograph of a surface blasted by steel grit

(c)   Profilograph of a surface blasted by brown corundum

(d)   Profilograph of a surface blasted by demetallized steelmaking slag

FIGURE 2.42   Profiles of surfaces blasted with different blasting materials (middle)–(a) steel shot, (b) steel grit, (c) brown corundum, (d) demetallized steel slag; BAC1: bearing area curve (right); magnifications of profiles in horizontal and vertical directions are indicated by arrows (left).

Differences in the microgeometries of surfaces blasted by an abrasive with different shapes (rounded and sharp-edged) are concisely expressed by *profilographs* and *bearing area curves* (also known as *material ratio curves* or *Abbott–Firestone curves*) (Figure 2.42). These curves can graphically describe the distribution of material within the profile height. By comparing the bearing area curves of the assessed profile, it is possible to explain the difference between surfaces blasted by sharp-edged and rounded blasting materials. Surfaces blasted by sharp-edged abrasives show a greater material ratio in the upper 40% of the profile, which creates more suitable conditions for good anchoring of subsequently applied adhesive.

The shape of valleys on surfaces blasted by sharp-edged blasting materials also plays a noticeable role (formation of wedge or anchor profile) in anchoring the adhesive into more rugged and more heterogeneous reliefs [31–34].

In the complex appraisal of the microgeometry of blasted surfaces, an evaluation in terms of a set of chosen parameters of roughness is necessary. According to the analysis [1,17,21], the most suitable roughness parameters are:

- R_a: Arithmetical mean deviation of the assessed profile
- R_z: Maximum height of the profile
- R_{Sm}: Mean width of the profile elements (eventually R_{Pc}: peak count per unit length)
- The bearing area curve of the profile

By a combination of these parameters, it is possible to distinguish the differences in the microgeometries of blasted surfaces.

From the microgeometry parameters' dependency on various blasting parameters, it is found that the roughness of the blasted surface increases with increasing abrasive grain size, increasing the impact angle to a value of 75°, and with an increase of grain speed.

2.2.1.6.1.2   3-D Evaluation of Microgeometry of the Blasted Surfaces   Measuring and evaluation of a surface in 3-D, using spatial parameters, provides very valuable and practically usable information about relations between geometric characteristics of a surface and its functional properties. Fast development of measuring techniques and the corresponding software nowadays allows the application of the advantages of 3-D evaluation of a surface texture. Spatial measurement of the surface microgeometry in comparison with 2-D evaluation of a single profile assures more objective presentation of the whole surface. Individual spatial parameters of the texture are calculated from a considerably larger amount of data; thus, the values of the measured parameters are more precise. Spatial display of the surface profile also reduces the danger of lapse of some important surface property [1,35–37].

It is possible to obtain the data for spatial evaluation of the surface texture either by contact measurement (scanning of a set of normally parallel profiles) or by the optical technique. Optical devices use scanning rays that monitor the surface similarly to a contact scanner or a microscope.

Nowadays, a variety of devices are available (contact and optical), and by using these it is also possible to measure the surface texture. However, both systems of measurement are comparable only for the measurement of surfaces with simple profile geometry.

ISO 25178 is an international standard for the spatial texture of a 3-D surface. Measuring and processing of noticeably larger amounts of data that describe the spatial surface profile provide a huge amount of information, required for real reconstruction of the evaluated surface.

Spatial analysis of a blasted surface structure also offers a graphical display of a surface in axonometric projection (Figure 2.43a), a topographic map (Figure 2.43b), and a wireframe model (Figure 2.44). It is also possible to obtain 3-D surfaces by using various types of microscopes with the support of modern software (Figure 2.45).

2.2.1.6.2   Surface Volume

The surface volume (V_S) is a volume related to a unit nominal surface area. It is a volume between a plain pass through the highest peak and the real surface of the substrate. The surface volume can be determined on the sample by the so-called *drop method*. The principle consists in applying a contrasting substance with defined volume on a pretreated surface, which is spread by a glass plate to the level of the highest peaks under the action of constant reproducible loading. The size of the picture created by the liquid is then determined by a planimeter and, on the basis of this area. The surface volume is calculated for a unit area (Figure 2.46).

If a spatial surface display is available, it is possible with the help of appropriate software (e.g., Surfer) to determine surface volume when the intersecting plane crosses the highest peak of the surface. The surface volume is a variable that plays a role in the formation of adhesive joints. It presents a "dead" volume that has to be filled by the adhesive to a level of the highest peaks of pretreated surface.

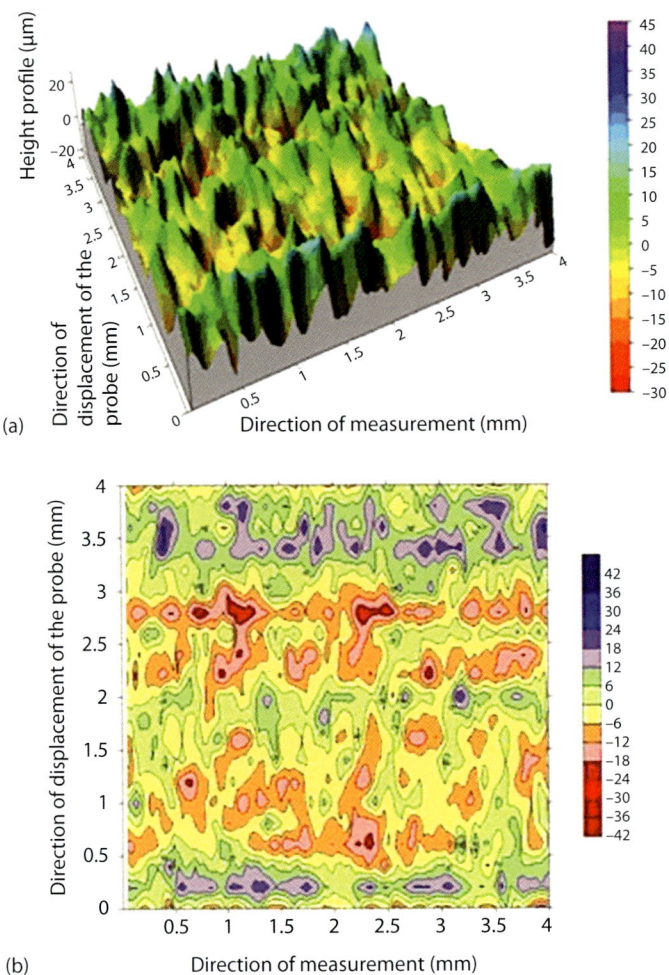

FIGURE 2.43 (a) 3-D map of blasted surface, (b) topographic map of blasted surface. (From Brezinová, J. et al., *Abrasive Blast Cleaning and its Application*, Trans Tech Publications, Pfäffikon, Switzerland, 2015.)

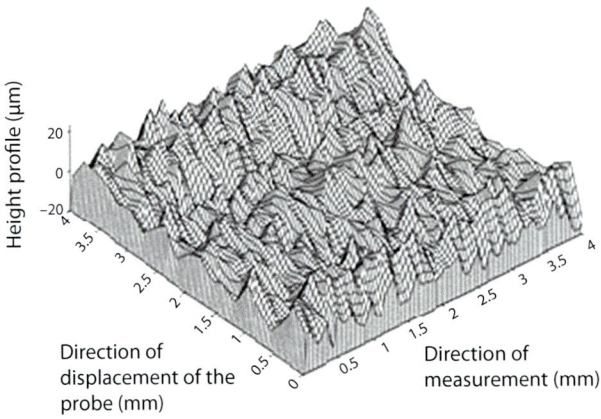

FIGURE 2.44 Wireframe model of blasted surface.

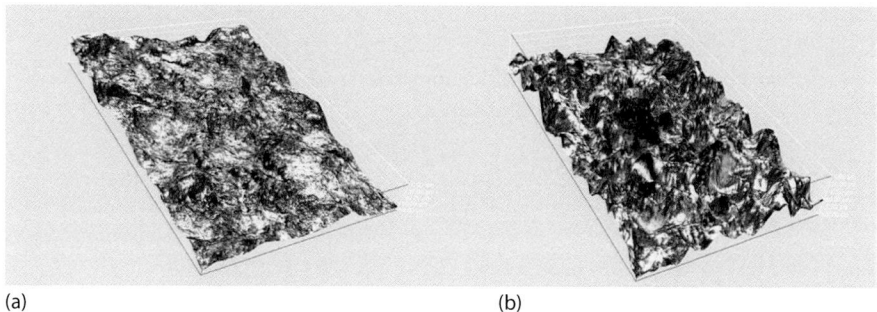

(a)   (b)

FIGURE 2.45 3-D pictures of surface blasted by (a) steel shot, (b) grit (confocal microscope). (From Jankura, D. et al., *Materials in Mechanical Engineering and Technology of their Finalization*, Technical University of Košice, Košice, Slovakia, 2011.)

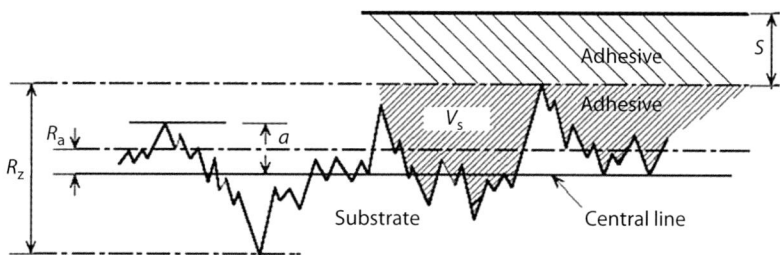

FIGURE 2.46 Schematic representation of cross-section of blasted surface with adhesive. s: thickness of adhesive; V_S: surface volume; R_z: maximum height of the profile. (From Brezinová, J. et al., *Abrasive Blast Cleaning and its Application*, Trans Tech Publications, Pfäffikon, Switzerland, 2015.)

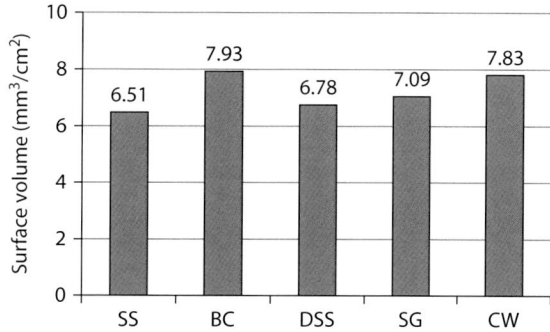

FIGURE 2.47 Values of surface volumes of surfaces blasted by various blasting abrasives. SS: steel shot; BC: brown corundum; DSS: demetallized steelmaking slag; SG: steel grit; CW: cut wire. (From Brezinová, J. et al., *Abrasive Blast Cleaning and its Application*, Trans Tech Publications, Pfäffikon, Switzerland, 2015.)

Comparing the values of surface volumes for different types of blasting materials, Figure 2.47 shows that with application of a rounded blasting material—the shot—the lowest surface volume is achieved.

2.2.1.6.3 Real Surface Area

The real surface area is the area of blasted surface profile stretched out into a plain. Experimentally, it is possible to determine the size of the blasted surface area by galvanic nickeling. The principle consists in a galvanic deposition of a nickel layer with a certain weight on the blasted surface and deposition of the same weight of nickel on a nonblasted smooth sample. On the basis of the determination of the thickness of the nickel layer on both samples (on the metallographic cross-section), it is possible to determine the real size of the blasted surface area.

It is possible to determine the real size of the surface area by using a number of software products.

2.2.1.6.4 Cleanliness of the Blasted Surface

Contaminants on the blasted surface can include:

- Residues of scales or corrosion products
- Metallic particles (dust) that are created due to abrasion or fragmentation of blasting grains
- Particles of blasting grains embedded into the surface of the substrate.

In the process of descaling by blasting, it is important to consider to what extent it is possible to remove scale. An absolutely clean surface in real conditions is very hard to achieve, and its achievement would increase the economic costs of the whole process. Therefore, in practice, it is possible to tolerate insignificant amount of contaminants; this can be expressed by the level of descaling, or by the level of a blasted surface cleanliness. To determine surface cleanliness, it is first necessary to determine the grade of surface contamination for the initial surface: the mill scale and rust grades. The ISO 8501-1 standard defines four rust grades of steel (Figure 2.48):

- A: Steel surface largely covered with adhering mill scale but little, if any, rust
- B: Steel surface that has begun to rust and from which the mill scale has begun to flake
- C: Steel surface on which the mill scale has rusted away or from which it can be scraped, but with slight pitting visible under normal vision
- D: Steel surface on which the mill scale has rusted away and on which general pitting is visible under normal vision

Levels of surface cleanliness achievable by blasting (marked by the Sa symbol) can have the following values (Figure 2.48):

- Sa 1: *Light blast cleaning*: When viewed without magnification, the surface shall be free from visible oil, grease, and dirt, and also from poorly adhering mill scale, rust, paint coatings, and foreign matter.
- Sa 2: *Thorough blast cleaning*: When viewed without magnification, the surface shall be free from visible oil, grease, and dirt and from most of the mill scale, rust, paint coatings, and foreign matter. Any residual contamination shall be firmly adhering.
- Sa 2½: *Very thorough blast cleaning*: When viewed without magnification, the surface shall be free from visible oil, grease, and dirt, and also from mill scale, rust, paint coatings, and foreign matter. Any remaining traces of contamination shall show only as slight stains in the form of spots or stripes.
- Sa 3: *Blast cleaning to visually clean steel*: When viewed without magnification, the surface shall be free from visible oil, grease, and dirt, and also from mill scale, rust, paint coatings, and foreign matter. It shall have a uniform metallic color.

After blasting (if nonmetallic abrasives are used), so-called *secondary contamination* is present on the surface, created by the residues of the blasting materials. These are either embedded in the surface or settled on the surface in the form of dust particles (Figure 2.49).

Determining the amount of contaminants on a metallic surface is notably difficult, because of their heterogeneous structure. This mainly concerns corrosion products or residues of mill scale, particles of blasting grains embedded in the surface of the substrate, greases of organic and inorganic

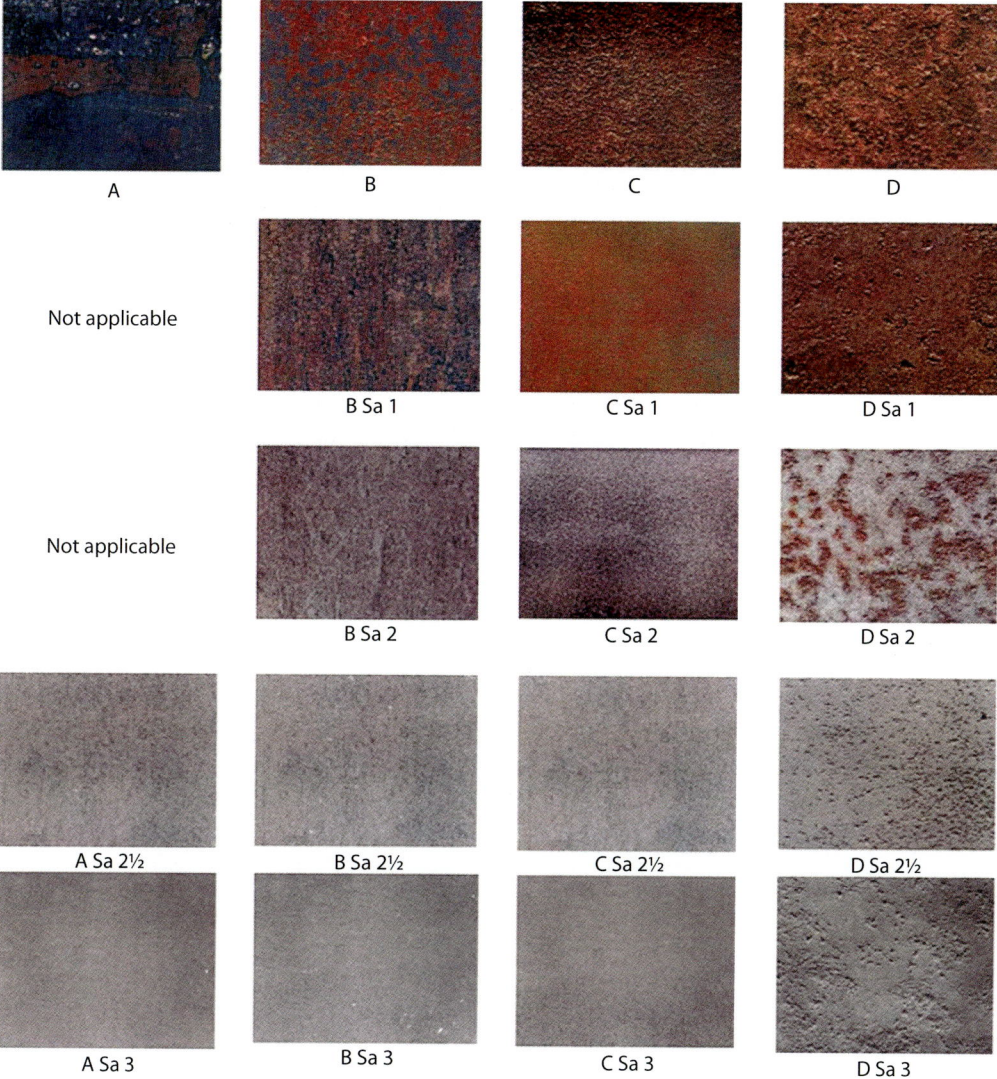

FIGURE 2.48 Four rust grades of steel (A, B, C, D). S1: light blast cleaning; Sa 2: thorough blast cleaning; Sa 2½: very thorough blast cleaning; Sa 3: blast cleaning to visually clean steel.

origin, residues of alkaline waste products after incomplete rinsing, and so on. This leads to oxidation, due to a high affinity of metal to oxygen.

Nowadays, the test for assessment of surface cleanliness using a pressure-sensitive tape according to ISO 8502-3 is used. Pressure-sensitive tape, in the set conditions, is stuck to a pretreated steel surface, then it is removed and visual appraisal (dust size particles, dust quantity rating) is made (Figure 2.50).

Images of the tapes with dust particles are also possible to evaluate by image analysis software (Figure 2.51). The method is based on a comparison of the reflectance of light from a white sheet of paper with reflectance from the tape removed from blasted substrate. The reflectance is expressed as a percentage.

In Figure 2.52, embedded particles stuck to the tape are highlighted by a red color. Software calculates the percentage coverage of the surface by abrasive particles.

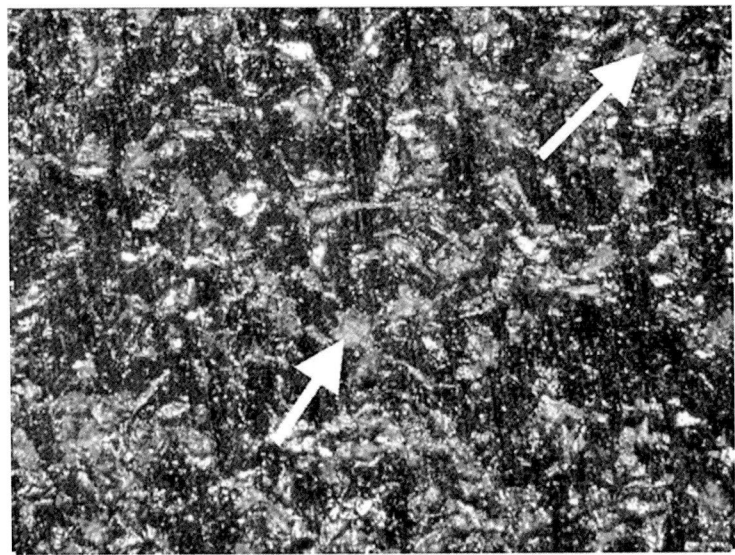

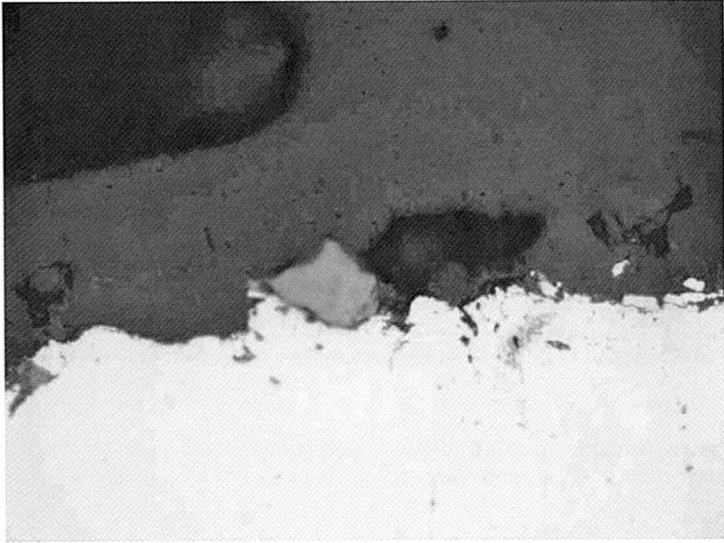

FIGURE 2.49 Blasted substrate; embedded grains of AG in top view (left, mag. 50×) and in cross-section (right, mag. 100×). (From Brezinová, J. et al., *Abrasive Blast Cleaning and its Application*, Trans Tech Publications, Pfäffikon, Switzerland, 2015; Jankura, D. et al., *Materials in Mechanical Engineering and Technology of their Finalization*, Technical University of Košice, Košice, Slovakia, 2011.)

The image in Figure 2.52 was processed by software for phase analysis. Dust particles represent 7.62% of the evaluated area (in red).

2.2.1.7   Surface Preparation Specifications Prior to Bonding

2.2.1.7.1   Preparation Before Blasting

It is necessary to visually check for the presence of oil, grease, salt, and other contaminants. These contaminants must be removed by degreasing or washing, and a check carried out to determine whether these contaminants have been removed. If this step is omitted, the grease will be transferred to the surface of the blasting abrasive, which could consequently contaminate cleaned surfaces. Contaminated blasting abrasive should not be used. It is appropriate to remove thick, strongly

Dust quantity rating

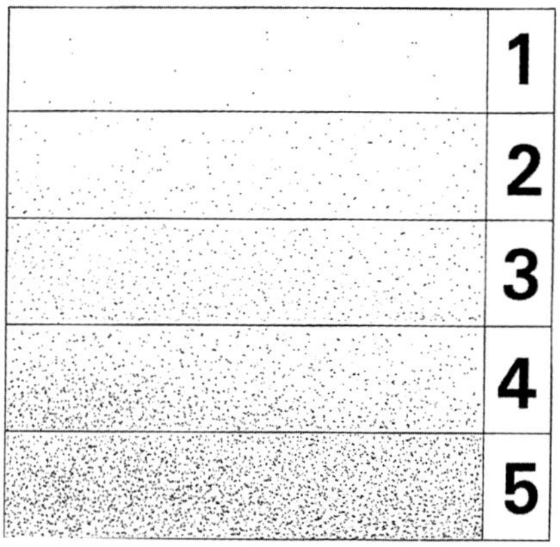

Dust size classes

0	Particles not visible under × 10 magnification
1	Particles visible under × 10 magnification but not with normal or corrected vision (particles usually are less than 50 μm in diameter)
2	Particles just visible with normal or corrected vision (usually particles are between 50 μm and 100 μm in diameter)
3	Particles clearly visible with normal or corrected vision (particles are up to 0.5 mm in diameter)
4	Particles between 0.5 mm and 2.5 mm in diameter
5	Particles larger than 2.5 mm in diameter

FIGURE 2.50   Assessment of surface cleanliness by pressure-sensitive tape.

adhering corrosion products and mill scale using powerful tools. Areas not to be blasted should be masked. In wet blasting, a degreasing agent can be added to the water used, which will degrease simultaneously with abrasive cleaning.

2.2.1.7.2 After Blasting

After dry blasting, it is necessary to remove dust or remaining weakly adhering particles of abrasive or contaminants by suction, brushing, or compressed air free of oil and moisture. To reduce the rest of the soluble contaminants, the surface can be washed with steam, hot water, solvent, or other suitable cleaner (followed by rinsing with pure water), and dried. After wet blasting, it is necessary to wash all surfaces with pure water to remove weakly adhering particles of blasting abrasives or remaining contaminants. The water may contain a corrosion inhibitor. Drying of surfaces prior to adhesive bonding can be realized by compressed air that is free of oil and moisture, or by other media (e.g., hot air). So-called *flash corrosion* may appear on the prepared surfaces. Therefore, adhesive bonding should be realized as soon as possible after cleaning the surface.

2.2.1.7.3 Assessment of Cleaned Surface

It is necessary to assess a cleaned surface according to ISO 8501 (preparation grades) and ISO 8502 (surface cleanliness), and also in terms of achieved surface roughness to ensure compliance with the requirements of the standards/agreement. If the surface does not meet the agreed requirements, the preparation process should be repeated.

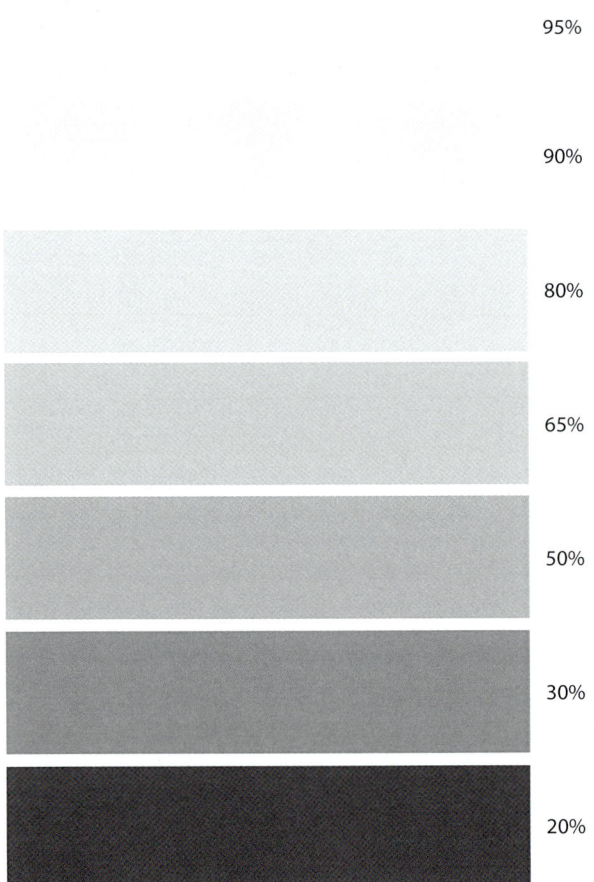

FIGURE 2.51   Scale of color shade and corresponding reflectance.

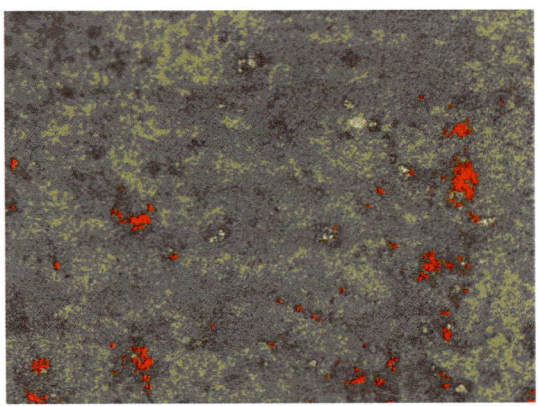

FIGURE 2.52   Embedded particles after blasting by Zirblast, 7.62%. (From Brezinová, J. et al., *Abrasive Blast Cleaning and its Application*, Trans Tech Publications, Pfäffikon, Switzerland, 2015.)

2.2.2 SURFACE CLEANING BY GRINDING

Grinding is one of the ways to prepare a product surface before applying coatings or adhesive bonding. Grinding is used mainly for removing imperfections or unevenness of the material surface; for example, seams on castings and forgings, pores, scales, corrosion products, the reinforcement of welds, and so on. In principle, grinding is comparable to milling (i.e., grinding with grinding wheels) or planing (i.e., grinding with abrasive belts). By pressing the metallic surface against a grinding wheel, a large number of abrasive grains with different geometries are in contact with the substrate. Each abrasive grain acts similarly to the cutting edge of a mill: it gradually cuts a chip of small section. On the grinding wheel, a large number of grains cut a large number of microscopic chips (Figure 2.53). The shape of the individual chips corresponds to the size and shape of the abrasive grain and the contact force between the grain and the surface.

Grinding moves from a larger to a smaller grain size. A larger grain size has a high capability of removing material. When changing the grain size of the grinding wheel, it is necessary to change the grinding direction to remove the grooves from the previous grinding. Grinding pastes can be used to improve surface quality.

The sequence of grinding operation involves:

- Rough grinding: The basic grinding operation, grit number 25–80; a rigid felt wheel or belt, always dry operation.
- Rough smoothing: A felt or cloth wheel, grit number 100–160; for hard materials, grinding pastes can be used.
- Fine smoothing: Surface preparation for electroplating, but not prior to bonding; grit number 200–250, using lubricants.
- Prepolishing: Grit number 315–400; wheels lubricated with paste.
- Fine polishing: The last grinding operation; grit number 400–600.

Some grinding operations may be omitted. The number of operations depends on the initial state of the surface and on the requirements for surface preparation prior to adhesive bonding.

The tools for grinding operations are *grinding wheels* or *abrasive belts.*

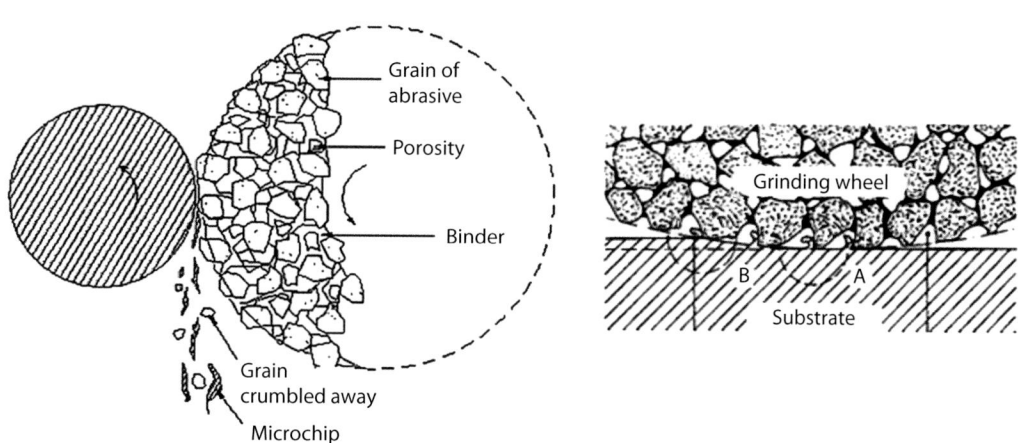

FIGURE 2.53 Schematic of grinding (left), contact between grinding wheel and substrate in detail (right). (From Brezinová, J. et al., *Abrasive Blast Cleaning and its Application*, Trans Tech Publications, Pfäffikon, Switzerland, 2015.)

2.2.2.1 Grinding Wheels

The ability of a wheel to remove material and the final surface quality after grinding are affected by the stiffness of the grinding wheel. The contact stiffness between the grinding wheel and the workpiece depends on the number of abrasive grains in contact with the workpiece and the stiffness of a single abrasive grain.

- *Felt wheels*: Are produced from animal hair. The felt wheel serves as a carrier for abrasive grains that are applied to the wheel as polishing paste, liquid, or glue.
- *Cloth wheels*: Are made up of multiple layers of suitable cloth that are loosely stacked together or stitched. The closer the stitch, the stronger the grinding wheel. The wheel serves as a carrier for abrasive grains, which are applied to the wheel as polishing paste, liquid, or glue.
- *Solid wheels*: Are made of a set of suitable abrasive grains glued together by a binder. Abrasives used: diamond, natural corundum, synthetic corundum, silicon carbide, and so on.

2.2.2.2 Abrasive Belts

An abrasive belt is a layer of abrasive adhering to a paper or textile carrier. Compared with a grinding wheel, it has the following advantages:

- Three or four times greater grinding area.
- An electrostatically applied abrasive is embedded in the binder (resin); its ability to remove material and cutting power are thereby improved.
- Lower temperatures during grinding, good surface quality.
- Light, sufficiently wide, easily replaceable.
- Can be prepared on endless belts.

2.2.2.3 Advanced Structured Abrasives

Structured abrasives consist of a series of unique 3-D structures (pyramids) that contain multiple layers of abrasive, distributed evenly across the backing material (Figure 2.54 [right]). Over a period of time, they wear away to continually and consistently expose fresh abrasives to the surface of the workpiece, delivering consistent, predictable finishes at high, even cut rates, with a longer belt life.

The resulting surface roughness after grinding depends on the grinding parameters, mainly on the grit number. The surface after grinding is oriented, characterized by parallel grooves. In the

(a)   (b)

FIGURE 2.54 Arrangement of abrasive grains in a standard (a) and structured abrasive tool (b) [38].

case where the surface is prepared prior to bonding, the grooves should be perpendicular to the loading of the adhesive joint.

2.2.3 SURFACE CLEANING BY BRUSHING

Surface preparation technology using rotating brushes is a process in which surfaces of parts are cleaned from weakly bonded impurities, corrosion products, and so on. The shape of the brushes is similar to grinding and polishing wheels. They are made primarily of steel wires, synthetic and natural bristles, cotton yarn, vegetal fibers, natural hair, and so on. The individual fibers of the brush penetrate the surface microprofile, and thus a smooth and clean surface is produced. Brushing can reduce the stress arising from the sharp edges. This fact leads to a reduction of crack formation on the surface of components during the operation, together with an increase in the strength of the material. The advantage of brushing is its simple principles, together with high productivity. The appearance of the surface after brushing is matt. Brushing can be carried out for the following purposes:

1. Cleaning metal surfaces by removing corrosive products, scale, and other contaminants
2. Deburring, smoothing sharp edges and corners on machined mechanical components
3. Smoothing scratches and traces after machining on the surface of highly stressed components
4. Polishing and treatment of metal and coated parts

There are a number of different types of brushes. They differ in shape and material of wire bristles (Figure 2.55). For a particular application, the appropriate type of brush is chosen according to the shape of the object, accessibility to cleaned areas, and initial and desired surface states.

The brush material used most is steel wire, with a diameter from 0.1 to 1.2 mm. For brushing of copper alloys, bronze or brass wire bristles are used. Thinner fibers (0.3–0.4 mm) are used for fine surface treatment; thicker fibers (1.2 mm) are more suitable for coarse contaminants and descaling.

In round wire brushes, rigid steel wire is used. During the rotation of the wire brush, centrifugal forces O_1, O_2, O_3 and cutting resistances R_1, R_2, R_3 act on the wire bristles (Figure 2.56).

The greater the weight of the wire (depending on material, length, and diameter), the greater the centrifugal force. The brushing wheel diameter and speed are important. It follows that at a given length and diameter of the wire, centrifugal force increases with speed; therefore, a brush will appear to be harder. The brush wire in contact with the surface breaks up dirt and leaves grooves and scratches on the cleaned surface. The depth and width of the grooves depend on the diameter of the wire (Figure 2.57). The shorter the diameter of the brush wire, the lower the resulting roughness of the cleaned surface.

FIGURE 2.55 Round wire brushes.

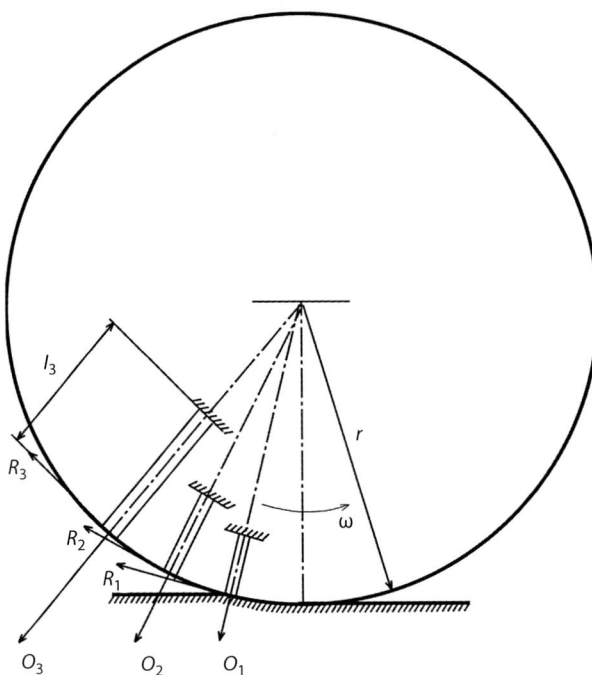

FIGURE 2.56 Forces acting during brushing. ω: angular speed of brush; r: radius of brush; R: cutting resistance of substrate against brushing; O: centrifugal force. (From Brezinová, J. et al., *Abrasive Blast Cleaning and its Application*, Trans Tech Publications, Pfäffikon, Switzerland, 2015.)

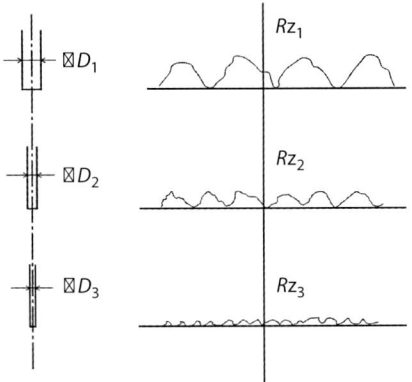

FIGURE 2.57 Influence of wire diameter ϕD on resulting surface roughness R_z. (From Brezinová, J. et al., *Abrasive Blast Cleaning and its Application*, Trans Tech Publications, Pfäffikon, Switzerland, 2015.)

2.3 SUMMARY

The chapter deals with three technologies of metallic surface mechanical treatment: abrasive blasting, grinding, and brushing. These technologies implement the principle of mechanical disruption of contaminants (corrosion products, scale, dust, old coatings) bonded to the substrate surface.

In an abrasive blasting technology, the selection of a suitable abrasive for the specific purpose of the treatment is very important. It is possible to choose metal abrasives with a higher cost as well as a higher durability, or nonmetallic abrasives based on minerals, slags, plastics, and so on, with a lower price and lower durability also. The different predispositions of particular abrasives to fragmentation and embedding into the substrate should also be considered, as embedded dust particles are able to reduce the strength of adhesively bonded joints. The material of the abrasive, as well as its grain size, controls the kinetic energy of the blasting grain, which affects the resulting cleanliness and roughness of the substrate. The shape of the abrasive used affects the mechanism of scale layer removal: Round grains (shot) disrupt the integrity of the scale not only at the point of impact, but also in neighboring areas; sharp angular grains (grit, cut wire) break the scale layer, while the scale near the edge of notches formed peels off partially, but can also be pushed into the notches. To achieve marked surface relief, it is necessary to use a monodisperse abrasive and narrow grain size distribution. The final cleanliness and roughness of the surface blasted also depends on the blasting parameters: grain speed/air pressure, blasting distance, and blasting angle [39]. To achieve appropriate surface preparation prior to bonding, it is necessary to ensure complete surface coverage by determining the necessary amount of abrasive applied per unit area or optimal blasting time. A prepared surface can be characterized by 2-D or 3-D roughness parameters, a surface volume that presents a "dead" volume that has to be filled by the adhesive to a level of the highest peaks of the pretreated surface, and achievable levels of surface cleanliness.

Surface cleaning by grinding uses grinding wheels or abrasive belts with randomly arranged or structured abrasive for contamination removal. It can be compared to microcutting processes. The ground surface is characterized by parallel grooves that should be oriented perpendicularly to the expected direction of bonded joint loading.

Surface cleaning by brushing uses the kinetic energy of a rotating wire brush for removal of contaminants. Surface contaminants are removed from the substrate by brushing as a result of centrifugal force acting on the wire of the brush and the cutting resistance of the material acting toward the wire movement. The resulting surface is oriented; the final surface roughness depends on the wire speed and the diameter of the wire on the brush. In grinding, as well as in brushing, the brush should be oriented so that the resulting grooves are in a direction perpendicular to the expected direction of loading the bonded joint.

REFERENCES

1. J. Brezinová, A. Guzanová and D. Draganovská. *Abrasive Blast Cleaning and its Application*, 1st edn, Trans Tech Publications, Pfäffikon, Switzerland (2015).
2. A. Momber. *Blast Cleaning Technology*, 1st edn, Springer-Verlag, Heidelberg (2008).
3. D. Jankura, J. Brezinová, J. Ševčíková, D. Draganovská and A. Guzanová. *Materials in Mechanical Engineering* and *Technology of their Finalization*, 1st edn, Technical University of Košice, Košice, Slovakia (2011).
4. H. Tobben. Back to Basics: Choosing the Right Media. https://www.theshotpeenermagazine.com/wp-content/uploads/tsp23no2.pdf. *The Shot Peener* 23, 18–20 (2009).
5. V. Schulze. *Modern Mechanical Surface Treatment, States, Stability, Effects*, 1st edn, Wiley-VCH, Weinheim, Germany (2006).
6. A. V. Pocius. *Adhesion and Adhesives Technology*, 3rd edn, Carl Hanser Verlag, Munich (2012).
7. S. Ebnesajjad. *Surface Treatment of Materials for Adhesion Bonding*, William Andrew Publishing, Norwich, NY (2006).
8. L. F. M da Silva, A. Öchsner and R. Adams (Eds.). *Handbook of Adhesion Technology: Volume 1*, Springer-Verlag, Heidelberg (2011).
9. R. F. Wegman and J. Van Twisk. *Surface Preparation Techniques for Adhesive Bonding*, 2nd edn, Elsevier, Oxford (2013).
10. C. V. Cagle, H. Lee and K. Neville. *Handbook of Adhesive Bonding*, McGraw Hill, New York (1973).
11. H. M. Clearfield, D. K. McNamara and G. D. Davis. Adherend surface preparation for structural adhesive bonding. In: *Adhesive Bonding*, L. H. Lee (Ed.), pp. 203–237, Springer US, New York (1991).
12. K. Tosha and K. Iida. Affected layer produced by grit blasting for austenitic stainless steel. *Int. J. Japan Soc. Precis. Eng.* 29, 46–47 (1995).
13. H. Y. Miao, D. Demers, S. Larose and M. Levesque. Experimental study of shot peening and stress peen forming. *J. Mater. Process. Technol.* 210, 2089–2102 (2010).

14. F. Otsubo, K. Kishitake, T. Akiyama and T. Terasaki. Characterization of blasted austenitic stainless steel and its corrosion resistance. *J. Thermal Spray Technol.* 12, 555–559 (2003).

15. A. Kalendová, D. Veselý, M. Kohl and J. Stejskal. Anticorrosion efficiency of zinc-filled epoxy coatings containing conducting polymers and pigments. *Prog. Org. Coat.* 78, 1–20 (2015).

16. B. Nasilowska, Z. Bogdanowicz and M. Wojucki. Shot peening effect on 904 L welds corrosion resistance. *J. Constr. Steel Res.* 115, 276–282 (2015).

17. K. Kambham, S. Sangameswaran, S. Potana and B. Kura. Productivity and consumption variation in dry abrasive blasting with coal slag. *J. Ship Prod.* 22, 1–8 (2006).

18. D. Draganovská, G. Ižaríková, A. Guzanová and J. Brezinová. The study of parameters of surface roughness by the correlation analysis. *Mater. Sci. Forum* 818, 15–18 (2015).

19. S. Kalpakjian and S. R. Schmid. *Manufacturing Engineering and Technology,* 7th edn, Pearson Education South Asia, Singapore (2014).

20. A. Guzanová, J. Brezinová, D. Draganovská and F. Jaš. A study of the effect of surface pre-treatment on the adhesion of coatings. *J. Adhes. Sci. Technol.* 28, 1–18 (2014).

21. V. Bačová and D. Draganovská. Analyses of the quality of blasted surfaces. *Mater. Sci.* 40, 125–131 (2004).

22. D. Kniewald, A. Guzanová and J. Brezinová. Utilization of fractal analysis in strength prediction of adhesively-bonded joints. *J. Adhes. Sci. Technol.* 22, 1–13 (2008).

23. S. Amada and H. Yamada. Introduction of fractal dimension to adhesive strength evaluation of plasma-sprayed coatings. *Surf. Coat. Technol.* 78, 50–55 (1996).

24. M. F. Bahbou, P. Nylén and J. Wigren. Effect of grit blasting and spraying angle on the adhesion strength of plasma-sprayed coatings. *J. Thermal Spray Technol.* 13, 508–514 (2004).

25. K. A. Soady, B. G. Mellor, J. Shackleton, A. Morris and P. A. S. Reed. The effect of shot peening on notched low cycle fatigue. *J. Mater. Sci. Eng.* 528, 8579–8588 (2011).

26. A. Moridi, M. H. Gangaraj, S. Vezzu and M. Guagliano. Fatigue behavior of cold spray coatings: The effect of conventional and severe shot peening as pre-/post-treatment. *Surf. Coat. Technol.* 283, 247–254 *(2015).*

27. K. Miková, S. Bagherifard, M. Guagliano and L. Trško. Fatigue behavior of X70 microalloyed steel after severe shot peening. *Int. J. Fatigue* 55, 33–42 (2013).

28. L. Trško, O. Bokuvka, F. Nový and M. Guagliano. Effect of severe shot peening on ultra-high-cycle fatigue of a low-alloy steel. *Mater. Des.* 57, 103–113 (2014).

29. R. Browning. An alternative solution to shot peening challenges. *The Shot Peener* 27, 6–10 (2013).

30. C. Diviani, C. Nouguier–Lehon, D. Hertz, H. Zahouani, M. Zarwel and T. Hoc. Surface impact analysis in shot peening process. *Wear* 302, 1058–1063 (2013).

31. S. Bagherifard, R. Ghelichi and M. Guagliano. Numerical and experimental analysis of surface roughness generated by shot peening. *Appl. Surf. Sci.* 258, 6831–6840 (2012).

32. S. Lou, X. Jiang and P. J. Scott. Correlating motif analysis and morphological filters for surface texture analysis. *Measurement* 46, 993–1001 (2013).

33. S. Bagherifard, R. Ghelichu and M. Guliano. A numerical model of severe shot peening (SSP) to predict the generation of a nanostructured surface layer of material. *Surf. Coat. Technol.* 204, 4081–4090 (2010).

34. A. B. Edward, P. S. Heyns and F. Pietra. Shot peening modeling and simulation for RCS assessment. *Procedia Manuf.* 7, 172–177 (2017).

35. A. Mannelqist and M. R. Groth. Comparison of fractal analyses methods and fractal dimension for pre-treated stainless steel surfaces and the correlation to adhesive joint strength. *Appl. Phys. A* 73, 347–355 (2001).

36. M. Taro, T. Chaise and D. Nelias. A methodology to predict the roughness of shot peened surfaces. *J. Mater. Process. Technol.* 217, 65–76 (2015).

37. X. J. Jiang and D. Whitehouse. Technological shifts in surface metrology. *CIRP Anna. Manuf. Technol.* 61, 815–836 (2012).

38. 3M™ Trizact™ Abrasives. http://www.3m.com/3M/en_US/trizact-us/ (2016).

39. T. Kim, L. Hyungyil, H. C. Hyun and S. Jung. Effects of Rayleigh damping, friction and rate-dependency on 3D residual stress simulation of angled shot peening. *Mater. Des.* 46, 26–37 (2013).

40. Toyo Seiko Co., Ltd. Coverage checker. http://www.toyoseiko.co.jp/english/product/product08.html (2017).

3 Plasma Surface Treatment to Enhance Adhesive Bonding

Nitu Bhatnagar

CONTENTS

3.1 INTRODUCTION

Major industrial sectors such as automotive, aerospace, and others are increasingly employing polymer composites in their load-bearing structural parts, which offers benefits such as being lightweight, good thermomechanical bulk properties, low cost, and recycling [1–4]. Apart from this, the rationale behind the choice of polymeric composites lies in their superior physical, mechanical, and chemical performances, along with higher resistance to corrosion and structural weakening from radiation [5]. The use of adhesively bonded joints in these load-bearing structures is of great interest to the aerospace and automotive industries as compared with mechanical methods (welding, riveting), which generally lead to stress and failure problems.

Structural and specialty adhesives account for about 30% of the total adhesive and sealant market, with uses in many industries [6]. This is because of time and cost savings, high corrosion and fatigue resistance, crack retardance, and good damping characteristics. Adhesive bonds are more effective in assembling composite structures than other mechanical joining methods, as they provide more uniform stress distribution, lower stress concentration, and better fatigue life and corrosion resistance [7]. Any adhesive joint can distribute the applied load over the entire bonded area and is suitable for joining dissimilar materials with low manufacturing cost [8].

It has been established that for successful application of polymeric composite materials to form structural parts using adhesive bonding, they need to have special surface properties such as polarity and high surface free energy (SFE) [9]. A basic requirement for successful bonding is the spreading of the adhesive on the surface of the adherend or substrate [10]. This will occur if the SFE of the adhesive is lower than that of the adherend, but the typical values of SFE are 30–50 mJ/m^2 for adhesives and about 12 mJ/m^2 to about 70 mJ/m^2 for polymers [11]. This shows that although these polymers have found wide suitability for high-end applications, they still have some limitations in achieving a good structural adhesive bond. Several conventional methods such as chemical treatment, thermal treatment, and mechanical treatment are used to modify polymeric surfaces, but they suffer from problems of uniformity, reproducibility, and cost-effectiveness. The need for adequate adhesion by activating the polymeric surface without affecting the bulk properties of the polymer has resulted in the development of plasma surface modification techniques [12], which offer a uniform, reproducible, economic, and environmentally friendly alternative [13]. It has been observed that the polar component of the SFE leading to increased total SFE of the polymer increases significantly due to generation of new functional groups as a result of surface oxidation by reactive species in the plasma [14]. The changes in the physicochemical characteristics and surface morphology of the polymer brought about by plasma result in strong adhesion [15].

3.2 ADHESIVES

The automotive, aerospace, building, biomedical, and other industries have been investigating the use of various types of adhesives and sealants for numerous applications in these industries.

An adhesive or glue is a mixture in a liquid or semiliquid state that bonds items together. Adhesives cure (harden) by either evaporating the solvent or by chemical reactions that occur between two or more constituents [16,17]. The adhesives are classified into different types. A detailed description of different types of adhesives is given in Table 3.1.

The choice of proper adhesive is critical to produce strong and durable bonds [24]. Unlike thermoplastics, thermosetting adhesives do not melt or flow on heating, but become rubbery and lose strength. The molecular chains present in thermosetting adhesives undergo irreversible cross-linking during curing [25]. The most familiar thermosetting adhesives are the family of epoxies. The limitations of other adhesives for the joining of different materials have been solved by epoxy adhesives, and presently these adhesives are widely used.

It is evident from the literature that thermoplastic resins can withstand temperatures comparable to many aerospace graphite–epoxy prepregs (177°C). A common example of an adhesive system found in the automotive industry is a paint coating applied to a polypropylene (PP) bumper bar. However, it requires some pretreatments. Adhesives that have been used successfully in aerospace are, typically, modified epoxy film adhesives with brand names such as FM300 (Cytec Industries Inc.) for both poly(phenylene sulfide) (PPS) [26] and poly(ether ether ketone) (PEEK) [27,28], FM 377 [29], and FM 87 [30]. The building industries require such sealing materials in the form of sealants that have the ability to withstand thermal expansion and contraction while still bonded to the substrate [4,31,32]. Kodokian and Kinloch [33] described the use of a wide range of epoxy- and acrylic-based adhesives for bonding applications: Hysol 9309.3, FM73M, Permabond F241, F245, F246, and V501, and Bostik M890 and M896. An extensive review of adhesives for PEEK–graphite composite bonding is also given in [34], and it was found that FM300 recorded a close second-highest lap shear strength value of 23 MPa to Scotch-Weld structural adhesive film (AF-163-2K) (24.5 MPa) when tested at 25°C. When it comes to the biomedical industry, a study of the adhesion mechanism between human hepatoma cell lines and polymers such as polystyrene, poly(methyl methacrylate), and polycarbonate (PC) is of much importance [35–38]. Researchers [38,39] have made an important observation that the major factor responsible for cell adhesion to polymer substrates is the SFE of the polymer, irrespective of whether the surface has been covered by a protein layer.

TABLE 3.1
Classification of Adhesives

Type	Description	Examples
	General Classification	
Thermosetting	Cross-linked by strong covalent intermolecular bonds, forming one giant molecule	Epoxies, polyesters, polyimides, phenolics
Thermoplastic	Molecules of most of thermoplastics combine into long polymer chains alternating with monomer units	Polyamides, cyanoacrylates, polyacrylates, poly(vinyl acetate) (PVAc)
Elastomeric	These consist of long lightly cross-linked molecules	Natural rubbers, silicones, acrylonitrile butadiene (nitrile), neoprene, polyurethane, styrene–butadiene
	On the Basis of Function	
Structural adhesives	Used to bond structural materials and have the ability to withstand significant loads or stresses	Polyurethanes, polysulfides, polyesters, acrylics, special formulations of cyanoacrylates, anaerobics, epoxy adhesives
Nonstructural adhesives	Also called *holding adhesives*, these are not required to support substantial loads but merely to hold lightweight materials in place [18]	Pressure-sensitive tapes, packaging adhesives
	On the Basis of Curing Method	
One-Part Adhesives		
Heat-activated curing adhesives	Adhesives of this type consist of a ready mixture of two components and harden from melt for thermoplastics	Epoxies, urethanes, polyimides
Light/UV-activated curing adhesives	Light-activated adhesives are cured under a visible or UV light of appropriate wavelength. Adhesives of this type usually contain photoinitiators to enhance curing reaction	Acrylics, cyanoacrylates, urethanes
Moisture-activated curing adhesives	Cured on reacting with moisture present on the substrate surface or in the air	RTV silicones, cyanoacrylates, urethanes
Pressure-sensitive adhesives (PSAs)	Adhesives of this type do not cure. These materials bond to adherend surfaces at room temperature as low pressure is applied [18]	Self-stick tapes
Two-Part Adhesives		
	A two-part adhesive is cured when its two (or more) components are mixed. The components react chemically, forming cross-links of the polymer molecules	Epoxies, urethanes, acrylics, silicones

Description of Common Adhesives

Name	Description	Advantages	Disadvantages
Polyurethanes	Formed by cross-linking highly reactive isocyanates with various polyols and are frequently used for joining	Possess low viscosity, good wettability, and penetrating ability—which help the adhesives to spread over the substrate surface—good cohesive strength, and high impact resistance; good flexibility at low temperatures	Pose processing problems that arise due to their reaction with water and their gaseous nature. Limited temperature resistance, sensitive to moisture in cured and uncured states [19]

(Continued)

TABLE 3.1 (CONTINUED)
Classification of Adhesives

Polysulfides	Obtained by the reaction of sodium polysulfide with organic dichlorides such as dichlorodiethyl formal and ethylene dichloride	Adhere well to almost all adherends	Require higher humidity for faster cure, show poor elastic recovery; also result in poor creep resistance [20]
Polyesters	Reaction products of dibasic acids with polyfunctional hydroxyl materials	Often used as hot-melt adhesives, have good resistance to water, nuclear radiation, and biodeterioration [20]	Generally have poor resistance to chemicals. Other undesirable properties include high flammability and tendency to brittleness and crazing [18]
Acrylics	Synthesized from a wide selection of acrylic and methacrylic ester monomers	Capable of adhering to most metals and polymers [21]. This is due to their low viscosity, good wettability, and penetrating ability that help the liquid to spread over the substrate surface	Possess poor cohesive and shear strengths, low elevated temperature strength, high flammability, and bad odor of uncured adhesive [19]
Cyanoacrylates	Known as *instant adhesives* or *superglues*, cyanoacrylates are also from the acrylic family tree, but having a completely different cure system	Very rapid curing adhesives; possess excellent lap shear strength and good shelf life [19]	Bond to skin instantly, posing safety risks. Unpleasant fumes, poor peel strength, poor resistance to impact loads, limited gap cure, poor durability on glass, low-temperature resistance (lose strength at about 85°C), and poor solvent resistance [22]
Anaerobics	Derived from methacrylates (Plexiglas), the basic ingredient in anaerobic adhesives is a special liquid of small molecules that can combine chemically to form a polymer or a group of molecules	Very versatile and are used in a wide variety of applications	Brittle, limited gap filling, cure rate is surface dependent, and very expensive [23]

3.3   ADHESION

Adhesion is a multidisciplinary topic that involves surface chemistry, physics, rheology, polymer chemistry, stress analysis, polymer physics, and fracture analysis [40]. The ASTM [41] defines adhesion as "the state in which two surfaces are held together by interfacial forces which consist of valence forces or interlocking actions or both." When the adhesion between two surfaces is caused by valence forces, this is called specific adhesion, while when they are held together by interlocking action, this is called mechanical adhesion. Both of these play important roles in understanding the effect of surface preparation on adhesion. These various types of intrinsic forces that may operate across the adhesive–substrate interface are referred to as the mechanisms of adhesion [42].

3.3.1 MECHANISM OF LIQUID–SOLID ADHESION

During liquid–solid adhesion, there are usually three interfaces involved: the solid–gas interface, the liquid–gas interface, and the solid–liquid interface. There are two conditions that are required to be met when a liquid spreads over a solid surface. Firstly, the SFE of the solid–gas interface must be greater than the combined interfacial energies of the liquid–gas and the solid–liquid interfaces. Secondly, the interfacial free energy of the solid–gas interface must exceed the interfacial free energy of the solid–liquid interface. The adhesion mechanisms between liquid adhesive and solid substrate have been explained on the basis of mechanical interlocking, adsorption, diffusion, and electrostatic theories. These mechanisms have been known to be dependent on the surface characteristics of the materials being used for the automotive and aerospace industries. These theories can be used, solely or in combination with each other, to describe almost any kind of adhesion phenomenon [43].

The mechanical interlocking theory of adhesion essentially proposes the interlocking of the adhesive into the irregularities of the substrate surface as the major source of intrinsic adhesion [42]. According to this theory, good adhesion occurs only when a liquid penetrates into the pores, holes, crevices, and other irregularities of the surface of the substrate, and locks mechanically into the texture of the substrate [44]. Pretreatment methods applied on a surface enhance adhesion [45], as they result in microroughness on the adherend surface, which can improve adhesion strength by providing mechanical interlocking. Beyond mechanical interlocking, the enhancement of adhesion strength is due to the roughening of the adherend substrate. Further, surface roughness increases surface area, improves the kinetics of wetting, and increases plastic deformation for better adhesion [46,47]. The size and shape of the special features created on the surface have an influence on adhesion, providing a tortuous path that prevents separation of the adhesive from the adherend [48]. However, this theory cannot explain the attainment of good adhesion strength between smooth surfaces [42].

The diffusion theory is based on the assumption that the adhesion strength of polymers to themselves (*autohesion*) or to one another is due to mutual diffusion of macromolecules across the interface, thus creating an interphase [49]. This implies that the molecular chains or chain segments are sufficiently mobile and mutually soluble [50]. To describe the self-diffusion phenomenon of polymers, several theories have been proposed: entanglement coupling [51], cooperativity [52], and reptation [53]. The reptation model has been applied to study tack, green strength, healing, and welding of polymers [54]. Some studies have shown that the interdiffusion phenomenon exists in mobile and compatible polymers and increases intrinsic adhesion. The diffusion theory, however, has found limited application where the polymer and adherend are not soluble or the chain movement of the polymer is constrained by its highly cross-linked, crystalline structure, or when it is below its glass transition temperature [55].

The electrostatic theory of adhesion, as primarily proposed by Deryaguin and coworkers, states that an electron transfer mechanism between the substrate and adhesive, having different electronic band structures, can occur to equalize the Fermi levels [56]. The result is the creation of an electrical double layer (EDL) at the interface [57]. These electrostatic forces at the interface are responsible for resistance to separation of the adhesive and the substrate. Thus, the adhesion depends on the magnitude of the potential barrier at the substrate–adhesive interface [56]. However, this kind of potential barrier does not exist in some cases [58]. This theory could not be widely accepted, because the EDL could not be identified without separating the adhesive bond. Also, as argued by many researchers [59,60], the effect of the EDL on the adhesive bond strength was exaggerated.

The adsorption theory states that adhesion results from intimate intermolecular contact between two materials, and involves surface forces that develop between the atoms in the two surfaces [61]. These forces may be due to physical adsorption—mainly van der Waals forces or secondary forces. These adhesion forces may also be due to acid–base interaction [62,63] or hydrogen bonds [64,65].

To obtain good adsorption, intimate contact must be reached, so that van der Waals interaction or acid–base interaction or both can occur; hence, good wetting is essential. Thus, criteria for good adhesion essentially become the same criteria as for good wetting, although this is a necessary but not sufficient condition [55].

3.3.2 Wetting

The wetting ability of a liquid is a function of the SFEs of the solid–gas interface, the liquid–gas interface, and the solid–liquid interface. The adhesion forces between the liquid and the second material will compete against the cohesive forces of the liquid. Liquids with weak cohesive bonds and a strong attraction to another material (or the desire to create adhesive bonds) will tend to spread over the material. Interfacial interactions are an important factor in the prediction of adhesion [66]. Liquids with strong cohesive bonds and weaker adhesion forces will tend to bead up or form a droplet when in contact with another material [67]. One way to quantify a liquid's surface wetting characteristic is to measure the contact angle of a drop of liquid placed on the surface of a substrate.

According to Young's equation, the interfacial tensions (liquid–vapor: γ_{LV}, solid–liquid: γ_{SL}, and solid–vapor: γ_{SV}) at the three-phase contact are related to the equilibrium contact angle θ through:

$$\gamma_{SV} = \gamma_{SL} + \gamma_{LV}\cos\theta \tag{3.1}$$

The contact angle (θ), as seen in Figure 3.1, is the angle at which the liquid–vapor interface meets the solid–liquid interface. The contact angle is determined by the resultant of the adhesion and cohesive forces.

The SFE and the dispersion (γ_S^D) and polar (γ_S^P) components for any polymer are calculated using the equation shown:

$$\left(1+\cos\theta\right)\gamma_{LV} = 2\left(\gamma_S^D\gamma_{LV}^D\right)^{\frac{1}{2}} + 2\left(\gamma_S^P\gamma_{LV}^P\right)^{\frac{1}{2}} \tag{3.2}$$

Firstly, the contact angle of deionized water θ is measured on the surface. Therefore, in this equation, θ is the measured contact angle of deionized water where the surface tension of deionized water γ_{LV} and its two components, the dispersion γ_{LV}^D and the polar γ_{LV}^P, are known and the two unknowns are γ_S^D and γ_S^P for the surface. Secondly, the contact angle θ of formamide is measured on the surface and, consequently, θ is the measured contact angle of formamide where surface tension of formamide γ_{LV} and its two components, the dispersion γ_{LV}^D and the polar γ_{LV}^P are known and the two unknowns are γ_S^D and γ_S^P for the surface. Thus, by solving these two equations, the unknowns γ_S^D and γ_S^P for the surface are calculated. Finally, the total SFE γ_S is determined by Equation 3.3:

$$\gamma_S = \gamma_S^D + \gamma_S^P \tag{3.3}$$

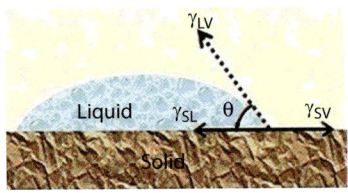

FIGURE 3.1 Contact angle on a solid surface. (From http://home.iitk.ac.in/~kbalani/vl-kb/wetting.html. With permission)

A deeper understanding of the adhesion process is achieved by calculating the thermodynamic solid–liquid work of adhesion (W_{SL}) required to separate a unit area of the two phases in contact [69]. The combination of the Dupré expression [70], which assumes that there is negligible liquid surface area change on adhesion to a solid surface [71], and Young's equation [72] leads to an equation that allows prediction of the bonding characteristics of a surface through two experimentally measurable parameters, the liquid surface tension and its contact angle on the solid surface [73]:

$$W_{SL} = \gamma_{SV} + \gamma_{LV} - \gamma_{SL} = \gamma_{LV}\left(1 + \cos\theta\right) \tag{3.4}$$

where:
W_{SL} is the solid–liquid work of adhesion
γ_{LV} is the surface tension or free energy of the liquid
θ is the liquid contact angle on the solid surface

One important factor that influences adhesion strength is the ability of the liquid to spread uniformly on the substrate [74].

For spontaneous wetting to occur:

$$\gamma_{SV} \geq \gamma_{SL} + \gamma_{LV} \tag{3.5}$$

However, this is an ideal case and is not possible with all polymers.

The tendency of a drop to spread over a flat solid surface increases as the contact angle decreases. Thus, the contact angle provides an inverse measure of wettability [75], as shown in Table 3.2.

3.3.3 POLYMER–POLYMER ADHESIVE JOINT

The science of a polymeric adhesive joint is concerned with two steps: the formation of the adhesive bond and the physical strength of the adhesive bond. As the forces of attraction between the polymeric chains are mainly van der Waals forces, the SFE of a polymer is low. A basic requirement for successful bonding is the spreading of the adhesive on the surface of the adherend or substrate. This will occur if the SFE of the adhesive is lower than that of the adherend, but the typical values of SFE are 30–50 mJ/m^2 for adhesives and about 12 mJ/m^2 to about 70 mJ/m^2 for polymers [11]. Since the SFE of the adhesive can hardly be altered, attempts are made to increase the SFE of the adherend by employing different surface modification techniques. When the polymers are modified by different treatments, polar groups are formed on the polymeric surface. The presence of polar groups on the polymer surface bridges the forces of physical adsorption and chemical adsorption by the formation of covalent bonds with the adhesive, where the covalent bond is the primary force responsible for adhesion. This covalent bond plays a significant role in producing adhesive bonds that are even stronger than the cohesive strength of polymeric materials. Thus, surface treatment is critical in any adhesive bonding operation.

3.4 SURFACE MODIFICATION OF POLYMERS

Adhesive bonds are dependent on the chemistry of the interface, and thus control of surface chemistry is critical to bond quality [42,76]. Thus, the most important step in adhesive bonding is surface modification, which not only prevents or removes contaminants that can adversely affect bonding, but also helps in creating chemically active sites on the surface to maximize bond strength. High-performance polymers very often do not possess the desired surface properties for strong adhesive bonding. They are hydrophobic in nature, and in general exhibit insufficient adhesive bond strength due to relatively low SFE [77].

TABLE 3.2
Relationship of Contact Angle with Wettability

| Contact Angle | Degree of Wetting | Strength of | |
		Solid–Liquid Interactions	Liquid–Liquid Interactions
$\theta = 0$	Perfect wetting	Strong	Weak
$0 < \theta < 90°$	High wettability	Strong	Strong
		Weak	Weak
$90° \leq \theta < 180$	Low wettability	Weak	Strong
$\theta = 180°$	Perfect nonwetting	Weak	Strong

Several conventional methods such as chemical treatment, thermal treatment, and mechanical treatment have been used to modify polymeric surfaces. The different treatment methods such as (1) chemical treatment, (2) thermal treatment, (3) mechanical treatment, and (4) electrical treatment, along with their merits and demerits, are discussed in the following subsections.

3.4.1 Chemical Treatment

The influence of chemical treatment on polymers includes surface alteration as well as other morphological changes in the surface layer, such as surface roughness [78,79]. The wetting characteristics of a polymeric surface are enhanced by the introduction of polar groups on the surface when it is treated with a heavily oxidative liquid chemical such as chromic anhydride with tetrachloroethane, chromic acid with acetic acid, or potassium dichromate with sulfuric acid. The polymer surface is then rinsed clean of the etching chemicals and dried. Oxidation may also be carried out with nitrogen oxide, cycloalkyl chromate, potassium permanganate, and sodium hypochlorite. Chlorosulfonation and flame treatment in the presence of a halogen could also be used. Bajpai et al. [80] have reported surface modification of the poly(methyl methacrylate)–poly(vinylidene fluoride) (PMMA–PVDF) blend system under the chemical environment of some organic liquids, namely, benzene, toluene, xylene, and acetone. Comyn et al. [81] reported that although the sulfuric acid–orthophosphoric acid–water treatments were ineffective in treating the surface of PEEK, the addition of 1% potassium permanganate, a strong oxidizing agent, resulted in strong bonds with failure predominantly away from the interfacial zone.

Carbon fiber (CF)-filled PEEK was chemically etched by Davies et al. [82] using a composition of $7\,g\ K_2Cr_2O_7 + 12\,g\ H_2O + 150\,g\ H_2SO_4$. This resulted in an increase of adhesive bond strength. Chromic acid etching increases the bondability of a plastic by introducing reactive sites, such as hydroxyl, carbonyl, carboxylic, and $-SO_3H$ groups, to the plastic surface and forming root-like cavities that provide sites for mechanical interlocking. The effect of this treatment varies from substrate to substrate. For example, increasing the etch time and temperature increases only the etch depth when etching PP. On the other hand, both the degree of oxidation and etch depth increase with treatment time for polyethylene (PE). All these reactions take place in vessels under very strict time–temperature–pressure conditions that are difficult to handle and are not feasible for use on an industrial scale. Moreover, they require expensive waste disposal and may also pollute the environment.

3.4.2 Thermal Treatment

Thermal treatment plays an important role in modifying the nature and concentration of surface functional groups [83]. Thermal treatments have been used to produce activated carbons with basic character, and such carbons were effective in the treatment of some organic hydrocarbons [84]. Heat treatment of carbon in inert atmospheres (hydrogen, nitrogen, or argon) could increase carbon hydrophobicity by removing hydrophilic surface functionalities, particularly various acidic groups [85–88].

Thermal treatment also oxidizes the surface of polymers by introducing polar groups [89] such as carbonyl, carboxyl, or amide when the plastics are exposed to a blast of hot air. It also utilizes a free radical mechanism accompanied by chain scission and some cross-linking, which helps in improving the wettability of the surface. Ansari and Wallace [90] reported that the redox properties of polypyrroles were improved with mild heating but were degraded at higher temperature. Lafdi et al. [91] used heat treatment to alter the physical properties of a nanofiber-reinforced epoxy composite. It was reported that heat treatment up to 1800°C resulted in improved flexural modulus and strength of the nanocomposite, while heat treatment to higher temperatures demonstrated a slight decrease in mechanical properties, likely due to the elimination of potential bonding sites caused by the elimination of the truncated edges of the graphene layers.

3.4.3 Mechanical Treatment

The mechanical abrasion process has been used for surface pretreatment of polymers, because it provides an effective and reliable way of ensuring improvement in adhesion of polymer surfaces [92]. It serves to increase the surface area of the material by roughening the exposed areas prior to adhesive bonding. Surface roughening increases bondability by dramatically increasing the number of mechanical interlocking sites. Mechanical abrasion is carried out by dry blasting, wet blasting, or hand/machine sanding. However, for low SFE plastics, the opposite effect can occur. Silverman and Griese [27] examined a joining method using mechanical fasteners in conjunction with FM300 adhesive for the AS4 CF–PEEK composite. Their findings revealed a comparatively low lap shear strength value of around 17 MPa, when a chromic acid-etch surface treatment was used prior to bonding and aluminum rivets were used to mechanically fasten the joint.

Bhowmik et al. [93] investigated the effect of mechanical polishing prior to surface modification of high-density polyethylene (HDPE) and PP sheets by exposure to dc glow discharge on the SFE and their adhesive joint strength to steel. Mechanical polishing of the HDPE and PP sheets by abrading with 800-grade emery paper prior to glow discharge treatment increased the adhesive joint strengths over those observed in the case of unpolished polymers exposed to glow discharge only. However, the use of prior mechanical polishing increases the joint strength only by a little more than 10%, compared with a five-to-seven-times increase in strength observed as a consequence of exposure to glow discharge of untreated samples. Mechanical processes are operator dependent and labor intensive, produce dust, and are usually only employed in situations with low production volumes. The abrasive materials that are used to remove particulates or residues are hazardous and require disposal [94].

3.4.4 Electrical Treatment

While these traditional procedures are well tested and reliable, they are also time-consuming manual processes. If not properly administered, these techniques can yield poor bond strengths, and can even result in damage to composites that are fabricated with high-modulus fibers [95,96]. They also suffer from problems of uniformity, reproducibility, and cost-effectiveness. The need for adequate adhesion by activating the polymeric surface without affecting the bulk properties of the polymer has resulted in the development of the plasma surface modification industry [12], which offers uniform, reproducible, economic, and environmentally friendly alternatives [13]. Electrical treatment under atmospheric-pressure plasma (*APP* or *normal plasma*) [97] and low-pressure plasma (LPP) (glow discharge) [98,99] are popular techniques that have been employed to modify the surfaces of polymers.

3.4.4.1 Electrical Treatment by Generation of Plasma

Electrical treatment by generation of ionized plasma is a suitable technique that is used to modify polymer surfaces without affecting their bulk properties. The term *plasma* was coined by Irving Langmuir. A distinct fourth state of matter, it is broadly defined as partially or wholly ionized gas with approximately equal numbers of positively and negatively charged particles. *Ionized* means that at least one electron is not bound to an atom or molecule, converting the atoms or molecules into positively charged ions. Plasma contains active species, such as electrons, ions, radicals, photons, and so on, that initiate chemical and physical modifications on the polymer surface [100,101] by making the surface electrically conductive and strongly responsive to electromagnetic fields. Ions and electrons present in plasma break polymer chains due their high kinetic energy [102]. At the same time, free radicals in plasma modify the chemical properties of polymers by introducing functional groups on the polymer surface [103,104]. The interactions of the polymer with the plasma also alter the molecular weight of the surface layers by scissoring, branching, and cross-linking. Thus, when the material whose surface is to be modified is placed in the plasma chamber, these energetic particles collide with the surface of the material and cause molecular disruptions. This leads to a drastic modification of the structure and properties of the surface [105], although this depends on the composition of the surface and the gas used. There are mainly two types of plasmas: (1) thermal plasma and (2) cold plasma.

Thermal plasma is used to destroy solid, liquid, and gaseous toxic halogenated and hazardous substances or to generate anticorrosion, thermal barrier, antiwear coatings, and so on. Cold plasmas are used for surface modifications of materials, ranging from simple topographical changes to the creation of surface chemistries and coatings that are radically different from the bulk material [106]. Cold plasmas are generated by glow discharges at reduced pressures of 0.01–10 torr, and pressure of around 1 torr is sufficient for surface modification of polymers [107]. LPP using glow discharge and APP using corona discharge are convenient methods for surface modification of polymers enhance their adhesion characteristics.

3.4.4.1.1 Low-Pressure Plasma (Glow Discharge)

A glow discharge is an ionized gas consisting of equal concentrations of positive and negative charges and a large number of neutral species. It is generated by applying a potential difference (of a few hundred volts to a few kilovolts) between two electrodes that are inserted in a cell filled with a gas (an inert gas or a reactive gas) at a pressure ranging from a few millitorr to 10 torr [108]. Due to the potential difference, electrons emitted from the cathode by the cosmic radiation are accelerated away from the cathode, and give rise to collisions with the gas atoms or molecules (excitation, ionization, dissociation). The excitation collisions give rise to excited species, which can decay to lower levels by the emission of light. The ionization collisions create ion–electron pairs. The ions are accelerated toward the cathode, where they release secondary electrons. These electrons are accelerated away from the cathode and can give rise to more ionization collisions. In its simplest form, the combination of secondary electron emission at the cathode and ionization in the gas gives rise to self-sustained plasma [109], as shown in Figure 3.2.

Due to the various collision processes occurring in the plasma, a large number of different plasma species are generated: electrons, atoms, molecules, several kinds of radicals, several kinds of (positive and negative) ions, excited species, and so on. These active species activate the surface of the polymer and cause the surface layer to be oxidized or cross-linked, enabling it to become significantly active for participation in adhesion. There are different variants of glow discharge plasma:

- Direct current (dc)
- Radio-frequency (RF)
- Microwave (MW)

In dc glow discharge, a continuous potential difference is applied between cathode and anode, giving rise to a constant current. However, this setup gives problems when one of the electrodes is

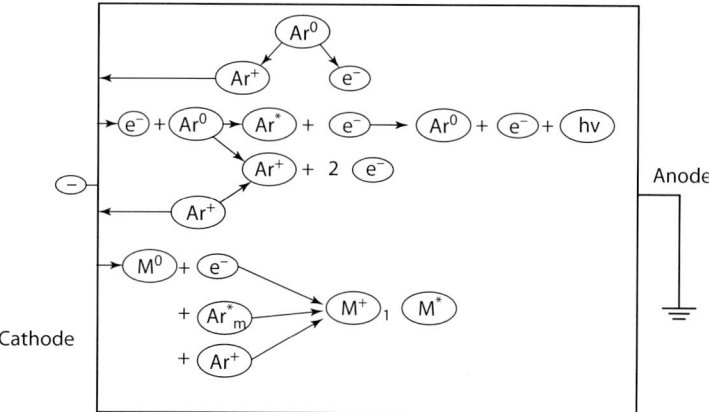

FIGURE 3.2 Schematic overview of the basic plasma processes in a glow discharge. When a potential difference is applied between two electrodes (target material M is the cathode and the substrate is attached to the anode), Ar gas will break down into electrons and positive ions. The latter can cause secondary electron emission at the cathode. The emitted electrons give rise to collisions in the plasma; for example, excitation (which is often followed by deexcitation with emission of radiation) and ionization (which creates new electrons and ions, and therefore makes the glow discharge a self-sustaining plasma). Besides the argon ions, fast argon atoms bombarding the cathode can also give rise to sputtering, which is important for several applications. (From Bogaerts, A. et al., *Spectrochimica Acta Part B*, 57, 109–119, 2002.)

nonconducting as, due to the constant current, the electrodes become charged up, which may lead to burnout of the glow discharge. This problem is overcome by applying an alternating voltage between the two electrodes, as in the capacitively coupled radio-frequency (CCRF) glow discharge. The charge accumulated during one half of the cycle is neutralized by the opposite charge accumulated during the next half-cycle in the RF glow discharge. RF plasmas can be sustained at lower gas pressure, and are characterized by higher ionization efficiency as compared with dc plasmas.

The majority of plasma processing has been carried out at low pressure in a vacuum chamber and is viewed as a necessary processing requirement. In principle and practice, however, APPs provide a critical advantage over widely used LPPs, as they do not require expensive and complicated vacuum systems. Without a vacuum system, the costs of materials processing are reduced substantially and materials issues related to vacuum compatibility are not of concern. Therefore, the use of APPs is beginning to greatly expand the current scope of materials processing.

3.4.4.1.2 Atmospheric-Pressure Plasma

APPs [97] are gaining greater acceptance than other plasma techniques, because they are easy to integrate into existing production lines and can treat specific parts of a substrate selectively [110,111]. Also, in contrast to most corona treatments (and dielectric barrier discharges) [112], APPs are not limited to flat and thin substrates, but can also be used for large three-dimensional objects. APP shows significant potential for improving interfacial adhesion. It minimizes downstream wastes from subsequent surface preparation. Since it can be automated, it reduces process variability while increasing reliability and processing rates [113].

APP is the name given to the special case of plasma in which the pressure approximately matches that of the surrounding atmosphere—the so-called *normal pressure*. It is generated by alternating current (ac) excitation (corona discharge and plasma jets). By means of high-voltage discharge (5–15 kV, 10–100 kHz) in the plasma jet, a pulsed electric arc is generated. A process gas, usually oil-free compressed air flowing past this discharge section, is excited and converted to the plasma state. This plasma then passes through a jet head to arrive on the surface of the material to be treated. The jet head is normally at earth potential and, in this way, the potential-carrying part of the plasma stream is largely held back [114]. In addition, it determines the geometry of the emergent beam. It is observed that during APP discharge, ion bombardment physically and chemically removes oxides and reducible compounds from surfaces and many other contaminants are vaporized. In addition, gas molecules are accelerated to an excited state, releasing active free radicals and ultraviolet (UV) energy [115]. Free radicals activate chemical reactions on surfaces, inducing intermolecular cross-linking. When compared with corona discharges, APPs produce significantly more homogeneous and uniform surface activation across material surfaces, and increase the microroughness of surfaces, with introduction of active species.

3.5 EFFECT OF PLASMA TREATMENT ON PHYSICOCHEMICAL PROPERTIES OF POLYMERS

Many researchers have documented well that plasma treatment of polymer surfaces not only increases the polarity of the polymer surfaces, but also results in surface roughening [116–118]. These surface properties, such as polarity and roughness, play an important role in successful application of polymers to form structural parts using adhesive bonding [116,119].

The presence of polar groups is readily established by determining SFE through contact angle measurements [120]. According to Williams et al. [121], adhesion is correlated with the fraction of the polymer surface sites that are oxidized and converted into active functional groups. This physicochemical change brought about on the polymer surface is determined by x-ray photoelectron spectroscopy (XPS) and infrared (IR) spectroscopy. The change in surface morphology on the polymer surface is measured by atomic force microscopy (AFM) analysis. Further, it is worthwhile to note that the physicochemical change on the surface of polymers plays an important role in contribution to adhesion strength.

3.5.1 CONTACT ANGLE MEASUREMENT AND SURFACE FREE ENERGY DETERMINATION

Pelagade et al. [122] studied the effect of low-energy (300 eV) argon plasma treatment on the surface of PC. They observed that the SFE increased from 38.5 mJ/m^2 to 74.9 mJ/m^2 on increasing the treatment time, as shown in Figure 3.3. The corresponding water contact angle decreased from 63° to 17°.

Pandiyaraj et al. [98] studied the effect of dc glow discharge air plasma on poly(ethylene terephthalate) (PET) films. It was observed that the contact angle on the untreated PET surface was

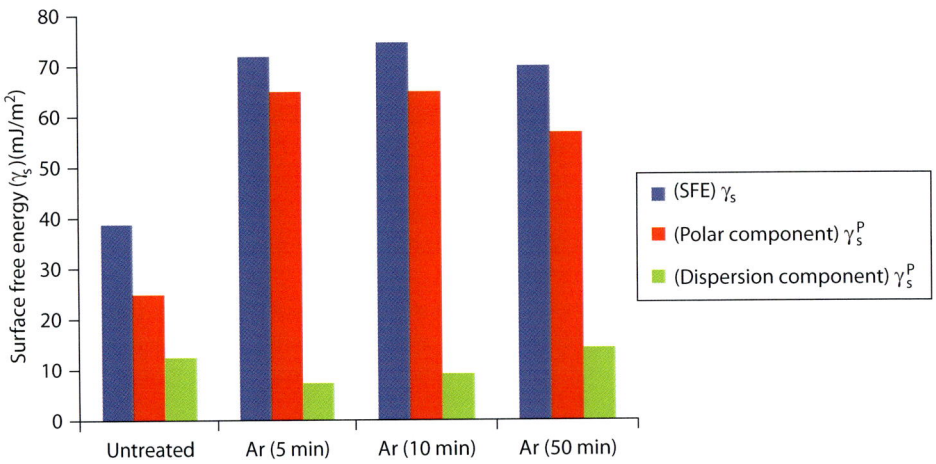

FIGURE 3.3 Comparison of surface free energy (SFE) and its polar and dispersion components before and after the Ar plasma treatment. (From Pelagade, S. et al., *J Phys.: Conf. Ser.*, 208, Paper ID: 012107, 1–8, 2010.)

86.2° for distilled water and decreased by 50% after plasma treatment. The SFE of plasma-modified PET films increases on increasing the exposure time, as shown in Figure 3.4a. It reaches a maximum value at 15 min of exposure time, and thereafter tends to saturate. This may be due to lack of any further change in the oxygen content incorporated onto the surface. Figure 3.4a clearly demonstrates that the total SFE of the PET film attained a higher value when the film was exposed to

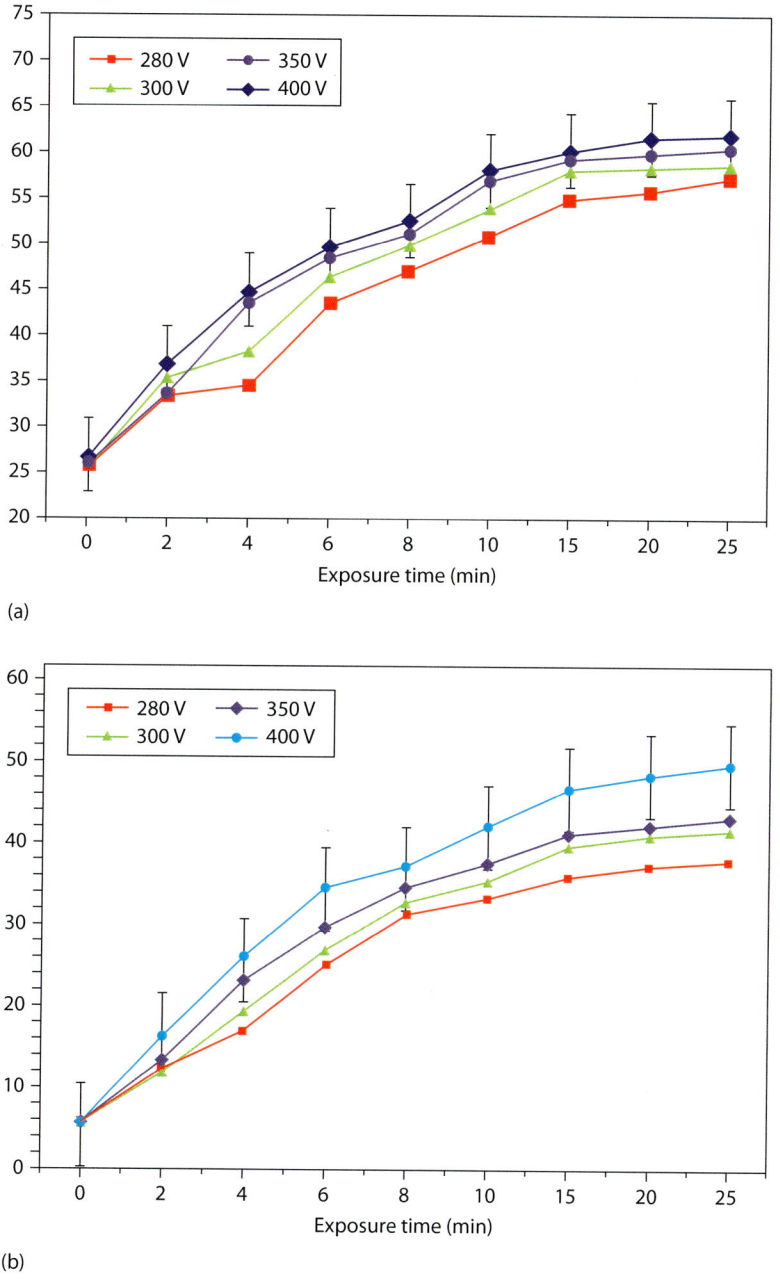

(a)

(b)

FIGURE 3.4 Variation of the (a) total SFE and (b) polar component of PET film with time of exposure to dc glow discharge plasma treatment at different discharge potentials. (From Pandiyaraj, K.N. et al., *J Phys.: Conf. Ser.*, 208, Paper ID: 012100, 1–7, 2010.)

higher discharge potential compared with lower potentials. A similar kind of trend was observed in the increase in polar components, as shown in Figure 3.4b.

Kitova et al. [123] studied the RF plasma treatment of PC substrates, and found that all RF plasma treatments led to an increase in the polar component of SFE of PC. Narushima and Ikeji [102] used oxygen plasma to improve the adhesion strength at Cu–PEEK interfaces. It was observed that the water contact angle on the untreated PEEK surface was 98°, which decreased to 62° and 66°, respectively, after pulsed oxygen plasma and conventional oxygen-plasma treatments. They concluded that both pulsed oxygen plasma and conventional oxygen plasma introduced hydrophilic oxygen functional groups that resulted in an increase in SFE of the polymer.

Iqbal et al. [124] investigated the adhesion performance of CF- and glass fiber (GF)-reinforced PPS after surface modification with APP. It was observed that surface modification by APP resulted in a substantial increase in the polar component of SFE of these materials, as shown in Table 3.3. In the case of CF-reinforced PPS composite, the increase in SFE was 52% after the plasma treatment.

Lommatzsch et al. [125] used an APP jet to activate PE samples. The presence of polar functional groups on the polymer surface was revealed by contact angle measurements, where it was found that the water contact angle on an untreated PE surface was 93.3°, which decreased to 21.6° after air-plasma treatment.

Arpagaus et al. [126] found that a plasma exposure time of 0.14 s effectively improved the wettability of HDPE and copolyamide (Co-PA) powders. The water contact angle was reduced from more than 90° (untreated powder) down to 65° for HDPE and to 76° for Co-PA powder as a consequence of O_2–Ar-plasma treatment, indicating a more hydrophilic surface. The total SFEs increased by a factor of 1.5, showing an increase from 31.2 mJ/m² to 47.9 mJ/m² for HDPE and to 44.9 mJ/m² for Co-PA.

Bhatnagar and coworkers [116,127] used low-pressure air–N_2 plasma under RF glow discharge as well as APP to alter wetting, chemical, and morphological characteristics of PEEK. It was observed that the contact angles of deionized water and formamide on the PEEK surface decreased significantly when the surface of the PEEK was modified by LPP and it decreased more with APP treatment.

It was also interesting to note that the primarily polar component of the SFE of PEEK increased on its exposure to both LPPs and APPs, while the change in the dispersion component of the SFE of PEEK was relatively small. Thus, an increase in the polar component of SFE results in an increase in the total SFE of PEEK. It was observed that the SFE of the as-received PEEK was 51.1 mJ/m², and increased after surface modification to the corresponding highest value of 70.7 mJ/m² obtained in this study when PEEK was exposed to APP for 30 s, as shown in Table 3.4. However, exposure to LPP increased the SFE only to 60 mJ/m² for an exposure time of 120 s.

Consequently, the increase in the total SFE of PEEK is dominated by the polar component (Table 3.4). It is, therefore, apparent that APP is capable of increasing SFE more as compared with LPP.

TABLE 3.3
Surface Free Energies of Substrates with Polar and Dispersion Components before and after Plasma Treatment

Substrate Material	γ_S^p(mJ/m²)	γ_S^p(mJ/m²)	γ_S (mJ/m²)
Untreated CF–PPS composite	24.7	22.9	47.6
Untreated GF–PPS composite	26.3	17.7	44.0
Plasma-treated CF–PPS composite	55.9	16.6	72.5
Plasma-treated GF–PPS composite	52.9	17.6	70.5

Source: Iqbal, H.M.S., et al., *Polym. Eng. Sci.*, 50, 1505–1511, 2010 with permission.

TABLE 3.4

Surface Free Energy Values and Its Polar and Dispersion Components of PEEK

Sample	Treatment Time (s)	γ_S^P(mJ/m²)	γ_S^D (mJ/m²)	γ_S (mJ/m²)
Untreated PEEK	0	4.3	46.8	51.1
LPP-treated PEEK	30	24	29	53
	60	30	26.2	56.2
	120	36	24	60
	240	34	23	57
	480	31	20	51
APP-treated PEEK	30	54	16.7	70.7
	60	52	16	68

Source: Jha, S. et al., *J. Appl. Polym. Sci.*, 118, 173–179, 2010 with permission.

3.5.2 X-Ray Photoelectron Spectroscopy Analysis

XPS analysis of O_2–Ar-plasma-treated HDPE powders revealed the formation of oxygen functional groups on the surface, such as C–O–C (alcohol or ether), O–C=O (acid or ester) and C=O (aldehyde or ketone). These polar groups enable interaction with water through hydrogen bonds, and are responsible for higher powder wettability [126].

XPS analysis of untreated PET, and that treated under RF glow discharge using pulsed argon and oxygen plasmas, shows that the C1s spectra could be decomposed into three components: C–H at 285.0 eV, C–O at 286.5–286.8 eV, and C=O at 288.9–289.1 eV. But, the pulsed Ar and O_2 plasma treatments led to a decrease in the relative concentration of the C–H group and an increase in the relative concentration of the C–O and C=O groups [128].

Narushima and Ikeji [102] compared the effects of pulsed oxygen plasma and conventional oxygen plasma on the surface of PEEK for its surface metallization with Cu. The C1s spectrum of PEEK film can be resolved into three main components: 285.0 eV (C–H group), 286.4 eV (C–O group), and 287.4 eV (C=O group) [129]. The relative intensities of the C–H, C–O, and C=O components of untreated PEEK film are 75%, 25%, and 5%, respectively. Both pulsed and conventional oxygen-plasma treatments decrease the relative intensity of the C–H group and result in the formation of a new functional group, COO, with relative intensities of 1% and 5%, respectively, as listed in Table 3.5. This is in accord with an increased oxygen:carbon (O:C) ratio.

Kim et al. [130] exposed a polyimide (PI) surface to Ar–N_2 plasmas after KOH treatment. The change in the surface functionalities of the PI was studied by XPS analysis. The chemical composition of the untreated sample was determined to be 73.1% C, 20.6% O, and 6.3% N from the integrated peak signal intensities, as shown in Figure 3.5a. After the KOH and Ar–N_2 plasma treatments, the chemical composition of the PI surface changed to 62.4% C, 21.4% O, and 12.5% N (Figure 3.5c), indicating an increase in formation of polar groups on the surface.

TABLE 3.5

Percentage Content of Each Component

Sample	C–H	CO	C=O	COO	O:C ratio
Untreated PEEK	75	25	5	—	0.17
Conventional oxygen plasma-treated PEEK	68	26	6	5	0.23
Pulsed plasma-treated PEEK	69	25	6	1	0.24

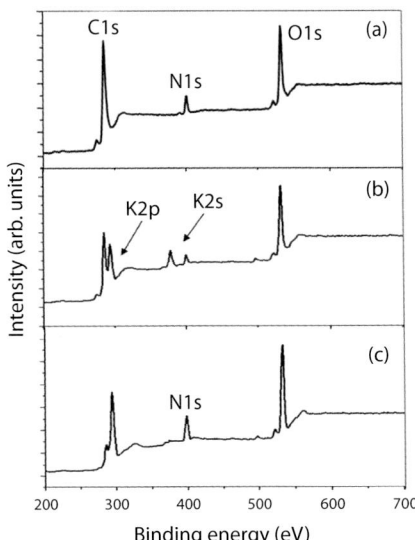

FIGURE 3.5   The XPS survey spectra of PI at: (a) the initial state, (b) after alkaline treatment, and (c) after Ar and N_2 plasma treatment. (From Kim, C. et al., *J. Korean Phys. Soc.*, 54, 621–627, 2009.)

Surface studies on PET film exposed to dc glow discharge air plasma indicate that polar groups are created on the treated surface [98]. This was analyzed from the XPS C1s high-resolution spectra for both the untreated and dc glow discharge air plasma-treated samples, as shown in Figure 3.6a and b. The spectrum of untreated PET is composed of four main components at 285 eV (component C1), 286.43 eV (component C2), 288.95 eV (component C3), and 291.37 eV (component C4), which may be attributed to C–C, C–O, O=C–O, and π–π* shake-up satellite in phenyl groups, respectively (Figure 3.6a) [117,131,132]. The spectrum of plasma-treated PET film (Figure 3.6b) also shows components C1, C2, and C3, and additional peaks at 286.76 (component C5), 288.10 eV (component C6), and 286.0 (component C7), which may be attributed to C–OH and/or C–O–O–H, –C=O, and C–N groups [131].

The XPS C1s spectra (Figure 3.7) show significantly higher oxygen peaks when PEEK is exposed to APP using air as the process gas as compared with LPP using air–N_2 as the process gas, indicating extensive oxidation of the polymer surface [116]. The XPS C1s spectrum (Figure 3.7b) of LPP-treated PEEK shows a new peak at 289.0 eV (COO group) (which is not present in the untreated PEEK), along with peaks at 285.0 eV (C–H group), 286.5 eV (C–O group), and 287.9 eV (C=O group). The C1s peak of APP-treated PEEK (Figure 3.7c) can also be deconvoluted into four components: 285.0 eV (CH group), 286.5 eV (C–O group), 287.9 eV (C=O group), and 289.0 eV (COO group), but with higher concentrations of all the three functional groups, C–O, C=O, and COO, as is evident from Table 3.6.

3.5.3   Atomic Force Microscopy Analysis

According to some investigators [133–135], changes in surface morphology on polymer surfaces also play an important role in improving adhesion through a process known as *mechanical interlocking*. Various researchers have studied polymer surfaces by AFM to investigate the topographical changes on the surface of polymers when they are exposed to plasma.

The variation in surface roughness after the surface treatment of PI films was determined using AFM [130].

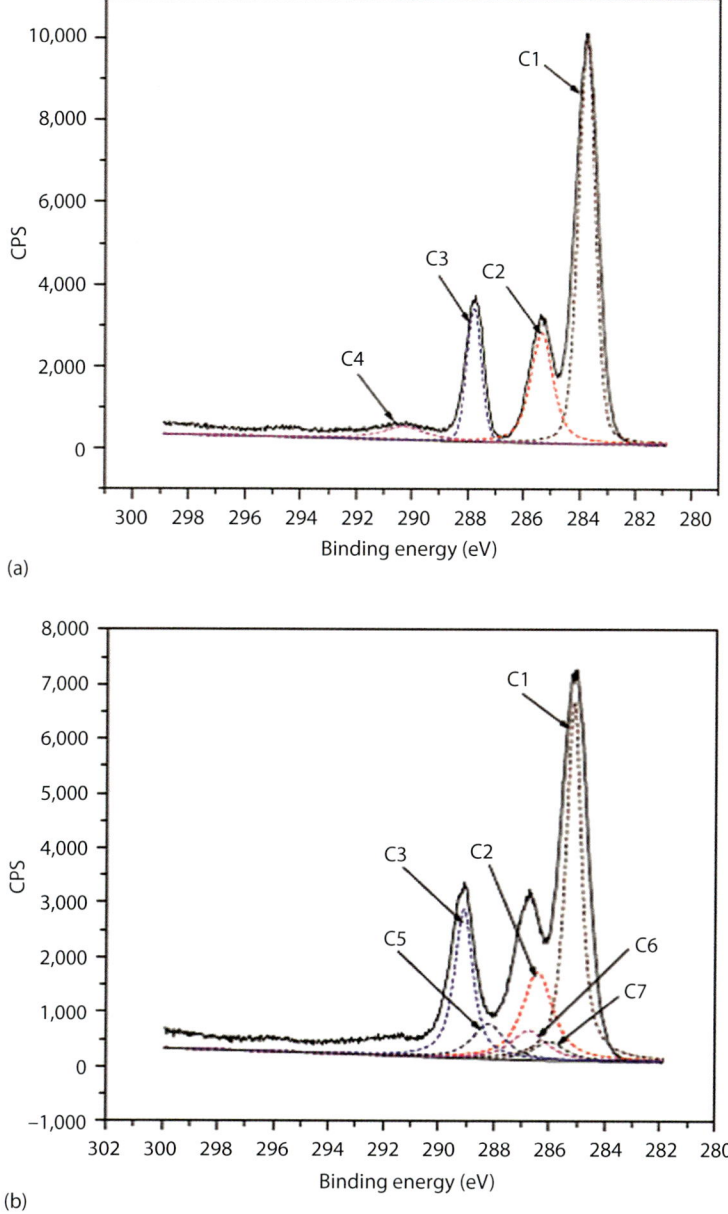

FIGURE 3.6 XPS spectra of (a) untreated PET and (b) plasma-treated PET. (From Beamson, G., and Briggs, D., *High Resolution XPS of Organic Polymers: The Scienta ESCA300 Database*, John Wiley, Chichester, UK, 1992.)

The untreated surface of the PI showed a root mean square surface roughness (R_{rms}) of 1.42 nm, as shown in Figure 3.8a. After KOH + Ar–N$_2$ plasma treatment, root mean square surface roughness rose to 28.52 nm, as shown in Figure 3.8b.

Pandiyaraj et al. [117] examined the topography of untreated PET films and of those exposed to dc glow discharge air plasma by AFM. It was observed that the surface of untreated PET film was relatively smooth (Figure 3.9a), but after plasma treatment, the surface of the PET film showed rough morphology, as shown in Figure 3.9b.

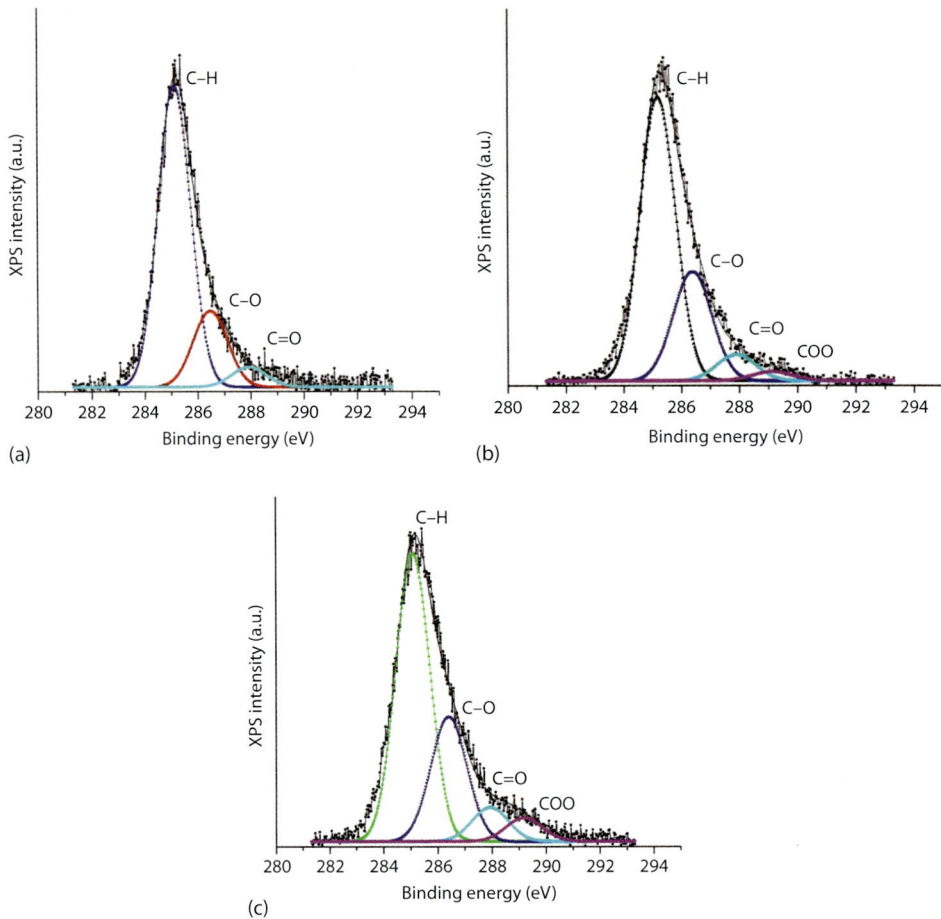

FIGURE 3.7 XPS C1s spectra: (a) untreated PEEK, (b) low-pressure plasma (LPP) treated PEEK, (c) atmospheric-pressure plasma (APP) treated PEEK. (From Bhatnagar, N. et al., *Surf. Eng. Appl. Electrochem.*, 48, 117–126, 2012.)

TABLE 3.6
Percentage Content of Each Component

Sample	C–H	CO	C=O	COO
Untreated PEEK	75	19.8	5.2	—
LPP-treated PEEK	64.2	27.0	6.5	2.3
APP-treated PEEK	59.6	27.6	7.6	5.2

When PC is exposed to low-energy Ar plasma, the roughness of the PC sample increases, as shown in Figure 3.10b, which helps in the enhancement of its adhesion property [122].

O'Kell et al. [136] have examined by AFM the topography of untreated PE surfaces and those exposed to RF glow discharge under air and nitrogen gases at 13 MHz. It was observed that both plasmas caused the polymer surface to roughen, with nitrogen plasma treatment having a more pronounced effect. AFM studies by Gupta et al. [132] on RF glow discharge-exposed polyester films

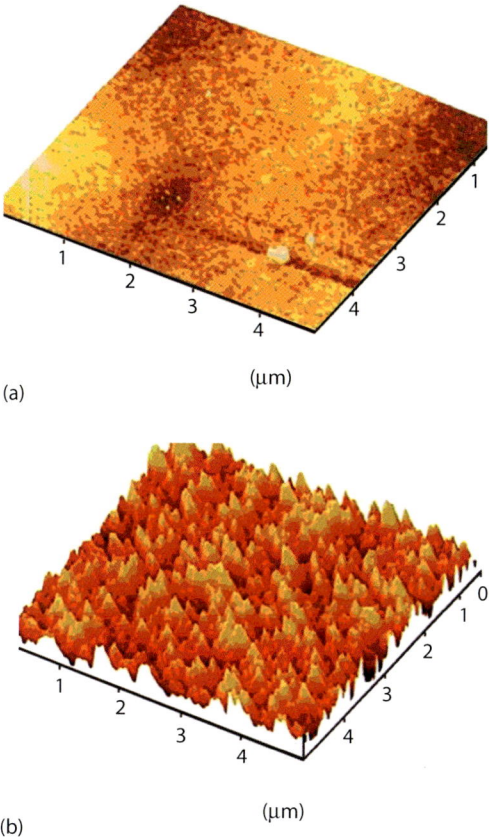

(a)

(μm)

(b)

(μm)

FIGURE 3.8 Variation in surface roughness of (a) bare PI films: $R_{rms} = 1.42$ nm, (b) KOH + Ar 100 W + N$_2$ 60 W: $R_{rms} = 28.52$ nm. (From Kim, C. et al., *J. Korean Phys. Soc.*, 54, 621–627, 2009.)

revealed that the polymer surface became progressively rougher on exposure to glow discharge Ar plasma followed by O$_2$ exposure, while the virgin film had a relatively smooth surface.

Studies using AFM carried out by Bhatnagar et al. [116] also show that exposure to plasma results in significant transformation on the polymer surface. After the LPP treatment with air–N$_2$, the surface of the PEEK showed a rough morphology, as shown in Figure 3.11b, and after APP treatment using air, there was a further increase in surface roughness, as is evident from Figure 3.11c. This apparent increase in surface roughness results in improvement in wettability and bond strength.

3.5.4 Adhesion Characteristics of Surface-Modified Polymers

The work of adhesion increases with the increase in SFE, which results in higher adhesive joint strength [10].

Williams et al. [121] observed that the lap shear strength and crack delamination resistance of CF–epoxy composites increased by 50% for surfaces prepared by helium–oxygen plasma activation instead of solvent wiping and abrasion. Zhang et al. [137] used air plasma to modify PPS film to improve interfacial adhesion and mechanical properties of PPS–GF cloth composites. They observed that tensile strength and notched impact strength of plasma-treated PPS–GF cloth (50:50) composite increased by 11% (from 248.49 MPa to 275.70 MPa).

Iqbal et al. [124] compared the tensile lap shear strength of PPS–CF-bonded joints before and after APP treatment with neat epoxy adhesive and nanofilled epoxy adhesive. The joint strength

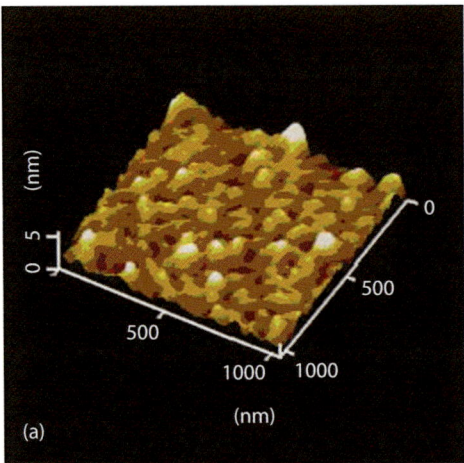

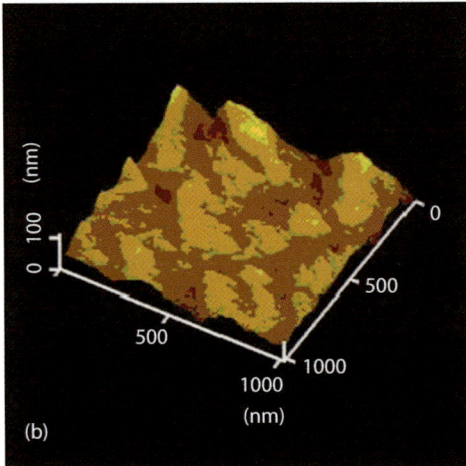

FIGURE 3.9   AFM micrographs of (a) untreated PET film and (b) plasma-treated PET film. (From Pandiyaraj, K.N. et al., *Surf. Coat. Technol.* 202, 4218–4226, 2008.)

increased from 6.1 to 21.5 MPa after APP treatment, resulting in a more than three times improvement in the adhesion properties of the composite joint after the plasma treatment. Analysis of failed surfaces of untreated PPS–CF composites revealed that the failure was essentially interfacial between composite and adhesive, but APP treatment resulted in cohesive failure of the substrate material.

Bhatnagar and coworkers [127] observed that the tensile lap shear strength of PEEK bonded with epoxy adhesive was significantly influenced by surface modification of PEEK under APP. The result revealed that the tensile lap shear strength of adhesive joints increased from 0.5 MPa to 4.00 MPa after LPP treatment using air–N_2 gas and further increased to 8.00 MPa after APP treatment with air. The fractured surfaces of untreated, LPP-treated, and APP-treated PEEK revealed that APP treatment resulted in a shift of locus of failure from the adhesive–substrate interface to within the adhesive or substrate material. This is in accord with the 15 times increase in the joint strength as compared with untreated PEEK.

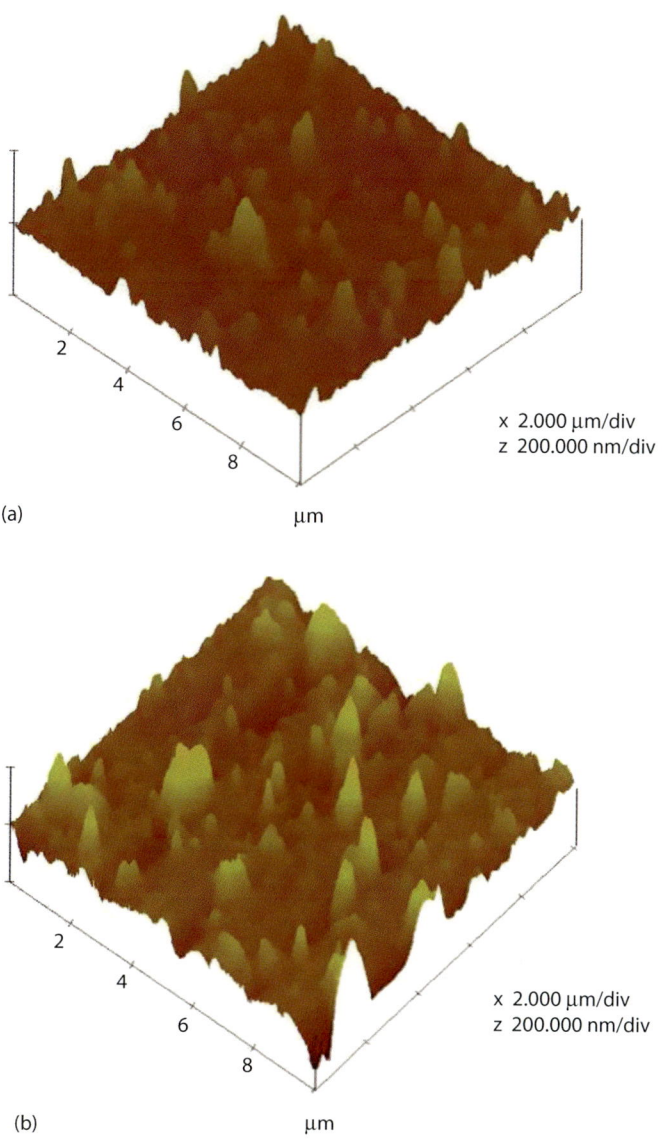

FIGURE 3.10 AFM photomicrograph of (a) untreated PC and (b) Ar plasma-treated PC. (From Pelagade, S. et al., *J Phys.: Conf. Ser.* 208, Paper ID: 012107, 1–8, 2010.)

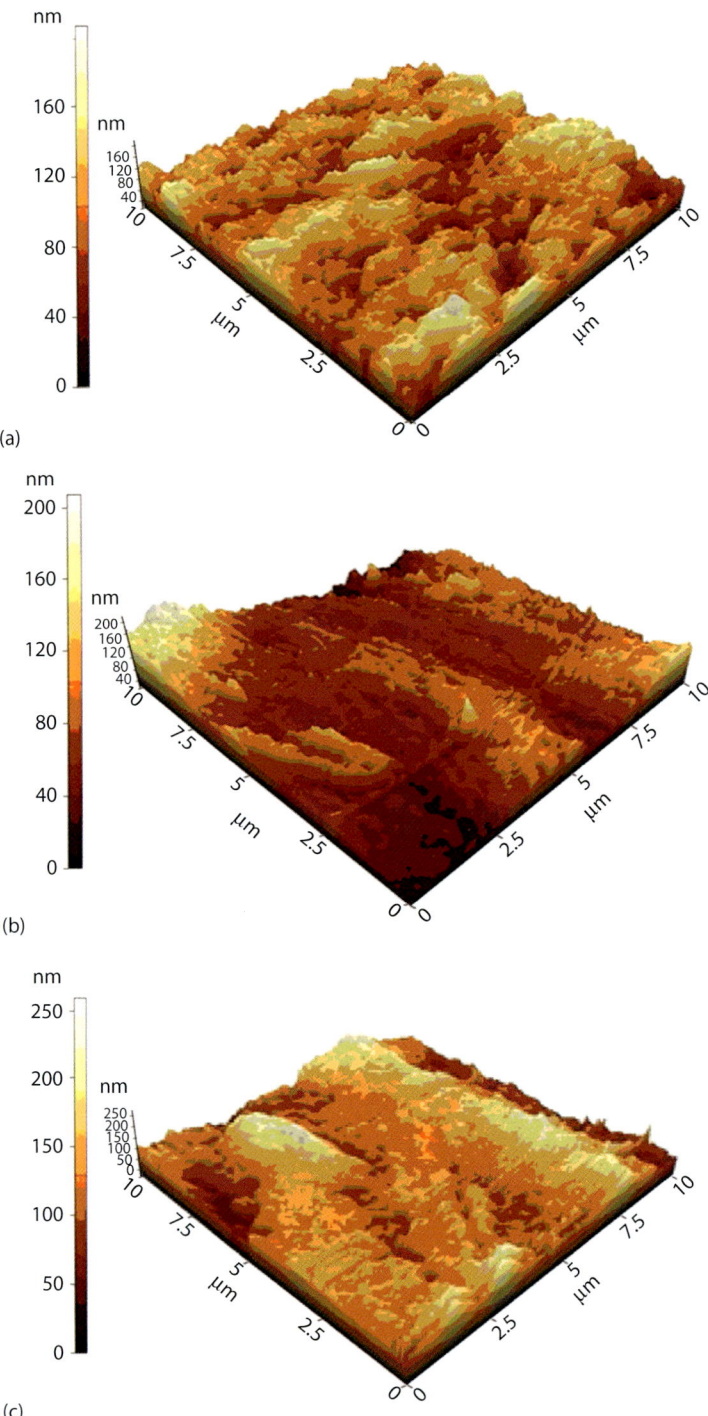

FIGURE 3.11 AFM micrographs of (a) untreated PEEK, (b) low pressure plasma-treated PEEK, (c) atmosphere pressure plasma treated PEEK [116].

3.6 SUMMARY

Surface modification of polymers on exposure to plasma results in the formation of polar groups on the polymer surfaces. This leads to an increase in the polar component of SFE, which further results in an increase in total SFE. As a result, the wettability of the polymer surface increases, resulting in better spreading of the liquid resin (adhesive) and increasing work of adhesion due to polar groups, which could form covalent bonds with the resin. The resin solidifies by cross-linking occurring due to hardener. Also, the high temperature generated during exposure of the polymer surface to plasma and, in particular, to APP, raises the surface temperature to a level that exceeds the glass transition temperature (T_g) of the polymer, which could create micropits on the polymer surface and make the surface rough. Thus, both SFE and surface roughness, which play a significant role in increasing the joint strength, have been found to increase manifold by modifying the polymer surface on exposure to plasma, especially APP.

ACKNOWLEDGMENTS

This book chapter is based on my PhD thesis. I am thankful to my supervisors, Prof. Sangeeta Jha and Prof. Shantanu Bhowmik for suggesting my PhD thesis topic, guiding me to successful completion of my PhD. I am grateful to the authorities of BIT Mesra, Ranchi and SMIT, Sikkim for permitting me to use their lab facilities. I am also thankful to TU Delft, the Netherlands and KAIST, South Korea for providing analytical data used in this chapter.

REFERENCES

1. A. H. Mourad. Thermo-mechanical characteristics of thermally aged polyethylene/polypropylene blends. *Mater. Design* 31, 918–929 (2010).
2. M. Friedman and G. Walsh. High performance films: Review of new materials and trends. *Polym. Eng. Sci.* 42, 1756–1788 (2002).
3. J. P. Sargent. Durability studies for aerospace applications using peel and wedge tests. *Int. J. Adhesion Adhesives* 25, 247–253 (2005).
4. A. R. Hutchinson and S. Iglauer. Adhesion of construction sealants to polymer foam backer rod used in building construction. *Int. J. Adhesion Adhesives*, 26, 555–556 (2006).
5. H. W. Bonin, J. R. Vantine and V. T. Bui. A container based on polymer composite materials for the ultimate disposal of spent nuclear fuel and radioactive waste. *Nuclear Technol.* 169, 150–179 (2010).
6. G. McGrath, Adhesives: An introduction. http://www.azom.com/article.aspx?ArticleID=189 (2001).
7. Z. A. Aga and E. Woldesenbet. Surface preparation effect on performance of adhesively-bonded graphite/epoxy composites subjected to impact. *J. Adhesion Sci. Technol.* 21, 51–65 (2007).
8. J. Custódio, J. Broughton and H. Cruz. A review of factors influencing the durability of structural bonded timber joints. *Int. J. Adhesion Adhesives*, 29, 179–185 (2009).
9. S. Bhowmik, H. W. Bonin, V. T. Bui and R. D. Weir. Modification of high-performance polymer composite through high-energy radiation and low-pressure plasma for aerospace and space applications. *J. Appl. Polym. Sci.* 102, 1959–1967 (2006).
10. K. L. Mittal. The role of the interface in adhesion phenomena. *Polym. Eng. Sci.* 17, 467–473 (1977).
11. A. V. Pocius. Adhesives. In: *Physical Properties of Polymers Handbook*, Part V, J. E. Mark Ed.) pp. 479–486, Springer, New York (2007).
12. M. Strobel, C. S. Lyons and K. L. Mittal (Eds.). *Plasma Surface Modification of Polymers: Relevance to Adhesion*, CRC Press, Boca Raton, FL (1994).
13. M. Goldman, A. Goldman and R. S. Sigmond. The corona discharge, its properties and specific uses. *Pure Appl. Chem.* 57, 1353–1362 (1985).
14. N. Bhatnagar, S. Jha and S. Bhowmik. Energy dispersive spectroscopy study of surface modified PEEK. *Adv. Mater. Lett.* 2, 52–57, (2011).
15. N. Bhatnagar. Effect of plasma on adhesion characteristics of high performance polymers: A critical review. *Rev. Adhesion Adhesives* 1, 397–412 (2013).
16. R. H. Todd, D. K. Allen and L. Alting. *Manufacturing Processes Reference Guide*, Industrial Press, New York (1994).

17. J. H. Lau, C. P. Wong, N. C. Lee and S. W. R. Lee. *Electronics Manufacturing: With Lead-free, Halogen-free, and Conductive-adhesive Materials*, McGraw-Hill Professional, New York (2002).

18. S. Ebnesajjad and A. H. Landrock. *Adhesives Technology Handbook*, 2nd edn, pp. 63–135, William Andrew, Norwich, NY (2008).

19. F. C. Campbell (Ed.). *Joining: Understanding the Basics*, ASM International (2011).

20. A. T. Wolf. Construction sealants. In: *Handbook of Sealant Technology*, K. L. Mittal and A. Pizzi (Eds.), pp. 312–388, CRC Press, Boca Raton, FL (2009).

21. A. J. Kinloch. *Structural Adhesives: Developments in Resins and Primers*, Elsevier Applied Science Publishers, pp. 7–11 (1986).

22. http://www.bemrc.org/bemrc-build-class/ca-glue-article.pdf

23. J. D. David. *Adhesives and Sealants: Technology, Applications and Markets*, Rapra Market Report, Smithers Rapra Publishing, Shawbury, UK (2003).

24. R. Vodicka. Thermoplastics for Airframe Applications: A Review of the Properties and Repair Methods for Thermoplastic Composites. A report from DSTO Aeronautical and Maritime Research Laboratory, Melbourne, Australia (1996).

25. G. C. Mays and A. R. Hutchinson. *Adhesives in Civil Engineering*, p. 333, Cambridge University Press (1992).

26. B. R. Bonazza. Adhesive bonding of improved PPS thermoplastic composites. *Proc. 35th International SAMPE Symposium*, pp. 859–870 (1990).

27. E. M. Silverman and R. A. Griese. Joining methods for graphite/PEEK thermoplastic composite. *SAMPE J.* 25, 34–38 (1989).

28. S. Y. Wu. Adhesive bonding of thermoplastic composites. *Proc. 35th International SAMPE Symposium*, 846–858 (1990).

29. D. K. Kohli. Development of a high toughness heat resistant 177°C (350°F) curing 907 film adhesive for aerospace bonding applications: FM 377 adhesive. *Proc. 35th International SAMPE Symposium*, pp. 907–922 (1990).

30. J. W. Powers and W. J. Trzaskos. Recent developments in adhesives for bonding advanced thermoplastic composite. *Proc. 34th International SAMPE Symposium*, 1987–1998 (1989).

31. R. Mahlberg, H. E. M. Niemi, F. Denes and R. M. Rowell. Effect of oxygen and hexamethyldisiloxane plasma on morphology, wettability and adhesion properties of polypropylene and lignocellulosics. *Int. J. Adhesion Adhesives* 18, 283–297 (1998).

32. R. M. Paroli, K. K. Y. Liu and T. R. Simmons. Thermoplastic polyolefin roofing membranes. *Construction Technology Update No. 80*, Institute for Research in Construction, National Research Council of Canada (1999).

33. G. K. A. Kodokian and A. J. Kinloch. The adhesive fracture energy of bonded thermoplastic-fibre composites. *J. Adhesion* 29, 193–218 (1989).

34. D. C. Goeders and J. L. Perry. Adhesive bonding PEEK/IM-6 composite for cryogenic applications. *Proc. 36th International SAMPE Symposium*, pp. 348–361 (1991).

35. W. Alexander and G. Eric. UV-based patterning of polymeric substrates for cell culture applications. *Biomed Microdevices* 4, 33–41 (2002).

36. M. Lampin, R. Warocquier-Clérout, C. Legris, M. Degrange and M. F. Sigot-Luizard. Correlation between substratum roughness and wettability, cell adhesion, and cell migration. *J. Biomed. Mater. Res. Part B* 36, 99–108 (1997).

37. K. L. Ohashi and R. H. Dauskardt. Effects of fatigue loading and PMMA precoating on the adhesion and subcritical debonding of prosthetic-PMMA interfaces. *J. Biomed. Mater. Res.*, 51, 172–183 (2000).

38. Q. Zhao, Y. Liu and E. W. Abel. Effect of temperature on the surface free energy of amorphous carbon films. *J. Colloid Interface Sci.* 280, 174–183 (2004).

39. J. M. Schakenraad, H. J. Busscher, C. R. Wildevuur and J. Arends. Thermodynamic aspects of cell spreading on solid substrata. *Cell Biophys.* 13, 75–91 (1988).

40. F. Awaja, M. Gilbert, G. Kelly, B. Fox and P. J. Pigram. Adhesion of polymers. *Prog. Polym. Sci.* 34, 948–968 (2009).

41. ASTM D907. *Standard Definitions of Terms Relating to Adhesives*, Annual Book of ASTM Standards Vol. 15, 06 (1984).

42. A. J. Kinloch. *Adhesion and Adhesives: Science & Technology*, Chapman and Hall, New York (1987).

43. A. Nihlstrand. Plasma modification of polypropylene materials for improved adhesion. Doctoral thesis, Chalmers University of Technology, Gothenburg, Sweden (1996).

44. K. W. Allen. Theories of adhesion. In: *Handbook of Adhesion*, 2nd edn, D. E. Packham (Ed.), John Wiley & Sons (2005).

45. H. M. Clearfield, D. K. McNamara and G. D. Davis. Adherend surface preparation for structural adhesive bonding. In: *Adhesive Bonding*, L. H. Lee (Ed.), pp. 203–235, Plenum Press, New York (1991).

46. J. R. Evans and D. E. Packham. Adhesion of polyethylene to metals: The role of surface topography. *J. Adhesion*, 10, 177–191 (1979).

47. C. W. Jennings. Surface roughness and bond strength of adhesives. *J. Adhesion* 4, 25–41 (1972).

48. L. W. Fisher. *Selection of Engineering Materials and Adhesives*, CRC Press, Boca Raton, FL (2005).

49. D. Tabor and R. H. S. Winterton. The direct measurement of normal and retarded van der Waals forces. *Proc. Roy. Soc.* A312, 435–450 (1964).

50. S. S. Voyutskii. *Autohesion and Adhesion of High Polymers*, Wiley Interscience, New York (1963).

51. J. Klein. The self-diffusion of polymers. *Contemp. Phys.* 20, 611–629 (1979).

52. S. F. Edwards and J. W. V. Grant. The effect of entanglements on the viscosity of a polymer melt. *J. Phys. A* 6, 1186–1195 (1973).

53. F. Brochard and P. G. de Gennes. Polymer-polymer interdiffusion. *Europhys. Lett.* 1, 221–224 (1980).

54. S. Wu, H. K. Chuang and C.D. Han. Diffuse interface between polymers: Structure & kinetics. *J. Polym. Sci. Polym. Phys. Edition* 24, 143–159 (1986).

55. J. Schultz and M. Nardin. Theories and mechanisms of adhesion. In: *Handbook of Adhesive Technology: Revised and Expanded*, A. Pizzi and K. L. Mittal (Eds.), pp. 53–68, Marcel Dekker, New York (2003).

56. B. V. Deryaguin, N. A. Krotova and V. P. Smilga. *Adhesion of Solids*, Plenum Press, New York (1978).

57. K. W. Allen. At forty cometh understanding: A review of some basics of adhesion over the past four decades. *Int. J. Adhesion Adhesives* 23, 87–93 (2003).

58. J. Krupp and W. Schnabel. Light modulated electrostatic double layer adhesion. *J. Adhesion* 5, 269–296 (1973).

59. A. D. Roberts. Surface charge contribution in rubber adhesion and friction. *J. Phys. D: Appl. Phys.* 10, 1801–1819 (1977).

60. W. Possart. Experimental and theoretical description of the electrostatic component of adhesion at polymer/metal contacts. *Int. J. Adhesion Adhesives* 8, 77–83 (1988).

61. G. Ramarathnam, M. Libertucci, M. M. Sadowski and T. H. North. Joining of polymers to metal. *Welding J.* 71, 483–490 (1992).

62. K. L. Mittal and H. R. Anderson, Jr. (Eds.). *Acid-Base Interactions: Relevance to Adhesion Science and Technology*, CRC Press, Boca Raton, FL (1991).

63. K. L. Mittal (Ed.). *Acid-Base Interactions: Relevance to Adhesion Science and Technology*, Volume 2, CRC Press, Boca Raton, FL (2000).

64. D. L. Allara, F. M. Fowkes, J. Noolandi, G. W. Rubloff and M. V. Tirrell. Bonding and adhesion of polymer interfaces. *Mater. Sci. Eng.* 83, 213–226 (1986).

65. F. M. Fowkes. Role of acid-base interfacial bonding in adhesion. *J. Adhesion Sci. Technol.* 1, 7–27 (1987).

66. N. Encinas, J. Abenojar and M. A. Martínez. Development of improved polypropylene adhesive bonding by abrasion and atmospheric plasma surface modifications. *Int. J. Adhesion Adhesives* 33, 1–6 (2012).

67. S. J. Park and M. K. Seo (Eds.). *Interface Science and Composites*, p. 852, Elsevier, New York (2011).

68. http://home.iitk.ac.in/~kbalani/vl-kb/wetting.html

69. C. J. Lee, S. K. Lee, D. C. Ko, D. J. Kim and B. M. Kim. Evaluation of surface and bonding properties of cold rolled steel sheet pretreated by Ar/O2 atmospheric pressure plasma at room temperature. *J. Mater. Process Technol.* 209, 4769–4775 (2009).

70. A. W. Adamson. *Physical Chemistry of Surfaces*, 4th edn, Chapter 10, Wiley-Interscience, New York (1982).

71. J. Lyklema. *Fundamentals of Interface and Colloid Science, Liquid–Fluid Interfaces*, Vol. 3, Academic Press, London (2000).

72. F. W. Fowkes. Additivity of intermolecular forces at interfaces. I. Determination of the contribution to surface and interfacial tensions of dispersion forces in various liquids. *J. Phys. Chem.* 67, 2538–2541 (1963).

73. W. Su. *Polymer Interface and Adhesion*, Marcel Dekker, New York (1982).

74. F. A. Al-Zahrani, A. A. Al-Masmoom and U. A. Khashaba. Impact of polymers and polymeric composites on the development of new designs in mechanical, electrical, and civil engineering: A review. *MASAUM J. Reviews Surveys* 1, 184–195 (2009).

75. E. G. Shafrin and W.A. Zisman. Constitutive relations in the wetting of low energy surfaces and the theory of the retraction method of preparing monolayers. *J. Phys. Chem.* 64, 519–524 (1960).

76. L. J. Hart-Smith, D. Brown and S. Wong. Surface preparations for ensuring that the glue will stick in bonded composite structures. Paper MDC 93K0126. In: *Proceedings of 10th DoD/NASA/FAA Conference on Fibrous Composites in Structural Design* (1993).

77. M. J. Shenton, M. C. Lovell-Hoare and G.C. Stevens. Adhesion enhancement of polymer surfaces by atmospheric plasma treatment. *J. Phys. D: Appl. Phys.* 34, 2754–2760 (2001).

78. G. L. Lashkov, A. V. Veniaminov and O. B. Ratner. Study of diffusion of anthracene structure compounds in PMMA by holographic relaxometry. *Polym. Sci.* (USSR) 28, 487–492 (1986).

79. E. N. Krupenya, A. Yu Kagan, I. D. Simonov-Emel Yanov, V. N. Kuleznev and L. L. Mukhina, Influence of plasticisers on the strength of multi-layer materials based on polyvinylchloride. *Int. Polym. Sci. Technol.* 16, T/55–T/57 (1989).

80. R. Bajpai, V. Mishra, P. Agrawal and S. C. Datt. Surface modification on PMMA: PVDF polyblend: Hardening under chemical environment. *Bull. Mater. Sci.* 25, 21–23 (2002).

81. J. Comyn, L. Mascia, B. M. Parker, S. J. Shaw and G. Xiao. Surface treatments for PEEK. In: *Proceedings of IOM Adhesives Section Meeting on Surface Preparation for Adhesive Bonding*, pp. 17–26 (1997).

82. P. Davies, C. Courty, N. Xanthopoulos and H. J. Mathieu. Surface treatment for adhesive bonding of carbon fibre-poly(etherether ketone) composites. *J. Mater. Sci. Letters* 10, 335–338 (1991).

83. W. Shen, Z. Li and Y. Liu. Surface chemical functional groups modification of porous carbon. *Recent Patents on Chem. Engg.* 1, 27–40 (2008).

84. L. R. Radovic, I. F. Silva, J. I. Ume, J. A. Menendez, C. A. Leony Leon and A. W. Scaroni. An experimental and theoretical study of the adsorption of aromatics possessing electron-withdrawing and electron-donating functional groups by chemically modified activated carbons. *Carbon* 35, 1339–1348 (1997).

85. J. A. Menendez, M. J. Illan-Gomez, C. A. Leony Leon and L. R. Radovic. On the difference between the isoelectric point and the point of zero charge of carbons. *Carbon* 33, 1655–1657 (1995).

86. J. A. Menendez, J. Phillips, B. Xia and L. R. Radovic. On the modification and characterization of chemical surface properties of activated carbon: In the search of carbons with stable basic properties. *Langmuir*, 12, 4404–4410 (1996).

87. J. A. Menendez, L. R. Radovic, B. Xia and J. Phillips. Low-temperature generation of basic carbon surfaces by hydrogen spillover. *J. Phys.Chem.* 100, 17243–17248 (1996).

88. S. Shin, J. Jang, S. H. Yoon and I. Mochida. A study on the effect of heat treatment on functional groups of pitch based activated fiber using FTIR. *Carbon* 35, 1739–1743 (1997).

89. T. C. Uzomah and S. C. O. Ugbolue. Time and temperature effects on the ultimate properties of thermally oxidized polypropylene films. *J. Appl. Polym. Sci.* 66, 1217–1226 (1997).

90. R. Ansari and G. G. Wallace. Effect of thermal treatment on the electrochemical properties of conducting polypyrrole polymers. *Polymer* 35, 2372–2377 (1994).

91. K. Lafdi, W. Fox, M. Matzek and E. Yildiz. Effect of carbon nanofiber heat treatment on physical properties of polymeric nanocomposites: Part I. J Nanomater. Article ID 52729, http://dx.doi.org/10.1155/2007/52729 (2007).

92. P. I. F. Niem, T. L. Lau and K. M. Kwan. The effect of surface characteristics of polymeric materials on the strength of bonded joints. *J. Adhesion Sci. Technol.* 10, 361–372 (1996).

93. S. Bhowmik, P. K. Ghosh and S. Ray. Surface modification of HDPE and PP by mechanical polishing and DC glow discharge and their adhesive joining to steel. *J. Appl. Polym. Sci.* 80, 1140–1149 (2001).

94. E. Finson and S. L. Kaplan. Surface treatment. http://www.plasmatechsystems.com/about/pubs/surface-treatment.asp (1997).

95. R. J. Zaldivar, H. I. Kim, G. L. Steckel, J. P. Nokes and B. A. Morgan. Effect of processing parameter changes on the adhesion of plasma-treated carbon fiber reinforced epoxy composites. *J. Composite Mater.* 44, 1435–1453 (2010).

96. M. Davis and D. Bond. Principles and practices of adhesive bonded structural joints and repairs. *Int. J. Adhesion Adhesives* 19, 91–105 (1999).

97. M. Thomas and K. L. Mittal (Eds.). *Atmospheric Pressure Plasma Treatment of Polymers: Relevance to Adhesion*, Wiley-Scrivener, Beverly, MA (2013).

98. K. N. Pandiyaraj, V. Selvarajan and R. R. Deshmukh. Effects of operating parameters on DC glow discharge plasma induced PET film surface. *J. Phys.: Conf. Ser.* 208, 1–7 (2010).

99. D. P. Subedi, D. K. Madhup, K. Adhikari and U. M. Joshi. Plasma treatment at low pressure for the enhancement of wettability of polycarbonate. *Indian J. Pure Appl. Phys.* 46, 540–544 (2008).

100. J. Larrieu, B. Held, H. Martinez and Y. Tison. Ageing of atactic and isotactic polystyrene thin films treated by oxygen DC pulsed plasma. *Surf. Coat. Technol.* 200, 2310–2316 (2005).

101. C. M. Weikart and H. K. Yasuda. Modification, degradation, and stability of polymeric surfaces treated with reactive plasmas. *J. Polym. Sci., Part B: Polym. Phys.* 38, 3028–3042 (2000).

102. K. Narushima and H. Ikeji. Plasma surface modification of poly (aryl ether ether ketone) and surface metallization using copper metal. *Sen'I Gakkaishi* (translated into English) 65, 127–131 (2009).

103. K. Narushima, H. Ikeji, Y. W. Park, Y. Isono and Y. R. Islam. Plasma surface modification of aromatic liquid crystal polymers and its resultant improvement in adhesive strength with copper thin films. *Sen'I Gakkaishi* (translated into English) 63, 287–294 (2007).

104. N. Inagaki. Modification of polymer surfaces by plasma. *Shinku* (translated into English) 43, 856–862 (2000).

105. S. L. Kaplan, E. S. Lopata and J. Smith. Plasma processes and adhesive bonding of polytetrafluoroethylene. *Surf. Interf. Anal.* 20, 331–336 (1993)

106. G. Bonizzoni and E. Vassallo. Plasma physics and technology: Industrial applications. *Vacuum*, 64, 327–336 (2002).

107. E. M. Liston, L. Martinu and M. R. Wertheimer. Plasma surface modification of polymers for improved adhesion: A critical review. *J. Adhesion Sci. Technol.* 7, 1077–1089 (1993).

108. A. Sparavigna. Plasma Treatment Advantages for Textile, Cornell University Library, https://arxiv.org/ftp/arxiv/papers/0801/0801.3727.pdf, pp. 1–16 (2008).

109. A. Bogaerts, L. Wilken, V. Hoffmann, R. Gijbels and K. Wetzig. Comparison of modeling calculations with experimental results for rf glow discharge optical emission spectrometry. *Spectrochimica Acta Part B* 57, 109–119 (2002).

110. A. Schutze, J. Y. Jeong, S. E. Babayan, J. Park, G. S. Selwyn and R. F. Hicks. The atmospheric-pressure plasma jet: A review and comparison to other plasma sources. *IEEE Trans. Plasma Sci.* 26, 1685–1694 (1998).

111. C. Tendero, C. Tixier, P. Tristant, J. Desmaison and P. Leprince. Atmospheric pressure plasmas: A review. *Spectrochimica Acta Part B*, 61, 2–30 (2006).

112. H.-E. Wagner, R. Brandenburg, K. V. Kozlov, A. Sonnenfeld, P. Michel and J. F. Behnke. The barrier discharge: Basic properties and applications to surface treatment. *Vacuum* 71, 417–436 (2003).

113. M. A. (Tony) Belcher, K. L. Krieg, P. J. Van Voast and K. Y. Blohowiak. Nonchemical surface treatments using atmospheric plasma systems for structural adhesive bonding. In: *Proc. SAMPE* 2013, Long Beach, CA (2013).

114. N. Jiang, A. Ji and Z. Cao. Atmospheric pressure plasma jets beyond ground electrode as charge overflow in a dielectric barrier discharge setup. *J. Appl. Phys.* 108, 033302 (2010).

115. R. A. Wolf. New atmospheric plasma and photografting approach for permanent surface tension and coating adhesion. Report by Enercon Industries Corp., Milwaukee, WI (2007).

116. N. Bhatnagar, S. Jha, S. Bhowmik, G. Gupta, J. B. Moon and C. G. Kim. Physico-chemical characteristics of high performance polymer modified by low and atmospheric pressure plasma. *Surf. Eng. Appl. Electrochem.* 48, 117–126 (2012).

117. K. N. Pandiyaraj, V. Selvarajan, R. R. Deshmukh and M. Bousmina. The effect of glow discharge plasma on the surface properties of poly (ethylene terephthalate) (PET) film. *Surf. Coat. Technol.* 202, 4218–4226 (2008).

118. A. Qureshi, S. Shah, S. Pelagade, N. L. Singh, S. Mukherjee, A. Tripathi, U. P. Deshpande and T. Shripathi. Surface modification of polycarbonate by plasma treatment. *J. Phys.: Conf. Ser.* 208, Paper ID: 012108, 1–6 (2010).

119. M. Noeske, J. Degenhardt, S. Strudthoff and U. Lommatzsch. Plasma jet treatment of five polymers at atmospheric pressure: Surface modifications and the relevance for adhesion. *Intl. J. Adhesion Adhesives*, 24, 171–177 (2004).

120. F. M. Etzler. Determination of the surface free energy of solids: A critical review. *Rev. Adhesion Adhesives* 1, 3–45 (2013).

121. T. S. Williams, H. Yu and R. F. Hicks. Atmospheric pressure plasma activation of polymers and composites for adhesive bonding: A critical review. *Rev. Adhesion Adhesives* 1, 46–87 (2013).

122. S. Pelagade, N. L. Singh, S. Shah, A. Qureshi, R. S. Rane, S. Mukherjee, U. P. Deshpande, V. Ganesan and T. Shripathi. Surface free energy analysis for bipolar pulsed argon plasma treated polymer films. *J Phys.: Conf. Ser.* 208, Paper ID: 012107, 1–8 (2010).

123. S. Kitova, M. Minchev and G. Danev. RF plasma treatment of polycarbonate substrates. *J. Optoelectron. Adv. Mater.* 7, 2607–2612 (2005).

124. H. M. S. Iqbal, S. Bhowmik, J. A. Poulis and R. Benedictus. Effect of plasma treatment and electron beam radiations on the strength of nanofilled adhesive-bonded joints. *Polym. Eng. Sci.* 50, 1505–1511 (2010).

125. U. Lommatzsch, D. Pasedag, A. Baalmann, G. Ellinghorst and H. E. Wagner. Atmospheric pressure plasma jet treatment of polyethylene surfaces for adhesion improvement. *Plasma Process. Polym.* 4, S1041–S1045 (2007).

126. C. Arpagaus, A. Rossi and P. R. von Rohr. Short-time plasma surface modification of polymer powders in a downflowing tube reactor. *Surf. Coat. Technol.* 200, 525–528 (2005).

127. S. Jha, S. Bhowmik, N. Bhatnagar, N. K. Bhattacharya, U. Deka, H. M. S. Iqbal and R. Benedictus. Experimental investigation into the effect of adhesion properties of PEEK modified by atmospheric pressure plasma and low pressure plasma. *J. Appl. Polym. Sci.* 118, 173–179 (2010).

128. N. Inagaki, S. Tasaka, K. Narushima and H. Kobayashi. Surface modification of PET films by pulsed argon plasma. *J. Appl. Polym. Sci.* 85, 2845–2852 (2002).

129. K. Narushima, S. Tasaka and N. Inagaki. Surface modification of poly(aryl ether ether ketone) by pulsed oxygen plasma. *Jpn. J Appl. Phys* 41, 6506–6516 (2002).

130. C. Kim, D. Jeong, J. Hwang and H. Chae. Argon and nitrogen plasma surface treatments of polyimide films for electroless copper plating. *J. Korean Phys. Soc.* 54, 621–627 (2009).

131. G. Beamson and D. Briggs. *High Resolution XPS of Organic Polymers: The Scienta ESCA300 Database*, John Wiley, Chichester, UK (1992).

132. B. Gupta, J. Hilborn, C. Hollenstein, C. J. G. Plummer, R. Houriet and N. Xanthopoulos. Surface modification of polyester films by RF plasma. *J. Appl. Polym. Sci.* 78, 1083–1091 (2000).

133. A. N. Gent and C. W. Lin. Model studies of the effect of surface-roughness and mechanical interlocking on adhesion. *J. Adhesion* 32, 113–125 (1990).

134. R. J. Zaldivar, H. I. Kim, G. L. Steckel, D. Patel, B. A. Morgan and J. P. Nokes. Surface preparation for adhesive bonding of polycyanurate-based fiber-reinforced composites using atmospheric plasma treatment. *J. Appl. Polym. Sci.* 120, 921–931 (2010).

135. W. S. Kim, I. H. Yun and J. J. Lee. Evaluation of mechanical interlock effect on adhesion strength of polymer-metal interfaces using micro-patterned surface topography. *Intl. J. Adhesion Adhesives* 30, 408–417 (2010).

136. S. O'Kell, T. Henshaw, G. Farrow, M. Aindow and C. Jones. Effects of low-power plasma treatment on polyethylene surfaces. *Surf. Interface Anal.* 23, 319–327 (1995).

137. S. Zhang, G. Huang, X. Wang, Y. Huang, J. Yang and G. Li. Effect of air plasma treatment on the mechanical properties of polyphenylene sulfide/glass fiber cloth composites. *J. Reinforced Plast. Composites* 32, 786–793 (2013).

4 Applications of Nanoparticles in Adhesives
Current Status

Farid Taheri

CONTENTS

4.1 INTRODUCTION

Adhesively bonded joints (ABJs) are gaining increasing popularity as the preferred joining method in various structural applications. Their applications have been noticeably increasing in recent years, especially in assembling fiber-reinforced polymer (FRP) composites structural components [1]. ABJs are generally used to mate at least two components (adherends) with a layer of adhesive in between. Therefore, the adhesive layer is considered as a critical part of a joint system, as it is the constituent that transfers the load from one adherend to another. There are many reasons and advantages for the increasing popularity of ABJs compared with conventional mechanically fastened joints. Some of the advantages of adhesive bonding over alternative assembly techniques, especially when joining FRP composites, are listed here [2]:

- More even distribution of stresses over the entire bond region.
- Lower magnitudes of localized stress concentrations.
- Superior fatigue resistance in comparison with that of mechanically fastened joints.
- Excellent resistance to mechanical vibration.
- Less need for machining and better compliance to critical tolerances can be achieved compared with mechanical fastening methods.
- Leak-free joints.
- Relatively lighter weight and lower cost.

In addition to these advantages, the most practical reasons for selecting an ABJ over mechanical fastening are: (1) stress concentration around holes in FRP can go as high as seven times as the far-field stress, depending on the FRP's layup sequence; (2) hole perforation in FRP adherends, especially in carbon and boron epoxies, is much costlier when compared with metallic adherends; (3) delamination around holes caused by drilling of FRP adherends is very likely, while this issue is nonexistent in the case of metallic adherends. To alleviate and manage such potential issues, *smart adhesives*, which are another class of adhesives [3,4], have been developed. Nonetheless, in recent years, one of the most dynamic areas of research related to ABJs has been concerned with improving the performance of adhesives by inclusion of various nanoparticles (NPs).

To be pragmatic, however, despite the numerous advantages of ABJs and the exponential increase in their applications, there still remain several challenges when adhesives are considered as the bonding/joining agent. Long curing time, relatively poor temperature and humidity resistance, elaborate surface preparation in some cases (e.g., joining aluminum adherends), and difficulties in disassembly are some of the drawbacks of using adhesives for mating structural components. As a result, the attempts to modify the properties of adhesives to overcome such drawbacks have been the focus of a large number of studies [5–7]. One of the most attempted approaches in recent years, opted for by several researchers, has been the use of NPs to reinforce adhesives. The use of various types of NPs (e.g., various types of nanocarbons, nanosilica and nanoalumina, and nanoclay, nanosilver, and nanocelullose, to mention a few) has been demonstrated to be one of the most promising approaches for enhancing the mechanical response of adhesives and resolving some of the abovementioned challenges.

This chapter attempts to present a brief review of some of the relevant works that have been conducted in recent years relative to the inclusion of NPs for enhancing the mechanical properties of adhesives. In fact, this chapter expands the work of the author presented recently in a review article [8].

While various NPs have been used to enhance various properties of adhesives, such as their conductivity or resistance to various environmental elements, the works reviewed in this chapter will be limited to those related to the enhancement of mechanical response of adhesives. Issues such as those related to mixing of NP in adhesives, improving their distribution within adhesives, and improving the interface bonding characteristics of the particles to adhesive will also be discussed and reviewed.

4.2 NANOPARTICLES

In the context of the present discussion, an NP is considered to be a particle with one of its dimensions being less than 100 nm. In other words, an NP is considered as a bridge between atomic and bulk scales. Such NPs provide the following key advantages [9]: large specific surface area, high surface energy, reduced numbers of structural imperfections, and distinctively different physical properties from those of bulk materials. Unlike bulk materials, whose properties are fairly size independent, NPs exhibit size dependency. This size dependency can be attributed to the significant increase in the ratio of the number of atoms on the surface of the particle to the total number of atoms forming the particle, as the NP's dimensions approach zero.

As briefly mentioned, there are several types of NPs that have been used to reinforce adhesives. Each type of NP exhibits one or more unique physical properties that cannot be achieved at macroscale. These unique properties have encouraged many scientists to devote their efforts in tailoring the desired properties of adhesives (including thermal conductivity, hygrothermal stability, mechanical properties, etc.) by utilizing different types of NPs. In doing so, several factors must be considered. For instance, the compatibility of the adhesive and NPs, the functionality of the NPs, and the service conditions must be carefully examined. In addition, any approaches that could improve the compatibility and uniform distribution of the NPs should also be explored. The following section briefly discusses some of the most commonly used NPs to reinforce adhesives.

4.2.1 CARBON-BASED NPs

A group of the most widely used NPs for reinforcing adhesives, with the aim to improve the mechanical performance and fracture toughness of adhesives, is carbon-based NPs. Perhaps one of the major reasons for their wide popularity is that these particles have relatively large aspect ratios, which allow them to offer excellent mechanical bonding to the adhesive and also bridging action in an event of a crack development. As a result, a large number of studies have focused on enhancing the properties that these NPs can offer to their host resins or matrices. Carbon nanofillers come in three commonly used configurations. These configurations are: (1) spherical NPs (e.g., nanodiamond [ND] particles); (2) the most widely used cylindrically shaped carbon nanotubes (CNTs), multiwalled carbon nanotubes (MWCNTs), and carbon nanofibers (CNFs); and (3) the two-dimensional configuration, known as *graphene nanoplatelets* (GNPs). The first configuration is primarily used to provide outstanding tribological properties [10]; therefore, they are widely used to enhance surface morphology of various materials. In addition, it has been shown that inclusion of CNTs and CNFs into polymers leads to a notable increase in the mechanical properties [11,15–18], electrical properties [19,22,23], and thermal properties [20–23] of the resulting nanocomposites. Besides, the nanocomposites reinforced with GNPs are postulated to offer remarkable thermal [24], mechanical [18,25], and electrical properties [18,26].

All these impressive features have encouraged the researchers to explore the reinforcing effect of nanomaterials when included in different polymer matrices. The main disadvantages of carbon nanomaterials are known to be: (1) the inconsistency in the quality of carbon NPs, especially in their mass-produced form; and (2) their cost, with the exception of GNPs, which are relatively considerably less expensive than other types of nanocarbon particles.

4.2.2 METAL-BASED NPs

For many years, metallic nanoparticles (MNPs) have occupied the attention of engineers and scientists. One of the most significant usages of MNPs has been in the area of biomedical sciences and engineering [27]. MNPs can be synthesized and modified to obtain various chemical functionalities. MNPs can be formed literally from any metallic element found in the periodic table. Nanogold, nanosilver, and metal oxides, such as alumina (Al_2O_3) or zirconia (ZrO_2) NPs, are some examples of the commonly used MNPs. The major applications of MNPs, besides their usage to enhance the mechanical properties of polymers [28], have been to enhance the electrical [29], optical [30], and magnetic properties [31] of polymers.

4.2.3 POLYHEDRAL OLIGOMERIC SILSESQUIOXANES

Although not as popular and widely used as nanocarbon particles, or metal-based nanoparticles, polyhedral oligomeric silsesquioxanes (POSS) nanoparticles have attracted considerable attention in recent years. They are nanostructures with the empirical formula $RSiO_{1.5}$, where R may be a hydrogen atom or an organic functional group [32]. POSS nanostructures have diameters in the range 1–3 nm. POSSs are primarily used mainly owing to their small dimension and easy incorporation into polymeric materials. As a result, they have been shown to provide excellent reinforcing attributes to polymers. POSSs have also been reported to be effective as a facilitator for dispersing other types of NPs [33].

4.2.4 NANOCLAYS

Nanoclays are a relatively inexpensive group of NPs that have been widely used to enhance the fire-retardant properties of polymers [34]. This rather unique enhancing property of nanoclays does not preclude their ability to also enhance the mechanical properties of polymers [35]. Nanoclays have also been incorporated into polymers to enhance their gas permeation barrier properties [36]. Researchers have also explored other interesting enhancing properties offered by nanoclay NPs, such as the enhancement gained in the rheological properties [37] and liquid infusion-resistant properties [38]. Nanoclays also come in so-called *organoclays*, which are an enticing class of hybrid organic–inorganic NPs. This class of NPs is used in several applications: as rheological modifiers, oil and gas adsorbents, and drug delivery carriers.

4.2.5 NANOSILICA (SiO₂)

This class of NPs is divided into two categories, depending whether they are porous (referred to as the *P-type*), or whether they are spherically shaped (hence, *S-type*). The P-type has relatively greater specific surface area in comparison with the S-type [39]. Nanosilicas are the basis for a great deal of biomedical research due to their stability, low toxicity, and ability to be functionalized with a range of molecules and polymers. Nanosilicas have porous and relatively large surfaces that could accommodate a large number of hydroxyl groups and unsaturated residual bonds. Nanosilicas have proven to be an effective reinforcement for polymers to enhance their strength [40], flexibility [41], and durability [42]. Moreover, nanosilicas have been used as an additive to improve the workability and strength of high-performance and self-compacting concrete [43].

4.3 DEVELOPMENT OF POLYMER-BASED NANOCOMPOSITES

As briefly stated, one of the factors that has impeded a wider usage of NPs within polymers has been the cost associated with mixing and dispersing them within a given polymer. Therefore, development of suitable strategies for their manufacturing processes has also attracted considerable efforts in recent years. It should be noted that there are several factors that should be considered when NPs are used within polymers. For instance, in selecting an effective and efficient dispersion technique, one should be concerned that the process would not damage the NPs, and that it would yield consistent results and be suitable for mass production. In the next section, the manufacturing methods commonly used to generate nanocomposites are discussed.

Within the parameters that affect manufacturing of polymer-based nanocomposites (i.e., the so-called *end product*) is the attainment of a uniform dispersion of NPs in the polymer matrix. One of the earlier sources in which the constitutive theories and equations of suspensions are discussed is the textbook authored by Shenoy [44]. In this book, the effects of shape, concentration, dimensions, and size distributions of the particles have been discussed. An entire chapter is also dedicated to the mixing/blending/compounding of fillers into polymer melts. In addition, Huang et al. [45], have also presented a nice survey of the procedures used by several researchers in regard to mixing various nanocarbon particles in resins. The comparison of the various methods reported in their paper is extracted and presented in Table 4.1.

TABLE 4.1

Fabrication Methods of Carbon-Based Polymer Composites

Method	Filler	Polymer	Description	Filler Dispersion	Filler Fraction (%)	Ref.
Resin infiltration	Aligned CNT arrays	epoxy	Aligned CNTs were mechanically densified and then epoxy was infused into the CNT forest by capilarity-driven wetting force.	∥[b]	20 vol	[13]
Hot-press molding	Aligned CNT sheets	epoxy	Epoxy resin prepreg was impregnated into a horizontally aligned CNT sheet by hot press then the CNT/epoxy prepreg was cured under a pressure.	═[c]	21.4 vol	[26]
Resin infiltration	Aligned CNT sheets	epoxy	CNT arrays were first converted to horizontally aligned CNT sheets, then the CNT sheets were soaked in epoxy solution. After drying, the soaked buckypapers were cured at a high temperature.	═[c]	27 vol	[27]
Resin infiltration	Aligned CNT buckypapers	epoxy	The buckypapers were impregnated with epoxy. After drying, the impregnated buckypapers were cured at a high temperature.	═[c]	50 vol	[28]
Latex mixing	SWCNT	PS[a]	Mixing latex and single-walled CNT.	Random	30 wt	[29]
Layer by layer	SWCNT	epoxy	Building up layered composite film by depositing CNT solution on the top of epoxy. After removing the solvent by irradiation, the composites were prepared by hot press of stack of composite layers.	═[c]	39.1 wt	[30, 31]
Resin infiltration	CNT buckypapers	epoxy	The CNT buckypapers are immersed into epoxy/curing solution, then removing the solvent and performing curing.	═[c]	49.2 wt	[32, 33]
Resin infiltration	CNT buckypapers	epoxy	The CNT buckypapers are immersed into epoxy solution. The impregnated buckypaper was then cured under a hot press.	═[c]	60 wt	[34]
Filtration	SWCNT buckypapers	PS[a]	The dispersion of solvent, CNT and polymer were vaccum filtered through a Teflon filter paper.	═[c]	82 wt	[35]
Casting	Carbon sheets	epoxy	Mixing the carbon sheets with polymer in solution and then casting.	═[c]	72 vol	[36]
Mixing	Graphite nanoplatelets	epoxy	Preparing graphite nanoplatelets from natural graphite and then mixing them with epoxy, molding, degassing, and curing at vaccum oven.	Random	25 vol	[25]

Source: Huang, X. et al., *J. Phys. Chem. C,* 116, 23812–23820, 2012.

[a] PS: polystyrene. [b] ∥: perpendicular to the in-plane direction. ═: parallel to the in-plane direction.

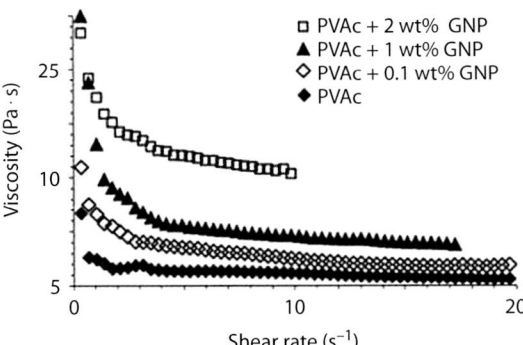

FIGURE 4.1   Variation in the Brookfield viscosity for PVAc hosting different contents of GNP as a function of shear rate. (From Jalili, M.M. et al., *Prog. Org. Coatings*, 59, 81–87, 2007.)

The issue that arises during mixing and dispersion of nanoparticles, which is also referred to as *agglomeration of NPs*, is essentially due to the small dimensions and relatively large surface area of NPs. As a result, the intermolecular forces attract individual particles to one another; consequently, they tend to form large clumps or agglomerates. To promote the efficiency of nanocomposites, it is vital to minimize the formation of such agglomerates and ensure a uniform dispersion of NPs. The task of uniformly dispersing particles becomes a greater challenge as the volume content of the NPs in polymers increases, which, in turn, increases the viscosity of the host polymer, and as a result, the mix becomes less workable. The graph shown in Figure 4.1, generated by Pinto et al. [46], reports the increase in the viscosity of polyvinyl acetate (PVAc) as a function of GNP content and shear rate.

Uniform dispersion of NPs can be achieved in two stages. First, NPs are mixed in resin with some sort of mechanical stirrer. Subsequently, a more elaborate physical method is used to further disperse the NPs in the resin. The dispersion can be further facilitated by chemical means (or functionalization). Otorgust et al. [47] have stated that more uniform dispersion and interfacial interactions of the nanocomposites containing CNTs could be achieved by covalent and noncovalent modifications. It has been demonstrated that the dispersion of tungsten sulfide nanoparticles (WS_2) in the polymer medium can further enhance dispersion of CNTs within the polymer. They stated that WS_2 particles can be dispersed in polymers using simple techniques such as shear mixing and sonication, without a requirement for any surface modifications. The readers are directed to references [50–51] for a critical review of the effect of the inclusion of carbon and WS_2 nanotubes and fullerene-like nanostructures in thermoset adhesives.

In the following section, first, the most commonly used techniques for the dispersion of nanomaterials are briefly discussed. Subsequently, the chemical enhancement methods will be briefly summarized.

4.3.1 SONICATION

This physical method entails the dispersion of NPs with the aid of high-energy sound waves, often projected at frequencies greater than 20 kHz. The resulting cavitation, which is developed at the end of the physical probe, generates high-speed liquid jets. These jets have very large velocities, as high as 1000 km/h, which causes breakage of the contact within the NP agglomerates, thereby separating them from one another (see Figure 4.2). In addition, the NPs that are accelerated by the liquid jets may also crash into one another, resulting in further dispersion [48]. It should be noted that the temperature of the liquid elevates as a result of this physical action. The higher temperature, in turn, decreases the viscosity of the mixture during the process, further facilitating the dispersion process. However, overheating or burning of the polymer could occur if one does not monitor the process carefully. Moreover, an additional advantage and outcome of sonication is the degasification of the mixture. High-speed

FIGURE 4.2 Formation of cavitation phenomenon occurring during ultrasonication produced by two differently shaped probes. (From Hielscher, T., *Ultrasonic Lab Devices and Industrial Processors*, Ultrasonic Technology, Germany, 2014. ©2017 www.hielscher.com)

mechanical stirring produces a large number of gas bubbles within the mixture, which in some cases could even cause brightening of the mixture's color. In such an event, the sound waves used in the ultrasonication would force the bubbles out of the liquid, thus degasifying the mixture.

Two types of sonication methods are generally used for the purpose of processing nanocomposites. These are: (1) the bath method, and (2) the probe sonication method. In the bath method, the ultrasonic waves are propagated into the liquid through the bath walls. On the other hand, in the probe method, concentrated ultrasonic waves exit from a bar-shaped probe. The resulting ultrasonic cavitation is usually powerful enough to break NP agglomerates, especially when they are in bulk volume; thus, the probe method is more commonly used for this purpose. The shortfall of this method is that the material is not uniformly exposed to the sound waves' energy in its entirety. This issue can be addressed by continuous mechanical stirring of the liquid during sonication.

Different factors are associated with an efficient sonication process, including the power, time, and cycles of sonication. Montazeri and Chitsazzadeh [49] and Suave et al. [50] investigated the effect of these factors on the dispersion and structural damages of CNTs. They showed that exposure of the mixture to a little amount of ultrasonic energy would lead to residual agglomerates in the system, while overusing sonication could cause severe damage to NPs. The scanning electron microscopy (SEM) micrographs shown in Figure 4.3, taken from the nanocomposites manufactured utilizing different time intervals of the sonication process, illustrate these phenomena. It can be seen that after 15 min of sonication (Figure 4.3a), there are still large agglomerates in the system [52]. After 45 min, the desired state of dispersion is obtained and the structural integrity of NPs is also maintained. However, after 135 min of sonication, NPs are severely damaged (i.e., broken into small pieces); as a result, they would not deliver any positive attributes to the host matrix.

4.3.2 CALENDERING (USING THREE-ROLL MILL)

Another effective physical method used for the dispersion of NPs is calendering, which requires slightly more expensive equipment than a sonicator. The equipment used for this process is a three-roll mill machine. This equipment exerts shear forces to the matrix through three adjacent rollers, with a tiny gap between them. The rollers rotate in opposite directions, at progressively increasing speeds, to break down NP agglomerates, thus dispersing them in the matrix. Figure 4.4 shows an actual three-roll mill machine and the schematic of the calendering mechanism. The material is loaded between the feed and center rollers. Because of the narrow gaps (at the nanometer scale)

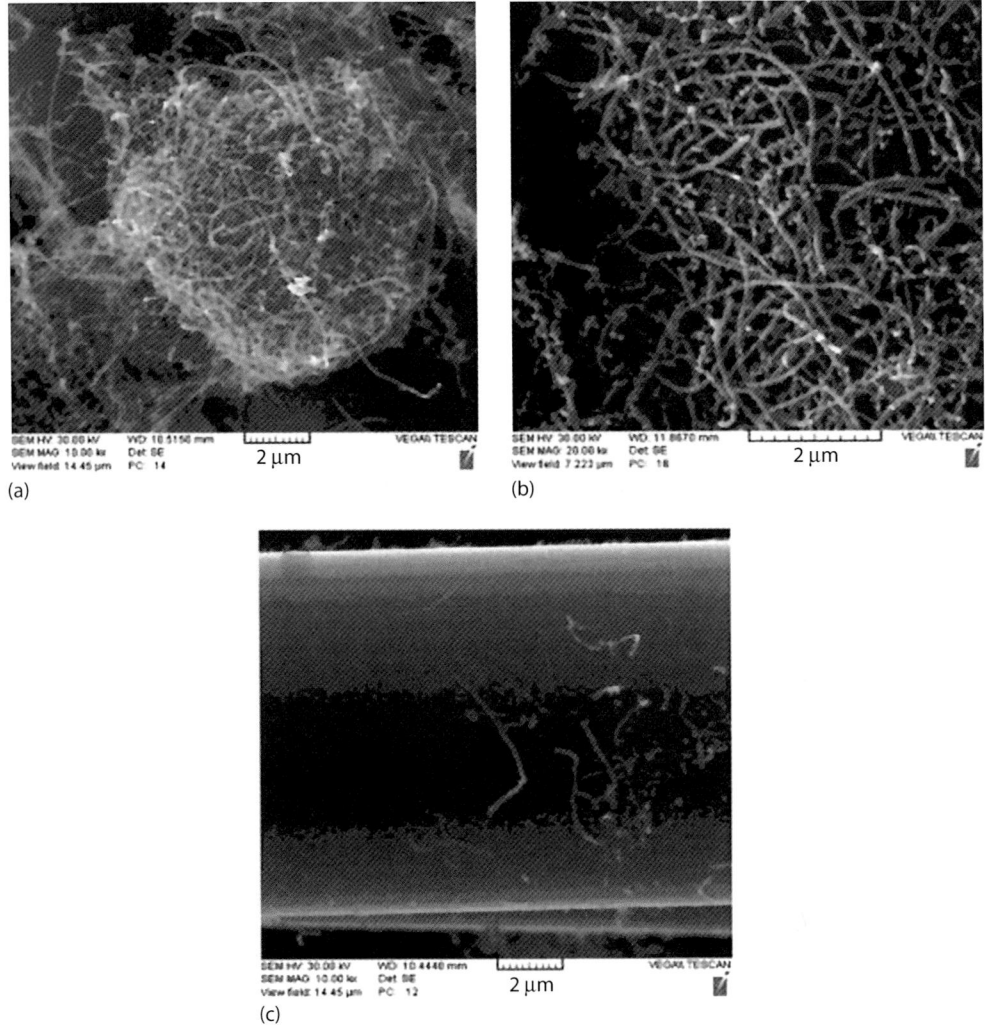

FIGURE 4.3 SEM micrographs of CNT/epoxy nanocomposites after the burn-off test, sonicated at 100 W, after (a) 15 min, (b) 45 min, and (c) 135 min. (From Shenoy, A.V., *Rheology of Filled Polymer Systems*, Kluwer Academic Publishers, the Netherlands, 1999.)

between the rollers, essentially all the material remains within the feed region. The mixture makes its way through the rollers, experiencing very large shear force magnitudes that, in turn, disperse NPs in the matrix. As the resin comes out of the other side of the rolls, the material, which remains on the center roll (as a very thin layer), goes through the gap between the center roller and apron roller, experiencing even larger shear forces that result due to the higher rolling speed. The sharp knife-edge plate located immediately adjacent to the last roller collects the processed material from the apron roller. The three-roll milling process may be repeated for a number of cycles, until the material is perfectly dispersed. Calendering has proven to be a very effective means for dispersing different types of NPs, especially carbon-based NPs in resins [51–53].

Although, comparatively, the damaging effect of calendering is much less than that exerted by sonication, an investigation recently carried out by Ahmadi-Moghadam and coworkers [54] demonstrated that depending on the type and dimension of NPs, very small roller gap distances could lead to breakage of NPs and reduce their effective length. Therefore, to obtain the desired results, the number of calendering cycles and the gap distances should be carefully established. The duration of

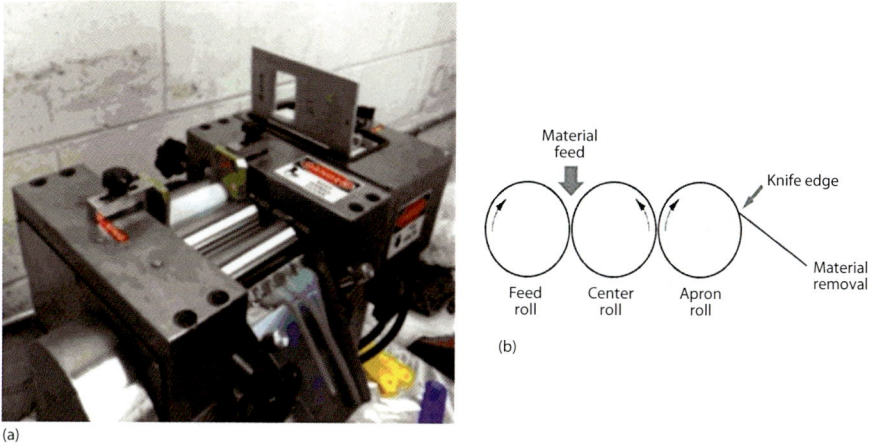

FIGURE 4.4 Three-roll mill equipment (a); schematic of calendering mechanism (b).

each cycle is significantly dependent on gap distance between the rollers. Consequently, depending on the gap distance between the rollers, calendering could be accomplished more quickly or more slowly in comparison with sonication. The other advantage of calendering over sonication is that the entire resin–NP mixture is uniformly subjected to the applied energy, thus leading to more consistent results.

4.3.3 HIGH-SHEAR MIXING

High-shear mixers have also been used to disperse NPs within resins. These mixers consist of a vertically driven shaft together with a high-shear disk type blade. The blade rotates at very high speed, some up to 20,000 rpm, which creates a radial flow pattern in a stationary vessel. A typical high-shear mixer and the schematic of the vortex created by its rotating blades, which, in turn, pulls in the material in the vessel toward the blades' sharp edges, are shown in Figure 4.5. The blades' surfaces then mechanically tear apart the particles, thereby reducing their size, and at the same time dispersing them throughout the resin [55]. Many researchers have used high-shear mixing as a dispersion technique for processing nanocomposites [56]. It should be noted that a higher speed does not necessarily translate into a more uniformly dispersed nanocomposite. For example, Al-Qadhi et al. [57], who exfoliated clay nanoparticles in an epoxy resin using speeds up to 8000 rpm, discovered that 6000 rpm generated the best particle dispersion.

4.4 MECHANICAL PROPERTIES OF NANOREINFORCED ADHESIVES

Development on NP-reinforced adhesives encompasses nearly three decades of intensive research. A great majority of the studies conducted in this field have had their main driving objective set on achieving higher mechanical properties of ABJs. This includes improvements in peel and shear strengths, as well as the fracture toughness of ABJs. In fact, the number of studies in these contexts has become enormous in recent years. Tables 4.2 and 4.3 provide a summary of some of the studies conducted, targeting the improvements in peel and shear strengths of ABJs gained by the inclusion of various NPs. The gain in fracture toughness, as a function of percentage by weight inclusion of GNPs attained by various researchers, is shown in the chart illustrated in Figure 4.6. It should be stated that the data presented in the tables and figure are merely a snapshot of some of the recent studies on these topics. A more elaborate compilation of improvement in fracture toughness of polymers gained by inclusion of various NPs can be found in the concise literature survey presented by Domun et al. [58].

As can be seen, a wide range of results have been reported on the influence of NPs on the shear and peel strengths of adhesives. The reported results range from a remarkable improvement of mechanical properties by 757% to a degradation of as much as 30%. This enormous range in gain

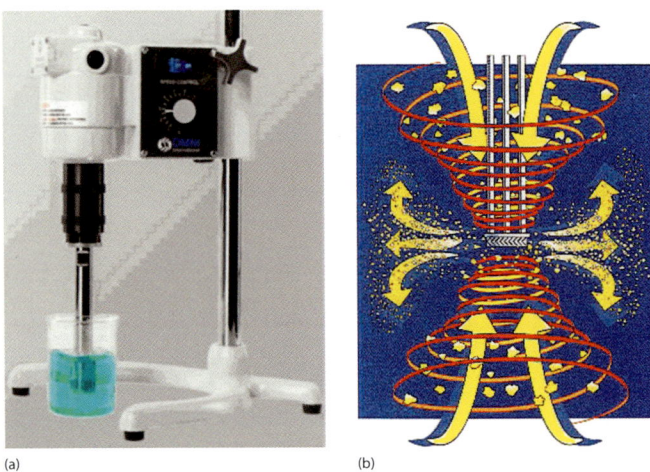

FIGURE 4.5   Typical high-speed shear mixer (a); schematic of vortex created by rotating blades (b). (From Suave, J. et al., *J. Mater. Sci. Eng. A*, 509, 57–62, 2009.)

or loss of enhancement is attributed to a large number of parameters that influence the outcome of inclusion of NPs in adhesives or matrices in general. Certain factors, including manufacturing-related parameters, quality of NPs, particle size, and functionality of NPs, to mention a few, all contribute to either enhancement or degradation of mechanical properties of adhesives reinforced by NPs. More consistent results could be expected by elimination of the aforementioned undesirable factors. In the next section, a review of some of the parameters that could induce either enhancement or degradation of properties of adhesives is presented.

It should be noted that the degradation of ABJ strength as a result of inclusion of NPs, which has been reported in some of the studies, is in concert with those reported for polymers hosting various types of NPs. Perhaps the most prevalent factor that could instigate the degradation of mechanical properties of polymers as a result of inclusion of NPs is the agglomeration of NPs; however, other factors such as trapped gases in the adhesive mixture and damage to NPs structure could also contribute. Some researchers have also postulated that the interference and negative effect of NPs on the curing process of host polymers could impede the enhancement expected from NP reinforcement. For instance, Dorigato et al. [59] reported reduction in the glass transition temperature of the host polymer

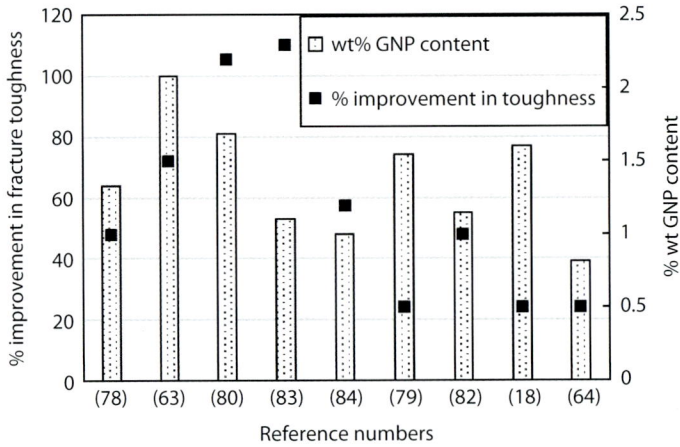

FIGURE 4.6   Improvement gained in fracture toughness based on percentage weight content of GNP. Note that the 100% gain was possible by combining GNP and CNT.

at relatively high NP content. This could be ascribed to the opposing effects blocking the formation of polymeric chains and thereby reducing the evolution of cross-linking. On the other hand, Gkikas et al. [60] incorporated CNTs into an epoxy at 0.5 wt%, and observed the following: (1) an enhancement of the storage modulus and glass transition temperature; and (2) a decrease in the rate of deterioration of these properties after the nanocomposite was subjected to thermal shock and hygrothermal exposure.

4.4.1 Effect of NPs on Adhesive Stiffness

ABJ performance is significantly affected by the stiffness of the overlap region. This region's stiffness is comprised of the stiffness of the adherends, as well as that of its adhesive layer. A large

TABLE 4.2
Influence of NPs on the Peel Strength of the Host Adhesives

Adhesive	NPs	Adherends	NP (wt%)	Strength Increase (%)	Reference
PVAc	CNT	Glass/steel	0.1	7.9	[161]
			0.3	26.3	
			1	5.3	
Epoxy	CNT	White iron	1	29	[162]
Epoxy	SiO$_2$	Steel/aluminum	1	66	[163]
			8	30	
			23	−6	
Polyurethane	SiO$_2$	Cast polypropylene	0.5	212	[88]
			1	462	
			1.5	375	
			2	337	
Epoxy	POSS (with various functionalities; see column 4)	Aluminum	0.3 (aminoethyl)	100	[164]
			4 (epoxycyclohexyl)	42	
			1 (glycidoxypropyl)	68	
			5 (isocyanatopropyl)	757	
			10 (octaphenyl)	126	
Epoxy	Tungsten disulfide	Aluminum	0.3	6	[165]
			0.5	73	
			1	21	
Polystyrene	Fe$_2$O$_3$	Iron	5	236	[166]
			10	276	
			25	260	
Epoxy	Al$_2$O$_3$	Steel	1	75	[167]
			2	350	
			5	125	
Epoxy	Al$_2$O$_3$	Steel	2	116	[168]
Epoxy	Al$_2$O$_3$	Aluminum/composite	5	23	[169]
			10	49	
Epoxy	SWCNTs	CFRP	1	44	[183]
Polyurethane	Nanoclay	Aluminum	1	30	[170]
			3	10	
			5	10	

TABLE 4.3
Influence of NPs on the Shear Strength of the Host Adhesives

Adhesive	NPs	Adherends	NP	Strength Increase (%)	Reference
Epoxy	CNT	Copper	0.8	−19	[171]
Polyimide	MWNT	Steel	0.5	8	[172]
			1	23	
			1.5	4	
Epoxy	MWNT	Carbon/epoxy	1	24	[173]
			5	43	
PVAc	Graphene	Wood	0.75	36	[174]
			1.5	136	
			3	309	
PVAc	Graphene	Beech veneer	0.1	41	[46]
			0.15	52	
			0.3	49	
			0.5	38	
			1	15	
			2	8	
Epoxy	Graphene	Epoxy	1	61	[164*]
Epoxy	Carbon black	Glass/epoxy	0.5	17	[105]
			1	40	
			1.5	46	
			2	38	
			3	9	
Epoxy	Carbon black	Glass/epoxy composite	0.5	14	[106]
			1	28	
			1.5	13	
			2	9	
			3	2	
Epoxy	CNF	Carbon fiber/epoxy	0.25	−3	[89]
			0.5	5	
			1	−6	
Epoxy	SiO_2	Steel	2	15	[163]
			3	28	
			5	35	
			10	22	
			25	−28	
Silylated polyether	SiO_2 (stemming from tetraethyl orthosilicate)	Glass	5	28	[175]
			8	72	
			10	109	
			13	83	
Epoxy	Nanoclay	Glass/epoxy	1	7	[176]
			3	−11	
			5	−18	
PVAc	Nanoclay	Wood	1	11	[177]
			2	20	
			4	25	

(Continued)

TABLE 4.3 (CONTINUED)
Influence of NPs on the Shear Strength of the Host Adhesives

Adhesive	NPs	Adherends	NP	Strength Increase (%)	Reference
Polyurethane	Nanoclay	Aluminum	1	−6	[164,170]
			3	6	
			5	68	
Epoxy	Al_2O_3	Epoxy-steel	1	8	[178]
			2	9	
			5	2	
			10	−1	
			50	−6	
PVAc	Al_2O_3	Wood	1	4	[179]
			2	10	
			4	7	
Epoxy	Zirconia	Aluminum	0.5 (vol%.)	30	[59]
			1 (vol%.)	60	
			1.5 (vol%.)	4	
Phenol–formaldehyde	CuO	Polywood	5.2	14	[180]
Soy protein	$CaCO_3$	Polywood	1	194	[181]
			2	176	
			3	179	
			5	209	
			8	200	
			10	150	
			15	121	
Epoxy	POSS (with various functionalities; see column 4)	Aluminum	0.3 (aminoethyl)	−10	[164]
			4 (epoxycyclohexyl)	−24	
			1 (glycidoxypropyl)	−14	
			5 (isocyanatopropyl)	14	
			10 (octaphenyl)	14	
Epoxy	Nanoelasto-meric copolymer	Aluminum	10	3	[182]
			20	91	
			30	32	
		Copper	10	26	
			20	94	
			30	31	
		Steel	10	1	
			20	15	
			30	11	
Polyurethane	Nanosilica	Aluminum	2	55	[62]
Epoxy	SWCNTs	CFRP	1	−15	[183]
Epoxy	Tungsten disulfide	Aluminum	0.3	5	[165]
			0.5	9	
			1	5	
			3	1	
			5	14	

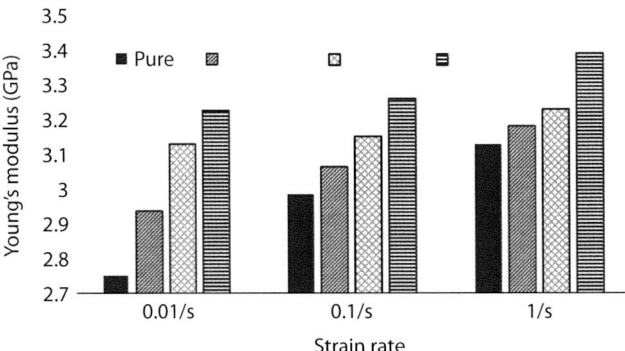

FIGURE 4.7 Comparison of the Young's modulus for the neat epoxy and epoxy–GNP nanocomposites evaluated at various strain rates. (From Shadlou, S. et al., *Mater. Des.*, 59, 439–447, 2014.)

number of the studies, some of which are reported in Tables 4.2 and 4.3, have reported enhancement in the gain in stiffness of polymers that could be attained by addition of NPs. Some direct examples of research are those reported in references [18,61–67]. It is believed that the increase in stiffness of polymers is caused due to the constraint imposed on molecular mobility of the polymer chains by the NPs. NPs, which usually have much greater stiffness than their host polymers, become nested within the spaces that exist among the polymer chains, thus reducing the chains' flexibility [68]. Shadlou et al. [69] studied the effect of graphene nanoplatelets percentage by weight content and loading rate on the tensile and compressive properties of epoxy nanocomposites. The values of the tensile Young's modulus of their nanocomposites measured at various loading rates are shown in Figure 4.7. The results depict a noticeable increase in the Young's modulus of the nanocomposite as the loading rate increases. Moreover, comparison of the results at a given loading rate indicates that the addition of GNPs improved the stiffness of their nanocomposites. The results also reveal the influence of the loading rate on the efficiency gained through the use of GNPs; in other words, as seen, the efficiency of the GNP content decreases with increasing loading rate.

The effects of different NPs and loading rates on the ultimate strength of adhesively bonded single-lap joints (SLJs) have also been reported by Soltannia and Taheri [70]. As the results of their experiments shows (see Figure 4.8), joints formed by GNP-reinforced epoxy exhibit significant increases in their stiffness and strength. The average ultimate shear strength of SLJs with carbon adherends was increased as

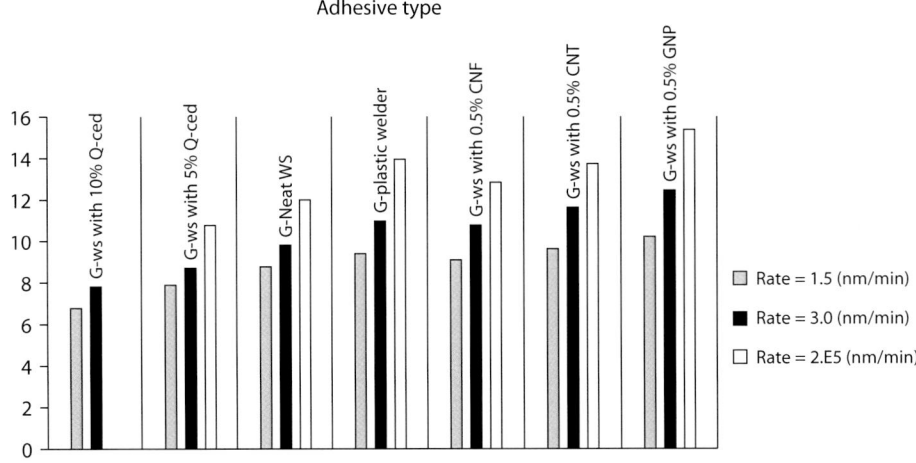

FIGURE 4.8 Effect of loading rate on nanoreinforced adhesively bonded single-lap joints with graphite/epoxy adherends. (From Soltannia, B. and Taheri, F., *Int. J. Composite Mater.*, 3, 181–190, 2013.)

much as 32% (relative to the neat adhesive), when SLJs were subjected to the highest loading rate tried in their experiments; the increase was also significant (26% on average) when SLJs were tested under the quasi-static loading rate. A large amount of information related to the change in the stiffness, and the entire response of an epoxy resin reinforced with various NPs adhering to both glass- and carbon-reinforced epoxy adherends has also been reported by Soltannia and Taheri [70].

Considerable efforts have been expended by researchers with the aim of identifying or developing accurate and reliable models to predict the stiffness of nanocomposites. The Halpin–Tsai model is one of the most commonly used models to predict the Young's modulus of the composites. According to this model, the maximum attainable Young's modulus of a composite with uniform fiber distribution and perfect fiber/matrix bond is given by:

$$E_C = \left(\frac{3}{8} \frac{1 + 2(\lambda)\eta_l V_f}{1 - \eta_l V_f} + \frac{5}{8} \frac{1 + 2\eta_T V_f}{1 - \eta_T V_f} \right) E_m \tag{4.1}$$

in which

$$\eta_l = \frac{E_f/E_m - 1}{E_f/E_m + 2(\lambda)}, \quad \eta_T = \frac{E_f/E_m - 1}{E_f/E_m + 2}, \quad \text{and} \quad \lambda = l_f/d_f \tag{4.2}$$

where:

E_C is the Young's modulus of the composite
l_f, d_f, and E_f are the length, average diameter, and Young's modulus of the reinforcement
E_m is Young's modulus of the matrix
V_f is the reinforcement volume content

As seen, there is a noticeable difference between the experimental and theoretically predicted results (Figure 4.9). This is mainly due to the underlying assumptions used in developing the model, that is, uniformity in distribution of reinforcement, perfect bond between matrix and reinforcement, and the absence of any void in the matrix. As a result, various researchers have proposed modifications to the model to improve its predictive accuracy when applied to nanocomposites [69,71,72].

In another study, Soltannia et al. [74] investigated the effects of nanoreinforcement on the mechanical response of a nanoreinforced resin, subjected to different strain rates. The loading rates considered were 1.5, 15, 150, and 1500 mm/min (strain rates of 10^{-3}, 10^{-2}, 10^{-1}, and 1 s^{-1}). They also conducted a parametric study to ascertain the influences of various parameters (i.e., the particle type, their percentage by weight) on the mechanical properties of the nanocomposite. The results obtained through their modeling technique compared well with those obtained experimentally.

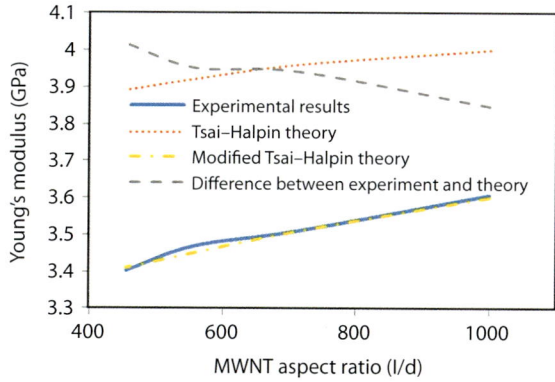

FIGURE 4.9 Comparison of experimental results and theoretical predictions of the Young's modulus of MWNT reinforced nanocomposites. (From Ayatollahi, M.R. et al., *Polym. Test.*, 30, 548–556, 2011.)

Their solution starts with the representation of the inelastic strain components expressed in terms of the deviatoric stress components as follows:

$$\dot{\varepsilon}^I_{ij} = 2D_0 \exp\left[-\frac{1}{2}\left(\frac{Z}{\sigma_e}\right)^{2n}\right]\left(\frac{S_{ij}}{2\sqrt{J_2}} + \alpha\delta_{ij}\right) \tag{4.3}$$

where:

$\dot{\varepsilon}^I_{ij}$ is the inelastic strain rate tensor, which can be defined as a function of deviatoric stress

J_2 is the second invariant of the deviatoric stress tensor

Z and α are the state variables

D_0 is the maximum inelastic strain rate

n is the rate dependency of the material

σ_e is the equivalent (or effective) stress

Note that α controls the level of hydrostatic stress' effect as a state variable. Moreover, D_0 and n are material constants. The equivalent (or effective) stress (σ_e) can be defined as a function of the mean stress, such that the summation of the normal stress components, σ_{kk}, is three times the mean stress, as follows:

$$\sigma_e = \sqrt{3J_2} + \sqrt{3\alpha\sigma_{kk}} \tag{4.4}$$

Soltannia et al. [74] represented this equation in the following incremental forms for it to be solved numerically:

$$d\varepsilon^I_{ij} = \left(2D_0 \exp\left[-\frac{1}{2}\left(\frac{Z}{\sigma_e}\right)^{2n}\right]\left(\frac{S_{ij}}{2\sqrt{J_2}} + \alpha\delta_{ij}\right)\right)dt \tag{4.5}$$

$$de^I_{ij} = d\varepsilon^I_{ij} - \frac{d\varepsilon^I_{kk}}{3}\delta_{ij} \tag{4.6}$$

$$de^I_e = \sqrt{\frac{2}{3}de^I_{ij}de^I_{ij}} \tag{4.7}$$

$$dZ = q(Z_1 - Z)de^I_e \tag{4.8}$$

$$d\alpha = q(\alpha_1 - \alpha)de^I_e \tag{4.9}$$

where:

$\dot{e}^I_e$ is the effective deviatoric inelastic strain rate

q is a material constant representing the hardening rate

Z_1 and α_1 are constants representing the maximum values of Z and α

The hardening rate is determined through trial and error, based on the inelastic shear strain attaining a plateau, or the tensile strain corresponding to the saturation region of the stress–strain curve. Note that the initial value of Z is the magnitude of the stress at the point where the shear stress–strain curve becomes nonlinear. In summary, all the required material constants Z_1, Z_0, α_1, α_0, n, and D_0 of the model can be determined using the conventional stress–strain curves obtained under constant strain rates on the neat polymer under tension, compression, and shear loading states.

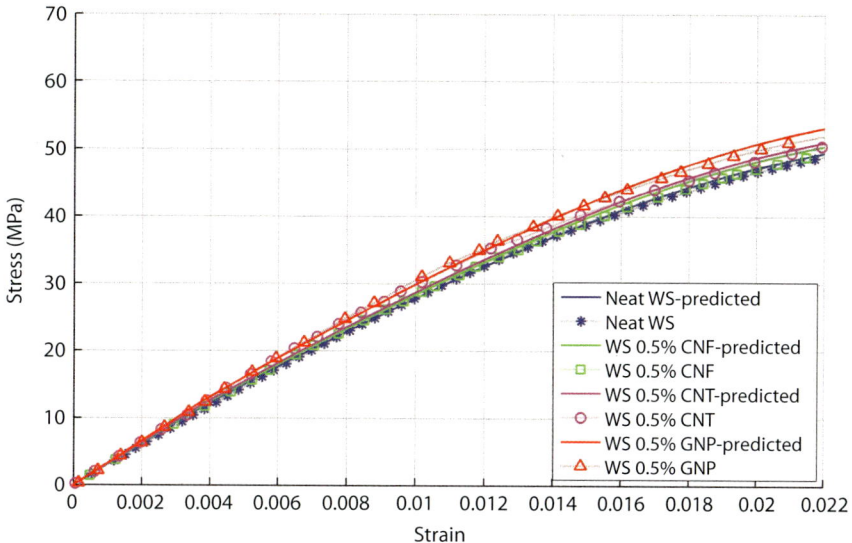

FIGURE 4.10 Comparison of the predicted and experimental stress–strain curves of epoxy West-System epoxy resin reinforced with various types of nanoparticles obtained by the model developed by Soltannia et al. [74]. (From Soltannia, B. et al., *Nanomaterials*, 2016, Article ID 9841972, 2016.)

For instance, the values of n and Z_1 can be identified using the shear stress–strain curves constructed under various strain rates.

On evaluation of the constants of the model, Equation 4.3 can be solved numerically, using the four-step Runge–Kutta (R–K–4) method. The results obtained by their model compared with experimental results are illustrated in Figure 4.10. As can be seen, very good agreement is obtained.

Multiscale modeling is another technique for prediction of the stiffness of nanocomposites, which is more complicated as compared with semiempirical theories; however, such models usually lead to more accurate results [75,76]. Multiscale modeling refers to a modeling approach in which multiple models at different scales are used simultaneously to describe a system [77].

4.4.2 EFFECT OF NPS ON FRACTURE TOUGHNESS

When considering fracture failure modes in ABJs, there are essentially three different modes by which ABJs can fail. The first one is referred to as the *cohesive mode*, which is the desirable failure mode. In this mode, a crack would initiate and propagate within the bulk adhesive layer (see Figure 4.11a). The second one is the *interfacial failure mode*; in this mode, a crack would initiate at the adhesive–adherend interface and subsequently propagate within the interface (see Figure 4.11b). The third mode, which is referred to as a *mixed-mode failure*, occurs when a crack initiates at the

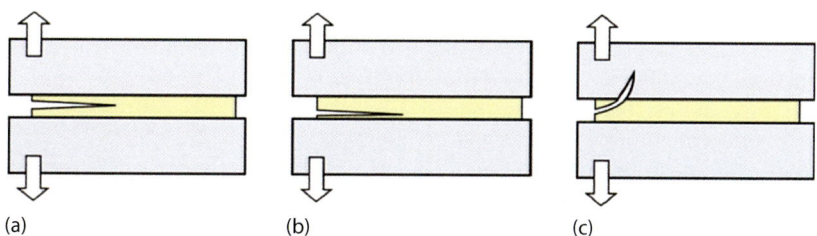

(a) (b) (c)

FIGURE 4.11 Failure mechanisms in ABJs: (a) cohesive failure (bulk adhesive failure), (b) interfacial failure, (c) crack running into substrate. (From Wikipedia: https://en.wikipedia.org/wiki/Adhesive.)

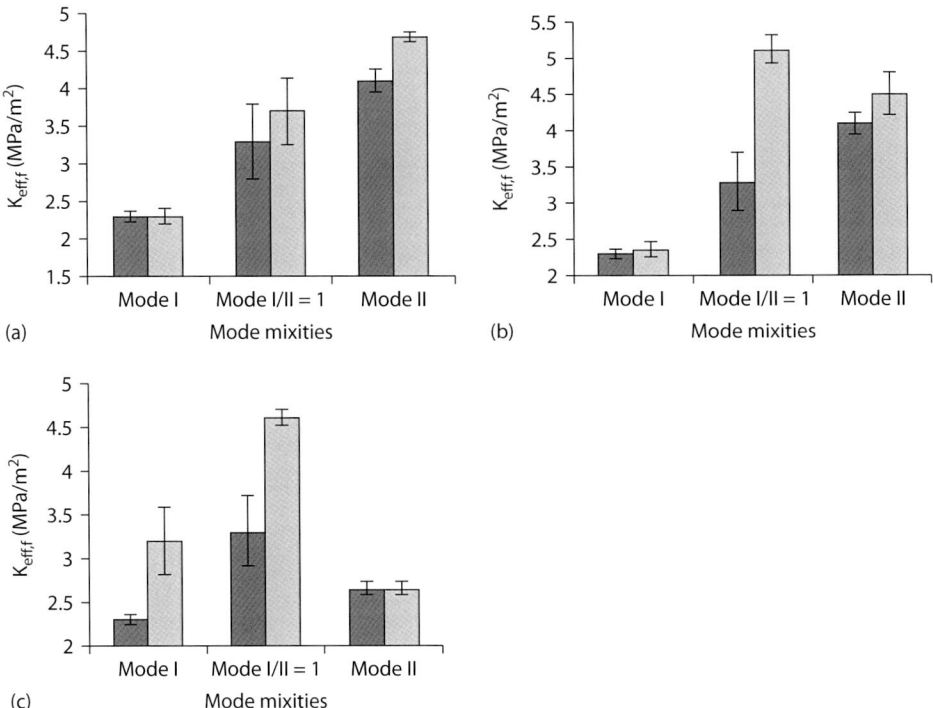

FIGURE 4.12 Effective fracture resistance, K_{eff}, for pure epoxy and for nanocomposites reinforced with (a) ND, (b) CNF, and (c) GO. Dark-gray bars refer to neat epoxy, and light-gray bars refer to NP-doped epoxy. (From Shadlou, S. et al., *Compos. Struct.*, 95, 577–581, 2013.)

interface boundary, but instead of growing within the interface layer, it propagates into the adherend (see Figure 4.11c). As stated, the cohesive mode is the preferred mode, because the interfacial mode often initiates due to improper surface treatment of adherends or faulty procedure during fabrication of the joint. Therefore, the fracture toughness of the adhesive should be evaluated through cohesive failure mode. This parameter is the key factor that governs the mechanical strength and durability of ABJs, in particular if a microcrack is present in the adhesive.

The use of NPs for enhancement of the fracture toughness of adhesives has been reported by several researchers (see, e.g., [15,20,64,78–84,85]). Shadlou et al. [15] investigated the effect of inclusion of different carbon-based NPs on the fracture toughness of an epoxy resin under different fracture mode mixities. Figure 4.12 shows the experimental results obtained for pure epoxy and the three types of nanocomposites. It is seen that the effective stress intensity factor, K_{eff}, increases from Mode I to Mode II in pure epoxy and also for nanodiamond/epoxy nanocomposites. However, by adding the CNF and GNP reinforcements, the maximum value of K_{eff} occurs in a mixed-mode state, and then, as the fracture state moves from the mixed-mode to the Mode II state, the enhancement of K_{eff} in nanocomposites reinforced with CNF and GNP decreases gradually.

Ahmadi-Moghadam et al. [18] also investigated the enhancement in fracture toughness of epoxy resin using functionalized GNPs. Both oxidation and functionalization processes were utilized to treat the GNPs. They observed remarkable increase in fracture toughness of the resin as a result of functionalizing GNPs with amino (NH_2). A remarkable increase of nearly 100% was achieved when 0.5 wt% of the functionalized GNP was used. Shokrieh and coworkers [64] investigated the influence of GNPs on Mode I fracture toughness of an epoxy system by evaluating the critical stress intensity factor of the nanocomposite adhesive. They also compared their findings with the performance of the same epoxy that was reinforced with graphene nanosheets (GNSs). Figure 4.13 illustrates a summary of their findings. It can be seen that, similarly to the findings of Ahmadi-Moghadam et al.

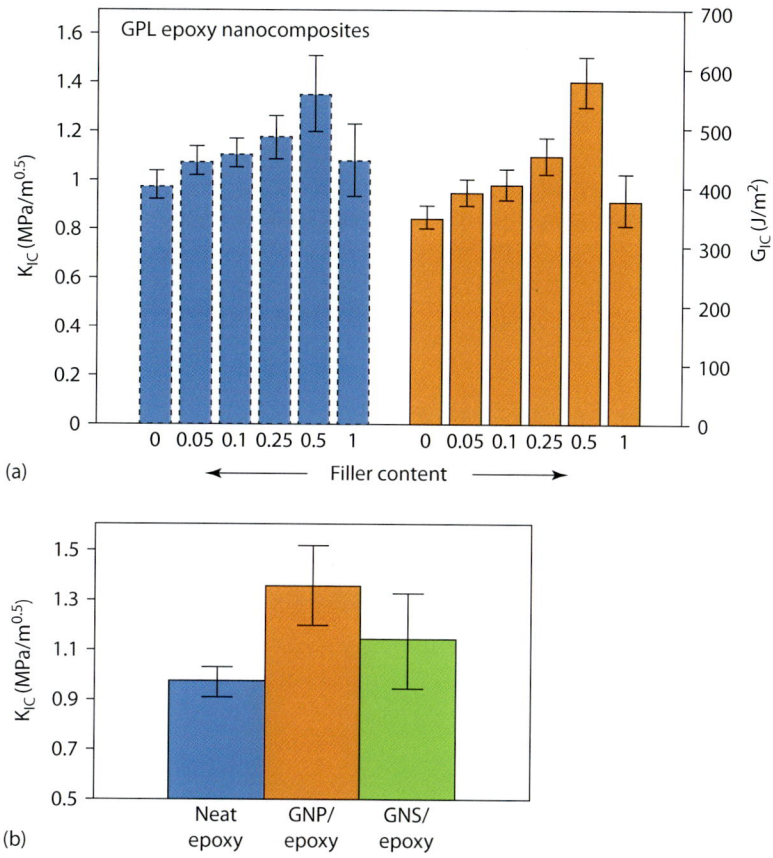

FIGURE 4.13 (a) Influence of percentage weight GNP content on fracture toughness (K_{IC}) and fracture energy (G_{IC}) of GNP-reinforced epoxy; (b) comparison of the fracture toughness of the epoxy reinforced with 0.5 wt% GNP and GNS. (From Shokrieh, M.M. et al., *Fatigue Fract. Eng. Mater. Struct.*, 37, 1116–1123, 2014.)

[18], Shokrieh et al. also observed 0.5 wt% inclusion of GNP to produce the greatest enhancement in toughening the epoxy. Moreover, the higher contents of GNP decreased the resin's toughness by a significant margin (likely due to agglomeration).

In another study, Ahmadi-Moghadam and Taheri [17] conducted a systematic investigation to assess the fracture responses of E-glass fiber-reinforced epoxy laminates, with their midplane resin layer modified with 0.5 wt% of different types of GNPs (i.e., unmodified and functionalized GNP). They evaluated Modes I and II fracture toughness of their GNP-reinforced composite. To evaluate Mode III fracture toughness, they employed a newly introduced torsion test. SEM examination of the fracture surfaces was also carried out to examine the enhancing effect of GNPs with respect to the energy release rate of the laminates and their failure modes. The enhancement in fracture toughness values for the three modes are illustrated in Figure 4.14. In the figure, UG refers to untreated GNP, OG refers to oxidized GNP, G-SI refers to silane-treated GNP, and G-NH2 refers to ammonia-treated GNP.

Principally two mechanisms are responsible for the enhancement in fracture toughness gained by the inclusion of NPs in polymers. These are: (1) crack deviation; and (2) crack bridging. Either of these mechanisms may be dominant over the other one, depending on the geometry of the NPs. For instance, in nanocomposites hosting CNTs of relatively long cylindrical shape, the main energy-dissipating mechanism is crack bridging. This is because CNTs have very small diameters compared with their lengths and, thus, the crack deviation may take a secondary role [68]. The crack-deviation

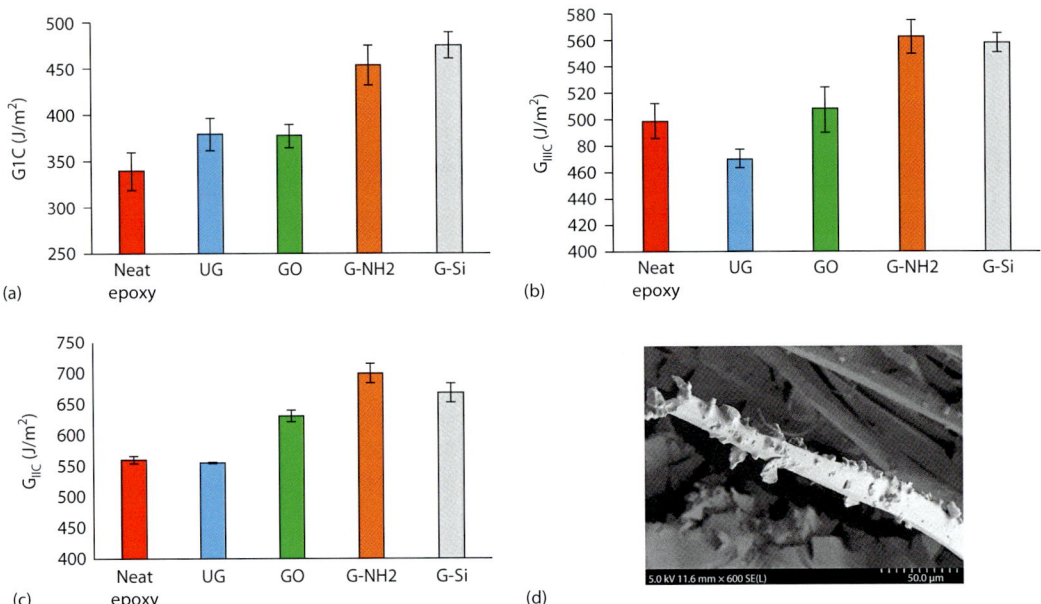

FIGURE 4.14 Influence of GNP inclusion (as treated with various chemicals) on the fracture toughness of epoxy; (a) Mode I, (b) Mode II, (c) Mode III; (d) the SEM image illustrates the excellent adhesion of GNP as a result of functionalization of GNPs with silane. (From Ahmadi-Moghadam, B. and Taheri, F., *Eng. Fract. Mech.*, 143, 97–107, 2015.)

mechanism can be identified by investigation of the roughness of the fracture surfaces of nanocomposites. That is to say, the higher the roughness of the fracture surface, the more the crack would deviate from its original plane. Figure 4.15 shows the increase in fracture surface roughness in the neat epoxy, and the epoxy containing 4 wt% of A–P–Fe_2O_3 NPs. In this figure, atomic force microscopy (AFM) images were taken with a 500 nm × 500 nm scan area from specimens prepared by Sun et al. [67]. In the images, the full height scale is 100 nm. As can be seen, the surface is relatively smooth in the case of neat epoxy. The so-called *river line* seen in the image (see Figure 4.15a), indicates the characteristic brittle fracture behavior of the epoxy. In contrast, the fracture surface of the epoxy resin containing A–P–Fe_2O_3 NPs is somewhat rougher than that seen in the neat epoxy (see Figure 4.15b). Moreover, careful examination of the image shown in Figure 4.15b reveals curled and entangled regions, indicating that the A–P–Fe_2O_3 NPs were embedded and tightly held by the epoxy matrix and were broken instead of being pulled out during the test. This further confirms the presence of a strong interfacial bond between the two constituents.

The bridging mechanism plays an important role in improving the fracture toughness of composites [86]. When a NP bridges the two fracture surfaces, depending on the embedded length,

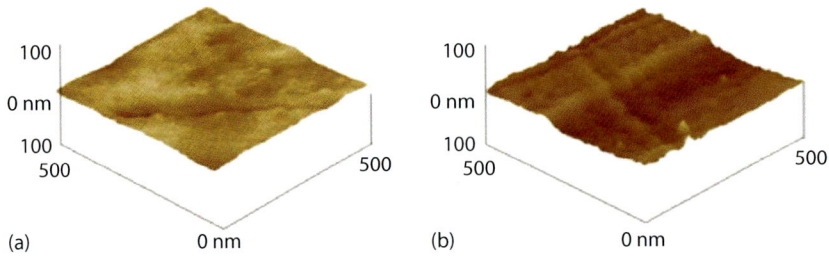

FIGURE 4.15 Comparison of roughness of fracture surfaces of (a) neat epoxy and (b) epoxy containing 4 wt% of aminated PVP-modified Fe_2O_3 NPs. (From Sun, T. et al., *Mater. Des.*, 87, 10–16, 2015.)

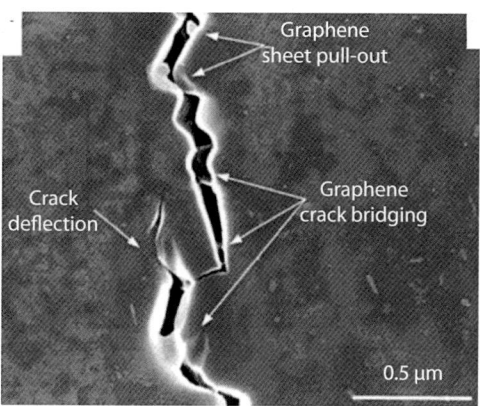

FIGURE 4.16 Micrographs showing the interaction of GNPs in bringing and deflecting cracks (note also the pulled-out GNPs on crack-front). (From Koratkar, N. A., *Graphene in Composite Materials: Synthesis, Characterization and Applications*, p. 148, Destech Publishers, Lancaster, PA, 2013.)

interfacial strength, mode of fracture, and flexibility of the NP, it may either fracture or pull out from one of the surfaces. To see this bridging mechanism, one requires high magnification micrographs. A micrograph clearly illustrating the bridging mechanism in a GNP-reinforced Si_3N_4 composite is illustrated in Figure 4.16.

4.4.3 EFFECT OF NPs ON SURFACE WETTABILITY

As stated previously, the preferred failure mode of ABJs is the cohesive fracture failure, that is, within the adhesive layer. Under this failure mode, NPs can exhibit their highest potential for bringing and deflecting cracks. To ensure such mechanism, one has to ensure strong interfaces between the adhesive layer and its adherends. The interfacial strength is, therefore, one of the key factors governing the satisfactory performance of ABJs. A weak interfacial strength would lead to interfacial failure of ABJs, as schematically illustrated in Figure 4.11b. The adhesive interfacial strength is largely affected by the surface preparation of adherends, but is also related to the wettability of the adhered substrate by the adhesive. The wettability can be quantified in terms of the interfacial tension. In other words, the higher the interfacial tension, the poorer the wettability [87]. The interfacial tension can be calculated using the following empirical formula [88]:

$$\gamma_{SL} = \frac{(\gamma_S^{0.5} - \gamma_L^{0.5})^2}{1 - 0.015(\gamma_S\gamma_L)^{0.5}} \tag{4.10}$$

where γ_S and γ_L are the surface free energy of the substrate and the surface tension of the nanocomposite slurry, respectively, and γ_{SL} represents the interfacial tension between the nanocomposite slurry and the substrate. Figure 4.17 shows the variation in the surface tension of nano-SiO_2-reinforced nanocomposite slurry reported by Fu et al. [88]. The wettability can also be directly determined by measuring the static contact angle. A smaller contact angle evidences better wettability of the adhesive. Dorigato et al. [59] measured the epoxy–water contact angle. Representative images of a water droplet on pure epoxy, on the epoxy filled with 0.5 vol%. zirconia, and on the aluminum substrate are illustrated in Figure 4.18. The measured values for all types of materials are shown in Figure 4.19. The results show that the presence of zirconia NPs results in a noticeable decrease of the equilibrium contact angle values. In fact, for the pure epoxy sample, an equilibrium contact angle of 82° was measured, while a mean angle of 71.5° was measured for the epoxy reinforced with 0.5 vol%. zirconia. These values indicate that the water contact angle values on the nanocomposite are closer to that displayed by the aluminum substrate (59.4°). Dorigato et al. [59] associated the

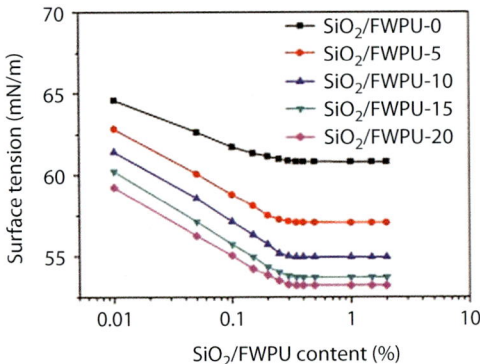

FIGURE 4.17 Surface tension of the SiO$_2$/FWPU nanocomposite slurry (FWPU: fluorinated waterborne polyurethane). (From Fu, H. et al., *J. Ind. Eng. Chem.*, 20, 1623–1632, 2013.)

subsequent increase of contact angle observed in specimens with relatively high NP contents to the degradation in dispersion uniformity of their NPs, possibly due to the formation of agglomerates. They further concluded that the inclusion of zirconia NPs led to a better interfacial wettability and chemical compatibility between the adhesive and substrate, positively contributing to the shear resistance of the joints.

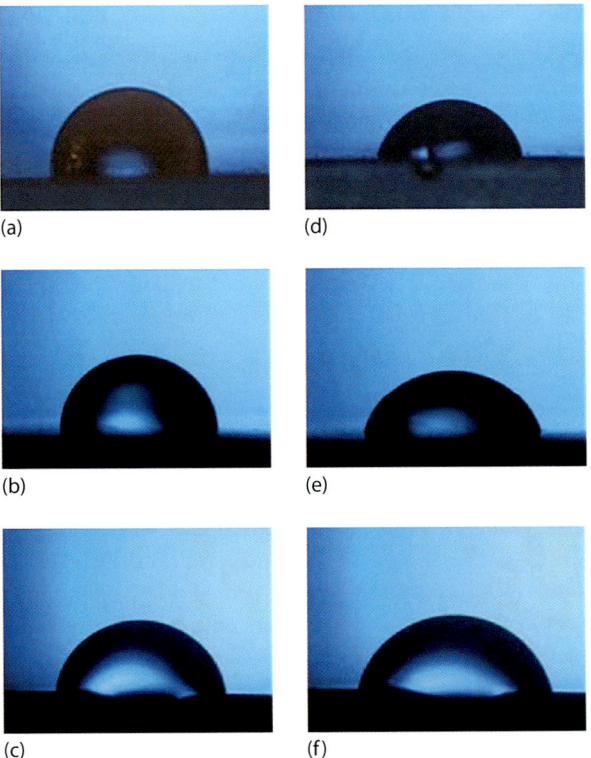

FIGURE 4.18 Images of epoxy–water contact angles; (a), (b), and (c) represent static condition; (d), (e), and (f) represent the equilibrium (vibrated) conditions; (a) and (d) are epoxy, (b) and (e) are epoxy-B-0.5 (epoxy modified with calcined zirconia), and (c) and (f) are aluminum substrates. (From Dorigato, A. et al., *Compos. Interfaces*, 17, 873–892, 2010.)

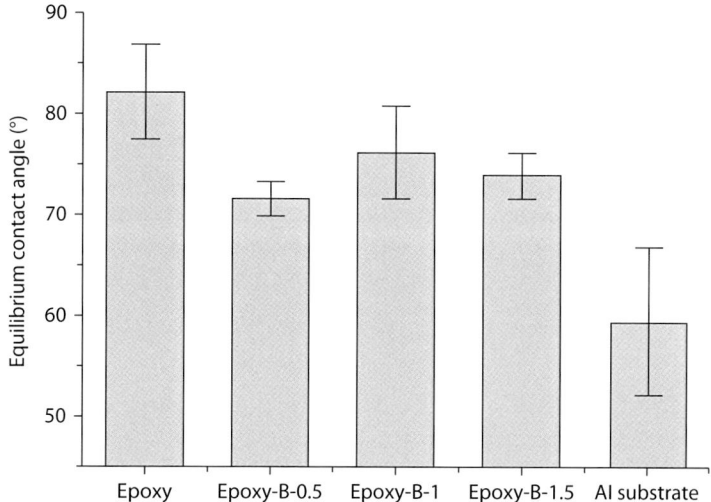

FIGURE 4.19 Substrate–water equilibrium contact angles for epoxy-B-x (epoxy-zirconia) nanocomposites and aluminum substrate. (From Dorigato, A. et al., *Compos. Interfaces*, 17, 873–892, 2010.)

Pinto et al. [46] also measured equilibrium contact angles of ultrapure water drops on PVAc films hosting various GNP loadings (see Figure 4.20). As can be seen, the measured contact angle is approximately 43° on PVAc, while the angle increased to approximately 56° for PVAc specimens hosting 0.1 wt% GNPs. The higher hydrophobicity of the film has been attributed to the presence of partially exposed GNPs on the specimen surface.

Prolongo et al. [89] investigated the effect of different surface treatments and also different contents of CNFs on the contact angle of epoxy adhesive. The results are depicted in Figure 4.21. It should be noted that the plasma technique noted in the figure is another means to prepare the surface of composite adherends for bonding. In this technique, the adherend surfaces are bombarded by ionized gas, generated by radio-frequency energy. The highest values of contact angle were measured on the as-received adherends (i.e., those without any surface treatment), and no notable differences were observed when a treatment was applied. They further observed that the contact angle for the neat epoxy adhesive decreased when CNF content increased. They ascribed this trend to the

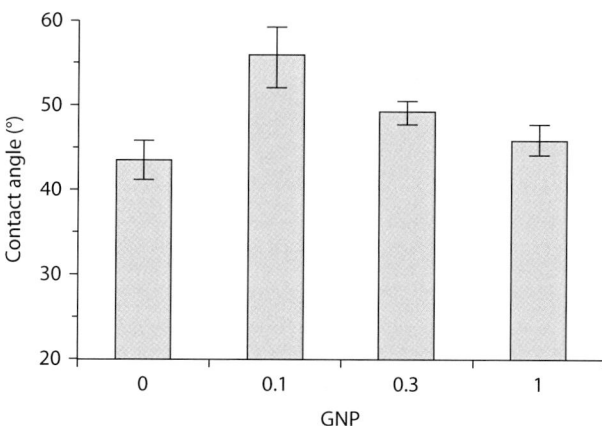

FIGURE 4.20 Equilibrium contact angle of ultra-pure water drops on PVAc films with different GNP loadings. Bars represent standard deviation. (From Pinto, A. M. et al., *Polym. Int.*, 62, 928–935, 2013.)

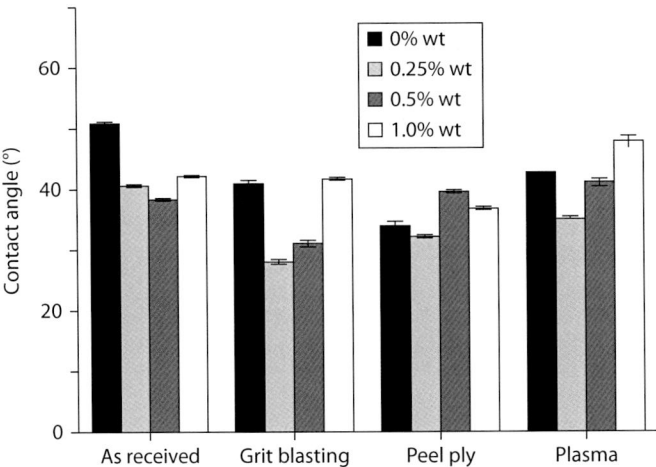

FIGURE 4.21   Contact angle of neat epoxy adhesive and adhesives reinforced with 0.25%, 0.5%, and 1 wt% CNF on carbon fiber–epoxy laminates (untreated and treated with grit blasting, peel ply, and atmospheric plasma). (From Prolongo, S.G. et al., *J. Adhesion*, 85, 180–199, 2009.)

nanoscale size of CNF and the higher chemical compatibility between the carbon fiber-reinforced epoxy composite substrates and the nanoreinforced epoxy adhesive.

It is worth mentioning that the interfacial tension can also be calculated by using the measured contact angles, for which different approaches have been proposed [90–93]. For instance, γSL can be calculated using Young's equation [94,95]:

$$\gamma_S = \gamma_L \cos\theta + \gamma_{SL} \tag{4.11}$$

where is the measured contact angle of nanocomposite slurry on the polyolefin films.

An alternative method to the traditional surface preparation methods, which are generally time-consuming manual processes, is the atmospheric-pressure plasma activation method [96]. This approach has been gaining acceptance as a desirable method of surface preparation prior to bonding [97]. It has been noted that while the traditional procedures have been well-tried and accepted, nevertheless, they yield lower bond strengths, and can even result in damage to adherends when high-modulus FRP adherends are to be mated [96]. Moreover, the traditional approaches of abrading surfaces generate dust, which is an environment, health, and safety concern. Consequently, the atmospheric-pressure plasma activation method has been developed as an alternative, environmentally friendly approach, by which the adherend surfaces are activated without abrasion and any damage to the bulk material. In this procedure, after exposure of the bond surfaces to sufficient plasma afterglow, the surfaces become fully activated when bonded and cured with epoxy adhesives, thus facilitating perfect and strong bonding. It has been claimed that failure of ABJs prepared by this technique would be 100% in the desirable cohesive failure mode. As a result, the lap shear strength and crack delamination resistance (GIC) of the ABJs increase by severalfold in comparison with those produced by the traditional approaches of either solvent wiping or abrasion [98]. The readers are directed to the work of Williams et al. [98] for a detailed review of this technique. They have also discussed the efficacy of this technique in terms of the mechanical properties, the water contact angle (i.e., surface energy), the surface roughness, and the surface composition.

4.4.4   EFFECTS OF NPS ON RESIDUAL STRESSES

Structural components made of various polymers and ABJs that are formed by dissimilar materials become subjected to significant residual stresses during their curing process. The magnitude of the residual stresses becomes even more significant in ABJs that are formed by dissimilar adherends. Not only do residual stresses in ABJs develop when the joint is undergoing its curing process, but they become even more exacerbated when the joint assembly is subjected to thermal and hygrothermal loading conditions during its service life. The magnitude of residual stresses becomes greater when the mismatch in the coefficient of thermal expansion (CTE) of the adhesive and adherends is large. In ABJs formed by hot-cured adhesives, the residual stress would start developing during their bonding process, which would then remain in the assembly throughout their service lives; on the other hand, ABJs made with cold-cured adhesives would not retain a substantial level of residual stress. However, a significant amount of thermal stresses could develop when the joint is exposed to thermal load. The magnitude of residual stresses becomes more significant in ABJs mating to metallic adherends. This is because, in general, there is a greater mismatch in the CTE values of metals and adhesives in comparison with that between FRPs and adhesives. For instance, the CTE of typical epoxies hovers around 60×10^{-6} m/mK, while for aluminum it is 22×10^{-6} m/mK, for steel it is around 13×10^{-6} m/mK, and it would vary considerably for FRPs, depending on the FRP's layup and fiber orientation (a typical value for epoxy reinforced with unidirectional E-glass fibers would be 0.66×10^{-5} m/mK, while values of -0.533×10^{-6} m/mK and 1.592×10^{-6} m/mK have been reported for cross-ply and quasi-isotropic graphite–epoxy laminates [C6000/PMR-15, from Celanese Corp., Irving, TX], respectively [99]).

In general, and compared with their host polymers, NPs have a very small CTE. For instance, the CTE of a graphene sheet is about -0.5×10^{-6} m/mK (valid between 80°C and 197°C) [100]; thus, addition of small contents of NPs to polymer can reduce the polymers' CTE by a noticeable margin. Chow [101] proposed an equation for predicting the CTE of nanocomposites. He extended the *mean field theory*, which was originally developed for evaluating the elastic modulus of composites, to evaluate the CTE of composite materials. He suggested the following equations for calculating the longitudinal, $_L$, and transversal, $_L$, CTEs of a composite.

$$\theta_L = \theta_m + \frac{k_f}{k_m} \frac{(\gamma_f - \gamma_m)G_1 V_f}{2K_1 G_3 + K_3 G_1} \tag{4.12}$$

$$\theta_T = \theta_m + \frac{k_f}{k_m} \frac{(\gamma_f - \gamma_m)G_3 V_f}{2K_1 G_3 + K_3 G_1} \tag{4.13}$$

where:
θ_m is the matrix linear CTE
k_f and k_m are the bulk moduli of nanofiller and matrix, respectively
V_f is the filler volume fraction
γ_f and γ_m are the bulk thermal expansions of nanofiller and matrix, respectively
K_i is the stress intensity factor
G_i is the fracture toughness
K_i and G_i are calculated by Equations 4.14 and 4.15, respectively:

$$K_i = 1 + \left(\frac{k_f}{k_m} - 1\right)[(1 - V_f)\alpha_i + V_f] \tag{4.14}$$

$$G_i = 1 + \left(\frac{\mu_f}{\mu_m} - 1\right)[(1 - V_f)\beta_i + V_f] \tag{4.15}$$

where μ is the shear modulus, and α_i and β_i are functions of the GNP characteristic ratio (i.e., thickness to diameter (t/D)) and Poisson's ratio of the matrix, respectively [100].

Several researchers have reported the reduction in CTE of host polymers that results from inclusion of various types of NPs [7,30,102–104]. An example of such research can be found in [105], in which the effect of carbon black NPs on the CTE of its host epoxy has been reported. The variation

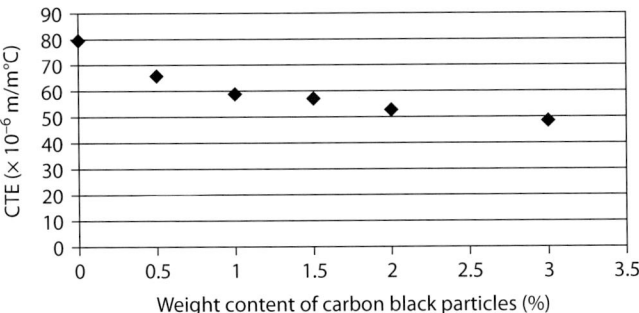

FIGURE 4.22   Coefficient of thermal expansion (CTE) of epoxy adhesive with respect to the weight content of carbon black particles. (From Park, S.W. et al., *J. Adhesion Sci. Technol.*, 23, 95–113, 2009.)

in CTE of the epoxy adhesive with respect to the carbon black content is shown in Figure 4.22. As seen, the CTE decreases as the carbon black content is increased; this is because of the lower value of CTE of carbon black. The minimum CTE was reported as 48.4 10^{-6} m/mK when the carbon black content was 3.0 wt%, which is 40% less than that of the neat epoxy. Park and Lee [106] also used the finite element method to investigate the influence of carbon black on the maximum principal strain in their ABJs (see Figure 4.23). The graphs illustrate that the value of maximum principal strain decreased at higher contents of carbon black.

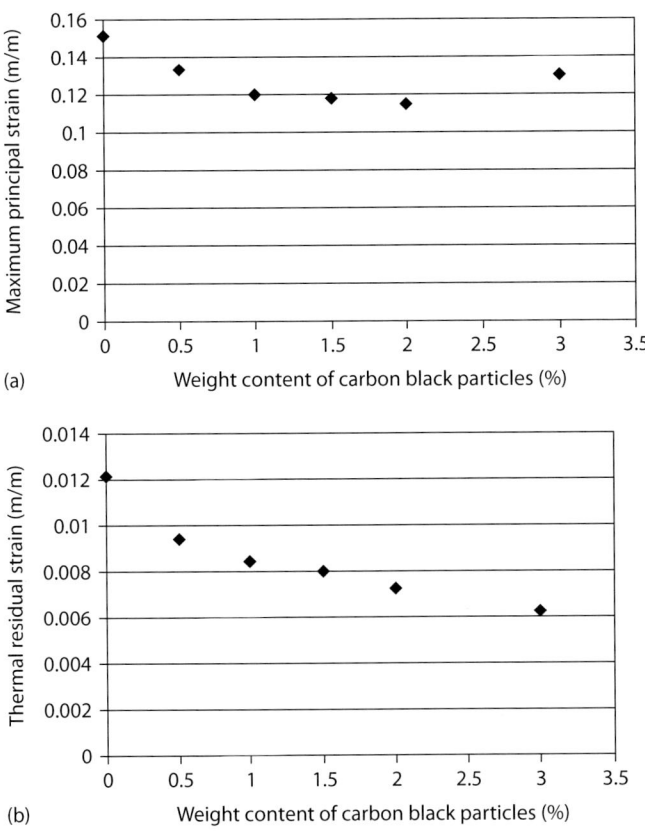

FIGURE 4.23   (a) Maximum principal strain in ABJ with respect to the weight content of the carbon black particles subjected to tensile load of 16 kN; (b) thermal residual strain. (From Park, S.W. et al., *J. Adhesion Sci. Technol.*, 23, 95–113, 2009.)

4.4.5 EFFECT OF NPS ON DURABILITY OF ADHESIVES

Durability of polymer composites has been shown to be significantly affected by thermal and hygro-thermal loadings [107]. The durability of ABJs made of composite adherends are shown to be significantly susceptible to such loadings as well. Exposure of ABJs to especially varying (cyclic) moisture and temperature can cause detrimental outcomes to the durability of ABJs [108]. The absorbed fluids may plasticize and induce relaxation of adhesives, and degrade the adhesives' mechanical properties; the latter is considered to be one of the primary causes of failure of ABJs [108,109]. In this section, a summary of some of the notable investigations into the effect of NPs on the durability of ABJs is reported.

The durability of a CNT-reinforced epoxy adhesive bonded aluminum alloy joints, utilizing the Boeing wedge test [110], was investigated by Yu et al. [111]. The schematic of the test specimen is shown in Figure 4.24.

They also studied the effect of the CNT content on the durability and failure mode of the joints. The wedge test creates a relatively high stress concentration at or near the interface, thus increasing the joint's sensitivity to environmentally caused degradation. Therefore, the test is usually used to provide quantitative durability data for ABJs. A tensometer was inserted into the wedge, which was subjected to a constant load and speed into the bondline of the flat specimens. This process creates a tensile stress in the region surrounding the crack tip. After the specimens were allowed to equilibrate at the ambient conditions for several hours, the initial crack length was recorded. Subsequently, while the wedge in each specimen was subjected to constant load (and the crack), the specimens were immersed in 60°C water. They then measured the resulting crack propagation for a period of up to 90 h, to establish the relative bond durability performance in a humid environment. The results are depicted in Figure 4.25.

The results shown in Figure 4.25 indicate that the initial crack length, which was measured immediately after the wedge test specimen was immersed in water (i.e., at time 0 h), was very different from one specimen to another. For the specimens bonded with CNT-reinforced epoxy, the initial crack length depended highly on the CNT weight fraction. The initial crack length decreased when CNT weight fraction was increased from 0% to 1 wt%, but it then increased only slightly as the CNT weight fraction increased from 1% to 5 wt%. The initial crack length for the joint with epoxy filled with 0.5 wt% CNTs was about 70.3% lower in comparison with that of the joint made with neat epoxy. However, the best results were obtained for the joint made with epoxy containing 1 wt% CNT. The results clearly indicate that the addition of CNTs into the epoxy significantly improved the bond strength of the ABJs.

Another interesting and notable investigation is that conducted by Gkikas et al. [60]. They investigated the performance of CNT-modified epoxy resins and their hybrid composites after hygrothermal exposure and thermal shock conditions. They observed that the diffusion coefficient and water uptake at equilibrium significantly reduced after the addition of CNTs to their carbon fiber-reinforced polymers (CFRPs). They attributed the gain to the reduction in the free volume and

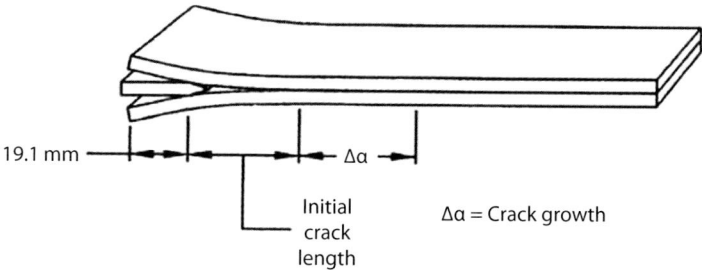

FIGURE 4.24 Configuration of the wedge-test specimen. (From ASTM-D3762-03, American Society for Testing and Materials, Philadelphia, 2010.)

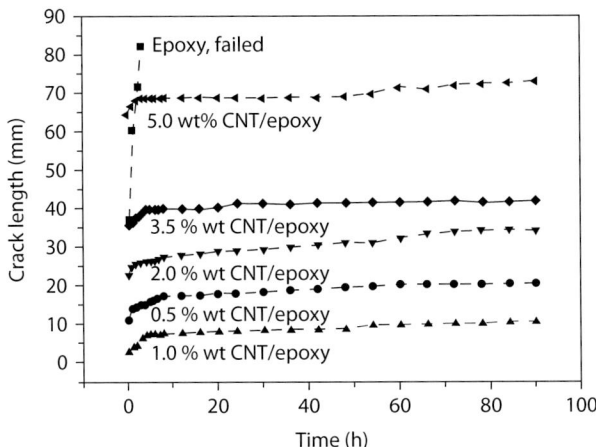

FIGURE 4.25 Crack length of the CNT-filled epoxy adhesive joints as a function of immersion time in 60°C water. (From Yu, S. et al., *J. Appl. Polym. Sci.*, 111, 2957–2962, 2009.)

creation of a tortuous path created by the inclusion of CNTs that, in turn, hindered the diffusion of water. Interestingly, however, the interlaminar shear strength of their unexposed specimens was slightly reduced with the addition of CNTs. Moreover, the interlaminar shear strength (ILSS) of the specimens subjected to hygrothermal exposure was reduced by 50%–60%; this observation was similar when considering both their neat and doped systems.

The reader is also directed to the work of Heshmati et al. [112], who have presented a thorough review of the current state of research work conducted on the environmental durability of adhesively bonded steel–FRP joints, also identifying the research needed in this field. In their paper, they have identified the most important factors that influence the environmental durability of adhesively bonded steel–FRP joints. They also presented the main aging mechanisms that have been observed. It should be noted that the review presented by Heshmati and his coworkers focuses on factors that affect the durability of ABJs used mainly in civil engineering applications.

4.5 OTHER ADVANTAGES OF NANOREINFORCED ADHESIVES

As stated in Section 4.1, the main goal of this chapter was to provide a summary of the enhancement of the mechanical properties of adhesives and ABJs that have been attained by inclusion of NPs in the adhesive, as shown by various researchers. Although enhancements in the strength and stiffness of adhesives are the most important attributes that NPs could impart to ABJs, these are not the only positive attributes NPs offer. The other positive attributes that one may obtain by inclusion of NPs in an adhesive are briefly discussed in the subsequent sections.

4.5.1 ELECTRICAL PROPERTIES

In general, polymers that are used to form structural adhesives are not electrically conductive; however, some types of NPs, especially carbon NPs, are superconductive. In fact, some are piezoelectric. Therefore, addition of an appropriate amount of such NPs to adhesives could significantly decrease their electrical resistance, also referred to as the *percolation threshold* (see [20,54]). Generally, there are two different types of thresholds, namely, static and kinetic thresholds. The static percolation threshold refers to a situation when randomly distributed filler particles form the percolating paths. In the kinetic percolation, the particles are free to move, thereby forming a conducting network at much lower particle concentrations [113]. Different parameters, however, could affect the resulting electrical conductivity, including the fabrication method, matrix wettability, and filler dimensions.

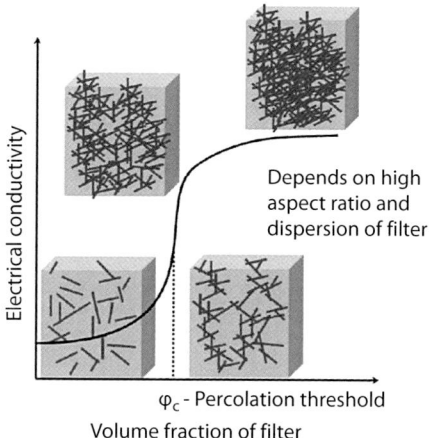

FIGURE 4.26 Schematic illustrating the influence of NPs dispersion on the electrical resistivity of nano-composites. (From Chandrasekaran, S., Development of nanoparticle modified polymer matrices for improved fiber reinforced composites. PhD thesis, Department of Mechanical Engineering, Technische Universität Hamburg-Harburg, Germany, 2014.)

It is noteworthy to mention that while an effective method of NP dispersion would result in better mechanical properties, it may, however, lower nanocomposites' electrical conductivity [114]. The percolation threshold is affected by the structural denseness of the NPs in the matrix and the available multiconductive paths; indeed, it has been shown that the conductivity value reaches saturation due to lesser influence of the links between the conducting particles. Based on these arguments, the electrical percolation curve has been represented in the form of an S-curve, as shown in Figure 4.26. This percolation threshold is strongly influenced by the geometry and aspect ratio of the filler [115]. This phenomenon is also supported by the test data reported by Ahmadi-Moghadam et al. [54] reproduced in Figure 4.27.

To develop an advanced electrical-conductive adhesive (ECA), Amoli et al. [20] dispersed high-aspect-ratio silver nanobelts (NBs) in an epoxy resin. They achieved an outstanding level of enhancement in the electrical conductivity of 1300% of their epoxy resin. This was made possible by incorporating a small amount of the Ag NBs into a conventional ECA with 60 wt% Ag

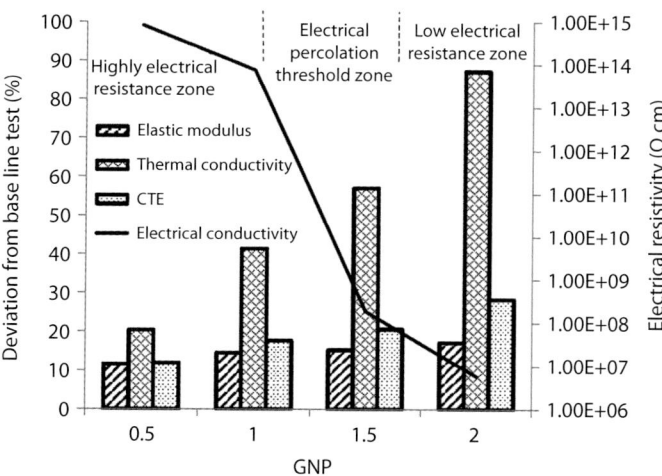

FIGURE 4.27 Test data illustrating the influence of NPs loading on mechanical, thermal and electrical properties. (From Ahmadi-Moghadam, B. and Taheri, F., *J. Mater. Sci.*, 49, 6180–6190, 2014.)

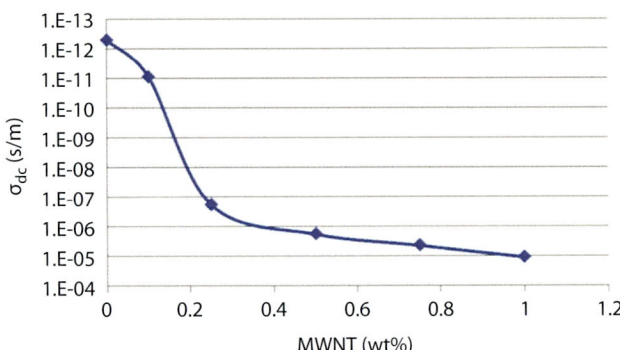

FIGURE 4.28 Variation of electrical resistivity (σ_{dc}) of epoxy/MWNT nanocomposites. (From Ayatollahi, M. et al., *Polym. Test.*, 30, 548–556, 2011.)

microflakes. The addition of 2 wt% of the NBs into a conventional ECA with 80 wt% Ag flakes reduced the bulk resistivity to a value comparable to that of a typical eutectic solder.

Chandrasekaran [115] prepared epoxy nanocomposites reinforced with three different types of carbon nanofillers (thermally reduced graphene oxide [TRGO], graphite nanoplatelets [GNPs] and multiwall carbon nanotubes [MWCNTs]). These nanocomposites were formed by dispersing the fillers in epoxy matrix using a three-roll mill. The effect of filler type on the mechanical, electrical, and thermal properties of the nanocomposite is studied. Graphene-based nanocomposites outperformed MWCNT/epoxy composites in terms of mechanical properties. However, discovered although the electrical conductivity of the GNP–epoxy composite reached 5.8×10^{-3} S/m at 2.0 wt%, the percolation threshold was measured to be 0.3 wt%.

Ayatollahi et al. [22] measured the DC electrical conductivity of CNT–epoxy for different NP contents. As seen in Figure 4.28, the resistivity decreases dramatically as the filler content approaches the value corresponding to the electrical percolation threshold, and the conductive network of the CNTs is formed at about 0.25 wt%.

Similar results have been reported by Yu et al. [116] and Wang et al. [161] for the surface electrical resistivity of epoxy adhesives. In addition, several analytical and semiempirical models have been developed to predict the electrical conductivity of nanocomposites. Some examples of the predictive models are the *excluded volume model* developed and proposed by Celzard et al. [117], Lu and Mai's model [118], and the more recently proposed model by Li and Kim, which is based on the average interparticle distance [119].

There are many applications in which electrically conductive adhesives are utilized. Some of the examples are paints with magnetic properties, circuit boards, damage sensing, and solar cells [120,121]. In addition to the applications mentioned, the use of conductive adhesives for structural health monitoring purposes has recently attracted considerable attention. Lim et al. [122] used a conductive CNT-modified adhesive (in addition to an acoustic emission sensor) to monitor the effect of adherend surface treatments on the failure mechanism of an SLJ. The results are shown in Figure 4.29. For their untreated adherend, the electrical resistance increased in a step-like manner, which represents a sudden failure in the adhesive. A relatively lower overall shear strength in an ABJ could result due to the potential weak interaction between the steel and adhesive, while the change in the resistance (i.e., jumps in the electrical resistance) can signify the onset of damage long before the specimen reaches its ultimate shear strength. On the other hand, in case of saline-treated adherends, the resistance increases gradually throughout the loading regime, which would indicate that a progressive damage event is occurring during the loading cycle, possibly due to damage in the adhesive layer.

In another research work [123], the effect of GNPs on the electrical conductivity of epoxy resin was assessed, and the potential for detection of damage by monitoring the change in the electrical

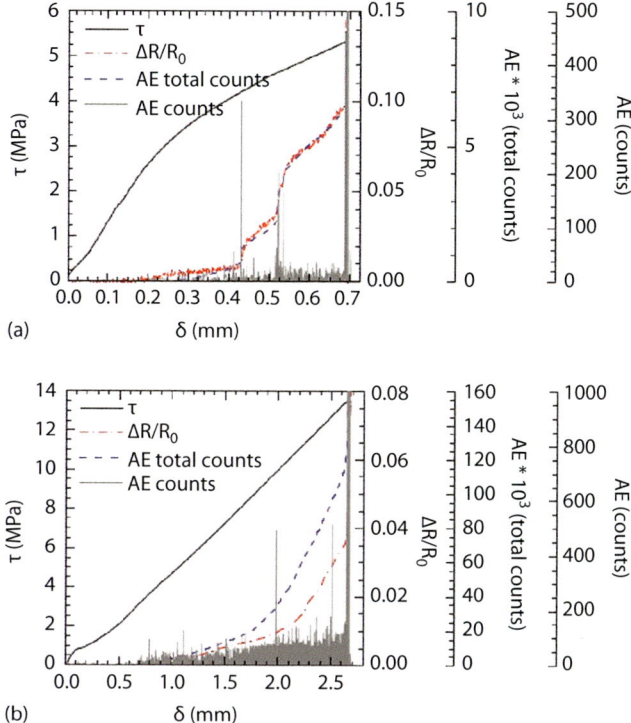

FIGURE 4.29 Mechanical, electrical, and acoustic emission responses of (a) untreated specimens (b) saline-treated specimens (where is the shear stress and is the displacement). Note the step-like variation in the electrical resistance in untreated adherend (left) versus its gradual increase in the saline-treated adherend (right). (From Lim, A. S. et al., *Compos. Sci. Technol.*, 71, 1183–1189, 2011.)

conductivity was demonstrated. In that study, the GNP-doped epoxy resin was prepared and used to laminate unidirectional carbon-epoxy plates. A small 10 mm diameter thin Teflon patch was used to create delamination within the midplane of the six-layer laminate. The invisible flaw was detected by monitoring the electrical resistance change in CFRP laminates made of GNP-reinforced resin. Their results indicated that, firstly, the electrical conductivity of CFRP laminates that were prepared by the GNP-reinforced epoxy resin dramatically decreased when the weight fraction of GNP exceeded 1.5 wt%. Secondly, their electrical resistance change (ERC); the ERC method was demonstrated to have successfully distinguished the invisible delamination within the laminates. It was also shown that the presence of GNPs increased the sensitivity of CFRPs to the electrical resistivity change resulting from delamination between the laminae and microcracks present within the matrix. The measured change in electrical conductivity of their GNP-reinforced laminates was found to be 27% higher than that of the laminate without GNPs (see Figure 4.30).

4.5.2 Thermal Properties

Just as the CTE of adhesives would be affected by the inclusion of NPs, so would their thermal conductivity. Some NPs have extremely large values of thermal conductivity. For instance, the thermal conductivity of CNTs at room temperature can be as high as 6600 W/mK. Therefore, the thermal conductivity of the host polymer reinforced with these NPs would be significantly altered. The main applications of thermally conductive adhesives are in heat sink bonding, potting/encapsulating sensors, and in fabrication of power semiconductors.

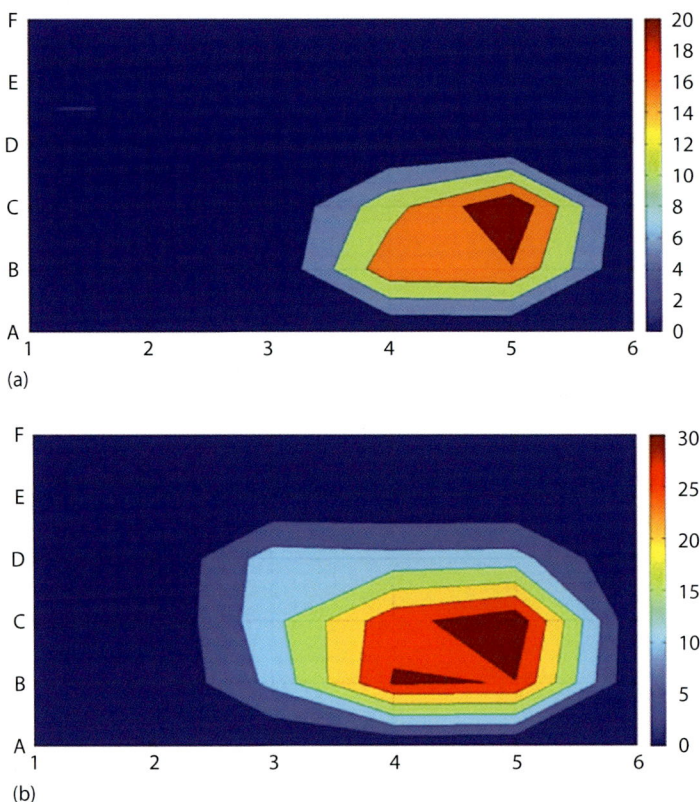

(a)

(b)

FIGURE 4.30 Contour variation in electrical conductivity signifying the presence of a hidden delamination (a) in a graphite–epoxy specimen composite without GNPs; (b) in a graphite–epoxy composite specimen hosting GNPs. (From Ahmadi-Moghadam, B. and Taheri, F., *Proceedings of the 19th International Conference on Composite Materials [ICCM19]*, Montreal, Quebec, Canada, 2013.)

Cui et al. [124] developed a new high-performance phase change graphene and polyolefin hot-melt pressure-sensitive adhesive with high thermal conductivity and appropriate hardness, without the overflow (or running) problem. They showed that the inclusion of an appropriate amount of graphene increased both the thermal contact conductivity and hardness of the adhesive. They found the best weight percentage content of graphene that developed the lowest thermal contact resistance to be 6 wt%. They claimed their adhesive film could perform better than most commercially available thermal interface materials, asserting that the performance could also be enhanced further by the application of greater temperature and pressure.

With the aim of improving the thermal conductivity of epoxy composites, Gao and Zhao [125] prepared nanocomposites with various morphologies [i.e., nano-AlN particles, AlN–graphene nanohybrids (AlN/GE) and AlN–carbon nanotubes nanohybrids (AlN/CNTs)]. They observed that the epoxy composite with AlN–GE nanohybrids had the highest enhancement in its thermal conductivity in comparison with their neat epoxy. Moreover, the density of epoxy composites with the synthesized nanofillers was decreased and the corresponding thermal stability was enhanced.

Qiao et al. [126] investigated the effect of nanosilver particles on the thermal conductivity of an epoxy-based adhesive, and the results are shown in Figure 4.31. An obvious increase in the thermal conductivity value is observed (i.e., an increase from 1.057 W/mK (70 wt%) to 1.70 W/mK (83 wt%). This indicates that the thermal conductivity increases with an increase in the nanosilver content. Their results further reveal that the thermal conductivity is not as sensitive to agglomeration of NPs as is the mechanical strength.

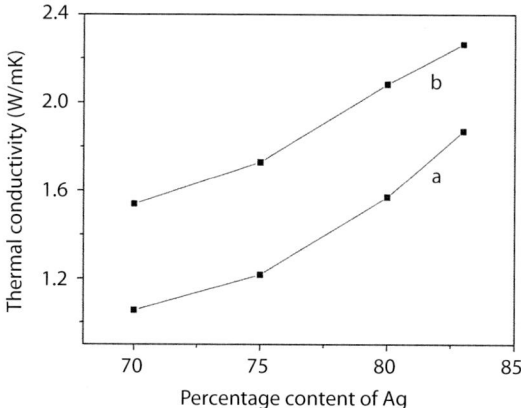

FIGURE 4.31 Relationship between thermal conductivity of epoxy and percentage weight content of nanosilver (a) spherical silver and (b) mixed spherical and flaky silver. (From Qiao, W. et al., *Int. J. Adhesion Adhesives*, 48, 159–163, 2014.)

4.5.3 GAS PERMEATION BARRIER PROPERTIES

Despite all the advantages that are offered by polymers, because of their relatively large molecules, their inherent permeability to gases and vapors, including oxygen, carbon dioxide, and organic vapors has impeded their use in certain applications, including the food industry [127]. Inclusion of NPs, however, can significantly minimize the aforementioned shortfall of polymers. The review of the state-of-the-art literature conducted by Cui et al. [127] concentrates on the influence of inclusion of graphene on the gas barrier performance of reinforced polymers. A comprehensive review on modeling aspects of gas barrier performance of NP-reinforced polymers is also provided in the same article. Table 4.4 reports a summary of more notable research works in this area, extracted from Cui et al. [127]. The number in brackets seen in the first column of the table refers to the references cited in their manuscript (see [127]).

As for the optimum amount load of NPs to minimize gas permeability, it has been shown that, in fact, the most effective reduction in gas permeability may be attained by the inclusion of relatively lower volume fractions of NPs. The lower volume would also enable the nanocomposite to retain the optical clarity of the neat resin, which is desired mostly in packaging applications [36,128].

In regard to adhesives, the use of nanoclay to enhance gas permeation properties of polyurethane adhesive was investigated by Osman et al. [129]. The dependency of the water vapor transmission rate through the nanocomposites as a function of the volume fraction of inorganics is illustrated in Figure 4.32. It can be seen that the transmission rate decreased asymptotically with an increasing volume fraction of nanoclays in all types of nanocomposites. It was also observed that maximum efficiency could be obtained if the nanoclay particles were oriented parallel to one another and perpendicular to the gas penetration direction.

4.5.4 BIOADHESIVES

Many naturally formed materials with glue-like properties are loosely termed as *bioadhesives*. However, many researchers use this term in the context of a synthetic material that is designed to adhere to biological tissue. Indeed, the use of bioadhesives and their ability to control bleeding and wound closure dates back to ancient times. Researchers have reported successful application of different types of bioadhesives containing POSS [130], SiO_2 [131], nanoclays [132,133], silica [134], zirconia [135], nanosilver [136], to mention a few. In addition to increasing the mechanical strength of the base matrix, incorporation of various NPs could create other properties that

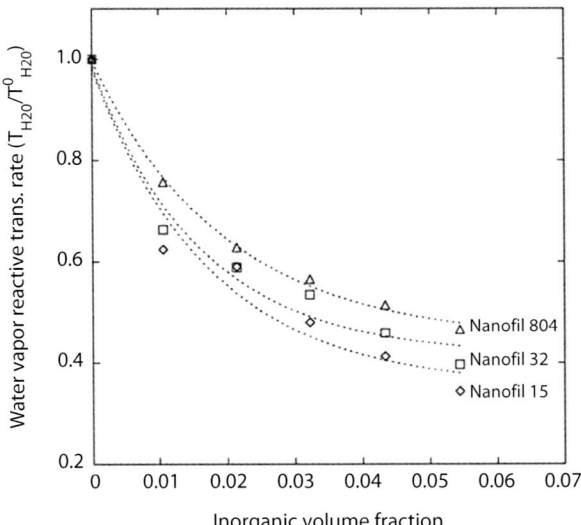

FIGURE 4.32 Dependence of the water vapor transmission rate through the PU–nanocomposites on the inorganic volume fraction. The dotted lines are guides for the eye (Nanofil 15, Nanofil 32, and Nanofil 804 are trade names for various types of organoclays by Sud-Chemie [Moosburg, Germany]). (From Osman, M. A. et al., *Macromolecules*, 36, 9851–9858, 2003.)

could be of interest. For instance, Kassaee et al. [137] showed that the addition of silver NPs to an acrylic resin rendered a dental composite that exhibited strong antibacterial activity against *Escherichia coli*.

As another example, it is important for the adhesives used in dentistry to be easily distinguishable by x-raying teeth. The development of radiopaque adhesives using Ta_2O_5/SiO_2 NPs has been reported by Schulz et al. [131], and the improvement in radiopacity is illustrated in Figure 4.33. As seen, the increase in Ta_2O_5 content increased the composite radiopacity by a notable margin. In fact, the nanocomposites containing 40% mol (83 wt%) Ta_2O_5 exhibited higher radiopacity than dentin and enamel. However, the surface-functionalized particles exhibited slightly lower radiopacity than the untreated ones. A recent article authored by Mehdizadeh and Yang [138] is also a good source for gaining an insight into various bioadhesives.

4.5.5 BIOINSPIRED ADHESIVES

There exist another classification of adhesives that has been relatively recently identified as *bioinspired* adhesives. This class of adhesives has also emerged as a result of recent advances in nanotechnology. As a result, a significant amount of effort has been expended by various researchers to (1) characterize a broad array of adhesive mechanisms imparted by utilizing biological organisms in their natural environments; and (2) mimic performance of such adhesives through synthetic platforms [139]. The performance of these adhesives has been observed to be remarkable; this class of adhesives provides effective solutions to stringent requirements especially required by the health industry. These adhesives are required to have high adhesion strength, perform reliably during their intended service lives, and essentially be able to adhere to any surface, including in aquatic and fluid environments.

The efforts thus far have led to several interesting applications in various industries, such as [140]:

- In the wood industry: Adhesives made of environmentally friendly materials used in manufacturing plywood.

TABLE 4.4
Gas Permeability of Graphene/Polymer Nanocomposites

Polymer	Filler	Filler Loading	Processing*	Gas	Permeability
PLA [6]	GONS	1.37 vol%.	Solution	O_2	1.145×10^{-14} cm^3 cm/cm^2 s^{-1} Pa^{-1}
				CO_2	1.293×10^{-14} cm^3 cm/cm^2 s^{-1} Pa^{-1} (50% RH)
Cellulose [18]	GNPs	5 wt%	Solution	O_2	$\sim 0.8 \times 10^{-18}$ m^3 m/m^2 s Pa
				CO_2	$\sim 1.15 \times 10^{-18}$ m^3 m/m^2 s Pa
PS [61]	Graphene	2.27 vol%.	Solution	O_2	1.84 barrer[a]
IIR [62]	TRG	5 phr	Solution	O_2	28.4 ml/m^2.24 h
PMMA [67]	Graphene oxide	1 wt%	Solution	O_2	~ 1.25 ml/[m^2.day.atm] (RH = 50%)
HDPE [96]	DA-GO, DA-RGO	1 wt%	Solution	O_2	1.75×10^{-14} cm^3 cm/(cm^2 s Pa)
LLDPE [135]	DA-G	1 wt%	Solution	O_2	19.5 fm/Pa.S
				N_2	5.7 fm/Pa.S
PLA [136]	Graphene oxide, GNP	0.4 wt%	Solution	O_2	1.2×10^{-18} m^2/s Pa
				N_2	0.25×10^{-18} m^2/s Pa
PET [137]	fGO	3 wt%	Solution	O_2	1.14×10^{-3} barrer
PVA [138]	GONS	0.72 vol%	Solution	O_2	0.24×10^{-15} cm^3 cm cm^{-2} s^{-1} Pa^{-1}
EP [139]	Fe$_3$O$_4$/GNPs	1 wt%	Solution	He	2.2×10^{-60} Pa m^3/s
PU [140]	Graphene oxide	1 wt%	Solution	He	0.970 ± 0.027 barrer
PANI [141]	Graphene	0.5 wt%	Solution	O_2	0.1056 barrer
PND [142]	AFG	5 wt%	Solution	O_2	~ 25 cc/m^3 day
PPC [143]	EFG	5 wt%	Solution	O_2	51.8 cm^3 m^{-2} day^{-1}
EVOH [144]	TRG	0.5 wt%	Solution	O_2	8.517×10^{-15} cm^3 cm cm^{-2} s^{-1} Pa^{-1}
PVA [145]	Graphene oxide, RGO	0.3 wt%	Solution	O_2	5.14×10^{-15} mol s^{-1} m^{-1} Pa^{-1} (60% RH)
TPU [106]	iGO, TRG	3 wt%	Solution, *In situ.* Melt	N_2	N/A
PVA [146]	Graphene oxide	0.07 vol%	Solution, Recrystallization	O_2	$<5.0 \times 10^{-20}$ cm^3 cm cm^{-2} Pa^{-1} s^{-1}
PET [147]	GNPs	1.5 wt%	Melt	O_2	0.1 cc/m^2/day/atm
PU [148]	RGO	2 wt%	Melt	O_2	4.13×10^{-12} cm^3(STP)/cm-s-cmHg
PEN [149]	Graphite & FGS	10 wt% 4 wt%	Melt	H_2	0.76 barrer 0.61 barrer
PP [150]	Exfoliated GNPs	3 vol%	Melt	O_2	~ 175 cc mil/m^2 atm
Nylon [151]	FG	0.3 wt%	Melt	O_2	10.1 cc/(m^{3}day) (100% RH)
PC [152]	FGS	3 wt%	Melt	He	8.8 barrer
				N_2	0.2 barrer
PS [153]	Graphene oxide	2 wt%	*In situ*	O_2	2.24 barrer
				N_2	0.43 barrer
PEI [154]	RGO	12.5 wt%	*In situ*	H_2	~ 46 cc/m^2 day atm
PI [155]	RGO	30 wt%	*In situ*	O_2	26.07 cm^3 cm^{-2} 24 h^{-1} atm^{-1}
PMMA [156]	Graphene	0.5 wt%	*In situ*	O_2	0.81 barrer
PAN [157]	Expanded graphite	4 wt%	*In situ*	O_2	~ 0.27 l/cm^2/min
BPEI[36]	Graphene oxide	N/A	LbL	O_2	<0.05 cm^3 cm^{-2} day-1
PEI [158]	Graphene	N/A	LbL	O_2	0.05 cc/m^2 day

(Continued)

TABLE 4.4 (CONTINUED)
Gas Permeability of Graphene/Polymer Nanocomposites

Polymer	Filler	Filler Loading	Processing*	Gas	Permeability
PDDA, SPVDF [159]	Graphene oxide	N/A	LbL	H_2	3.1 cc mm/m² day atm
PEI [160]	Graphene oxide	91 wt%	LbL	O_2	0.12 cc/m² . atm. day
				H_2	158.1 cc/m² . atm. day
				CO_2	<1 cc/m² . atm. day
XNBR [111]	Graphene oxide	1.9 vol%	Latex co-coagulation	O_2	~1.5 × 10⁻¹⁷ m² Pa⁻¹ s⁻¹
SBR [161]	Graphene	7 phr	Latex compounding	O_2	N/A

Source: Extracted from Y. Cui et al., *Carbon*, 98, 313–333, 2016. The number in brackets seen in the first column refers to the references cited in this reference.

a melt, solution, and *in situ* in this column refer to melt mixing, solution mixing, and *in situ* polymerization, respectively.

b 1 barrer = 10⁻¹⁰ cm³ (STP) cm/cm² s cmHg [137], [162], [163], [164] and [165] = 7.5005 × 10⁻¹⁸ m²/s Pa [166] and [167].

- For drug delivery: A bioinspired adhesive system based on microspheres has been shown to significantly increase the systemic absorption of conventional drugs and polypeptides across the nasal membrane, without the need for use of absorption enhancing agents.
- In wound-healing dressing: A variety of bioinspired surgical adhesives have been developed or are under development with applications in homeostasis, skin closure, and sealing of the colon and blood vessels.
- In the military: Bioinspired adhesives extracted from geckos are controllable, reversible, and adhere to surfaces with a strength that approaches the strength of geckos' toe pads.

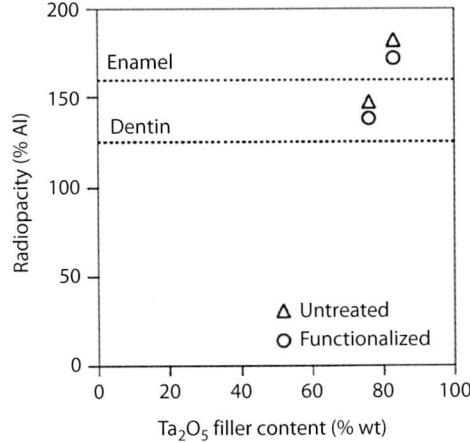

FIGURE 4.33   Radiopacity of nanocomposites containing 20 wt% filler with 35%–83 wt% Ta_2O_5 content normalized with respect to that of equally thick aluminum (particles were functionalized with γ-methacryloxypropyltrimethoxysilane [MPS]). (From Schulz, H. et al., *J. Dentistry*, 36, 579–587, 2008.)

4.6 MODELING OF ABJs

Prediction of the response of ABJs by various methods has been considered extensively by researchers over the past few decades. Analytical, experimental, and numerical methods of various capabilities have been developed and proposed. Most analytical models have aimed at analyzing the adhesive layer shear and peel stresses, and the deformation of the overlap region in ABJs, as well as predicting the fracture response of ABJs, using various fracture mechanics principles. However, most analytical models, despite the fact that they can encompass the intricacies of ABJs made of advanced composites, are only capable of predicting the response of simple ABJ configurations. Even the more recently developed rigorous solutions, such as that proposed by Shahin et al. [141], fall short of being applicable to many practical configurations of ABJs. As a result, the application of the finite element method (FEM), in modeling the response of ABJs, has consumed significant efforts since the 1970s. Taheri et al. [142] provide a detailed review of the computational methods used to characterize delamination and disbonding in adhesives. In that article, they also cover the pertinent structural health monitoring and damage detection techniques.

Most FEM-based works conducted until relatively recently have been based on modeling ABJs as a continuum. However, the more recent advances in numerical methods have led to development and inclusion of various more robust methods, such as the cohesive zone method (CZM) and the extended finite element method (XFEM). Various forms of these techniques are now available in some of the more reputable commercial finite element codes, such as ABAQUS and ANSYS.

In the following sections, the concept of computational simulation of ABJs using the CZM will be briefly discussed. The section will be followed by a brief introduction to the most recent advancement in the area of numerical modeling, namely, XFEM.

4.6.1 COHESIVE ZONE METHOD

The idea of describing the fracture phenomenon in materials as a function of separation of two surfaces was originally introduced by Dugdale [143] and subsequently by Barenblatt [144]. This notion is the foundation of the so-called *cohesive zone* or *damage zone* models. The cohesive zone (CZ) refers to a surface on which displacement discontinuities occur, thus requiring additional constitutive description as compared with a continuum. A CZM describes the variation (or jump) in the normal and tangential components of the displacement across the cohesive surfaces (see Figure 4.34), with suitable traction functions (see, e.g., [145]). The CZM approach is developed under the assumption that the damage mechanisms leading to fracture are located within a thin layer of material in

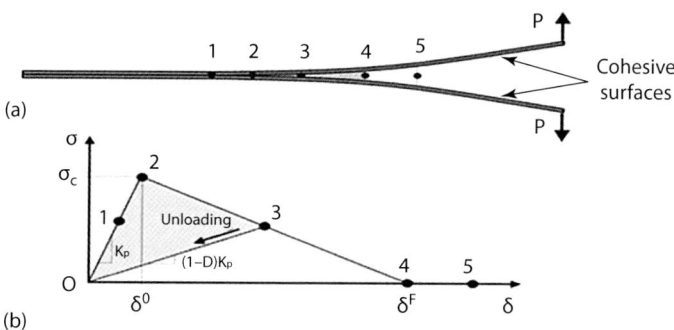

FIGURE 4.34 Schematic model of two separating cohesive surfaces (a); the bilinear cohesive zone constitutive model (b) (numbered stages correspond to the positions identified on [a]). (From Camanho, P.P. and Dávila, C.G., Mixed-mode decohesion finite elements for the simulation of delamination in composite materials, NASA Technical Report/TM, 2002.)

a region ahead of the crack tip, referred to as the *failure process zone (FPZ)*. The material behavior within the FPZ is characterized using a traction–separation law. CZMs are essentially based on the energy principle and a traction–separation law that idealizes the two separating surfaces. To combine damage and fracture mechanics concepts, the area under the traction–separation law is equated to the critical strain energy release rate of fracture, G_c. Furthermore, in the case of an isotropic material, the model can be represented by three parameters: (1) the critical energy release rate (G_c), (2) the critical limiting maximum stress, and (3) the shape of the curve that describes the traction–separation response of the material. CZMs are also categorized based on the basic functions used to define them (e.g., bi- and multilinear, polynomial, trigonometric, and exponential). Moreover, CZMs are widely used in the finite element simulations of crack tip plasticity and creep, crazing in polymers, response of ABJs, and delamination in composites and multilayered materials.

Among the CZM models, the linear elastic/linear softening (bilinear) model is the simplest and also the most commonly used. Figure 4.34 illustrates a typical bilinear constitutive model calibrated for Mode I (separation mode) fracture. The graph illustrates the stress variation as a function of separation displacement, as well as identifying the key points, while the separation of two fractured surfaces is in progress. In the graph, point 1 represents the response of the material in the elastic range, where the material experiences no damage. Point 2 corresponds to the stress state at the onset of damage. The material softening response occurs after this stage, as the damage propagates. The shaded triangular area represents the energy dissipated in a partially damaged state. The unloading that takes place at point 3 would not be recordable during testing, thus, no damage parameter could be accounted for within this stage. At point 4, the separation of the two surfaces commences (see [146–148]).

Alfano [149] conducted one of the earliest studies in which the CZM concept was used to characterize the response of ABJs. He identified the influence of the interface law on algorithmic numerical performance and the resulting accuracy. He identified the trapezoidal law as the least capable means for modeling ABJ response, from both the perspective of numerical stability and convergence of the solutions. From the perspective of accuracy, the exponential law was identified as the optimal one, while the bilinear law was proved to yield the best compromise between the solution time and accuracy. Later, Alfano et al. [150] investigated Mode I fracture in ABJs by modeling the response of the double-cantilever beam (DCB) specimen, combined with the concept of intrinsic piece-wise linear cohesive surface relation. They also established the influence and sensitivity of the CZM's parameters (i.e., fracture strength and critical energy release rate) in predicting the overall performance.

In another study, Alfano et al. [151] explored the performance of various CZMs in modeling cleavage type fracture in ABJs. They tailored a CZM to model Mode I fracture in ABJs that correlated the tensile traction to the displacement jump (as a result of crack opening) along the fracture process zone. They also discussed the difficulties experienced when attempting to directly measure the required fracture parameters. They further demonstrated the procedure to establish the required parameters for their CZM, which was achieved by adjusting the CZM parameters such that the numerical results converged to the experimental ones.

Li et al. [152] used a CZ approach to model the Mode I fracture of adhesive joints made from a polymer–matrix composite. It was concluded that there is a distinction between the characteristic strength of the interface associated with the toughness and the intrinsic cohesive strength at the interface. The characteristic strength of the interface strength (or the peak strength of the triangular region shown in Figure 4.34). Their observations of fiber bridging indicated that, as with any bulk composite material, a second strength parameter could be defined, which is the strength associated with the intrinsic cohesive strength of the interface bondline. This is generally greater than the characteristic strength. They further stated that one would require the value of this intrinsic cohesive strength to accurately analyze some geometries that have very small characteristic dimensions or crack lengths. Also, Li et al. [153] presented a two-parameter model for CZM with a characteristic toughness and a characteristic strength to predict the fracture of notched or cracked specimens.

They determined that the engineering behavior, in terms of strength, deformation, and energy dissipation, were well described by their two-parameter model; however, when considering laminated composites, additional parameters would be required (e.g., the matrix-cracking strength).

It is postulated that the same concept as discussed here can be extended to capture the response of ABJs formed by nanoreinforced adhesives. However, studies considering this topic are very scarce.

4.6.2 EXTENDED FINITE ELEMENT METHOD

Despite the most recent efforts expended to advance and improve the predictive capabilities of the CZM technique, the results obtained by it have been shown to be quite sensitive to the mesh used to discretize an ABJ's bonded region. Moreover, cohesive modeling as a phenomenological continuum theory is associated with some disadvantages, such as an uncertain range of validity and use of constitutive parameters that may not have clear physical meaning, which could be difficult to evaluate experimentally [154].

To resolve the abovementioned issues, a novel numerical method, referred to as the *extended finite element method (XFEM)*, was developed by Melenk and Babuška [155] with the original aim of accurately and effectively capturing the fracture of composites. The basic idea behind the method is based on the mathematical foundation of the partition of unity finite element method (PUFEM) [154,156]. Subsequently, Belytschko and Black [157] developed a minimal remeshing finite element method by using the discontinuous enrichment function, which could account for the presence of a crack. Later, the aforementioned method was improved by Dolbow et al. [158], and since then it has been referred to as XFEM. The advantage of the XFEM method is that it can predict the entire crack propagation within the body independent of the mesh. This means that in XFEM, the crack can initiate and propagate on any location within the finite element mesh. This becomes possible by incorporation of the *enrichment functions*. These functions define the continuous fields to simulate the singular asymptotic material behavior ahead of the crack tip. Therefore, simulation in this case allows displacement jumps along with crack propagation. The enrichment functions have been proven to mimic the response of the region close to the crack tip with reasonable accuracy.

A few researchers have used XFEM to simulate delamination in composite materials. For instance, Sosa and Karapurath [159] investigated the integrity of XFEM in simulating delamination in a cantilever fiber-reinforced composite beam. The XFEM results were compared against the results produced by standard FEM and experiment results. They concluded that XFEM could be considered as a promising technique for the failure analysis of composite structures. Moreover, Moreno et al. [160] numerically and experimentally analyzed the initiation and propagation of crack in a cruciform specimen subjected to a quasi-static biaxial tensile loading. The XFEM was used to study the linear elastic fracture problem with the finite element (FE) software ABAQUS. They obtained a good agreement between the numerical and experimental results. However, in comparison with the applications of CZM for modeling the response of ABJs, the use of XFEM to model such response has been very scarce thus far.

4.7 SUMMARY

This chapter provided a summary of the recent advances made in nanoparticles and their applications in adhesives and ABJs. There are numerous applications in which nanoreinforced adhesives are used, mainly due to their multifunctional properties. In comparison with bulk materials, NPs possess relatively large specific surface areas, high surface energy, a reduced number of structural imperfections, and distinctively different physical properties than the bulk form. Various types of NPs produced from a variety of materials, and their attributes were discussed. As stated, each NP type exhibits one or more unique physical properties that cannot be offered by the same material in its bulk form.

The materials presented in various sections in this chapter demonstrated that one of the biggest challenges in manufacturing polymeric nanocomposites is the uniform dispersion of NPs in a given matrix. Due to their extremely small dimensions, NPs tend to form large bundles or agglomerates. The investigation of various dispersion methods, especially those suitable for mass production of nanocomposites, has been the objective of many ongoing investigations.

Moreover, despite the fact that most researches in regard to the inclusion of nanomaterials in adhesives have concentrated on enhancing the strength and stiffness properties of adhesives, nonetheless, there has been a large number of studies focusing on other properties of adhesives, such as their thermal, electrical, and gas permeability properties.

Finally, the challenges in characterizing the performance of ABJs by analytical means were discussed. The notion of utilizing numerical techniques, which overcome the limitations of analytical methods, was briefly discussed. The discussion included the more recent advancements in the area of modeling of the response of ABJs, namely, the use of CZM and XFEM.

ACKNOWLEDGMENTS

Some of the works that were conducted by the author, and cited throughout this chapter, were made possible by financial supports obtained from the Natural Sciences and Engineering Research Council of Canada and the Auto21 Network of Centres of Excellence. The financial supports are gratefully acknowledged and appreciated.

REFERENCES

1. R. Hosseinzadeh. Analysis and enhancement of tubular bonded joints subject to torsion. PhD thesis, Dalhousie University, Halifax, Nova Scotia, Canada (2009).
2. http://www.azom.com/. Polymer adhesives—advantages of adhesive bonding over alternative assembly techniques by master bond (2011).
3. J. Cheng and F. Taheri. A smart single-lap adhesive joint integrated with partially distributed piezoelectric patches. *Int. J. Solids Struct.* 43, 1079–1092 (2006).
4. J. Cheng and F. Taheri. A novel smart adhesively bonded joint system. *Smart Mater. Struct.* 14, 971–981 (2005).
5. R. Hosseinzadeh and F. Taheri. Non-linear investigation of overlap length effect on torsional capacity of tubular adhesively bonded joints. *Compos. Struct.* 91, 186–195 (2009).
6. R. A. Esmaeel and F. Taheri. Stress analysis of tubular adhesive joints with delaminated adherend. *J. Adhesion Sci. Technol.* 23, 1827–1844 (2009).
7. J. Cheng, F. Taheri and H. Han. Strength improvement of a smart adhesive bonded joint system by partially integrated piezoelectric patches. *J. Adhesion Sci. Technol.* 20, 503–518 (2006).
8. S. Shadlou, B. Ahmadi-Moghadam and F. Taheri. Nano-enhanced adhesives: A critical review. *Rev. Adhesion Adhesives* 2, 371–412 (2014).
9. G. Korotcenkov. *Chemical Sensors: Fundamentals of Sensing Materials. Volume 2: Nanostructured Materials*, Momentum Press, New York (2010).
10. M. R. Ayatollahi, S. Doagou-Rad and S. Shadlou. Nano-/microscale investigation of tribological and mechanical properties of epoxy/MWNT nanocomposites. *Macromol. Mater. Eng.* 297, 689–701 (2012).
11. A.S. Artemov. Polishing nanodiamonds. *Phys. Solid State* 46, 687–695 (2004).
12. M. Ayatollahi, E. Alishahi, S. Doagou and S. Shadlou. Tribological and mechanical properties of low content nanodiamond/epoxy nanocomposites. *Compos. Part B* 43, 3425–3430 (2012).
13. H. Wan, C. Lei, H. Zhou, J. Chen and Y. Jia. Effects of nano-LaF_3 on the friction and wear behaviors of PTFE-based bonded solid lubricating coatings under different lubrication conditions. *Appl. Surf. Sci.* 382, 73–79 (2016).
14. N. Tiruvenkadam, P. R. Thyla, M. Senthilkumar and M. Bharathiraja. Development of optimum friction new nano hybrid composite liner for biodiesel fuel engine. *Transp. Res. Part D Transp. Environ.* 47, 22–43 (2016).
15. S. Shadlou, E. Alishahi and M. Ayatollahi. Fracture behavior of epoxy nanocomposites reinforced with different carbon nano-reinforcements. *Compos. Struct.* 95, 577–581 (2013).

16. J. Gao, W. Li, H. Shi, M. Hu and R. K. Li. Preparation, morphology, and mechanical properties of carbon nanotube anchored polymer nanofiber composite. *Compos. Sci. Technol.* 92, 95–102 (2014).

17. B. Ahmadi-Moghadam and F. Taheri. Influence of graphene nanoplatelets on modes I, II and III interlaminar fracture toughness of fiber-reinforced polymer composites. *Eng. Fract. Mech.* 143, 97–107 (2015).

18. B. Ahmadi-Moghadam, M. Sharafimasooleh, S. Shadlou and F. Taheri. Effect of functionalization of graphene nanoplatelets on the mechanical response of graphene/epoxy composites. *Mater. Des.* 66, Part A 142–149 (2015).

19. J. Lin, R. Wei, D. C. Bitsis and P. M. Lee. Development and evaluation of low friction TiSiCN nanocomposite coatings for piston ring applications. *Surf. Coatings Technol.* 298, 121–131 (2016).

20. B. M. Amoli, E. Marzbanrad, A. Hu, Y. N. Zhou and B. Zhao. Electrical conductive adhesives enhanced with high-aspect-ratio silver nanobelts. *Macromol. Mater. Eng.* 299, 739–747 (2014).

21. M. H. Gabr, N. T. Phong, K. Okubo, K. Uzawa, I. Kimpara and T. Fujii. Thermal and mechanical properties of electrospun nano-celullose reinforced epoxy nanocomposites. *Polym. Test.* 37, 51–58 (2014).

22. M. Ayatollahi, S. Shadlou, M. Shokrieh and M. Chitsazzadeh. Effect of multi-walled carbon nanotube aspect ratio on mechanical and electrical properties of epoxy-based nanocomposites. *Polym. Test.* 30, 548–556 (2011).

23. R. L. Poveda and N. Gupta. Electrical properties of carbon nanofiber reinforced multiscale polymer composites. *Mater. Des.* 56, 416–422 (2014).

24. C. Li, L. Miao, X. Tan, M. Han and J. Jiang. Thermal conductivity of graphene nanoribbons with regular isotopic modification. *J. Comput. Theor. Nanosci.* 11, 348–352 (2014).

25. E. Alishahi, S. Shadlou, S. Doagou and M. R. Ayatollahi. Effects of carbon nanoreinforcements of different shapes on the mechanical properties of epoxy-based nanocomposites. *Macromol. Mater. Eng.* 298, 670–678 (2013).

26. H. Zhang, G. Lee, C. Gong, L. Colombo and K. Cho. Grain boundary effect on electrical transport properties of graphene. *J. Phys. Chem. C* 118, 2338–2343 (2014).

27. V. V. Mody, R. Siwale, A. Singh and H. R. Mody. Introduction to metallic nanoparticles. *J. Pharm. Bioallied Sci.* 2, 282–289 (2010).

28. C.-W. Chou, S.-H. Hsu, H. Chang, S.-M. Tseng and H.-R. Lin. Enhanced thermal and mechanical properties and biostability of polyurethane containing silver nanoparticles. *Polym. Degrad. Stab.* 91, 1017–1024 (2006).

29. S. K. Sardana, V. S. Chava, E. Thouti, N. Chander, S. Kumar, S. Reddy and V. K. Komarala. Influence of surface plasmon resonances of silver nanoparticles on optical and electrical properties of textured silicon solar cell. *Appl. Phys. Lett.* 104, 073903 (2014).

30. A. Evlyukhin, A. Stepanov, A. Dmitriev, A. Akhmanov, V. Bagratashvili and B. Chichkov. Influence of metal doping on optical properties of Si nanoparticles. *Opt. Commun.* 316, 56–60 (2014).

31. P. Kucheryavy, J. He, V. T. John, P. Maharjan, L. Spinu, G. Z. Goloverda and V. L. Kolesnichenko. Superparamagnetic iron oxide nanoparticles with variable size and an iron oxidation state as prospective imaging agents. *Langmuir* 29, 710–716 (2013).

32. E. Ayandele, B. Sarkar and P. Alexandridis. Polyhedral oligomeric silsesquioxane (POSS)-containing polymer nanocomposites. *Nanomaterials* 2, 445–475 (2012).

33. W. Yu, J. Fu, X. Dong, L. Chen and L. Shi. A graphene hybrid material functionalized with POSS: Synthesis and applications in low-dielectric epoxy composites. *Compos. Sci. Technol.* 92, 112–119 (2014).

34. G. Ji and G. Li. Effects of nanoclay morphology on the mechanical, thermal, and fire-retardant properties of vinyl ester based nanocomposite. *Mater. Sci. Eng. A* 498, 327–334 (2008).

35. J. H. Park and S. C. Jana. The relationship between nano- and micro-structures and mechanical properties in PMMA–epoxy–nanoclay composites. *Polymer* 44, 2091–2100 (2003).

36. G. Choudalakis and A. D. Gotsis. Permeability of polymer/clay nanocomposites: A review. *Eur. Polym. J.* 45, 967–984 (2009).

37. V. Pistor and A. J. Zattera. Thermal and rheological properties of poly (ethylene-co-vinyl acetate) (EVA) nanoclay. In: *Handbook of Polymer Nanocomposites Processing, Performance and Application*, pp. 129–152, J. K. Pandey, K. R. Reddy, A. N. Mohanty and M. Misra (Eds.), Springer, Berlin, Heidelberg (2014).

38. T. Mohan and K. Kanny. Water barrier properties of nanoclay filled sisal fibre reinforced epoxy composites. *Compos. Part A* 42, 385–393 (2011).

39. http://www.us-nano.com/

40. M. Z. Rong, M. Q. Zhang, Y. X. Zheng, H. M. Zeng and K. Friedrich. Improvement of tensile properties of nano-SiO$_2$/PP composites in relation to percolation mechanism. *Polymer* 42, 3301–3304 (2001).

41. I. Javni, W. Zhang, V. Karajkov, Z. Petrovic and V. Divjakovic. Effect of nano-and micro-silica fillers on polyurethane foam properties. *J. Cell Plast.* 38, 229–239 (2002).

42. M. M. Jalili, S. Moradian, H. Dastmalchian and A. Karbasi. Investigating the variations in properties of 2-pack polyurethane clear coat through separate incorporation of hydrophilic and hydrophobic nano-silica. *Prog. Org. Coatings* 59, 81–87 (2007).

43. B. B. Mukharjee and S. V. Barai. Influence of nano-silica on the properties of recycled aggregate concrete. *Constr. Build. Mater.* 55, 29–37 (2014).

44. A. V. Shenoy. *Rheology of Filled Polymer Systems*, Kluwer Academic Publishers, Dordrecht, the Netherlands (1999).

45. X. Huang, C. Zhi and P. Jiang. Toward effective synergetic effects from graphene nanoplatelets and carbon nanotubes on thermal conductivity of ultrahigh volume fraction nanocarbon epoxy composites. *J. Phys. Chem. C* 116, 23812–23820 (2012).

46. A. M. Pinto, J. Martins, J. A. Moreira, A. M. Mendes and F. D. Magalhães. Dispersion of graphene nanoplatelets in poly(vinyl acetate) latex and effect on adhesive bond strength. *Polym. Int.* 62, 928–935 (2013).

47. G. Otorgust, A. Sedova, H. Dodiuk, S. Kenig and R. Tenne. Carbon and tungsten disulfide nanotubes and fullerene -like nanostructures in thermoset adhesives: A critical review. *Rev. Adhesion Adhesives* 2, 311–363 (2015).

48. T. Hielscher. *Ultrasonic Lab Devices and Industrial Processors*, Ultrasonic Technology, Germany (2014). https://www.hielscher.com/products.htm (Retrieved July 2017).

49. A. Montazeri and M. Chitsazzadeh. Effect of sonication parameters on the mechanical properties of multi-walled carbon nanotube/epoxy composites. *Mater. Des.* 56, 500–508 (2014).

50. J. Suave, L. A. F. Coelho, S. C. Amico and S. H. Pezzin. Effect of sonication on thermo-mechanical properties of epoxy nanocomposites with carboxylated-SWNT. *Mater. Sci. Eng. A* 509, 57–62 (2009).

51. N. A. Koratkar. *Graphene in Composite Materials: Synthesis, Characterization and Applications*, p. 148, Destech Publishers, Lancaster, PA (2013).

52. A. Yasmin, J. L. Abot and I. M. Daniel. Processing of clay/epoxy nanocomposites by shear mixing. *Scripta Mater.* 49, 81–86 (2003).

53. B. Ahmadi-Moghadam, B. Soltannia and F. Taheri. Interlaminar crack detection in graphene nanoplatelet/CFRP composites using electric resistance change. In: *Proc. 19th International Conference on Composite Materials*, pp. 3597–3607 (2013).

54. B. Ahmadi-Moghadam and F. Taheri. Effect of processing parameters on the structure and multi-functional performance of epoxy/GNP-nanocomposites. *J. Mater. Sci.* 49, 6180–6190 (2014).

55. H. Charles. *High Speed Dispersers*, Ross & Son Company, New York (2014).

56. R. Andrews, D. Jacques, M. Minot and T. Rantell. Fabrication of carbon multiwall nanotube/polymer composites by shear mixing. *Macromol. Mater. Eng.* 287, 395–403 (2002).

57. M. Al-Qadhi, N. Merah, Z. Khan, K. Mezghani, Z. Gasem and M. J. Adinoyi. Effect of sonication and high shear mixing parameters on nanoclay dispersion in epoxy. In: *Proceedings of 15th European Conference on Composite Materials: Composites at Venice, ECCM* 2012 (2012).

58. N. Domun, H. Hadavinia, T. Zhang, T. Sainsbury, G. H. Liaghat and S. Vahid. Improving the fracture toughness and the strength of epoxy using nanomaterials: A review of the current status. *Nanoscale* 7, 10294–10329 (2015).

59. A. Dorigato, A. Pegoretti, F. Bondioli and M. Messori. Improving epoxy adhesives with zirconia nanoparticles. *Compos. Interfaces* 17, 873–892 (2010).

60. G. Gkikas, D. D. Douka, N. M. Barkoula and A. S. Paipetis. Nano-enhanced composite materials under thermal shock and environmental degradation: A durability study. *Compos. Part B* 70, 206–214 (2015).

61. L. Guadagno, M. Sarno, U. Vietri, M. Raimondo, C. Cirillo and P. Ciambelli. Graphene-based structural adhesive to enhance adhesion performance. *RSC Adv.* 5, 27874–27886 (2015).

62. G. Jia-Hu, L. Yu-Cun, C. Tao, J. Su-Ming, Q. Hui, Q. Ning, Z. Hua, Y. Tao and H. Wei-Ming. Synthesis and properties of a nano-silica modified environmentally friendly polyurethane adhesive. *RSC Adv.* 5, 44990–44997 (2015).

63. P.-N. Wang, T.-H. Hsieh, C.-L. Chiang and M.-Y. Shen. Synergetic effects of mechanical properties on graphene nanoplatelet and multiwalled carbon nanotube hybrids reinforced epoxy/carbon fiber composites. *Nanomaterials* (2015), Article ID 838032, 9 pages (2015).

64. M. M. Shokrieh, S. M. Ghoreishi, M. Esmkhani and Z. Zhao. Effects of graphene nanoplatelets and graphene nanosheets on fracture toughness of epoxy nanocomposites. *Fatigue Fract. Eng. Mater. Struct.* 37, 1116–1123 (2014).

65. M. R. Ayatollahi, S. Shadlou and M. M. Shokrieh. Mixed mode brittle fracture in epoxy/multi-walled carbon nanotube nanocomposites. *Eng. Fracture Mech.* 78, 2620–2632 (2011).

66. M. M. Shokrieh, A. R. Kefayati and M. Chitsazzadeh. Fabrication and mechanical properties of clay/epoxy nanocomposite and its polymer concrete. *Mater. Des.* 40, 443–452 (2012).

67. T. Sun, H. Fan, Z. Wang, X. Liu and Z. Wu. Modified nano Fe_2O_3-epoxy composite with enhanced mechanical properties. *Mater. Des.* 87, 10–16 (2015).

68. M. R. Ayatollahi, M. M. Shokrieh, S. Shadlou, A. R. Kefayati and M. Chitsazzadeh. Mechanical and electrical properties of epoxy/multi-walled carbon nanotube/nanoclay nanocomposites. *Iran. Polym. J.* 20, 835–843 (2011).

69. S. Shadlou, B. Ahmadi-Moghadam and F. Taheri. The effect of strain-rate on the tensile and compressive behavior of graphene reinforced epoxy/nanocomposites. *Mater. Des.* 59, 439–447 (2014).

70. B. Soltannia and F. Taheri. Static, quasi-static and high loading rate effects on graphene nano-reinforced adhesively bonded single-lap joints. *Int. J. Composite Mater.* 3, 181–190 (2013).

71. M. R. Ayatollahi, S. Shadlou, M. M. Shokrieh and M. Chitsazzadeh. Effect of multi-walled carbon nanotube aspect ratio on mechanical and electrical properties of epoxy-based nanocomposites. *Polym. Test.* 30, 548–556 (2011).

72. A. Montazeri, J. Javadpour, A. Khavandi, A. Tcharkhtchi and A. Mohajeri. Mechanical properties of multi-walled carbon nanotube/epoxy composites. *Mater. Des.* 31, 4202–4208 (2010).

73. M. Salahinejad. Nano-QSPR modelling of carbon-based nanomaterials properties. *Curr. Top. Med. Chem.* 15, 1868–1886 (2015).

74. B. Soltannia, I. Haji Gholami, S. Masajedian, P. Mertiny, D. Sameoto and F. Taheri. Parametric study of strain rate effects on nanoparticle-reinforced polymer composites. *Nanomaterials* 2016, Article ID 9841972, 9 pages (2016).

75. M. M. Shokrieh and R. Rafiee. Prediction of mechanical properties of an embedded carbon nanotube in polymer matrix based on developing an equivalent long fiber. *Mech. Res. Commun.* 37, 235–240 (2010).

76. M. Ayatollahi, S. Shadlou and M. Shokrieh. Multiscale modeling for mechanical properties of carbon nanotube reinforced nanocomposites subjected to different types of loading. *Compos. Struct.* 93, 2250–2259 (2011).

77. E. Weinan and J. Lu. Multiscale modeling. *Scholarpedia* 6, 11527 (2011).

78. D. R. Bortz, E. G. Heras and I. Martin-Gullon. Impressive fatigue life and fracture toughness improvements in graphene oxide/epoxy composites. *Macromolecules* 45, 238–245 (2012).

79. S. Chandrasekaran, N. Sato, F. Tölle, R. Mülhaupt, B. Fiedler and K. Schulte. Fracture toughness and failure mechanism of graphene based epoxy composites. *Compos. Sci. Technol.* 97, 90–99 (2014).

80. S. Chatterjee, F. Nafezarefi, N. H. Tai, L. Schlagenhauf, F. A. Nüesch and B. T. T. Chu. Size and synergy effects of nanofiller hybrids including graphene nanoplatelets and carbon nanotubes in mechanical properties of epoxy composites. *Carbon* 50, 5380–5386 (2012).

81. M. S. Goyat, S. Suresh, S. Bahl, S. Halderand and P. K. Ghosh. Thermomechanical response and toughening mechanisms of a carbon nano bead reinforced epoxy composite. *Mater. Chem. Phys.* 166, 144–152 (2015).

82. M. A. Rafiee, J. Rafiee, I. Srivastava, Z. Wang, H. Song, Z.-Z. Yu and N. Koratkar. Fracture and fatigue in graphene nanocomposites. *Small* 6, 179–183 (2010).

83. L.-C. Tang, Y.-J. Wan, D. Yan, Y.-B. Pei, L. Zhao, Y.-B. Li, L.-B. Wu, J.-X. Jiang and G.-Q. Lai. The effect of graphene dispersion on the mechanical properties of graphene/epoxy composites. *Carbon* 60, 16–27 (2013).

84. I. Zaman, T. T. Phan, H.-C. Kuan, Q. Meng, L. T. Bao La, L. Luong, O. Youss and J. Ma, Epoxy/graphene platelets nanocomposites with two levels of interface strength. *Polymer* 52, 1603–1611 (2011).

85. R. Benzaid, J. Chevalier, M. Saâdaoui, G. Fantozzi, M. Nawa, L. A. Diaz and R. Torrecillas. Fracture toughness, strength and slow crack growth in a ceria stabilized zirconia–alumina nanocomposite for medical applications. *Biomaterials* 29, 3636–3641 (2008).

86. F. H. Gojny, M. H. G. Wichmann, B. Fiedler and K. Schulte. Influence of different carbon nanotubes on the mechanical properties of epoxy matrix composites: A comparative study. *Compos. Sci. Technol.* 65, 2300–2313 (2005).

87. K. L. Mittal. The role of the interface in adhesion phenomena. *Polym. Eng. Sci.* 17, 467–473 (1977).

88. H. Fu, C. Yan, W. Zhou and H. Huang. Nano-SiO_2/fluorinated waterborne polyurethane nanocomposite adhesive for laminated films. *J. Ind. Eng. Chem.* 20, 1623–1632 (2013).

89. S. G. Prolongo, M. R. Gude, J. Sanchez and A. Ureña. Nanoreinforced epoxy adhesives for aerospace industry. *J. Adhesion* 85, 180–199 (2009).

90. P. J. Dynes and D. H. Kaelble. Surface energy analysis of carbon fibers and films. *J. Adhesion* 6, 195–206 (1974).

91. A. W. Neumann, R. J. Good, C. J. Hope and M. Sejpal. An equation-of-state approach to determine surface tensions of low-energy solids from contact angles. *J. Colloid Interface Sci.* 49, 291–304 (1974).

92. K. Johnson, K. Kendall and A. Roberts. Surface energy and the contact of elastic solids. *Proc. R. Soc. London A* 324, 301–313 (1971).

93. F. M. Etzler. Determination of the surface free energy of solids. *Rev. Adhesion Adhesives* 1, 3–45 (2013).

94. P. R. Waghmare and S. K. Mitra. Contact angle hysteresis of microbead suspensions. *Langmuir* 26, 17082–17089 (2010).

95. G. Ertl, H. Lüth and D. L. Mills (Eds.). *Surface Science Techniques*, Springer Series in Surface Science, Springer, Heidelberg (2013).

96. M. Davis and D. Bond. Principles and practices of adhesive bonded structural joints and repairs. *Int. J. Adhesion Adhesives* 19, 91–105 (1999).

97. R. J. Zaldivar, H. I. Kim, G. L. Steckel, J. P. Nokes and B. A. Morgan. Effect of processing parameter changes on the adhesion of plasma-treated carbon fiber reinforced epoxy composites. *J. Compos. Mater.* 44, 1435–1453 (2010).

98. T. S. Williams, H. Yu and R. F. Hicks. Atmospheric pressure plasma activation of polymers and composites for adhesive bonding: A critical review. *Rev. Adhesion Adhesives* 1, 46–87 (2013).

99. S. S. Tompkins. *NASA Technical Memorandum* 87572. NASA-TM-87572 19850023836 (1985).

100. J-W. Jiang, J-S. Wang and B. Li. Thermal expansion in carbon nanotubes and graphene: Nonequilibrium Green's function approach. *Phys. Rev. B* 80, 205429 (2009).

101. T. S. Chow. The effect of particle shape on the mechanical properties of filled polymers. *J. Mater. Sci.* 15, 1873–1888 (1980).

102. L.-N. Tsai, Y.-T. Cheng and W. Hsu. Nanocomposite effects on the coefficient of thermal expansion modification for high performance electro-thermal microactuator. In: *Proc 18th IEEE International Conference on MEMS*, pp. 467–470 (2005).

103. P. Yoon, T. Fornes and D. Paul. Thermal expansion behavior of nylon 6 nanocomposites. *Polymer* 43, 6727–6741 (2002).

104. Y. Rao and T. N. Blanton. Polymer nanocomposites with a low thermal expansion coefficient. *Macromolecules* 41, 935–941 (2008).

105. S. W. Park, B. C. Kim and D. G. Lee. Tensile strength of joints bonded with a nano-particle-reinforced adhesive. *J. Adhesion Sci. Technol.* 23, 95–113 (2009).

106. S. W. Park and D. G. Lee. Strength of double lap joints bonded with carbon black reinforced adhesive under cryogenic environment. *J. Adhesion Sci. Technol.* 23, 619–638 (2009).

107. S. Eslami, F. Taheri-Behrooz and F. Taheri. Long-term hygrothermal response of perforated GFRP plates with/without application of constant external loading. *Polym. Compos.* 33, 467–475 (2012).

108. M. M. Abdel Wahab. Fatigue in adhesively bonded joints: A review. *ISRN Mater. Sci.* 2012, Article ID 746308, 25 (2012).

109. G. D. Davis. Durability of adhesive joints. In: *Handbook of Adhesive Technology*, 2nd edn, pp. 273–291, A. Pizzi and K. L. Mittal (Eds.), CRC Press, Boca Raton, FL (2003).

110. ASTM-D3762-03. Standard test method for adhesive-bonded surface durability of aluminum (wedge test). American Society for Testing and Materials, Philadelphia (2010).

111. S. Yu, M. N. Tong and G. Critchlow. Wedge test of carbon-nanotube-reinforced epoxy adhesive joints. *J. Appl. Polym. Sci.* 111, 2957–2962 (2009).

112. M. Heshmati, R. Haghani and M. Al-Emrani. Environmental durability of adhesively bonded FRP/steel joints in civil engineering applications: State of the art. *Compos. Part B* 81, 259–275 (2015).

113. W. Bauhofer and J. Z. Kovacs. A review and analysis of electrical percolation in carbon nanotube polymer composites. *Compos. Sci. Technol.* 69, 1486–1498 (2009).

114. F. H. Gojny, M. H. G. Wichmann, B. Fiedler, I. A. Kinloch, W. Bauhofer, A. H. Windle and K. Schulte. Evaluation and identification of electrical and thermal conduction mechanisms in carbon nanotube/epoxy composites. *Polymer* 47, 2036–2045 (2006).

115. S. Chandrasekaran. Development of nano-particle modified polymer matrices for improved fibre reinforced composites. PhD thesis, Department of Mechanical Engineering, Technische Universität Hamburg-Harburg, Germany (2014).

116. S. Yu, M. N. Tong and G. Critchlow. Use of carbon nanotubes reinforced epoxy as adhesives to join aluminum plates. *Mater. Des.* 31, S126–S129 (2010).

117. A. Celzard, E. McRae, C. Deleuze, M. Dufort, G. Furdin and J. F. Marêché. Critical concentration in percolating systems containing a high-aspect-ratio filler. *Phys. Rev. B* 53, 6209–6214 (1996).

118. C. Lu and Y. W. Mai. Influence of aspect ratio on barrier properties of polymer-clay nanocomposites. *Phys. Rev. Lett.* 95, 088303 (2005).

119. J. Li and J.-K. Kim. Percolation threshold of conducting polymer composites containing 3D randomly distributed graphite nanoplatelets. *Compos. Sci. Technol.* 67, 2114–2120 (2007).

120. Y. Li and C. P. Wong. Recent advances of conductive adhesives as a lead-free alternative in electronic packaging: Materials, processing, reliability and applications. *Mater. Sci. Eng. R.* 51, 1–35 (2006).

121. R. Gomatam and K. L. Mittal (Eds.). *Electrically Conductive Adhesives*, CRC Press, Boca Raton, FL (2008).

122. A. S. Lim, Z. R. Melrose, E. T. Thostenson and T.-W. Chou. Damage sensing of adhesively-bonded hybrid composite/steel joints using carbon nanotubes. *Compos. Sci. Technol.* 71, 1183–1189 (2011).

123. B. Ahmadi-Moghadam and F. Taheri. Detection in grapheme nanoplatelet/CFRP composites using electric resistance change. In: *Proceedings of the 19th International Conference on Composite Materials ICCM19)*, Montreal, Quebec, Canada (2013). Available online: http://confsys.encs.concordia.ca/ICCM19/AllPapers/FinalVersion/AHM80835.pdf

124. T. Cui, Q. Li, Y. Xuan and P. Zhang. Preparation and thermal properties of the graphene–polyolefin adhesive composites: Application in thermal interface materials. *Microelectron. Reliab.* 55, 2569–2574 (2015).

125. Z. Gao and L. Zhao. Effect of nano-fillers on the thermal conductivity of epoxy composites with micro-Al_2O_3 particles. *Mater. Des.* 66, Part A, 176–182 (2015).

126. W. Qiao, H. Bao, X. Li, S. Jin and Z. Gu. Research on electrical conductive adhesives filled with mixed filler. *Int. J. Adhesion Adhesives* 48, 159–163 (2014).

127. Y. Cui, S. I. Kundalwal and S. Kumar. Gas barrier performance of graphene/polymer nanocomposites. *Carbon* 98, 313–333 (2016).

128. H. Kim, Y. Miura and C. W. Macosko. Graphene/polyurethane nanocomposites for improved gas barrier and electrical conductivity. *Chem. Mater.* 22, 3441–3450 (2010).

129. M. A. Osman, V. Mittal, M. Morbidelli and U. W. Suter. Polyurethane adhesive nanocomposites as gas permeation barrier. *Macromolecules* 36, 9851–9858 (2003).

130. H. Dodiuk-Kenig, Y. Maoz, K. Lizenboim, I. Eppelbaum, B. Zalsman and S. Kenig. The effect of grafted caged silica (polyhedral oligomeric silesquioxanes) on the properties of dental composites and adhesives. *J. Adhesion Sci. Technol.* 20, 1401–1412 (2006).

131. H. Schulz, B. Schimmoeller, S. E. Pratsinis, U. Salz and T. Bock. Radiopaque dental adhesives: Dispersion of flame-made Ta_2O_5/SiO_2 nanoparticles in methacrylic matrices. *J. Dentistry* 36, 579–587 (2008).

132. M. Atai, L. Solhi, A. Nodehi, S. M. Mirabedini, S. Kasraei, K. Akbari and S. Babanzadeh. PMMA-grafted nanoclay as novel filler for dental adhesives. *Dental Mater.* 25, 339–347 (2009).

133. L. Solhi, M. Atai, A. Nodehi and M. Imani. A novel dentin bonding system containing poly (methacrylic acid) grafted nanoclay: Synthesis, characterization and properties. *Dental Mater.* 28, 1041–1050 (2012).

134. S. Kasraei, M. Atai, Z. Khamverdi and S. K. Nejad. The effect of nanofiller addition to an experimental dentin adhesive on microtensile bond strength to human dentin. *J. Dentistry Tehran Univ. Med. Sci.* 6, 36–41 (2009).

135. U. Lohbauer, A. Wagner, R. Belli, C. Stoetzel, A. Hilpert, H. D. Kurland, J. Grabow and F. A. Muller. Zirconia nanoparticles prepared by laser vaporization as fillers for dental adhesives. *Acta Biomater.* 6, 4539–4546 (2010).

136. A. Akhavan, A. Sodagar, F. Mojtahedzadeh and K. Sodagar. Investigating the effect of incorporating nanosilver/nanohydroxyapatite particles on the shear bond strength of orthodontic adhesives. *Acta Odontol. Scand.* 71, 1038–1042 (2013).

137. M. Kassaee, A. Akhavan, N. Sheikh and A. Sodagar. Antibacterial effects of a new dental acrylic resin containing silver nanoparticles. *J. Appl. Polym. Sci.* 110, 1699–1703 (2008).

138. M. Mehdizadeh and J. Yang. Design strategies and applications of tissue bioadhesives. *Macromol. Biosci.* 13, 271–288 (2013).

139. C. E. Brubaker and P. B. Messersmith. The present and future of biologically inspired adhesive interfaces and materials. *Langmuir* 28, 2200–2205 (2012).

140. P. M. Favi, S. Yi, S. C. Lenaghan, L. Xia and M. Zhang. Inspiration from the natural world: From bio-adhesives to bio-inspired adhesives. *J. Adhesion Sci. Technol.* 28, 290–319 (2014).

141. K. Shahin, G. Kember and F. Taheri. An asymptotic solution for evaluation of stresses in balanced and unbalanced adhesively bonded joints. *Mech. Adv. Mater. Struct.* 15, 88–103 (2008).

142. F. Taheri, S. Shadlou and R. A. Esmaeel. Computational modelling of delamination and disbond in adhesively bonded joints and the relevant damage detection approaches: A critical review. *Rev. Adhesion Adhesives* 1, 413–458 (2013).

143. D. Dugdale. Yielding of steel sheets containing slits. *Mech. Phys. Solids* 8, 100–104 (1960).

144. G. I. Barenblatt. The mathematical theory of equilibrium cracks in brittle fracture. *Adv. Appl. Mech.* 7, 55–129 (1962).

145. A. Needleman. A continuum model for void nucleation by inclusion debonding. *Appl. Mech.* 54, 525–31 (1987).

146. C. G. Dávila and P. P. Camanho. Decohesion elements using two and three-parameter mixed-mode criteria. In: *American Helicopter Society Conference*, Williamsburg, VA, Citeseer (2001). Available online at: http://ntrs.nasa.gov/archive/nasa/casi.ntrs.nasa.gov/20020010916.pdf

147. M. J. Loikkanen. *Guidelines for Modeling Delamination in Composite Materials using LS-DYNA*, Livermore Software Technology Corporation, Livermore, CA (2011).

148. P. P. Camanho and C. G. Dávila. Mixed-mode decohesion finite elements for the simulation of delamination in composite materials, NASA Technical Report/TM-(2002). Available online: http://ntrs.nasa.gov/archive/nasa/casi.ntrs.nasa.gov/20020053651.pdf

149. G. Alfano, On the influence of the shape of the interface law on the application of cohesive-zone models. *Compos Sci Technol* 66, 723–730 (2006).

150. M. Alfano, F. Furgiuele, A. Leonardi, C. Maletta and G. Paulino. Fracture analysis of adhesive joints using intrinsic cohesive zone models. *Key Eng. Mater.* 348, 13–16 (2007).

151. M. Alfano, F. Furgiuele, A. Leonardi, C. Maletta and G. Paulino. Mode I fracture of adhesive joints using tailored cohesive zone models. *J. Fracture* 157, 193–204 (2009).

152. S. Li, M. Thouless, A. Waas, J. Schroeder and P. Zavattieri. Use of mode-I cohesive-zone models to describe the fracture of an adhesively-bonded polymer-matrix composite. *Compos. Sci. Technol.* 65, 281–293 (2005).

153. S. Li, M. Thouless, A. Waas, J. Schroeder, P. Zavattieri. Use of a cohesive-zone model to analyze the fracture of a fiber-reinforced polymer–matrix composite. *Compos. Sci. Technol.* 65, 537–49 (2005).

154. A. Needleman. Some issues in cohesive surface modeling. *Procedia IUTAM* 10, 221–246 (2014).

155. J. M. Melenk and I. Babuška. The partition of unity finite element method: Basic theory and applications. *Comput. Methods Appl. Mech. Eng.* 139, 289–314 (1996).

156. C. A. Duarte and J. T. Oden. An h-p adaptive method using clouds. *Comput. Methods Appl. Mech. Eng.* 139, 237–262 (1996).

157. T. Belytschko and T. Black. Elastic crack growth in finite elements with minimal remeshing. *Int. J. Numer. Methods Eng.* 45, 601–620 (1999).

158. J. Dolbow, N. Moës and T. Belytschko. An extended finite element method for modeling crack growth with frictional contact. *Comput. Methods Appl. Mech. Eng.* 190, 6825–6846 (2001).

159. J. C. Sosa and N. Karapurath. Delamination modelling of GLARE using the extended finite element method. *Compos. Sci. Technol.* 72, 788–791 (2012).

160. M. S. Moreno, J. Curiel-Sosa, J. Navarro-Zafra, J. M. Vicente and J. L. Cela. Crack propagation in a chopped glass-reinforced composite under biaxial testing by means of XFEM. *Compos. Struct.* 119, 264–271 (2015).

161. T. Wang, C. H. Lei, A. B. Dalton, C. Creton, Y. Lin, K. A. S. Fernando, Y.-P. Sun, M. Manea, J. M. Asua and J. L. Keddie. Waterborne, nanocomposite pressure-sensitive adhesives with high tack energy, optical transparency, and electrical conductivity. *Adv. Mater.* 18, 2730–2734 (2006).

162. H. S. Hedia, L. Allie, S. Ganguli and H. Aglan. The influence of nanoadhesives on the tensile properties and mode-I fracture toughness of bonded joints. *Eng. Fracture Mech.* 73, 1826–1832 (2006).

163. S. Sprenger, C. Eger, A. Kinloch, J. H. Lee, A. Taylor and D. Egan. Nano-modified ambient temperature curing epoxy adhesives. *Adhaesion Kleben Dichten* 17–21 (2004).

164. H. Dodiuk, S. Kenig, I. Blinsky, A. Dotan and A. Buchman. Nanotailoring of epoxy adhesives by polyhedral-oligomeric-sil-sesquioxane (POSS). *Int. J. Adhesion Adhesives* 25, 211–218 (2005).

165. A. Buchman, H. Dodiuk-Kenig, A. Dotan, R. Tenne and S. Kenig. Toughening of epoxy adhesives by nanoparticles. *J. Adhesion Sci. Technol.* 23, 753–768 (2009).

166. X.-R. He, R. Zhang, Q. Chen, Y.-Q. Rong and Z.-Q. Yang. Different surface functionalized nano-Fe_3O_4 particles for EVA composite adhesives. *Int. J. Adhesion Adhesives* 50, 128–135 (2014).

167. L. L. Zhai, G. P. Ling and Y. W. Wang. Effect of nano-Al_2O_3 on adhesion strength of epoxy adhesive and steel. *Int. J. Adhesion Adhesives* 28, 23–28 (2008).

168. L. Zhai, G. Ling, J. Li and Y. Wang. The effect of nanoparticles on the adhesion of epoxy adhesive. *Mater. Lett.* 60, 3031–3033 (2006).

169. E. N. Gilbert, B. S. Hayes and J. C. Seferis. Nano-alumina modified epoxy based film adhesives. *Polym. Eng. Sci.* 43, 1096–1104 (2003).

170. H. Dodiuk, I. Belinski, A. Dotan and S. Kenig. Polyurethane adhesives containing functionalized nanoclays. *J. Adhesion Sci. Technol.* 20, 1345–1355 (2006).

171. J. Li and J. K. Lumpp. Electrical and mechanical characterization of carbon nanotube filled conductive adhesive. In: *Proc. 2006 IEEE Aerospace Conference*, p. 6 (2006).

172. M. B. Saeed and M.-S. Zhan. Adhesive strength of nano-size particles filled thermoplastic polyimides. Part-I: Multi-walled carbon nano-tubes (MWNT)–polyimide composite films. *Int. J. Adhesion Adhesives* 27, 306–318 (2007).

173. K.-T. Hsiao, J. Alms and S. G. Advani. Use of epoxy/multiwalled carbon nanotubes as adhesives to join graphite fibre reinforced polymer composites. *Nanotechnology* 14, 791–793 (2003).

174. U. Khan, P. May, H. Porwal, K. Nawaz and J. N. Coleman. Improved adhesive strength and toughness of polyvinyl acetate glue on addition of small quantities of graphene. *ACS Appl. Mater. Interfaces* 5, 1423–1428 (2013).

175. Y. Zhang, B. You, H. Huang, S. Zhou, L. Wu and A. Sharma. Preparation of nanosilica reinforced waterborne silylated polyether adhesive with high shear strength. *J. Appl. Polym. Sci.* 109, 2434–2441 (2008).

176. S. M. R. Khalili, M. Tavakolian and A. Sarabi. Mechanical properties of nanoclay reinforced epoxy adhesive bonded joints made with composite materials. *J. Adhesion Sci. Technol.* 24, 1917–1928 (2010).

177. A. Kaboorani and B. Riedl. Effects of adding nano-clay on performance of polyvinyl acetate (PVA) as a wood adhesive. *Compos. Part A* 42, 1031–1039 (2011).

178. Y. A. Gorbatkina, V. G. Ivanova-Mumzhieva and T. M. Ul'yanova. Adhesiveness of an epoxy oligomer filled with aluminum oxide powders. *Polym. Sci. Series C* 49, 131–134 (2007).

179. A. Kaboorani and B. Riedl. Nano-aluminum oxide as a reinforcing material for thermoplastic adhesives. *J. Ind. Eng. Chem.* 18, 1076–1081 (2012).

180. W. Gao and G. Du. Curing kinetics of nano cupric oxide (CuO)-modified PF resin as wood adhesive: Effect of surfactant. *J. Adhesion Sci. Technol.* 27, 2421–2432 (2013).

181. D. Liu, H. Chen, P. R. Chang, Q. Wu, K. Li and L. Guan. Biomimetic soy protein nanocomposites with calcium carbonate crystalline arrays for use as wood adhesive. *Bioresource Technol.* 101, 6235–6241 (2010).

182. S. Khoee and N. Hassani. Adhesion strength improvement of epoxy resin reinforced with nanoelastomeric copolymer. *Mater. Sci. Eng. A* 527, 6562–6567 (2010).

183. M. B. Jakubinek, B. Ashrafi, Y. Zhang, Y. Martinez-Rubi, C. T. Kingston, A. Johnston and B. Simard. Single-walled carbon nanotube–epoxy composites for structural and conductive aerospace adhesives. *Compos. Part B* 69, 87–93 (2015).

Part 2

Adhesive Classes

5 Protein Adhesives

Charles R. Frihart* and Linda F. Lorenz

CONTENTS

5.1 INTRODUCTION

Nature uses a wide variety of chemicals for providing adhesion internally (e.g., cell to cell) and externally (e.g., mussels to ships and piers). This adhesive bonding is chemically and mechanically complex, involving a variety of proteins, carbohydrates, and other compounds. Consequently, the effect of protein structures on adhesive properties is only partially understood and creates an inherent difficulty in understanding protein structure–property relationships. This complexity has limited researchers in the past to mainly depend on empirical studies for insight, with a few exceptions.

For most protein adhesives, bonding of wood has not been a problem under dry use conditions, as the strength of the adhesives is sufficient to cause wood failure, but the adhesive bonds are weaker under wet-use conditions. Because of its economic importance, protein bonding has been widely studied in the adhesion of mussels and barnacles to marine surfaces. The bioadhesion of organisms

---

* Indicates corresponding author

in marine environments is not covered here, because these proteins have not been used for formulating wood adhesives, but they have led to biomimicry and bioinspired research on modified and synthetic proteins that could be used as adhesives, as covered in Section 5.9.

Although nature has used proteins as adhesives for a wide variety of purposes, man-made protein adhesives have been used mainly for wood and paper bonding, because of the substrate's ability to transport the water solvent away from the bondline. Although protein adhesives have been greatly surpassed by synthetic adhesives for most bonding applications [1], wood bonding is one area in which there has been some increase in the use of protein adhesives in recent years [2]. Figure 5.1 shows the many types of wood products that have been made with protein adhesives. If sufficient wet cohesive strength can be developed, the adhesives can be used in an assortment of applications because of their ability to bond to a variety of wood species. However, the waterborne nature of protein adhesives has limited their use with most nonwood substrates.

Synthetic polymers have two main advantages over natural polymers. The first is that synthetic polymers can be designed to have specific properties, because the physical properties, functionality, and molecular weights can be controlled for peak adhesive performance. With natural polymers,

FIGURE 5.1 Wood products made with soy adhesives (except for the bottom product, made with a casein adhesive). From top to bottom: decorative plywood, hardboard, medium density fiberboard, particleboard, oriented strandboard, engineered wood flooring, and glulam.

utility can only be obtained from what nature has synthesized for another purpose, unless the bio-polymers can be properly modified. The second advantage is that synthetic materials can be made repeatedly with the same structural properties, whereas natural polymers have considerable variability in their properties.

The two main drivers leading to an increase in research on protein adhesives are the interest in materials made from natural sources and concerns about health hazards from the monomers used in synthetic wood adhesives. Wood is already a green material, but using natural polymer adhesives makes these products even greener. The health hazard concerns about synthetic polymers have been important because the most commonly used wood adhesive, urea–formaldehyde (UF), is made with formaldehyde, which is considered carcinogenic. UF adhesives can continue to release significant amounts of formaldehyde under conditions of high heat and humidity for the life of the product [3–5]. Formaldehyde emissions from products made with UF adhesives have been a concern for a long time [6,7]. The easiest to manufacture products are also those UF adhesives with the greatest strength. However, these products are made with higher formaldehyde-to-urea (F/U) ratios, which result in the highest formaldehyde emissions [5,8]. Thus, the trend has been to use lower F/U ratios, make changes to the synthesis conditions, and add scavengers to provide products with lower formaldehyde emissions [9]. The lower emission products are driven by new regulations, companies' desires to be green, or concern about long-term legal liability for producing products with higher formaldehyde emissions [10–13]. Although some applications allow higher formaldehyde emissions, the trend has been to lower the formaldehyde emissions to the level that naturally occurs in wood. Processing can increase the formaldehyde emissions of some wood products, requiring adhesives that can absorb this formaldehyde and are free of added formaldehyde [14]. The desire for formaldehyde-free adhesives has been an important driver for the resurgence of soy adhesives, because they retain low emission levels even under high temperature and humidity conditions [4]. The problem concerning formaldehyde emissions resides solely with UF adhesives, since the phenol-, resorcinol-, and melamine–formaldehyde adhesives generally do not have formaldehyde emission issues.

Although the emphasis has been on emissions from the final wood product as used by the customer, there are also concerns about hazardous emissions during the production processes for adhesives and bonded wood products [14,15]. Given that adhesives generally need to react to form the adhesive bond, there is always a concern that adhesives may react with human proteins, nucleic acids, and so forth impacting worker health. Certainly, this has been an issue with some of the formaldehyde-containing adhesives. Concern surrounding isocyanate adhesive use in wood bonding processes has also been raised [16]. Generally, wood adhesives have the advantage of being water-borne; thus, they normally have a low amount of volatile organic compounds relative to solvent-borne adhesives. Although workplace exposure to the adhesive can be minimized by good manufacturing practices, some facilities do not even consider using isocyanate adhesives for fear of worker exposure.

Many market forces, including the facts of being biobased, being potentially free of added formaldehyde, and having better recyclability, favor proteins as wood adhesives if they can provide sufficient product performance. The most available source of proteins is plants, primarily seed crops. These plants are separated into oil seeds (harvested mainly for their oil) and cereal grains (harvested mainly for their carbohydrate content). Table 5.1 shows the approximate compositions of a variety of seed materials. The soybean is clearly the best source, given the high protein content compared with other oil seeds, and soybeans are much higher in protein than cereal grains. Even though the high-value human food market is the driving force behind the growing and processing of these crops, industrial chemicals provide valuable outlets for portions of these products that are often used as animal (cattle and pig) feeds. Finding additional value for the by-products can improve the overall economics of growing and processing these plants. This was especially true in the past for many industries that converted waste materials into valuable adhesives, especially in animal and plant processing. Collagen from cattle and horse hooves and hides and fish skins was important for glues in the paper and furniture industries. Additionally, blood glues from pigs and cattle were important for the wood bonding industry. The use of proteins in pet foods, gardening aids, and other higher-value products has led to

TABLE 5.1

Comparison of Protein Contents and Other Compounds in Compositions of Various Oilseeds and Cereal Grains in Weight Percentages (Average Data Taken from a Number of Sources)

Source	Protein	Fat	Starch	Fiber	Ash
Oil Seeds, Harvested Primarily for Oils					
Soybean	51	20	0	6	6
Rapeseed	26	46	0	5	8
Canola	22	41	22	10	5
Cottonseed	22	20	35	19	5
Linseed	24	43	29	8	3
Peanuts	30	50	14	3	3
Safflower	21	41	15	19	5
Sunflower	21	55	18	2	3
Cereal Grains, Harvested Primarily for Carbohydrates					
Wheat	12	2	72	2	2
Corn/maize	10	5	80	2	1
Oats	11	6	56	11	3
Rice	8	1	76	1	1
Rye	12	2	72	2	2

Source: Lusas, E.W, Oilseeds and oil-bearing materials, in: *Handbook of Cereal Science and Technology*, K. Kulp and G. Points, Jr. (Eds.), pp. 297–362, Marcel Dekker, New York, 2000; Salunkhe, D. K. et al., *World Oilseed Chemistry, Technology, and Utilization*, Van Nostrand Reinhold, New York, 1991; Wolf, W. J., Soybeans and other oilseeds, in: *Kirk-Othmer Encyclopedia of Chemical Technology*, 3rd edn, Vol. 21, pp. 417–442, Wiley, New York, 1983; Lasztity, R., *Chemistry of Cereal Proteins*, CRC Press, Boca Raton, FL, 1995. Modified from X. S. Sun. Overview of plant polymers: Resources, demands, and sustainability. In: *Bio-Based Polymers and Composites*, R. P. Wool and X. S. Sun (Eds.), pp. 1–14, Elsevier-Academic Press, Burlington, MA, 2005.

the near extinction of most protein adhesives. Although animal-derived protein adhesives are less important, some of the plant materials are more likely to be used as adhesives in the future.

To think effectively about the adhesive performance of proteins, it is important to understand the structure–property relationships of proteins. This is very important, because proteins are very different in structure from most polymers, especially synthetic polymers, and thus have very different structure–property relationships. In this chapter, after an overview of protein structure, protein adhesives from specific sources are discussed. Unfortunately, because of the wide variety of protein properties, there is no real way with current knowledge to relate performance characteristics with specific structural elements for these proteins in normal plant products. Hydrophobic and polar interactions, colloidal properties, and the aggregation state of the protein are important parameters that influence adhesive properties. Protein colloidal properties are influenced by the hydrophobic and hydrophilic properties of the surface and the ionic charge of the protein [21]. Thus, an understanding of the hydrophobic properties of the protein is important, as well as distribution of the polar groups in the protein. The aggregation of the protein, through both hydrophobic and hydrophilic attractions, suggests that the native protein properties are a summation of a number of individual protein chains of different compositions and structures [22]. These specific native quaternary aggregates often exist even if some of the individual proteins have denatured secondary structures [22,23]. Developing an understanding of the properties of these protein aggregates is important, with much of the information coming from studies for food applications [24]. This information is an important addition to the adhesive literature in understanding protein rheological and strength properties, especially for the properties in the wet state, but it provides less information about dry and rewetted state properties.

Proteins are important contributors to the texture properties of food, which depend on both the chemical and physical properties of the proteins as they interact with other food ingredients. Food proteins are characterized by their solubility, viscosity, water binding, gelation, cohesion–adhesion, elasticity, emulsification, foaming, and fat and flavor bonding [21]. Many of the proteins' properties are related to their surface tension; this is commonly determined by measuring the ease of aqueous foam formation and its stability, and by the gelation properties. The good surfactant behavior of proteins suggests that they should be good adhesives for wetting bonding surfaces. Gelation, cohesion, and elasticity are dependent on gel formation properties, which are measured by indentation tests after heating aqueous dispersions followed by cooling them to form a gel. These properties would, hopefully, carry over to their cohesive strength properties as adhesives.

Because the protein adhesive formulations widely used in the past are not used much nowadays, only a brief coverage of these adhesive formulations is provided and references are made to the best literature sources for more detailed application information. More importantly, this chapter contains information on protein structures and their related chemical and physical properties, which influence current and future applications of proteins as adhesives. The amino acid composition of each type of protein is given, but the literature data are for a single sample and do not reflect the natural variability in composition. Yet, understanding the compositions can assist in developing specific modifications or coreactants for future research work, such as providing information on the amount of side-chain functional groups potentially available for modification. However, because of the folding of the protein, many of these functional groups are not available for reaction if they are buried inside the protein.

5.2 PROTEIN STRUCTURE

It is difficult to make sense of a protein's properties without understanding its structural complexity involving specific hierarchical orders. The protein chain is composed of 21 naturally occurring amino acids. The amino acids are linked in a linear fashion through condensation reactions between the amino and carboxylic acid groups. The protein's backbone imposes some restrictions on the chain flexibility, especially in the case of the proline and hydroxyproline amino acids, which cause a pronounced kink in the chain because of their cyclical structure. The side chains impose additional restrictions on chain conformations. The sequence of the amino acids in the protein chain is called the *primary structure* [21,22]. Sometimes, a change in sequence can alter the protein's function, especially in enzymes, and other times it may not, such as in storage proteins. For a particular plant protein, amino acid composition has some minor natural variation in its composition caused by plant and growth condition diversity.

As shown in Figure 5.2, the different amino acids are either aliphatic, aromatic, or heterocyclic with many types of functional groups. Thus, a wide variety of interactions are possible between the hydrophobic and hydrophilic (acidic, basic, hydroxyl, and thiol) side groups on the amino acids. Different proteins vary in the percentages of amino acids and in amino acid sequences. Given that typical proteins are more than 200 amino acids long and each amino acid is selected from one of the 21 amino acids, this means that there can be a huge number of protein sequences.

The primary sequence of amino acids determines the secondary structure, which, in turn, influences the tertiary structure of a protein and the protein's physicochemical properties. Certain sequences of amino acids can lead to either α-helix or β-sheet structures, which are called the *secondary structures* [22]. These secondary structures form as the protein is being synthesized, and are important for developing the native structure of the protein because they control the folding of the protein. These structures are intrachain elements and can be disrupted by heat and various chemical additives. They should not be confused with the interchain helix formed by collagen proteins and the interchain sheet crystallites that form in synthetic polymers [25].

Most proteins, including those used in adhesives, are globular proteins that have enough hydrophobic portions or weakly hydrophilic portions to cause the protein to fold inward in an aqueous

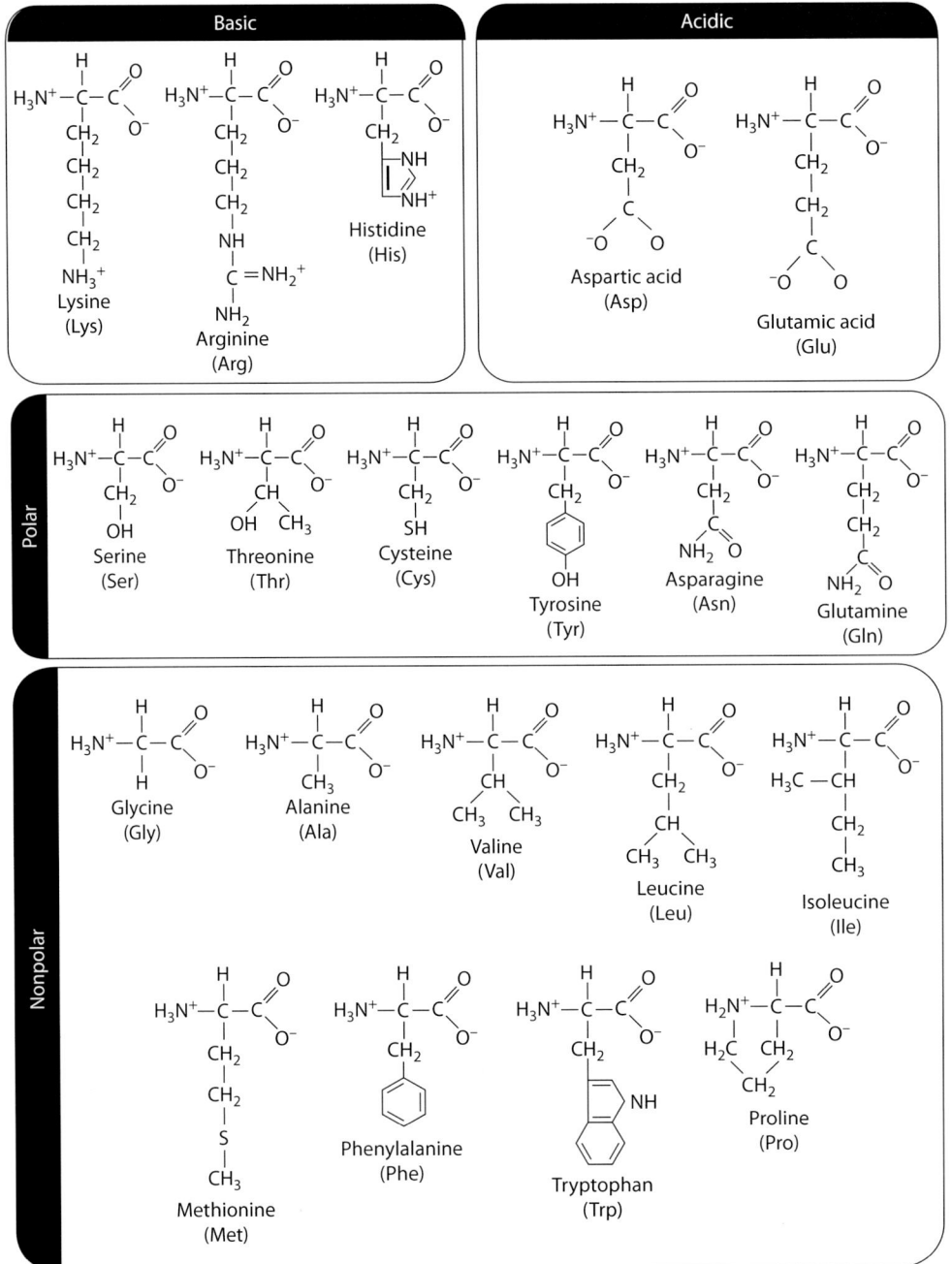

FIGURE 5.2 Common amino acids functional groups.

environment (Figure 5.3). During this hydrophobic collapse, intramolecular association of hydrophobic amino acids moves them mainly to the inside of the globular structure and the polar amino acids preferentially to the outside. However, the primary and secondary structures limit the ability of particular amino acids to move to these separate regions. The polar groups on the inside of the globule are stabilized by associating with other polar groups on the chain. These types of interactions include not only hydrogen bonds, but also acid–base salt bridges and disulfides from the thiol

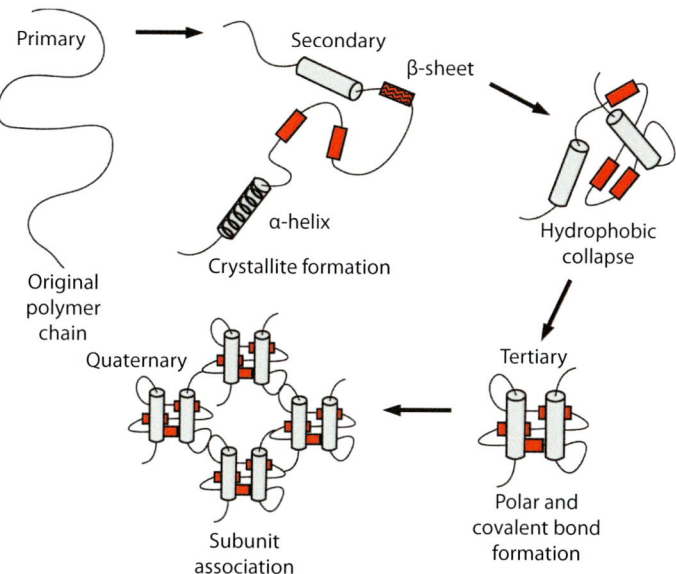

FIGURE 5.3   Protein folding as discrete steps for illustration purposes. In reality, the primary, secondary, and tertiary structures occur mainly at the same time.

groups. Some of the hydrophobic amino acids end up on the outside of the structure. This folding of the protein is referred to as its *tertiary structure*.

As the protein is being synthesized, the secondary and tertiary structures form to provide the native state of the protein. This native state is only a localized energy minimum, and thus, the native state may not be the most stable state [23,26]. As shown in Figure 5.4, there are many energy minimums (B, C, etc.) in addition to the native state (A) that may not be very different in energy from the native state, although there are some larger troughs for particular conformations. The small activation energies mean that the protein can readily change its conformation depending on the external environment. The similar energy minimums represent changes that may occur in the protein colloid with changes in the electrical double layer with alterations of the aqueous environment, such as the addition of salts, urea, or other chemicals. The larger activation energy maximums and deeper troughs represent the unfolding and refolding of the protein chain within the protein globule during thermal denaturation of the proteins. Thus, it can be very misleading to assume that there is a single

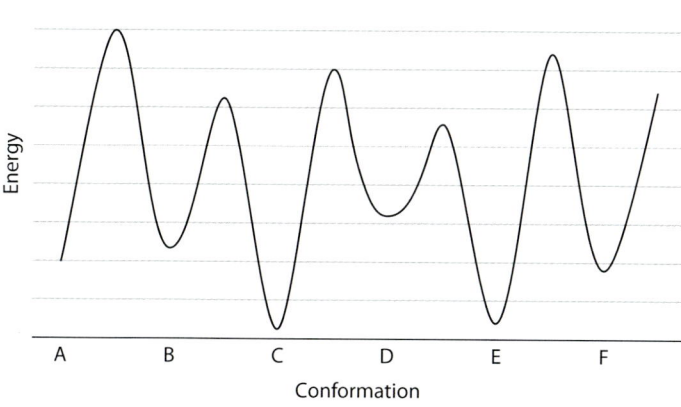

FIGURE 5.4   Typical protein energy versus conformation, where differences are a few calories per mole between the native state A and denatured states B, C, and so forth.

conformation that cannot be changed by external conditions. It can also be misleading to think of unfolding as going into an elongated chain, rather than just an intermediate step of the globular protein in changing its folding conformation.

Although some of the hydrophobic groups end up on the surface because of folding restrictions, these groups can minimize their energy by associating with surface hydrophobic regions on other proteins. Association of proteins is very common in native environments and isolated proteins. This association of protein globules is referred to as the *quaternary structure*. Although hydrophobic attractions hold most protein complexes together, the proteins can also be held together by polar bonds and specific chemical bonds such as disulfides involving thiol groups on separate chains. The association of proteins is often considered a balance between hydrophobic attraction and electrostatic repulsion [21]. As the pH of the dispersion increases, the electrostatic repulsion increases enough to overcome the hydrophobic attraction, allowing the proteins to separate and to unfold in a way that opens the structure.

Proteins are often characterized under very dilute conditions. Because the external environment can influence the secondary, tertiary, and quaternary structures, caution needs to be used in extrapolating the available information from these dilute studies to other environments. In fact, a prominent area of study is the effect of protein crowding on the protein properties. Although this research is concerned with the crowded environment in cells [27], it is also an important factor in adhesives because of the need to have the solids content of the adhesives as high as possible. The crowding can significantly impact protein–protein and protein–carbohydrate interactions compared with the characterization of proteins under very dilute conditions.

Given the complex structural nature of proteins, when discussing adhesive performance, the history of the protein in prior processing needs to be understood. It is also important to remember that most natural sources contain many different proteins and, in many cases, the proteins are intimately associated with other materials, such as carbohydrates, which can influence their properties. The natural state of the protein is called the *native state*, which is a singular structure for each protein, and any transformation from the native state is called a *denatured state*. Obviously, there are many denatured states, whereas there is only one native state. Some disruptions can be measured to understand what has happened, such as the loss of secondary structure, but detailed changes in the tertiary structure are hard to determine. Many methods have been developed to measure the properties of proteins [28].

Although the discussion of proteins ends at times at the quaternary structure, there are often higher-order structures of a macromolecular nature. These can be due to larger protein agglomerates [29], or they can also be proteins associating with carbohydrates [30]. In fact, after synthesis of the amino acid backbone, proteins can be modified by attachment of carbohydrates or phosphates. An example of this is the native colloids in milk, which are a mixture of proteins, calcium phosphate, and fats, as discussed in Section 5.5.

The association of proteins becomes very important during the aqueous gelation process in water, which links the proteins together to form a large macromolecular structure [31]. Heat gelation is very important in the food industry and has been well studied. Gelation is probably driven more by hydrophobic interactions, because stable polar interactions are less likely to form in an aqueous environment. A useful model for globular proteins is a string of beads with branch points to give a three-dimensional gel that traps a lot of water [29,31]. These larger aggregates probably provide the cohesive strength for protein adhesives. Hydrogen, disulfide, and acid–base bonds may also provide additional strength between the individual protein molecules for the adhesive bond strength.

There is often confusion about proteins for more than one reason. The first is that although the literature often considers proteins to be a single entity or a single type of protein, a protein source such as soybean or blood contains many proteins that show a great variety of different structures and properties. Additional composition variation exists because of differences in the specific source and processing of the proteins. The processing can influence the proteins' bonding properties. as illustrated by the differences in chemical and physical properties, which are shown in Table 5.2 for

TABLE 5.2

Functional Characteristics of Various Isolated Soy Proteins and How They Are Rated

Soy	Solubility*	pH	Viscosity	Dispersibility*	Gelation*	Emulsification*	Water Binding*
A	7[a]	7	7	2	7	6	7
B	6	6	1	3	1	7	1
C	4	5	4	5	4	3	4
D	7	7	5	4	5	6	6
E	7	6	3	1	3	7	3
F	2	3	2	6	2	2	2
G	7	6	2	3	2	7	2
H	1	1	1	7	1	1	1

Source: Reproduced from Egbert, W. R., Isolated soy protein: Technology, properties, and applications, in *Soybeans as Functional Foods and Ingredients*, K. Liu (Ed.), pp. 134–162, AOCS Press, Champaign, IL, 2004. With permission.

*Solubility: Percentage of protein staying in the aqueous phase; Dispersibility: How fast the protein solubilizes; Gelation: The ability to form a rigid gel after heating and cooling; Emulsification: The ability to mix water and fat; Water binding: The ability of the protein to hold water in a food product.

[a]*Rating system:* 7 = very high, 4 = moderate, 1 = very low. pH range reported as 7 = high (approx. 7.5) and 1 = low (approx. 4.5).

different soy protein isolates [32]. The second reason is that there are very few isolated proteins found in natural products; proteins are typically found as structures containing different amino acid sequences. In native form, these complexes are usually characterized, but once proteins have been denatured, the variety of structures present grows dramatically. Thus, a researcher needs to be cautious in making generalized claims about a specific protein being a good adhesive. The importance of understanding the source and processing of proteins in ascribing mechanisms to adhesive performance cannot be overstated.

There are also several other issues with proteins that can lead to confusion. The terms *solubility* and *dispersibility* are used for commercial products. Most proteins are not truly soluble, because of their high molecular weight and low polarity; thus, proteins are dispersions in water. The term dispersibility refers to how quickly the protein solids are wetted when brought into contact with water, whereas solubility refers to how much of the protein stays dispersed after mechanical agitation of the protein–water mixture is stopped. The term *unfolded protein* is often used to describe changes in the protein. "Unfolded" does not mean that the protein goes into an extended polymer structure in water, but that the organized areas in the polymer become disrupted by denaturation conditions through unfolding and refolding while still in the globular state. In addition, the protein is altered by swelling with water, conferring a more plastic behavior while still remaining in a globular state. It is important to remember that many of the proteins used for wood bonding are globular in nature and their strength comes from the interactions of these globules. Thus, colloid science is important in understanding the behavior of protein structures as adhesives.

Because proteins from a variety of sources make good wood adhesives, understanding how they are similar and different is valuable. Proteins have been categorized based on a variety of properties. One category is based on the type of protein. Simple proteins consist of only the polymerized amino acids. Examples of these proteins include some albumins, collagens, keratins, and the glycinin in soy. Conjugated proteins, also called *heteroproteins*, include glycoproteins (protein plus carbohydrates are quite common, such as the conglycinin in soy) and phosphoproteins (phosphate groups added to the proteins in egg yolk and caseins). Proteins can be divided into two groups based on shape: fibrous (such as collagen and α-keratins) and globular (such as grain and legume storage proteins). Another major differentiation of proteins is albumins, which dissolve in water but not in salt solutions, compared with globulins, which dissolve in salt solutions.

5.3 SOYBEAN

Soybeans were historically, and are again, the dominant source of industrial protein adhesives, and have been used for commercial bonded wood products. Soy proteins are mainly storage proteins used to provide the growing plant with nutrients. Although many other protein adhesives were used long before soybeans, most are fairly expensive or limited in quantity. Soy proteins appeared to be a good replacement for the other proteins given cost and availability, but considerable effort was needed to develop soybean flour into a useful adhesive for wood bonding [33]. Although soy flour adhesives did not achieve the same degree of moisture resistance as some other proteins, such as casein and blood, they were suitable for interior plywood, leading to a significant expansion of that industry [1,34]. Soy protein's recent return to the wood bonding market has been almost exclusively in interior products, with the aid of coreactant components in the adhesive formulation to enhance water resistance.

Since a number of soy protein products are available, it is important to understand how each one is derived, because the processing affects the physical and chemical properties of the protein. After the soybeans have been dried and solvent-extracted to remove the valuable oil, the solvent is removed and the soybeans are ground to produce flour [35]. Soybean flour is relatively inexpensive, available in very large quantities without disrupting the human food supply, and fairly consistent in composition. It contains about equal amounts of protein and carbohydrates, along with lipids and salts. The carbohydrates are about half soluble, consisting mainly of sucrose, raffinose, and stachyose, and half insoluble, consisting of rhamnose, arabinose, galactose, glucose, xylose, and mannose polymers [36]. The protein components have many types of functionalized side chains (Figure 5.2). The individual soy proteins are globular and readily form aggregates that constitute the 7S (β-conglycinin) and 11S (glycinin) fractions, with conglycinin plus glycinin making up about 80% of the total protein content of the flour. A 2S fraction consists of low molecular mass polypeptides in the range of 8–20 kDa and is mainly the soybean trypsin inhibitors [37,38]. The 15S protein is probably a dimer of glycinin. As the data shown in Table 5.3, there are considerable differences in composition between the conglycinin and glycinin. As with most vegetable proteins, the weight percentages of aspartate plus aspartic acid, and glutamate plus glutamic acid, are large for both proteins, but the conglycinin has more arginine and lysine than the glycinin. The glycinin is a dodecamer with two coplanar rings of protein with a molecular mass of 300–380 kDa, with each ring composed of three acidic polypeptides, each with a molecular mass of about 35 kDa, and three basic polypeptides, each with a molecular mass of ≈20 kDa. The acidic and basic polypeptides are linked together by a single disulfide bond in addition to the ionic attraction. The β-conglycinin is a trimeric glycoprotein with a molecular mass of 150–200 kDa, and consists of three types of subunits: α' (72 kDa), α (68 kDa), and β (52 kDa). Information is provided here in compact form, and the reader is referred to the following references for additional information on soy protein structure [31,37,39–41].

It is natural to expect that with all these reactive moieties in the soy proteins (Table 5.3), it would be easy to develop covalent bonds of the protein subunits and agglomerates with cross-linkers. However, many of these reactive groups are not exposed on the surface, but are buried inside the globule, making reactions difficult. Furthermore, the polar groups inside the protein form interactions with other polar groups to stabilize the protein structure, making reactions even more difficult with these polar side chains.

Of all the commercial protein adhesives, those using soy flour are the only ones with a high carbohydrate:protein ratio. Commercially, the flour comes in three varieties, based on the protein dispersibility index (PDI), with 90 PDI having the lowest level of thermal treatment and the highest amounts of the native proteins and 70 and 20 PDI having increasing levels of thermal treatment and increased amounts of denatured proteins. The PDI is the percentage of protein (as determined by nitrogen content) in the flour that can form a stable dispersion in water. The flour can be further

TABLE 5.3

Amino Acid Contents in Weight Percentage for Soy Protein, Soy Glycinin, and Soy Conglycinin

Amino Acid	Soy Protein	Soy Glycinin (11S)	Soy Conglycinin (7S)
Alanine	6.0	5.2	3.7
Arginine	4.7	5.8	8.8
Aspartate + aspartic acid	16.9	20.0	20.5
Cysteine	1.0	1.4	0
Glutamate + glutamic acid	11.6	11.9	14.1
Glycine	9.7	7.5	2.9
Histidine	5.0	1.9	1.3
Isoleucine	2.9	4.2	6.4
Leucine	8.0	7.0	10.3
Lysine	5.9	4.4	7.0
Methionine	1.2	1.0	0.3
Phenylalanine	3.9	3.8	7.4
Proline	5.7	6.9	4.3
Serine	7.5	6.7	6.8
Threonine	4.0	3.9	2.8
Tryptophan	NA[a]	0.8	0.3
Tyrosine	2.6	2.8	3.6
Valine	3.4	4.8	5.1

Source: Sun, X. S., Isolation and processing of plant materials, in: *Bio-Based Polymers and Composites*, R. P. Wool and X. S. Sun (Eds.), pp. 33–55, Elsevier-Academic Press, Burlington, MA, 2005.

[a]*NA:* not available.

processed by removal of the soluble carbohydrates to form a protein concentrate or by removal of both soluble and insoluble carbohydrates to yield a protein isolate.

There are several ways to make soy protein concentrate, but they all involve denaturation of the proteins to decrease their solubility to allow soluble carbohydrates to be separated from the protein. The most common method uses an aqueous ethanol extraction of the soluble carbohydrates and some low molecular weight proteins, and involves an ethanol denaturation and insolubilization of the main proteins [42]. Concentrates can have different properties, because some are produced directly from the extraction process, whereas others go through a jet cooking process that involves a short, high-temperature treatment of the aqueous soy, followed by rapid cooling. Jet cooking is often called *functionalization*, because the soy functions better in food applications. This jet cooking greatly alters the protein, although the altered structure is not well understood [32]. It is important to understand the processing history used to make the concentrate, because processing history can dramatically alter the properties of soy concentrate and complicate the interpretation of the data. Because of the alteration of the protein in the extraction process, concentrates cannot be simply viewed as the soy flour with the soluble carbohydrates removed.

The purest commercial soy protein available is the isolate, which is produced by dissolving the native soy flour in water to allow the insoluble carbohydrates to be removed by centrifugation, followed by precipitation of the proteins by acidification to the isoelectric point and centrifugation. This process is used to prepare the native soy protein isolate (SPI) [35]. However, commercial SPI is far different from the conditions and equipment used to prepare the isolate in the laboratory, and is intentionally altered to improve its functionality in food applications. These functionalized proteins are produced through jet cooking and sometimes by enzyme treatments [32]. Although native SPI has the same differential scanning calorimetric glass transition peaks as those in native flour, and water dispersion is relatively low in viscosity, neither of these is true of commercial SPIs. In addition, commercial SPI is orders of magnitude better in providing wood bonds with good wet strength than the laboratory SPI [2]. Thus, the use of commercial SPI as a model for protein in soy flour is often technically quite misleading.

5.3.1 High-pH Soy Flour Adhesives

Although plywood was developed using other adhesives, such as animal protein, it gained commercial significance using soy flour with high pH (>12) conditions [1,33,34,43]. Much of the literature emphasizes the use of basic conditions with sodium salts for solubility and calcium salts for water resistance. In the current theories of proteins, basic conditions cause the polymer chains to be separated and somewhat uncoiled, because of the electrostatic repulsion overcoming the hydrophobic attraction. This can certainly be true for monovalent cations, but divalent calcium cations should hold the proteins together better. One report indicated that commercial formulations were more complicated, including the use of carbon disulfide [33]. These adhesives were the main adhesives for interior plywood until UF adhesives took over the market in the 1940s and 1950s. As with any new adhesive, UF formulations and curing conditions needed to be optimized to produce a consistent product. The transfer to UF required several changes from the high-pH soy adhesives, including the use of hot presses to replace the cold presses used with soy. Another change for safety and economic reasons required the UF adhesive to be made in a supplier's factory and shipped to the plywood plant, where the curing catalyst and filler were added just prior to application of the adhesive. The soy adhesives had been made in the plywood plant and, thus, the formulation could be modified for changes in wood furnish and plant conditions, such as temperature and humidity.

Another commercial application of soy adhesives was in finger-jointed studs using a honeymoon adhesive with soy hydrolysate applied to one wood surface and a phenol–resorcinol–formaldehyde (PRF) to the other wood surface [44,45]. This allowed successful finger jointing of green (undried) pieces to form nominal 2 × 4 in. (actually 38 × 89 mm [1.5 × 3.5 in.] after drying) studs [46]. This process was developed to use an environmentally friendly soy adhesive that substantially decreased adhesive costs by decreasing the amount of PRF needed and cut processing costs by not having to dry the wood first.

Given that the soy adhesives were manufactured under basic conditions, as are phenol–formaldehyde (PF) adhesives, it occurred to some researchers to combine soy with PF to decrease the cost [47–52]. However, a significant problem was that the soy formulations were much more viscous in combined adhesives compared with the PF alone. Two ways to address this problem were to keep the amount of soy low or to hydrolyze the soy flour to decrease its molecular weight. However, neither of these methods was useful, because they did not allow incorporation of much soy flour into the PF adhesive. Later research showed that the high viscosity was not as much of a problem as had been previously thought, because the protein dispersions are shear thinning. Thus, although high viscosity for a regular PF would lead to application problems, because it is nearly Newtonian in viscosity, a high-viscosity soy adhesive worked under the high shear conditions involved with most application equipment. Thus, it is not necessary to match the viscosity of a phenolic resin with that of a soy adhesive. With this approach, up to 50% soy can be used in an adhesive resin that is useful as an oriented strandboard (OSB) face resin. An interesting discovery was that the pH of an alkaline soy–PF adhesive could be lowered to a neutral pH to make a soy–PF dispersion that is both a light-colored, low-viscosity adhesive and one that provides good bond strength [53]. Normally, PF resole adhesives are dark colored and need to be kept at high pH to prevent precipitation, unless a surfactant is used. Most proteins are surface-active agents, and this property is widely used in the food and other industries but has not been used much in the chemical industry. Although soy can be used as a substitute for a large portion of the PF, this process has not been commercialized.

5.3.2 MODERATE-pH SOY FLOUR ADHESIVES

In the previous section, we covered adhesives made using soy flour with pH values of 12 or higher to denature the soy protein. This section discusses soy adhesives made using pH values of 6–10. Lower pH values had not been used, because they approach the isoelectric point of the proteins (pH 4–6 depending on the protein), which causes the protein to become more compact and precipitate. A pH below the isoelectric point is not in a useful range for a wood adhesive, because wood degrades with time at low pH values (<3). Soy protein with a wide variety of functional groups can react with a number of other chemicals [2,60]. For example, soy protein can react with formaldehyde and other aldehydes, mainly under more basic conditions, but this depends on the aldehyde. At a lower pH <7, the reaction products are often weak gels. Most of the studies on the higher pH systems above 12 were for adhesives used in lamination, such as plywood, but a number of investigations have also involved binder applications [60].

The discovery of polyamidoamine–epichlorohydrin (PAE) resin as a coreactant with soy by Li and others [54,55] led to a successful line of commercial Soyad adhesives by Solenis of Wilmington, DE (previously by Ashland and Hercules) for interior wood products, with a pH between 5 and 9 [56]. The concept is that PAE can react with both soy and wood, forming a bridge between wood surfaces. Although the original disclosure was with commercial SPI, the reaction with PAE also works well with soy flour. This technology was welcomed commercially, because it is a no-added formaldehyde (NAF) interior adhesive, and the California Air Resources Board (CARB) has decreased formaldehyde emissions from interior wood products in response to formaldehyde being declared a carcinogen. Columbia Forest Products (Greensboro, NC) wanted all their interior plywood to be NAF and developed the PureBond product line. Although soy with PAE is a major adhesive in this application in the United States, ultralow-formaldehyde-emitting UF is still a major player in the world market. These soy adhesives are also used for engineered wood flooring and particleboard.

Another soy technology that has been used commercially involves soy and magnesium oxides, in which the soy probably serves as a binder for the oxide and wood [57,58]. This technology has successfully worked in plywood plants. Both soy flour and SPI have been used as binders, along with other additives in this type of product.

Numerous other ways to modify soy proteins for adhesive applications are not reviewed here, because good reviews have been published [2,41,59,60]. Only very limited studies exist in the literature on these processes, and to our knowledge, none were examined commercially. Sun has reviewed research on modifying native SPI showing that urea and surfactant addition can improve bond strength [59]. Sun has also reviewed aspects of soy structure and different ways to modify soy for adhesive applications [37,41]. Vnučec and others have carried out a more thorough review of soy protein modification for soy adhesives [60]. Frihart and Birkeland have addressed the problem of comparing different soy products for wood bonding [2]. The major problem in assessing these studies is that bonding is often done with each study using its own wood species and only one bonding condition. With the large effects of soy type, wood species, bonding temperature, rheology, and type of strength test, relating the strength tests from different studies is tenuous.

5.4 OTHER PLANT PROTEINS (WHEAT GLUTEN, RAPESEED/ CANOLA, LUPINE, COTTONSEED)

Other plant proteins have not been as widely explored as soy or used commercially in the past, probably because of limited availability and cost. However, because of the tremendous advances in the amount of plant materials grown per acre through genetic selection and engineering, fertilization and irrigation, and mechanical improvements at all stages of plant production, plant protein has become by far the least expensive protein source. Although soybeans are an inexpensive plant source in the United States, Argentina, Brazil, and China, they do not grow well in colder climates. This has led to research on using other plant protein sources in other parts of the world. There has also been examination of plants that are more useful for their chemical content than for food or clothing uses, but so far there has been little published research on these plants for their utility as adhesives. This may change if the biofuels market becomes more profitable, because the interest would be more in the oils and carbohydrates and not in the proteins. The main outlet for excess proteins has been primarily as animal feed and not as industrial chemicals.

5.4.1 WHEAT GLUTEN

Although wheat straw has been examined and commercialized as a replacement for wood in making composites for home buildings and furnishings, the proteins in the seeds have been examined on a limited extent as the main adhesive. Wheat gluten certainly plays a major role in food products, given its cohesive and adhesive nature. Nonetheless, the protein content of wheat is low compared with that for other protein sources. Additionally, understanding wheat proteins is complex because of the many varieties of wheat. Besides the main divisions of hard grain (durum) wheat and soft grain wheat, there are many genetic variations, different growing conditions, and a variety of processing conditions [61]. The hard grain wheat gives strong doughs used in pizza making, whereas the soft grain wheat gives weak, extensible doughs used in cookie making. Breads use grains in the middle range.

With more than 100 different proteins in wheat, they are grouped into categories based on their structure and properties [61]. Analysis of the proteins by size exclusion high-performance liquid chromatography shows a continuum of peaks from monomeric to polymeric, making separation of pure proteins difficult. Among the monomeric (single-chain) proteins are the gliadin and albumin groups. The gliadins are monomeric storage proteins with molecular weights of 30–80 kDa, whereas the monomeric albumins have weights of 20–30 kDa, with many of them being enzymes (amino acid composition shown in Table 5.4). Compared with soy, wheat proteins are high in glutamate plus glutamic acid and proline, and low in aspartame plus aspartic acid and lysine among

TABLE 5.4
Amino Acid Content in Weight Percentage of the Main Wheat Proteins

Amino Acid	Glutenin	Gliadin	Albumin, Monomeric
Alanine	4.4	3.3	8.4
Arginine	3.0	2.0	5.7
Aspartate + aspartic acid	3.7	2.8	7.6
Cysteine	2.6	3.3	8.1
Glutamate + glutamic acid	28.9	34.6	10.8
Glycine	7.5	3.1	8.3
Histidine	1.9	1.9	0
Isoleucine	3.7	4.3	1.7
Leucine	6.5	6.9	7.6
Lysine	2.0	0.6	5.0
Methionine	1.4	1.2	2.6
Phenylalanine	3.6	4.3	0.1
Proline	11.9	16.2	7.5
Serine	6.9	6.1	6.4
Threonine	3.4	2.4	2.4
Tryptophan	1.3	0.4	3.0
Tyrosine	2.5	1.8	3.4
Valine	4.8	4.8	11.3

Source: MacRitche, F. and Lafiandra, D., Structure and functionality of wheat proteins, in: *Food Proteins and Their Applications*, S. Damodaran and A. Paraf (Eds.), pp. 293–325, Marcel Dekker, New York, 1997.

the reactive amino acids. The other major group is the polymeric (aggregated) proteins containing three groups (glutenins, high molecular weight albumins, and triticins). The glutenins make up about 85% of the polymeric proteins, and are similar to the gliadins in amino acid composition, but glutenins are made up of many different types of aggregates with a wide range of molecular weights. The gliadins are high in proline and low in lysine, as is the glutenin. The high molecular weight albumins are mainly amylases, whereas the triticins are globular proteins. These proteins are unusually low in glutamate plus glutamic acid and low in aspartate plus aspartic acid compared with other plant proteins.

Although a lot of research has been done on separation of proteins using a variety of analytical techniques, the main evaluation of protein properties has been in dough mixing and property tests [61]. The ability of gluten to hold dough together indicates that these proteins may have some utility as an adhesive. Wheat flour contains 75%–85% starch and only 10%–15% proteins (wheat gluten) [61]. Therefore, the flour is of limited value as an adhesive unless there is a coreactant that decreases the water sensitivity of the starch. The natural occurrence is for the starch to be in granules and the protein more in the continuous phase. Although the protein is only a small part of the composition, being part of the continuous phase, it has a large impact on the physical properties of the dough.

Due to these limitations, there has been research on using wheat flour as an adhesive. By far the biggest use has been to provide tack in wood adhesives, such as PF, because wheat gluten imparts the stickiness in doughs. Therefore, it is not surprising that it provides tack to wood adhesives. Some adhesives, such as phenolics and isocyanates used in wood bonding, are often not tacky enough to hold the layers together in plywood, fibers in fiberboard, particles in particleboard, or strands in OSB prior to hot pressing and curing. Gluten has been examined as an adhesive for particleboard [62], wood panels [63], and two-ply veneer bonds as an alternative to SPI under basic conditions [64]. Characterization of wheat glutens in relation to bond strength of two-ply veneer bonds was also carried out [65]. Prior work has not shown gluten to be a superior adhesive to soy in these applications.

5.4.2 Rapeseed and Canola

Although rapeseed readily grows in many areas, its use as an agricultural product has been limited because of the toxic nature of the plant [66,67]. The amino acid content of rapeseed, given in Table 5.5, is not very different from the soy protein in most functionalized amino acids in comparison with the data in Table 5.2. Although the proteins are not influenced by the toxic components, these components have limited the processing of the plant for commercial applications. High erucic acid content limited the use of the oils, because erucic acid has been associated with cardiac lesions, and the carbohydrates contain glucosinolates responsible for metabolism disruption in cattle and pigs. The oil is still useful for lubrication and other purposes, and the plant is used for winter ground cover.

As with many oilseeds, rapeseed is dried, crushed, and defatted to produce rapeseed meal. The meal is treated to decrease toxicity and used as animal feed. Although the toxins limit the food use of rapeseed, these toxins do not hinder the use of the meal or flour as an adhesive. Rapeseed proteins can be isolated from the meal, using a process similar to other oilseeds. Dispersing the meal in water at pH 11–12 dissolves the proteins, allowing them to be separated from insoluble meal components using centrifugation [67,70]. For rapeseed, sodium sulfite is added to limit oxidation of the phenolic compounds, which may then react with the proteins [71]. The supernatant is first acidified to pH 3.5–4 with dilute acid to precipitate the protein and then separated from the soluble carbohydrates and other soluble compounds by centrifugation. Given the limited or nonexistent human consumption, rapeseed flour is not readily available in the United States, and instead, most workers have used the meal of rapeseed or canola because of its availability. Rapeseed has also been treated with formaldehyde as a plywood adhesive [72].

Research in Canada led to the production of canola in 1979, a cultivar of rapeseed with oil low in erucic acid and meal with a low content of glucosinolates [66,67]. Additional cultivars have also been developed. The ability to use all parts of canola has supported an increase in total rapeseed

TABLE 5.5
Amino Acid Content in Weight Percentage of Rapeseed Protein and Canola Protein

Amino Acid	Rapeseed Protein	Canola Protein
Alanine	5.2	4.4
Arginine	10.2	5.8
Aspartate + aspartic acid	13.8	7.3
Cysteine	1.5	2.4
Glutamate + glutamic acid	21.2	18.1
Glycine	6.0	4.9
Histidine	2.8	3.1
Isoleucine	4.8	4.3
Leucine	8.2	7.1
Lysine	5.8	5.6
Methionine	2.3	2.0
Phenylalanine	3.3	4.4
Proline	5.9	3.8
Serine	5.6	4.0
Threonine	5.3	4.4
Tryptophan	NA[a]	1.3
Tyrosine	NA[a]	3.2
Valine	6.2	5.5

Source: Josefsson, E., *J. Sci. Food Agric.*, 21, 99–103, 1970; Canola Council of Canada, http://www.canolacouncil.org/oil-and-meal/canola-meal/nutrient-composition-of-canola-meal/protein-and-amino-acids/.

[a]*NA:* not available.

production from 19.2 million tn. in 1985 to 72.5 million tn. in 2013. Rapeseed oil is used for bio-diesel fuel and lubricants, as well as in food in the form of canola oil. Because canola is a variant of rapeseed, the proteins are similar to those in rapeseed and are mainly globular [67]. As shown in Table 5.5, the amino acid composition is similar between rapeseed and canola, but there is less aspartate plus aspartic acid and glutamate plus glutamic acid in canola. Acid- and base-hydrolyzed canola meal was examined as an additive for bonding fiberboard [73]. Canola protein isolate was used to bond cherrywood veneer at 190°C to give good wet bond strength in two-ply wood bonds. The addition of sodium bisulfite to decrease the dimercaptan bonds and viscosity did not greatly decrease bond strength [74]. Given the amounts of rapeseed and canola grown, and that these are a ready source of protein, it is surprising that more research has not been done on them.

5.4.3 Lupine

Lupine is a legume that has been used for food for thousands of years. It is a common plant in North and South America, around the Mediterranean Sea, and in Australia [75]. In some places, it is often used for its decorative flowers, but it is also harvested in a number of countries as a food in various forms. Although it has not received the attention that the soybean has, it is used as flour in South American and European countries. The ability of lupine to grow in soils and climate unfavorable to soybeans has led to studies on improving its yield. Selection and modern agriculture practice has increased its protein and oil content, and made it worth considering economically [76]. Lupine has been compared with soy because of the similar properties of high protein content and protein

TABLE 5.6
Amino Acid Content in Weight Percentage of Lupine Proteins

Amino Acid	Lupine Flour	Lupine Globulins	Lupine Albumins
Alanine	3.1	2.7	3.6
Arginine	11.1	11.8	12.6
Aspartate + aspartic acid	10.8	10.6	12.6
Cysteine	2.2	1.5	1.7
Glutamate + glutamic acid	25.1	27.4	26.6
Glycine	4.0	3.7	2.0
Histidine	2.1	1.8	2.1
Isoleucine	4.3	4.5	3.8
Leucine	7.3	7.4	7.6
Lysine	4.4	3.9	5.4
Methionine	0.5	0.3	0.5
Phenylalanine	3.9	3.9	3.2
Proline	4.1	4.4	4.1
Serine	5.2	5.4	5.3
Threonine	3.9	3.1	5.0
Tryptophan	0.4	0.5	0.2
Tyrosine	4.6	5.6	1.8
Valine	3.4	3.2	3.6

Source: Duranti, M. and Cerletti, P., *J. Agr. Food Chem.*, 27, 977–978, 1979.

composition (Table 5.6), except that the glutamate plus glutamic acid content is higher than aspartate plus aspartic acid in lupine, whereas the opposite is true for soy (Table 5.3). The proteins in lupine are mainly globulins with some albumins [76].

Lupine seed or bean proteins are 87% globulins and 13% albumins [78]. Of the globulins, 44% are vicilin-like, 33% are legumin-like, and 23% are others [76]. The vicilin proteins are storage globulins, have molecular weights between 143 and 335 kDa, and are all glycoproteins, whereas the legumin-like globulins are larger in size and do not have attached carbohydrates. The albumins have been less characterized.

Limited work has been done on bonding wood with lupine adhesives, but two studies indicate that they are about equal to soy adhesives in bond strength with lower dispersion viscosities [79,80].

5.4.4 COTTONSEED

Cotton is grown and used for its fiber, but the seeds need to be removed first; Whitney's cotton gin, one of the most famous inventions, removes the seeds from the cotton, which eliminated hand separation. Since the seed is mainly a low-value but abundant by-product, research has been carried out to find a use for the cotton seed protein that is of higher value than being used in animal feeds. As with many other plant proteins, it has been investigated as an adhesive for wood bonding. Cottonseed meal proteins are fairly typical in composition for plant proteins, but have a higher arginine content (Table 5.7).

He and coworkers have looked at the performance of cottonseed protein isolate in plywood bonding [82,83], whereas Cheng et al. looked at the performance of the meal for producing wood composites [84]. There are insufficient data available to compare the cottonseed proteins with the soy proteins as wood adhesives.

TABLE 5.7
Amino Acid Content in Weight Percentage of Cotton Seed Meal Proteins

Amino Acid	Cottonseed Meal	Amino Acid	Cottonseed Meal
Alanine	4.5	Lysine	4.4
Arginine	11.5	Methionine	1.8
Aspartate + aspartic acid	10.1	Phenylalanine	5.7
Cysteine	1.6	Proline	4.3
Glutamate + glutamic acid	21.7	Serine	4.7
Glycine	4.9	Threonine	3.6
Histidine	2.8	Tryptophan	NA[a]
Isoleucine	3.4	Tyrosine	3.4
Leucine	6.6	Valine	5.1

Source: O'Mara, F.P. et al., *J. Anim. Sci.*, 75, 1941–1949, 1997.
[a]*NA:* not available.

5.5 MILK (CASEIN AND WHEY)

5.5.1 CASEIN

Milk is an important source of proteins with some industrial uses, but mainly food uses. Although early recorded history does not indicate much use of milk as an adhesive, it has been used as a binder in milk paint. This type of binder has been used for thousands of years and is still commercially available. Although milk itself is not used as an adhesive, because of the high fat and sugar content, the proteins from it have had good adhesive utility. Generally, the proteins are separated from milk into casein and whey protein fractions, with the casein being more commercially important, whether it is separated as the isolated casein or used for cheese and yogurt production. Casein and whey proteins have different amino acid compositions (Table 5.8), but the real differences are molecular weights, protein conformation, and phosphorylation of the casein protein. Native casein is a complex micelle that has two purposes in milk besides the nutritional value of the protein, which are binding and dispersing calcium phosphate and butterfat in milk [85]. These two functions are located in different portions of the casein micellar structure. Calcium binding appears to be associated with phosphorylated serine residues in localized domains [85]. The second function involves having the polar groups localized in the protein sequences so that the rest of the chain is quite hydrophobic to hold the butterfat in the milk. The large casein micelles give milk its opaqueness and provide stability to the milk at different temperatures and pH conditions. The casein is known to have low crystallinity (lack of α-helices and β-sheets) because of the high content of proline and hydroxyproline (not normally part of the protein analysis) (Table 5.8). These data counter the idea that crystallinity in soy is important for its adhesion. The main whey components are β-lactoglobulin, α-lactalbumin, and other low molecular weight proteins that provide low dispersion viscosities.

The cow milk protein that has been used commercially as a wood adhesive is the casein-type adhesive from skim milk. The casein protein has been produced only in New Zealand for many years, because of excess milk production and low cheese manufacturing there [88]. The dispersed casein micelle consists of casein proteins, calcium phosphate, and fat. The casein protein is isolated from milk by acid precipitation, using either inorganic acids (hydrochloric or sulfuric) or lactic acid produced from lactose by enzymatic reaction. The casein proteins are not just a source of amino acids for the growing calf, but also provide calcium that is bonded to the phosphorylated serine residues in the casein protein micelle. Sensitivity to acid is built into the casein micelles, allowing the

TABLE 5.8
Amino Acid Content in Weight Percentage of Cow's Milk, Casein Protein, and 80% Whey Protein Concentrate

Amino Acid	Milk	Casein	Milk Whey 80% Concentrate
Alanine	2.3	2.76	5.55
Arginine	3.8	3.35	2.71
Aspartate + aspartic acid	5.8	7.57	9.18
Cysteine	0.3	**Part of methionine value**	
Glutamate + glutamic acid	21.7	21.85	15.84
Glycine	0.4	1.73	5.32
Histidine	2.3	2.54	0.78
Isoleucine	6.0	4.59	4.97
Leucine	10.8	8.89	10.66
Lysine	6.9	7.75	8.81
Methionine	2.9	3.20	7.97
Phenylalanine	5.5	10.14	5.82
Proline	9.8	9.33	6.66
Serine	5.4	5.54	5.30
Threonine	4.4	4.05	6.87
Tryptophan	6.0	1.04	1.73
Tyrosine	1.2	**Part of phenylalanine value**	
Valine	6.6	5.64	1.84

Source: Macy, I.G. et al., The composition of milks. A compilation of the comparative composition and properties of human, cow, and goat milk, in: *Colostrum and Transitional Milk*, National Academy of Sciences and National Research Council, Washington, DC, 1953; Sindayikengera, S. and Xia, W.-S., *Jhejiang Univ. Sci. B*, 7 (2), 90–98, 2006.

complex to break apart in stomach acid, freeing up the components. Although the isolation process irreversibly breaks up the protein micelle, it leaves protein aggregates that have both a large hydrophobic side and a polar side with phosphorylated groups. This aggregate is made up of four types of casein proteins, with three of them containing substantial amounts of serine phosphate, along with a negative charge at pH 6.6, and the fourth one being glycosylated [85].

The precipitated acidic casein is converted into an adhesive using a combination of sodium hydroxide and calcium oxide [1,89]. The pH is raised from the isoelectric point of about 4.5 to greater than 12, usually using a mixture of sodium hydroxide and lime [85,89]. The sodium provides greater solubility, but is poor in water resistance, with the opposite being true for the calcium. The slowly dissolving lime is added shortly before use to the sodium caseinate, to prevent viscosity from getting too high prior to application. Among the additives for casein adhesives were soy to lower cost and blood to give better wet strength [43,89,90]. Except for blood proteins, the caseinate adhesives produced the most water-resistant protein adhesive of the twentieth century. This adhesive was used for making exterior plywood and propellers for wooden airplanes, especially in World War I, often with blood mixed in. It was also used in early glulam beams, because of its strength and good water resistance. In fact, it was recently reported that casein-bonded glulam beams were used in an unheated and non-air-conditioned building for 75 years before the building was disassembled and the beams tested for strength [91]. Most of the designed strength was retained.

The high price and limited availability of casein has held back any further development in this area. Casein adhesives continue to be used in certain applications such as the construction of fire-resistant doors.

5.5.2 WHEY

After precipitating the casein from milk, whey is processed to separate the proteins from the lactose, fats, and minerals. However, most of the whey comes not from casein precipitation from low-fat milk, but from cheese or yogurt production using high-fat or low-fat milk. The whey proteins have been divided into three classes, based on the conditions used to separate them from the casein portion. The first is from hard cheese production, which uses rennet coagulation of milk at pH 6.6. The second is from soft cheese production, which uses acid coagulation of casein, and cottage cheese production. A newer area is the acidic whey from yogurt production. Unless the protein is concentrated, these whey products have more lactose than protein, because most of the lactose goes into the whey. Conversely, the fat content of the whey fraction is not very high, because most of the fat goes into the cheese or yogurt portion.

Whey concentrate has most of the nonprotein materials removed, and the isolate is even purer (its amino acid composition is given in Table 5.8). The concentrates are generally made by ultrafiltration of the whey (<25% protein) to allow the low molecular weight organics, especially lactose and fat, and inorganics, to be removed [92]. However, the range of products is quite varied in protein content (30%–80%), depending on the degree of filtration. Further purification leads to a whey protein isolate that is 90%–95% protein. In addition to ultrafiltration, ion exchange has been used to purify the whey proteins.

Although whey contains many proteins with diverse functions, all of the proteins are globular, with the β-lactoglobulin and the α-lactalbumin accounting for about 70% of the protein weight [93]. The β-lactoglobulin has a molecular mass of 18.3 kDa with 43%–50% of proteins in β-sheets, 10%–15% in α-helices, and 15%–20% in β-turns. It is a compact dimer molecule with two disulfide bridges. The α-lactalbumin has a low content of organized sheets of 30% α-helix and 9% β-sheets, with one bound calcium ion and four disulfide bridges. These proteins also contribute to most of the hydration, gelling, and surface-active properties of the whey proteins. The immunoglobulins are a heterogeneous family of glycoproteins from 148 to 1000 kDa and tend to be heat sensitive. The other main protein is bovine serum albumin with 582 amino acids. Other minor proteins can play an important role in the properties of whey [93].

Whey products are used in a wide variety of food products, including animal feed [93]. Whey has also been tested as a wood adhesive with an aqueous polymer isocyanate for glulam [94].

5.6 EGG WHITE

All parts of the egg contain proteins with concentrations of 3% of the shell, 11% of the white, and 17.5% of the yolk. For a long time, egg whites were used as binders for paints, but there was no record of them being used as an adhesive. Although eggs are mainly used as a food source, the whites can also be a source for nonfood, industrial applications. The whites do not completely drain out of the eggshells in industrial egg processing, and the remaining egg whites washed from the eggshells cannot be used for food applications because of health concerns. This leaves these whites available for industrial uses.

Ovalbumin is a high percentage of the protein in egg whites, and is also easy to separate from the other egg components in a very pure form [95,96]. The ovalbumin is a monomeric phosphoglycoprotein with four cysteine and one cystine amino acid in 385 amino acids [95,97]. The different ovalbumins vary in the amount of phosphorylation at the two specific serine sites. The more stable ovalbumin forms have a 92.5°C denaturation temperature, compared with 84.5°C for a less stable form. Although most analytical methods do not show a difference, the more stable ovalbumin is slightly more compact. The four cysteine residues in the native protein are not very accessible, preventing them from joining together as disulfides. This could change after denaturation. Table 5.9 shows that the ovalbumin, egg white, and egg yolk do not have large differences in amino acids, except that the glutamate plus glutamic acid and cysteine are greater in the ovalbumin than in the egg white and greater in the egg white than in the egg yolk.

Gelation in water is important to food properties, and shows the development of structure caused by heat. In some ways, it is similar to the structure that provides water resistance to protein adhesives after hot bonding of the adhesives. Gel properties of the ovalbumin have been extensively studied and shown to be sensitive to the gelation conditions [95]. Heat gelation is apparently driven mainly by hydrophobic interactions between protein molecules, rather than by hydrogen bond or disulfide formation. At higher pH conditions, in which the electrostatic repulsion is greater, the gel is similar to a closely packed string of beads with each protein molecule being a bead. At low concentrations, sol structures form, whereas at higher concentrations, cross-links develop to form a three-dimensional network, with both types of structures being transparent. As the pH is decreased near the isoelectric point or as the ionic strength of the aqueous phase increases, the gel structure becomes random aggregates and is less linear, with these gels being opaque. The opaque gels have greater strength than the transparent gels.

Despite the extensive studies in the last century on egg proteins for food uses and some uses of egg as a binder, the wood adhesive properties of egg proteins had not been reported until recently [99]. Compared with most other proteins, dispersions of ovalbumin are low in viscosity. This could be caused by the monomeric nature of ovalbumin rather than polymeric protein aggregates of most of the proteins. Although their viscosities are low, ovalbumin dispersions provide good wet and dry bond strengths without any coreactant [99]. The aqueous dispersions of dried egg white are almost as good as those of the ovalbumin, but those of the dried egg yolk are much poorer, and those of the dried whole egg are in between the whites and yellow portions of eggs. These differences in adhesive strength are not obvious if one solely depends on making predictions from the amino acid contents in Table 5.9.

TABLE 5.9
Amino Acid Content in Weight Percentage of Ovalbumin, Egg White, and Egg Yolk

Amino Acid	Ovalbumin	Egg White	Egg Yolk
Alanine	5.7	5.9	5.1
Arginine	5.9	5.7	7.0
Aspartate + aspartic acid	9.2	7.7	8.5
Cysteine	8.0	2.4	1.7
Glutamate + glutamic acid	15.7	13.8	12.2
Glycine	3.2	3.6	3.6
Histidine	2.4	2.1	2.3
Isoleucine	7.1	6.3	6.2
Leucine	9.9	8.6	8.5
Lysine	6.4	5.9	6.7
Methionine	5.4	3.8	2.6
Phenylalanine	7.5	6.3	4.5
Proline	3.8	3.7	4.5
Serine	8.5	6.8	8.2
Threonine	5.7	4.4	5.2
Tryptophan	NA[a]	1.5	1.5
Tyrosine	3.8	4.1	4.7
Valine	6.1	7.6	7.0

Source: Lewis, J. C. et al., *J. Biol. Chem.*, 186, 23–35, 1950.
[a]*NA:* not available.

5.7   COLLAGEN

5.7.1   Animal (Bones and Hides)

Animal glues are the earliest recorded adhesives used in wood bonding. Ancient Egyptian drawings demonstrated veneer bonding with a hot-melt adhesive such as collagen [1]. Seed, milk, or egg proteins provide essential nutrients to young organisms for growth, but collagen-based animal proteins are used for structural purposes during the entire life span of the animal. Instead of individual protein chains being coiled globules, the collagen protein forms elongated triple-stranded helical coils [100]. These coiled proteins that make up the collagen fibrils provide not only good strength, but also good flexibility. The high contents of proline and hydroxyproline prevent the protein chain from forming a globular shape, and the many polar groups provide good interaction between the chains. Therefore, coils are helically linked together [25,100]. The different protein contents for animal and fish collagen are given in Table 5.10.

It is important to understand how different these protein adhesives are from most of the other proteins discussed in this chapter. The amino acid composition in Table 5.10 is highly nonpolar [25,103]. Thus, it would appear that the protein should be globular in nature to minimize water interaction, yet the high contents of proline and hydroxyproline not only prevent the protein from folding into a globular structure, but instead generate a helical structure. Some globular proteins can be converted into a more fibrous structure under certain processing conditions [104], but do not exist that way in nature. Although globular proteins are generally dispersible without modification, fibrous collagens need to be broken apart by chain separation with some depolymerization to be dispersed in water. The degree of chain separation and depolymerization generally has a larger effect than the source on the performance of these adhesives, leading to the different grades mainly determined by viscosity and gel hardness values [100]. These extended proteins behave similarly

TABLE 5.10
Amino Acid Content in Weight Percentage for Animal and Fish Collagen

Amino Acid	Animal Hide Collagen	Fish Collagen
Alanine	11.0	10.4
Arginine	8.8	9.1
Aspartate + aspartic acid	6.7	7.5
Cysteine	0	0
Glutamate + glutamic acid	11.4	11.4
Glycine	27.5	28.2
Histidine	0.8	9.1
Hydroxyproline	14.1	8.3
Isoleucine	1.7	1.6
Leucine	3.3	3.2
Lysine	4.5	3.7
Methionine	0.9	2.3
Phenylalanine	2.2	2.0
Proline	16.4	12.4
Serine	4.2	7.9
Threonine	2.2	3.3
Tryptophan	NA[a]	NA[a]
Tyrosine	0.3	0.6
Valine	2.6	2.3

Source: Eastoe, J. E., *Biochem. J.*, 61, 589–600, 1955; Eastoe, J. E., *Biochem. J.*, 65, 363–368, 1957.
[a]*NA:* not available.

to a typical synthetic polymer with greater interchain interactions and greater influence by a wide variety of additives [25,100,103]. These interchain interactions are not only polar in some domains but should also be hydrophobic in other domains, given the high concentrations of the nonpolar glycine, proline, alanine, and other nonpolar amino acids [25]. The protein chains are still in a helical shape, and these helices tend to reform intertwined structures [100]. The extended structure allows for a greater ability to formulate products with a wider variety of performance characteristics than is possible with the globular proteins. Thus, it is important to consider the effect of the primary structure, or polypeptide backbone, on the properties of the protein (globular or extended) and the ability to modify the protein properties for the adhesives.

There is tremendous variability in the collagen, depending on its source (skins, connective tissue, cartilage, or bones) and the age and type of animal source (type and raising of cattle, type of pigs). The main source has been the hides and bones of beef cattle [103]. The bones and hides are processed using different procedures, because of the need to separate the collagen from other different materials. After cleaning to remove most of the inorganic and other organic molecules, the fibrils are disentangled and hydrolyzed using an inorganic acid and heat [103]. The source of the collagen and the hydrolysis conditions determine the properties of the adhesive, although the difference in source becomes less critical as the chains are shortened to lower the molecular weight of coiled proteins [100]. These proteins are mainly sold on the basis of viscosity, color, and gel strength measurement. The average molecular weight of these hydrolyzed proteins is in the range of 29–250 kDa. Increasing the molecular weight generally increases the viscosity and adhesive strength [100], with the higher molecular weights yielding gels that need to be heated for application.

The original animal glues were sold in the dry state and then dispersed in hot water to provide a thick solution. In the solution state, the adhesive is not stable for extended periods of time. Therefore, the adhesive is made up just prior to using it. This is basically a hot-melt adhesive, because it needs to be heated for application, and then the initial strength is obtained as the adhesive cools. The initial strength is provided by a high propensity of these coiled polymers to associate, forming a gelled network that provides high initial strength by cooling for holding the substrates together. The strength rapidly increases as some of the water is lost to evaporation or absorption into the substrate. Further strength is provided as the gel forms a more entangled network and additional water moves away from the bondline [100,103]. Strong bonds form in the dry state, because of the intertwined protein chains and extensive hydrogen and hydrophobic bonds between the coiled proteins. There can also be polar bond formation between the acidic and basic groups, with a reported tensile strength of up to 69 MPa (10,000 lb./in.^{2}) [25]. The dried film forms a continuous, noncrystallizing film that resists most solvents and oils.

To avoid the necessity of dissolving the dry glue in hot water, liquid animal glues have been developed using gel-depressant chemicals. These chemicals limit the association of the proteins in water, allowing the gelled state to be postponed [100,103]. As the water departs from the glue, the gel-depressant nature of these additives disappears and a strong bond is formed. The addition of glycerin plasticizers results in jelly adhesives that provide products with permanent flexibility [103]. For nonwarping applications such as bookbinding, the jelly glues are mixed with plasticizers, extenders, and fillers to provide a fast-setting product with limited shrinkage after drying.

The affinity of animal glues for water leads to poor water resistance, which can be either a disadvantage or an advantage. Certainly, the adhesives' tendency to soften and, in some cases, to delaminate is usually a disadvantage. However, this is useful when a recyclable adhesive is desired; the adhesive is strong in the dry state but softens for adhesion with water treatment and dissolves away with more extensive water exposure. Moisture sensitivity can be decreased by precipitating the protein with inorganic sulfates, borates, or formaldehyde [103]. These adhesives were used with paper products, gummed tapes, and coated abrasives, as well as with wood for furniture construction and woodworking industries. These adhesives were initially displaced by poly(vinyl acetate) adhesives or white glues. Now, there are a variety of synthetic polymers, such as cross-linking poly(vinyl acetate) and polyurethane, types that are easier to use and provide greater bond durability. The main use of animal adhesives today is in historic preservation and by hobbyists who want to use traditional methods [100].

5.7.2 FISH

Fish glues are, in many ways, similar to animal glues. but are often in a separate category because of property and performance differences. Fish glues are liquid at room temperature. This provides an advantage compared with animal glues, which originally were solids that needed to be dissolved in hot water and applied hot to prevent gelation prior to application [105]. Although more recent versions of animal glue are made in liquid form, there is still a difference between animal and fish glues [100]. Liquid fish glue can bond a wide variety of substrates (metal, rubber, glass, cement, cork, wood, and paper) and will not soften up to 260°C. However, they do soften in water, which allows them to be used in self-sticking stamps and as a temporary adhesive.

The adhesive comes from the skins and bones of fish, especially cod skins, by extraction with hot water, and is concentrated to about 45% solids [105]. Cleanliness of the processing near the fish processing plant and use of added deodorant helps keep the product odor low. The glues are between 30 and 60 kDa in molecular weight, although some gelatin versions are higher in molecular weight [100]. The fish glues can be more linear than animal glues, leading to greater water solubility. After drying, the glues have internal stresses because of volume shrinkage; this shrinkage can increase with higher humidity conditions, which leads to both greater interchain helix formation and increased brittleness. Fish amino acid contents from bones and skin are similar to that of animal collagens in having unusually high amounts of proline and hydroxyproline (Table 5.10).

Because the protein is amphoteric, it has a certain buffering capability, but pH levels below 3 and above 9 should be avoided because of degradation of the protein. Fish glue has shown strengths of 32 MPa (4641.2 lb./in.^{2}) with 50% wet wood failure in ASTM D905 wood shear testing (ASTM International, West Conshohocken, PA). Their liquid nature and high strength led to their use in furniture. However, the brittle aspect of the glue can lead to failure with time. Fish glue's water solubility can be decreased by adding polyvalent salts, such as aluminum or ferric sulfates, oxidation with chromates, or adding aldehyde cross-linkers, such as formaldehyde, glyoxal, or glutaraldehyde [105]. Because of their compatibility with many waterborne polymer dispersions, fish glue has been added to these dispersions to increase their tackiness. As with other collagen adhesives, fish glues are still used today by hobbyists and furniture restorers.

5.8 BLOOD

Not surprisingly, blood is one of the wood adhesives that goes back to the beginning of recorded history, because of its ability to dry into a strong material with adequate water resistance. Blood became a common adhesive for wood bonding in the twentieth century because it provided good exterior durability. Fresh blood was used for a long time, because blood does not keep well under ordinary conditions. The use of blood expanded greatly between 1910 and 1925 with the development of a low-temperature method of drying the blood without decreasing its solubility and the need for water-resistant plywood for military aircraft [1]. Most blood adhesives require heat curing, limiting their use in plywood, which was often cold-cured. These adhesives were generally displaced by phenolics in the 1930s, but the need for more water-resistant interior plywood led to a revival of blood adhesives as wet-strength enhancers. Because of their high cost and variability, blood proteins were often used with other proteins. The use of blood in wood adhesives peaked at 22.7 Gg (50 million lb.) in the 1960s and steadily declined to near zero. Although the data on amino acid content in Table 5.11 are for guinea pig blood, the data for other blood sources are not very different. The lysine content is high while the combined aspartate and glutamate are low compared with soy protein.

TABLE 5.11

Amino Acid Content in Weight Percentage of Blood Proteins from Guinea Pig

Amino Acid	Blood	Hemoglobin	Plasma
Alanine	7.6	9.0	6.5
Arginine	4.5	2.6	6.1
Aspartate + aspartic acid	10.2	10.0	10.3
Cysteine	2.1	NA[a]	3.8
Glutamate + glutamic acid	12.5	7.1	16.9
Glycine	3.8	4.1	3.5
Histidine	6.5	9.1	4.4
Isoleucine	3.4	3.2	3.6
Leucine	6.8	12.5	2.3
Lysine	9.9	9.5	10.2
Methionine	1.4	0.9	1.8
Phenylalanine	7.3	5.0	6.7
Proline	5.0	3.0	6.5
Serine	5.5	5.7	5.4
Threonine	3.7	6.8	1.2
Tryptophan	NA[a]	NA[a]	NA[a]
Tyrosine	3.2	0.8	5.0
Valine	6.8	7.7	6.1

Source: Chang, Y.-Y.H. and Judson, C.L., *Compr. Biochem. Physiol.*, 62A, 753–755, 1979.
[a]*NA:* not available.

There are three general categories of dried blood [107]. The most important as an adhesive is soluble blood, which is the most dispersible protein for good water-resistant bond strength after heat curing. The heat is minimized in the drying of the blood by using a spray or vacuum dryer. For inclusion in this category, more than 80% of the protein needs to be soluble. The other extreme is insoluble heat-dried blood that has low solubility (less than 20%) and leads to gritty dispersions; this is the most common form today and used by gardeners. The remaining category of partially soluble blood is between these extremes. Blood was mainly used to enhance the water-resistance properties of other protein adhesives. The casein–blood adhesives were considered the most durable for exterior plywood [90,107] until PF adhesives were eventually formulated to provide a better adhesive. Blood was also used with soy adhesives to improve their properties. For many years, blood was added to PF adhesives to improve their properties, including foamability and tack. Although blood improved the moisture resistance of protein adhesives, it had to be dried to prevent degradation. Not only was there a large difference between pig and cattle blood, but the blood varied with the breed and time of the year. This difference mainly showed up in the rheology of the adhesive, which needed formulation adjustments with new blood batches.

Although blood has not been used much for adhesives, there could be some availability in the future as by-products when blood is used as the source of specific chemicals [108]. Although this would not be a large volume, blood was generally added in small amounts to other proteins and synthetic adhesives to improve tack and, in some cases, water resistance.

5.9 SYNTHETIC AND MODIFIED PROTEINS

One study has shown that specific sequences of synthetic polypeptides can provide better adhesives than other sequences [109]. Given the difficulty of making these proteins, this type of process does not currently have commercial viability. However, testing hypotheses of protein interactions with synthetic proteins could be very interesting scientifically.

Although protein glues have always been derived from natural sources, this does not mean that they cannot be improved, for example, by genetic manipulation. Genetic modification is a significant area now for many crops to improve resistance to various insects, diseases, and pesticides used to treat the crop. Other genetic modifications have provided the canola variant of rapeseed and high oleic content soybeans. Although these changes have not been directed toward changing the proteins, changing other properties of the crop can lead to inadvertent changes in the proteins, some of which may be beneficial.

Biomimicry and bioinspired are two newer approaches for product development. The difference between them is that biomimicry attempts to carry out a function in the same manner as is done in nature, whereas bioinspired approaches use principles learned from natural systems but with different chemistry or structure. An interesting biomimetic approach is to incorporate 3,4-dihydroxy-L-phenylalanine (L-DOPA) amino acid into the protein, so that when exposed to oxygen, the dihydroxybenzene group forms the hydroquinone. Two of these hydroquinones on different chains can then couple to cross-link the polyamide chains. The L-DOPA can also form bridges by chelating metals. For a biomimetic approach, Liu and Li successfully grafted dopamine to a commercial soy protein isolate (CSPI) via an amide linkage imparting DOPA-like phenolic functional groups found in marine adhesive protein, a strong and water-resistant adhesive [110]. For a bioinspired approach, they also successfully grafted cysteine onto CSPI by amide linkages, increasing the content of the −SH group significantly [111]. The free mercapto group content in soy protein could greatly increase the strength and water resistance of wood composites bonded with the modified soy protein. Mussel protein contains a high amount of mercapto-containing cysteine.

5.10 PROSPECTS

Some protein sources that were used as wood adhesives in the past are not likely to be available in suitable volume and with viable cost. Many of these proteins have found other uses in human or animal food or specialty chemicals. Although the technology developed in the early twentieth century for protein adhesives was suitable at the time for making bonded wood products, this technology is not capable of meeting today's production or performance criteria for wood products. This emphasizes the need for novel technology for using proteins. Professor Li at Oregon State University has developed the most new routes to modify soy proteins [48,54,55,57,80,110,111]. Additional creativity along these lines, especially if it is tied in with colloidal nature proteins, may hopefully develop more protein adhesives that can better compete with synthetic adhesives.

Although protein adhesives provide good dry strength, many are deficient in wet strength. Today, much more is known about protein structure and more ways are known for modifying proteins. Unfortunately, chemical modifications are often not well coordinated with what is known about protein structure. The work with soy (e.g., the much higher strength for the commercial isolate compared with the native isolate) and egg whites (e.g., the much lower viscosity of egg protein at high solids compared with soy) shows that there is untapped potential in proteins that may hopefully be converted into stronger adhesives by properly controlling the colloidal and coalescence properties. The polyamidoamine–epichlorohydrin coreactant demonstrates that a good coreactant can make soy proteins commercially viable adhesives. Hopefully, with a better understanding of the proteins' structures and the colloidal properties, the desired properties can be achieved.

5.11 SUMMARY

Proteins serve many purposes in nature, such as adhesives, enzymes, structural tissue, and amino acid sources for growing organisms. Both plants and animals have been sources of proteins for wood adhesives. Although proteins had dominated the wood bonding market for centuries, starting in the 1940s fossil fuels became the dominant source of adhesives for wood bonding because of their durability, ease of use, and relatively low cost. This trend may be reversing for a variety of reasons, with concern about formaldehyde emissions from UF being chief among them. The concern about hydrocarbon cost and the desire for higher biobased-content products are additional drivers.

Although some of the most effective protein adhesives, such as casein and blood adhesives, are no longer available at reasonable prices and volumes, some other protein sources have good availability and prices. Principal among these more abundant proteins are plant-based proteins that have an infrastructure in place to grow, gather, and process the crop. In many cases, protein is the lower-value portion of the crop, with the oil of oilseeds being more valuable than the meal or flour.

However, native proteins by themselves have not shown sufficient strength in most cases when the bonded samples are exposed to wet conditions. Proteins will need to be physically or chemically modified or used with a coreactant to provide sufficient cohesive strength to be able to withstand wet conditions. The improved products can be developed empirically, as in the past, or systematically by understanding the colloidal nature of proteins. Taking leads from food chemistry and adopting the knowledge developed by bioscientists, who study natural protein products, will be critical to allow the work to be done more systematically.

A lot of development was needed to make phenolic, amino, and isocyanate types of adhesives suitable for wood bonding. The original poly(vinyl acetate) adhesives had good adhesion but poor moisture resistance. The cross-linked versions were developed later to provide the adhesives with much improved water and heat resistance. Hence, the protein adhesives need to be considered as starting materials, not finished products. The success with Soyad adhesives from Solenis and the Purebond wood products made by Columbia Forest Products are signs that the right chemistry can help soy adhesives overcome their deficiencies.

ACKNOWLEDGMENTS

The authors thank Jim Wescott of Heartland Resource Technologies, Chris Hunt of the Forest Products Laboratory, Mike Birkeland of AgriChemical Technologies, Russ Egbert of Archer Daniels Midland, Mike Porter of Cargill, and consultant Dani Zhu for their helpful comments on proteins.

REFERENCES

1. A. L. Lambuth. Protein adhesives for wood, in *Handbook of Adhesive Technology*, 2nd edn, A. Pizzi and K. L. Mittal (Eds.), pp. 457–478, CRC Press, Boca Raton, FlL (2003).
2. C. R. Frihart and M. J. Birkeland. Soy properties and soy wood adhesives, *ACS Symp. Seri.* 1178, 167–192 (2014).
3. G. E. Myers. Effects of temperature and humidity on formaldehyde emission from UF-bonded boards: A literature critique. *Forest Prod. J.* 35 (9), 20–31 (1985).
4. C. R. Frihart, J. M. Wescott, T. L. Chaffee and K. M. Gonner. Formaldehyde emissions from urea-formaldehyde- and no-added-formaldehyde- bonded particleboard as influenced by temperature and relative humidity. *Forest Prod. J.* 62 (7–8), 551–558 (2012).
5. A. Pizzi. Urea-formaldehyde adhesives. In: *Handbook of Adhesive Technology*, 2nd edn, A. Pizzi and K. L. Mittal (Eds.), pp. 635–652, CRC Press, Boca Raton, FL (2003).
6. G. E. Myers. Formaldehyde emission from particleboard and plywood paneling: Measurement, mechanism, and product standards. *Forest Prod. J.* 33 (5), 27–37 (1983).
7. G. E. Myers and M. Nagaoka, Emission of formaldehyde by particleboard: Effect of ventilation rate and loading on air-contamination levels. *Forest Prod J* 31 (7), 39–44 (1981).
8. G. E. Myers. How mole ratio of UF resin affects formaldehyde emission and other properties: A literature critique. *Forest Prod. J.* 34 (5), 35–41 (1984).
9. S. Zhang, J. Li, J. Zhang, Y. Li and Q. Gao. Study on properties of modified low molar ratio urea-formaldehyde resins (I), *Adv. Mater. Res* 113–114, 2016–2020 (2010).
10. EPA. Formaldehyde. February 26. http://www.epa.gov/formaldehyde. Accessed March 4, 2016 (2016).

11. ARB. Composite Wood Products ATCM 2016. Formaldehyde emissions in wood products. http://www.arb.ca.gov/toxics/compwood/compwood.htm. Accessed March 4, 2016 (2016).

12. Chimar-Hellas. Update on Formaldehyde Emissions from Wood-based Panels. http://www.chimarhellas.com/wp-content/uploads/2008/07/formaldehyde_2008.pdf. Accessed February 12, 2016 (2008).

13. Intertek. ATCM: Reducing Formaldehyde Emissions from Composite Wood Products. https://www.google.com.br/?gfe_rd=cr&ei=Y2bbVtn2IcHK8geo85SgCA&gws_rd=ssl#q=formaldehyde+emissions+wood+products. Published June 22, 2010. Accessed December 20, 2015.

14. M. J. Birkeland, L. Lorenz, J. M. Wescott and C. R. Frihart. Determination of native (wood derived) formaldehyde by the desiccator method in particleboards generated during panel production. *Holzforschung* 64 (4), 429–433 (2010).

15. J. B. Wilson. Life-cycle inventory of particleboard in terms of resources, emissions, energy and carbon. *Wood Fiber Sci.* 42 (CORRIM Special Issue), 90–106 (2010).

16. HSE. P43 Using Isocyanate-based Adhesives. COSHH Essentials for Printers 05/07. http://www.hse.gov.uk/pubns/guidance/p43.pdf. Accessed March 22, 2016 (2007).

17. E. W. Lusas. Oilseeds and oil-bearing materials. In: *Handbook of Cereal Science and Technology*, K. Kulp and G. Points, Jr. (Eds.), pp. 297–362, Marcel Dekker, New York (2000).

18. D. K. Salunkhe, J. K. Chavan and R. N. Adsule. *World Oilseed Chemistry, Technology, and Utilization*, Van Nostrand Reinhold, New York (1991).

19. W. J. Wolf. Soybeans and other oilseeds. In: *Kirk-Othmer Encyclopedia of Chemical Technology*, 3rd edn, Vol. 21, pp. 417–442, Wiley, New York (1983).

20. R. Lasztity. *Chemistry of Cereal Proteins*, CRC Press, Boca Raton, FL (1995).

21. S. Damodaran. Food proteins: An overview. In: *Food Proteins and Their Applications*, S. Damodaran and A. Paraf (Eds.), pp. 1–24, Marcel Dekker, New York (1997).

22. T. E. Creighton. *Proteins: Structure and Molecular Principles*, W.H. Freeman and Company, New York (1984).

23. J. I. Boye, C.-Y. Ma and V. R. Harwalkar. Thermal denaturaton and coagulation of proteins. In: *Food Proteins and Their Applications*, S. Damodaran and A. Paraf (Eds.), pp. 25–56, Marcel Dekker, New York (1997).

24. S. Damodaran and A. Paraf (Eds.). *Food Proteins and Their Applications*, Marcel Dekker, New York (1997).

25. C. L. Pearson. Animal glues and adhesives. In: *Handbook of Adhesive Technology*, 2nd edn, A. Pizzi and K. L. Mittal (Eds.), pp. 283–298, Marcel Dekker, New York (1994).

26. Y. V. Wu and G. E. Inglett. Denaturation of plant proteins related to functionality and food applications, a review. *J. Food Sci.* 39, 218–225 (1974).

27. Y. Wang, M. Sarkar, A. E. Smith, A. S. Krois and G. J. Pielak. Macromolecular crowding and protein stability. *J. Am. Chem. Soc.* 134, 16614–16618 (2012).

28. P. K. Nandi. Chemical and physical methods for the characterization of proteins. In: *Food Proteins and Their Applications*, S. Damodaran and A. Paraf (Eds.), pp. 597–602, Marcel Dekker, New York (1997).

29. K. Nishinari, Y. Fanga, S. Guo, and O. G. Phillips. Soy proteins: A review on composition, aggregation and emulsification. *Food Hydrocolloids* 39, 301–318 (2014).

30. V. B. Tolstoguzov. Protein-polysaccharide interactions. In: *Food Proteins and Their Applications*, S. Damodaran and A. Paraf (Eds.), pp. 171–198, Marcel Dekker, New York (1997).

31. D. Oakenfull, J. Pearce and R. W. Burley. Protein gelation. In: *Food Proteins and Their Applications*, S. Damodaran and A. Paraf (Eds.), pp. 111–142, Marcel Dekker, New York (1997).

32. W. R. Egbert. Isolated soy protein: Technology, properties, and applications. In: *Soybeans as Functional Foods and Ingredients*, K. Liu (Ed.), pp. 134–162, AOCS Press, Champaign, IL (2004).

33. I. F. Laucks. Little journeys into the chemurgic industries: Soybean adhesives. *The Chemurgic Digest* 2 (20) 173–176 and (21) 185–188 (1943).

34. A. L. Lambuth. Soybean glues. In: *Handbook of Adhesives*, 3rd edn, I. Skeist (Ed.), pp. 172–180, Van Nostrand Reinhold, New York (1990).

35. X. S. Sun. Isolation and processing of plant materials. In: *Bio-Based Polymers and Composites*, R. P. Wool and X. S. Sun (Eds.), pp. 33–55, Elsevier-Academic Press, Burlington, MA (2005).

36. E. M. Bainy, S. M. Tosh, M. Corredig, V. Poysa and L. Woodrow. Varietal differences of carbohydrates in defatted soybean flour and soy protein isolate by-products. *Carbohydrate Polym.* 72, 664–675 (2008).

37. X. S. Sun. Thermal and mechanical properties of soy proteins. In: *Bio-Based Polymers and Composites*, R. P. Wool and X. S. Sun (Eds.), pp. 292–326, Elsevier-Academic Press, Burlington, MA (2005).

38. S. Utsumi, Y. Matsumura and T. Mori. Structure-function relationships of soy proteins. In: *Food Proteins and Their Applications*, S. Damodaran and A. Paraf (Eds.), pp. 257–291, Marcel Dekker, New York (1997).

39. D. Fukushima. Recent progress of soybean protein foods: Chemistry, technology, and nutrition. *Food Rev. Int.* 7 (3), 323–351 (1991).

40. J. E. Kinsella, S. Damodaran and B. German. Physiochemical and functional properties of oilseed proteins with emphasis on soy proteins. In: *New Protein Foods: Seed Storage Proteins*, A. M. Altschul and H. L. Wilcke (Eds.), pp. 108–179, Academic Press, New York (1985).

41. X. S. Sun. Soy protein polymers and adhesion properties. *J. Biobased Mater. Bioenergy* 5, 1–24 (2011).

42. Agriculture and Consumer Protection. Soybean Protein Concentrate. http:www.fao.orrg/docrep/to532e/t0532e06.htm. Accessed October 1, 2015 (2006).

43. C. Bye. Casein and mixed protein adhesives. In: *Handbook of Adhesives*, 3rd edn, I. Skeist (Ed.), pp. 135–152, Van Nostrand Reinhold, New York (1990).

44. P. H. Steele, R. E. Kreibich, P. J. Steynberg and R. W. Hemingway. Finger jointing green southern yellow pine with a soy-based adhesive. *Adhes. Age* 41 (10),49–56 (1998).

45. M. Sterley. Adhesive systems for green gluing, finger jointing of green Scots pine. In: *Green Gluing of Wood: Process – Products – Market*, COST E-34, Borås, Sweden, B. Källander (Ed.), SP Swedish National Testing and Research Testing Institute, Stockholm, Sweden (2005).

46. M. Lipke. Green glued fingerjoint wall studs: Industrial experience of Hampton Lumber Mills, Inc. In: *Green Gluing of Wood: Process – Products – Market*. COST E-34, Borås, Sweden, B. Källander (Ed.), SP Swedish National Testing and Research Testing Institute, Stockholm, Sweden (2005).

47. N. S. Hettiarachchy, U. Kalapathy and D. J. Myers. Alkali-modified soy protein with improved adhesive and hydrophobic properties. *J. Am. Oil Chem. Soc.* 72, 1461–1464 (1995).

48. J. Huang and K. Li. A new soy flour-based adhesive for making interior type II plywood. *J. Am. Oil Chem. Soc.* 85, 63–70 (2008).

49. N. Ji, J. Long, H. Y. Zheng and J. Q. He. Phenolic resin modified soybean protein adhesive. *Adv. Mater. Res.* 236–238, 3–7 (2011).

50. U. Kalapathy, N. S. Hettiarachchy, D. Myers and K. C. Rhee. Alkali-modified soy proteins: Effect of salts and disulfide bond cleavage on adhesion and viscosity. *J. Am. Oil Chem. Soc.* 73, 1063–1066 (1996).

51. U. Kalapathy, N. S. Hettiarachchy and K. C. Rhee. Effect of drying methods on molecular properties and functionalities of disulfide bond-cleaved soy proteins. *J. Am. Oil Chem. Soc.* 74, 195–199 (1997).

52. M. Kuo, D. Adams, D. Myers, D. Curry, H. Heemstra, J. L. Smith and Y. Bian. Properties of wood/agricultural fiberboard bonded with soybean-based adhesives. *Forest Prod. J.* 48 (2), 71–75 (1998).

53. J. Wescott and C. R. Frihart. Water-resistant vegetable protein adhesive dispersion compositions. U.S. Patent 7345136, assigned to Heartland Resource Technologies LLC and U.S. Department of Agriculture (2008).

54. K. Li, S. Peshkova and X. Geng. Investigation of soy protein-kymene® adhesive systems for wood composites. *J. Am. Oil Chem. Soc.* 81, 487–491 (2004).

55. K. Li. Formaldehyde-free lignocelluosic adhesives and composites made with the adhesives. U.S. Patent 7252735, assigned to Oregon State University (2007).

56. C. R. Frihart, M. J. Birkeland, A. J. Allen and J. M. Wescott. Soy adhesives that can form durable bonds for plywood, laminated wood flooring, and particleboard. In: *Proceedings of the International Convention of Society of Wood Science and Technology and United Nations Economic Commission for Europe – Timber Committee*, Geneva, Switzerland, WS-20 (2010).

57. Y. Jang and K. Li. An all-natural adhesive for bonding wood. *J. Am. Oil Chem. Soc.* 92, 431–438 (2015).

58. M. J. Birkeland, R. Wu and S. S. Weaver. Adhesive materials and methods of forming lignocellulosic composites using such adhesive materials. U.S. Patent Appl. 20150151449 A1, assigned to Design Adhesives Inc. (2014).

59. X. S. Sun. Soy protein adhesives. In: *Bio-Based Polymers and Composites*, R. P. Wool and X. S. Sun (Eds.), pp. 327–368, Elsevier-Academic Press, Burlington, MA (2005).

60. D. Vnučec, A. Kutnar and A. Goršek. Soy-based adhesives for wood-bonding: A review. *J. Adhes. Sci. Technol.* 31, (8) 910–931 (2017). http://dx.doi.org/10.1080/01694243.2016.1237278. Accessed September 26, 2016.

61. F. MacRitche and D. Lafiandra. Structure and functionality of wheat proteins. In: *Food Proteins and Their Applications*, S. Damodaran and A. Paraf (Eds.), pp. 293–325, Marcel Dekker, New York (1997).

62. S. Khosravi, F. Khabbaz, P. Nordqvist and M. Johansson. Protein-based adhesives for particleboards. *Ind. Crop. Prod.* 32, 275–283 (2010).

63. H. Lei, A. Pizzi, P. Navarrete, A. Redl and A. Wagner. Gluten protein adhesives for wood panels. In: *Wood Adhesives*, A. Pizzi and K. L. Mittal (Eds.), pp. 353–366, CRC Press, Boca Raton, FL (2010).

64. P. Nordqvist, F. Khabbaz and E. Malmström. Comparing bond strength and water resistance of alkali-modified soy protein isolate and wheat gluten adhesives. *Int. J. Adhesion Adhesives* 30, 72–79 (2010).

65. P. Nordqvist, P. E. Johansson, F. Khabbaz and E. Malmström. Characterization of hydrolyzed or heat treated wheat gluten by SE-HPLC and 13C NMR: Correlation with wood bonding performance. *Ind. Crop. Prod.* 51, 51–61 (2013).

66. Wikipedia. Rapeseed. (2016). https://en.wikipedia.org/wiki/Rapeseed. Accessed March 2, 2016.

67. S. H. Tan, R. J. Mailer, C. L. Blanchard and S. O. Agboola. Canola proteins for human consumption: Extraction, profile, and functional properties. *J. Food Sci.* 76, R16-R27 (2011).

68. E. Josefsson. Glucosinolate content and amino acid composition of rapeseed (*brasszca napus*) meal as affected by sulphur and nitrogen nutrition. *J. Sci. Food Agric.* 21, 99–103 (1970).

69. Canola Council of Canada. Protein and Amino acids. (2016). http://www.canolacouncil.org/oil-and-meal/canola-meal/nutrient-composition-of-canola-meal/protein-and-amino-acids/. Accessed June 1, 2016.

70. G. Mieth, J. Bruckner, J. Kroll and J. Pohl. Rapeseed: Constituents and protein products part 2: Preparation and properties of protein-enriched products. *Nahrung* 27 (8), 759–801 (1983).

71. Y. M. Tzeng, L. L. Diosady and L. J. Rubin. Production of canola protein materials by alkaline extraction, precipitation, and membrane processing. *J. Food Sci.* 55, 1147–1151 (1990).

72. D. Narayanamurti, V. Ranganathan and D. C. Roy. Rapeseed protein-formaldehyde dispersions as plywood adhesives. Indian Forest Leaflet No. 58, Forest Research Institute, Dehra Dun, India (1943).

73. I. Yang, G.-S. Han, S. H. Ahn, I.-G. Choi, Y.-H. Kim and S. C. Oh. Adhesive properties of medium-density fiberboards fabricated with rapeseed flour-based adhesive resins. *J. Adhes.* 90, 279–295 (2014).

74. N. Li, G. Qi, X. S. Sun, M. J. Stamm and D. Wang. Physicochemical properties and adhesion performance of Canola protein modified with sodium bisulfite. *J. Am. Oil Chem. Soc.* 89, 897–908 (2012).

75. Wikipedia, Lupinus. (2015). https://en.wikipedia.org/wiki/Lupinus. Accessed March 2, 2016.

76. M. Duranti. The structure of lupine seed proteins. *Nahrung* 30 (3–4),221–227 (1986).

77. M. Duranti and P. Cerletti. Amino acid composition of seed proteins of *lupinus albus*. *J. Agr. Food Chem.* 27, 977–978 (1979).

78. P. Cerletti. Lupin seed proteins. In: *Developments in Food Proteins*, B. J. Hudson (Ed.), pp. 133–171, Applied Science Publishers, London (1983).

79. M. Vidal, E. Vergara, M. Núñez, A. Ballerini and E. Ramírez. Evaluation of lupin flour (LF)-based adhesive for making sustainable wood materials. In: *Proceedings of the International Convention of Society of Wood Science and Technology and United Nations Economic Commission for Europe – Timber Committee*, Geneva, Switzerland. Society of Wood Science and Technology, Monona, WI (2010).

80. K. Li and J. Huang. Development and characterization of a formaldehyde-free adhesive from lupine flour, glycerol, and a novel curing agent for particleboard (PB) production. *Holzforschung*, 70 (10), 927–935 (2016).

81. F. P. O'Mara, J. J. Murphy and M. Rath. The amino acid composition of protein feedstuffs before and after ruminal incubation and after subsequent passage through the intestines of dairy cows. *J. Anim. Sci.* 75, 1941–1949 (1997).

82. Z. He, D. C. Chapital and H. N. Cheng. Comparison of the adhesive performances of soy meal, water washed meal fractions, and protein isolates. *Modern Appl. Sci.* 10 (5), 112–120 (2016).

83. Z. He, D. C. Chapital, H. N. Cheng and M. K. Dowd. Comparison of adhesive properties of water- and phosphate buffer-washed cottonseed meals with cottonseed protein isolate on maple and poplar veneers. *Int. J. Adhes. Adhes.* 50, 102–106 (2014).

84. H. N. Cheng, M. K. Dowd and Z. He. Investigation of modified cottonseed protein adhesives for wood composites. *Ind. Crop. Prod.* 46, 399–403 (2013).

85. D. G. Dalgleish. Structure-function relationships of caseins. In: *Food Proteins and Their Applications*, S. Damodaran and A. Paraf (Eds.), pp. 199–224, Marcel Dekker, New York (1997).

86. I. G. Macy, H. J. Kelly and R. E. Sloan. The composition of milks. A compilation of the comparative composition and properties of human, cow, and goat milk. In: *Colostrum and Transitional Milk*, National Academy of Sciences and National Research Council, Washington, DC(1953).

87. S. Sindayikengera and W.-S. Xia. Nutritional evaluation of caseins and whey proteins and their hydrolysates from Protamex. *Jhejiang Univ. Sci. B* 7 (2), 90–98 (2006).

88. C. R. Southward. Casein products. *III-Dairy-E-Casein*, 13 (1998). http://nzic.org.nz/ChemProcesses/dairy/3E.pdf. Accessed June 15, 2016.

89. H. K. Salzberg. Casein glues and adhesives. In: *Handbook of Adhesives*, 3rd edn, I. Skeist (Ed.), pp. 158–171, Van Nostrand Reinhold, New York (1990).

90. W. D. Detlefsen. Blood and casein adhesives for bonding wood. In: *Adhesives from Renewable Resources*, R. W. Hemingway, A. H. Conner and S. J. Branham (Eds.), pp. 447–452, American Chemical Society, Washington, DC(1989).

91. D. R. Rammer, J. de Melo Moura and R. J. Ross. Structural performance of the second oldest glued-laminated structure in the United States. In: *Proceedings of the Structures Congress 2014*, Boston, MA, pp. 1233–1243 (2014).

92. H. M. Jayaprakasha and Y. C. Yoon. Production of functional whey protein concentrate by monitoring the process of ultrafiltration. Asian-Australasian. *J. Anim. Sci.* 18, 433–438 (2005).

93. P. Cayot and D. Lorient. Structure-function of whey proteins. In: *Food Proteins and Their Applications*, S. Damodaran and A. Paraf (Eds.), pp. 225–256, Marcel Dekker, New York (1997).

94. Z. Gao, W. Wang, Z. Zhao and M. Guo. Novel whey protein-based aqueous polymer-isocyanate adhesive for glulam. *J. Appl. Polym. Sci.* 120, 220–225 (2011).

95. E. Doi and N. Kitabatake. Structure and function of egg proteins. In: *Food Proteins and Their Applications*, S. Damodaran and A. Paraf (Eds.), pp. 325–340, Marcel Dekker, New York (1997).

96. A. C. Awade, S. Moreaua, D. Mollk, G. Brulkb and J.-L. Maubois. Two-step chromatographic procedure for the purification of hen egg white ovomucin, lysozyme, ovotransferrin and ovalbumin and characterization of purified proteins. *J. Chromatogr. A* 677, 279–288 (1994).

97. M. Zemser, M. Friedman, J. Katzhendler, and L. L. Greene. Relationship between functional properties and structure of ovalbumin. *J. Protein Chem.* 13, 261–274 (1994).

98. J. C. Lewis, N. S. Snell., D. J. Hirschmann and H. Fraenkel-Conra. Amino acid composition of egg proteins. *J. Biol. Chem.* 186, 23–35 (1950).

99. C. R. Frihart, H. Satori, Z. Rongxian and M. J. Birkeland, Ovalbumin as a wood adhesive. In: *Proceedings of the 37th Annual Meeting of The Adhesion Society*, San Diego, CA (2014).

100. N. C. Schnellmann,. Animal glues: A review of their key properties relevant to conservation. *Rev. Conserv.* 8, 55–66 (2007).

101. J. E. Eastoe. The amino acid composition of mammalian collagen and gelatin. *Biochem. J.* 61, 589–600 (1955).

102. J. E. Eastoe. The amino acid composition of fish collagen and gelatin. *Biochem. J.* 65, 363–368 (1957).

103. J. R. Hubbard. Animal glues. In: *Handbook of Adhesives*, 3rd edn, I. Skeist (Ed.), pp. 139–151, Van Nostrand Reinhold, New York (1990).

104. H.-C. Huang. The Production of Textile Fibers from Soy Proteins. Iowa State University, Ames, IA. Retrospective Theses and Dissertations, Paper 10480. http://lib.dr.iastate.edu/cgi/viewcontent.cgi?article=11479&context=rtd. Accessed December 20, 2016 (1994).

105. R. E. Norland. Fish glues. In: *Handbook of Adhesives*, 3rd edn, I. Skeist (Ed.), pp. 152–157, Van Nostrand Reinhold, New York (1990).

106. Y.-Y. H. Chang and C. L. Judson. Amino acid composition of human and guinea pig blood proteins, and ovarian proteins of the yellow fever mosquito *aedes aegypti*; and their effects on the mosquito egg production. *Compr. Biochem. Physiol.* 62A,753–755 (1979).

107. A. L. Lambuth. Blood glues. In: *Handbook of Adhesives*, 3rd edn, I. Skeist (Ed.), pp. 181–191, Van Nostrand Reinhold, New York (1990).

108. C. S. F. Bah, A. E. D. A. Bekhit, A. Carne and M. A. McConnell. Slaughterhouse blood: An emerging source of bioactive compounds. *Compr. Rev. Food Sci. Food Saf.* 12, 314–331 (2013).

109. X. Mo, Y. Hiromasa, M. Warner, A. N. Al-Rawi, T. Iwamoto, T. S. Rahman, X. Sun, and J. M. Tomich. Design of 11-residue peptides with unusual biophysical properties: Induced secondary structure in the absence of water. *Biophys. J.* 94, 1807–1817 (2008).

110. Y. Liu and K. Li. Chemical modification of soy protein for wood adhesives. *Macromol. Rapid Commun.* 23, 739–742 (2002).

111. Y. Liu and K. Li. Modification of soy protein for wood adhesives using mussel protein as a model: The influence of a mercapto group. *Macromol. Rapid Commun.* 25, 1835–1838 (2004).

6 Rubber-Based Adhesives

*A.A. Shybi, Siby Varghese, Hanna J. Maria, and Sabu Thomas**

CONTENTS

---

* Indicates corresponding author

6.1 INTRODUCTION

Substances that are termed *adhesives* perform the function of joining two materials together by certain interfacial interactions. The increasing demands for adhesive bonding for various purposes compared with other joining techniques are due to the special advantages and properties offered by adhesive materials. Advantages such as low cost, compatibility with adherend materials, simple equipment requirements, and ease of application make adhesives more acceptable compared with other joining techniques. A large number of materials are used as adhesive raw materials; however, rubber-based adhesives show some special properties compared with other generic classes of adhesives. This is due to the availability of various types and forms such as natural, synthetic, rubber derivatives, reclaimed rubbers, and rubber latex that can be used in adhesive preparation. Thus, the role of rubber-based adhesives in the adhesive industry is very significant. In rubber-based adhesives, the basic material is rubber and the elasticity of the basic rubber compound influences the properties of the final adhesive product. However, the main property associated with rubber-based adhesives is tack, and natural rubber-based adhesives are well known for this tackiness. The properties of rubber-based adhesives can be modified by the addition of various compounding ingredients. A review of the earlier research carried out in the field of rubber-based adhesives is presented in this chapter.

6.2 THEORIES OF ADHESION

Adhesion refers to the state in which two surfaces are held together by interfacial forces arising from the interactions of the atoms, molecules, or ions of the two surfaces [1]. Adhesion is part of *zygology*, the science of joining materials; welding, riveting, soldering, and all other forms of mechanical bonding techniques are some other areas that are included in this science [2]. One can call a substance an adhesive only if it can perform the function of joining two adherends or substrates. Adherends are those materials on which the adhesive is applied to obtain a strong bond. In general, there are two ways to study the science of adhesion: One is based on the chemistry and physics of surfaces and interfaces, and the second is based on the mechanics of adhesive joints [3]. A large number of theories regarding adhesion have been proposed, but no one theory alone can explain all adhesion behaviors; instead, they can explain only certain aspects regarding the phenomenon of adhesion. The main theories relating to adhesion are mechanical theory, electrostatic theory, diffusion theory, and adsorption theory [4,5]. A brief discussion of the various theories of adhesion was given by DeLollis [6], and a review on the mechanical interlocking and adsorption interactions was reported by Allen [7].

6.2.1 MECHANICAL THEORY OF ADHESION

This theory of adhesion considers the phenomenon of adhesion as a purely mechanical one, and bond formation is by mechanical interlocking of the adhesive into the surface features of adherends. According to this theory, surface of any material contains some irregularities such as valleys and ridges, even though this cannot be seen at the macroscopic level. The adhesive will flow and penetrate into the valleys and ridges of the adherends and then solidifies and hardens with the formation of a mechanical bond. The mechanical concept of adhesion was first reported by McBain and Hopkins in 1925 [8,9]. According to them, joints are of two types: mechanical and specific. Mechanical joints are possible only with porous materials, whereas specific adhesion occurs with smooth nonporous materials. The mechanical theory of adhesion gives importance to the necessity of adherends' surface roughness and cleanliness before adhesive application. The theory explains that increasing surface roughness increases the surface area of contact for adhesion. The presence of impurities or any other foreign substances on the adherend surface will greatly affect the actual adhesive joint formed.

6.2.2 ELECTROSTATIC THEORY OF ADHESION

The electrostatic theory of adhesion is based on the concept of formation of an electrical double layer at the interface between the adhesive and adherends while bonding them. Derjaguin and Smilga [10] proposed the electrostatic theory of adhesion, which states that at the interface between adhesive and adherend, transfer of electrons takes place and this electron transfer generates positive and negative charges. The attraction of these charges to one another holds the materials together.

6.2.3 ADSORPTION THEORY OF ADHESION

This theory explains that the phenomenon of adhesion between adherend and adhesive is due to the adsorption of adhesive onto the substrate, and causes formation of bonds between them. The formation of bonds can be due to many weak and strong forces such as dispersion forces, van der Waals forces, hydrogen bonds, or strong chemical forces arising from ionic, covalent, or metallic bonds. The concept of the *contact angle* between the adhesive and adherend surface and its role in adhesive performance is well explained by this theory. The significance of the adhesive's ability to wet and spread on the adherend surface is given much importance. When the contact angle (θ) between the adhesive and adherend (substrate) is zero or less than 90°, the force of attraction between them is high, and results in a strong adhesive bond. If the contact angle between them is large (greater than 90°), the bond formed will be very weak. Surface free energy of various materials is determined by contact angle measurements. Quantitative measurement of the polar acid–base interactions can be carried out by determining the acidic and basic strengths of the functional sites of solvents and various polymers. These measurements give an idea about the wettability, solubility, adsorption, and adhesion properties of adhesive materials [11]. The adhesion properties greatly depend on the acid–base interactions between adhesive and adherend materials [12]. The overall strength of adhesive joints depends on several parameters such as wetting, thermodynamic work of adhesion, and also interfacial tension. The role of these parameters and role of the interface in adhesion was discussed by Mittal [13].

6.2.4 DIFFUSION THEORY OF ADHESION

The reason for adhesion between adhesive and adherend surface according to diffusion theory is the diffusion of both adherend and adhesive across the interface. This theory is applied mainly to the adhesive action of polymer-type molecules, for which molecular entanglements are possible. For adhesion to occur by interdiffusion, both adhesive and adherend should be compatible with each other. This theory of adhesion was suggested by Voyutskii [14]. According to this theory, when the adhesive applied is in the form of solution, the adherend swells or dissolves in it and will diffuse into the adhesive. The diffusion theory of adhesion explains the effect of contact time on adhesion, and reports that the force of adhesion between the polymeric adhesive and the substrate increases with time; this increase is at first rapid, then slows down, and finally reaches a certain limiting value. The reason for this, according to this theory, is the slow penetration of adhesive molecules into the substrate. This theory also explains the effect of bonding temperature on adhesion, and concludes that adhesion is high at elevated temperatures due to the increased thermal motion of macromolecules or their segments at elevated temperatures. The effect of molecular weight, molecular shape, and polarity of the adhesive on adhesion properties is also explained by the diffusion theory.

6.3 CLASSIFICATION OF ADHESIVES

Adhesives can be classified in a number of ways, depending on their origin, basic chemical composition, purpose or end use of application, method of application, physical form, and so on.

6.3.1 Classification Based on Origin

Adhesives are classified according to their origin as natural or synthetic. Natural adhesives are obtained from natural origins such as natural rubber, starch glue, soybean glue, fish glue, animal glues, casein glue, and so on. Synthetic adhesives are made synthetically and include synthetic rubbers and synthetic polymers; for example, cyanoacrylates, poly(vinyl acetate), epoxies, polyurethanes, and so on.

6.3.2 Classification Based on Physical Form

Adhesives can be classified according to their physical form as liquid adhesives, solid adhesives, paste, films, tapes, and so on.

6.3.3 Classification Based on Basic Raw Materials

This classification includes the main raw materials used in the adhesives, such as rubber-based adhesives, epoxy adhesives, poly(vinyl acetate) adhesives, cyanoacrylates, starch adhesives, cellulose, and so on.

6.3.4 Classification Based on the Method of Application

Based on the method of application, adhesives are classified as contact adhesives, hot-melt adhesives, and pressure-sensitive adhesives. With contact adhesives, the adhesive is applied to both adherends and are allowed to dry before bonding. With hot-melt adhesives, the adhesive hardens on cooling, while pressure-sensitive adhesives adhere to the substrate by the application of a small amount of pressure.

6.3.5 Classification Based on Bonding Temperature

The temperature at which adhesives form bonds with the substrate is an important parameter. Based on this, some adhesives form bonds at room temperature, while others need some heat for bond formation or curing, and some others are cold-setting adhesives.

6.3.6 Classification Based on the Load-Carrying Ability of Adhesives

This includes structural adhesives, which can withstand extremely high loads and are applied usually below their glass transition temperature. Examples include adhesives based on epoxies, acrylics, urethanes, and cyanoacrylates. Depending on the load-carrying ability, this classification also includes semistructural adhesives and nonstructural adhesives.

6.4 ORIGIN OF RUBBER-BASED ADHESIVES

Historically, naphtha solutions of rubber were used for coating, waterproofing, and laminating textiles [15,16]. The use of rubber latex adhesives was first started in the middle of the nineteenth century. With the advancement of synthetic rubber production, a large number of synthetic elastomers has been prepared and tested in the adhesive industry.

6.4.1 Classification of Rubber-Based Adhesives as Latex-Based or Dry Rubber-Based

Rubber-based adhesives are products that contain natural rubber or synthetic rubbers and are capable of joining together two surfaces. Rubber-based adhesives are also called *elastomeric adhesives* and are widely used in industrial and household applications. The main properties associated with rubber-based adhesives are their tack and adhesion to various substrates. This property allows rubber-based adhesives to be in high demand in the adhesive industry [17]. Natural or synthetic rubbers are polymeric materials that are elastic and undergo long deformation by the application of an external force and return almost to the original shape with the release of the deforming force [18–23]. For a bonding technique that requires flexibility, resilience, and strength, rubber-based adhesives are most suitable [24]. Some rubber-based adhesives require vulcanization to produce adequate strength. Rubber-based adhesives are prepared either using rubber latex or by dissolving dry solid rubber in a suitable solvent [15,24]. Latex-based adhesives are less toxic, inexpensive, and more resistant to water than the dry rubber-based adhesives. Dry rubber-based adhesives are highly viscous and more resistant toward chemicals and solvents compared with latex adhesives.

6.4.1.1 Latex-Based Adhesives

In the area of rubber-based adhesives, latex-based adhesives occupy a unique position and are in great demand because of their special properties such as high solids content, wide control of viscosity, comparatively low toxicity, low flammability, and also low production cost [25–27]. Latex is a stable colloidal dispersion of a polymeric substance in an aqueous or nonaqueous medium [28]. Latex-based adhesives are produced either from natural rubber latex obtained from *Hevea brasiliensis* or from various other synthetic rubber latices. Latex adhesives differ essentially from dry rubber forms both in application and preparation. Dry rubber-based adhesives require mastication or milling, which reduces their aging resistance, whereas latex-based adhesives do not require mastication. Both the adhesion and cohesion properties associated with latex-based adhesives are major factors in the bonding of similar or dissimilar materials, and are mainly used for bonding porous substances such as leather, textiles, paper, and wood. The limitations associated with latex adhesives are their low resistance toward various aging factors, their tendency to coagulate on storage, and low wetting power for oily and greasy surfaces.

6.4.1.1.1 Natural Rubber Latex-Based Adhesives

Natural rubber latex is a milky liquid that consists of extremely small particles of rubber suspended or dispersed in an aqueous medium. Although it is obtained from a variety of trees, *Hevea brasiliensis* is the commercial source of natural rubber. Natural rubber latex is a stable colloidal dispersion of cis-1,4-polyisoprene of high molecular mass. Natural rubber latex is excellent for various adhesive applications, due to its self-tack property [29]. When latex adhesives are applied to porous adherend surfaces, an immediate tack development is observed [30]. Latex that is oxidized by thiols under hot conditions was used in the production of excellent adhesives. They usually give a sticky film by simple drying at room temperature [31]. The use of natural rubber latex as an adhesive for bonding rayon fibers to rubber in tires was reported by Gardner and Williams [32]. In their study, a mixture of natural rubber latex, casein, borax, and dispersed reclaimed rubber was used for adhesive preparation, and it was reported that an optimum level of adhesion was obtained when the casein:solid rubber ratio was 1:2. Formulations of natural rubber latex-based adhesives for various applications were reported [29,33]. In these reports, formulations for self-seal envelopes, adhesives for leather shoes, bags, travel goods, and tire cord adhesives using natural rubber latex were described. Jayasinghe [34] also reported the use of natural rubber field latex, centrifuged concentrated latex, creamed latex, vulcanized latex, and prevulcanized latex in adhesive applications. Blends of centrifuged formaldehyde stabilized natural rubber latex with urea–formaldehyde and melamine–formaldehyde were found to be efficient adhesives for plywood bonding [35]. Natural rubber latex is excellent for making pressure-sensitive adhesive tapes. The inherent tackiness associated with natural rubber is an important property for the preparation of pressure-sensitive adhesives. The use of natural rubber latex in pressure-sensitive adhesives and the factors affecting the development of tack were reported in many works [36,37]. The application of low-protein natural rubber latex for surgical bandage preparation was investigated by Bonnet [38]. Low-protein natural rubber latex was compounded with various additives such as tackifiers, plasticizers, and so on and used as an adhesive for surgical bandages. An adhesive based on natural rubber latex for bonding wood was prepared by John and Joseph [39]; this adhesive showed higher water resistance compared with commercial ones. In this work, the effect of prevulcanization and the effect of various ingredients on adhesive properties were also studied. Blends of natural rubber latex and styrene-vinyl acetate copolymer latex were used for pressure-sensitive adhesive application [40]. The adhesive properties of styrene-based, deproteinized natural rubber latex were studied by Neoh et al. [41]. The tensile and adhesion properties of the rubber latex increased with an increase in the styrene content. Chlorinated, epoxidized natural rubber latex-based adhesive showed that changes in the vulcanization system can affect adhesive properties and strength [42].

6.4.1.1.2 Synthetic Rubber Latex-Based Adhesives

Synthetic rubber latices show better aging resistance than natural rubber latex, whereas the inherent tackiness associated with synthetic latex is lower when compared with natural rubber latex. A number of synthetic latices are used in adhesive application, among these the important ones are polychloroprene-based, styrene–butadiene-based, and nitrile rubber-based adhesives.

6.4.1.1.2.1 Polychloroprene Latex-Based Adhesives Due to the environmental issues created by solvents used in adhesives, waterborne contact adhesives based on polychloroprene latex have attracted considerable attention. These are widely used in the adhesive industry when quick, high-strength, and permanent bonds are needed. Polychloroprene latex adhesives have wide applications in bonding porous materials such as leather and textiles; and also in wood bonding and rubber-to-metal bonding. These are also used for several other applications such as bonding furniture and automobile parts, and applications where good weather and oil resistance is needed [43–45]. Polychloroprene rubber latex prepared by dissolving vinyl acetate–ethylene copolymer in chloroprene monomer was used for bonding EVA material [46]. Using the adhesive prepared from polychloroprene latex and resorcinol–formaldehyde, polyester cord can be bonded to rubber [47]. In the work by Perlinski et al. [48], carboxylated polychloroprene latex adhesives were prepared using polychloroprene latex, an epoxy resin prepared by the reaction of p-aminophenol with epichlorohydrin, a metallic oxide stabilizer (mainly zinc oxide), tackifying resins selected from natural and modified rosins, and antioxidants. The role of the metallic oxide used was as an acid acceptor, neutralizing the hydrochloric acid released on aging of the polychloroprene latex. In this work, it was reported that the properties of conventional polychloroprene latex adhesives such as adhesion to metals, heat, and water resistance were enhanced by the addition of an epoxy resin. A sprayable adhesive based on nonionic polychloroprene latex was reported by Carnahan [49]. The nonionic latex adhesive consisted of chloroprene latex, which is a copolymer of chloroprene and methacrylic acid, dispersed and stabilized with poly(vinyl alcohol) along with a nonionic surfactant. The adhesive composition also contained a hydrocarbon resin, zinc oxide, and hindered bisphenol. The drying time of the sprayable adhesive prepared using nonionic latex was highly reduced compared with conventional polychloroprene latex. It was observed that the nonionic latex adhesive could be used as an aerosol-sprayable adhesive, free from the tendency to agglomerate and coagulate. Polychloroprene latex adhesive for bonding foamed polymeric materials to fabric was reported by Christell and Tabibian [50]. Wakayama et al. [51] reported that chloroprene rubber latex mixed with tackifier resins and metal oxides was found to be an adhesive with excellent room-temperature and high-temperature bonding strengths and water resistance. A contact adhesive of polychloroprene latex blended with styrene acrylate emulsion was reported by Zhang et al. [52]. Boric acid was added for easier mixing of the latex and styrene acrylate emulsion, and also for accelerating the drying rate of the adhesive. Good adhesive properties, and also good shelf stability, were obtained with 40% wt of styrene acrylate emulsion and 1.25% wt of boric acid.

6.4.1.1.2.2   Styrene–Butadiene Rubber Latex-Based Adhesives   Styrene–butadiene rubber (SBR) latices are used extensively in tire cord adhesive applications and for bonding fibrous materials. These latex-based adhesives increase the fatigue resistance of the bond formed [53]. SBR latex is also used in pressure-sensitive adhesive applications [54]. The use of SBR latex in pressure-sensitive adhesives was reported by Takemoto and Morrison [55]. The adhesive was prepared by mixing carboxylated SBR emulsion and tackifying resin. An adhesive dip for bonding different types of tire cords to rubber was prepared from an aqueous mixture of styrene–butadiene latex, tris(dihydroxybenzyl)phenol, and an alkali. It was observed that the adhesive thus prepared showed good adhesion strength for bonding polyester, nylon, and rayon tire cords [56]. Carpet backing adhesive was prepared using SBR latex mixed with saponified tall oil pitch. Addition of tall oil pitch improved both the initial tack and final adhesion of the carboxylated styrene–butadiene latex. Replacing 15% of styrene–butadiene latex with tall oil pitch was reported to improve the final adhesive strength [57]. A blend of SBR and acrylic latex emulsion was used as a water-based adhesive with superior water resistance [58]. A formulation of SBR vulcanizing-type latex adhesive is also reported [59]. A mixture of SBR latex and poly(vinyl alcohol) was used as an adhesive for various bonding applications [60]. An adhesive system based on latex of hydrogenated styrene–butadiene with an aqueous solution of a half-ester of maleinized liquid polybutadiene resin was also reported [61].

6.4.1.1.2.3   Nitrile Rubber Latex-Based Adhesives   Nitrile rubbers are rubber-like copolymers of unsaturated nitriles, such as acrylonitrile, with dienes (mainly butadiene). The oil resistance and adhesive properties of nitrile rubber increase with an increase in nitrile content [62]. Adhesives based on nitrile rubber latex and resorcinol–formaldehyde latex to bond nitrile rubber to glass fibers and organic synthetic fibers were prepared by Hisaki et al. [63]. The adhesive formulation they used also contained carbon black as filler, and a vulcanizing agent. It was found that the adhesive gave better initial adhesion strength than conventional adhesives for bonding fibers. An adhesive from hydrogenated nitrile rubber latex blended with chlorinated paraffin wax was used to coat textiles [64]. An adhesive composition based on hydrogenated nitrile–butadiene rubber latex with resorcinol–formaldehyde resin and ethylene urea compound showed excellent degradation resistance and high tackiness, and was used to bond rubber and glass [65]. To produce a toothed belt with excellent wear and water resistance, Isshiki et al. prepared an adhesive comprised of a self-cross-linking carboxylated unsaturated nitrile rubber latex with resorcinol–formaldehyde [66].

6.4.1.1.2.4   Butyl Rubber Latex-Based Adhesives   Butyl rubber is a copolymer of isobutylene with isoprene. Butyl rubber latex-based adhesives are mainly used for tire cord dipping [67].

6.4.1.1.2.5   Butadiene Rubber Latex-Based Adhesives   Butadiene rubber latex is used in the adhesive industry mainly for bonding rubbers to metals. To bond natural and synthetic rubbers to metals, adhesives of polybutadiene latex and poly-C-nitroso compound with various fillers and viscosity modifiers were prepared by Auerbach and Berry [68].

6.4.1.2   Dry Rubber-Based Adhesives

In the rubber-based adhesive industry, most of the adhesives are based on the solid form or dry form of the rubber. Dry rubber-based adhesives include rubber solutions in which the dry rubber is dissolved in suitable solvents. Dry rubber-based adhesives were also prepared using rubber resin solutions in which the properties of simple solutions were modified by the addition of various resins [30]. Compared with latex-based adhesives, dry rubber-based adhesives provide higher adhesion strength in most applications. But, the addition of solvents creates environmental problems. While considering solid rubber-based adhesives, both natural and synthetic rubbers are used in adhesive preparation. Elastomeric adhesion and adhesives were reviewed by Voyutskii [69]. Elastomeric modification of structural adhesives was reviewed by Pocius [70]. The chemistry and physical properties of the rubber-modified structural thermosetting polymers used as adhesives were reported in this review. The addition of elastomers increases the resistance of the adhesive to crack propagation. A rubber-based adhesive was also used for the preparation of pressure-sensitive surgical tapes [71].

6.4.1.2.1 Natural Rubber-Based Adhesives

The main property associated with natural rubber-based adhesives is tack, which can be modified by the addition of some other materials. In many situations where natural rubber-based adhesives offer good strength, this is mainly due to this self-tackiness [30,72]. Natural rubber-based solvent adhesives possess intrinsically greater adhesion to a substrate than latex adhesives. An adhesive tape using natural rubber mixed with natural resins dissolved in a solvent was reported. The rubber used was a natural rubber pale latex crepe (for a lighter color) that was sliced and masticated; after that resins were added to it. The resultant adhesive was then dissolved in petroleum solvent and used to make good adhesive tapes [73]. Pressure-sensitive adhesives can be prepared using unvulcanized elastomers of natural rubber along with various tackifying agents. Pressure-sensitive adhesives have the ability to stick instantly, and can be removed cleanly when needed [74]. Sherriff et al. [75] reported that with pressure-sensitive adhesives made from natural rubber, the tackifying resin was added to modify the viscoelastic properties of the adhesive. They found that the tack of the adhesive depended on the viscoelastic properties and glass transition temperature. They reported that the tackifier resin may form either a one- or two-phase system with natural rubber, and the two-phase system is not necessary for good tack properties of the adhesive. In another work, the surface structure of a pressure-sensitive adhesive consisting of natural rubber and pentaerythritol ester of hydrogenated resin was studied, and the authors reported that maximum tack was obtained at a resin content of 40%–60% wt [76]. In the work reported by Vitek [77], aliphatic and aromatic peroxides were used to cross-link natural rubber in the adhesive preparation, and the adhesive prepared was used as a hot-melt adhesive. Viscoelastic properties of pressure-sensitive adhesives made from a 1:1 ratio of natural rubber and resin were dependent on the structure of the resin used. The viscoelastic properties of the adhesive were also dependent on the concentration of the resin added [78,79]. Blends of natural rubber and pentaerythritol ester of hydrogenated rosin were used in the work carried out by Hino and Hashimoto [80]. The dynamic behavior of the deformation process of adhesives during peeling was studied using optical microscopy.

When a surgical adhesive is in contact with the skin, it absorbs skin secretions and moisture, causing a change in the rheological properties of the adhesive. These changes can be monitored by measuring the glass-to-rubber transition temperature of the adhesive. The glass transition temperature of natural rubber-based surgical adhesives was studied by Andrews et al. [81]. A pressure-sensitive tape based on natural rubber and poly(n-butyl acrylate) and the effect of various model fillers on its adhesive property were reported by Bhowmick [82]. The addition of filler generated an additional mechanism of energy dissipation during deformation of the adhesive. A natural rubber-based adhesive composition using natural rubber and bis-imide was used for bonding various elastomeric layers in tire retreading, which showed excellent blow-out properties [83]. An efficient adhesive system to bond rubbers to steel was developed by John and Joseph [84]. The adhesive system consisted of a solid rubber compound strip and a primer. In this work, the effects of carbon black and phenol–formaldehyde resin content on the adhesive strength were tested. Salt-water resistance and aging resistance were also tested, and it was found that this system showed better storage properties. The effect of miscibility on the peel strength of natural rubber-based pressure-sensitive adhesives blended with different tackifying resins such as rosin, terpenes, and petroleum resins was reported by Fujita et al. [85]. In this work, the miscibilities of all the adhesive blend systems used were reported using phase diagrams. The authors reported that when the pressure-sensitive adhesive system was completely immiscible, the adhesive peel strength decreased as compared with a miscible system. The use of acrylic tackifier resins in the adhesive preparation of natural rubber was studied by blending acrylic tackifiers in different amounts with the rubber, and adhesive properties were measured [86]. The carboxy-terminated liquid natural rubber was prepared by photochemical reaction of natural rubber with maleic anhydride [87], and the authors used this carboxy-terminated liquid natural rubber-based adhesive to bond rubber to rubber and rubber to metal. The role of zinc oxide in natural rubber-based pressure-sensitive adhesives was investigated by Poh and Chow [88]. In their study, toluene was used as the solvent and the tackifying resin used was coumarone–indene. They reported that with an increase in zinc oxide, both the tack and viscosity were increased, and the optimum adhesion strength was obtained at 40 phr zinc oxide and 80 phr coumarone–indene resin. The effect of silica filler on the adhesive properties of a natural rubber-based pressure-sensitive adhesive was investigated in another work [89]. In this study, natural rubbers used were ENR-50 and ENR-25 grades with coumarone–indene resin as the tackifier resin. The silica loading was varied from 10 to 50 parts and the resin added was 40 phr; toluene was used as the solvent. The authors reported that viscosity of the adhesive increased with an increase in silica loading, due to the reinforcement effect of the filler. For ENR-25-based adhesives, maximum loop tack and shear strength were obtained at 40 phr of silica loading. Another natural rubber-based, pressure-sensitive adhesive using coumarone–indene resin and petro resin was investigated by Poh and Firdaus [90]. They studied the viscosity, shear strength, and peel strength of the adhesive and showed that the viscosity and shear strength of the adhesive decreased with increasing petro resin and optimum peel strength was obtained at a resin content of 60 phr. A review of natural rubber-based, pressure-sensitive adhesives was reported by Khan and Poh [91]. In this review, factors affecting adhesion properties such as tack, shear, and peel were examined. In the work by Poh and Cheong [92], natural rubber cross-linked by benzoyl peroxide was used as an adhesive using toluene and coumarone–indene resin. They observed that a maximum of 2 phr benzoyl peroxide was optimum for both loop tack and peel strength. An epoxidized skim rubber that was obtained from saponified low-protein skim rubber was used for pressure-sensitive adhesive preparation in the work by Riyajan et al. [93]. They blended the skim rubber with poly(vinyl alcohol) containing tackifier and reported that optimum adhesive properties were observed at 20 phr poly(vinyl alcohol).

6.4.1.2.2 Synthetic Rubber-Based Adhesives

Solid or dry forms of synthetic rubbers are also extensively used in the adhesive industry. However, synthetic rubbers lack the inherent tackiness associated with natural rubber. So, resins and other materials are usually mixed with synthetic rubbers in adhesive preparation. Compared with natural rubber-based adhesives, for some applications, synthetic rubber-based adhesives show improved adhesive properties.

6.4.1.2.2.1 Polychloroprene Rubber-Based Adhesives   Polychloroprene rubber-based adhesive cements are widely used in footwear, automobiles, and other industrial areas [43,94]. Polychloroprene-based adhesives are found to be as efficient as natural rubber adhesives in most applications. In the work reported by Evans and Will [95], an adhesive based on polychloroprene rubber was used to coat a heat-recoverable sleeve. They used blends of aliphatic and aromatic solvents; for example, methyl ethyl ketone and toluene. To improve adhesion strength and heat resistance, tertiary butyl phenolic resins and antioxidants were added. Development of initial bond strength in solvent-based polychloroprene contact adhesives at a particular rubber:resin ratio in adhesive films was discussed by Kozakiewicz et al. [96], and explained the complex relations between the initial bond strength, solvent retention, and resin content in the adhesive film. In this work, they modified polychloroprene adhesive with butyl phenolic resin. The results showed that solvent retention was the main factor affecting the initial bond strength in solvent-based contact adhesives. A polychloroprene-based contact adhesive was used as a primer for ethylene–propylene diene monomer rubber (EPDM) roofing [97]. This adhesive was comprised of polychloroprene rubber with an admixture of thermoplastic block copolymers, halogenated butyl rubber, an aromatic hydrocarbon reinforcing resin, and toluene as solvent. Polychloroprene rubber adhesives develop rapid bond strength, good tack, and high resistance to degradation by oils and chemicals. To bond rubber to metal, an adhesive containing a primer component and an overcoat component was reported by Mowrey and Pontore [98]. The

primer component was made of polychloroprene rubber, a phenolic resin, and a metal oxide such as zinc or magnesium oxide. This adhesive system was very flexible. The effect of a piperylene–styrene copolymer content on the surface properties of polychloroprene adhesive was studied by Zukiene et al. [99], and they reported that addition of the piperylene–styrene copolymer increased the adhesive strength, and no chemical interaction between the piperylene–styrene and chloroprene adhesive could be observed. Technology to improve the performance of polychloroprene rubber-based adhesives was reported by De Silva et al. [100]. Molecular orientation and crystallization of polychloroprene rubber–synthetic cis-1,4-polyisoprene rubber blends on metallic surfaces to produce high-performance adhesive systems were studied by Kardan [101]. The results showed that when the polychloroprene level exceeded 50% in the blend, a parallel orientation of the methylene chains to the surface occurred and enhanced the adhesion of polymer chains to metallic surfaces. Contact adhesives based on polychloroprene rubber show good adhesive properties compared with most other contact adhesives. Polychloroprene rubber in combination with phenolic resin was reported to have good adhesive properties. For bonding aluminum to aluminum and SBR to SBR, a polychloroprene rubber–phenolic resin adhesive was prepared using a mixture of phenol and cardanol as phenolic resin. The effects of varying the solids content, resin content, and cross-linking agents were also studied and reported. The results showed that cardanol–phenol–formaldehyde resin is an effective ingredient in an adhesive for bonding aluminum to aluminum and SBR to SBR [102]. A continuation of the same work was carried out by varying the phenol:cardanol ratio in the formulation [103]. Compatibility improvement between chlorinated thermoplastic rubber and polychloroprene adhesive was reported by Mercedes Pastor Blas [104]. The polychloroprene adhesive consisted of polychloroprene rubber and a thermoreactive resin. Terpene phenolic resin, glycerol ester of wood rosin, and an aromatic hydrocarbon resin were used for the study. The author reported that the nature of the resin greatly influenced the viscoelastic properties of the adhesive. It was observed that the polychloroprene adhesive with no resin and with resins of different nature did not produce suitable adhesive joints between the chlorinated rubber and the leather. However, an increase in the elasticity and improvement in viscoelastic properties were observed in the presence of thermoreactive phenolic resin.

6.4.1.2.2.2 Styrene–Butadiene Rubber-Based Adhesives Styrene–butadiene rubber-based solvent adhesives have applications mainly in the tire industry. The adhesion properties of a pressure-sensitive adhesive prepared from blends of styrene–butadiene rubber and natural rubber were studied by Poh and Ong [105]. The results showed that the peel strength of the adhesive was optimum at 60% styrene–butadiene rubber in the blend. The maximum value for loop tack was obtained at 20% styrene–butadiene rubber and then decreased with increasing styrene–butadiene rubber content in the blend. The effect of kaolinite clay on the adhesion properties of styrene–butadiene rubber adhesive on an aluminum surface was investigated by Alwaan [106]. It was observed that addition of kaolinite clay decreased the adhesion strength of the styrene–butadiene rubber adhesive.

6.4.1.2.2.3 Nitrile Rubber-Based Adhesives Nitrile rubber-based adhesives generally have good shelf stability and high-temperature properties. These are used mainly for coating fabrics, thereby giving them oil and wear resistance [62]. For aluminum–aluminum bonding, an adhesive based on a self-vulcanizable rubber blend of carboxylated nitrile rubber and chlorobutyl rubber with silica filler was tested. The joint peel strength was found to depend on the filler loading, the state of cure, the molding temperature, and the adhesive film thickness. Maximum peel strength was obtained at a filler loading of 10 phr. The high peel strength in the filled adhesive system was due to the reinforcement effect of the filler added [107]. Another adhesive for aluminum–aluminum bonding was reported using a blend of carboxylated nitrile rubber and epichlorohydrin rubber, with carbon black as the filler. In the same work, an adhesive based on a blend of carboxylated nitrile rubber and chlorobutyl rubber using silica filler for aluminum bonding was also studied. Both adhesives showed good water resistance, and the adhesive based on epichlorohydrin carboxylated nitrile rubber showed good heat resistance also. It was found that the heat resistance of the nitrile rubber–butyl rubber blend was improved by the addition of silane primer [108]. Adhesives based on blends of nitrile rubber and phenolic resin containing various amounts of p-cresol and phenol were reported on by Achary and Ramaswamy [109]. Incorporation of p-cresol in phenolic resin was found to improve the thermal stability of the nitrile rubber–phenolic resin blend adhesive; also, adhesive strength increased with p-cresol content. They explained that interaction between phenolic resin and rubber increased toughness, and that improved toughness was the reason for enhancement of adhesive strength. The effect of silica filler on the adhesive blend was also tested and it was found that silica filler could act both as the reinforcing filler and also as a surface compatibilizer. In another work, natural rubber was bonded to metal using an adhesive made of nitrile rubber and phenolic resin. The phenolic resin used was p-cresol phenol–formaldehyde resin and the solvent used was methyl ethyl ketone. In this work, an adduct, toluene diisocyanate-nitrosophenol, was added, which increased the chemical interactions between the rubber and the adhesive, thereby increasing the peel strength [110]. An adhesive made of a mixture of nitrile rubber and phenolic resin was used to make automotive gaskets and bonding various surfaces of tin and asbestos [111]. Addition of phenolic resin imparted heat resistance and adhesion strength to nitrile rubber.

6.4.1.2.2.4 Butyl Rubber-Based Adhesives Butyl rubber adhesives show excellent tack strength. An adhesive contact cement prepared from a mixture of sulfonated derivative of butyl rubber and phenol–formaldehyde resin was found to be useful for bonding both porous and nonporous materials and showed good peel strength, tensile strength, and water impermeability [112]. For mirror assembly in vehicles, butyl rubber-based pressure-sensitive adhesive is used [113]. Butyl rubber-based adhesive to bond cured EPDM rubber was investigated by Chmiel and Young [114]. The adhesive was made of cross-linked halogenated butyl rubber, low molecular weight, high-softening point thermoplastic aliphatic hydrocarbon resin, and a solvent. This adhesive exhibited superior heat aging and hot-water aging resistance properties. Cured butyl rubber-based adhesives for membrane roofing materials such as EPDM and neoprene were made using butyl rubber. Curing agents such as quinone or phenolic resin and tackifiers such as polyisobutylene and hydrocarbon resin were used. The high-temperature stability of the adhesive composition and, hence, the resistance to aging were found to be greatly improved by the addition of zinc oxide and sulfur [115]. For EPDM roofing material, an adhesive based on butyl rubber that was cross-linked after polymerization was used. The adhesive composition comprised butyl rubber, halogenated butyl rubber, tackifying resin, and an isocyanate partially reacted with a diamine to obtain strong bond resistance to heat and humidity [116]. The addition of nanoclay into maleic anhydride-grafted butyl rubber enhanced the adhesive properties of butyl rubber [117].

6.4.1.2.2.5 Polybutadiene Rubber-Based Adhesives Polybutadiene rubber-based adhesives are also used in many bonding applications. For bonding tires and conveyor belts, butadiene rubber-based adhesive was found to be very useful. The adhesive showed excellent adhesion properties before and after vulcanization [118].

6.4.2 Reclaimed Rubber-Based Adhesives

Reclaimed rubber is also used for various adhesive applications [119]. Reclaimed rubber is obtained from used rubber tires and waste rubber products by the application of mechanical, thermal, and chemical agents. Reclaimed rubber adhesives give good adhesion strength for most applications and possess tack similar to natural rubber. Reclaimed rubber adhesives are cheaper than both natural and synthetic rubber-based adhesives. Reclaimed rubber adhesive of a plastic nature was described in the work by Zimmerli and Havenhill [120]. In this work, reclaimed rubber mixed with tackifiers such as ester gums or resins was used for adhesive bonding. Reclaimed rubber was also used to prepare water-dispersed rubber adhesives, which, on drying, formed tacky, smooth tenacious adhesive films. Ester gum was used as the tackifier. The adhesives prepared were used for bonding metals, wood, and glass [121]. An adhesive layer comprised of virgin butyl rubber with reclaimed butyl rubber and a tackifying agent was used to coat a steel pipe by wrapping it with a tape consisting of polyolefin backing and an adhesive layer. A cross-linking agent (p-dinitrosobenzene) and an activator (zinc oxide) were also used in the preparation [122]. Reclaimed rubber latex-based adhesives for bonding natural rubber vulcanizates and a natural rubber–butadiene rubber blend vulcanizate were studied by Job and Joseph [123]. They pointed out that reclaimed latex could be used instead of natural rubber in adhesive production, either partially or completely. The results showed that the adhesive retained its bond strength even after five weeks of aging. Vulcanized reclaimed rubber compounded with accelerators, resins and, antioxidants was used as an adhesive for some applications [124]. Reclaimed rubber from natural rubber or EPDM rubber, together with other additives such as paraffin oil or aromatic oil as the softener, and calcium carbonate, talc, or cement as the filler and tackifier showed adhesive properties for waterproofing asphalt sheets and high-density polyethylene sheets [125].

6.4.3 Importance of Rubber Derivatives in Rubber-Based Adhesives

Derivatives of both natural and synthetic rubbers are used in adhesive preparation, including chlorinated rubber, rubber hydrochloride, and cyclized rubber [44,94]. Chlorinated rubber-based adhesives are mostly used to bond rubber to metals and rubber to rubber. Chlorinated rubber also functions as an additive in chloroprene rubber and nitrile rubber adhesives. A chlorinated rubber-based adhesive used to bond metals was studied by Bradley and Dum [126]. In this work, polyalkylene polyamine was added to the chlorinated rubber for adhesive preparation. The results showed that the adhesive was able to form strong and tenacious bonds with a variety of substrates. For the preparation of adhesive for metal bonding using chlorinated ethylene–propylene rubber, nickel dithiocarbamate was added. The nickel dithiocarbamate acts as an antioxidant in this preparation [127]. An adhesive composed of chlorinated natural rubber with an aromatic p-nitroso compound, a polymaleimide compound, and zinc oxide was reported as a strong adhesives with excellent environmental resistance. Vulcanizing agents, phenolic epoxy resin, and carbon black were also added to the formulation [128]. For bonding wood, leather, and so on, chlorinated rubber adhesives are also used. The application of rubber hydrochloride in adhesives is also known, but is limited. To laminate a sheet of transparent rubber hydrochloride with a sheet of paper, a laminating adhesive prepared using a solution of rubber hydrochloride was used [129].

6.5 VARIOUS COMPOUNDING INGREDIENTS FOR THE PREPARATION OF RUBBER-BASED ADHESIVES

Rubber-based adhesives are adhesives that contain rubber as the base material. For the preparation of rubber-based adhesives, some compounding ingredients are also added. Not all rubber-based adhesives contain the same compounding ingredients; it depends on the purpose of application or, in other words, on the adhesive's utility. Depending on the compounding ingredients, the properties of the adhesive will vary. Rubber-based adhesives are either latex-based adhesives or dry rubber-based adhesives; the ingredients added in the latex are not always needed in dry rubber adhesives formulation. Besides rubber, the other main compounding ingredients present in rubber-based adhesives are tackifiers or resins, antioxidants, vulcanizing or curing agents, accelerators, fillers, solvents, viscosity modifiers or thickeners, sequestering agents, plasticizers or softeners, and some other compounding ingredients that are particularly applied depending on the desired final property of the adhesive [26,44].

6.5.1 Tackifiers and Resins

The role of tackifiers and resins in an adhesive formulation is to increase the tack and adhesion properties of the basic adhesive system [26,44]. Tack is the property of a material to form a bond of measurable strength immediately after contact with another surface under light pressure. So, tack differs from the ultimate strength of the adhesive. A tacky material shows some stickiness when lightly touched with fingers. The main roles of tackifiers and resins are in pressure-sensitive adhesive preparation. The important tackifiers used in the rubber-based adhesive industry are hydrocarbon resins, rosin and its derivatives, phenolic resins, and terpene resins and their derivatives. Some tackifiers provide good long-term tack, while some give short-term tack or only initial tack. Natural rubber has some inherent tack associated with it, but for some adhesive applications, this tack becomes insufficient, so that tackifiers are also added to the compound preparation. Addition of tackifier improves the wettability of rubber-based adhesives. Since most of the synthetic rubbers have poor tack, tackifiers are often used in the adhesive formulation. A pressure-sensitive adhesive using a 50:50 blend of natural rubber and styrene–butadiene was prepared using a tackifier resin, where the resin was unequally partitioned and was more compatible with the natural rubber phase [130]. Another work was carried out to analyze rosin esters and modified rosin esters in a number of rubber-based pressure-sensitive adhesives. Glyceryl rosinate, glyceryl disproportionated rosinate, pentaerythrityl rosinate, and pentaerythrityl hydrogenated rosinate were investigated during the study [131]. The use of coumarone–indene resin tackifier in natural rubber-based pressure-sensitive adhesives was investigated by Khan and Poh [132]. The miscibility of the tackifier and rubber was confirmed by differential scanning calorimetry and Fourier transform infrared spectroscopy. Raja et al. [133] studied the properties of natural rubber latex-based pressure-sensitive adhesives using three different aliphatic hydrocarbon tackifiers with different softening points at two different resin addition levels (25% and 50%). Adhesive formulated with 50% resin addition level showed good adhesive behavior when tested on both high- and low surface energy substrates. The purpose of adding tackifier is to hold the adherends in place while the adhesive bond strength develops. Due to the excellent solubility of terpene resins in the elastomers, these are often used in rubber-based adhesive preparation. The use of phenolic resin derived from a mixture of cardanol (a metasubstituted naturally occurring substance that is the main ingredient of cashew nut shell liquid) and phenol was reported by Varghese and Thachil [102,103]. The tackifying action of alkyl phenolic resin was studied by Durairaj [134]. In this work, he pointed out that when polar alkyl phenolic resins were mixed with nonpolar rubbers, the polar substance tends to move toward the surface of the compound. The action of polar groups at the surface of the rubber compound brings the surfaces together. Tack properties were observed to increase with an increase in the molecular weight of the alkyl phenolic resin and also with an increase in the softening point of the resin.

6.5.2 Vulcanizing Agents and Accelerators

Curing agents or vulcanizing agents are sometimes added to the rubber-based adhesives to vulcanize the adhesive. The addition of a vulcanizing (cross-linking) agent imparts some advantages to rubber-based adhesives, such as aging resistance to the bond developed. Rubber-based adhesives can be used without the use of cross-linking or vulcanizing agents in many applications. The addition of a vulcanizing agent is sometimes minimized; otherwise, the adhesive properties will be seriously affected. With rubber-based adhesives, the main vulcanizing agents used are sulfur, peroxides, and some organic halides. Accelerators are substances used to accelerate the rate or speed of vulcanization. The use of a quinolic ester as an accelerator for a natural rubber-based adhesive to repair rubber articles was reported. It was pointed out that the quinolic ester was a good structure-forming agent and adhesion promoter [135].

6.5.3 Fillers

Fillers are compounding ingredients added to reduce the cost of the adhesive product and, in some cases, as a reinforcing material for the adhesive. But, the addition of fillers to a high level will lower adhesion strength in most cases. Fillers generally used in rubber-based adhesives are carbon black, clays, silicas, and zinc oxide. In rubber-based adhesive technology, carbon black is added to increase the cohesive strength of the product and also to impart conductivity. The effect of carbon black as a filler for natural rubber-based adhesives was discussed by Kardan [136]. Eight different grades of carbon black were used in the study and the level of carbon black used was 5 phr. Results showed that reinforcing carbon black did not form any cross-link with the elastomer network. John and Joseph reported the effect of carbon black on a rubber compound strip adhesive system [84]. The effects of various fillers such as glass beads, Teflon powder, and glass fillers treated with alkyl-chlorosilane on the adhesion strength of natural rubber-based adhesives and poly(n-butyl acrylate) tapes were investigated by Bhowmick [82]. Adhesion strength measurements showed that adhesion strength increased for both treated and untreated glass bead-filled adhesives. According to the findings, it was concluded that the addition of fillers introduced an additional mechanism of energy dissipation during deformation of the adhesive and greater energy would be expended due to debonding of the filler. The effect of silica filler on the adhesion properties of a self-vulcanizable rubber blend based on carboxylated nitrile rubber and chlorobutyl rubber for bonding aluminum to aluminum was studied by Bhattacharya et al. [107]. The results showed that the joint peel strength depended on the filler loading, and the higher peel strength in the filled adhesive system was due to filler reinforcement, resulting in tear path deviation and the formation of Si–O–Al linkage at the aluminum–adhesive interface. At filler loading of 10 phr, maximum peel strength was obtained. In another study by Wang et al. [137], montmorillonite clay was used as a filler for a silicone rubber-based adhesive system. The clay was both reinforcing and compatible with the silicone rubber. The effects of nanoclay on the adhesion and physicomechanical properties of a polysulfide elastomer-based adhesive were discussed by Pradhan et al. [138]. They reported that the incorporation of nanoclay increased adhesion strength due to the interaction of the nanoclay with the liquid elastomer surface. For packaging tapes, instead of zinc oxide, other fillers like calcium carbonate, aluminum hydrate, or clay can be used. The use of magnesium oxide as a filler for an adhesive prepared from epoxidized natural rubber was discussed by Poh and Gan [139].

6.5.4 ANTIOXIDANTS

The function of antioxidants in adhesive compounding is to decrease the aging of the adhesive compound and adhesive bond, and also to protect the bond against deterioration. Most of the adhesives involve the addition of antioxidants, but some exclude their use. The antioxidants that are mainly used in rubber-based adhesives are aromatic amine antioxidants and substituted phenolic antioxidants. In some adhesives, zinc diethyl dithiocarbamate is also added as an antioxidant. In the preparation of self-sealing envelopes, zinc diethyl dithiocarbamate served both as a fungicide and an antioxidant [29].

6.5.5 SOLVENTS

Solvents are an important part of rubber-based adhesive preparation, especially in the case of dry rubber-based solvent cements or solution adhesives. The solvents employed are mainly naphtha, toluene, benzene, carbon tetrachloride, methyl ethyl ketone, methyl isobutyl ketone, and so on. During the selection of the solvent for an adhesive preparation, care must be given so that the solvent is nontoxic, cheap, and has the required rate of evaporation. The choice of solvent should be such that it is appropriate for the base rubber, with minimum hazards, and should possess a suitable drying rate. In the case of solvent adhesives, the viscosity of the adhesive varies with the solvent. Some solvents provide high viscosity, while others may give a low or medium viscosity.

6.5.6 PLASTICIZERS AND SOFTENERS

The purpose of adding softeners and plasticizers into rubber-based adhesives is to make the processing of the rubber easier. The flexibility of the adhesive film can be improved by the addition of plasticizers in rubber-based adhesives. The plasticizers applied are mainly mineral oils, stearic acid, lanolin, and zinc laurate.

6.5.7 VISCOSITY MODIFIERS OR THICKENERS

Viscosity modifiers are mainly used in latex-based adhesives to increase the viscosity of the adhesives, and also to impart some rheological properties. In latex-based adhesives, common viscosity modifiers used are water-soluble cellulose such as methyl cellulose, hydroxyethyl cellulose, and some proteinaceous substances such as casein [26]. For some latex adhesive preparations, thickeners like karaya gum, tragacanth, and other synthetic materials such as poly(vinyl alcohol) and poly (acrylic acid) are also used.

6.5.8 COMPOUNDING INGREDIENTS PARTICULARLY APPLICABLE TO LATEX-BASED ADHESIVES

Besides the abovementioned compounding ingredients, rubber latex-based adhesives may contain some other ingredients such as colloid stabilizers, preservatives, emulsifiers, antifoaming agents, antifreezing agents, and so on [26]. These are present only in small quantities; otherwise, the adhesive properties will decrease.

6.6 METHODS OF PREPARATION OF RUBBER-BASED ADHESIVES

Rubber-based adhesives are prepared either as latex-based adhesives or as dry rubber-based adhesives. The method employed for latex-based adhesive preparation is entirely different from that for dry rubber-based adhesives.

6.6.1 PREPARATION OF LATEX-BASED ADHESIVES

Since rubber latex is a colloidal dispersion of rubber particles in an aqueous or nonaqueous medium, the compounding ingredients added should be in aqueous or in dispersion form. The compounding ingredients are added into the latex in such a way that their addition will not affect the stability of the latex particles. If the compounding ingredients added to the latex are water-insoluble powders, they should be added as water dispersions. The dispersions are prepared using a ball mill or colloidal mill [140]. The total solids content, dry rubber content, viscosity, and so on should be adjusted depending on the purpose of application and also on the method of application. While other ingredients such as antifoaming agents are added to the latex, the pH of the latex should be in the correct range, otherwise it will also change the properties of the latex adhesive. The rubber film deposited from the latex is stronger and has good aging properties compared with that obtained from rubber solution, because the latex is not subjected to mastication or milling.

6.6.2 PREPARATION OF DRY RUBBER-BASED ADHESIVES

The methods of preparation of dry rubber-based adhesives or of the solid form of rubber-based adhesives are totally different from those of latex-based adhesives. The first step in the preparation of almost all dry rubber-based adhesives is mastication or, in other words, milling [44,119]. Dry rubber, whether natural or synthetic, is subjected to mastication, so a reduction in the molecular weight of the polymer will occur. As a result, the masticated rubber is able to dissolve in suitable solvents and can be processed easily. Mastication or milling is a mechanical process of the elastomer; the rubber is passed between two rollers and the size of the gap between the two rollers can be adjusted. The compounding ingredients are added or blended to the dry rubber, either during or after the mastication of the rubber; the choice depends on the compounder as to which method is suitable for a particular application. As in the case of latex-based adhesives, here, also, care must be taken for the thorough mixing of the compounding ingredients. An incomplete or improper mixing will result in variations in viscosity, solids content, and adhesion strength. These changes will also affect the ultimate adhesive properties of the product prepared. In the case of solvent-based rubber adhesives, the addition of solvent is usually carried out after mastication of the rubber compound and, thus, the masticated rubber is dissolved in suitable solvents. The main difference between latex-based adhesives and dry rubber solvent-based adhesives is that the solvent cements are highly viscous and are of low solids content when compared with latex-based adhesives and, also, the use of solvents may cause some kind of toxicity and fire hazard during the use of solvent-based adhesives. The method of preparation of chloroprene rubber-based contact adhesive is described in an Indian patent assigned to Krishnamurti et al. [141]. To a reaction kettle containing toluene and hexane, a mixture containing polychloroprene rubber and phenolic resin dissolved in toluene were added, followed by the addition of magnesium oxide. The adhesive was finally prepared by adding an antioxidant to this solution, and the adhesive was found to be superior in adhesion strength. Self-vulcanizing adhesives are another type of adhesive, where the adhesive solution is a mixture of two parts, one part containing the vulcanizing agent and the second part containing the accelerator. Some typical formulations of rubber-based adhesives are given in Tables 6.1 through 6.3 [29,33].

TABLE 6.1
Self-Seal Envelope Adhesive Formulation Using Natural Rubber Latex

Ingredients	Parts by Weight (Dry Weight)
60% natural rubber latex (ammonia preserved)	100
10% stabilizer (potassium hydroxide) solution	0.2
50% accelerator (zinc diethyl dithiocarbamate) dispersion	0.5

TABLE 6.2
Formulation for Tire Cord Adhesive

Ingredients	Parts by Weight (Wet Weight)
Natural rubber latex (60%)	125
Vinyl pyridine terpolymer latex (40%)	62.5
Resorcinol–formaldehyde resin solution	265
Water	507

TABLE 6.3
Formulation for Pressure-Sensitive Packaging Tape Adhesive

Ingredients	Parts by Weight
Natural rubber	100
Zinc oxide	50
Tackifying resin (ester gum)	175
Softener (lanolin)	25
Antioxidant	1

6.7 METHODS OF APPLICATION OF RUBBER-BASED ADHESIVES

The methods of application of rubber-based adhesives include spraying, knife coating, and using stiff brushes and roll coaters. Roll coaters are used when the adherends are flat sheets and when continuous, uniform coatings are required. The method of application depends on the particular adhesive and its purpose of application; or, in other words, it can be said that the preparation of an adhesive also depends on the method used for its application. The method of application of a rubber-based adhesive also depends on the type of adherends to be joined. The adherends used should be clean and free from all kinds of contaminants, otherwise they will affect the adhesion strength properties. To obtain maximum adhesion strength properties, the adhesive should be in contact with the adherend at all points; thus, a good contact between the adhesive and adherend should occur. For this, it is important to have a properly prepared adherend. Surface roughness of some adherends such as leather will increase the surface area for adhesive application. Depending on the adherend, the surface preparation method will vary. For metal adherends, which are usually associated with oil and other contaminants, surface preparation is very important and is carried out with the help of chemicals or degreasing techniques. After surface preparation, the adhesive is applied via the preferred method. While applying the adhesive to the adherend, proper care is needed to obtain uniform thickness of the adhesive film. Variations in film thickness have a marked effect on strength properties. Improper joining of the adherends will cause variations in the strength of bonded joints. Improper joining occurs when the adherends are joined before proper drying of the adhesive, or after a long period of drying of the adhesive. Some adhesive bonds require curing for final adhesive strength development. These include curing under pressure, temperature, chemical, or electrical methods of curing.

6.8 METHODS OF TESTING RUBBER-BASED ADHESIVES

Testing of adhesives gives us an idea about their qualitative and quantitative aspects, such as adhesion strength, viscosity, storage life, and so on. ASTM standards for testing of adhesives are given in Table 6.4.

6.8.1 VISCOSITY

Most rubber-based adhesives contain various compounding ingredients in addition to the basic rubber. To determine their consistency and flow properties, viscosity measurements are carried out [142]. Viscosity is an important property of rubber-based adhesives, since it gives us an idea about adhesives' wetting and flow properties. Some adhesive applications require high viscosity, whereas others need low or medium viscosity. Thus, measurement of the viscosity of an adhesive

TABLE 6.4
Some ASTM Standards for Testing Adhesives

Designation	Title
D 1084-16	Standard Test Method for Viscosity of Adhesives
D 4800-94 (2015)	Standard Guide for Classifying and Specifying Adhesives
D 1144-99 (2016)	Standard Practice for Determining Strength Developments in Adhesive Bonds
D 1337-10 (2016)	Standard Practice for Storage Life of Adhesives by Viscosity and Bond Strength
D 6862-11 (2016)	Standard Test Method for 90 Degree Peel Resistance of Adhesives
D 3163-01 (2014)	Standard Test Method for Determining Strength of Adhesively Bonded Rigid Plastic Lap Shear Joints in Shear by Tension Loading
D 6195-03 (2011)	Standard Test Method for Loop Tack

is important, as it tells us about its performance. The viscometers used for adhesive viscosity measurements include the rotational viscometer, the Ford-type efflux viscosity cup, the Stormer-type viscometer, and so on.

6.8.2 Storage Life

The storage life of an adhesive is a measure of the time period up to which an adhesive can be used properly. This is done by testing the adhesive properties for a particular time period of storage at a particular temperature (usually the adhesive application temperature) and comparing the values obtained for bond strength and the consistency between them.

6.8.3 Adhesion Strength Measurements: Tack, Peel, and Shear

The tack or stickiness of an adhesive is the ability of instantaneous bonding of one adhesive-coated surface to another. Tack is measured as the stress required to break the bonds between adherend surfaces that are in contact for a relatively short time. Measurement of tack is important in the case of rubber-based, pressure-sensitive adhesives. The main standard methods for tack measurement are the loop tack, probe tack, quick stick, and rolling ball tack.

Peel testing is an important method for testing the adhesion strength of rubber-based adhesives. Peel tests are actually destructive methods for testing adhesion strength, in which two specimens or adherends that are adhesively joined are peeled off under loading. To carry out peel tests, one or both of the adherends should be flexible, and the test involves pulling the flexible adherend at an angle of 90° or 180°. Rubber-based adhesives usually have excellent peel strength but low shear strength, and their resiliency provides good fatigue and impact properties. Since rubbery adhesives are viscoelastic in nature, the peel strength of rubbery adhesives depends on the rate of detachment and the test temperature. The effect of nanoclay on the peel strength of adhesives made of ammonium dichromate cured liquid polysulfide elastomer was described by Pradhan et al. [138]. The results of a 180° peel test showed that nanoclay addition increased the adhesive joint strength of aluminum–aluminum bonds. The addition of nanoclay improved the adhesive ability to dissipate a greater amount of energy during the debonding process of the peel test. Some adhesives are termed *peel-up-type adhesives*, which help the adherend and adhesive to be peeled off from the substrate without breakage of the adherend substrate and without any adhesive remaining on the substrate. The preparation of a peel-up-type adhesive with high adhesion strength against peel force and shear stress was reported based on natural rubber latex [143]. The adhesive prepared in this study was film forming at room temperature and had high viscosity; for the testing of the adhesive, a 180° peel test was used. In another work, the concept of peel stress relaxation during 180° peel testing and the relation between adhesive joint strength and the relaxation phenomenon was studied by Bhattacharya and De [144]. In this work, an adhesive made of a self-vulcanizable rubber blend of chlorobutyl rubber and carboxylated nitrile rubber was used for aluminum–aluminum joints. The authors reported that peel stress relaxation depends on molding time, silica filler loading, test temperature, and peel rate. They also observed that the peel stress relaxation mechanism depends on whether the bond failure is interfacial or cohesive. The effect of an external electric field on the adhesion and cohesion properties of chlorine-containing rubber, such as a chlorinated natural rubber-based adhesive for bonding of rubber to metal, was tested by Potapov et al. [145]. To bond chloroprene rubber-based vulcanizates, modified chloroprene rubber adhesives formulated with some adhesion modifiers such as epoxy resin and residue from aniline preparation were used [146,147]. The results showed that the increase in adhesion was due to the increase in the concentration of polar groups on adding modifiers, and also due to the formation of additional chemical bonds that increased the density of the three-dimensional network. To bond various parts of a speaker made of molded polypropylene, carboxylated synthetic chloroprene rubber-based adhesive containing chlorinated polypropylene was used, and the adhesive showed good adhesion strength [148]. The adhesive performance of a water-based contact adhesive made of natural rubber latex grafted with n-butyl acrylate and methyl methacrylate was studied via a 90° peel test [149].

Measurement of shear strength determines the ability of the adhesive to distribute the load under shear loading. In many adhesive applications, the shear strength plays a vital role in the final properties of the adhesive. Adhesion properties of chloroprene rubber adhesives containing phenolic resin (a mixture of phenol and cardanol) showed that for aluminum–aluminum joints, the optimum shear strength was observed at a phenol:cardanol ratio of 80:20 [103].

6.9 SUMMARY

In adhesive bonding technology, rubber-based adhesives have a vital role due to many unique properties associated with them. Both rubber latex and dry solid adhesives are found to function as important adhesive raw materials. Combinations of two or more rubber blends can result in enhanced adhesive properties with high weathering resistance. In the rubber-based adhesive field, the selection of rubber for a particular application is an important parameter. The modification of basic rubber with compounding ingredients can improve both the adhesion strength and, also, the overall adhesive properties. Due to environmental safety regulations, solvent-based adhesives should be free from toxic solvents. So, solvents used in rubber-based adhesives should be free from toxicity and hazard, but at the same time the required adhesive properties should be attained. Due to this, water-based adhesives are in great demand nowadays. Latex-based adhesives are less toxic and more environmentally friendly than solvent-based ones. But, the adhesion properties and applications of waterborne adhesives are limited compared with those of solvent-based ones. In this chapter, an attempt was made to review the works carried out in the field of rubber-based adhesives. Natural and synthetic rubber latex-based and dry rubber-based adhesives were included. In addition, some of the earlier works carried out in the field of reclaimed rubber and modified rubber adhesives were reviewed. Various theories relating to adhesion were also covered. Short descriptions of the compounding ingredients required for rubber adhesives preparation and the methods of testing of adhesives were also included in this review.

REFERENCES

1. Standard Terminology of Adhesives. ASTM D 907 - 08b (2008).
2. V. Vohralik and W. C. Wake. Theories of adhesion and joint strength. In: *Industrial Adhesives and Sealants*, B. S. Jackson (Ed.), pp. 30–48, Hutchinson Benham, London, UK (1976).
3. H. F. Mark, N. M. Bikales, C. G. Overberger and G. Menges. *Encyclopedia of Polymer Science and Engineering*, Vol. 1, 2nd edn, pp. 476–546, John Wiley & Sons, New York (1985).
4. J. Schultz and M. Nardin. Theories and mechanisms of adhesion. In: *Handbook of Adhesive Technology*, A. Pizzi and K. L. Mittal (Eds.), pp. 19–33, Marcel Dekker, New York (1994).
5. W. C. Wake. Theories of adhesion and uses of adhesives: A review. *Polymer* 19, 291–308 (1978).
6. N. J. DeLollis. Theory of adhesion, mechanism of bond failure, and mechanism for bond improvement. *Rubber Chem. Technol.* 43, 229–243 (1970).
7. K. W. Allen. Some reflections on contemporary views of theories of adhesion. *Intl. J. Adhesion Adhesives* 13, 67–72 (1993).
8. J. W. McBain and D. G. Hopkins. On adhesives and adhesive action. *J. Phys. Chem.* 29, 188–204 (1925).
9. D. E. Packham. The mechanical theory of adhesion: A seventy year perspective and its current status. In: *First International Congress on Adhesion Science and Technology: Invited Papers (Festschrift in Honour of K.L. Mittal)*, W. J. van Ooij and H. R. Anderson, Jr. (Eds.), pp. 81–108, CRC Press, Boca Raton, FL (1998).
10. B. V. Derjaguin and V. P. Smilga. Electronic theory of adhesion. *J. Appl. Phys.* 38, 4609–4616 (1967).
11. F. M. Fowkes. Quantitative characterization of the acid-base properties of solvents, polymers, and inorganic surfaces. In: *Acid-Base Interactions: Relevance to Adhesion Science and Technology*, K. L. Mittal and H. R. Anderson, Jr. (Eds.), pp. 93–115, CRC Press, Boca Raton, FL (1991).
12. W. V. Chang and X. Qin. Repulsive acid-base interaction: Fantasy or reality. In: *Acid-Base Interactions: Relevance to Adhesion Science and Technology*, Vol. 2, K. L. Mittal (Ed.), pp. 4–53, CRC Press, Boca Raton, FL (2000).

13. K. L. Mittal. The role of the interface in adhesion phenomena. *Polym. Eng. Sci.* 17, 467–473 (1977).
14. S. S. Voyutskii. The diffusion theory of adhesion. *Rubber Chem. Technol.* 33, 748–756 (1960).
15. F. H. Wetzel. Introduction to rubber-based adhesives. In: *Handbook of Adhesives*, I. Skeist (Ed.), pp. 188–208, Reinhold Publishing, Chapman & Hall Ltd., London (1962).
16. R. E. Bennett. Pressure sensitive adhesive tape technology. In: *Rubber Products Manufacturing Technology*, A. K. Bhowmick, M. M. Hall and H. A. Benarey (Eds.), pp. 855–866, Marcel Dekker, New York (1994).
17. L. E. Puddefoot, Rubber adhesives in industry. *Trans. Institution Rubber Industry* 24, 199–210 (1948).
18. P. K. Freakley and A. R. Payne. *Theory and Practice of Engineering with Rubber*, pp. 3–23, Applied Science Publishers, London (1978).
19. IRI. *Rubber Engineering*, pp. 48–113, Tata Mc-Graw Hill, New Delhi (1998).
20. A. N. Gent. Elasticity. In: *Engineering with Rubber*, A. N. Gent (Ed.), pp. 33–65, Hanser, Munich (1992).
21. S. Blow. *Handbook of Rubber Technology*, pp. 25–60, Galgotia, New Delhi (1998).
22. C. M. Blow. An outline of rubber technology. In: *Rubber Technology and Manufacture*, C. M. Blow (Ed.), pp. 20–54, Newnes-Butterworths, London (1971).
23. W. J. S. Naunton. Synthetic rubber. In: *The Applied Science of Rubber*, W. J. S. Naunton (Ed.), pp. 199–235, Edward Arnold Publishers, London (1960).
24. L. C. Bateman. Rubber and modern adhesives. *Rubber Dev* 2, 13–15 (1949).
25. T. Gorton. Latex product manufacturing technology. In: *Rubber Products Manufacturing Technology*, A. K. Bhowmick, M. M. Hall and H. A. Benarey (Eds.), pp. 823–844, Marcel Dekker, New York (1994).
26. D. C. Blackley. *Polymer Latices*, Vol. 3, pp. 474–543, Chapman & Hall Ltd., London (1997).
27. R. J. Noble. *Latex in Industry*, pp. 641–664, Palmerton Publishing, New York (1953).
28. D. C. Blackley. *Polymer Latices*, Vol. 1, pp. 1–32, Chapman & Hall Ltd., London (1997).
29. W. C. Wake. Latex adhesives. *NR Technol.* 5, 69–74 (1974).
30. H. J. Northeast. Production methods and uses of rubber-based adhesives. *Rubber Dev.* 12, 57–62 (1959).
31. M. Conte. Plasticization of rubber in the form of latex. *Rubber Chem. Technol.* 27, 271–276 (1954).
32. E. R. Gardner and P. L. Williams. Latex-reclaim mixtures for rubber-cord fabric adhesion. In: *Proceedings of 2nd Rubber Technology Conference*, p. 478 (1948).
33. Some natural rubber adhesive formulations. *NR Technol.* 3, 1–6 (1972).
34. P. Jayasinghe. The future of natural rubber in adhesives. *Bull. Rubber Res. Inst. Sri Lanka* 8, 42–43 (1973).
35. W. G. Weeraratne and M. Nadarajah. The use of natural rubber latex-resin blends as an adhesive for plywood. *Rubber Res. Inst. Ceylon-Quarterly J.* 49, 37–48 (1972).
36. K. F. Gazeley. Natural rubber latex in pressure-sensitive adhesives. *NR Technol.* 12, 85–92 (1981).
37. W. B. Griffith, Jr., A. G. Bunn, I. E. Uhl and K. S. Ho. Pressure sensitive adhesive tape containing natural rubber latex. US Patent 6489024 B2, assigned to Rohm & Haas Company (December 3, 2002).
38. F. Bonnet. Auto-adhesive compositions. European Patent 1229943 A1, assigned to National Starch and Chemical Investment (August 14, 2002).
39. N. John and R. Joseph. Studies on wood-to-wood bonding adhesives based on natural rubber latex. *J. Adhesion Sci. Technol.* 11, 225–232 (1997).
40. M. M. Magida, Y. H. Gad and H. H. El-Nahas. The use of compatible blend of styrene-vinylacetate copolymer/natural rubber latex in pressure-sensitive adhesive applications by using irradiation and chemical initiation. *J. Appl. Polym. Sci.* 114, 157–165 (2009).
41. S. B. Neoh, X. M. Lee, A. R. Azura and A. Hashim. Effect of in situ polymerization of styrene onto natural rubber on adhesion properties of styrene-natural rubber (SNR) adhesives. *J. Adhesion* 86, 859–873 (2010).
42. A. Thitithammawong, N. Ruttanasupa and C. Nakason. Preparation and properties of chlorinated epoxidised natural rubber latex and its latex-based adhesive. *J. Rubber Res.* 15, 19–34 (2012).
43. L. S. Bake, Neoprene cements. In: *Handbook of Adhesives*, I. Skeist (Ed.), pp. 268–285, Reinhold Publishing, New York (1962).
44. SBP. *Handbook of Adhesives*, pp. 96–108, Small Business Publishing, Delhi (1979).
45. H. F. Mark, N. M. Bikales, C. G. Overberger and G. Menges. *Encyclopedia of Polymer Science and Technology*, Vol. 1, pp. 547–577, John Wiley & Sons, New York (1964).
46. K. Watanabe and M. Ose. Method for production of polychloroprene latex and composition. US Patent 20070043165, assigned to Denki Kagaku Kogyo (February 22, 2007).
47. E. C. Atwell. Bonding of polyethylene terephthalate fibers to certain rubbers. US Patent 3060078, assigned to Burlington Industries (October 23, 1962).
48. W. Perlinski, I. J. Davis and J. F. Romanick. Neoprene latex contact adhesives. US Patent No 4485200 A, assigned to National Starch & Chemical Company (November 27, 1984).

49. O. W. Carnahan. Sprayable nonionic neoprene latex adhesive and method of preparation. US Patent 5444112 A, assigned to Cj's Distributing (August 22, 1995).

50. A. L. Christell and R. M. Tabibian. Polychloroprene adhesive latex composition. US Patent 5527846 A, assigned to DuPont (June 18, 1996).

51. H. Wakayama, T. Hayashi and Y. Harada. Chloroprene rubber latex for adhesive, process of production thereof, and adhesive composition using the same. US Patent 5977222, assigned to Tosoh (November 2, 1999).

52. K. Zhang, H. Shen, X. Zhang, R. Lan and H. Chen. Preparation and properties of a waterborne contact adhesive based on polychloroprene latex and styrene-acrylate emulsion blend. *J. Adhesion. Sci. Technol.* 23, 163–175 (2009).

53. J. F. Anderson and H. P. Brown. Styrene-butadiene rubber adhesives. In: *Handbook of Adhesives*, I. Skeist (Ed.), pp. 248–253, Reinhold Publishing, Chapman & Hall Ltd., London (1962).

54. A. D. Hickman. Styrene/butadiene latex-based adhesives. In: *Handbook of Pressure Sensitive Adhesive Technology*, 2nd edn, D. Satas (Ed.), pp. 295–316, Reinhold, New York (1989).

55. S. G. Takemoto and O. J. Morrison. Pressure-sensitive adhesives based on carboxylated SBR emulsion. US Patent 4189419 A, assigned to Avery International Corporation (February 19, 1980).

56. T. T. Wun. Adhesive dip comprising a rubber latex and tris(dihydroxybenzyl) phenol. US Patent 3578613 A, assigned to Koppers Co. Inc. (May 11, 1971).

57. C. G. Force. Treatment of styrene-butadiene rubber. US Patent 4272419 A, assigned to Westvaco (Jun 9, 1981).

58. F. Silvers and M. D. Whitaker. Water based adhesives. US Patent 7427644 B2, assigned to Interlock Industries (September 23, 2008).

59. R. F. Mausser. *The Vanderbilt Latex Handbook*, 3rd edn, pp. 225–229, R.T. Vanderbilt Company, Norwalk, CT (1987).

60. S. Sakurada, Y. Miyazaki, T. Hattori, M. Shiraishi and T. Inoue. Adhesive composition consisting of polyvinylalcohol solution or polyvinylacetate latex modified with hydrophobic solution of isocyanate compound. US Patent 3931008 A, assigned to Kuraray Company (January 6, 1976).

61. D. A. Pelton. Adhesive composition and method for adhering textiles to EPDM rubber. Canadian Patent CA 2477685 A1 (September 18, 2003).

62. H. P. Brown and J. F. Anderson. Nitrile rubber-adhesives. In: *Handbook of Adhesives*, I. Skeist (Ed.), pp. 229–249, Reinhold Publishing, Chapman & Hall Ltd., London (1962).

63. H. Hisaki, O. Mori, A. Okamura, M. Oyama, M. Sekiguchi and H. Tanabe. Adhesive for bonding rubber to fibers. US Patent 5017639 A, assigned to Nippon Zeon (May 21, 1991).

64. D. K. Parker and R. F. Roberts. Hydrogenated nitrile rubber latex composition. Canadian Patent CA 2209442 A1, assigned to Goodyear Tire and Rubber Company (March 16, 1998).

65. Y. Kageyama. Adhesive for bonding rubber and glass fiber. US Patent 5847033 A, assigned to Toyota (December 8, 1998).

66. S. Isshiki, M. Sato, M. Hirayama, O. Mori, Y. Ozawa and S. Takeuchi. Adhesive composite comprising rubber and canvas, and toothed belt. Patent WO 1999025784, assigned to Nippon Zeon (May 27, 1999).

67. K. F. Richards. Butyl rubber and polyisobutylene. In: *Handbook of Adhesives*, I. Skeist (Ed.), pp. 21–228, Reinhold Publishing, Chapman & Hall Ltd., London (1962).

68. R. A. Auerbach and D. B. Berry. Adhesive compositions. US Patent 5036122 A, assigned to Lord Corporation (July 30, 1991).

69. S. S. Voyutskii. Elastomeric adhesion and adhesives. *Rubber Chem. Technol.* 34, 1188–1189 (1961).

70. A. V. Pocius. Elastomer modification of structural adhesives. *Rubber Chem. Technol.* 58, 622–636 (1985).

71. C. W. Bemmels. Adhesive tapes. In: *Handbook of Adhesives*, I. Skeist (Ed.), pp. 584–592, Reinhold Publishing, Chapman & Hall Ltd., London (1962).

72. S. P. Tilley. Adhesives in industry. *Rubber Developments* 15, 79–84 (1962).

73. Self adhesive tape: A boon to home and industry. *Rubber Dev.* 20, 95–98 (1967).

74. Adhesive plasters: Cool, clean and comfortable. *Rubber Dev.* 21, 55–58 (1968).

75. M. Sherriff, R. W. Knibbs and P. G. Langley. Mechanism for the action of tackifying resins in pressure-sensitive adhesives. *J. Appl. Polym. Sci.* 17, 3423–3438 (1973).

76. K. Hino, T. Ito, M. Toyama and H. Hashimoto. Morphological studies on wettability and tackiness of pressure-sensitive adhesives. *J. Appl. Polym. Sci.* 19, 2879–2888 (1975).

77. R. Vitek. Hot-melt adhesive compound and method for the production of the same. US Patent 4091195 A, assigned to Kores Holding (May 23, 1978).

78. J. B. Class and S. G. Chu. The viscoelastic properties of rubber–resin blends. I. The effect of resin structure. *J. Appl. Polym. Sci.* 30, 805–814 (1985).

79. J. B. Class and S. G. Chu. The viscoelastic properties of rubber–resin blends. III. The effect of resin concentration. *J. Appl. Polym. Sci.* 30, 825–842 (1985).

80. K. Hino and H. Hashimoto. Morphological studies on the adhesion mechanism of pressure-sensitive adhesives. *J. Appl. Polym. Sci.* 30, 3369–3376 (1985).

81. F. H. Andrews, T. A. Khan, J. K. Rieke and J. F. Rudd. Adhesion to skin III. Glass transition temperatures of surgical adhesives with added sebum. *Clin. Mater.* 1, 205–213 (1986).

82. A. K. Bhowmick. Effect of model fillers on the strength of pressure-sensitive tapes. *J. Adhesion Sci. Technol.* 3, 371–381 (1989).

83. R. N. Majumdar, T. E. Duncan, R. M. D'Sidocky, J. R. Herberger, L. T. Lukich and B. G. Dunn. Elastomeric laminates containing a solventless elastomeric adhesive composition. US Patent 5503940 A, assigned to Goodyear Tire and Rubber (April 2, 1996).

84. N. John and R. Joseph. Rubber-to-steel bonding studies using a rubber compound strip adhesive system. *J. Adhesion Sci. Technol.* 12, 59–69 (1998).

85. M. Fujita, M. Kagiyama, A. Takemura, H. Ono, H. Mizumachi and S. Hayashi. Effects of miscibility on peel strength of natural-rubber-based pressure-sensitive adhesives. *J. Appl. Polym. Sci.* 70, 777–784 (1998).

86. Y. C. Leong, M. S. Lee and S. N. Gan. The viscoelastic properties of natural rubber pressure-sensitive adhesive using acrylic resin as a tackifier. *J. Appl. Polym. Sci.* 88, 2118–2123 (2003).

87. G. Dileep and S. A. Avirah. The use of carboxy terminated liquid natural rubber (CTNR) as an adhesive in bonding rubber to rigid and non-rigid substrates. *J. Elastomers Plastics* 35, 227–234 (2003).

88. B. T. Poh and S. K. Chow. Effect of zinc oxide on the viscosity, tack, and peel strength of ENR 25-based pressure sensitive adhesives. *J. Appl. Polym. Sci.* 106, 333–337 (2007).

89. I. Khan and B. T. Poh. Effect of silica on viscosity, tack, and shear strength of epoxidized natural rubber-based pressure-sensitive adhesives in the presence of coumarone-indene resin. *J. Appl. Polym. Sci.* 118, 3439–3444 (2010).

90. B. T. Poh and S. Z. Firdaus Syed Putra. Viscosity, shear strength, and peel strength on poly(ethylene terephthalate) of (natural rubber)-based adhesives containing hybrid tackifiers. *J. Vinyl Additive Technol.* 17, 209–212 (2011).

91. I. Khan and B. T. Poh. Natural rubber-based pressure-sensitive adhesives: A review. *J. Polym. Environ.* 19, 793–811 (2011).

92. B. T. Poh and S. K. Cheong. Adhesion behavior of natural rubber-based adhesives crosslinked by benzoyl peroxide. *J. Appl. Polym. Sci.* 124, 1031–1035 (2012).

93. S. Riyajan, N. Phupewkeaw, S. Maneechay and A. Kowalczyk. An emulsion from green epoxidized skim rubber blended with poly(vinyl alcohol) for use as a bioadhesive. *Int. J. Adhesion Adhesives* 45, 84–89 (2013).

94. J. Shields. *Adhesives Handbook*, pp. 1–6, Butterworths, London (1970).

95. J. H. Evans and G. W. Will. Wraparound closure sleeve. US Patent 3770556 A, assigned to Raychem (November 6, 1973).

96. J. Kozakiewicz, B. Kujawa-Penczek, P. Penczek and K. Puton. Factors affecting initial bond strength in solvent-based contact adhesives. *J. Appl. Polym. Sci.* 26, 3699–3705 (1981).

97. D. P. Miller and L. G. Dammann. Primer for use on EPDM roofing materials. US Patent 4897137 A, Ashland Oil (January 30, 1990).

98. D. H. Mowrey and N. M. Pontore. Polychloroprene-based adhesive system. US Patent 5093203 A, Lord Corporation (March 3, 1992).

99. K. Zukiene, V. Jankauskaite and K. V. Mickus. Adhesion studies on piperylene-styrene copolymer modified polychloroprene adhesive: Phase morphology and surface composition. *J. Adhesion* 84, 601–618 (2008).

100. K. J. K. de Silva, A. H. L. Renuka Nilmini and D. Dayaratne. Improvement in performance of polychloroprene rubber based adhesive. *J. Rubber Res. Inst. Sri Lanka* 84, 50–62 (2001).

101. M. Kardan. Adhesive and cohesive strength in polyisoprene/polychloroprene blends. *Rubber Chem. Technol.* 74, 614–621 (2001).

102. L. A. Varghese and E. T. Thachil. Studies on the adhesive properties of neoprene–phenolic blends. *J. Adhesion Sci. Technol.* 18, 181–193 (2004).

103. L. A. Varghese and E. T. Thachil. Adhesive properties of blends of phenol/cardanol–formaldehyde copolymer resin with polychloroprene rubber. *J. Adhesion. Sci. Technol.* 18, 1217–1224 (2004).

104. M. M. Mercedes Pastor Blas. Compatibility improvement between chlorinated thermoplastic rubber and polychloroprene adhesive. *Rubber Chem. Technol.* 82, 18–36 (2009).

105. B. T. Poh and L. N. Ong. Adhesion properties of styrene-butadiene rubber (SBR)/Standard Malaysian Rubber (SMR L)-based adhesives in the presence of phenol formaldehyde resin. *Express Polym. Lett.* 1, 654–659 (2007).

106. I. M. Alwaan. A study on the adhesion of styrene-butadiene rubber with red kaolinite on aluminum surface. *ISRN Chem. Eng.* Article ID: 212567 (2014).

107. T. Bhattacharya, B. K. Dhindaw and S. K. De. Effect of silica filler on aluminum-aluminum bonding by a self-vulcanizable blend based on chlorobutyl rubber and carboxylated nitrile rubber. *J. Adhesion Sci. Technol.* 6, 537–556 (1992).

108. T. Bhattacharya, T. B. Ghosh and S. K. De. Durability studies of aluminium/aluminium joints bonded by chlorobutyl rubber-carboxylated nitrile rubber and epichlorohydrin rubber-carboxylated nitrile rubber blends. *Intl. J. Adhesion Adhesives* 12, 233–239 (1992).

109. P. S. Achary and R. Ramaswamy. Reactive compatibilization of a nitrile rubber/phenolic resin blend: Effect on adhesive and composite properties. *J. Appl. Polym. Sci.* 69, 1187–1201 (1998).

110. P. S. Achary, C. Gouri and R. Ramaswamy. Reactive bonding of natural rubber to metal by a nitrile–phenolic adhesive. *J. Appl. Polym. Sci.* 81, 2597–2608 (2001).

111. N. Krishnamurti. A process for the preparation of nitrile rubber-based adhesives for bonding automotive gasket. Indian Patent 216095 (March 21, 2008).

112. J. J. Higgins and C. P. Ofarrell. Adhesive cements containing a sulfonated derivative of butyl rubber and laminates therefrom. US Patent 3867247 A, assigned to Exxon Research (February 18, 1975).

113. D. E. Lawson. Side view mirror assembly with butyl rubber adhesive. US Patent 4200359 A, assigned to Lawson David (April 29, 1980).

114. C. T. Chmiel and D. A. Young. Adhesive for bonding cured EPDM rubber containing a crosslinked halogenated butyl rubber. US Patent 4851462 A, assigned to Uniroyal Plastic (July 29, 1989).

115. T. J. Chiu. Adhesive of butyl rubber, curing agent, c-black and tackifier. US Patent 5095068 A, assigned to Ashland Oil (March 10, 1992).

116. L. G. Dammann, M. C. Clingerman, J. W. Fieldhouse and H. Tsai. Adhesive for bonding EPDM rubber roofing membrane. Patent WO 1993003914 A1, assigned to Ashland Oil (March 10, 1993).

117. O. Akhtar and R. Krista. Butyl adhesive containing maleic anhydride and optional nanoclay. European Patent 1892277 A2, assigned to Lanxess (February 27, 2008).

118. Y. Nagai and K. Yamaguchi. Butadiene rubber adhesive composition with specified low syndiotactic-butadiene content. US Patent 6187855B1, assigned to Bridgestone Corporation (February 13, 2001).

119. S. Palinchak and W. J. Yurgen. Natural and reclaimed rubber adhesives. In: *Handbook of Adhesives*, I. Skeist (Ed.), pp. 209–220, Reinhold Publishing, Chapman & Hall Ltd., London (1962).

120. W. F. Zimmerli and R. S. Havenhill. Adhesive composition and method of making same. US Patent 1892123 A, assigned to Goodrich (December 27, 1932).

121. S. N. Henry. Water dispersed rubber adhesive. US Patent 2411905 A, assigned to Minnesota Mining and Manufacturing Company (December 3, 1946).

122. G. M. Harris and S. J. Thomas. Pipe wrapping adhesive tape method. US Patent 4268334 A, assigned to Kendall (May 19, 1981).

123. L. Job and R. Joseph. Studies on the adhesives for rubber to rubber bonding. *J. Adhesion Sci. Technol.* 9, 1427–1434 (1995).

124. L. C. Lima. Method of manufacture of paint and adhesive rubber from vulcanized rubber. US Patent 5607981 A, assigned to Relastomer (March 4, 1997).

125. S. K. Lee, J. Y. Lee, I. K. Cho and J. S. Kim. Non-curing adhesive waterproof material using recycled rubber, and manufacturing method thereof. Patent EP 2489713 A1, assigned to Renew System (August 22, 2012).

126. H. P. Bradley and J. L. Dum. Metal bonded by means of chlorinated rubber adhesives containing a poly-alkylene polyamine. US Patent 2459742, assigned to Firestone Tire (January 18, 1949).

127. T. Iwasa and H. Yokoi. Rubber adhesive. US Patent 4686122 A, assigned to Toyoda Gosei Company (August 11, 1987).

128. D. H. Mowrey. One-coat rubber-to-metal bonding adhesive. Patent EP 0705314 A1, assigned to Lord Corporation (April 10, 1997).

129. A. M. George. Container and method of manufacture. US Patent 2333330 A (November 2, 1943).

130. B. C. Copley. Tackification studies of natural rubber/styrene-butadiene rubber blends. *Rubber Chem. Technol.* 55, 416–427 (1982).

131. Y. Kumooka. Analysis of rosin and modified rosin esters in adhesives by matrix-assisted laser desorption/ionization time-of-flight mass spectrometry (MALDI-TOF-MS). *Forensic Sci. Intl.* 176, 111–120 (2008).

132. I. Khan and B. T. Poh. Effect of molecular weight and testing rate on adhesion property of pressure-sensitive adhesives prepared from epoxidized natural rubber. *Mater. Des.* 32, 2513–2519 (2011).

133. P. R. Raja, A. G. Hagood, M. A. Peters and S. G. Croll. Evaluation of natural rubber latex-based PSAs containing aliphatic hydrocarbon tackifier dispersions with different softening points: Adhesive properties at different conditions. *Intl. J. Adhesion Adhesives* 41, 160–170 (2013).

134. R. B. Durairaj. Superior tackifying resin. *Rubber Sci.* 26, 148–157 (2013).

135. S. Y. Kukushkin, L. R. Lyusova, V. A. Glagolev and E. E. Potapov. Development of the processing of natural rubber adhesives for repairing rubber articles. *Intl. Polym. Sci. Technol.* 30, 60–61 (2003).

136. M. Kardan. Carbon black reinforcement in natural rubber based adhesives and sealants. *Intl. Polym. Sci. Technol.* 31, 7–10 (2004).

137. J. Wang, Y. Chen and Q. Jin. New organic montmorillonite: Application to room-temperature vulcanized silicone rubber adhesive system. *J. Adhesion* 82, 389–405 (2006).

138. S. Pradhan, P. K. Guchhait, K. D. Kumar and A. K. Bhowmick. Influence of nanoclay on the adhesive and physico-mechanical properties of liquid polysulfide elastomer. *J. Adhesion Sci. Technol.* 23, 2013–2029 (2009).

139. B. T. Poh and F. C. Gan. Viscosity and peel strength of magnesium oxide-filled adhesive prepared from epoxidized natural rubber (ENR 25). *Polymer-Plastic Technol. Eng.* 49, 191–194 (2010).

140. D. C. Blackley. *Polymer Latices*, Vol. 3, pp. 1–33, Chapman & Hall Ltd., London (1997).

141. N. Krishnamurti, B. S. Sitaramam, A. R. Sastry and D. B. Rohinikumar. A formulation useful as contact adhesives for leather, rubber, and PVC surfaces. Indian Patent 215683 (March 21, 2008).

142. M. Petronio. Properties, testing, specification and design of adhesives. In: *Handbook of Adhesives*, I. Skeist (Ed.), pp. 61–80, Reinhold Publishing, Chapman & Hall Ltd., London (1962).

143. T. Sugimura, Y. Ozawa and M. Takahashi. Peel-up type adhesives. US Patent 4350723 A, assigned to Nippon Zeon (September 21, 1982).

144. T. Bhattacharya and S. K. De. Peel stress relaxation of aluminium-aluminium joints bonded by chlorobutyl rubber-carboxylated nitrile rubber blend. *J. Adhesion Sci. Technol.* 6, 861–877 (1992).

145. E. E. Potapov, Y. D. Nebratenko and G. G. Salych. Influence of external electrical action on the adhesion properties of rubber adhesive composites. *Intl. Polym. Sci. Technol.* 28, 12–17 (2001).

146. A. N. Keibal, N. S. Bondarenko and F. V. Kablov. Amine containing modifier for polychloroprene based adhesives. *Intl. Polym. Sci. Technol.* 32, 69–71 (2005).

147. A. N. Keibal, N. S. Bondarenko, V. F. Kablov and N. G. Sergeev. Polychloroprene adhesives with increased adhesion to polychloroprene rubber based vulcanisates. *Intl. Polym. Sci. Technol.* 32, 29–30 (2005).

148. K. Okuzawa, Y. Baba and K. Wakabayashi. Synthetic chloroprene rubber adhesive composition and process for producing speaker by using the same. US Patent 7001480 B1, assigned to Natsushita Electrics (February 21, 2006).

149. P. Chumsamrong and J. Mondobyai. Preparation, adhesive performance and stability of natural rubber latex grafted with N-butyl acrylate (BA) and methyl methacrylate (MMA). *Suranaree J. Sci. Technol.* 14, 269–276 (2007).

7 Elastic Adhesives

Johann Klein* and Christina Despotopoulou

CONTENTS

7.1 INTRODUCTION

During the last few years, the trend in bonding seems to have changed from rigid bondlines to elastic joints and bondlines. The main benefit of elastic bonding is its ability to absorb any kind of stress or strain that can arise between two different substrates such as, for example, mechanical stress or stress due to different thermal expansion coefficients. Thus, by counteracting naturally occurring strains, elastic bonding ensures a strong, secure, and durable bond. Apart from this mechanical advantage of elastic bonding, there are also a number of secondary advantages, such as damping of vibrations or the easy removal/peel off of the adhesive, that add significantly to its success [1].

The popularity of elastic adhesives is undoubtedly strongly connected with the progress that has been made in utilizing novel reactive polymers. The main technology segments include polyurethane

---

* Indicates corresponding author

(PU) chemistry, long-chain polyethers, and silane curing organic prepolymers. Unlike the previously used elastic adhesives that were based on conventional elastomers and cured only through a physical process (e.g., solvent-based adhesives and hot-melts), the new generation reactive systems are more user-friendly and offer tailor-made solutions to any customer requirement.

7.2 MECHANICAL PROPERTIES

Shanks and Kong define elastomers as materials that exhibit a rapid and large reversible strain in response to a stress. Translating this definition into structural features means that polymers with minimal intermolecular forces are required. Hydrogen bonding or polar functional groups should be missing from the ideal elastomer structure, which is also the reason why most elastomers are high molecular weight polymers lacking molecular complexity [2].

Such elastomeric materials are the most common ingredients of sealants. Due to their properties, they fulfill the sealing requirements perfectly, with a low tensile strength of <1 MPa and a high elongation, typically in the range of 250%–600%. However, elastic adhesives, though they very often use the same curing and polymer chemistry as the elastic sealants, show a very different property profile. With a characteristic tensile strength ranging from 1 to 10 MPa and an elongation of 70%–300%, they place themselves in the middle of the application spectrum between the aforementioned sealants and the structural adhesives that typically display tensile strengths of 10–30 MPa and a minimal elongation of 0%–70% [3].

In each adhesive class, polymers with a certain behavior are chosen as the main component of each formulation. Their properties and behavior are based on their chemical structure, chemical bond flexibility, molecular weight and molecular weight distribution, and, most importantly, their glass transition temperature (T_g), which characterizes them. This behavior can be either elastic, plastic, or rigid [4].

A material shows elastic behavior if it regains its original configuration on release of stress. If, however, the material does not display a high resilience and stays deformed even after the applied load has ceased, then this behavior is characterized as plastic. Stress–strain curves of polymers can immediately show how plastic or elastic each material is. The behavior of the material also influences the behavior of the resulting bondline.

The advantages of elastic adhesives can be shown very well through photoelastic recordings of transparent materials. Bonding transparent poly(methyl methacrylate) (PMMA) or polycarbonate substrates shows that brittle bondlines can be recognized through stress peaks on the overlapping ends. Such bondlines can very easily crack, and ultimately lead to adhesion failure on application of load. In contrast, elastic bondlines seem to evenly absorb the stress and distribute it across the whole bonded surface. Thus, by increasing the bonded surface, a greater stress can be absorbed, despite the fact that the mechanical strength of elastic adhesives (cohesion) is significantly lower compared with structural adhesives [5].

One further property that directly influences the behavior of a material is its glass transition temperature. The T_g value of a polymer refers to the temperature at which the material transitions from a hard, glassy state to a soft, rubbery state. This process is time-dependent. Hence, for a polymer to be an effective elastomer, its in-service temperatures should be well above its T_g temperature. Most polymers that show elastic behavior at ambient temperatures have a T_g that lies well in the range of −20°C to −70°C [2].

Based on the characteristics and properties mentioned, a number of polymers qualify for use in elastic bonding. The properties of each polymer are further enhanced in a formulation through addition of fillers, adhesion promoters, catalysts, and a number of other components to ensure ease of use and reliable performance, tailored for each desired application. In the next sections, such polymeric materials are described. Depending on the chemical reactivity of each polymer, two different application systems are distinguished by how to attach the elastic adhesive to the surface. If adhesion occurs through a chemical reaction of the polymer with the surface, the adhesive class is called *reactive elastic adhesives*, whereas if the polymer does not change chemically during the application and the attachment to the surface, the process is called *physically setting* and adhesion is achieved through sufficient wetting and subsequent solvent evaporation. This interpenetration of the polymer results in mechanical strength. Each of these two categories described here utilizes different polymers that are also differently formulated to produce the final adhesive product. In the next sections, the chemistry and properties of these elastic adhesive categories are described in more detail.

7.3 PHYSICALLY SETTING ELASTIC ADHESIVES

In physically setting elastic adhesives, polymers and elastomers are used that do not change chemically on application and, hence, do not react with the surface of the substrate. The process of setting depends mainly on the wetting of the surface, and the properties of the resulting bond also depend on the polymer used. For this reason, polymers and elastomers are preferred that are easily dispersed, dissolved, or melted in different liquids to ensure sufficient wetting. After application, the adhesive is hardened by a physical process involving most commonly either drying (dispersion adhesives) or cooling (hot-melt adhesives). Such nonreactive elastic adhesives generally offer cost-efficient solutions, as they are easy to apply and require low equipment expenditure. Some drawbacks of this technology, however, are the shrinkage associated with evaporation of the solvent, the chemical hazards often associated with the use of organic solvents, and the point that this process can be reversed by exposing the bondline to solvents or heat [6]. Nevertheless, such adhesive systems are used in a number of elastic applications, especially in the automotive industry. Naturally occurring rubbers, as well as synthetic polymers, are the more commonly used materials in these formulations and are responsible for the properties of the resulting joints. The main polymer structures used for these types of adhesives are highlighted in the next sections.

7.3.1 ALIPHATIC AND AROMATIC HYDROCARBON ELASTOMERS

7.3.1.1 Natural Rubber

Natural rubber or latex is a naturally occurring elastomer that is made from the sap of the rubber tree and consists predominantly of the polymer of *cis* 1,4-isoprene with molecular weights in the range of 100,000–1,000,000 g/mol (Figure 7.1).

Natural rubber can also be subjected to the process of vulcanization to increase the degree of cross-linking through disulfide bonds. The majority of rubber produced each year is used in the automotive industry in tires. Nevertheless, rubber also finds application in water- and solvent-based elastic adhesives, especially in applications where abrasion, wear resistance, shock absorption, and electrical resistance are required.

7.3.1.2 Styrene–Butadiene Rubber

Styrene–butadiene copolymers (also known as *SBR* and *Buna S*) are the synthetic equivalent of natural rubber. They consist of hard and soft segments, with the hard styrene blocks imparting thermoplastic properties and the softer butadiene midblocks giving the material its elastomeric properties (Figure 7.2) [7,8].

Styrene–butadiene rubbers are formulated in a number of pressure sensitive adhesives such as tape adhesives, footwear adhesives, and nonstructural construction applications.

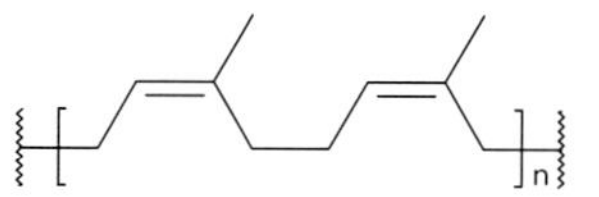

FIGURE 7.1 Structure of 1,4-polyisoprene.

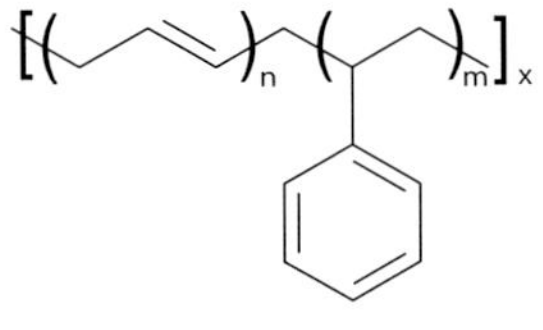

FIGURE 7.2   Structure of styrene–butadiene rubber (SBR).

7.3.1.3 Polyolefins

Polyolefins, similarly to the polymers of ethylene propylene diene monomer (EPDM rubber) building blocks, are a type of synthetic rubber resulting from the copolymerization of ethylene, propylene, and a diene component. Currently, the most commonly used dienes in the manufacture of EPDM rubbers are dicyclopentadiene, ethylidene norbornene, and vinyl norbornene (Figure 7.3).

The ethylene content in these rubbers ranges from 45% to 85%. A higher ethylene content improves the mixing and extrusion properties of the polymer, allowing a higher loading. The dienes make up 2.5%–12 wt% of the composition, and act as cross-linkers in further curing of the polymer with sulfur and/or resins.

Similarly to the other nonpolar elastomers, EPDM rubbers show good electrical resistance, as well as heat and weather resistance.

7.3.2 Halogen Substituted Elastomers

7.3.2.1 Polychloroprene

The main representative of the halogenated class of elastomers is polychloroprene, also known under the DuPont trademark neoprene (Figure 7.4). Different grades of polychloroprene can be found in the market with various degrees of cross-linking and vulcanization.

Polychloroprene is particularly famous for displaying a good combination of desired properties such as good chemical and weather resistance, good flexibility even at lower temperatures, good adhesion to substrates, and low flammability. Apart from its use as protective clothing, it is also used in gasketing and sealing applications in the automotive industry and as a raw material in solvent- and water-based contact adhesives.

7.3.3 Nitrile Rubber

Nitrile functionalized rubber, also known as *Buna N rubber*, can be synthesized from the copolymerization of butadiene and acrylonitrile (Figure 7.5). Its properties can be tuned by adjusting the level of acrylonitrile in the composition. Typical nitrile contents in these rubbers range from 15% to 50%. Higher levels of nitrile in the polymer lead to higher strength and chemical resistance

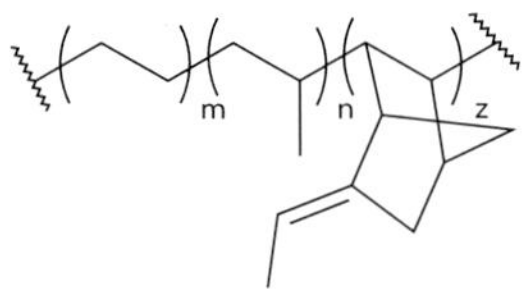

FIGURE 7.3   Structure of ethylene propylene diene monomer (EPDM rubber).

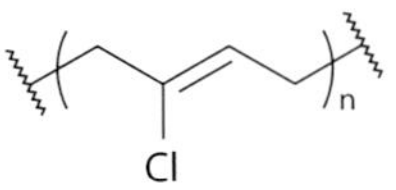

FIGURE 7.4   Structure of polychloroprene polymer.

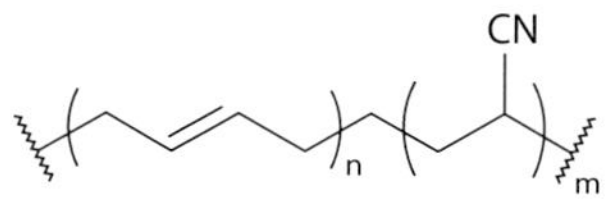

FIGURE 7.5   Structure of Buna N rubber.

and lower permeability to gases. However, the increase in nitrile content also leads to a higher T_g and, hence, to a more brittle and less flexible polymer with higher application temperatures. One of the rubber's main advantages is its excellent resistance to oil fuels and heat, and that is why its main application is in the automotive industry for sealing and gasketing, especially in contact with hot oil [2].

7.3.4 SULFIDE ELASTOMERS

Organic polysulfide rubbers refer to materials with the general structure R_2S_x that are synthesized by copolymerization of sulfur and hydrocarbon compounds. The process of vulcanization, discovered by C. Goodyear, is one of the main routes to cross-link hydrocarbons with polysulfides that, depending on the degree of cross-linking, results in more rigid polymers with excellent fuel resistance, good dielectric properties, and low moisture vapor and gas transmission.

7.3.5 REACTIVE POLYMERS IN PHYSICALLY SETTING ADHESIVES

This section has, so far, focused on nonreactive elastomers that are dissolved or dispersed in a liquid (water or organic solvent) to form a physically setting elastic adhesive that adheres to the surface after drying (or, in the case of hot-melts, after cooling). For certain applications, however, it can be desirable to employ multiple setting mechanisms (e.g., physical and chemical), so that the properties of the resulting adhesive film can be varied or enhanced between the different setting steps. Reactive polymers such as PUs and silicones can, therefore, also be formulated in a solvent-based system and, thus, create a dual system, where first setting occurs on drying, followed by additional curing on reaction of the reactive groups of the polymer with the surface.

Such systems have a number of beneficial properties. As most of these adhesives are water dispersions, they have one major advantage; namely, the molecular weight of the polymer has no significant effect on the final viscosity, which would have a negative influence on the setting speed. They also carry no hazard pictograms according to the Globally Harmonized System of Classification and Labelling of Chemicals (GHS) and are more user- and environmentally friendly [6].*

However, due to polarity and compatibility issues between the water and the nonpolar organic polymer, such systems are more difficult to formulate and require a clever use of emulsifiers and stabilizing components. Such reactive polymers are, therefore, mostly found as 100% systems (systems where the polymer is formulated without the use of a solvent), where they are able to unfold their full performance potential.

7.4 REACTIVE ELASTIC ADHESIVES

7.4.1 SILICONES

Silicones are the first polymers that come to mind when talking about elastic sealants and adhesives. They have a low T_g, an excellent thermal and weathering stability, are resistant against discoloration, and can also introduce different degrees of cross-linking in the polymer matrix through a number

---

* The GHS system uses nine pictograms for labeling of containers and for workplace hazard warnings, to alert users of the chemical hazards to which they may be exposed. The pictograms can be viewed and downloaded from the official United Nations (UN) web page.

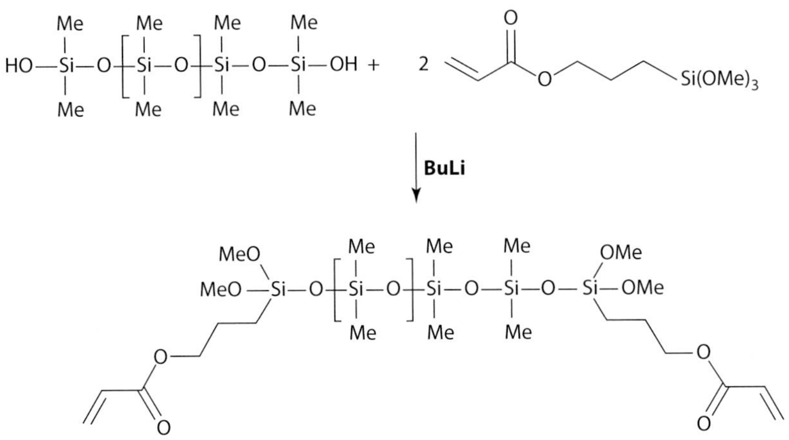

FIGURE 7.6 Synthesis of dual-cure silicone polymer.

of different ways. This unique combination of properties makes silicones a very attractive polymer class; however, they also suffer from certain drawbacks. Due to the low surface tension of these polymers and the resulting weak adhesion, silicones are suitable for only a limited number of substrates, such as glass, for example. Furthermore, evaporation of low molecular weight components (cyclics), which can occur and thus contaminate the substrates, causes a number of users, especially from the automotive and electronics industries, to avoid the use of elastic silicone-based adhesives.

Silicone polymers can be cured, depending on the application, using a number of different curing mechanisms. The most common ones are (1) the moisture cure of poly(dimethylsiloxane) (PDMS) polymers terminated with reactive silanes (RTV1); (2) the addition cure of hydride-terminated PDMS polymers that react with vinyl-terminated ones (RTV2); and (3) the radical cure of silicones that use radical initiators. Recently, silicone-based hot-melts as well as ultraviolet (UV) curing systems have also been reported. Different curing mechanisms can also be combined in one dual-cure system, whereby a silanol Si–OH moiety of the PDMS polymer is end-capped with an acrylate functionalized silane (Figure 7.6) [9].

Contrary to silicone-based sealants, where polymers with molecular weights of approximately 100,000 g/mol are needed, silicone adhesives normally use short-chain PDMS polymers in the range of 2,000–20,000 g/mol. To increase the cross-link density and the E-modulus, silicone resins are also mixed in the formulation. The more recent patents describe the addition of silsesquioxanes [10].

One further additive that plays a significant role in silicone-based elastic adhesives is pyrogenic silica gel. Its high surface area, which is enriched with pendent silanol (Si–OH) moieties, enables an easy incorporation of the silica into the adhesive matrix and acts as a reinforcing filler. Such pyrogenic silica gels are available in different forms and can be used according to the degree of toughness and viscosity one wants to achieve. Hydrophobic silicas with a high specific surface area are typically used to achieve higher rigidity. These types of materials also bring with them, however, a higher final viscosity, due to the respective Si–OH interactions (in the form of hydrogen bonding) and, thus, limit the amount of silica gel that can be incorporated into the adhesive formulation. Other types that have a lower specific surface or have different silanes grafted on their surfaces are used to balance viscosity and toughness in the final product.

Finally, for such an adhesive formulation to be effective and for reaction with the substrate surface to occur, a catalyst is needed to initiate the curing. The catalyst is first chosen based on the curing mechanism for which the polymer is designed. Two-component (2c) silicones that utilize the addition cure mechanism need platinum-based catalysts to cure, whereas the most prominent curatives for RTV1 systems are organotin and organotitanium catalysts. Radically curing silicone systems make use of radical initiators such as dicumyl peroxide, *tert*-butyl peroxybenzoate, 2,5-dimethyl-bis-(2,5-*tert*-butylperoxy)-hexane and bis-(2,4-dichlorobenzoyl) peroxide.

The components described here should provide a brief overview of the broad formulation spectrum of the silicone adhesives. Through the right choice of polymer, resin, filler, and catalyst the basic network of the adhesive is built. This network can then be enhanced and tuned according to the intended application through addition of adhesion promoters (e.g., organofunctionalized silanes, functionalized silicone resins), plasticizers (e.g., high molecular weight silicone oils and mineral oils), and nonreactive fillers (e.g., chalk, talcum).

7.4.2 Polyurethanes

When Otto Bayer first developed PUs in 1937, he could not have imagined how much influence this class of materials would have in shaping our modern world. The main reason for their significant influence is their unique diversity, based on the various combination possibilities of isocyanate and polyol components that result, each time, in a different polymer system with different properties. One characteristic example of this polymer technology is PU foams. In 1940, PU foams were already used in the construction of airplanes, while today their everyday use in mattresses, construction foams, sandwich panels, and automotive parts is irreplaceable. A number of thermoplastic materials used in fibers, floors, and textiles are based on PU chemistry. Furthermore, the development of nonyellowing aliphatic isocyanates has enabled the technology to further expand in the coatings and paint market. In 1941, the first PU systems were applied as adhesives for rubbers, metals, and glass in the automotive industry [11]. Finally, the application of PU adhesives in the bonding of glass windshields by General Motors in 1963 led the way toward the development of a new class of adhesives: elastic adhesives.

7.4.2.1 Two-Component PU Systems

Elastic adhesives based on two-component PUs are very widespread and well established, as they are one of the simplest and cheapest categories of reactive elastic adhesives. They generally consist of one resin component (OH or NH functionalized) and one hardener component (NCO functionalized). As these two components need to be mixed before the curing step, such PU systems are mainly used in industrial production processes where the necessary mixing equipment is in place and the severe toxicity of the isocyanate component can be handled with the appropriate engineering controls.

Through the versatility of the PU chemistry, and by wisely choosing the corresponding OH and NCO components, a number of different polymers can be designed with different physical properties tailored to each application. Suitable hardener components can be chosen from a pool of various di- and multifunctional isocyanates that are well known in the PU field (Figure 7.7).

When formulating the resin component, aromatic isocyanates are preferred, due to their increased reactivity and corresponding faster curing speed. Technical-grade methylene diphenyl diisocyanates (MDI) types are also often used, because of their very attractive price and low volatility. Polyfunctional isocyanates present in the technical-grade diisocyanates can also be utilized to achieve the desired cross-linking degree and, hence, the desired elasticity. Diisocyanates are reacted in excess with diols and diamines to form a *prepolymer* that is used to adjust the NCO content and viscosity of the hardener component to the desired level.

The resin component usually consists of polyols with sufficiently high molecular weight to ensure that the T_g of the polymer will not exceed the application temperature of the adhesive. Typically, conventional poly(propylene glycol)s (PPGs) are used with molecular weights ranging from 500 to 2000 g/mol. The elasticity can then be adjusted by mixing in tri- and tetrafunctional PPGs. However, most conventional polyether diols that have been synthesized through base catalysis always contain a certain amount of monools that can act as a chain stopper. The precise content

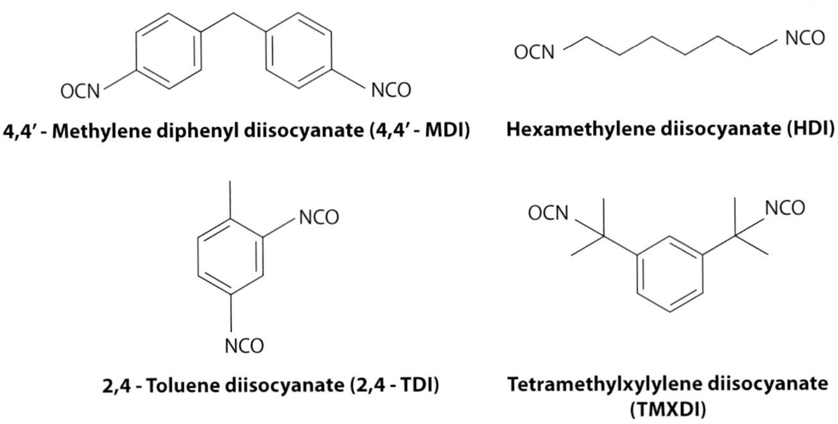

FIGURE 7.7 Structures of commonly used isocyanates.

of monools, therefore, needs to be known prior to the formulation, so that higher functionalized polyols can be added to compensate for the monool content and thus achieve the same degree of cross-linking. This way, the desired mechanical properties of the final formulation are retained in spite of the monool impurities.

Apart from PPGs, linear amorphous polyester diols can also be used in similar molecular weight ranges. Polybutadiene diols are excellent alternatives for reducing the T_g and increasing the adhesive's moisture resistance. Furthermore, by adding solid components with a melting point equal to or below 60°C, a so-called *high position tack* can be achieved (this hot-melt-type hardening of the hotly applied adhesive renders a mechanical fixation of the substrates unnecessary). These solid components can also contain reactive groups, such as one-component (1c) PU prepolymers based on crystalline diols (e.g., polycaprolactones or polycarbonate diols) [12].

Renewable raw materials such as castor oil, partially dehydrated castor oil [13], or products from the ring-opening reaction of epoxidized soybean oil (ESO) with carboxylic acids, can also be added to increase the sustainability of the formulations.

Similarly to the silicone formulations, an indispensable component of all adhesive formulations is the catalyst. The same applies also to the two-component PU adhesives in the reaction between the OH moiety and the NCO functionality to form the urethane link. This reaction can either be catalyzed with a Lewis acid, by activating the carbonyl functionality, or with a Lewis base, by activating the alcohol. The resulting urethane groups can also act as a Lewis base, so that the use of the catalyst can be minimized or even eliminated [14]. Examples of commonly used Lewis bases are tertiary amines such as 1,8-diazabicyclo[5.4.0] undec-7-en (DBU), 1,4-diazabicyclo[2.2.2]octane (DABCO), 2,2-dimorpholinodiethylether (DMDEE), or triethanolamine. The tertiary amine is frequently incorporated into the polyol component, so as to prevent migration of the catalyst out of the finished bond.

Examples of commonly used Lewis acids are metal salt-based catalysts such as tin-, lead-, bismuth-, or zinc carboxylates, organotin structures such as dibutyltin- or dioctyltin-based catalysts, iron-based catalysts such as iron acetylacetonate, as well as titanium or zirconium complexes. Before choosing the right catalyst, all possible side reactions need to be considered. Such reactions include the dimerization of isocyanates to form allophanates, or their trimerization to isocyanurates. Furthermore, the presence of water in the polyol part can lead to the formation of urea groups that could, in turn, react further with isocyanates to form biuret structures. These side reactions would result in an increase in the cross-link density of the final polymer that would significantly reduce the elasticity of the adhesive. For this reason, catalysts that can selectively catalyze only the urethane synthesis should be preferred. Unfortunately, not many catalytic systems currently available offer such selectivity. Recent developments show, however, that zinc complexes with ketoamide ligands are very selective [15].

Apart from the selectivity, one other important criterion in the choice of catalyst is its speed and curing profile. If the adhesive is intended for application on large surfaces, then the catalyst must be chosen so that the pot-life of the product is sufficiently long to allow the adhesive to be fully applied before the curing starts. In these cases, the catalysts are often encapsulated and can be activated either through heat [16], or through microwave irradiation, in the case of encapsulated ferromagnetic particles [17].

Depending on the adhesive requirements, plasticizers (e.g., phthalic esters, benzoic esters etc.) and fillers (e.g., quartz powder, talcum, and chalk) can be further incorporated in the formulation to tune its properties. Rheological additives such as hydrogenated castor oil or polyamides are also added to adjust the final viscosity of the mixed components. Such additives are mostly added to the less sensitive polyol component that needs to be thoroughly dried afterward to remove volatile components and water. In some cases, drying agents such as zeolite-based pastes can also be added to ensure that no residual moisture is left in the component [18].

7.4.2.2 One-Component PU Systems

The 1c elastic PU adhesives are prepared by reacting diols with an excess of diisocyanates to form NCO-terminated prepolymers (Figure 7.8).

Curing occurs then through reaction of the isocyanate groups in the polymer with moisture to form urea groups on release of carbon dioxide (Figure 7.9).

By adjusting the raw material ratios, polymers with different molecular weights and different NCO contents are produced. Similarly to the two component (2c) PUs, long-chain diols (M_w 2000–6000 g/mol) are also used in the 1c systems. However, the incorporation of multifunctional polyols to achieve a higher cross-linking degree is less commonly utilized in this system, as it is directly associated with a significant increase in the viscosity of the prepolymer that makes the compounding and product application more difficult. The choice of the isocyanate component in the case of 1c systems is limited to reactive isocyanates such as MDI or TDI; otherwise, moisture curing occurs very slowly. The molecular weight of the elastic PU prepolymers lies in the range of 5,000–15,000 g/mol, with an NCO content of 1%–3%. Based on the current synthesis routes for the preparation of the 1c PU prepolymers, residual monomeric diisocyanate is always present in the mixture, something

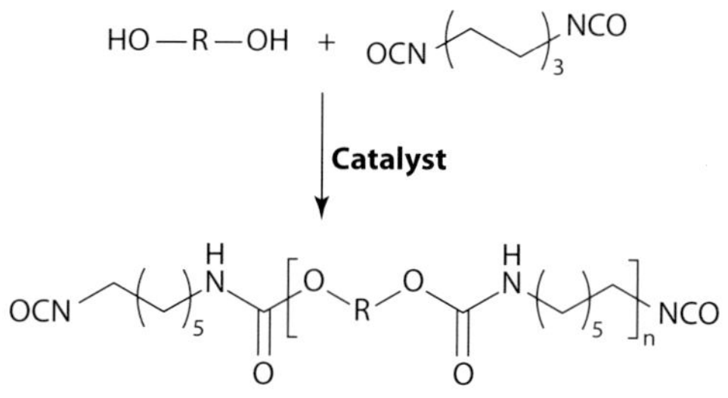

FIGURE 7.8 Reaction of polymer diols (R=polymer) with diisocyanates to form PUs.

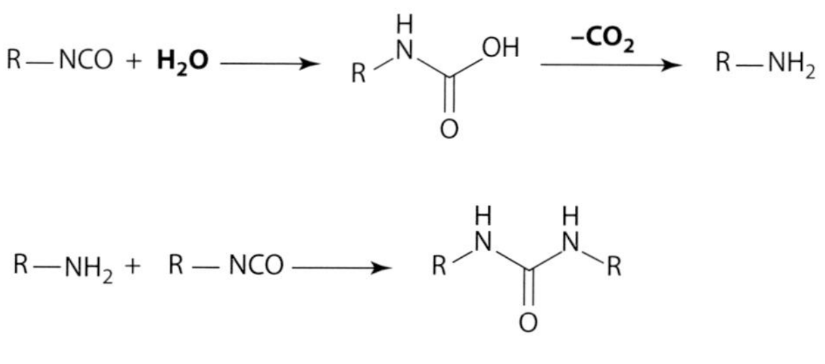

FIGURE 7.9 Reaction of isocyanate-terminated polymer with moisture.

that leads to significant health hazards. Companies synthesizing these prepolymers are currently try-ing to minimize the issue, by using various distillation processes to remove most of the monomeric diisocyanate and keep the NCO content minimal [19]. The choice of catalyst in the case of the 1c polyurethanes is similar to that described in Section 7.4.2.1 for the 2c systems. The same rules apply to these formulations as well, with the only exception that the right choice of catalyst is more critical in the 1c systems, as it directly affects the end viscosity of the prepolymer. In particular, the trimer-ization side reaction leading to the isocyanurates needs to be inhibited by all means. As the synthesis of the prepolymer occurs at elevated temperatures, catalysts are chosen with respective reaction temperatures in the range of 50°C–100°C, so as to ensure a fast reaction and a high selectivity for urethane formation. Such catalysts are mostly Lewis acids, with the most prominent example being dibutyltin laurate (DBTL), which shows a remarkable selectivity at higher temperatures. To control the resulting viscosity, chain-terminating agents such as benzoyl chloride can be used [20].

Finally, catalysts are not only needed for the synthesis of the prepolymer, but also for the final curing of the NCO functionalities with moisture. Fortunately, the same catalytic systems are also effective in this step, so that organotin catalysts are the general catalyst choice.

The formulation components of a 1c PU adhesive are very similar to the ones already described for the 2c systems. Here, too, fillers, plasticizers, and rheological additives are used to tune the adhesive physical, mechanical, and chemical properties. However, layered fillers, such as carbon black and talcum, are preferred, as they not only increase the final toughness of the adhesive, but also suppress the formation of carbon dioxide bubbles, which leads to weak bonds.

7.4.2.3 Pseudo 1c PU Systems

The formation of gas bubbles during curing of polyurethanes leading to nonuniform surfaces is a significant problem with this class of materials. The PU and paint industry, therefore, invested early enough in the development of latent systems whereby the NCO-terminated prepolymers are mixed with latent hardeners such as oxazolidine [21], polyketimines, and polyaldimines [22]. On contact with moisture, these materials release hydroxyamines and polyamines that subsequently react with the prepolymers to further cross-link the system (Figure 7.10).

One significant advantage of these systems is that, due to the increased reactivity of the hydroxy-amines and polyamines, PU prepolymers that are end-capped with the toxicologically less severe aliphatic isocyanates can be used. However, the resulting by-products after curing, such as formal-dehyde, aldehydes, and ketones, are again associated with toxicity issues, low vapor pressures, and unwanted smells, so that they negate their original advantage.

Further developments in this field have recently resulted in nonvolatile aldehydes as by-products that remain in the system without any negative effects, making this again very attractive and user-friendly [23].

7.4.2.4 Surface Treatment for PU Adhesives

Prior to each application of PU adhesives, the surfaces need to be particularly clean and free from moisture, dust, and grease. Depending on the surface, a number of different processes are described in the literature, from simple treatment with solvent to more complicated cleaning systems. The trend here is also shifting to more environmentally friendly solutions such as the safer water-based

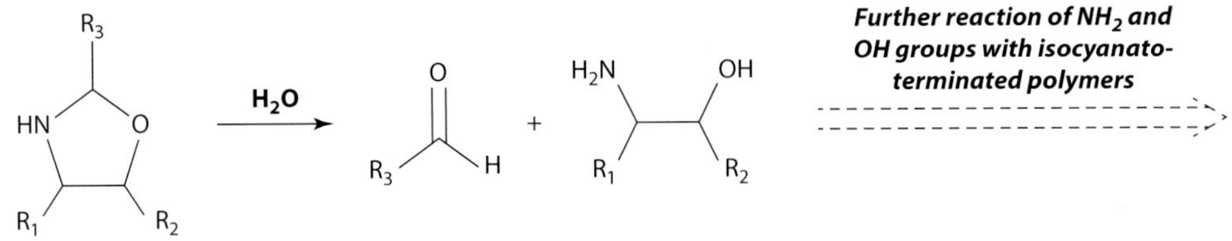

FIGURE 7.10 Hydrolysis of latent hardener oxazolidine with moisture to produce hydroxyamines.

systems. These systems are, in principle, alkaline solutions containing grease- and dust-dissolving surface-active substances.

Another important pretreatment step to enhance adhesion of elastic adhesives is the chemical treatment of the surface through the use of a primer. These primers usually consist of an adhesion promoter that displays good film-forming property and a good surface wettability to the respective substrate. They are mostly dissolved in a liquid or solvent (predominantly organic solvents) in low concentration and applied directly on the substrate surface to form a very thin film. The role of the adhesion promoter in the primer is, on the one hand, to act as an anchor to the surface and, on the other hand, to build up a strong interaction with the adhesive layer on top. The type of primer required depends strongly on the nature of the substrate that needs to be bonded.

The most common primers are based on organofunctional silanes. The silane moiety acts like an anchor and adheres to the metal surface, whereas the organofunctionality forms the link to the adhesive layer. Some examples of such silanes are depicted below (Figure 7.11) [24].

Also, there is a clear trend to move away from solvent-based primers to more environmentally safe water-based systems. Water-based primers are, usually, silanes that have been precondensed in slightly acidic water to become soluble [25].

A different type of primer has also recently been used in the bonding of glass with PU adhesives. So-called *black primers* are used that include, apart from a film-forming polymer, carbon black, which aids in blocking out the light and increasing the UV stability of the bond formed. These primers contain mostly isocyanate components and are also applied as a solvent-based mixture of low concentration [26]. In this type of primer, the role of the anchor to the substrate is assumed by different adhesion-promoting functional groups such as, for example, heterocyclic moieties [27]. Isocyanato-based primers can also be partially cross-linked with hydrogenated polyisoprene diols, to form polyisocyanato primers. These types of primers are mostly used on nonpolar substrates [28].

7.4.3 SILANE-MODIFIED POLYMERS

In the last few decades, silane-modified polymer (SMP) technology has entered the field of elastic adhesives with an impressive momentum. Organic polymers (polyethers, PUs, and acrylates) are functionalized with alkoxysilane groups that enable them to cure on contact with moisture, similarly to the RTV1 systems in silicone chemistry, to release the respective alcohol (most commonly methanol) (Figure 7.12).

A significant advantage of such systems is that the cross-linking points needed to ensure good elasticity are formed during the curing mechanism of the adhesive, and do not need to be inserted in the prepolymer, as is the case in the 1c PUs. This way, high viscosity and the associated difficult workability are successfully avoided.

In addition, SMPs are also capable of bubble-free curing, good adhesion to a broad spectrum of substrates, and have no considerable health concerns, making them very attractive for most users.

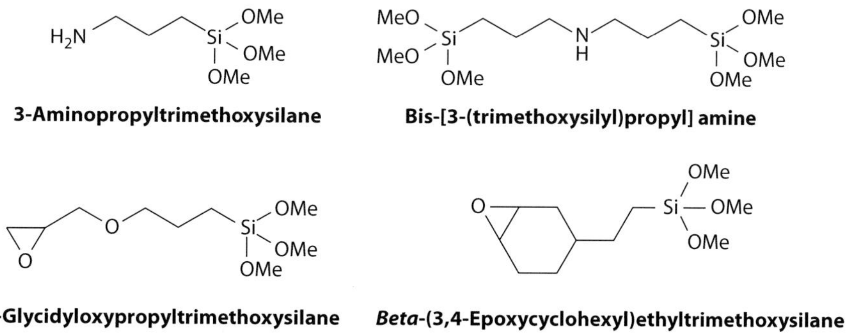

3-Aminopropyltrimethoxysilane   **Bis-[3-(trimethoxysilyl)propyl] amine**

3-Glycidyloxypropyltrimethoxysilane   *Beta*-**(3,4-Epoxycyclohexyl)ethyltrimethoxysilane**

FIGURE 7.11 Examples of structures of amino- and epoxysilanes used as primers.

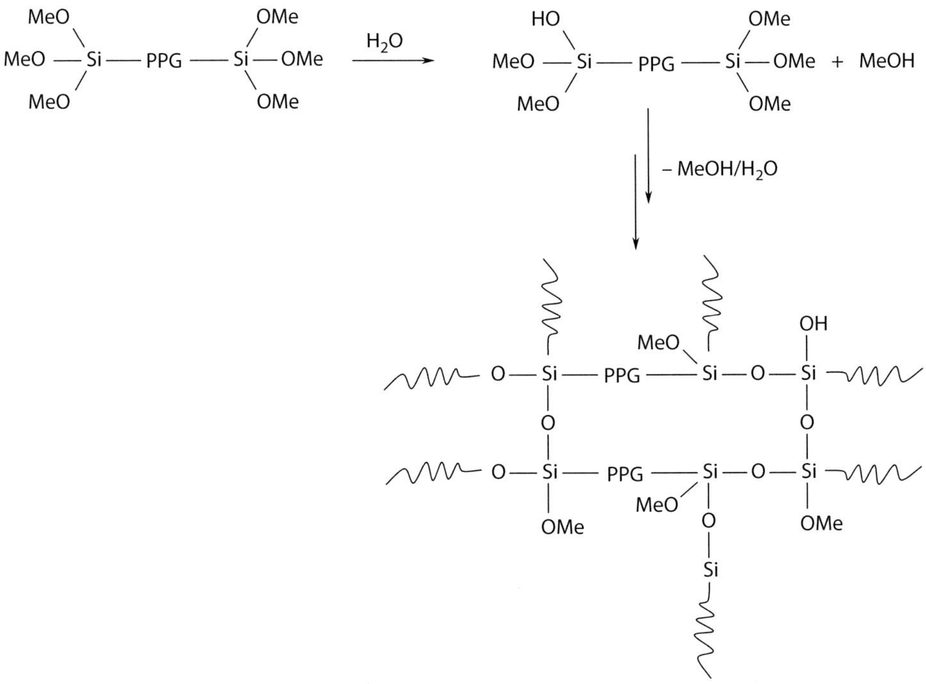

FIGURE 7.12 Curing reaction of a polyether SMP (PPG backbone) with moisture.

The only drawback is the release of small amounts of methanol in the atmosphere during curing, and that, compared with polyurethanes, the SMP prepolymers are more expensive. The higher cost of the SMPs is due to their slightly more elaborate synthesis and the higher cost of raw materials.

7.4.3.1 Silane-Terminated Polyethers

In 1976 Kaneka disclosed a process to convert propylene oxide to allyl terminated PPGs. These PPG polymers could then be reacted with methyldichlorosilane through hydrosilylation and subsequently with methanol to yield silane modified polymers (Figure 7.13) [29].

These first SMPs were used in the 1990s as binding agents for sealants in the construction of earthquake-proof high-rise buildings in Asia. Elastic adhesives containing branched silane- terminated polyethers (STPEs) were introduced to the market much later. By combining these SMPs with side-chain-functionalized acrylates, novel elastic adhesives could be prepared displaying a higher modulus and a better weather resistance [30].

SMP synthesis, involving propoxylation, a chain-extension step, etherification with allyl chloride, and a final hydrosilylation step, has the advantage that the polymerization of propylene oxide can also, in principle, occur using conventional basic catalysts. The allyl-terminated monools that are formed as by-products via the rearrangement of propylene oxide to allyl alcohol are converted directly after chain extension to the desired allyl-terminated polymer, so that at least the etherification step can be partly avoided. The development of more efficient catalysts for this process, such as double metal cyanide (DMC) catalysts, also enable the preparation of higher molecular weight PPG polymers with a shorter reaction time. The long-chain PPG polymers that became

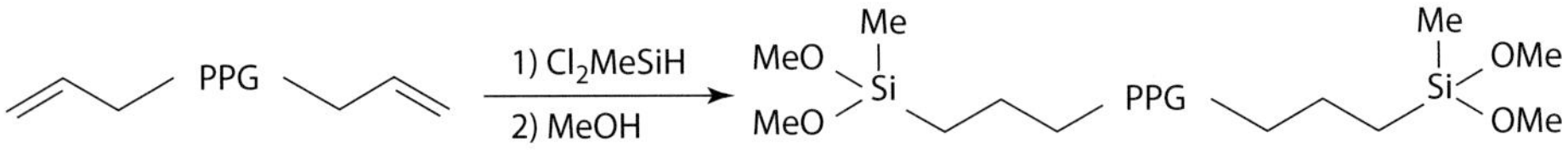

FIGURE 7.13 Hydrosilylation reaction to convert allyl-terminated poly(propylene glycol) to silane-terminated polymers.

available through DMC catalysis toward the end of the 1990s, as well as the availability of isocyanatopropyltrimethoxysilane, helped the development of SMPs significantly [31].

The DMC catalysts that sped up the progress in the SMP field had been known since the 1960s [32]. They are prepared through the reaction of $K_2Co(CN)_6$ and $ZnCl_2$ in the presence of special ligands and have the structure shown in Figure 7.14.

The reactivity and selectivity of the DMC catalyst can be enhanced by replacing the originally used glyme ligand with *tert*-BuOH, whereby PPGs with a molecular weight of up to 20,000 g/mol can be successfully prepared. Through this process, the amount of monool impurities is also kept significantly low.

At approximately the same time as these advancements, isocyanatopropyltrimethoxysilane also became available in the market. This very sensitive compound can be synthesized through thermolysis of carbamatosilane, which, in turn, is prepared either by reacting aminosilane with dimethyl carbonate or, alternatively, through the reaction of chloropropyltrimethoxysilane with potassium isocyanate (Figure 7.15).

The thermolysis of carbamatosilane occurs at temperatures of 250°C–350°C and is a technically challenging process, as the competing reaction of methanol back to the carbamatosilane needs to be inhibited. Alternatively, the so-called *hot oil process* offers a solution to this problem by adding the carbamatosilane dropwise to the hot carrier and directly removing the methanol through distillation [33]. In the more recently developed processes, the thermolysis step can occur at 500°C in a plug flow reactor [34]. A milder possibility is to carry out the thermolytic step in a thin film evaporator in the presence of catalysts, to minimize the time that the thermally sensitive silane group is exposed to heat [35].

Isocyanatosilane can then react with the long-chain PPG through a simple PU reaction to form a urethane link and, thus, the desired STPE. Such silane-modified polyethers are readily available in the market in a molecular weight range of 12,000–18,000 g/mol. Nevertheless, only a handful of companies are able to produce such SMPs, the reason being the high cost associated with the elaborate synthesis of the isocyanatosilane, as well as its difficult handling due to its toxicity.

In a slightly modified procedure, PPG diols of medium chain length with low unsaturation can first be chain extended with a diisocyanate to OH-terminated PU and then end-capped with an isocyanatosilane [36]. These additional PU groups in the chain allow for a higher modulus in the final

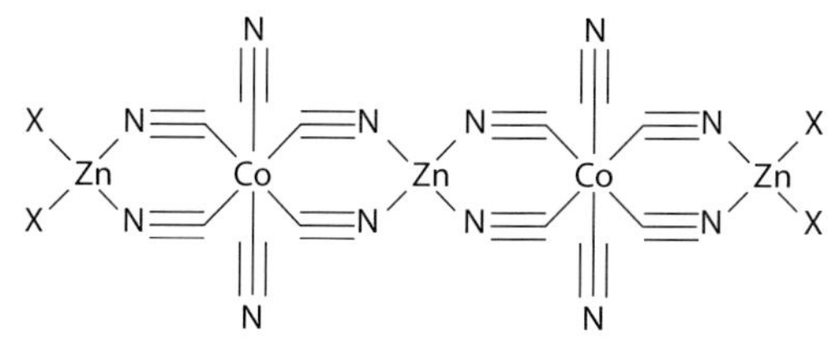

FIGURE 7.14   General structure of DMC catalyst (X = ligand).

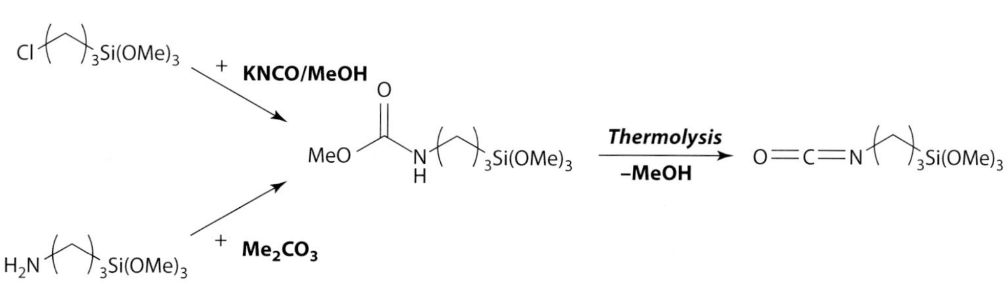

FIGURE 7.15   Synthesis routes to isocyanatopropyltrimethoxysilane.

elastic adhesive; the prepolymer is, however, also associated with higher viscosity, due to hydrogen bond interaction.

The next leap in the advancement of SMPs happened with the development of isocyanatomethyl-dimethoxymethylsilane. This isocyanatosilane is synthesized through photochlorination of dimethyldichlorosilane [37] (one of the most important silicone raw materials) to dichloro(chloromethyl) methylsilane. Esterification of this precursor with methanol to chloromethylsilane, followed by reaction with potassium cyanate in the presence of methanol to carbamatosilane, and after pyrolysis with one of the processes mentioned, leads to the desired silane (Figure 7.16).

By end-capping PPG polyethers using this type of silane, a new class of silane-terminated polymers are obtained that seem to cure with moisture significantly faster. This observation is called the *alpha-silane effect* [38] (Figure 7.17).

Nevertheless, the progress of SMP materials prepared through the hydrosilylation route did not remain stagnant either. A number of trimethoxysilyl materials with improved curing speed are already in the market and, judging from the developments described in the recent patent literature, methoxymethyldimethoxy-terminated polyether, meant to cure even faster, should also be available soon [39].

Finally, reactive silane groups can be attached to the polymer, not only via hydrosilylation and reaction with isocyanatosilane, but also via a third route, the copolymerization of epoxysilanes. By employing DMC catalysis, 3-glycidoxypropyltriethoxysilane can, for example, be reacted with propylene oxide [40] (Figure 7.18).

To concentrate the reactive silyl groups at the ends of the polymer chain, a two-step process is used. First, the PPG is homopolymerized, and in a second step a mixture of propylene oxide and epoxysilane is grafted onto the chain. One advantage of this type of SMP is the release of the less toxic ethanol on curing of the polymer, as opposed to methanol.

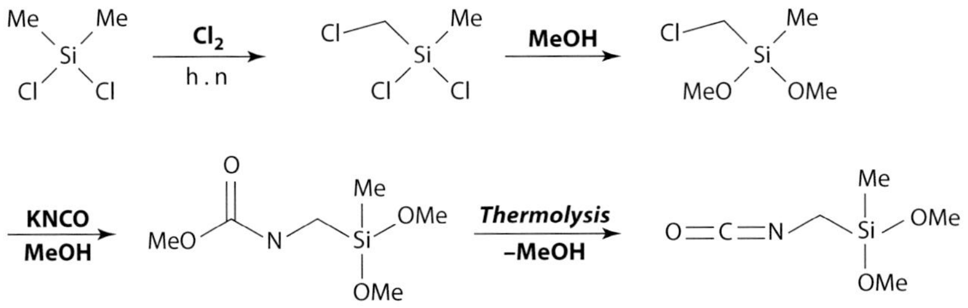

FIGURE 7.16   Synthesis of isocyanatomethyldimethoxymethylsilane.

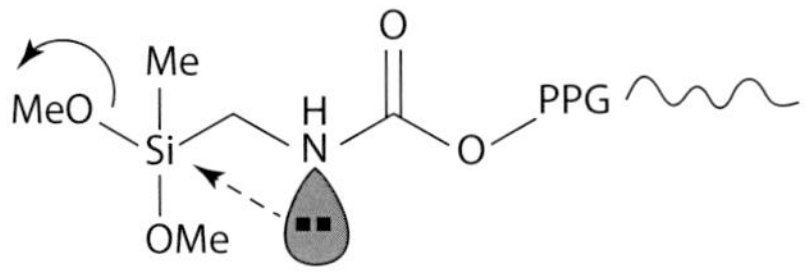

FIGURE 7.17   Alpha-silane effect.

FIGURE 7.18   Copolymerization of propylene oxide and 3-glycidoxypropyltriethoxysilane for the synthesis of multi-silane-functionalized polyethers.

7.4.3.2 Silane-Terminated PUs

In 1968, Union Carbide described for the first time the synthesis of silane-terminated PUs (SPUs) by reacting NCO-terminated PU prepolymers with aminopropyltrialkoxysilane (Figure 7.19).

At that time, only conventional polyols were available for SMP synthesis; the same polyols that were also used for the 1c PUs. The limitations in the possible molecular weights, as well as the high content of impurities that inhibit chain propagation present in those polyols, made the development of novel polymers with interesting properties very difficult. Furthermore, to completely end-cap the short-chain, low M_w NCO prepolymers and large amounts of the expensive aminosilane had to be used, which made these first-generation silane-terminated polymers very costly. The result was that in those first years, these polymers only found application in niche markets that were able to absorb the cost such as, for example, primers and high-priced coatings.

Only in 1992 did Henkel describe a mixing process that enabled the synthesis of higher molecular weight polymers with better elastic properties. By mixing diols and triols, an average functionality of 2 could be achieved [41]. Similarly to the STPEs, in the case of the SPUs huge progress was made once DMC-based polyols became available. The synthesis of higher molecular weight SPU polymers led to the formulation of products with superior mechanical properties [42]. These properties could be further enhanced by incorporating hard sections into the polymer structure; for example, poly(tetrahydrofuran) polymer segments [43].

One main drawback of the this process for synthesizing SPUs is the observed high viscosity that results from the numerous urea links in the polymer and the associated hydrogen bonding. In the last few decades, a number of concepts have been devised to solve this viscosity issue. One very elegant approach is to reduce the hydrogen bonding in the system by using secondary aminosilanes to only produce substituted urea links. The most commonly used secondary amine (one of the first ones to be patented) is bis-3-aminopropyltrimethoxysilane [44]. Further examples of such structures include phenylaminopropyltrimethoxysilane [45], cyclohexylaminopropyltrimethoxysilane, and butylaminotrimethoxysilane. Substituted aminosilanes with branched alkyl chains can also be used [46].

Nevertheless, the easy access to 3-aminopropyltrimethoxysilane has led many companies to establish it as a raw material for the *in situ* synthesis of further substituted aminosilanes. In a very simple process, 3-aminopropyltrimethoxysilane is reacted via a Michael addition with maleic acid esters and subsequently used directly as an end-capping agent [47]. A further cyclization to hydantoin that can occur during this reaction can also lead to a further reduction of the hydrogen bonding and the resulting viscosity [48] (Figure 7.20).

The Michael addition step is also utilized in a different process, where primary aminosilanes are reacted with acrylates to form secondary aminosilanes that are then used to end-cap NCO-terminated prepolymers [49].

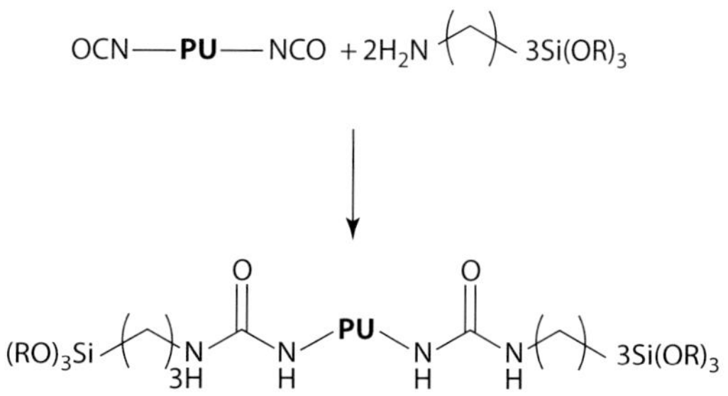

FIGURE 7.19 Synthesis of silane-terminated polyurethanes.

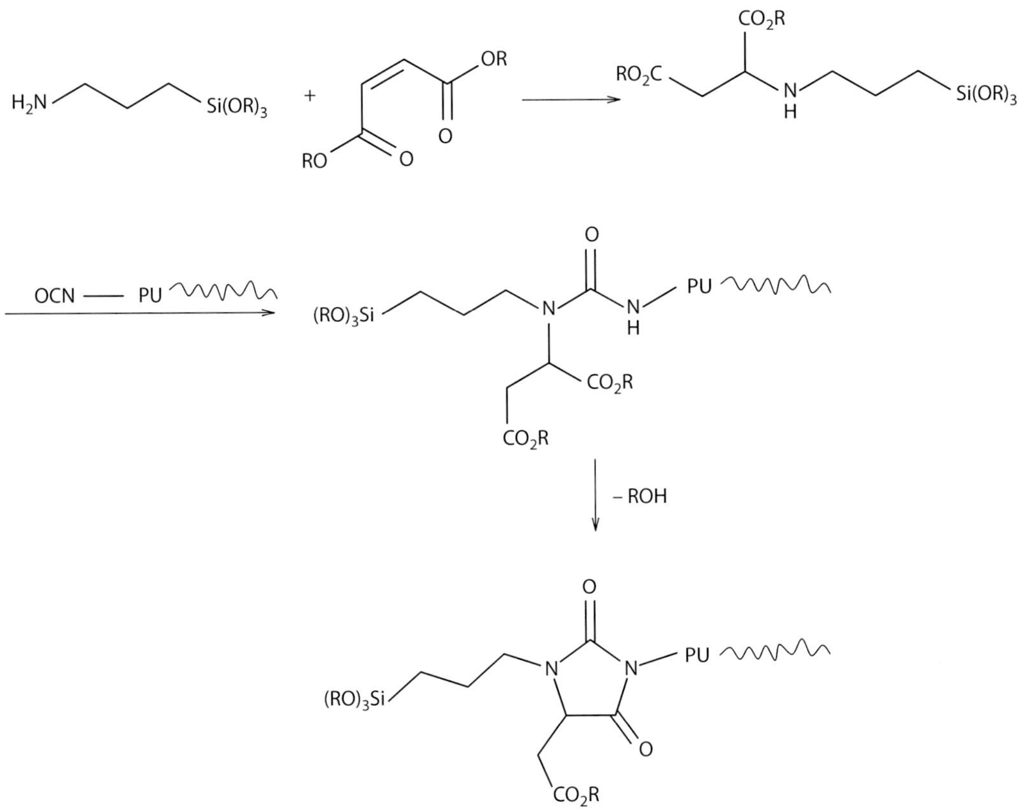

FIGURE 7.20 End-capping reaction of an isocyanate using a secondary amine and further cyclization to hydantoin.

7.4.3.3 Silane-Terminated Acrylates

Apart from the SPUs and STPEs described, silylated polyacrylates have also recently started gaining popularity. Acrylates are, in general, known for their good weathering resistance, low discoloration, and excellent adhesion. In the past, a number of different processes have been described for the preparation of vinyl polymers with statistically distributed silyl groups on the backbone. In one of the oldest processes, acrylate or methacrylate polymers with pendent OH groups were reacted with isocyanatosilane [50].

A different route describes the synthesis of such polymers through the radical copolymerization of acrylates with silane-functionalized acrylates; for example, methacryloxypropyltrimethoxysilane. However, the statistically uneven distribution of the silane groups in such prepolymers leads to material properties such as brittleness and tear propagation that render the cured polymer inapplicable for elastic bonding. Nevertheless, mixtures of such materials with more elastic prepolymers such as STPE display a number of positive effects [51].

Silane-terminated acrylates were described for the first time at the end of the 1990s. They were synthesized through atom-transfer radical polymerization (ATRP) of acrylates and other vinyl polymers to form OH-terminated acrylates that could then be reacted with isocyanatosilane, as described. The patent disclosing this method also describes the synthesis of similar acrylic polymers prepared through the ATRP process that are alkenyl terminated and are subsequently reacted via hydrosilylation with methyldimethoxysilane to obtain the respective silane terminated polymers [52].

7.4.3.4 SMP Formulations

Formulations of elastic adhesives that are based on SMP technology can be quite varied. The most important component after the polymer is the catalyst used. The catalyst is responsible for determining the hydrolysis and condensation rates of alkoxy (mostly methoxy) leaving groups and thus accelerates the curing. The most commonly used catalysts are tin organic complexes; for example, dibutyltin dilaurate. The catalyst type and the necessary concentration are directly dependent on the nature of the SMP. For instance, in the case of γ-methyldimethoxysilane groups, a high concentration of the catalyst is needed to compensate for the low reactivity of the end-capper. Apart from tin organic molecules, there are a number of other metal-based catalysts that can also be used in these formulations, such as, for example, titanium esters [53] and bismuth compounds. Also, Lewis acids such as BF_3 [54] and organic bases such as polycyclic amines, amidines, and guanidines [55] have also been reported.

Apart from the polymer and the catalyst, a typical SMP formulation also contains adhesion promoters, to enhance the adhesion to different substrates. As an example, aminosilanes were one of the first adhesion promoters for silane-modified PUs to be described. Finally, an elastic adhesive would not be complete without the addition of fillers and plasticizers to improve its rheology and mechanical properties and, of course, the use of stabilizers to ensure good stability, shelf life and UV resistance.

7.5 MARKET VALUATION AND APPLICATIONS

The global market for elastic adhesives in industrial applications was estimated for 2015 at approximately 527 kt (with a value of US$3.8 billion). North America is the market leader with a share of 200 kt (US$1.4 billion), followed by Europe (169 kt, US$1.34 billion) and Asia-Pacific (124 kt, US$0.82 billion) [56].

The oldest and most important industrial application of elastic adhesives is the bonding of front windshields in the automotive industry. This application is a prime example of how elastic bonding has replaced the previously used mechanical fixing of such glass shields. Elastic bonding has managed, in this application, to not only prolong the shelf life of the glass-to-car body joints, but has also shortened the assembly time, enhanced the noise-damping properties, and reduced the vibrations. This so-called *direct glazing process* was first introduced in this industry in 1963 by General Motors. In Europe, the process was first used by Audi in the 1980s to make the streamline optimized model Audi 100. Up until now, 1c PUs have mainly been used in this application, due to the reduced cost and the significant optimization of their mechanical performance and their production process (*high position tack*, fast curing) that has occurred in the last years. Nevertheless, in certain markets, SMP systems are also now becoming more popular due to their reduced chemical hazards compared with alternative technologies.

Windshields are not the only example of elastic bonding in the automotive industry. The side panels in buses and train cars are also bonded using PU adhesives, as well as headlights, spoilers, and various metal panels. In principle, elastic adhesives are mostly applied to join parts that are exposed to vibration and multiple mechanical strains. This is not only applicable in the automotive industry. In the white goods industry also, elastic adhesives are becoming more popular. Apart from the good adhesive function of these materials, their vibration and noise-damping properties play a significant role in their choice, especially in the assembly of refrigerators, washing machines, and dishwashers. Elastic adhesives based on SMP and PU formulations are also used in loudspeakers, where the joints need to withstand strong vibrations.

Apart from the industrial applications, an even bigger field for elastic adhesives is construction. The global market for construction applications was estimated for 2015 at approx. 780 kt (total value US$5.6 billion) [51]. Here, Europe holds the biggest market share with 263 kt (US$2 billion) followed by North America (227 kt, US$1.61 billion) and then Asia-Pacific (208 kt, US$1.37 billion).

The trend here is similar to the automotive applications, with structural glazing of complicated glass constructions being one of the most important fields. The constant developments of new materials open up enormous architectural possibilities [57]. However, tailor-made adhesives that are robust and able to withstand forces such as strong winds, heat, and earthquakes are continuously required. Silicone adhesives are the most dominant technology that addresses these issues, due to their flexibility and excellent heat and moisture resistance. In this field, physically setting adhesives, with formulations based on latex, polystyrene, and PUs also claim a significant portion of the market, especially in drywall, flooring, and roofing applications.

One further important application is the assembly of solar panels, by inserting silicon wafers into metallic constructions. For these applications, mostly SMP adhesives are used, due to their good and broad adhesion spectrum [58]. One of the recent innovations in this field are SMP systems with improved initial adhesion (semi-hot-melt). Another option is the cheaper, silicone-based systems that are used mainly in China.

Another field in construction applications where elastic adhesives are gaining popularity is flooring. Currently, in Europe, the application of parquet floors on large areas using elastic adhesives has become state of the art. SMP technology is almost exclusively being used for these flooring applications, particularly as it is a 1c, very easy to apply adhesive and, more importantly, it is nonhazardous and carries no GHS pictogram. SMP-based adhesives can now be found in great variety on the market, and their unique properties become especially apparent in flooring applications. Elastic parquet adhesives can absorb perfectly well any wood movement that may occur due to temperature and moisture differences, so that any cracking can be prevented. In addition, their noise-damping property aids significantly in noise regulation, especially in apartment buildings. In the North American and Asia-Pacific markets, elastic floor adhesives are not yet very dominant, mainly due to the already-established mechanical assembly processes. Small surface needs in flooring applications are then addressed with solvent-based PU adhesives. Nevertheless, a rise in SMP technology needs is expected even in North America, as the trend for more environmentally and user-friendly products grows.

Recently, SMP elastic adhesives have even been reported in tile applications [59]. Standard cheap cement-based products are normally most commonly used for such applications; however, elastic adhesives are superior in absorbing potential strains (e.g., bonding tiles to wooden floors).

Finally, assembly adhesives are one more application of elastic adhesives in the construction field. Such adhesives usually come in cartridges, and can be easily applied to fix panels, mirrors, baseboards (skirting boards), and so forth, where previously elaborate mechanical processes were needed to assemble them [60]. One-component systems based on PU, pseudo 1c PUs, and SMPs are used predominantly. In Europe, SMP systems dominate in this field, as they are nonhazardous and user-friendly and can be universally applied.

The SMP trend has not caught up yet in North America and Asia-Pacific, where the cheaper solvent-based PUs are still in use. In the last few years, transparent assembly adhesives have claimed a significant portion of the European market. These are used frequently with transparent substrates such as glass, acrylics, and polycarbonate, predominantly in the do-it-yourself (DIY) field.

7.6 PROSPECTS

Elastic bonding is expected in the coming years to gain significant importance. At the same time, the trend seems to be shifting from the well-established PUs to the emerging SMP adhesives, particularly because of their improved safety, user-friendliness, and broad application field in both professional and DIY circles. Nevertheless, these systems still need to improve in cost-effectiveness, as there is still a significant cost gap between SMPs and PUs. The higher formulation cost of these SMPs stems from the use of organofunctional silanes. Due to this limiting factor, it is expected that, in the near future, in industrial applications, PU-based adhesives are going to continue to be the preferred choice, as long as the necessary technical equipment is in place to ensure the safe handling and elaborate surface treatment required for this technology. In the field of glass constructions, silicone-based adhesives are also going to continue to play an important role, as their weather resistance is excellent.

To keep up with these trends, the research and development (R&D) efforts in the field of SMPs are, at the moment, focused on not only reducing the production costs of these adhesives, but also on improving their mechanical performance, so that they can enter structural applications in the future as well. Some very interesting approaches have already been described in this direction, where SMPs are combined with silicone resins for improved tensile strength [61]. Efforts to increase the water resistance of SMP adhesives are also currently underway, so that the technology can enter the field of elastic wood bonding as well.

In line with the global focus to reduce the amount of toxic and harmful materials, and the stricter EU regulations, a huge R&D initiative is currently focusing on replacing the organotin catalysts, currently present in the formulations in low levels, with less toxic materials. Under the same initiative, aiming at greener and more environmentally friendly products, a lot of work is currently being done to also reduce or eliminate the volatile organic content (VOC) issues currently being debated due to methanol release [62].

Finally, the increased patent activity in the field of SMPs in the last years with a significant output in innovation supports the prediction placing SMP technology as a key future technology in elastic applications.

7.7 SUMMARY

Within the past few years, the importance of elastic bonding has increased significantly, particularly due to its ability to absorb stress, noise, and vibrations. Four main polymer classes that seem to complement each other in terms of properties have provided, so far, the required elasticity for the current elastic applications, namely, solvent-based rubbers, polyurethanes, silicones, and the newly developed SMPs. This chapter has introduced the main chemistry of each technology, as well as its advantages and current limitations. There is a huge variety of differently designed polymers and formulations that the user can choose from to find the optimum adhesive, tailored for each application, substrate, and budget. Nevertheless, as the trend moves to more sustainable, easier to handle, and more robust products, these technologies also will evolve to bring more sophisticated solutions to the market. This chapter serves, therefore, not only as a summary of the current state-of-the-art elastic technologies, but also as a glimpse into the huge potential and versatility these chemistries are offering.

ACKNOWLEDGMENTS

We would like to thank Dr. Therese Hemery, Dr. Jan-Erik Damke, Dr. Horst Beck, Dr. Uwe Franken, Dr. Matthias Kohl, Dr. Sebastien Lanau, Dr. Henry Ashton, and Cormac Duffy for their input and valuable contributions to this chapter. We would also like to extend our sincere gratitude to William F. Harrington, whose very informative and interesting chapter on elastomeric adhesives in the second edition of the *Handbook of Adhesive Technology* served as the inspiration for our work here.

REFERENCES

1. M. Pröbster. *Elastisch Kleben: Aus der Praxis für die Praxis*, pp. 35–42, Springer Vieweg, Wiesbaden, Germany (2013).
2. R. A. Shanks and I. Kong. General purpose elastomers: Structure, chemistry, physics and performance. In: *Advances in Elastomers I: Blends and Interpenetrating Networks*, P. M. Visakh, S. Thomas, A. K. Chandra and A. P. Mathew (Eds.), p. 494, Springer Berlin, Germany (2013).
3. M. Pröbster. *Kompaktlexikon Dichtstoffen und Fugen*, pp. 49–51, Frauenhofer IRB Verlag, Stuttgart, Germany (2010).
4. G. Alliger and F. C. Weissert. Elastomers. *Indust. Eng. Chem.*, 59 (8), 80–90 (1967).
5. B. R. Burchardt and P. W. Merz. Elastic bonding and sealing in industry. In: *Adhesives and Sealants: General Knowledge, Application Techniques, New Curing Techniques*, Vol. 2, P. Cognard (Ed.), pp. 355–480, Elsevier Oxford, UK (2006).

6. H. Onusseit. *Adhesive Technology: Basic Principles*, 1st edn, pp. 17–31, DIN Deutsches Institut für Normung e.V. and Henkel AG & Co. KGaA (Eds.), Beuth Verlag GmbH, Berlin, Germany (2012).

7. British Plastics Federation. "Thermoplastic elastomers, TPE,TPR." http://www.bpf.co.uk/plastipedia/polymers/thermoplastic_elastomers.aspx

8. W. F. Harrington. Elastomeric adhesives. In: *Handbook of Adhesive Technology*, 2nd edn, A. Pizzi and K. L. Mittal (Eds.), pp. 515–530, Marcel Dekker, New York (2003).

9. L. D. Bennington, H. K. Chu and R. P. Cross. Radiation- and/or moisture-curable silicone compositions. EP 0772636, assigned to Loctite Corporation (1995).

10. F. Bohin, G. Joubert, O. Loubet, A. Pouchelon and D. Lorenzetti. Cross-linkable adhesive silicone composition and use of said composition for bonding various substrates. US Patent 6562180, assigned to Rhodia (1999).

11. Polyurethanes. Polyurethanes and their application. www.polyurethanes.org.

12. M. Pröbster and M. Schumann. Adhesive/sealant material. EP 705290, assigned to Henkel (1993).

13. J. Klein, H. Horskorte and B. Beuer. Verwendung partiell dehydratisierter Ricinusöle in 2-komponentigen Polyurethan-Klebemassen. DE 4114022, assigned to Henkel (1991).

14. L. Thiele and R. Becker. Catalytic mechanisms of polyurethane formation. In: *Advances in Urethane Science and Technology*, Vol. 12, K. C. Frisch and D. Klempner (Eds.), pp. 52–89, Technomic Publications, Lancaster, PA (1993).

15. R. Cannas and R. U. Burckhardt. Zink(II)-Komplexverbindungen als Katalysatoren für Polyurethan-Zusammensetzungen. EP 2604613, assigned to Sika (2013).

16. M. Kreyenschmidt and E. Jahns. Catalyst system containing catalysts encapsulated in wax. EP 1335944, assigned to BASF (2003).

17. C. Kirsten, G. Henke, C. Meckel-Jonas, L. Unger, F. Meier, T. Schmidt and T. J. Aquarius. Chemically reactive adhesive comprising at least one micro encapsulated component. EP 1313812, assigned to Henkel (2008).

18. U. Meier-Westhuis. *Polyurethanes Coatings Adhesives and Sealants*, pp. 248–251, Vincentz Network Verlag, Hanover, Germany (2007).

19. M. Krebs, K. Brosa, A. Brenger, U. Franken and C. Lohr. Reactive polyurethane compositions with low residual monomer content. EP 1434811, assigned to Henkel (2002).

20. M. Wintermantel, W. Meckel, M. Matner, H. Kraus and F. Kobelka. Polyurethane prepolymers with low viscosity based on 2,4'-MDI. EP 1619215, assigned to Bayer Material Science AG (2006).

21. M. Hajek and K. Wagner. Urethane oxazolidines. US Patent 4002601, assigned to Bayer AG (1977).

22. G. A. Haggis. Polymeric materials produced by interacting polyisocyanate and water in the presence of polyaldimine or polyketimine. US Patent 3420800, assigned to ICI (1964).

23. U. Burckhardt. Feuchtigkeitshärtende Polyurethanzusammensetzungen enthaltend Aldimin-haltige Verbindungen. EP1937741, assigned to Sika Technology AG (2008).

24. M. Huang and E. R. Pohl. Organofunctional silanes for sealants. In: *Handbook of Sealant Technology*, K. L. Mittal and A. Pizzi (Eds.), Chapter 2, CRC Press, Boca Raton, FL (2009).

25. P. A. Lorrach and E. Just. Aqueous silane systems based on tris(alkoxysilylalkyl)amines and the use thereof. US Patent 8772432, pp. 27–53, assigned to Evonik Degussa (2009).

26. M. Pröbster. *Elastisch Kleben: Aus der Praxis für die Praxis*, pp. 162–167, Springer Vieweg, Wiesbaden, Germany (2013).

27. M. Nakabayashi, M. Miyaji and Y. Kamatani. Primer for bonding polyester plastics. US Patent 4495020, assigned to Takeda Chemical Industries, Ltd. (1985).

28. P. W. Merz and S. Tsuno. Use as a Primer. EP 1149856, assigned to Sika (2005).

29. K. Isayama, I. Hatano. Vulcanizable silylether terminated polymer. US Patent 3971751, assigned to Kanegafushi Kagaku Kogyo Kabushiki Kaisha (1976).

30. T. Hirose, K. Isayama. Curing composition containing polyether having reactive silicone-containing group and a (meth)acrylate polymer. US Patent 4593068, assigned to Kanegafushi Kagaku Kogyo Kabushiki Kaisha (1983).

31. T. Doi, T. Watabe, T. Matsumoto, T. Onoguchi, K. Tsuruoka. Room temperature-setting compositions. Us Patent 6207766, assigned to Asahi Glass Company Ltd. (2001).

32. M. Ionescu. Synthesis of high-molecular weight polyether polyols with double metal cyanide catalysts. In: *Chemistry and Technology of Polyols for Polyurethanes*, 2nd edn, Vol. 1, pp. 177–196, Smithers Rapra Technology, Shawbury, Shrewsbury, Shropshire, UK (2016).

33. R. E. Sheridan and K. W. Hartman. Hot oil process for producing isocyanato organosilanes. US Patent 6008396, assigned to OSI Specialties (1998).

34. T. Korneck. Process for preparing organosilicon compounds containing isocyanate groups. US Patent 6979745, assigned to Wacker (2004).

35. V. Stanjek, F. Baumann and T. Frey. Process for preparing isocyanato organosilanes. US Patent 8158818, assigned to Wacker (2007).

36. R. Johnston and P. Lehmann. Compositions of silylated polymer and aminosilane adhesion promoters. EP 1216263, assigned to Crompton Corporation (2000).

37. R. Müller and R. Köhne. Über Silikone. LXIV. Elementares Fluor als Radikalspender bei der Chlorierung von Methylchlorsilanen. *J. Praktische Chemie* 21, 163–167 (1963).

38. Wacker Chemie AG. Geniosil®, Organofunctional silanes. http://www.wacker.com/cms/media/publications/downloads/6085_EN.pdf. accessed (2015).

39. K. Miyafuji, T. Fujimoto, K. Wakabayashi and T. Okamoto. Curable composition. EP 2604655, assigned to Kaneka (2015).

40. B. M. Brugger, F. Schubert and M. Lobert. Alkoxysilyl containing adhesive sealants having an increased rupture stress. WO 2012130674, assigned to Evonik (2012).

41. W. Emmerling and T. Podola. Alkoxysilane-terminated, moisture-hardening polyurethanes and their use in adhesives and sealing compositions. US Patent 4857623, assigned to Henkel (1987).

42. W. Klauck, L. Duhm and M. Majolo. Polyurethan und Polyurethanhaltige Zubereitung. EP 1093482, assigned to Henkel (2001).

43. T. Bachon, J. Lambertz, A. Ferencz, F. Koepnick and M. Majolo. Polymers with improved strength comprising mixed oxyalkyl units. EP 1678254, assigned to Henkel (2004).

44. M. H. Berger, W. P. Mayer and R. J. Ward. Silane-containing isocyanate terminated polyurethane polymers. US Patent 4374237, assigned to Union Carbide Corporation (1981).

45. T. M. Feng and G. R. Magrum. Arylaminosilane end-capped urethane sealants. EP 0676403, assigned to Witco (1995).

46. M. W. Huang and B. A. Waldman. Silane endcapped moisture curable compositions. EP 1204687, assigned to Crompton Corporation (2012).

47. B. A. Waldman, S. J. Landon and H. E. Petty. Curable silane-endcapped compositions having improved performance. EP 0831108, assigned to Crompton Corporation (1997).

48. G. Limbeck, R. Rettig and L. Schmalstieg. Polyurethane prepolymers containing alkoxysilane and hydantoin groups, process for their preparation and their use for the preparation of sealants. EP 0807649, assigned to Bayer (1998).

49. S. Sato and A. Sato. Process for the preparation of urethane resins and urethane resin compositions. EP 0919582, assigned to Konishi (1998).

50. S. Rizk, H. W. S. Hsieh and M. P. Mazzeo. Acrylic resin having pendant silane groups thereon, and methods of making and using the same. EP 182924, assigned to Essex (1984).

51. T. Kotani and Y. Y. Kanamori. Compositions of silicon containing (meth)acrylic acid ester polymers. EP 1277769, assigned to Kaneka (2002).

52. M. Fujita, M. Kusakabe and K. Kitano. Curable adhesive composition. EP 1000979, assigned to Kaneka (1998).

53. J. Baghdachi. Fast-cure polyurethane sealant composition containing titanium ester accelerators. US Patent 4889903, assigned to BASF (1988).

54. S. Mori, Y. I. Nomura and S. S. Kazuhiro. Resin composition and cold-setting adhesive. EP 1652891, assigned to Konishi (2004).

55. J. Baghdachi and K. K. Mahoney. Fast cure polyurethane sealant composition containing guanidine accelerators. EP 363685, assigned to Adco (1989).

56. Markets and Markets. Elastic bonding adhesives and sealants: Global forecasts to 2020. Maharashtra, India (2015).

57. B. Weller and I. Vogt. Adhesive joints in glass and solar engineering. *J. ASTM* 9, 124–151 (2012).

58. M. Kohl, M. Proebster and T. Fertig. Adhesives and sealants based on silane-terminated binders for bonding and sealing flexible solar films/photovoltaic modules. US Patent 20120055105, assigned to Henkel (2011).

59. I. Krügermann, L. Zander, A. Blaik, S. Auer and H. Loth. Fliesenkleber auf Basis Silan-modifizierter Polymere. WO 2011110384, assigned to Henkel (2010).

60. T. Stotten and M. Majolo. *Construction Adhesives: A Practical Guide*, pp. 15–23, Report from Henkel AG & Co. KGaA, Düsseldorf, Germany (2012).

61. V. Stanjek, B. Bachmeier and L. Zander. Crosslinkable compositions based on organyloxysilane terminated polymers. DE 102011081264, assigned to Wacker (2011).

62. M. W. Huang, A. Chaves, B. A. Waldman and S. J. Landon. Hydrolyzable silanes of low VOC-generating potential and resinous compositions containing same. US Patent 8088940, assigned to Momentive Performance Materials (2012).

Phenolic Resin Adhesives

Antonio Pizzi

CONTENTS

8.1 INTRODUCTION

Phenolic resins are the polycondensation products of the reaction of phenol with formaldehyde. Phenolic resins were the first true synthetic polymers to be developed commercially. Notwithstanding this, even now their structure is far from being completely clear, because the polymers derived from the reaction of phenol with formaldehyde differ in one important aspect from other polycondensation products. Polyfunctional phenols may react with formaldehyde in both the ortho and para positions to the hydroxyl group. This means that the condensation products exist as numerous positional isomerides for any chain length. This makes the chemistry of the reaction particularly complex and tedious to unravel. The result has been that although phenolic resins were developed commercially as early as 1908, were the first completely synthetic resins ever to be developed, have vast and various industrial uses today, and great strides have been made in both the understanding of their structure and technology and application, several aspects of their chemistry are still only partially understood.

It may be argued with some justification that such a state of affairs is immaterial, because satisfactory resins for many uses have been developed on purely empirical grounds during the last 80 years. However, it cannot be denied that gradual understanding of the chemical structure and mechanism of reaction of these resins has helped considerably in introducing commercial phenolic resins designed for certain applications and capable of performances undreamed of, formulations developed earlier by the empirical rather than the scientific approach. Knowledge of phenolic resin chemistry, structure, characteristic reactions, and kinetic behavior is an invaluable asset to the adhesive formulator in designing resins with specific physical properties. The characteristic that renders these resins invaluable as adhesives is their capability to deliver water-, weather-, and high-temperature resistance to the cured glueline of the joint bonded with phenolic adhesives, at relatively low cost.

8.2 CHEMISTRY

Phenols condense initially with formaldehyde in the presence of either acid or alkali to form a methylolphenol or phenolic alcohol, and then dimethylolphenol. The initial attack may be at the 2-, 4-, or 6-position. The second stage of the reaction involves methylol groups with other available phenol or methylolphenol, leading first to the formation of linear polymers [1] and then to the formation of hardened, highly branched structures.

Novolak resins are obtained by acid catalysis, in defect of formaldehyde. A novolak resin has no reactive methylol groups in its molecules and, therefore, without hardening agents it is incapable of condensing with other novolak molecules on heating. To complete resinification, further formaldehyde is added to cross-link the novolak resin. Phenolic rings are considerably less active as nucleophilic centers at an acid pH, due to hydroxyl and ring protonation.

(8.1)

However, the aldehyde is activated by protonation, which compensates for this reduction in potential reactivity. The protonated aldehyde is a more effective electrophile.

(8.2)

The substitution reaction proceeds slowly and condensation follows as a result of further protonation and the formation of a benzyl carbonium ion that acts as a nucleophile.

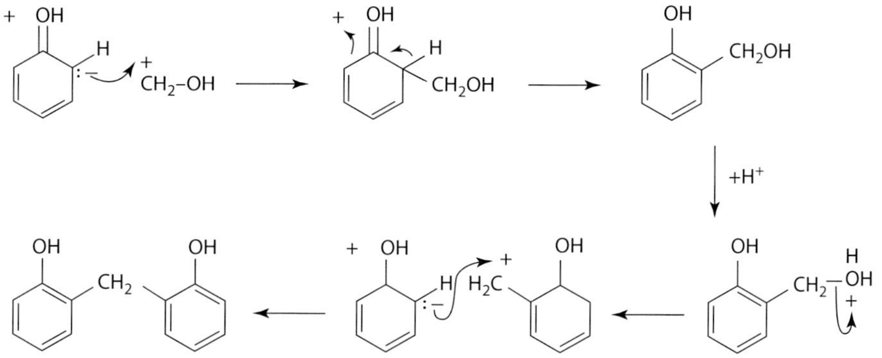

$$(8.3)$$

Resols are obtained as a result of alkaline catalysis and an excess of formaldehyde. A resol molecule contains reactive methylol groups. Heating causes the reactive resol molecules to condense to form large molecules, without the addition of a hardener. The function of phenols as nucleophiles is strengthened by ionization of the phenol, without affecting the activity of the aldehyde.

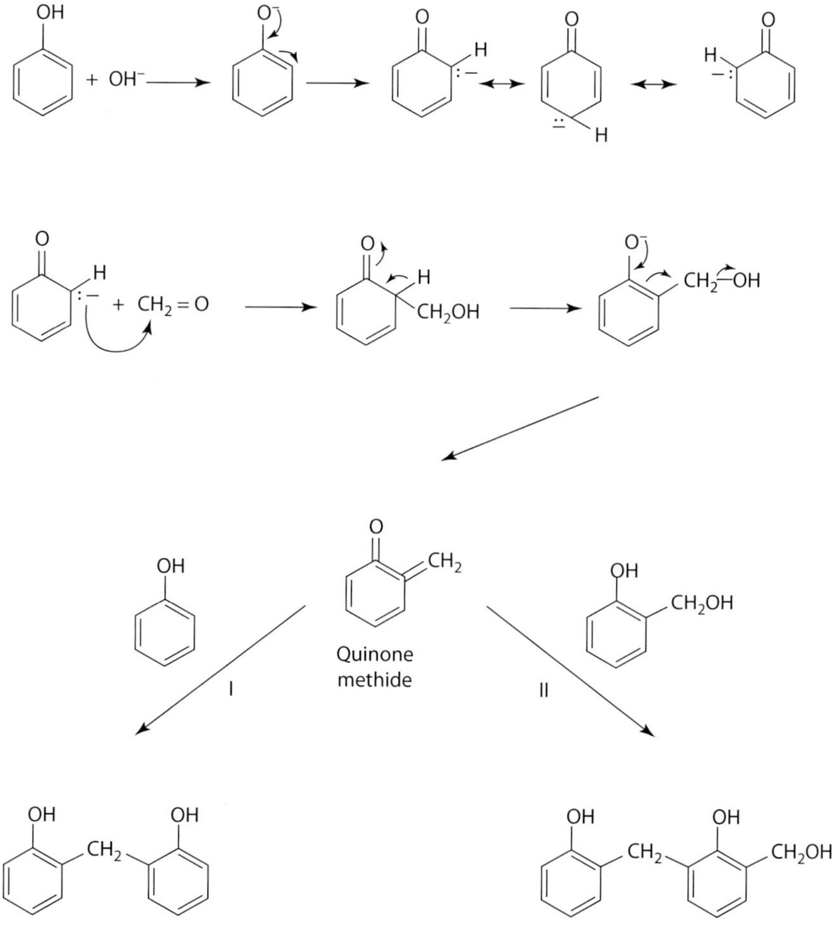

$$(8.4)$$

Quinone methide

I

II

Megson [2] states that Reaction II (in which resols are formed by the reaction of quinone methides with dimethylolphenols or other quinone methides) is favored during alkaline catalysis. A carbonium ion mechanism is, however, more likely to occur. Megson [2] also states that phenolic nuclei can be linked not only by simple methylene bridges, but also by methylene-ether bridges. The latter generally revert to methylol bridges if heated during curing with the elimination of formaldehyde.

The differences between acid-catalyzed and base-catalyzed processes are (1) in the rate of aldehyde attack on the phenol, (2) in the subsequent condensation of the phenolic alcohols, and to some extent (3) in the nature of the condensation reaction. With acid catalysis, phenolic alcohol formation is relatively slow. Therefore, this is the step that determines the rate of the total reaction. The condensation of phenolic alcohols and phenols forming compounds of the dihydroxydiphenylmethane type is, instead, rapid.

Dihydroxydiphenylmethanes are, therefore, predominant intermediates in novolak resins.

(8.5)

Novolaks are mixtures of isomeric polynuclear phenols of various chain lengths, with an average of five to six phenolic nuclei per molecule. They contain no reactive methylol groups and, consequently, cross-link and harden to form infusible and insoluble resins only when mixed with compounds that can release formaldehyde and form methylene bridges (such as paraformaldehyde or hexamethylenetetramine).

In the condensation of phenols and formaldehyde using basic catalysts, the initial substitution reaction (i.e., the formaldehyde attack on the phenol) is faster than the subsequent condensation reaction. Consequently, phenolic alcohols are, initially, the predominant intermediate compounds. These phenolic alcohols, which contain reactive methylol groups, condense either with other methylol groups to form ether links, or, more commonly, with reactive positions in the phenolic ring (ortho or para to the hydroxyl group) to form methylene bridges. In both cases, water is eliminated.

Mildly condensed liquid resols, which are the more important of the two types of phenolic resins in the formulation of wood adhesives, have an average of fewer than two phenolic nuclei in the molecule. The solid resols average three to four phenolic nuclei, but with a wider distribution of molecular size. Small amounts of simple phenol, phenolic alcohols, formaldehyde, and water are also present in resols. Heating or acidification of these resins causes cross-linking through uncondensed phenolic alcohol groups, and possibly also through reaction of formaldehyde liberated by the breakdown of the ether links.

As with novolaks, the methylolphenols formed condense with more phenols to form methylene-bridged polyphenols. The latter, however, quickly react in an alkaline system with more formaldehyde to produce methylol derivatives of the polyphenols. In addition to this method of growth in molecular size, methylol groups may interact with one another, liberating water and forming dimethylene-ether links ($-CH_2-O-CH_2-$). This is particularly evident if the ratio of formaldehyde to phenol is high. The average molecular weight of the resins obtained by acid condensation of phenol and formaldehyde decreases from over 1000 to 200, with increases in the molar ratio of phenol to formaldehyde from 1.25:1 to 10:1.

Thermomechanical analysis (TMA) on wood joints bonded with phenol–formaldehyde (PF) adhesives has shown that, frequently, the increase in joint modulus does not proceed in a single step but in two steps, yielding an increase of the modulus first derivative curve presenting two major peaks, rather than a single peak obtained for mathematically smoothed modulus increase curves [3]. This behavior has been found to be due to the initial growth of the polycondensation polymer leading, first, to linear polymers of critical length for the formation of entanglement networks. The reaching of this critical length is greatly facilitated by the marked increase in concentration of the PF polymer due to the loss of water on absorbent substrates, such as wood coupled to the linear increase of the average length of the polymer due to the initial phase of the polycondensation reaction. The combination of these two effects lowers markedly the level of the critical length needed for entanglement. Two modulus plateaus and two first derivative major peaks then occur, with the first due to the formation of linear PF oligomers entanglement networks, and the second one due to the formation of the final covalently cross-linked networks. The faster the reaction of phenolic monomers with formaldehyde, or the higher the reactivity of the PF resin, the earlier and at lower temperature the entanglement network occurs, and the higher is its modulus value in relation to the joint modulus obtained with the final, covalently cross-linked resin (Figure 8.1).

8.2.1 ACID CATALYSIS

Consideration must be given to the possibility of direct intervention by the catalyst in the reaction. Hydrochloric acid is the most interesting case of an acid catalyst, as is ammonia of an alkaline catalyst. When the PF reaction is catalyzed by hydrochloric acid, two mechanisms may come into play. A reaction route has been proposed that passes through the formation of bischloromethyl ether ($Cl–CH_2–O–CH_2–Cl$). Ziegler et al. [4] have suggested a route via the formation of a chloromethyl alcohol ($Cl–CH_2–OH$) as intermediate. The second route appears to be the more probable. Both hypotheses agree that chloromethylphenols are the principal intermediates. The chloromethylphenols have been prepared and isolated by various means. They are

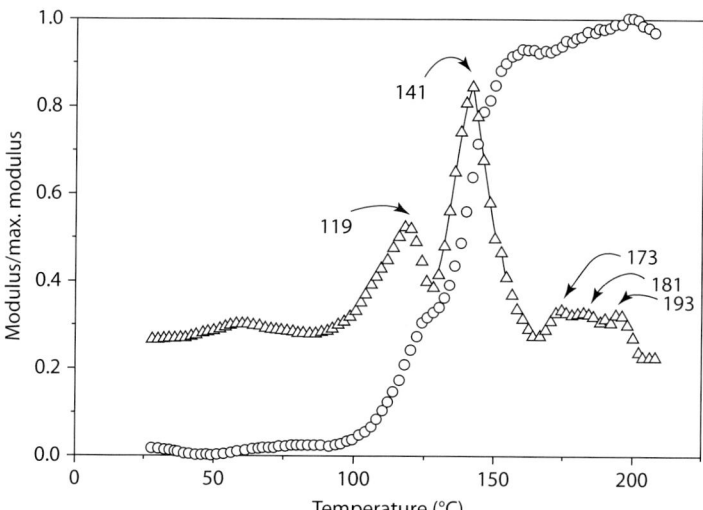

FIGURE 8.1 Thermomechanical analysis (TMA) of the hardening of a PF resin *in situ* in a wood joint. Increase of modulus of elasticity (MOE) of the joint as a function of temperature at a 10°C/min constant heating rate (O); first derivative (Δ).

highly reactive compounds that, with phenols, form dihydroxydiphenylmethanes and complex methylene-linked multiring polyphenols. Reaction is highly selective and takes place in the para position.

$$ \text{(8.6)} $$

8.2.2 ALKALINE CATALYSIS

Different mechanisms of alkaline catalysis have been suggested according to the alkali used. When caustic soda is used as the catalyst, the type of mechanism that seems most likely is that which involves the formation of a chelate ring similar to that suggested by Caesar and Sachanen [5]. The chelating mechanism was thought to initially cause the formation of a sodium–formaldehyde complex or of a formaldehyde–sodium phenate complex and is similar in concept to the mechanism advanced for metal ion catalysis of phenolic resins in the pH range 3–7. While cyclic metallic ion ring complexes have even been isolated [6], this is not the case for the sodium ring complex, evidence for its existence being rather controversial, and the predominant indication being that it does not form [7].

When ammonia is used as a catalyst, the resins formed are very different in some of their characteristics from other alkali-catalyzed PF resins. The reaction mechanism appears to be quite different from that of sodium hydroxide-catalyzed resins. An obvious deduction is that intermediates containing nitrogen are formed. Several such intermediates have been isolated from ammonia-catalyzed PF reactions [8,9] and hexamine-prepared resins [10–13] by various researchers. Similar types of intermediates are formed when amines or hexamethylenetetramine (hexamine) are used instead of ammonia. In the case of ammonia, the main intermediates are dihydroxybenzylamines and trihydroxybenzylamines, such benzylamine bridges having been shown to be much more temperature stable than previously thought and to impart particular characteristics to the resin [10–13].

$$ \text{(8.7)} $$

These intermediates contain nitrogen and have polybenzylamine chains. They react further with more phenol, causing splitting and elimination of the nitrogen as ammonia or producing, eventually, nitrogen-free resins. However, as benzylamine bridges have been shown to be much more temperature stable than previously thought, this requires a considerable excess of phenol and a high temperature or heating for a rather long time. With phenol–hexamethylenetetramine resins of molar ratio 3:1, the nitrogen content of the resin cannot be reduced to less than 7% when heated at 210°C. When the ratio is increased to 7:1, the nitrogen content on heating at 210°C can be reduced to less than 1%. Contrary to what was widely believed, it has been clearly demonstrated that in the preparation of PF resins starting from hexamethylenetetramine, the di- and trihydroxybenzylamine bridges that are initially formed are very stable and are able to tolerate for a considerable length of time a temperature as high as 100°C [10], yielding in certain aspects (only) resins of upgraded characteristics. This behavior is closely tied to the reaction characteristic of hexamethy-lenetetramine to form iminomethylene bases [11].

Ammonia-, amine-, and amide-catalyzed phenolic resins are characterized by greater insolubility in water than that of sodium hydroxide-catalyzed phenolic resins. The more the ammonia used, the higher the molecular weight and melting point obtained without cross-linking. This is probably due to the inhibiting effect of the nitrogen-carrying groups (i.e., $-CH_2-NH-CH_3$ or $-CH_2-NH_2$), which is caused by their slow rate of subsequent condensation and loss of ammonia. Ammonia, amines, and amides are sometimes used as accelerators during the curing of phenolic adhesives for wood products.

8.2.3 METALLIC ION CATALYSIS AND REACTION ORIENTATION

In the pH range 3–7, the higher rate of curing of phenolic resins prepared by metallic ion cataly-sis is due to preferential ortho methylolation [2,14] and, therefore, also to the high proportion of ortho–ortho links in the uncured phenolic resins prepared by metal ion catalysis. The faster curing rate of phenolic resins prepared by metallic ion catalysis is then due to the higher proportion of the free higher-reactive para positions available for further reaction during curing of the resin. The mechanism of the reaction [6] involves the formation of chelate rings between metal, formaldehyde, and phenols or phenol nuclei in a resin.

$$(8.8)$$

The rate of metal exchange in solution [6,15] and the instability of the complex formed deter-mine the accelerating or inhibiting effect of the metal in the reaction of phenol with formal-dehyde. The more stable Complex II is, the slower the reaction proceeds to the formation of Resin III. A completely stable Complex II should stop the reaction from proceeding to Resin III. If Complex II is not stable, the reaction will proceed to form PF resins of Type III. The rate of reaction is directly proportional to the instability or the rate of metal exchange in solution of Complex II. The acid catalysis due to the metal ion differs only in degree from that of the hydrogen ion [16].

The effect of the metal is stronger than that of hydrogen ions, because of higher charge and higher valence, since its interaction with donor groups is often much greater [16]. This allows phenolic resin adhesives to set in milder acid conditions. Most covalent metal ions accelerate the PF reaction. The extent of acceleration depends on the type of metal ion and its amount pres-ent. The capability of acceleration, in order of decreasing acceleration effectiveness, has been reported to be [6] Pb^2, Zn^2, Cd^2, $Ni^2 > Mn^2$, Mg^2, Cu^2, $Co^2 > Mn^3$, $Fe^3 \gg Be^2$, $Al^3 > Cr^3$, Co^3. The most important conclusion to be drawn is that the accelerating effect is indeed present in both the manufacture of PF resin and its curing. Therefore, the fast rate of curing of high-ortho phenolic resins can be ascribed only partially to the high proportion of para positions available. The other reason for the fast rate of curing is that the metallic ion catalyst is still present, and free to react, in the resin at the time of curing. In such a resin, a considerable number of ortho positions (especially of methylol groups in ortho positions to the phenolic hydroxyls) are still available for reaction and capable of complexing.

8.3 CHEMISTRY AND APPLICATIONS OF PHENOLIC RESIN ADHESIVES FOR WOOD

8.3.1 GENERAL PRINCIPLES OF MANUFACTURE

A typical phenolic resin is made in batches, in a jacketed, stainless steel reactor equipped with an anchor-type or turbine-blade agitator, a reflux condenser, vacuum distillation equipment, and heating and cooling facilities. Molten phenol, formalin (containing 37%–42% formaldehyde, or paraformaldehyde), water, and methanol are charged into the reactor in phenol:formaldehyde molar proportions between 1:1.1 and 1:2, and mechanical stirring is begun. To make a resol-type resin (such as that used in wood adhesives manufacture), an alkaline catalyst such as sodium hydroxide is added to the batch, which is then heated to 80°C–100°C. Reaction temperatures are kept under 95°C–100°C by applying a vacuum to the reactor, or by using cooling water in the reactor jacket. Reaction times vary between 1 and 8 h, depending on the pH, the phenol:formaldehyde ratio, the presence or absence of reaction retarders (such as alcohols), and the temperature of the reaction.

Since a resol can gel in the reactor, dehydration temperatures are kept well below 100°C, by applying a vacuum. Tests have to be done to determine, first, the degree of advancement of the resin and, second, when the batch should be finished and discharged from the reactor. Examples of such tests are measurements of the gel time of a resin on a 150°C hot plate or at 100°C in a water bath. Another method is to measure the turbidity point, that is, precipitating the resin in water or a solution of a certain concentration.

Resins that are water-soluble and have a low molecular weight are finished at as low a temperature as possible, usually around 40°C–60°C. It is important that the liquid, water-soluble resols retain their ability to mix with water easily when they are used as wood adhesives. Resols based on phenol are considered to be stable for 3–9 months. Properties of a typical resin are a viscosity of 100–200 cP at 20°C, a solids content of 55%–60%, a water mixability of a minimum of 2500%, and a pH of 7–13, depending on the application for which the resin is destined.

PF resins present lower reactivity at a pH of about 4. The accepted effect of the pH and of the phenol:formaldehyde molar ratio on the polymerization time and hardening time of phenolic resols is shown in Figure 8.2. However [7], the concepts expressed in the graph have been found to be only partially correct as regards the dependence of the PF adhesive rate of curing as a function of pH. Thus, at a pH higher than 7–8, the theory foresees continuous asymptotic acceleration of the polymerization reaction. This is not the case, because the formation of phenate ions has been shown not to be the only effect present. First, acceleration occurs, but after reaching a pH of approximately 8–9, the rate of hardening of the resin slows down considerably [7], contrary to conventional wisdom, as shown in Figure 8.2. There are several reasons for this behavior [7], the easier of these to accept being the formation of a ring involving phenol, a methylol group, and Na ions, which has been postulated already from the middle of the last century [5]. The existence of this ring has been shown to be untrue [7], and the persistence of the concept is due to the ease with which the behavior shown in Figure 8.2 can be explained. The reason for the acceleration, however, was ascribed and shown to be due to the existence of equilibria pertaining to quinone

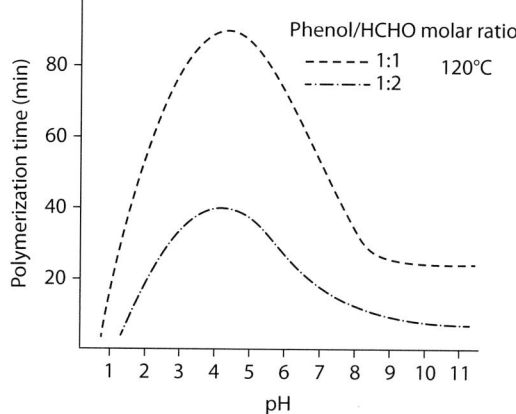

FIGURE 8.2   Polymerization time as a function of pH for phenolic resols of different molar ratios at 120°C. The theoretically derived progressive shortening of polymerization time at pHs higher than 9–10 is an incorrect concept.

methides [7,17]. The structure of the elusive oligomeric quinone methides in PF resins has also been elucidated [18].

(8.9)

The probable reason why the behavior in Figure 8.2 was not noticed earlier appears to be due to the long gel times of PF resins, which makes it very tedious to check reactivity effectively.

8.3.2   Curing Acceleration under Alkaline Conditions

8.3.2.1   α- and β-Set Acceleration

The so-called α- and β-*set acceleration* of curing for very alkaline PF resins for foundry core binders was pioneered in the early 1970s [19], although it was discovered in the early 1950s [19]. In this application, the addition of considerable amounts of such esters in liquid form (α-set) or as a gas (β-set), such as propylene carbonate, methyl formate, glycerol triacetate, and others was found to accelerate resin curing to extremely short times. This technique is now used extensively around the world for foundry core PF binders [19] and is being considered for wood adhesives [7] and rigid alkaline PF foams. The technique is applicable in the approximate pH range of 7–14. The mechanism that makes PF curing acceleration possible was initially explained [7] and different explanations exist (see the next paragraph) and are based on the carbanion behavior of the aromatic nuclei of phenate ions, leading to a more complex variant of the Kolbe–Schmitt reaction. The ester, or the residue of its decomposition, attacks the negatively charged phenolic nuclei, and the reaction is not limited to the ortho and para sites, the phenolic nuclei behaving then as a polycondensation reagent of functionality higher than 3, leading to much earlier gelling. Furthermore, polycondensation occurs not only according to the PF mechanism, but also according to a second reaction superimposed on it [7,20–22] (Figure 8.3).

Other explanations and mechanisms for this occurrence have also been advanced: Determination by TMA of the average number of degrees of freedom of polymer segments between cross-linking nodes of PF resin-hardened networks indicates that additives-accelerated PF resin polycondensation and hardening present several different acceleration mechanisms [20–22]. Some additives such as sodium carbonate appear to present a purely catalytic effect on the polycondensation reaction [20–22]. Other additives such as propylene carbonate present a catalytic effect, as well as inducing an increase in the average functionality of the system, due to alternate cross-linking reactions in which the accelerator itself does participate, leading to a tighter final network [20–22]. These alternate cross-linking reactions have, more recently, been shown to be of different natures, such as the propylene carbonate case in which the reaction appears to be related to a Kolbe–Schmitt reaction, or it could be similar to the acceleration effect due to the hydrolysis of formamide to formic acid and ammonia with the subsequent rapid reaction of the latter with two or more hydroxybenzyl alcohol groups of PF resols [20–22]. The rapid reaction of the $-NH_2$ group of formamide with two hydroxybenzyl alcohol groups of PF resols, a reaction that is also characteristic of urea and methylamine, also appears likely to occur. In some cases, such as in formamide, neither of the two acceleration mechanisms appears to be due to catalytic action only, but both appear to be related to additional cross-linking reactions. Both liquid and solid phase ^{13}C nuclear magnetic resonance (NMR) studies providing supporting evidence for the mechanisms proposed have been presented [20–22].

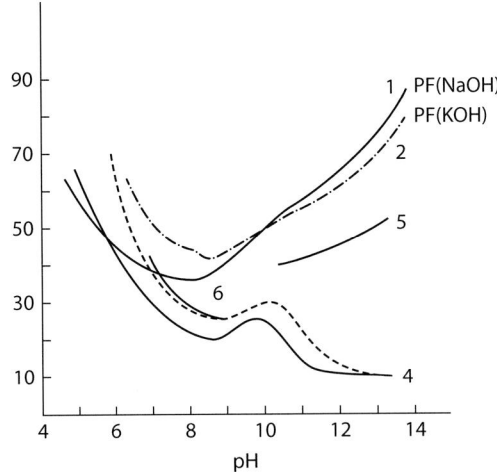

FIGURE 8.3 Cure retarding (in seconds) at high pH and ester acceleration effect of NaOH- and KOH-catalyzed PF resins (ester=propylene carbonate). Note Curve 5, the effect of 4 months aging of the PF resin of Curve 1 on the extent and starting pH of the retardation effect. Compare the start of acceleration for Curves 4 and 6, showing the differences between propylene carbonate and triacetin esters, and compare the starting point of acceleration at pH 5.5 and 7.1. The "bumps" on the curves at pH 8–11 are caused by methylene-ether formation, decomposition, and rearrangement. (From: Pizzi, A. and Stephanou, A., *J. Appl. Polym. Sci.*, 49, 2157–2160, 1993. With permission.)

Further proof of complex reactions between propylene carbonate and phenolic nuclei leading to compounds in which the carbonic acid has attacked the phenolic ring have been presented [20–22], based on the ^{13}C NMR spectrum of the product of the reaction of resorcinol with propylene carbonate, in the absence of formaldehyde. Resorcinol was chosen because its aromatic ring is a stronger nucleophile than phenol and, thus, if any reaction had occurred, this would be less elusive and much more easily observed [20–22]. The reaction products that appeared to be formed were carboxylic and dicarboxylic species. That they might be present was also derived from NMR studies [20–22]. It must be pointed out that such structures need only to be transitory and not permanent to obtain the same effects noted experimentally. Such a subsequent lability could be the reason why it is difficult to observe such linkages in the hardened resin, except for faster-reacting phenols, where it can be observed due to early immobilization of the network, which surely occurs.

It must also be remembered that in hot-temperature curing of phenolic resins, their polycondensation is accelerated particularly on a wood surface, first of all by a heterogeneous catalysis effect by the cellulose [23], and secondly by the substrates absorbing water and, thus, increasing the resin effective concentration, which is always a very significant effect in polycondensation reactions [3,24]. Under these conditions, the existence of the additional cross-linking mechanism will then be even more marked.

Once the nature of the accelerating mechanism induced by increased cross-linking and its existence through the determination of the increased tightness of the PF networks formed has been defined [7,20–22], it is necessary to address the nature of the other accelerating mechanism, which appears to be common to both sodium carbonate and propylene carbonate. The apparent failure by different analyses [25] such as ^{13}C NMR to find any trace of C=O bridges after purification of sodium carbonate-accelerated PF resins indicates quite clearly that the sodium carbonate effect may well be purely catalytic, or the C=O is transformed during the reaction to another group, or even more likely is that the C=O disappears from the system as CO_2 or precipitates completely as sodium hydrogen carbonate. The presence of C=O bridges has been clearly shown in nonpurified samples of accelerated PF resins [9,23] but, strictly speaking, this is only proof of the additional cross-linking mechanism discussed, or it could just be due to any carbonic acid salts still present in the system. That this mechanism exists is proven by the acceleration of hardening for high molar ratio (F/P $\geq$ 2.5) PF resins in which all available ortho and para sites on the phenol are blocked by methylene or methylol groups. In this case, a soft gel and no subsequent rapid hardening is obtained. The mechanism involved could then be one of the two proposed up to now, namely: (1) The mechanism based on a hydrogen carbonate ion intermediate activated complex [26], which presents inherent disadvantages that have been outlined [20]. This mechanism has been proposed without any evidence and direct evidence for it would be rather difficult to gather; and (2) the mechanism [26] based on a rapid transesterification reaction of the hydroxybenzyl alcohol group of a PF resol. This latter mechanism is based on the very facile transesterification of propylene carbonate with methanol, through which dimethyl carbonate is rapidly obtained [26]. The real mechanism involved and the existence of intermediate C=O bridges was finally proven by ^{13}C NMR in 2006 [21]. However, it must also be noted that even more recent research has shown that the reaction of dimethyl carbonate with the hydroxy groups of phenol is also very easy, leading to possible cross-linking through C=O bridges. This same reaction has been used as an intermediate to prepare polyurethane bridges without isocyanates [27–29].

It is interesting to remark that other reactive materials that will readily undergo transesterification analogous to that of propylene carbonate with methanol are trialkyl borates, tetraalkyl titanates, and trialkyl phosphites in alkaline environments, and that gas injection of methyl borate (and carbon dioxide) has been found to enhance as well the results of wood composites bonded with formaldehyde-based resins [30], just as the addition of propylene carbonate and glycerol triacetate has been shown to do in wood composites bonded with phenolic resins.

In the case of wood adhesives, firstly, glycerol triacetate (triacetin) and, secondly, guanidine carbonate are the accelerating esters of choice, yielding long pot-lives at ambient temperature and short cure times at higher temperature, and are used in proportions between 3% and 10% of adhesive resin solids. Propylene carbonate is unsuitable for wood adhesive application as it yields far too short pot-lives at ambient temperature. Methyl formate and other esters, including propylene carbonate, are instead used in foundry core binders, where sometimes the proportion of ester accelerator used is up to an equal amount of the resin solids; hence, the accelerator technology differs from one field of application to another. Most other esters are either much less effective accelerators at higher temperature, or they shorten the ambient temperature life of the resin to such an extent that in practice the resin cannot be used [31–36]. Triacetin gives long pot-lives and short cure times instead, due, among others, to its lower rate of hydrolysis at ambient temperature. Another series of compounds, some of which were finally found to yield sufficiently rapid acceleration at higher temperatures, still giving increased strength of the cured resin as well as sufficiently long shelf life at ambient temperature, were the salts of guanidine. Guanidine carbonate, guanidine hydrochloride, and guanidine sulfate were tried with positive results [33]. Guanidine carbonate appeared to be the best PF accelerator, and its accelerating capability remained acceptable, while the shelf life at ambient temperature of the PF and phenol–urea–formaldehyde (PUF) resins to which it had been added in different proportions and the performance in particleboard preparation were the same as triacetin [33]. Even in the case of some industrial higher condensation resins, the pot-life of the resins was as long as 3 weeks, with the guanidine carbonate already incorporated in the resin [33].

It has repeatedly been established that the activation energy of the polycondensation reaction of PF resins, and also of urea–formaldehyde (UF), melamine–formaldehyde (MF), and other resins, is markedly influenced by the presence of wood [21,37–43]. In the presence of wood as a substrate, the activation energy of the polycondensation reaction and, hence, of the hardening of PF and other resins, is considerably lowered. This implies that resin polymerization and cross-linking proceeds at a much faster rate when the resin is in molecular contact with one or more of the wood constituents [21,37]. It was indeed shown that catalytic activation of the hardening and advancement of PF and other polycondensation resins induced by the wood substrate did exist and was a rather marked effect. The reasons for this effect have been found to be due to the secondary attraction forces binding the resin to the substrate [21,37]; these cause variations in the strength of bonds and intensity of reactive sites within the PF oligomer considered, an effect well known in heterogeneous catalysis for a variety of other chemical systems [44]; bond cleavage and formation within a molecule are greatly facilitated by chemisorption onto a catalyst surface. This work also indicated which bonds in the PF resin were involved and what the extent of acceleration of the hardening reaction caused by such an effect was [37].

8.3.2.2 Urea Acceleration and Phenol–Urea–Formaldehyde (PUF) Exterior-Grade Resins

Low-condensation PF resins have been coreacted under alkaline conditions with up to 42% molar proportion of urea on phenol during resin preparation, to yield PUF resins capable of better performance and shorter hardening times than equivalent pure PF resins prepared under identical conditions [32,45]. The reason that urea reacts with relative ease with PF resols under alkaline reaction conditions can be ascribed to the relative reactivities toward methylol groups of urea and phenolic nuclei. A study has shown that there are definite pH ranges in which the reaction of unreacted urea –NH$_2$ and –NH– groups with formaldehyde in competition to phenol, or with the methylol groups carried by a PF resin, is more favorable than autocondensation of the PF resin itself [32,37,45] (Figure 8.4).

It is the urea that contributes to the acceleration of the resin cure by allowing the production of higher molecular mass linear oligomers needing fewer steps to cure, hence is faster. Copolymerization of up to a certain level of urea is a faster way to obtain mainly linear oligomers of higher molecular

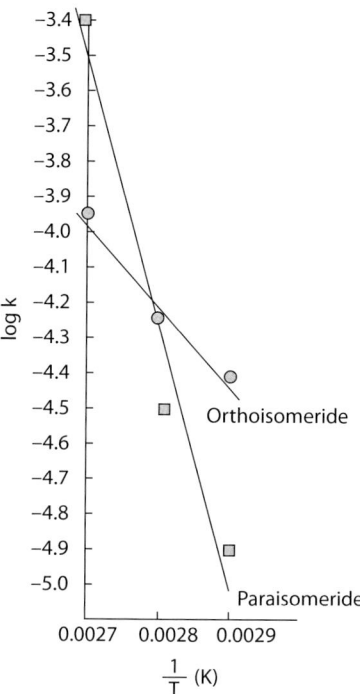

FIGURE 8.4 Log of reaction rate constant versus 1/T (T in kelvins) for the reaction of ortho- and para-hydroxybenzyl alcohol with urea, detailing the relative rates of condensation with urea of ortho- and para-phenolic methylols in a PF resin.

weight; in a sense, this is just a way to molecularly double very quickly the molecular weight of a PF resin while still maintaining the linearity of the higher molecular weight oligomers formed.

$$(8.10)$$

The extent of copolymerization, however, needs to be limited, otherwise the viscosity of the resin becomes unmanageable. However, the presence of free urea is able to control this trend (1) by the faster reaction of urea than phenol with the phenolic methylol group under alkaline conditions, and (2) by the possibility of reaction with phenol of the methylol groups formed by the reaction of HCHO with urea. This second reaction is an equilibrium that favors the formation of methylol ureas and subsequent products [32,33].

The water resistance of these PUF resins is comparable to pure PF resins when used as adhesives for wood particleboard. Part of the urea (between 18% and 24% molar on phenol at a phenol:HCHO 1:1.7 molar ratio, but higher at higher HCHO proportions) was found by ^{13}C NMR to be copolymerized to yield the alkaline PUF resin, while, especially at the higher levels of urea addition, unreacted urea is still present in the resin. Increase of the initial formaldehyde:phenol molar ratio decreases considerably the proportion of unreacted urea and increases the proportion of the PUF copolymer.

A coreaction scheme of phenolic and aminoplastic methylol groups with reactive phenol and urea sites based on previous work on model compounds has been proposed [32], with copolymerized urea functioning as a prebranching molecule in forming a hardened resin network.

$$(8.11)$$

The PUF resins prepared are capable of further noticeable curing acceleration by addition of ester accelerators, namely, glycerol triacetate (triacetin), to reach gel times as fast as those characteristic of catalyzed aminoplastic resins, but with wet-strength values characteristic of exterior PF resins [32,33]. Guanidine carbonate has also been shown to be an accelerator of PF resins. However, it gives slightly longer gel times than triacetin when simply added to a PF resin glue-mix. It is also capable of giving a glue-mix pot-life of the order of several days, hence, long enough to be premixed with the resin long before use [33]. Both triacetin and guanidine carbonate, used as simple glue-mix additives, increase the ultimate strength of the resin bond whatever the curing time used for the purpose, this being confirmed both by TMA as well as by application to wood particleboard. Synergy between the relative amounts of copolymerized urea and ester accelerator is very noticeable at their lower levels of addition to the resin, but this effect decreases in intensity toward the higher percentages of urea and triacetin. The relative performance of the different PUF resins prepared, under different conditions, allowed the preparation of wood particleboard bonded with accelerated PUF resins with the capability to achieve press times as fast as those of aminoplastic (UF and others) resins.

8.3.2.3 Physical Properties of Phenol–Formaldehyde Resins

Hardened PF resins have a specific gravity of approximately 1.2–1.3, a refractive index of 1.6, and a specific heat of 0.5. They are typically brown in color, and novolaks are lighter than resols. Resols are dark yellow, orange-reddish, or brownish even when made with pure raw materials. However, if the alkali is neutralized, resols become almost colorless. The best results were obtained with citric, lactic, and phosphoric acids. Pale-colored, hardened resins can be prepared with them. Phenolic resins are relatively stable up to about 200°C–250°C, although oxidative degradation at the methylene bridges produces substituted, dihydroxybenzophenones [46]. Above this temperature, they begin to char slowly, and at higher temperatures, charring is more rapid. At about 400°C, decomposition is rapid, yielding phenols and aldehydes and leaving a coke-like residue.

In the A-stage, simple PF resins are readily soluble in alcohols, esters, ketones, phenols, and some ethers, and insoluble in hydrocarbons and oils. As a class, resols tend to be more soluble in alcohols and water, and novolaks tend to be more soluble in hydrocarbons. In the early stages of condensation, resols are often soluble in water, owing to the presence of methylolphenols, especially polyalcohols. This is more pronounced with resols that are derived from phenol. Cresol-derived resols are less soluble, and xylenolic resols are almost insoluble in water. The solubility of A-stage resins in dilute aqueous sodium hydroxide or in mixtures of water and alcohols follows the same trend.

Solubility in alcohols and insolubility in hydrocarbons appears to go together. The alcohol and water solubility can be reduced only by using acetaldehyde, or other aldehydes, in place of formaldehyde, and by introducing hydrocarbon chains, particularly in the ortho or para position, in the aromatic ring. B-stage resins are soluble in only a few solvents, such as boiling phenols and acetone, aqueous sodium hydroxide, and deca- and tetrahydronaphthalenes. Resins in the hardened or C-stage are very resistant to most chemical reagents. They are unaffected by all ordinary organic solvents and water, although a few percentage points of water may be absorbed in a filled material, mainly by the filler, thus causing slight swelling. The C-stage resins dissolve slowly in boiling phenols such as naphthols. Resins from the simplest phenols can also be broken down and dissolved by hot, strong alkali solutions.

Simple PF resins are readily degraded by sodium hydroxide. However, cresol–formaldehyde and especially xylenol–formaldehyde resins are much less susceptible to alkaline degradation. Resins are often more resistant to strong alkaline solutions (i.e., 15%–20%) than to dilute solutions (i.e., 5%). The filler has a considerable influence on the chemical resistance of the resins. Inert mineral fillers have higher resistance than cellulosic fillers. C-stage resins are resistant to most acids, except sulfuric acid concentrated at more than 50%, formic acid, and oxidizing acids such as nitric and chromic acids. The insolubility of hardened resins in acetone is used to test the degree of cure of the resin. The curing temperature influences the amount that is insoluble in acetone after prolonged heating [24]. The higher the hardening temperature, the lower the amount of acetone extractives. The mechanical properties of hardened PF resins are greatly influenced by the moisture content. This applies more to resins containing fillers, plasticizers, and other ingredients. The rate of water absorption decreases with time, but thick samples may not reach an equilibrium even after several months in water. Therefore, in measuring the mechanical properties of resins, it is necessary to condition the test specimens under carefully controlled temperature and humidity prior to carrying out the tests. In many cases, the mechanical properties of hardened resins are largely dependent on the type of filler. This applies particularly to water absorption, tensile strength, and impact strength. It also applies to shear strength, but shear strength depends more on the adhesion to the substrate. The properties of the resin are more important than those of the filler in determining the compression strength.

8.4 APPLICATIONS

8.4.1 PF Wood Binders

Phenolic resins are used as binders for exterior-grade plywood and particleboard, which need the superior water resistance provided by these resins. In the manufacture of plywood, the phenolic resin adhesive is usually applied to the wood veneers by roller or extrusion coating. The coated veneer is then cross-grained, stacked, and cured in a multidaylight press for 5–10 min at 120°C–130°C and at a pressure of 11–16 kg/cm². In the manufacture of particleboard, the phenolic resin adhesives are applied to the wood chips by continuous blenders. The glued wood chips are formed in a mat and then pressed for 2–12 s/mm, depending on thickness, press temperature, and moisture content, at 190°C–230°C and 25–35 kg/cm². The only type of phenolic resin used commercially for this application are resol-type resins, which have the structure shown:

$$HOH_2C \quad \text{—} \quad OH \quad \text{—} \quad CH_2 \quad \left[\quad OH \quad \text{—} \quad CH_2 \quad \right]_n \quad OH \quad \text{—} \quad CH_2OH \tag{8.12}$$

n > 0

These are hardened by heating after the addition of small amounts of wax emulsion and insecticide solution in the case of particleboard, and of vegetable or mineral fillers and tackifiers in the case of plywood. Accelerators are sometimes added in both types of glue-mixes. The pH of these resins varies between 10 and 13.5 and is generally between 12 and 12.5.

In dealing with wood-related factors that affect glue bonds, it is important to remember that adhesion is at least 95% physicochemical in nature. The mechanical aspects of bond formation (such as keying cured adhesive solids in the wood surface) contribute negligibly to the bond strength or wood failure. The main chemical forces in thermosetting resins adhesion are covalent bonds and hydrogen bonds, plus secondary forces such as van der Waals forces and any other types of electrostatic and dipolar forces. It is, therefore, essential that the resin contains a significant number of functional groups and that the wood surface presents a significant number of reactive sites to enable the resin to bond. Any factors that limit resin functionality or block reactive sites on the wood structure necessarily impede adhesion.

8.4.1.1 Properties of Phenolic Adhesives for Plywood

Certain attributes of phenolic resins have been designed to give the strongest and most durable plywood bonds. Laboratory and field experiences have demonstrated that certain types of PF plywood resins perform significantly better on veneers than do others. These superior resins have several properties in common:

1. They are relatively low in alkali content, generally about 0.33 to no more than 0.5 mol.
2. They have a lower molecular weight for hardwood veneers than do phenolic resins designed for softwood gluing.
3. They are high in methylol group content. Alternatively, they may contain free formaldehyde or require a matching catalyst that contains paraformaldehyde.
4. Even in dried adhesive carrier films or powdered resins, phenolic resins for hardwoods share the B-stage characteristic of reliquefying briefly under heat and pressure to allow transfer and flow on the gluelines of a plywood panel. This liquefaction can occur without water. This is unlike all but the most recently developed softwood phenolic resins.
5. Phenolic resins for hardwoods have higher thermal softening points than those of many other conventionally prepared resins. This indicates a network that has more cross-links after final cure, and also greater durability.
6. They have a 40%–45% solids content and viscosity of 150–600 cP at 25°C.

In general, lower resin alkali content and lower molecular weight are associated with slower cure, which explains why adhesive resins for softwood and plywood are both more alkaline and more condensed. Adequate exterior-grade adhesion can be attained on softwood veneer if these considerably condensed phenolic resins, which also cure more quickly, are used. Conversely, the increased functionality of hardwood adhesive resins partially compensates for their inherently slower cure rate. Hardwood phenolic resins require about 30% longer press times for adequate cure.

Notwithstanding their good adhesion capability, phenolic resins for hardwood gluing carry one distinct disadvantage: They do not prepress as well as softwood phenolic resins. Their lower condensation and longer flow times are a disadvantage. Prepressing is important, because it is done to minimize face veneer losses and to reduce precure times on hot platens. Therefore, other means can be employed to obtain the required tackiness. Additives and adhesive formulations for this purpose are available. To increase the prepress capacity of low alkali, longer-flow-time phenolic resins suitable for hardwood gluing, small amounts of starch or poly(vinyl alcohol) can be added to the resin glue-mix just before use.

Finished resin viscosity is increased to allow for the thickening effect of the additives without reducing the average molecular weight of the phenolic resin. Water-soluble thickeners [such as hydroxyethyl cellulose, poly(ethylene glycol)s, and maleic anhydride copolymers] contribute to prepress tack. However, they cause a large increase in resin viscosity, so the amount added must be small. Consequently, the benefit from prepressing is limited. Most animal- and vegetable-based thickeners, such as gum arabic, are subject to hydrolysis in alkaline phenolic resins, and lose effectiveness in a matter of hours.

8.4.1.2 Additives

A number of additives and modifiers contribute useful properties to phenolic resins used for wood gluing. Multipurpose additives are the aminoresins and urea–formaldehyde and melamine–formaldehyde polymers. These include not only urea and melamine plywood adhesive resins, but also dimethylol urea, trimethylol melamine, and hexamethylol melamine. Added in amounts from 5% to 15% of phenolic resin solids, they improve resin tack and prepressing, increase long-assembly-time tolerance, shorten press times, and enhance resin functionality. This results in stronger bonds on wood veneers. As long as they are used in limited quantities, they have a negligible effect on long-term phenolic bond durability. They appear to be well protected from hydrolytic degradation by the cured phenolic polymer network. The dispersion of the aminoresin molecules in the alkaline medium of the phenolic resin glue-mix inhibits their curing reaction, which is acid-catalyzed. It causes them to function as methylolated cross-linking units for the PF polymer.

Urea can be used by itself in large amounts in the system of producing PF resins of very high molecular weight and such a high viscosity to be almost solid at ambient (but not higher) temperature. Such resin, while still hot, is "drowned" in up to 40% of urea (based on resin solids) to decrease its viscosity to a perfectly manageable level. These PFs in urea solutions have high reactivity and short curing time, due to the much higher degree of polymerization of the resin [32,47]. They show little copolymerization with the PF in their liquid state and some copolymerization in the hardened network, the majority of the urea still remaining free to leach out after curing. For these reasons, they must not be confused with copolymerized PUFs of high urea proportion, which are also operative in short hardening time applications. These are described more fully elsewhere [32,33].

Formaldehyde in liquid solution or solid form and formaldehyde-generating compounds are also phenolic resin additives that improve functionality and decrease curing times. Paraformaldehyde is used most frequently, but hexamethylenetetramine, formaldehyde–sodium bisulfite complexes, tris(hydroxymethyl)nitromethane, and glyoxal are also used. Significant effects are obtained when 3%–5% is added, based on phenolic resin solids. Further reduction in curing time is possible if 1%–2% resorcinol is added, or resorcinol-acting natural extractives, such as condensed (flavonoid) tannin extracts (wattle, mimosa, and others). The rescorinol-like materials can or cannot be mixed with paraformaldehyde and added to the liquid phenolic resin glue-mix. Equally, ester accelerators such as glycerol acetate (triacetin) and guanidine carbonate and other esters are also used as very effective cure accelerators and to improve final hardened strength (see Section 8.3.2.1).

Formaldehyde addition overcomes the effect of phenolic extractives in certain hardwood species, which prevent proper cure or adhesion of PF resins. Free formaldehyde appears to react rapidly with these phenolic extractives before they can interfere with the phenolic resin-curing mechanism. With certain wood species that are rich in extractives, this technique has been used to increase bond durability from interior-grade to true exterior-grade performance.

Natural phenolic compounds are used both as replacements for substantial portions of synthetic phenol in plywood adhesive resins and as glue-mix additives to improve performance; 4%–6% is added, based on phenolic resin solids. They bring about improvements in assembly-time tolerance and flow with no significant change in adhesion. Additions of wattle tannin or other condensed flavonoid tannin extracts with or without additional formaldehyde in the glue-mix produce faster hot-pressing cycles. However, some assembly-time tolerance and pot-life have to be sacrificed in the process, but full exterior-grade durability is retained.

The lignin residues from wood pulp production are another class of extractives currently receiving attention as phenolic resin additives. Substituted phenols, such as cresols and xylenols, have been used as glue-mix additives for phenolic adhesive resins to improve assembly-time tolerance. They are also used as solvents to remove oleoresinous deposits on the surface of pitchy softwood veneers. They can be used as flow promoters in phenolic hardwood adhesives. To avoid interference with the rate of resin curing, the amount added should not exceed 3%–4% of the phenolic resin solids content.

Complexing additives commonly include the soluble salts of boron, chromium, zinc, cobalt, lead, manganese, magnesium, and others. When added to phenolic resin adhesives, some of these compounds have been successful in reducing press times and in improving prepress performance [24]. Borax is widely used in North America to shorten the prepress cycles of phenolic plywood glues for softwoods. However, these compounds tend to increase the molecular weight of a phenolic resin, by complexing several molecules together through their phenolic hydroxyl groups. The gain in the resin molecular weight and prepress tack is sometimes accompanied by a reduction in assembly-time tolerance and the loss of the B-stage melt-flow behavior. In hardwood gluing, this is sometimes not advantageous, and the addition of complexing salts should be approached with caution.

Mixed borate salts are very effective as a treatment for the preservation of wood products against fungi and most insects. However, the boron salts, which become localized in high concentration on the veneer surface, tend to gel the phenol resin before it can reach the wood surface and bond to it. However, very dilute aqueous solutions of borates (i.e., 0.25%) applied to softwood veneers in their green state decrease their thermal degradation during high-temperature drying and preserve their reactive sites for bonding with phenolic adhesives.

8.4.1.3 Formulation of Plywood Glue-Mixes

The guiding principles for the preparation of plywood adhesive glue-mixes are:

1. To maintain the highest possible phenolic solids content in the glue-mix (preferably in the range 30%–40%).
2. To incorporate a cellulosic filler, such as nutshell flour of 200 mesh or finer, in the proportion of about 20%–40% of phenolic solids. Nonabrasive inorganic fillers may also be satisfactory. Alternatively, to add about half this amount of unrefined starchy material, such as wheat flour.
3. To add no alkali, or at the most only 1%–2%, to disperse and stabilize the starchy material.
4. To add only enough water to produce a glue viscosity that can be handled by the gluing equipment. The preferred viscosity range is 1500–2500 cP, measured at 25°.
5. To ensure proper wetting of the veneer by the glue film, a surface-active agent should be added (about 0.1%–0.25% of resin solids).

Examples of glue-mixes incorporating these principles are listed in Table 8.1.

8.4.1.4 General Observations on Particleboard Manufacture [24]

In the case of the application of phenolic adhesives to the manufacture of exterior-grade particleboard, the closest attention must be focused on the application of the resin rather than on its formulation. A good phenolic resin for plywood can be used successfully for the manufacture of

TABLE 8.1
Examples of Glue-Mixes

Material	Amounts in Parts		
40%–45% solids PF resin	100	100	100
300 mesh coconut (or walnut) shell flour	12	14	10
Industrial wheat flour	6	—	—
50% sodium hydroxide	2	—	—
Surfactant	0.1	0.1	0.1
Water	10	5	
Total parts	130.1	120.1	110.1
Phenolic solids	31%–35%	33%–37%	36%–41%

particleboard, once the various conditions of application have been understood. The press time of the board varies according to the type of adhesive, its reactivity, and the moisture content of the glued particles. In many cases, a light water spray is applied to the top surfaces of the board before prepressing to shorten the press time. The thin film of water covering the surface is vaporized when it comes in contact with the hot press platens and migrates from the surfaces toward the core of the panel, causing a faster increase in temperature and a faster cure.

The water spray prevents precuring of the adhesive on the surface of the board during closure of the press before contact with the hot top press platen. The wood undergoes a partially irreversible plastic deformation during pressing, caused by the combined action of pressure and heat. Different products can be obtained by varying the type of press cycle of the board. Different pressing procedures are available. A procedure for industrial three-layer particleboards may read as follows:

1. *Maximum pressure*: 23–27 kg/m^2 is reached as fast as possible after press closure (i.e., 35–50 s; other processes use pressures as high as 35 kg/cm^2).
2. *Contact with the gauge bars*: As a rule, contact is made after 60–120 s from the start of the press closure. The higher the density, the longer the time it takes. This, however, can be reduced by increasing the moisture content of the glued mat.
3. *Steam escape*: Expected to begin 1–3 min after making contact with the gauge bars.
4. *Pressure decrease*: After approximately 1.5–2 min of maximum pressure, the pressure is slowly decreased until the final pressure on the panel is as low as 2–3 kg/cm^2. This takes place toward the end of the cycle, just before press opening.

This pressing procedure produces a board with high-density face layers and the shortest possible press time, at a given temperature and a low power consumption. The main properties of panels with high-density face layers are the stiffness of the panel; better warp resistance; high dimensional stability; hard, glossy, and shockproof surfaces that need less adhesive for subsequent veneering; and narrow thickness tolerances.

Considerable differences in the properties of the final board can be obtained by varying the moisture contents of surface and core layers, and by using faster resins in the core layer and slower reacting resins in the surface layer, so as to increase the board core density and improve the density profile of the panel as a function of its thickness. This is one of the contributory factors to improve properties based on the adhesive application technology rather than on the characteristics of the adhesive itself. In wood particleboard manufacture, factors pertaining to the application/pressing technology contribute as much as 50% to the final performance, the rest being due to how good the adhesive is itself. This can also be achieved by varying the geometry and sizes of the wood chips, the density of the board, and so on. Small variations in the manufacture and characteristics of the phenolic resin used do not affect the property of the finished particleboard as much as the factors listed. Experiments [51] on the correlation of curing and bonding properties of particleboard glued with resol-type phenolic resins by differential scanning calorimetry show that resols tend to reach two endotherm peaks: the first at 65°C–80°C and the second at 150°C–170°C. Resols used for particleboard have been shown to begin curing at lower temperatures than those for novolak resins. Resol-glued particleboard shows no bond formation at 120°C. At 130°C, the resol-glued panels show internal bond strengths of 0.55–0.7 MPa. The internal bond strength in the wet tests increases as the board core temperature goes over 150°C during pressing. The normal temperatures for 12–13 mm thick board glued with phenolic adhesives are 170°C–230°C. The press time is 18–12 s/mm for standard PF resins but, today, PUFs [32], and esters-accelerated [32,33], tannin-accelerated [24], and urea-drowned PF resin [47] adhesives can reach press times as fast as 5 s/mm at 190°C–210°C in industrial application [31,32]. Typical results obtained using PF adhesives for particleboard are shown in Table 8.2.

8.4.1.5 Dry-Out Resistance

One of the more common difficulties in bonding pine veneers and chips is adhesive dry-out. Dry-out is associated with the high liquid absorbance of pine sapwood, and it appears especially during long assembly times. This problem can be overcome by using resins modified through reaction with alkylated phenols, especially 3,4-xylenol [48]. Another technique used to achieve similar results is the manipulation of synthesis procedures used in preparing a standard PF resin [48]. The dry-out resistance imparted by alkylated phenols is due to the initial semi-thermoplastic character of the resin. This is due to their monomer, which is, thus, only bifunctional to the linear polymer that is consequently formed.

If a linear and essentially non-cross-linking prepolymer is prepared from phenol and formaldehyde, it can be coreacted with a nonlinear and cross-linking prepolymer to form a resin. The latter resin will have some initial semi-thermoplastic or dry flow character, but will be primarily a thermosetting one. The product is an alkaline novolak–resol copolymer. Evaluation of this copolymer concept has shown that many resins possess a controlled initial semi-thermoplastic character that improves resistance to dry-out. Good dry-out resistance is achieved without loss of press-time efficiency or broad-range bonding ability. Such resins perform noticeably better than other types of resins that are resistant to dry-out.

Such a resin of the alkaline novolak–resol type can be prepared by coreacting a prepolymer, prepared by reacting formaldehyde and phenol in the molar ratio 2.6:1.0, and a prepolymer obtained by reacting formaldehyde and phenol in a molar ratio of 1:1.

The two prepolymers are then mixed in 50:50 proportions by mass and coreacted.

8.4.2 Foundry Sand Binders and Mineral Fiber Binders

Phenolic resins are also extensively used in the binding of foundry molds. Both resol and novolak resins are used for this application. The sand is coated with the phenolic resin at a load of 3%–4%. The PF resin can be used either as an organic solvent solution or in powder form. Coating of the

TABLE 8.2
Results Obtained Using PF Adhesives for Particleboard

Density (g/cm³)	Swelling after 2 h Boil in Water		Internal Bond Strength		
	Measured Wet (%)	Measured Dry (%)	Dry (kg/cm²)	After 2 h Boil (kg/cm²)	Cold-Water Swelling (%)
0.700	15	2–3	10–11	5–8	9–11

substrate can be carried out either at ambient or at higher temperature. In higher-temperature coatings, novolaks are the preferred resins, and in this application, waterborne resins (75% resin) can also be used. Hexamethylenetetramine, as well as wax, is added. Hexamine is often added separately from the resin to avoid precuring.

Another equally important field of application of phenolic resins is in the binding of mineral fibers such as glass fiber and rock wool. These are used for thermal and acoustic insulation at densities in the range 2.5–70 kg/m^3. Both powder and liquid resins, generally in water solution, are used for this purpose. Liquid resins are generally applied at about 10% concentration in water; the water evaporates, cooling the fiber and avoiding decomposition of the resins; and the resinated mat is then cured in a hot-air circulation oven at 175°C–200°C for 2–5 min.

Acid-setting (but not only these) PF resins are extensively used in these fields. These resins, however, cannot be used to bind wood or impregnate paper for laminates, due to the acid hydrolysis of the cellulose they would cause. However, even in the wood bonding field, self-neutralizing acid-setting PF resins have been developed, although these are not used industrially. Self-neutralizing systems for the hardened glueline are based on special hardeners that allow rapid curing of the resin and equally quick return to neutrality of the hardened glueline on joint cooling. A good joint bonded with a rapidly self-neutralizing PF resin shows high strength and high levels of adhesion, and this system shows some promise in some applications of wood bonding [37].

8.4.3 BINDERS FROM PF COPOLYMERS WITH OTHER RESINS

The characteristics of PF resins and the reactive chemical groups they present render them particularly suitable for the preparation of binders by coreaction with other resins. This is still a relatively young field, and the most interesting and relevant coresins that are used or explored in this respect are the aminoplastic resins, in particular UF and MF (the copolymerization with the latter being a somewhat older use) and diisocyanates.

While MF resins have been known for a long time to be able to form true copolymers with PF resins, this has not been the case for UF resins. Until quite recently, copolymerization between PF and UF resins or urea was not thought to be likely, the system curing as a polymer blend only, and applications of this type have really been shown to be useful also [32]. Such deduction was based on the lack of detection of any methylene bridge between phenolic nuclei and the amido group of urea. Recently, PUF resins of two different types and for two different purposes were shown to be able to copolymerize. First, it has been shown that copolymerization between PF and urea resins occurs under acid conditions. The driving force for this work was the aim to produce a PUF copolymer in which the methylol groups are on the amido group of urea and, thus, able to cure rapidly at very mild acidic pH values as a UF resin while retaining a level of water resistance of the resin. Second, in the alkaline pH range, PF resins were shown to be able to copolymerize rapidly with urea, doubling the PF linear degree of polymerization, while presenting complete water resistance of the cured resin. This latter approach was extended to copolymerize fairly high molar amounts of urea and to form a true PUF of excellent exterior performance, much lower cost, and much faster curing and press time while still curing at the alkaline pH characteristic of true PF resins [32,33]. The PUF resin showed higher strength of the finished network and increasing hardened strength when the proportion of urea was progressively increased. In this reported work, the proportion of urea capable of copolymerizing to form the PUF was shown to depend on the formaldehyde:phenol molar ratio of the resin; within certain limits, the higher the molar ratio, the higher the proportion of urea capable of copolymerizing. The upper proportion of urea was limited by the relative increase in viscosity and related pot-life shortening of the resin [32,33].

Both behaviors are easily explained by the relative rates of PF condensation and of urea hydroxymethylation and subsequent combination. Thus, in Figure 8.5, the relative gel times of PF and UF resins are shown, as well as taking into consideration the relative rate constants of PF autocondensation and urea hydroxymethylation and self-condensation, indicating quite clearly in which

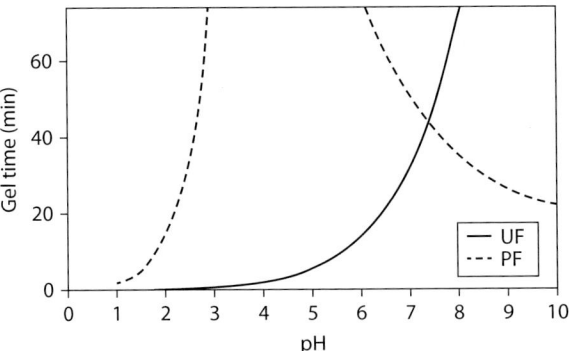

FIGURE 8.5   Relationship of gel time to pH for PF and UF resins.

pH ranges copolymerization is possible and with which species [49]. From Figure 8.5, urea and PF resin, with little free formaldehyde or methylol ureas, will easily copolymerize at a pH higher than 7; however, PF and UF resins copolymerize in the pH range 6–9. The types of reactions and mechanisms involved have been worked out and summarized in Formula 8.11, indicating all the main reactions occurring in the formation of PUF resins [32], even taking into consideration that the urea hydroxymethylation reaction is an equilibrium and that reversible reactions do occur.

PF resols in water solution have been shown [50] to react rapidly and readily with 4,4′-diphenyl methane diisocyanate (polymeric MDI [pMDI]) with minimal deactivation of the isocyanate groups by water. This peculiar behavior is based on the much faster rate of reaction of the isocyanate group with the PF methylol groups (hydroxybenzyl alcohols groups) than with water [56]. Such adhesives are now used industrially, to a limited extent, for bonding difficult-to-glue hardwood veneer species into exterior-grade plywood [51], and present exceptional adhesion, ease of bonding, adequately long pot-life, and very high joint strength. There is now more industrial interest in using these adhesives for particleboard and for wood panels other than plywood, since it has been perceived that they are an excellent and viable alternative to both pure isocyanates as well as pure phenolics. The reactions that lead the adhesive to network cross-linking with these copolymers are (proportions between 5% and 30% of isocyanate on PF resin solids are used when the two are mixed in the glue-mix and core-acted in the hot press while pressing the wood panel; 1%–2% are used when they are prereacted, but this latter approach is not, as yet, used industrially and might present some problems):

$$(8.13)$$

(1) the formation of urethane bridges derived by the reaction of the isocyanate group with the methylol groups of the PF resin [37,50] (Formula 8.13), which is, perhaps, one of the two dominant ones introducing better dissipation of energy at the interface and hence contributing to strength improvement; (2) the classical formation of methylene bridges in the PF resin only [37,50,51] (Formula 8.13); (3) the formation of polyurea networks, due to the reaction of the polymeric poly-isocyanate with the water of the PF resin solution [37,50,51]; and (4) the reaction of the PF methylol group to form methylene bridges on the aromatic nuclei of the pMDI [50].

$$(8.14)$$

This last reaction was also observed to occur between the methylol groups of lignin and the aromatic nuclei of pMDI [52–54]. These reactions are also observed in mixed PF, prereacted lignin, and pMDI [55].

Phenolic resins have also been coreacted with proteins, both to decrease their dependence on oil for deriving phenol and to render them more environmentally acceptable. One example of this is the cocondensation of phenol, wheat gluten protein hydrolysates, and formaldehyde to give mixed resins in which the coreaction of the protein and the phenol with the formaldehyde has been shown to produce mixed species by MALDI-TOF spectroscopy [56]. Up to 30% phenol can be substituted with some gluten protein hydrolysates. The coreaction is based on the reactivity of amido and amino groups of the protein with formaldehyde and the cocondensation of the methylolated protein with methylol phenol oligomers. Even adhesives in which gluten hydrolysates were used alone showed potential as natural interior adhesives for plywood [57].

Other natural materials in addition to proteins have been used to coreact with phenol and form-aldehyde to limit the proportion of oil-derived phenol in the resin. Recently, resins in which up to 70%–80% of phenol has been substituted in the resin preparation with an hydrolysable tannin, namely, chestnut tannin extract, have also been reported, with acceptable results for their applica-tion to wood panels [58].

Phenolic resins based on the reaction of phenol with aldehydes to obtain phenolic resins not dependent on formaldehyde have also been developed, especially with biobased aldehydes. Recently, a method was developed to transform lignin-based aromatic aldehyde precursors, that is, 4-hydroxybenzaldehyde and vanillin, into reactive and difunctional biobased aromatic aldehydes. These were used to prepare formaldehyde-free partially biobased phenolic resins with thermal properties competitive with those of PF resins [59]. Other researchers have concentrated on using 5-hydroxymethylfurfural (HMF) produced *in situ* from glucose reacted with phenol to form resins to bind glass fiber composites. This resin is reported to be heat curable using cross-linking agents such as tetraethylammonium chloride [60,61].

8.4.4 PREDICTION OF PROPERTIES

Only a small amount of work has been done up till now concerning the prediction of bond strengths and other properties based on the results of the analysis of the PF resin. Correlation equations to evaluate the chemical structures in various PF resins with different F/U molar ratios and different types of preparation, on the one hand, and the achievable internal bond as well as the subsequent formaldehyde emission, on the other hand, have been developed [62]. These equations are valid only for well-defined series of resins. The basic aim of such equations is the prediction of the properties of the wood-based panels based on the composition and the properties of the resins used. For this purpose, various structural components of the liquid resin are determined by ^{13}C NMR and their ratios related to board results and, thus, to the strength results of the hardened resin. Various papers in the chemical literature describe examples of such correlations, in particular for UF, MF, melamine-urea-formaldehyde (MUF), and PF resins [62–65]. For example, one type of equation correlating the dry internal bond (IB) strength (tensile strength perpendicular to the plane of the panel) of a particleboard bonded with PF adhesive resins is as follows [37,62]:

$$\text{Resin cross-linking} \approx IB = a \cdot A / (A + B + C) + b \cdot Mo / (A + B + C) + c \cdot Me / (A + B + C) \quad (8.1)$$

where:
- IB   strength is expressed in MPa
- A   is the sum of peak areas of phenolic ortho and para sites still free to react (110–122 ppm)
- B   is the sum of peak areas of phenolic meta sites (125–137 ppm)
- C   is the sum of peak areas of phenolic ortho and para sites already reacted (125–137 ppm)
- Mo   is the sum of peak areas of phenolic methylol groups (59–66 ppm)
- Me   is the sum of peak areas of methylene bridges connecting the phenolic nuclei (30–45 ppm)
  Coefficients a, b, and c are characteristics of the type of resin and depend on a variety of manufacturing parameters

Equation 8.1 is one of the simpler equations of this type, the equations for UF resins in particular being more complex. Similar equations correlating the level of crystallinity of hardened aminoplastic resin, the IB strength of the board prepared with it, the level of cross-linking of the resin, and the formaldehyde emission of the panel and resin with the ^{13}C NMR spectrum of the liquid resin have also been presented [37,63–66].

For certain boards, good correlation exists. However, it must be assumed that a general correlation for various resins and various panels will not exist and that, maybe, other correlation equations must be used. Nevertheless, these results are rather important, because they show that at least for a special combination of resin type and board type, correlation between analysis of the resin in its liquid state and the strength of the same resin in hardened form exists, and that forecast of performance can be made just based on the analysis of the liquid resin. Furthermore, the various variables used and the corresponding chemical and physical groups in the liquid resin will also be decisive for other resin combinations and manufacturing procedures, although the values of the coefficients for each individual equation may differ. However, it also must be considered that the range of molar ratios under investigation in the papers mentioned is rather broad. At the moment, it is not possible to use these equations for predictions within too narrow a range of molar ratios.

8.5 RESORCINOL ADHESIVES

8.5.1 INTRODUCTION

Resorcinol–formaldehyde (RF) and phenol–resorcinol–formaldehyde (PRF) cold-setting adhesives are used primarily in the manufacture of structural, exterior-grade glulam, fingerjoints, and other exterior timber structures. They produce bonds not only of high strength, but also of outstanding water and weather resistance when exposed to different climatic conditions [67,68]. PRF resins are prepared mainly by grafting resorcinol onto the active methylol groups of the low-condensation resols obtained by the reaction of phenol with formaldehyde. Resorcinol is the chemical species that gives to these adhesives their characteristic cold-setting behavior. At ambient temperature and on addition of a hardener, it provides accelerated and improved cross-linking not only to RF resins, but also to the PF resins onto which resorcinol has been grafted by chemical reaction during resin manufacture. Resorcinol is an expensive chemical, produced in very few locations around the world (to date, only three commercial plants are known to be operative: in the United States, Germany, and Japan), and its high price is the determining factor of the cost of RF and PRF adhesives. It is for this reason that the history of RF and PRF resins is closely interwoven, by necessity, with the search for a decrease in their resorcinol content, without loss of adhesive performance.

In the past decades, significant reductions in resorcinol content have been achieved, from pure RF resins to PRF resins in which phenol and resorcinol were used in equal or comparable amounts, to the modern-day commercial resins for glulam and fingerjointing in which the percentage, by mass, of resorcinol in liquid resins is of the order of 15%–18%. A step forward has also been the development and commercialization of the *honeymoon* fast-set system [69], composed either of just synthetic PRF resins or of a PRF resin coupled with the use of tannin extracts, that in certain countries is used to obtain PRFs of 8%–9% resorcinol content without loss of performance and with some other advantages (such as gluing of high-moisture-content timber). This was a system improvement, not an advance in the basic formulation of PRF resins.

8.5.2 Chemistry of RF Resins

The same chemical mechanisms and driving forces presented for PF resins apply to resorcinol resins. Resorcinol reacts readily with formaldehyde to produce resins that harden at ambient temperature if formaldehyde is added. The initial condensation reaction, in which A-stage liquid resins are formed, leads to the formation of linear condensates only when the resorcinol:formaldehyde molar ratio is approximately 1:1 [70]. This reflects the reactivity of the two main reactive sites (positions 4 and 6) of resorcinol [71]. However, reaction with the remaining reactive but sterically hindered site (2-position) between the two hydroxyl functions of resorcinol also occurs [67]. The proportion of 4- plus 6-linkages relative to 2-linkages is 10.5:1 on a molar basis. The difference in reactivity of the two types of sites (i.e., the 4- or 6-position relative to the 2-position) is then 5:1 [70]. Linear components always appear to form in preference to branched components in A-stage resins [70], that is, leading to the preferential formation of linear rather than branched condensates. This fact can be attributed to:

1. The presence of two reactive nucleophilic centers on the terminal units, as opposed to single centers of doubly bound units already in the chain.
2. The greater steric hindrance of the available nucleophilic center (nearly always at the 2-position) of the doubly bonded units, as opposed to the lower steric hindrance of at least one of the nucleophilic centers of the terminal units (a 4- or 6-position is always available). The former is less reactive as a result of the increased steric hindrance. The latter are more reactive.
3. The lower mobility of doubly bound units, which further limits their availability for reaction.

$$(8.15)$$

The absence of methylol (–CH$_2$OH) groups in all six lower molecular weight RF condensates that have been isolated [4] reflects the high reactivity of resorcinol under acid or alkaline conditions. It also shows the instability of its para-hydroxybenzyl alcohol groups and their rapid conversion to para-hydroxybenzyl carbonium ions or quinone methides. This explains why identical condensation products are obtained under acid or alkaline reaction conditions [4]. In acid reaction conditions, methylene ether-linked condensates are also formed, but they are highly unstable and decompose to form stable methylene links in 0.25–1 h at ambient temperature [72,73].

From a kinetic point of view, the initial condensation reaction to form dimers is much faster than the subsequent condensation of these dimers and higher polymers. The condensation reaction of resorcinol with formaldehyde on an equal molar basis and under identical conditions also proceeds at a rate that is approximately 10–15 times faster than that of the equivalent PF system [5]. The high reactivity of the RF system renders it impossible to have these adhesives in resol form. Therefore, only RF novolaks, that is, resins not containing methylol groups, can be produced. Thus, all the resorcinol nuclei are linked together through methylene bridges, with no methylol groups present and generally without any presence of methylene-ether bridges either.

The reaction rate of resorcinol with formaldehyde is dependent on the molar ratio of the two constituents, the concentration of the solution, pH, temperature, presence of various catalysts, and amounts of certain types of alcohols present [74–76]. The effect of pH and temperature on the reactivity and gel time of the RF system presents the same trend as for all PF reactions, with a minimum of reactivity at around a pH of 4 and the reaction becoming rapidly faster at progressively more alkaline and more acid pH values [77]. Methanol and ethanol slow down the rate of reaction. Other alcohols behave similarly, the extent of their effect being dependent on their structure. Methanol lengthens gel time more than other alcohols, higher alcohols being less effective. The retarding effect on the reaction is due to temporary formation of hemiacetals between the methanol (or other alcohols) and the formaldehyde. This reduces the reaction rate, because of the lower concentration of available formaldehyde [75,76]. Other solvents also affect the rate of reaction by forming complexes or by hydrogen bonding with the resorcinol [75,77].

In the manufacture of pure resorcinol resins, the reaction would be violently exothermic unless controlled by the addition of alcohols [70]. Because the alcohols perform other useful functions in the glue-mix, they are left in the liquid glue. PRF adhesives are generally prepared, firstly, by reaction of phenol with formaldehyde to form a PF resol polymer that has been shown to be in the highest percentage, and is often completely linear [1]. This can be represented as Structure I.

$$(8.16)$$

m ≥ 0 in integer numbers
0 ≤ n ≤ 2 in integer numbers

I

In the reaction that follows, the resorcinol chemical is added in excess, in a suitable manner, to Polymer I to react with the –CH$_2$OH groups to form PRF polymers in which the terminal resorcinol groups can be resorcinol chemical or any type of RF polymer.

$$(8.17)$$

P > 1 in integer numbers

In reality, as resorcinol chemical is expensive, the resin manufacturers tend to limit the amount of resorcinol grafted onto the PF resol. This is, then, generally lower than what is necessary to form structures as shown in Formula 8.17. Structures such as these do occur, but the majority of the polymers present in the PRF resin are of the type shown:

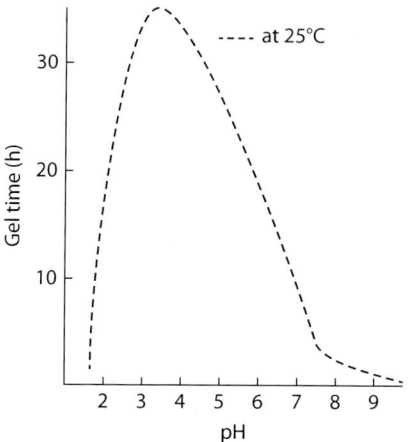

$$(8.18)$$

With n and m > 1 and integer numbers

where the residual third reactive site of the resorcinol that is still free is the site through which cross-linking takes place by reaction with the added formaldehyde hardener (Figure 8.6).

Where straight resorcinol adhesives are not suitable, resins can be prepared from modified resorcinol [77]. Examples of these types of resins are those used for tire cord adhesives, in which a pure RF resin is used; or, alternatively, alkylresorcinol or oil-soluble resins suitable for rubber compounding are obtained by prereaction of resorcinol with fatty acids in the presence of sulfuric acid at high temperature, followed by reaction with formaldehyde. Worldwide, more than 90% of resorcinol adhesives are used as cold-setting wood adhesives. The other most notable application is as tire cord adhesives, which constitute less than 5% of the total use.

8.5.3 WOOD LAMINATING AND FINGERJOINTING ADHESIVES

Various adhesive formulations can be used for the manufacture of laminated wooden beams and fingerjoints for structural purposes. Only those adhesive formulations that at some time or other have been used in industrial applications will be described. All these formulations are based totally or partially on resorcinol, and the hardening process is carried out at ambient temperature up to 50°C [24,78].

FIGURE 8.6 Effect of pH and temperature on the reactivity of the resorcinol–formaldehyde system.

8.5.3.1 Adhesive 1: Resorcinol–Formaldehyde Adhesive

An RF novolak is produced according to the schematic reaction shown:

$$n \geq 0 \tag{8.19}$$

If paraformaldehyde and fillers—generally wood and nutshell flours—are added, the resin becomes capable of setting in 2–3 h and curing in 16–24 h at ambient temperature.

8.5.3.2 Adhesive 2: Phenol–Resorcinol–Formaldehyde Adhesive and Powder, Liquid, or Sludge Hardener

A PF resol is prepared and resorcinol is grafted onto it according to the schematic reaction shown:

$$\tag{8.20}$$

The resin produced is capable of setting in 2–3 h and cures in 16–24 h at ambient temperature, once paraformaldehyde and wood and nut flour fillers have been added. This is the most commercially used type of resorcinol cold-setting adhesive, with variations mainly concerning the hardener. Thus, sometimes and especially in Europe, sludges formed by suspensions of organic or inorganic fillers in water mixed with liquid hardeners such as formalin and UF precursor concentrates are used as hardeners for this type of adhesive. As these sludges and liquid hardeners have a bad formaldehyde odor during usage, hardeners based on odorless oxazolidines are often used instead of formaldehyde-based ones in these liquid or sludge hardeners.

8.5.3.3 Adhesive 3: Urea–Resorcinol–Formaldehyde Adhesive

A UF resin is prepared and resorcinol is grafted onto it as both terminal units or as middle connecting units, leaving each resorcinol to present one or two still reactive sites through which to effect the final cross-linking and hardening of the resin. The schematic reaction is shown:

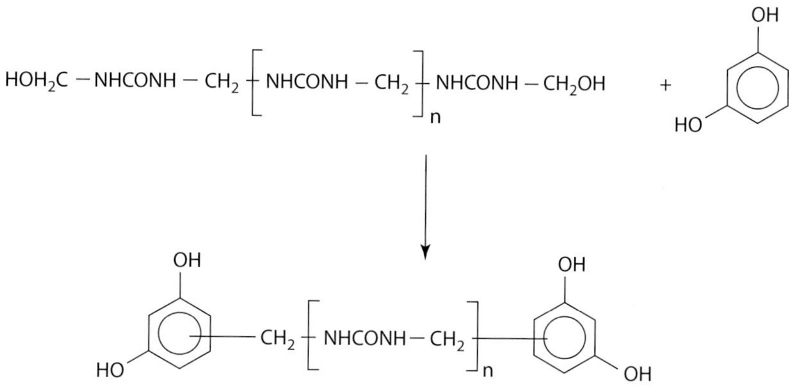

HOH2C – NHCONH – CH2 [NHCONH – CH2]n NHCONH – CH2OH
+
OH
HO
OH
HO
CH2 [NHCONH – CH2]n
OH
OH

(8.21)

The behavior of this adhesive is identical to that of Adhesive 2, although larger amounts of resorcinol are used. These adhesives have a higher urea:formaldehyde ratio, are water-resistant, and are capable of radio-frequency curing. They can also be used for plywood manufacture, although the high price of resorcinol renders them unsuitable for such an application.

8.5.3.4 Adhesive 4: Phenol–Resorcinol–Formaldehyde and Liquid Hardener Type 1

A PF resol is prepared that constitutes the resin. Resorcinol or RF novolaks, in aqueous or water–alcohol solutions, are used as hardeners according to the scheme shown:

OH
OH
OH
HOH2C
CH2
CH2
CH2OH
+
OH
OH
H2C
CH2
CH2
CH2
OH
OH
OH
CH2
CH2
OH
Cured resin

(8.22)

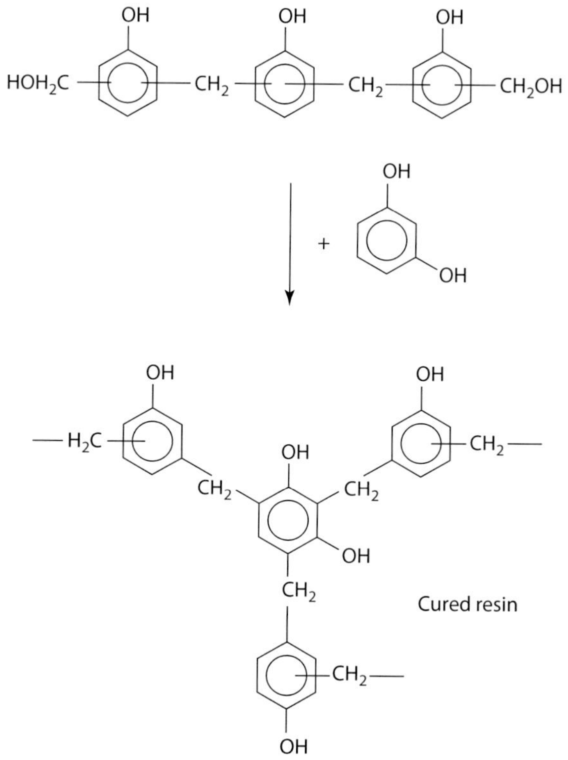

8.5.3.5   Adhesive 5: Phenol–Resorcinol–Formaldehyde and Liquid Hardener Type 2

A phenol–resorcinol–formaldehyde (PRF) similar to Adhesive 2 is prepared and a PF resol of the same type as Adhesive 4 is used as hardener. The schematic curing reaction is as follows:

$$n \geq 0$$
$$m \geq 0$$

(8.23)

Cured resin
(similar to adhesive 4)

    The PRF adhesives are always delivered as a liquid and must be blended with a hardener before use. PRF adhesives of Type 2 in which a liquid PRF adhesive is mixed with a powder hardener have been the most commonly used industrial systems, although in Europe the use of sludge hardeners of Type 2 is fairly common too. When using a powder hardener in adhesives of Types 1, 2, and 3, the liquid adhesive resin (50%–60% solids content) is mixed with the powder hardener in a 5:1 weight ratio before use. The powder hardener is generally a mixture of 10 parts paraformaldehyde and 10% fillers. It is comprised of 200-mesh wood flour or a mixture of wood flour and nutshell flour, also 200 mesh. Adhesives of Types 4 and 5 have a liquid resin:liquid hardener ratio of 1:1 by mass. This is so because the hardener is also a resin. Adhesives of Types 4 and 5 have been used quite extensively in the past in certain markets, but have now been superseded by adhesives of Type 2, which have several handling advantages.

    Adhesives of Type 3, although good, have not really caught on commercially, and were developed as an alternative to the PRF of Type 2 due to the ever-increasing price of oil-derived phenol. In this regard, adhesives of Type 2 in which the phenol has been completely substituted by a flavonoid tannin have also been developed, and have been used commercially in a few southern hemisphere countries now for more than 30 years. Their preparation and performance are detailed in Chapter 9 of this book. Pure RF adhesives (Type 1) were used extensively earlier (until some 25–35 years ago). They fell into disfavor because of the high price of the resorcinol chemical needed to make them, and also due to the shortage of resorcinol supply in 1960 and 1970 (oil crises). They are still used in some industrial applications, particularly at low-temperature curing and when difficult wood-gluing problems arise, but they constitute less than 1% of the total market by volume, except in tire cord adhesives, which consume, in general, between 1% and 12% of the resorcinol produced in the world (hence, about one-sixth of the amount used for wood adhesives). PRF adhesives together with liquid hardeners are used quite extensively in Europe, as they have several handling advantages for this market. These adhesives are mixed before use in a weight ratio of adhesive to liquid hardener from 5:1 to 2.5:1 or even 1:1 (depending on the type of hardener). The powder hardener is generally a mixture of equal parts of paraformaldehyde fine powder and fillers, these latter being comprised of 200-mesh wood flour or a mixture of wood flour and nutshell flour of 200–300 mesh. Adhesive types with a liquid hardener use formaldehyde, paraformaldehyde, or formurea (a prepolymer of urea and formaldehyde) as hardener. When an undue formaldehyde odor needs to be avoided, these types of adhesives use also oxazolidine as a hardener. In addition, these hardeners also contain fillers and thickening agents.

    All properly formulated resorcinol-based adhesives must have a viscosity low enough in aqueous–alcoholic solutions to flow with ease into all the interstices of the wood surface. Wetting ability is promoted by an alcohol. The paraformaldehyde used as hardener in powder hardener adhesives is an addition polymer composed of a few to over 100 formaldehyde monomers. It dissolves slowly in water by depolymerization to formaldehyde monomers. The rate of depolymerization depends on the degree of polymerization of the paraformaldehyde, the size of the paraformaldehyde powder particles, and the pH. Therefore, the working life or pot-life of a glue-mix can be adjusted by selecting the type of paraformaldehyde and the pH correctly. However, the ratio of resorcinol to phenol will have an influence on the pH value to be chosen for hardening. Fillers are added to give consistency to the glue-mix, to control viscosity and thixotropic characteristics, to form a fibrous reinforcement of the adhesive film, and to lessen the cost. Wood flour is used as a filler to obtain better gap filling properties where rough or uneven surfaces must be bonded, or where low bonding pressures must be used. Nutshell flours, such as coconut shell flour, walnut shell flour, peach pip shell flour, macadamia nutshell flour, or even olive stone flour, are used as fillers as well and to provide smooth-flowing powder mixtures. Inorganic fillers are also used, especially in Europe, although this is a practice to avoid, as saws wear out rapidly afterward. Clays and fumed silica can also be used in very small amounts to control the thixotropic consistency of the glue-mix.

As the formaldehyde reacts with the resorcinol-based resin, condensation occurs with the formation of high molecular weight polymers. There is considerable secondary forces interaction between the growing resorcinol polymers and the noncrystalline hemicellulosic and lignocellulosic molecules of the wood substrate. The highly polar methylol groups and the phenolic hydroxy groups link to cellulose and lignin groups by van der Waals, hydrogen, and electrostatic bonds. The growing adhesive polymers continue to interact to form colloidal particles, and then a gelatinous film. This mechanism depends strongly on the moisture content of the wood, which determines the rate of water and solvent absorption.

The advantage of ambient-temperature curing is that the moisture escapes gradually from the hard film formed on curing, inducing a minimum of residual stresses in the joint and allowing the glueline to assume the aspect of a molecularly porous solid. As a consequence, the hard film is able to transpire in the same way as wood, which minimizes cracking or crazing and allows the glued joint to survive exposure to extremes of humidity cycles. To shorten the curing time, heating chambers (40°C–60°C) are often used in the manufacturing of glulam, and an even faster curing can be achieved using radio-frequency curing presses. Typical levels of strength and wood failure results obtained in specific standard tests are shown in Table 8.3.

8.5.4 Special Adhesives with Reduced Resorcinol Content

8.5.4.1 Fast-Setting Adhesives for Fingerjointing and Glulam

Together with the more traditional fingerjointing adhesives that have just been discussed, a series of ambient-temperature fast-setting separate application systems have also been developed. These eliminate the long delays caused by the use of more conventional PRF adhesives, which require lengthy periods to set. These types of resorcinol adhesives are applied separately. They were first

TABLE 8.3

Typical Tensile Strength and Percentage Wood Failure Results Obtained with Synthetic PRF Resins

Dry Test		24 h Cold Water Soak		6 h Water Boil Test	
Strength (N)	Wood Failure (%)	Strength (N)	Wood Failure (%)	Strength (N)	Wood Failure (%)
3000–3500	90–100	2600–3200	75–100	2500–3000	75–100

Source: Pizzi, A., *Handbook of Adhesive Technology*, Marcel Dekker, New York, 2003.

developed in the United States [79,80] to bond large components where presses were impractical. Kreibich [80] describes these separate application or honeymoon systems as follows: Component A is a slow-reacting PRF resin with a reactive hardener. Component B is a fast-reacting resin with a slow-reacting hardener. When A and B are mated, the reactive parts of the components react within minutes to form a joint that can be handled and processed further. Full curing of the slow-reacting part of the system takes place over time. The m-aminophenol used for Component B is a very expensive chemical and, for this reason, these systems were discarded and not used industrially [24]. In their original concept, Component A was a traditional PRF cold-setting adhesive at its standard pH of 8–8.5 to which formaldehyde hardener was added. Flour fillers may be added or omitted from the glue-mix. Component B is a phenol–meta-aminophenol–formaldehyde resin with a very high pH (and therefore high reactivity) that contains no hardener, or only a very slow hardener.

More recently, a modification of the system described by Kreibich has been used extensively in industry with good success. Part A of the adhesive is again a standard PRF cold-setting adhesive with powder hardener added at its standard pH. Part B can either be the same PRF adhesive with no hardener and the pH adjusted to 12, or a 50%–55% tannin extract solution at a pH of 12–13, provided that the tannin is of condensed or flavonoid type, such as mimosa, quebracho, or pine bark extract, with no hardener [3,81]. The results obtained with these two systems are good, and the resin not only has all the advantages desired, but also, as a result of the use of vegetable tannins and of halving the resorcinol content of the entire adhesive system, it is considerably cheaper [3,78,79].

The adhesive works in the following manner. Once Component A of the glue-mix is spread on one fingerjoint profile and Component B on the other fingerjoint profile and the two profiles are joined under pressure, the reaction of Component B with the hardener of Component A is very fast. In 30 min at 25°C, fingerjoints prepared with these adhesives generally reach the levels of strength that fingerjoints glued with more conventional phenolic adhesives are able to reach only after 6 h at 40°C–50°C or in 16–24 h at 25°C [3,82]. The clamping time of laminated beams (glulam) bonded with these fast-setting honeymoon adhesives is, on average, only 3 h at ambient temperature, compared with the 16–24 h necessary with traditional PRF resins [81,82]. These adhesives also present two other advantages, namely, (1) they are capable of bonding without any decrease in performance at temperatures down to 5°C, and (2) they are able to bond green timber with high moisture content, a feat that has been used in industrial glulam bonding since their commercial introduction in 1981. Several variations on this theme exist, such as the *Greenweld system* from New Zealand, in which Component B is a solution just composed of a thickener and ammonia as a strong accelerator for Component A (comprising the PRF resin and the hardener). This system, however, appears to suffer from the presence of the odor of ammonia, which is unacceptable in some sophisticated markets.

8.5.4.2 Branched PRF Adhesives

Recently, another step forward has been taken in the formulation of PRF adhesives of lower resorcinol content. Liquid resorcinol or PRF resins appear to be mostly linear [70]. The original concept of *branching* erroneously maintained that if a chemical molecule capable of extensive branching (three or more effective reaction sites with an aldehyde) the PF and PRF resins is used before, during, or after the preparation of the PF resin, the polymer in the branched PRF adhesive has (1) higher molecular weight than in normal PRF adhesives where branching is not present and (2) higher viscosity in water or water–solvent solutions of the same composition and of the same resin solids content (concentration). It also requires a much lower resorcinol amount on total phenol to present the same performance as normal linear PRF adhesives. This can be explained schematically as follows:

Resorcinol $- CH_2 -\left[Phenol - CH_2 \right]_n$ Resorcinol

Resorcinol $- CH_2 -\left[Phenol - CH_2 \right]_n$ Resorcinol

Resorcinol $- CH_2 -\left[Phenol - CH_2 \right]_n$ Resorcinol

n in integer numbers

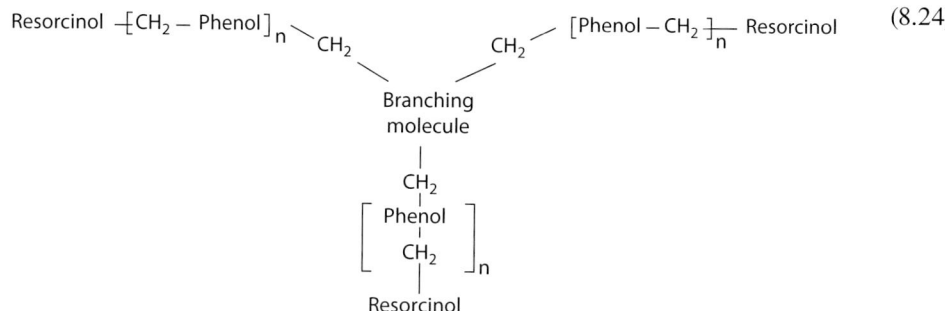

$$\text{Resorcinol} -\left[CH_2 - Phenol \right]_n CH_2 \qquad CH_2 \left[Phenol - CH_2 \right]_n \text{Resorcinol} \qquad (8.24)$$

Branching
molecule
|
CH_2
|
$$\left[\begin{array}{c} CH_2 \\ | \\ Phenol \\ | \\ CH_2 \end{array} \right]_n$$
|
Resorcinol

where $n > 1$ and is an integer number comparable to, similar to, or equal to n in the preceding scheme for the production of PRF resins (Formula 8.23).

When comparing linear and branched resins for every n molecules of phenol used, a minimum of two molecules of resorcinol are used in the case of a normal, traditional linear PRF adhesive, whereas only one molecule of resorcinol for n molecules of phenol is used in the case of a branched PRF adhesive. The amount of resorcinol has thus been halved, or approximately halved, in the case of the branched PRF resin. A second effect caused by the branching is a noticeable increase in the degree of polymerization of the resin. This causes a considerable increase in the viscosity of the liquid adhesive solution. Because PRF adhesives must be used within fairly narrow viscosity limits, to readjust the viscosity of the liquid PRF adhesive within these limits, the resin solids content in the adhesive must be lowered considerably, with a consequent further decrease in the total liquid resin of the amount of resorcinol and of the other materials except solvents and water. This decreases the cost of the resin further without decreasing its performance.

Thus, to conclude, the decrease of resorcinol by branching of the resin is based on two effects:

1. A decrease of resorcinol percentage in the polymer itself, hence in the resin solids, due to the decrease in the number of the PF terminal sites onto which resorcinol is grafted during PRF manufacture.
2. An increase in molecular weight of the resin, which requires a decrease in the percentage resin solids content to a workable viscosity, and thus decreases the percentage of resorcinol in liquid resin (not on resin solids).

It is clear that, in a certain sense, a branched PRF will behave as a more advanced, almost precured phenolic resin. While the first effect described is a definite advance on the road to better engineered PRF resins, the second effect can also be obtained with more advanced (reactionwise) linear resins. The contribution of the second effect to the decrease in resorcinol is no less marked than that of the first effect. It is, however, the second effect that accounts for the difference in behavior between branched and linear PRF adhesives.

Branching molecules that can be used may be resorcinol, melamine, urea, and others [83]. Urea is the favorite one, because it is much cheaper than the others and needs to be added in only 1.5%–2% of total resin. When urea is used as a brancher, the adhesive assumes an intense and unusual (for resorcinol resins) blue color after a few days; hence its nickname, *blueglue*. However, later work [84,85] has shown that tridimensional branching has very little to do with the improved performance of these low resorcinol content adhesives, with tridimensionally branched molecules contributing, at best, no more than 8%–9% to the total strength [84,85]. In reality, addition of urea causes the reaction as foreseen, but not in three-point branching. This is equivalent to saying that most of the resin doubles linearly in molecular weight and degree of polymerization, while the final effect, good performance at half the resin resorcinol content, is maintained [84,85]. This effect is based on the relative reactivity of urea and unreacted phenol sites in competition for phenolic methylols. Thus, while the final macroeffect is as wanted, at the molecular level it is only a kinetic effect due to the different relative reactivities of urea and phenol under the reaction conditions used. Thus:

$$\text{Resorcinol–CH}_2\text{–[–phenol–CH}_2]_n \text{ resorcinol}$$
$$\text{Resorcinol–CH}_2\text{–[–phenol–CH}_2]_n \text{ resorcinol}$$
$$\downarrow$$
$$\text{Resorcinol–CH}_2\text{–[–phenol–CH}_2]_n\text{– urea–CH}_2\text{–[–phenol–CH}_2]_n \text{ resorcinol}$$

The halving of the resorcinol content is still obtained, but between 90% and 100% of the polymers in the resin are still linear.

It is noticeable that the same degree of polymerization and the doubling effect cannot be obtained by lengthening the reaction time of a PF resin without the addition of urea [84,85]. These liquid resins then work at a resorcinol content of only 9%–11%, hence, considerably lower than that of standard commercial PRF resins. These resins can also be used with good results for honeymoon fast-setting adhesives in PRF–tannin systems, thus further decreasing the total content of resorcinol in the total resin system to a level as low as 5%–6%.

8.5.4.3 Cold-Setting PF Adhesives Containing No Resorcinol or Using Alternative Aldehydes

Due to the pressure from regulatory agencies and public awareness of formaldehyde, also in the case of resorcinol adhesives, alternative aldehydes have been tried for partial substitution of the formaldehyde content of the adhesive. Phenol–resorcinol–furfural resins have been successfully prepared in which all the formaldehyde used to prepare the resin itself has been substituted with furfural [86]. However, notwithstanding the decrease in formaldehyde, it still has to be used as a hardener to achieve the proper rate of curing at ambient temperature for these adhesives [86].

As the cost of cold-setting exterior-grade adhesives based on resorcinol is very high, due to the high cost of resorcinol itself, the tendency is to decrease the amount of resorcinol while maintaining the performance of the adhesive unaltered. This leads to the concept of exterior cold-setting phenolic adhesives of zero level resorcinol. As the rate of reaction at ambient temperature of alkaline PF resins is not sufficient to harden the adhesive, some modifications need to be introduced to overcome the absence of resorcinol in this type of resin. This can be done in several ways: (1) By using standard PF thermosetting resol resins and hardening them by increasing the glueline temperature by radio-frequency in fingerjointing and glulam manufacture. The system is expensive and needs considerably higher capital outlay and more careful handling of both the equipment and the joint, for results that are certainly not particularly exciting. (2) By using resins in which the PF resol of adhesives of Type 2 is grafted with a resorcinol substitute; for example, a natural polyflavonoid tannin [87]. This system is truly cold-setting and yields relatively good results, but is just on the lower limit of the standard requirements [87]. (3) By using self-neutralizing acid-setting PF resols. The term *acid-setting*, when used in the presence of a lignocellulosic substrate, makes wood technologists shudder, conjuring up visions of extensive acid-induced substrate degradation and early exterior joint failure. And this is indeed the case! In reality, some exterior aminoplastic resins do harden in the moderately acid range without any major substrate degradation problems. PF resins, however, while hardening very rapidly under acid conditions, do need very acid conditions to give a hardened strong network, and this elevated acidity is not really acceptable as regards long-term durability of the substrate. The damage due to the acid hydrolysis of cellulose and other wood carbohydrates is particularly aggravated and compounded by the long-term effect of the acid remaining in the glueline after resin hardening. However, the main negative effect due to acid-induced degradation of the substrate has been overcome by using acid-setting PF resins containing no resorcinol, but hardened by using a self-neutralizing catalyst [88]. According to this principle, the adhesive first becomes acid to allow the PF resin to cure, and after hardening, the hardened glueline self-neutralizes in a very short time [88]. The great majority of the effects of substrate degradation are thus avoided, and very strong and durable exterior wood joints are produced [88]. The system works well in radio-frequency cured joints, yielding much better results than the alkaline resols of point (1), and can work well under purely cold-setting conditions [88] (Figure 8.7).

Other rapid-setting adhesive systems containing no resorcinol are those based on MUF resins and on one-component polyurethanes, which are both described in Chapters 10 and 11, this volume [89,90].

8.6 FORMULATION OF PRF ADHESIVES

A basic formulation capable of giving more than adequate results is presented here, so that a starter in the field can get acquainted with these types of adhesives. The procedure for the preparation of this resin can be modified in many ways by varying the catalyst, concentration, molar ratio, and condensation conditions.

- Phenol, 110 parts by mass plus 22 parts water
- First formalin 37% solution, 49 parts by mass

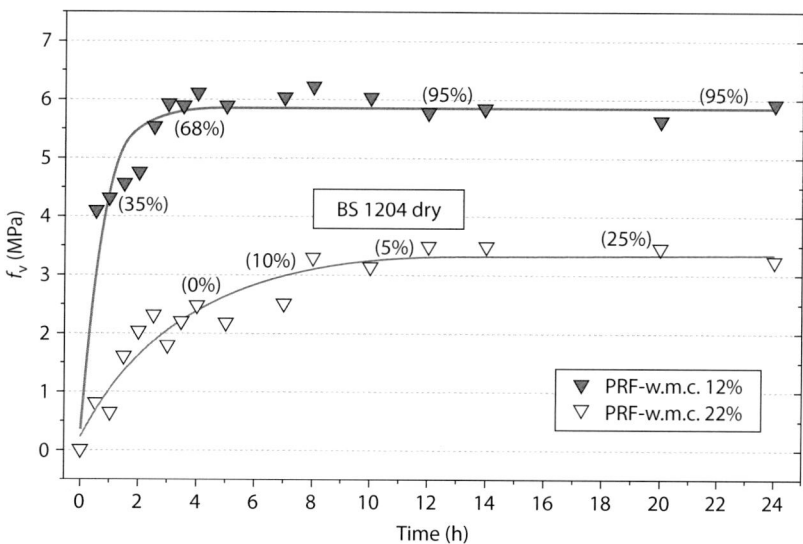

FIGURE 8.7 Typical tensile strength f_v and wood failure (in percent) increase as a function of time of a pure PRF honeymoon adhesive system at 12% and 22% wood moisture content (w.m.c.) [87–90]. Note that at the unusually high moisture content of 22%, the requirements of the standards are reached in less than 24 h as the timber starts to dry [87–90]. Test on beech according to British Standard BS 1204.

- H_2SO_4 10% solution, 22 parts by mass
- First NaOH 40% solution, 4.5 parts by mass
- Second NaOH 40% solution, 9.25 parts by mass
- Second formalin 37% solution, 90–93 parts by mass
- Methanol or methylated spirits, 30 parts by mass (at start of reaction)
- Resorcinol, 71 parts by mass
- Tannin extract, 19 parts by mass (only as a thickener; another thickener can also be used, as long as it is not excessively sensitive to water attack)

Phenol, water, methanol, and the first amount of formalin solution are charged in the reaction vessel and heated mildly until the phenol is dissolved. H_2SO_4 is added and the temperature increased to reflux under continuous mechanical stirring. The mixture is refluxed for 3.5–4 h (in general, approximately 4 h) under continuous mechanical stirring. It is cooled to 50°C–60°C, and the following are added: the two amounts of NaOH 40% solution (slowly) and the second amount of formalin solution, while maintaining the mix under continuous mechanical stirring. The mixture is then refluxed for 4.5–4.75 h, and then resorcinol is added. The mixture is refluxed for a further 30–50 min. Spray-dried mimosa tannin extract is added immediately before or during cooling to adjust the viscosity to the desired level. The pH must be adjusted to 8.5–9.5, depending on the pot-life required. The reaction must be carried out under continuous mechanical stirring throughout the whole reaction period. The hardener is a 50:50 mixture of 96% fine powder paraformaldehyde (usually a fast grade) and 180–200-mesh softwood flour (60:40 mass proportion), which is added to the liquid resin generally in proportion by weight of liquid resin:powder hardener = 100:20–100:25.

8.7   SUMMARY

Phenolic resins, both thermosetting and cold-setting, are the oldest synthetic resins developed, and they are extensively used for adhesives in large amounts and in a large number of different applications. Phenolic resin adhesives constitute approximately 50% of all the different applications of these resins. Their chemistry is complex, but their performance as exterior and weatherproof wood adhesives is excellent and unparalleled, especially considering their relatively affordable cost and their relative ease of manufacture. However, significant areas for investigation and improvement still exist, and progress in these adhesives is continuing. This chapter has outlined the role and most significant characteristics of these adhesives, as well as some more recent developments, although it does not pretend to be exhaustive.

REFERENCES

1. A. Pizzi, R. M. Horak, D. Ferreira, and D. G. Roux. Condensates of phenol, resorcinol, phloroglucinol and pyrogallol, as flavonoids A- and B-rings model compounds with formaldehyde, Part 2. *Cell. Chem. Technol.* 13, 753–762 (1979).
2. N. J. L. Megson. *Phenolic Resin Chemistry*, Butterworth, London (1958).
3. R. Garcia and A. Pizzi. Cross-linked and entanglement networks in thermomechanical analysis of poly-condensation resins. *J. Appl. Polym. Sci.* 70, 1111–1116 (1998).
4. E. Ziegler, I. Hontschik and L. Milowiz. Über die Kondensation von Phenolpseudohalogeniden mit Phenolen. *Monatsch*, 78, 334–335 (1948).
5. C. Caesar and A. N. Sachanen. Thiophene-formaldehyde condensation. *Ind. Eng. Chem.* 40, 922–928 (1948).
6. A. Pizzi. Phenol and tannin-based adhesive resins by reactions of coordinated metal ligands, Part 1: Phenolic chelates. *J. Appl. Polym. Sci.* 24, 1247–1255 (1979); Phenolic resins by reactions of coordinated metal ligands. *J. Polym. Sci., Polym. Lett.* 17, 489–491 (1979).
7. A. Pizzi and A-Stephanou. On the chemistry, behaviour and cure acceleration of phenol-formaldehyde resins under very alkaline conditions. *J. Appl. Polym. Sci.* 49, 2157–2160 (1993).

8. K. Hultzsch. Studien auf dem Gebiet der Phenol-Formaldehyd-Harze, XIV. Mitteil.:Über die Ammoniak-Kondensation und die Reaktion von Phenolen mit Hexamethylentetramin, *Chem. Ber.* 82, 16 (1949).

9. A. Zinke. The chemistry of phenolic resins and the processes leading to their formation. *J. Appl. Chem.* 1, 257–266 (1951).

10. S. Sojka, R. A. Wolfe and G. D. Guenther. Formation of phenolic resins: Mechanism and time dependence of the reaction of phenol and hexamethylenetetramine as studied by carbon-13 nuclear magnetic resonance and Fourier transform infrared spectroscopy. *Macromolecules* 14, 1539–1543 (1981).

11. F. Pichelin, C. Kamoun and A. Pizzi. Hexamine hardener behaviour – effects on wood glueing, tannin and other wood adhesives. *Holz. Roh. Werkst.* 57, 305–317 (1999).

12. C. Kamoun and A. Pizzi. Mechanism of hexamine as a non-aldehyde polycondensation hardener, Part 1: Hexamine decomposition and reactive intermediates. *Holzforschung Holzverwertung* 52, 16–19 (2000).

13. C. Kamoun and A. Pizzi. Mechanism of hexamine as a non-aldehyde polycondensation hardener, Part 2: Recomposition of intermediate reactive compounds. *Holzforschung Holzverwertung* 52, 66–67 (2000).

14. D. A. Fraser, R. W. Hall and A. J. L. Raum. Preparation of 'high-*ortho*' novolak resins I. Metal ion catalysis and orientation effect. *J. Appl. Chem.* 7, 676–689 (1957).

15. R. W. Kluiber. Inner complexes. III. Ring bromination of β-dicarbonyl chelates. *J. Am. Chem. Soc.* 82, 4839–4842 (1960).

16. A. E. Martell and M. Calvin. *Chemistry of Metal Chelate Compounds*, Prentice Hall, Englewood Cliffs, NJ (1952).

17. K. Lenghaus, G. G. Qiao and D. H. Solomon. Model studies of the curing of resole phenol-formaldehyde resins Part 1. The behaviour of ortho quinone methide in a curing resin. *Polymer* 41, 1973–1979 (2000).

18. G. G. Qiao, K. Lenghaus, D. H. Solomon, A. Reisinger, I. Bytheway and C. Wentrup. 4,6-Dimethyl-*o*-quinone methide and 4,6-dimethylbenzoxete. *J. Org. Chem.* 63, 9806–9811 (1998).

19. P. H. R. B. Lemon. An improved sand binder for steel castings. *Int. J. Mater. Prod. Technol.* 5, 25–47 (1990).

20. A. Pizzi, R. Garcia and S. Wang. On the networking mechanisms of additives accelerated PF polycondensates. *J. Appl. Polym. Sci.* 66, 255–266 (1997).

21. H. Lei, A. Pizzi, A. Despres, H. Pasch and G. Du. Esters acceleration mechanisms in phenol-formaldehyde resin adhesives. *J. Appl. Polym. Sci.* 100, 3075–3093 (2006).

22. J. Zhang, A. Pizzi, J. Li and W. Zhang. MALDI-TOF-MS analysis of the curing of phenol-formaldehyde (PF) wood adhesives induced by propylene carbonate. *Eur. J. Wood Prod.* 73, 135–138 (2015).

23. A. Pizzi, B. Mtsweni and W. Parsons. Wood-induced catalytic activation of PF adhesives autopolymerization vs. PF/wood covalent bonding. *J. Appl. Polym. Sci.* 52, 1847–1856 (1994).

24. A. Pizzi. Phenolic resins wood adhesives. In: *Wood Adhesives Chemistry and Technology*, Vol. 1, A. Pizzi (Ed.), pp. 105–178, Marcel Dekker, New York (1983).

25. S. Tohmura and M. Higuchi. Acceleration of the cure of phenolic resin adhesives. 6. Cure-accelerating action of propylene carbonate. *Mokuzai Gakkaishi* 41, 1109–1114 (1995).

26. A. Knop and L. Pilato. *Phenolic Resins*, Springer-Verlag, Berlin (1985).

27. M. Thebault, A. Pizzi, S. Dumarcay, P. Gerardin, E. Fredon and L. Delmotte. Polyurethanes from hydrolysable tannins obtained without using isocyanates. *Ind. Crops Prod.* 59, 329–336 (2014).

28. M. Thebault, A. Pizzi, H. Essawy, A. Baroum and G. Van Assche. Isocyanate free condensed tannin-based polyurethanes. *Eur. Polym. J.* 67, 513–523 (2015).

29. M. Thebault, A. Pizzi, F. M. Al-Marzouki, S. Abdalla and S. Al-Ameer. Isocyanate-free polyurethanes by coreaction of condensed tannins with aminated tannins. *J. Renewable Mater.* 5, 21–29 (2017).

30. R. L. Geimer, A. Leao, D. Ambruster and A. Pablo. Property enhancement of wood composites using gas injection. In: *Proceedings of the 28th International Particleboard Symposium*, Washington State University, Pullman, pp. 243–259 (1994).

31. A. Pizzi and A. Stephanou. Phenol-formaldehyde wood adhesives under very alkaline conditions, Part 2: Acceleration mechanism and applied results. *Holzforschung* 48, 150–156 (1994).

32. C. Zhao, A. Pizzi and S. Garnier. Fast advancement and hardening acceleration of low condensation alkaline PF resins by esters and copolymerized urea. *J. Appl. Polym. Sci.* 74, 359–378 (1999).

33. C. Zhao, A. Pizzi, A. Kühn and S. Garnier. Fast advancement and hardening acceleration of low condensation alkaline PF resins by esters and copolymerized urea. Part 2: Esters during resin reaction and effect of guanidine salts. *J. Appl. Polym. Sci.* 77, 249–259 (2000).

34. G. Du, H. Lei, A. Pizzi and H. Pasch. Synthesis-structure-performance relationship of cocondensed phenol-urea-formaldehyde resins by MALDI-TOF and ^{13}C NMR. *J. Appl. Polym. Sci.* 110, 1182–1194 (2008).

35. Y. Beng Hoong, A. Pizzi, L. Abd Chuah and J. Harun. Phenol-urea-formaldehyde resin co-polymer synthesis and its influence on Elaeis palm trunk plywood mechanical performance evaluated by ^{13}C NMR and MALDI-ToF mass spectrometry. *Int. J. Adhes. Adhes.* 63, 117–123 (2015).

36. H. Lei, G. Du, A. Pizzi, A. Celzard and Q. Fang. Influence of nanoclay on phenol-formaldehyde and phenol-urea-formaldehyde resins for wood adhesives. *J. Adhes. Sci. Technol.* 24, 1567–1576 (2010).

37. A. Pizzi. *Advanced Wood Adhesives Technology*, Marcel Dekker, New York (1994).

38. S. Proszyk and R. Zakrzewski. Activation energy of curing reaction of phenolic resins in the presence of some selected species of wood. *Folia Forestale Polonica, B* 23, 101–109 (1992).

39. R. Zakrzewski and S. Proszyk. Badanie energii aktywacji reakcji utwardzania zywyci mocznikovo-formaldehydowej U70 w obecnosci niektorych gatunkow drewna. *Roczniki Akad. Roln. Poznaniu (Ann. Acad. Agric. Poznan)* 117, 91–97 (1979).

40. H. Mizumachi. Activation energy of the curing reaction of urea resin in the presence of wood. *Wood Sci.* 6, 14–18 (1973).

41. H. Mizumachi and H. Morita. Activation energy of the curing reaction of phenolic resin in the presence of woods. *Wood Sci.* 7, 256–260 (1975).

42. S.-Z. Chow. A kinetic study of the polymerization of phenol-formaldehyde resin in the presence of cellulosic materials. *Wood Sci.* 1, 215–219 (1969).

43. M. V. Ramiah and G. E. Troughton. Thermal studies on formaldehyde glues and cellobiose-formaldehyde glue mixtures. *Wood Sci.* 3, 120–125 (1970).

44. G. C. Bond. *Heterogeneous Catalysis, Principles and Application*, Oxford Science Publishers, Clarendon Press, Oxford (1987).

45. A. Pizzi, A. Stephanou, I. Antunes and G. De Beer. Alkaline PF resins linear extension by urea condensation with hydroxybenzyl alcohol groups. *J. Appl. Polym. Sci.* 50, 2201–2207 (1993).

46. R. T. Conley and J. F. Bieron. A study of the oxidative degradation of phenol-formaldehyde polycondensates using infrared spectroscopy. *J. Appl. Polym. Sci.* 8, 103–117 (1963).

47. K. Oldorp and R. Marutzky. Untersuchungen an Spanplatten mit harnstoffmodifizierten PF-Harzen. *Holz Roh Werkst.* 55, 75–77 (1998).

48. C. M. Chen and J. T. Rice. Synthesizing phenol-formaldehyde resins for controlled dryout resistance. *Forest Prod. J.* 26(6), 17–21 (1976).

49. N. Meikleham. Scanning, development and application of industrial resins from SASOL phenosolvan pitch. M.Sc. thesis, University of the Witwatersrand, Johannesburg, South Africa (1993).

50. A. Pizzi and T. Walton. Non-emulsifiable, water-based diisocyanate adhesives for exterior plywood, Part 1: Novel reaction mechanisms and their chemical evidence. *Holzforschung* 46, 541–547 (1992).

51. A. Pizzi, J. Valenzuela and C. Westermeyer. Non-emulsifiable, water-based, diisocyanate adhesives for exterior plywood, Part 2: Industrial application. *Holzforschung* 47, 69–72 (1993).

52. A. Pizzi and A. Stephanou. Rapid curing lignins-based exterior wood adhesives, Part 1: Diisocyanates reaction mechanisms and application to panel products. *Holzforschung* 47, 439–445 (1993).

53. A. Pizzi and A. Stephanou. Rapid curing lignins-based exterior wood adhesives, Part 2: Acceleration mechanisms and application to panel products. *Holzforschung* 47, 501–506 (1993).

54. D. B. Batubenga, A. Pizzi, A. Stephanou, P. Cheesman and R. Krause. Isocyanate/phenolic wood adhesives by catalytic acceleration of copolymerysation. *Holzforschung* 49, 84–86 (1995).

55. N.-E. El Mansouri, A. Pizzi and J. Salvadó. Lignin-based polycondensation resins for wood adhesives. *J. Appl. Polym. Sci.* 103, 1690–1699 (2007).

56. M. C. Lagel, A. Pizzi and A. Redl. Phenol-wheat protein-formaldehyde adhesives for wood-based panels. *Pro Ligno* 10(3), 3–17 (2014).

57. H. Lei, A. Pizzi, P. Navarrete, S. Rigolet, A. Redl and A. Wagner. Gluten protein adhesives for wood panels. *J. Adhesion Sci. Technol.* 24, 1583–1596 (2010).

58. S. Spina, X. Zhou, C. Segovia, A. Pizzi, M. Romagnoli, S. Giovando, H. Pasch, K. Rode and L. Delmotte. Phenolic resin adhesives based on chestnut hydrolysable tannins. *J. Adhesion Sci. Technol.* 27, 2103–2111 (2013).

59. G. Foyer, B.-H. Chanfi, B. Boutevin, S. Caillol and G. David. New method for the synthesis of formaldehyde-free phenolic resins from lignin-based aldehyde precursors. *Eur. Polym. J.* 74, 296–309 (2016).

60. C. Xu, Y. Zhang and Z. Yuan. Formaldehyde-free phenolic resins, downstream products, their synthesis and use. US Patent application 20160355631, assigned to the University of Western Ontario (2016).

61. Y. Zhang, Z. Yuan and C. Xu. Engineering biomass into formaldehyde-free phenolic resin for composite materials, *AIChE J.* 61(4), 1275–1283 (2015). DOI:10.1002/aic.14716.

62. L. A. Panamgama and A. Pizzi. A ^{13}C NMR analysis method for phenol-formaldehyde resins strength and formaldehyde emission. *J. Appl. Polym. Sci.* 55, 1007–1015 (1995).

63. E. E. Ferg, A. Pizzi and D. Levendis. A ^{13}C NMR analysis method for urea-formaldehyde resin strength and formaldehyde emission. *J. Appl. Polym. Sci.* 50, 907–915 (1993).

64. L. A. Panamgama and A. Pizzi. A ^{13}C NMR analysis method for MUF and MF resins strength and formaldehyde emission. *J. Appl. Polym. Sci.* 59, 2055–2068 (1996).

65. A. T. Mercer and A. Pizzi. A ^{13}C NMR analysis method for MF and MUF resins strength and formaldehyde emission from wood particleboard: Part 1: MUF resins. *J. Appl. Polym. Sci.* 61, 1687–1696 (1996).

66. A. T. Mercer and A. Pizzi, A ^{13}C NMR analysis method for MF and MUF resins strength and formaldehyde emission from wood particleboard: Part 2: MF resins. *J. Appl. Polym. Sci.* 61, 1697–1702 (1996).

67. J. M. Dinwoodie. Properties and performance of wood adhesives. In: *Wood Adhesives Chemistry and Technology*, Vol. 1, A. Pizzi (Ed.), pp. 1–58, Marcel Dekker, New York (1983).

68. R. E. Kreibich. Exposure of glue lines to weather. In: *Wood Adhesives: Present and Future*, A. Pizzi (Ed.), pp. 1–18, Applied Polymer Symposium 40 (1984).

69. A. Pizzi, D. du T. Rossouw, W. Knuffel and M. Singmin. "Honeymoon" phenolic and tannin-based fast setting adhesive systems for exterior grade fingerjoints. *Holzforschung Holzverwertung*, 32, 140–151 (1980).

70. A. Pizzi. Resorcinol adhesives. In: *Handbook of Adhesive Technology*, 2nd edn, A. Pizzi and K. L. Mittal (Eds.), pp. 599–615, Marcel Dekker, New York (2003).

71. R. A. V. Raff and B. M. Silverman. Kinetics of the uncatalyzed reactions between resorcinol and formaldehyde. *Ind. Eng. Chem.* 43, 1423–1427 (1951).

72. D. du T. Rossouw, A. Pizzi, and G. McGillivray. The kinetics of condensation of phenolics polyflavonoid tannins with aldehydes. *J. Polym. Sci. Chem.* 18, 3323–3338 (1980).

73. A. Pizzi and P. van der Spuy. Kinetics of the metal catalyzed condensations of phenols and polyflavonoid tannins with formaldehyde. *J. Polym. Sci. Chem.* 18, 3447–3454 (1980).

74. R. A. V. Raff and B. H. Silverman. Influence of alcohols upon the uncatalyzed reaction between resorcinol and formaldehyde. *Canad. J. Chem.* 29, 857–862 (1951).

75. A. R. Ingram. Electronic interpretation of the uncatalyzed reaction between resorcinol and formaldehyde in alcohols and dioxane. *Canad. J. Chem.* 29, 863–870 (1951).

76. C. T. Liu and T. Naratsuka. Resorcinol resin. I. Influence of solvent upon the velocity of uncatalyzed resorcinol-formaldehyde reaction. *Mozukai Gakkaishi* 15, 79–83 (1969).

77. R. H. Moult. Resorcinolic adhesives. In: *Handbook of Adhesives*, 2nd edn, I. Skeist (Ed.), Reinhold, New York, pp. 417–423 (1977).

78. G. G. Marra. Development of a method for rapid laminating of lumber without the use of high frequency heat. *Forest Prod. J.* 6 (2), 97–100 (1956).

79. G. F. Baxter and R. E. Kreibich. Fast-curing phenolic adhesive system. *Forest Prod. J.* 23 (1), 17–22 (1973).

80. R. E. Kreibich. High speed adhesives for wood gluing industry. *Adhes. Age* 17, 26–30 (1974).

81. A. Pizzi and F. A. Cameron. Fast-set adhesives for glulam. *For. Prod. J.* 34 (9), 61–65 (1984).

82. A. Pizzi and F. A. Cameron. Fast setting adhesives for fingerjoints and glulam. In: *Wood Adhesives Chemistry and Technology*, Vol. 2, A. Pizzi (Ed.), pp. 229–306, Marcel Dekker, New York (1989).

83. A. Pizzi. Low resorcinol PRF cold set adhesives: The branching principle. In: *Wood Adhesives Chemistry and Technology*, Vol. 2, A. Pizzi (Ed.), pp. 190–210, Marcel Dekker, New York (1989).

84. E. Scopelitis and A. Pizzi. The chemistry and development of branched PRF wood adhesives of low resorcinol content. *J. Appl. Polym. Sci.*, 47, 351–360 (1993).

85. E. Scopelitis and A. Pizzi. Urea-resorcinol-formaldehyde adhesives of low resorcinol content. *J. Appl. Polym. Sci.* 48, 2135–2146 (1993).

86. A. Pizzi, E. Orovan and F. A. Cameron. The development of weather- and boil-proof phenol-resorcinol-furfural cold-setting adhesives. *Holz Roh Werkst.* 42, 467–472 (1984).

87. A. Pizzi, F. A. Cameron and E. Orovan. Cold-set tannin-resorcinol-formaldehyde adhesives of lower resorcinol content. *Holz Roh Werkst.* 46, 67–71 (1988).

88. A. Pizzi, R. Vosloo, F. A. Cameron and E. Orovan. Self-neutralizing acid-set PF wood adhesives. *Holz Roh Werkst.* 44, 229–234 (1986).

89. M. Properzi, A. Pizzi and L. Uzielli. Honeymoon MUF adhesives for exterior grade glulam. *Holz Roh Werkst.* 59, 413–421 (2001).

90. M. Properzi, A. Pizzi and L. Uzielli. Performance limits of pure MUF honeymoon adhesives for exterior grade glulam and fingerjoints. *Holzforschung Holzverwertung* 53, 114–117 (2001).

9 Natural Phenolic Adhesives derived from Tannins and Lignin

Antonio Pizzi

CONTENTS

9.1 INTRODUCTION

The loose terms *natural* or *renewable phenolic adhesives* have been used to identify polymeric compounds of natural vegetable origin that have been modified and/or adapted to the same use as some classes of purely synthetic adhesives [1]. At present, two classes of these adhesives exist: one already extensively commercialized in the southern hemisphere and the other on a slow path to commercialization. These two types of resins are tannin-based adhesives [2] and lignin adhesives [3–6]. Both types are aimed primarily at substituting synthetic phenolic resins. In some respects, such as performance, they closely mimic, or are even superior to, synthetic phenolic adhesives, while in others they behave in a vastly different manner from their synthetic counterparts. In this chapter, we focus primarily on tannin-based adhesives, because they have already been in extensive industrial use in the southern hemisphere, in certain fields of application, for the last 20 years. These adhesives are of some interest not only for their excellent performance in some applications, but also for their mostly environmentally friendly composition. Lignin adhesives are treated briefly here and in detail in another chapter.

9.2 TANNIN-BASED ADHESIVES

The word tannin has been used loosely to define two different classes of chemical compounds of mainly phenolic nature: hydrolyzable tannins and condensed tannins. The former, including chestnut, myrabolans (*Terminalia* and *Phyllantus* tree species), and divi-divi (*Caesalpina coraria*) extracts, are mixtures of simple phenols such as pyrogallol and ellagic acid and of esters of a sugar, mainly glucose, with gallic and digallic acids [2] linked together as pentagalloyl glucose chains [7] or gallic acid chains [8,9]. They can be and have been used, successfully, as partial substitutes (up to 50%) of phenol in the manufacture of phenol–formaldehyde (PF) resins [10,11]. Their chemical behavior toward formaldehyde is analogous to that of simple phenols of low reactivity, and their moderate use as phenol substitutes in the abovementioned resins does not present difficulties. They are not polymeric in their natural state after extraction; the low level of phenol substitution they allow, their low nucleophilicity, limited worldwide production, and higher price somewhat renders them of lower interest both chemically and economically for use as adhesives. However, effective adhesive formulations where, in the synthesis of a PF resin, phenol has been substituted up to 80% with a hydrolyzable tannin have also been developed [12].

Condensed tannins, on the other hand, constituting more than 90% of the total world production of commercial tannins (200,000 tn./year), are both chemically and economically more interesting for the preparation of adhesives and resins. Condensed tannins and their flavonoid precursors are known for their wide distribution in nature and particularly for their substantial concentration in the wood and bark of various trees. These include various *Acacia* (wattle or mimosa bark extract), *Schinopsis* (quebracho wood extract), *Tsuga* (hemlock bark extract), and *Rhus* (sumac extract) species, from which commercial tannin extracts are manufactured, and various *Pinus* bark extract species. Where bark and wood of trees were found to be particularly rich sources of condensed tannins, commercial development ensued through large-scale afforestation and/or industrial extraction, mainly for use in leather tanning. The production of tannins for leather manufacture reached its peak immediately after World War II and has since progressively declined. This decline of their traditional market, coupled with increased price and decreased availability of synthetic phenolic materials due to the advent of the energy crisis of the early 1970s, stimulated fundamental and applied research on the use of such tannins as a source of condensed phenolics.

9.2.1 CONDENSED TANNINS

The structure of the flavonoid constituting the main monomer of condensed tannins can be represented as in Figure 9.1.

The flavonoid unit in Figure 9.1 is repeated 2–11 times in mimosa tannin, with an average degree of polymerization of 4–5, and up to 12 times for pine tannins, with structures as shown in Figure 9.2 with an average degree of polymerization of 6–7 for their soluble extract fraction [13,14].

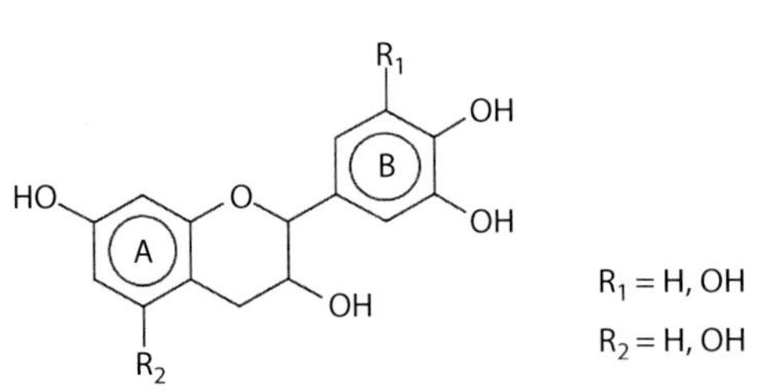

$$R_1 = H, OH$$
$$R_2 = H, OH$$

FIGURE 9.1   Structure of the flavonoid main monomer of condensed tannins.

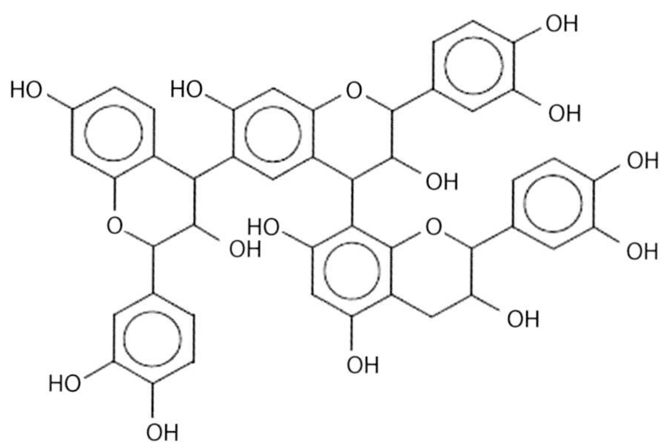

FIGURE 9.2   Typical sequence of flavonoid units in a mimosa tannin trimer.

The nucleophilic centers on the A-ring of a flavonoid unit tend to be more reactive than those found on the B-ring. This is due to the vicinal hydroxyl substituents, which cause general activation in the B-ring without any localized effects such as those found in the A-ring.

Formaldehyde reacts with tannins to produce polymerization through methylene bridge linkages at reactive positions on the flavonoid molecules, mainly the A-rings. The reactive positions of the A-rings are positions 6 or 8 (according to the type of tannin) of all the flavonoid units and both positions 6 and 8 of the upper terminal flavonoid units. The A-rings of mimosa and quebracho tannins show reactivity toward formaldehyde comparable to that of resorcinol [15–18]. Assuming the reactivity of phenol to be 1 and that of resorcinol to be 10, the A-rings have a reactivity of 8–9. However, because of their size and shape, the tannin molecules become immobile at a low level of condensation with formaldehyde, so that the available reactive sites are too far apart for further methylene bridge formation. The result may be incomplete polymerization and, therefore, a weak cross-linked network. Bridging agents with longer molecules should be capable of bridging the distances that are too long for methylene bridges to be formed. Alternatively, other techniques can be used to solve this problem.

In condensed tannins from mimosa bark, the main polyphenolic pattern is represented by flavonoid analogues based on resorcinol A-rings and pyrogallol B-rings. These constitute about 70% of the tannins. A secondary but parallel pattern is based on resorcinol A-rings and catechol B-rings [2,15]. These tannins represent about 25% of the total of mimosa bark tannin fraction. The remaining part of the condensed tannin extract are the *nontannins* [15]. They may be subdivided into carbohydrates, hydrocolloid gums, and small amino and imino acid fractions [2,15]. The hydrocolloid gums vary in concentration from 3% to 6% and contribute significantly to the viscosity of the extract, despite their low concentration [2,15]. Similar flavonoid A- and B-ring patterns also exist in quebracho wood extract (*Schinopsis balansae* and *lorentzii*) [16–18], but no phloroglucinol A-ring pattern, or probably a much lower quantity of it, exists in quebracho extract [18–20]. Similar patterns to wattle (mimosa) and quebracho are shown by hemlock and Douglas fir bark extracts. However, completely different patterns and relationships do exist in the case of pine tannins [21–23], which present instead only two main patterns. One is represented by flavonoid analogues based on phloroglucinol A-rings and catechol B-rings [24,25]. The other pattern, present in much lower proportion, is represented by phloroglucinol A-rings and phenol B-rings [21,23]. The A-rings of pine tannins thus possess only the phloroglucinol type of structure, much more reactive toward formaldehyde than a resorcinol-type structure, with important consequences in the use of these tannins for adhesives.

In condensed polyflavonoid tannin molecules, the A-rings of the constituent flavonoid units retain only one highly reactive nucleophilic center, the second reactive site being occupied by the interflavonoid bonds. Resorcinolic A-rings (wattle) show reactivity toward formaldehyde comparable to, though slightly lower than, that of resorcinol [24]. Phloroglucinolic A-rings (pine) behave instead as phloroglucinol [2]. Pyrogallol or catechol B-rings are comparatively unreactive and may be activated by anion formation only at relatively high pH [19]. Hence, the B-rings do not participate in the reaction except at high pH values (pH 10), where the reactivity toward formaldehyde of the A-rings is so high that the tannin–formaldehyde adhesives prepared have unacceptably short pot-lives [24]. In tannin adhesives, only the A-rings are generally used to cross-link the network. With regard to the pH dependence of the reaction with formaldehyde, it is generally accepted that the reaction rate of wattle tannins with formaldehyde is slowest in the pH range 4.0–4.5 [25]; for pine tannins, the range is between 3.3 and 3.9.

Formaldehyde is, generally, the aldehyde used in the preparation, setting, and curing of tannin adhesives. It is normally added to the tannin extract solution at the required pH, preferably in its polymeric form of paraformaldehyde, which is capable of fairly rapid depolymerization under alkaline conditions, and as urea–formalin concentrates. Hexamethylenetetramine (hexamine) may also be added to resins due to its potential formaldehyde-releasing action under heat. Hexamine is, however, unstable in acid medium [26] but becomes more stable with increased pH values. Hence, under alkaline conditions, the liberation of formaldehyde may not be as rapid and as efficient as wanted. Also, it has been fairly widely reported, with a few notable exceptions [27], that bonds formed with hexamine as hardener are not as boil resistant [28] as those formed by paraformaldehyde. The reaction of formaldehyde with tannins may be controlled by the addition of alcohols to the system. Under these circumstances, some of the formaldehyde is stabilized by the formation of hemiacetals [e.g., $CH_2(OH)(OCH_3)$] if methanol is used [2]. When the adhesive is cured at an elevated temperature, the alcohol is driven off at a fairly constant rate and formaldehyde is progressively released from the hemiacetal. This ensures that less formaldehyde is volatilized when the reactants reach curing temperature, and that the pot-life of the adhesive is extended. Other aldehydes have also been substituted for formaldehyde [2,25,27].

In the reaction of polyflavonoid tannins with formaldehyde, two competitive reactions are present:

1. The reaction of the aldehyde with tannin and with low molecular weight tannin– aldehyde condensates, which are responsible for the aldehyde consumption.
2. The liberation of aldehyde, available again for reaction. The latter reaction is probably due to the passage of unstable $-CH_2-O-CH_2-$ ether bridges initially formed to $-CH_2-$ linked compounds.

   In the case of some tannins, namely, quebracho tannin, a third reaction of importance is also present:
3. The simultaneous hydrolysis of some interflavonoid bonds—hence, a depolymerization reaction—slows down hardening [28–30]. Notwithstanding that the two major industrial polyflavonoid tannins that exist, namely, mimosa and quebracho tannins, are very similar and both are composed of mixed prorobinetinidins and profisetinidins, one could not explain this anomalous behavior of quebracho tannin. It has now been possible to determine by nuclear magnetic resonance (NMR) spectroscopy [28] and particularly by laser desorption mass spectrometry (MALDI-TOF) for mimosa and quebracho tannins and some of their modified derivatives [30] that: (1) Mimosa tannin is predominantly composed of prorobinetinidins while quebracho is predominantly composed of profisetinidins; (2) mimosa tannin is heavily branched due to the presence of considerable proportions of angular units in its structure, while quebracho tannin is almost completely linear [30]. This structural difference is the one that contributes to the considerable differences in viscosity of water solutions of the two tannins and that (3) induces the interflavonoid link of quebracho to be more easily hydrolyzable, due to the linear structure of this tannin, confirming NMR findings [26,28] that this tannin is subject to a polymerization/depolymerization equilibrium. This also showed that the decrease of viscosity, due to acid–base treatments to yield tannin adhesive intermediates, also depends on quebracho from a certain level of

hydrolysis of the tannin itself and not only of the carbohydrates present in the extract (see Section 9.4). This tannin hydrolysis does not appear to occur in mimosa tannin, in which the interflavonoid link is completely stable to hydrolysis.

It is interesting to note that while $-CH_2-O-CH_2-$ ether bridged compounds have been isolated in the PF [26] reaction, their existence for fast-reacting phenols such as resorcinol and phloroglucinol has been postulated, but they have not been isolated, as these two phenols have always been considered to be too reactive with formaldehyde. They are detected by a surge in the concentration of formaldehyde observed in kinetic curves due to methylene-ether-bridges decomposition [22].

When heated in the presence of strong mineral acids, condensed tannins are subject to two competitive reactions. One is degradative, leading to lower molecular weight products, and the other is condensative, as a result of hydrolysis of heterocyclic rings (p-hydroxybenzyl ether links) [19]. The p-hydroxybenzyl carbonium ions formed condense randomly with nucleophilic centers on other tannin units to form *phlobaphenes* or *tanner's red* [19,31–33]. Other modes of condensation (e.g., free radical coupling of catechol B-ring units) cannot be excluded in the presence of atmospheric oxygen. In predominantly aqueous conditions, phlobaphene formation or formation of insoluble condensates predominates. These reactions, characteristic of tannins, not of synthetic phenolic resins, must be taken into account when formulating tannin adhesives.

Sulfitation of tannin is one of the oldest and most useful reactions in flavonoid chemistry. Slightly sulfited water is sometimes used to increase tannin extraction from the bark containing it. The total effect of sulfitation affords the important advantages of higher concentration of tannin phenolics in adhesive applications, due to enhanced solubility and decreased viscosity, and of higher moisture retention by the tannin resins, allowing slower adhesive film dry-out, hence longer assembly times [34]. However, in certain types of adhesives, sulfitation also represents a distinct disadvantage, in that sulfonate groups promote sensitivity to moisture with adhesive deterioration and bad water resistance of the cured glueline, even with adequate cross-linking [34–37].

In recent years, the importance of the marked colloidal nature of tannin extract solutions has come to the fore [29,38–47]. It is the presence of polymeric carbohydrates in the extract, as well as the higher molecular fraction of the polyphenolic tannins, that determines the colloidal state of tannin extract solutions in water [26,36]. The realization of the existence of the tannin in this particular state affects many of the reactions that lead to the formation and curing of tannin adhesives. Thus, reactions not thought possible in solution become, instead, not only possible, but the favored ones [28,38], while reactions mooted to be of determinant importance when found on model compounds not in colloidal state have, in reality, been shown to be inconsequential to tannin adhesives and their applications [45,46].

9.3   TECHNOLOGY OF INDUSTRIAL TANNIN ADHESIVES

The purity of vegetable tannin extracts varies considerably. Commercial wattle bark extracts normally contain 70%–80% active phenolic ingredients. The nontannin fraction, consisting mainly of simple sugars and high molecular weight hydrocolloid gums, does not participate in the resin formation with formaldehyde. Sugars reduce the strength and water resistance in direct proportion to the amount added. Their effect is a mere dilution effect of the adhesive resin solids, with consequent proportional worsening of adhesive properties. The hydrocolloid gums, instead, have a much more marked effect on both original strength and water resistance of the adhesive [2,44]. If it is assumed that the nontannins in tannin extracts have a similar influence on adhesive properties, it can be expected that unfortified tannin–formaldehyde networks can achieve only 70%–80% of the performance shown by synthetic adhesives.

In many glued wood products, the demands on the glueline are so high that unmodified tannin adhesives are unsuitable. The possibility of refining extracts has proved fruitless, largely because the intimate association between the various constituents makes industrial fractionation difficult.

Fortification is, in many cases, the most practical approach to reducing the effect of impurities. Fortification generally consists of copolymerization of the tannin with phenolic or aminoplastic resins [24,44,48]. It can be carried out during manufacture of the adhesive resin, during glue-mix assembly, just before use, or during adhesive use. If added in sufficient quantity, various synthetic resins have been found to be effective in reducing the nontannin fraction to below 20% and in overcoming other structural problems [24]. The main resins used are PF and urea–formaldehyde resols with a medium to high methylol groups content. These resins can fulfill the functions of hardeners, fortifiers, or both. Generally, they are used as fortifiers between 10% and 20% of total adhesive solids, and paraformaldehyde is used as a hardener. Such an approach is the favorite one for marine-grade plywood adhesives. These fortifiers are particularly suitable for the resorcinolic types of condensed tannins, such as mimosa. They can be copolymerized with the tannins during resin manufacture, during use, or both [2,24,44,46]. Copolymerization and curing are based on the condensation of the tannin with the methylol groups carried by the synthetic resin. Since tannin molecules are generally large, the rate of molecular growth in relation to the rate of linkage is high, so that tannin adhesives generally tend to have lower gelling and curing times and shorter pot-lives than those of synthetic phenolic adhesives. From the point of view of reactivity, phloroglucinol tannins (procyanidins) such as pine tannins are much faster than mainly resorcinol tannins such as mimosa. The usual ways of slowing these down and, for instance, to lengthen adhesive pot-life are:

1. To add alcohols to the adhesive mix to form hemiacetals with formaldehyde and, therefore, act as retardants of the tannin–formaldehyde reaction.
2. To adjust the adhesive's pH to obtain the required pot-life and rate of curing.
3. To use hexamine as hardener, which under the current industrial pressing conditions gives a very long pot-life at ambient temperature but still a short curing time at higher temperatures.

The viscosity of bark extracts is strongly dependent on concentration. The viscosity increases very rapidly above a concentration of 50%. Compared with synthetic resins, tannin extracts are more viscous at the concentrations normally required in adhesives. The high viscosity of aqueous solutions of condensed tannins is due to the following causes, in order of importance:

1. Presence of high molecular weight hydrocolloid gums in the tannin extract [44,47]. The viscosity is directly proportional to the amount of gums present in the extract [44,47].
2. Tannin–tannin, tannin–gum, and gum–gum hydrogen bonds. Aqueous tannin extract solutions are not true solutions, but rather, colloidal suspensions in which water access to all parts of the molecules present is very slow. As a consequence, it is difficult to eliminate intermolecular hydrogen bonds by dilution only [44,47].
3. Presence of high molecular weight tannins in the extract [30,44,47].

The high viscosity of tannin extracts solutions has also been correlated with the proportion of very high molecular weight tannins present in the extract. This effect is not well defined. In most adhesive applications, such as in plywood adhesives, the viscosity is not critical and can be manipulated by dilution.

In the case of particleboard adhesives, decrease of viscosity is, instead, an important prerequisite. When reacted with formaldehyde, unmodified condensed tannins give adhesives having characteristics that do not suit particleboard manufacture; namely, high viscosity, low strength, and poor water resistance. The most commonly used process to eliminate these disadvantages in the preparation of tannin-based particleboard adhesives consists of a series of subsequent acid and alkaline treatment of the tannin extract, causing hydrolysis of the gums to simple sugars and some tannin structural changes (Figure 9.3), thus decreasing viscosity and improving strength and water resistance of the unfortified tannin–formaldehyde adhesive [2,46]. Furthermore, such treatments

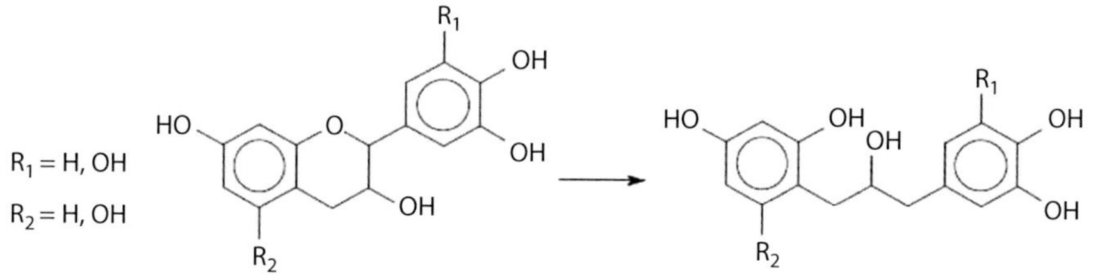

FIGURE 9.3 Reaction of opening of flavonoids heterocycle ring induced by acid–base treatment.

may cause partial rearrangement of the flavonoid molecules that cause liberation of some resorcinol *in situ* in the tannin, rendering it more reactive, allowing better cross-linking with formaldehyde, and ultimately yielding an adhesive that, without addition of any fortifier resins, gives excellent performance for exterior-grade particleboard [1,2,46].

This modification cannot be carried out too extensively, but only to a limited extent, to avoid precipitation of the tannin from solution by the formation of phlobaphenes.

Typical results obtained are shown in Table 9.1.

Particular gluing and pressing techniques have been developed for tannin particleboard adhesives [49,50] to achieve press times much shorter than those traditionally obtained with synthetic PF adhesives, although recent advances in synthetic PF resins have markedly limited such an advantage [51,52]. Press times of 7 s/mm of panel thickness have been achieved, and press times of 9 s/mm at 190°C to 200°C press temperature are in daily operation; these press times are becoming comparable to what is obtainable with urea–formaldehyde or melamine–formaldehyde resins at the same press temperatures (Table 9.1). The success of these simple types of particleboard adhesives relies heavily on industrial application technology, rather than just on the preparation technology of the adhesive itself [44,49,53]. A considerable advantage is the tolerance of much higher moisture content of the resinated chips with these adhesives than with any of the synthetic phenolic and aminoresin adhesives. In the case of wood particleboard and oriented strandboard (OSB) panels, the technology so developed allows hot pressing at moisture contents of around 24%, against values of 12% for traditional synthetic adhesives, and presents other advantages as well [44,53,54].

The best adhesive formulation for phloroglucinolic tannins such as pine tannin extracts is, instead, a comparatively new adhesive formulation that is also capable of giving excellent results when using resorcinolic tannins such as a wattle tannin extract [27,55–58]. The adhesive glue-mix consists only of a mix of an unmodified tannin extract 50% solution to which have been added paraformaldehyde and polymeric nonemulsifiable 4,4′-diphenylmethane diisocyanate (commercial pMDI) [27,55–58]. The proportion of tannin extract solids to pMDI can be as high as 70:30 based on mass, but can be much lower in pMDI content. This adhesive is based on the mechanism shown in Figure 9.4, by which the MDI, in water, is hardly deactivated to polyureas [56,58].

The properties of the particleboard manufactured with this system using pine tannin adhesives are listed in Table 9.2. The results obtained with this system are quite good and not too different

TABLE 9.1

Unfortified Tannin–Formaldehyde Adhesives Obtained by Acid–Alkali Treatment for Exterior-Grade Particleboard: Example of Industrial Board Results

	Swelling after a 2 h Boil				
Panel Density (g/cm³)	Measured Wet (%)	Measured Dry (Irreversible Swelling) (%)	Original Internal Bond (IB) Strength (kg/cm²)	IB after 2 h Boil (kg/cm²)	Cyclic test: Swelling after Five Cycles (%)
0.700	11.0	0.0	13.0	9.0	3.0

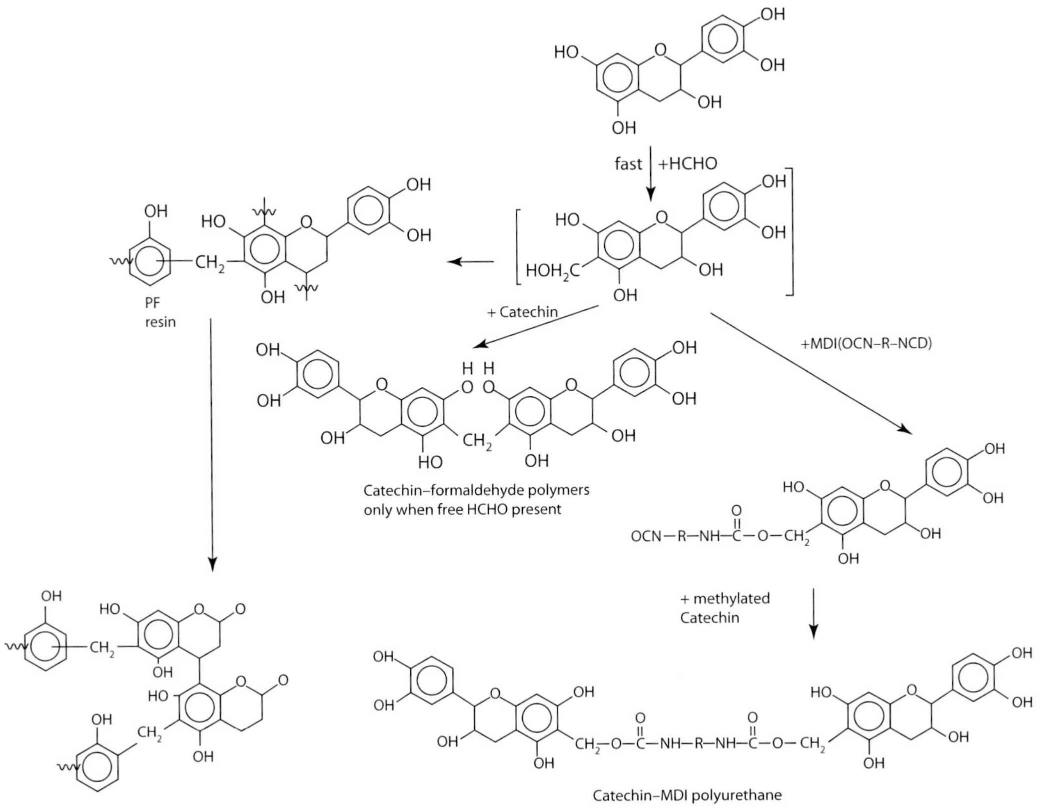

FIGURE 9.4 Joint reaction system of the joint reaction of aldehydes and polymeric isocyanates, leading to both methylene bridges and urethane bridges between flavonoid units of condensed tannins.

from the results obtained with some of the other tannin adhesives already described. In the case of the use of phloroglucinolic tannin extracts, no pH adjustment of the solution is needed. One point that was given close consideration is the deactivating effect of water on the isocyanate group of pMDI. It has been found that the amount of deactivation by water of this group when in a concentrated solution (50% or over) of a phenol is much lower than previously thought [55–58]. This is the reason that aqueous tannin extract solutions and pMDI can be reacted without substantial pMDI deactivation by the water present.

The quest to decrease or completely eliminate formaldehyde emission from wood panels bonded with adhesives, although not really necessary in tannin adhesives due to their very low emission (as most phenolic adhesives), has nonetheless promoted here, too, some research to further decrease formaldehyde emission. This has centered on two lines of investigations: (1) tannins autocondensation (see Section 3.4.1), and (2) the use of hardeners not emitting at all, simply because no aldehyde

TABLE 9.2

Properties of Particleboard Manufactured Using Pine Tannin Adhesives

	Swelling after a 2 h Boil				
Panel Density (g/cm³)	Measured Wet (%)	Measured Dry (Irreversible Swelling) (%)	Original Internal Bond (IB) Strength (kg/cm²)	IB after 2 h Boil (kg/cm²)	IB Retention after 2 h Boil (%)
0.690	15.0	4.3	8.4	4.3	51

has been added to the tannin. Methylolated nitroparaffins and, in particular, the simplest and least expensive of their class, namely, tris(hydroxymethyl)nitromethane [59,60], function well as hardeners for a variety of tannin-based adhesives while affording considerable side advantages to the adhesive and to the bonded wood joint. In panel products such as particleboard, medium density fiberboard (MDF), and plywood, the joint performance obtained is that of the exterior/marine-grade type. A considerable lengthening in glue-mix pot-life is also obtained, this being an advantage. Furthermore, the use of this hardener is coupled with such a marked reduction in formaldehyde emission from the bonded wood panel as to reduce emission to the level of formaldehyde emitted by just heating the wood (or slightly less, thus functioning as a mild depressant of emission from the wood itself). Furthermore, tris(hydroxymethyl)nitromethane can be mixed in any proportion with traditional formaldehyde-based hardeners for tannin adhesives. Its proportional substitution of such hardeners induce a proportionally marked decrease in the formaldehyde emission from the wood panel without affecting the exterior/marine-grade performance of the panel. MDF industrial plant trials confirmed all the properties reported, and the trial conditions and results have been reported [59,60]. A cheaper, but equally effective, alternative to hydroxymethylated nitroparaffins is the use of hexamine as tannin hardener. This sometimes causes problems of early agglomeration in some tannins [61], and a better solution proposed that overcame such problems was rather to use as hardener a mix of formaldehyde coupled with an ammonium salt. More recently, considerable interest has centered on the use of alternative, nonvolatile aldehydes. A number of research works on the use of glyoxal [62] have also appeared in the literature. This approach works well, but only with the more reactive types of condensed tannins, namely, procyanidins and prodelphinidins, such as those extracted from pine and spruce bark.

9.3.1 Corrugated Cardboard Adhesives

The adhesives developed for the manufacture of moisture-resistant corrugated cardboard are based on the addition of spray-dried wattle extract, urea–formaldehyde resin, and formaldehyde to a typical Stein Hall starch mix of 18%–22% starch content [63,64]. The wattle tannin–urea–formaldehyde copolymer formed *in situ* and any free formaldehyde left in the glueline are absorbed by the wattle tannin extract. The wattle extract powder should be added at a level of 4%–5% of the total starch content of the mix (i.e., carrier plus slurry). Successful results can be achieved in the range of 2%–12% of the total starch content, but 4% is the recommended starting level. The final level is determined by the degree of water hardness and desired bond quality. This wattle extract–urea–formaldehyde fortifier system is highly flexible and can be adopted to render moisture-resistant, starch-based glue-mixes.

9.3.2 Cold-Setting Laminating and Fingerjointing Adhesives for Wood

A series of different novolac-like materials are prepared by copolymerization of resorcinol with resorcinolic A-rings of polyflavonoids, such as condensed tannins [65–67]. The copolymers formed have been used as cold-setting exterior-grade wood adhesives, complying with the relevant international specifications [68]. Several formulations are used. The system most commonly used commercially relies on the simultaneous copolymerization of resorcinol and resorcinolic A-rings of the tannin, due to their comparable reactivities toward formaldehyde (Figure 9.5).

The final mixture of the products of this system is an adhesive that can be set and cured at ambient temperature by the addition of paraformaldehyde. Other cold-set systems exist and are described in the more specialized literature [2,62–64]. The typical results obtainable with these adhesives are shown in Table 9.3.

A particularly interesting system now used extensively in several southern hemisphere countries is the so-called *honeymoon* fast-setting, separate application system [69,70]. In this system, one of the surfaces to be mated in the joint is spread with a standard synthetic phenol–resorcinol–formaldehyde

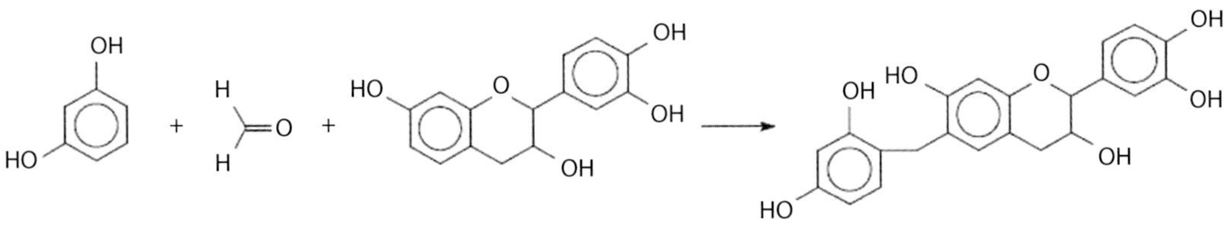

FIGURE 9.5 Coreaction of resorcinol and a condensed tannin flavonoid unit to form a resorcinol tannin–formaldehyde resin for wood structural cold-setting adhesives.

TABLE 9.3

Typical Results of Tannin–Resorcinol–Formaldehyde Cold-Setting Adhesives used on Beech Strips according to British Standard BS 1204

	Dry	After 24 h Cold-Water Soak	After 6 h Boil
Tensile strength (N)	3200–3800	2300–2900	2200–2800
Wood failure (%)	90–100	75–100	80–100

Source: British Standard BS 1204, 1965.

---

adhesive plus paraformaldehyde hardener. The second surface is spread with a 50% tannin solution at pH 12. When the two surfaces are joined, fingerjoints develop enough strength to be put into service within 30 min, and laminated beams (glulam) need to be clamped for only 2.5–3 h instead of the traditional 16–24 h, with a consequent considerable increase in factory productivity. This adhesive system also provides full weather- and boil-proof capabilities.

9.3.3 TIRE CORD ADHESIVES

Another application of condensed tannin extracts that has proved technically successful is as tire cord adhesives. Both thermosetting tannin formulations [71] and tannin–resorcinol–formaldehyde formulations have been experimented with successfully.

9.3.4 NEW CONCEPTS AND PRINCIPLES

As in the case of other formaldehyde-based resins, the interaction energies of tannins with cellulose obtained by molecular mechanics calculations [45] tend to confirm the effect of surface catalysis induced by cellulose also on the curing and hardening reaction of tannin adhesives. The considerable energies of interactions obtained can effectively explain weakening of the heterocycle ether bond, leading to accelerated and easier opening of the pyran ring in a flavonoid unit, as well as the facility with which hardening by autocondensation can occur. As with synthetic formaldehyde-based resins, the same effect explains the decrease in energy of activation of the condensation of polyflavonoids with formaldehyde, leading to exterior wood adhesives curing and hardening [72]. Furthermore, tannin–furfuryl alcohol adhesives, a technology derived from other tannin applications, has also been developed for adhesives and binders [73,74].

9.3.4.1 Hardening by Tannins Autocondensation

The autocondensation reactions characteristic of polyflavonoid tannins have only recently been used to prepare adhesive polycondensate hardening in the absence of aldehydes [75]. This autocondensation reaction is based on the opening under alkaline and acid conditions of the O1–C2 bond of the flavonoid repeat unit and the subsequent condensation of the reactive center formed at C2 with the free C6 or C8 sites of a flavonoid unit on another tannin chain [75–79]. Although this reaction may lead to considerable increase in viscosity, gelling does not generally occur. However, gelling occurs when the reaction occurs (1) in the presence of small amounts of dissolved silica (silicic acid or silicates) catalyst and some other catalysts [75–80], and (2) on a lignocellulosic surface [79]. In the case of the more reactive procyanidins and prodelphinidin-type tannins, such as pine tannin, cellulose catalysis is more than enough to cause hardening and to produce boards of strength satisfying the relevant standards for interior-grade panels [79]. In the case of the less reactive tannins, such as mimosa and quebracho, the presence of a dissolved silica or silicate catalyst of some type is essential to achieve panel strength as required by the relevant standards. Autocondensation reactions have been shown to contribute considerably to the dry strength of wood panels bonded with tannins, but are relatively inconsequential in contributing to the bonded panels' exterior-grade properties, which are rather determined by polycondensation reactions with aldehydes [79–81]. A combination of tannin autocondensation and reactions with aldehydes and a combination of radicals with ionic reactions have been used to decrease the proportion of aldehyde hardener used, as well as to decrease considerably the already low formaldehyde emission yielded by the use of tannin adhesives [79–81].

9.3.4.2 Polyurethane Adhesives from Polyphenolic Tannins

As regards the preparation of polyurethane adhesives starting from condensed or hydrolyzable tannins, two approaches have been taken: (1) Modification of the flavonoid tannin to render easier the reaction with isocyanates. This is due to the difficulty in reacting the flavonoid hydroxyl groups directly with isocyanates. In this type of application, the tannins are in direct competition with more suitable natural polyols, for which the literature abounds, to prepare semi-biosourced polyurethanes. (2) The use of a no-isocyanate approach to improve the acceptability and environmental friendliness of such adhesives.

For the first approach, benzoylation to reduce the number of hydroxyl groups of the tannin before reaction with diisocyanate was attempted a long time ago [82]. A more recent and more studied approach has been to prepare novel thermosetting tannin-based polyurethane adhesive resins using the hydroxypropyl and hydroxybutyl derivatives of purified condensed tannins from *Pinus pinaster* bark and other condensed tannin species by reaction with diisocyanates [83–86]. Hydroxypropyl tannins with a degree of substitution, increasing stepwise from 1 to 4, were cross-linked with either an aromatic (pMDI) or an aliphatic isocyanate (hexamethylene diisocyanate [HDI]), with good results. Hydroxypropylation and hydroxybutylation are one of the approaches used to react polyphenolic materials with isocyanates to obtain polyurethane adhesives and resins [87–89]. This also renders easier the reaction of flavonoid tannins with isocyanates, due to the introduction of much more approachable hydroxyl groups into the tannin structure, thus increasing reaction yield. A similar approach, that is, to introduce into the tannin structure more available –OH groups but through a reaction totally different from hydroxypropylation, consisted in reacting an aldehyde with the tannin and then using the hydroxyl groups formed by their addition onto the flavonoid structure, before their further condensation with other flavonoids, to react with an isocyanate [55,56,58,90]. Such a system has been used and is still used industrially for wood adhesives. More recently, the system has been adapted by eliminating formaldehyde and substituting it with glyoxal with good results [91], rendering the approach even more environmentally interesting.

For the second approach, the total elimination of the isocyanate in the preparation of polyurethane resins is based on the reaction of a phenol with a cyclic carbonate followed by reaction with a diamine. This has been proposed based exclusively on synthetic, not biosourced, materials [92–95]. The same reaction has been used with both a hydrolyzable tannin and condensed flavonoid tannins, by reacting them first with dimethyl carbonate, followed by reaction with hexamethylenediamine [96,97] (Figure 9.6). What was of further interest in this approach was that it was not necessary to use a purified tannin, as the 10%–12% carbohydrate fraction of commercial, industrial tannin extracts was also found to undergo the same two reactions, leading to urethane bridges. This was the first approach to obtain a biosourced polyurethane without using isocyanates; although here,

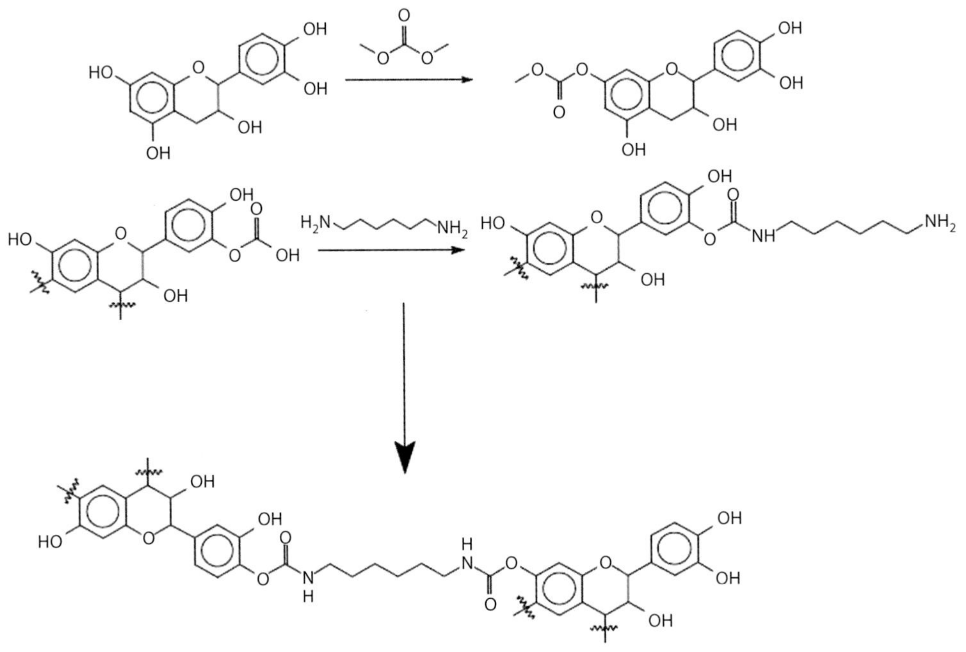

FIGURE 9.6 Sequence of reactions of dimethyl carbonate with flavonoid units, followed by the reaction with the diamine and subsequent diurethane bridge formed between flavonoid units of a condensed tannin.

too, as in approach (1), the percentage level of biosourced material was 45%–50%. As the amination of condensed tannins—and thus the conversion of their hydroxyl groups into amino groups—is an easy reaction [98], the aminated tannin was used to substitute the hexamethylene diamine, further improving to 70% or higher the percentage of biosourced material in the final polyurethane; this without using any isocyanate [99].

Biobased epoxy and epoxy–acrylic resins based on tannins have also been developed. They are described in Chapter 18 in this book.

9.3.4.3 New Cross-Linking Reactions

Procyanidin-type condensed tannins can produce adhesives for wood panels when using a natural nontoxic aldehyde, such as, for example, vanillin. An adhesive based on the reaction of a very fast-reacting procyanidin-type condensed tannin—namely, purified pine bark tannin—and food-grade nontoxic slow-reacting aldehydes derived from lignin was shown to satisfy well the relevant standards for bonding wood particleboard. Vanillin and a dialdehyde derivative from vanillin were the aldehydes used to yield an adhesive for wood panels containing only natural nontoxic materials [100]. The oligomers obtained and their distribution were determined by several systems of analysis.

The reaction of condensation and cross-linking of condensed flavonoid tannin extracts with triethyl phosphate (TEP) has also been found to lead to adhesives not containing any aldehyde and with no emissions [101]. Spectroscopy studies showed that reaction occurs mainly on the C3 of the flavonoid heterocycle ring and on the aromatic C4' and C5' carbons of the flavonoid B-ring, while TEP does not appear to react on the A-ring. The reaction appears to be dependent on the temperature. This cross-linking approach appeared to give stable, hardened bondlines at elevated temperature on metals such as aluminum or steel for use in engines, boilers, exhaust pipes, ovens, and others.

9.4 LIGNIN ADHESIVES

Lignin is a phenolic polymer that is one of the main polymeric constituents of wood. It is generally produced in great quantities as waste from paper pulp mills. It is composed of repeat phenylpropane units. Considerable research has been carried out on lignin adhesives and binders. While, for certain applications, such as binders for nontarred rough rural roads, lignin derivatives have been used for many years, in the main area of potential application—wood adhesives—industrial use has been lagging. A variety of effective lignin adhesive formulations exist and have already been reviewed extensively [3–6], some of these having been used for periods of up to 3 years in some particleboard or plywood mills. Many of these are not used today because they all always present two main problems: The formulation tends to be corrosive of the equipment in the plant, and the lignin in the formulation tends to slow down panel press time noticeably, with consequent loss of mill productivity. In North America, there are now encouraging indications that premethylolated lignin (prereacted with formaldehyde) can be added up to 20%–30% of synthetic phenolic resins for plywood without lengthening panel press times, and one or two mills already appear to have been using such a system for some time. Recent claims that industrial use of wood adhesive formulations containing up to 50% lignin has occurred for some years, but has now been discontinued purely on economic grounds, have proven to be true and reliable [4].

The only step forward that has found industrial application in the last 20 years is to prereact lignin with formaldehyde in a reactor to form methylolated lignin; thus, to carry out part of the reaction with formaldehyde first, and then add this methylolated lignin to PF resins at the 20%–30% level [102,103]. These resins have been used in some North American plywood mills [80]. Particularly in plywood mills, the press time is not the factor determining the output rate of the factory, and so one can afford to use relatively long press times with good results [102]. Similar prereactions with nonvolatile and nontoxic aldehydes such as glyoxal have also been used with good application results [104–108].

None of the many adhesive systems based on pure lignin resins—hence, without synthetic resin addition—has succeeded at an industrial level. Some were tried industrially, but for one reason or another—too long a pressing time, high corrosiveness of the equipment, and so forth—they did not meet with commercial success. Still notable among these is the Nimz system based on the self-cross-linking of lignin in the presence of hydrogen peroxide [46,109,110]. Only one system is still used successfully today in several mills worldwide, but this is only for high-density hardboard. This is the Shen system, based on self-coagulation and cross-linking of lignin by a strong mineral acid in the presence of some aluminum salt catalysts [109,111–113]. However, attempts to extend this system to the industrial manufacture of MDF are known to have failed.

Of interest in the MDF field is also the system of adding laccase enzyme-activated lignin to the fibers or activating the lignin *in situ* in the fibers, also by enzyme treatment [114,115], but this requires the addition of 1% pMDI to the board [94] to press at acceptably short press times, or else the extension of the pressing times to ridiculous lengths (100 s/mm board thickness, while industrial press times are of the order of 3–7 s/mm board thickness) [116]. In the former case, an adhesive had to be used, with the same result as obtained by pressing untreated hardboard—a 100-year-old process—hence just wasting expensive time and enzymes. The second case, instead, illustrates even more clearly where the problem lies and what breakthrough is necessary; lignin activation by enzymes works, but not fast enough. The breakthrough necessary is a new, strong catalyst for the enzymatic action capable of allowing press times of industrial significance. This has not been found, or even considered, as yet.

A promising new technology based on lignin use for wood adhesives is relatively recent and again uses premethylolated or preglyoxalated lignin in the presence of small amounts of a synthetic PF resin and pMDI [104–108,117–119]. The proportion of premethylolated or preglyoxalated lignin used is 65% of the total adhesive, the balance being 10%–15% PF resin and 20%–25% pMDI. The system is based on cross-linking caused by the simultaneous formation of methylene bridges and urethane bridges, overcoming with the latter the need for higher cross-link density that has been one of the problems that has stopped lignin utilization in the past. More interesting has been the recent development in which formaldehyde has been totally eliminated, by substituting it with a nontoxic, nonvolatile aldehyde, namely glyoxal [104–108]. In these formulations, lignin is preglyoxalated in a reactor, and the glyoxalated lignin obtained is mixed with tannin and pMDI, thus eliminating the need for any formaldehyde or formaldehyde-based resins [104–108]. This technology has also brought about the total elimination of synthetic resins in the adhesive, as described. Both technologies satisfy the requirements for industrially significant press times.

A lot of the literature on the subject of lignin adhesives just rehashes older systems, all based on the substitution of some phenol in PF resins. In general, these papers do not seem to be aware of the long press time problem and do not address it, perpetuating the myth of PF–lignin adhesives while repeating the same age-old errors. They lead new researchers in the field to believe they are doing something worthwhile, using press conditions that do not satisfy the requirements of the high press rate of the panel manufacturing industry.

Some new and rather promising technologies regarding lignin adhesives have, however, been developed recently. These are (1) adhesives for particleboard and other agglomerate wood panels based on a mix of tannin/hexamine with preglyoxalated lignin, and (2) similar formulations for high resin-content, high-performance agricultural fiber composites.

Mixed interior wood-panel tannin adhesive formulations were developed in which lignin is a considerable proportion (50%) of the wood-panel binder and in which no fortification with synthetic resins, such as the isocyanates and PF resins used in the past, was necessary to obtain results satisfying relevant standards. The wood panel itself is constituted of 99.5% natural materials, the 0.5% balance being composed of glyoxal, a nontoxic and nonvolatile aldehyde, for the preglyoxalation of lignin and hexamine, already accepted as a non-formaldehyde-yielding compound when in the presence of a condensed tannin. Both particleboard and two types of plywood were shown to pass the relevant interior standards with such adhesive formulations [105,106]. The mechanism of copolymerization and hardening can be understood from Figure 9.7.

Some new cross-linking reactions were recently reported, based on the use of unusual, unthought-of cross-linkers. Thus, as for tannins, the reaction of condensation and cross-linking of kraft and other types of lignin with TEP has also been found to lead to adhesives not containing any aldehyde and with no emissions [120]. This cross-linking approach appeared to give stable hardened bondlines at elevated temperature on metals such as steel for use in engines, boilers, exhaust pipes, ovens, and others.

Furthermore, the idea of mobilizing the lignin in wood itself as the adhesive to join wood and panels has been experimented with by just using citric acid as an additive on wood. Citric acid was first used to bond wood without the use of any adhesive in interior-grade particleboard panels

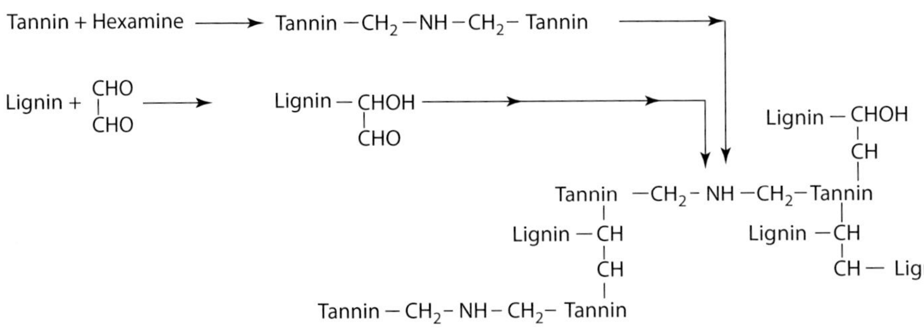

FIGURE 9.7 Scheme of hybrid adhesive system formed by coreacting tannin plus hexamethylenetetramine with preglyoxalated lignin.

[121]. Further studies showed that citric acid improved the resistance to water of wood joints of spruce along the end-grain-to-end-grain faces [122]. The use of 10%–20% citric acid as treatment has been shown to markedly improve both the mechanical performance and the water resistance of such joints. The changes observed by CP-MAS ^{13}C NMR of the spruce joints so obtained indicates that citric acid appeared to cause a catalytic effect, which led to rearrangements in the structure of lignin [122].

9.5 SUMMARY

Biobased adhesives, especially those based on tannins and lignin—abundant renewable raw materials—have been industrialized for the former and are well on the way to industrialization for the latter. Thus, due to their growth and the interest in them, biosourced adhesives can no longer be perceived as newcomers or outsiders in the adhesives market, but have the prospect of a sure growth in the medium-term future. As the research on their chemistry and formulation progresses, new applications are found and formulators are able to decrease their costs to levels comparable to those of synthetic oil-derived adhesives. Their growth must not be seen as in competition with the present, synthetic, oil-derived adhesives, but rather as complementary to these. It must also be seen as a substitution of synthetic adhesives in the medium- to long-term future, when the sources of the latter will start to decrease.

REFERENCES

1. A. Pizzi. An assessment of the future industrial prospects for panel adhesives from renewable natural materials. *Holzforschung Holzverwertung* 43, 83–87 (1991).
2. A. Pizzi. Tannin-based wood adhesives. In: *Wood Adhesives: Chemistry and Technology*, Vol. 1, A. Pizzi (Ed.), pp. 198–273, Marcel Dekker, New York (1983).
3. H. H. Nimz. Lignin-based wood adhesives. In: *Wood Adhesives: Chemistry and Technology*, Vol. 1, A. Pizzi (Ed.), pp. 275–312, Marcel Dekker, New York (1983).
4. E. Janiga. Masonite and lignosulphonate resins. In: *Proceedings of Wood Adhesives 2000*, Forest Products Society, Madison, WI (2000).
5. N. G. Lewis and T. R. Lantzy. In: *Adhesives from Renewable Resources*, Lignin in Adhesives ACS Symposium Series 385 R.W. Hemingway, A. Conner, and S.J. Branham (Eds.), pp. 13–26, American Chemical Society, Washington, DC (1989).
6. W. G. Glasser and S. Sarkanen (Eds.). *Lignin, Properties and Materials*, ACS Symposium Series 397, American Chemical Society, Washington, DC (1989).
7. H. Pasch and A. Pizzi. On the macromolecular structure of chestnut ellagitannins by MALDI-TOF mass spectrometry. *J. Appl. Polym. Sci.* 85, 429–437 (2002).
8. N. Radebe, K. Rode, A. Pizzi, S. Giovando and H. Pasch. MALDI-TOF-CID for the microstructure elucidation of polymeric hydrolysable tannins. *J. Appl. Polym. Sci.* 128, 97–107 (2013).
9. S. Giovando, A. Pizzi, H. Pasch and N. Pretorius. Structure and oligomers distribution of commercial Tara (*Caesalpina spinosa*) hydrolysable tannin. *ProLigno* 9(1), 22–31 (2013).
10. E. Kulvik. Chestnut wood tannin extracts in phenol-formaldehyde resins. *Adhesives Age* 18(3), 20–24 (1975).
11. E. Kulvik. Chestnut wood tannin extracts in plywood adhesives. *Adhesives Age* 19(3), 19–21 (1976).
12. S. Spina, X. Zhou, C. Segovia, A. Pizzi, M. Romagnoli, S. Giovando, H. Pasch, K. Rode and L. Delmotte. Phenolic resin adhesives based on chestnut hydrolysable tannins. *J. Adhesion Sci. Technol* 27, 2103–2111 (2013).
13. M. Fechtal and B. Riedl. Use of eucalyptus and *Acacia mollissima* bark extract-formaldehyde adhesives in particleboard manufacture. *Holzforschung* 47, 349–357 (1993).
14. D. Thompson and A. Pizzi. Simple ^{13}C NMR methods for the quantitative determination of polyflavonoid tannins characteristics. *J. Appl. Polym. Sci.* 55, 107–112 (1995).
15. D. G. Roux. *Modern Applications of Mimosa Extract*, pp. 34–41, Leather Industries Research Institute, Grahamstown, South Africa (1965).
16. A. Pizzi and H. Scharfetter. The chemistry and development of tannin-based wood adhesives for exterior plywood. *J. Appl. Polym. Sci.* 22, 1745–1761 (1978).

17. D. G. Roux and E. Paulus. Condensed tannins. 8: The isolation and distribution of interrelated heartwood components of *Schinopsis* spp. *Biochem. J.* 78, 785–789 (1961).

18. H. G. C. King, T. White and R. B. Hughes. The occurrence of 2-benzyl-2-hydroxycoumaran-3-ones in quebracho tannin extract. *J. Chem. Soc.* 3234–3239 (1961).

19. D. G. Roux, D. Ferreira, H. K. L. Hundt and E. Malan. Structure, stereochemistry, and reactivity of natural condensed tannins as basis of their extended industrial utilization. *Appl. Polym. Symp.* 28, 335–353 (1975).

20. J. W. Clark-Lewis and D. G. Roux. Natural occurrence of enantiomorphous leucoanthocyanidian: (+)-mollisacacidin (gleditsin) and quebracho(−)-leucofisetinidin. *J. Chem. Soc.* 1402–1406 (1959).

21. R. W. Hemingway and G. W. McGraw. Adhesives from southern pine bark. *Appl. Polym. Symp.* 28, 1349–1364 (1975).

22. D. T. Rossouw, A. Pizzi and G. McGillivray. The kinetics of condensation of phenolics polyflavonoid tannins with aldehydes. *J. Polym. Sci. Polym. Chem.* 18, 3323–3338 (1990).

23. L. J. Porter. Extractives of *Pinus radiata* bark. *N. Z. J. Sci.* 17, 213–218 (1974).

24. P. Navarrete, A. Pizzi, H. Pasch, K. Rode and L. Delmotte. MALDI-TOF and ^{13}C NMR characterisation of maritime pine industrial tannin extract. *Ind. Crops Prod.* 32, 105–110 (2010).

25. K. F. Plomley. Tannin-formaldehyde adhesives for wood, Paper 39, Division of Australian Forest Products Technology, pp. 1–16 (1966).

26. N. J. L. Megson. *Phenolic Resins Chemistry*, p. 300, Butterworth, Sevenoaks, UK (1958).

27. A. Pizzi. Tannin-based adhesives. *J. Macromol. Sci. Reviews Macromol. Chem.* C18(2), 247–294 (1980).

28. A. Pizzi and A. Stephanou. A ^{13}C NMR study of polyflavonoid tannin adhesives intermediates, Part 1: Non-colloidal, performance-determining rearranragements. *J. Appl. Polym. Sci.* 51, 2109–2124 (1994).

29. S. Garnier, A. Pizzi, O. C. Vorster and L. Halasz. Comparative rheological characteristics of industrial polyflavonoid tannin extracts. *J. Appl. Polym. Sci.* 81, 1634–1642 (2001).

30. H. Pasch, A. Pizzi and K. Rode. MALDI-TOF mass spectrometry of polyflavonoid tannins. *Polymer*, 42, 7531–7539 (2001).

31. R. Brown and W. Cummings. Polymerisation of flavans. Part II: The condensation of 4-methoxyflavan with phenols. *J. Chem. Soc.* 4302–4304 (1959).

32. K. Freudenberg and J. M. Alonso de Lama. Zur Kenntnis des Catechin-Gerbstoffe. *Annalen Chim.* 612, 78–93 (1958).

33. R. Brown, W. Cummings and J. Newbould. Polymerisation of flavans. Part IV: The condensation of flavan-4-ols with phenols. *J. Chem. Soc.* 3677–3682 (1961).

34. A. Pizzi. Sulphited tannins for exterior wood adhesives. *Colloid Polym. Sci.* 257, 37–40 (1979).

35. L. K. Dalton. Tannin-formaldehyde resins as adhesives for wood. *Aust. J. Appl. Sci.* 1, 54–70 (1950).

36. L. K. Dalton. Resins from sulphited tannins as adhesives for wood. *Aust. J. Appl. Sci.* 4, 136–146 (1953).

37. J. R. Parrish. Particleboard from wattle wood and wattle tannin. *J. S. African Forest Assoc.* 32, 26–31 (1958).

38. A. Pizzi and A. Stephanou. A ^{13}C NMR study of polyflavonoid tannin adhesives intermediates. Part 2: Colloidal state reactions. *J. Appl. Polym. Sci.* 51, 2125–2130 (1994).

39. A. Merlin and A. Pizzi. An ESR study of the silica-induced autocondensation of polyflavonoid tannins. *J. Appl. Polym. Sci.* 59, 945–952 (1996).

40. E. Masson, A. Merlin and A. Pizzi. Comparative kinetics of the induced radical autocondensation of polyflavonoid tannins, Part I: Modified and non-modified tannins. *J. Appl. Polym. Sci.* 60, 263–269 (1996).

41. E. Masson, A. Pizzi and A. Merlin. Comparative kinetics of the induced radical autocondensation of polyflavonoid tannins, Part III: Micellar reactions vs. cellulose surface catalysis. *J. Appl. Polym. Sci.* 60, 1655–1664 (1996).

42. E. Masson, A. Pizzi and A. Merlin. Comparative kinetics of the induced radical autocondensation of polyflavonoid tannins, Part 2: Flavonoid units effects. *J. Appl. Polym. Sci.* 64, 243–265 (1997).

43. R. Garcia, A. Pizzi and A. Merlin. Ionic polycondensation effects on the radical autocondensation of polyflavonoid tannins: An ESR study. *J. Appl. Polym. Sci.* 65, 2623–2632 (1997).

44. A. Pizzi. Wattle-based adhesives for exterior grade particleboard. *Forest Prod .J.* 28(12), 42–46 (1978).

45. A. Pizzi and A. Stephanou. A comparative ^{13}C NMR study of polyflavonoid tannin extracts for phenolic polycondensates. *J. Appl. Polym. Sci.* 50, 2105–2113 (1993).

46. A. Pizzi, *Advanced Wood Adhesives Technology*, pp. 1–300, Marcel Dekker, New York (1994).

47. S. Garnier, A. Pizzi, O. C. Vorster and L. Halasz. Comparative rheological characteristics of industrial polyflavonoid tannin extracts. *J. Appl. Polym. Sci.* 81, 1634–1642 (2001).

48. Specification for particleboard, Deutscher Normenausschuss V100 and V313, DIN 68761, Part 3 (1967).

49. A. Pizzi. Utilizing wattle-based adhesives in making particleboard. *Adhesives Age* 21(9), 32–35 (1978).

50. A. Pizzi. Hybrid interior particleboard using wattle tannin adhesives. *Holzforschung Holzverwertung* 31(4), 85–87 (1979).

51. A. Pizzi and A. Stephanou. Phenol-formaldehyde wood adhesives under very alkaline conditions. Part 2: Acceleration mechanism and applied results. *Holzforschung* 48, 150–155 (1994).

52. C. Zhao, A. Pizzi and S. Garnier. Fast advancement and hardening acceleration of low condensation alkaline PF resins by esters and copolymerized urea. *J. Appl. Polym. Sci.* 74, 359–378 (1999).

53. A. Pizzi, J. Valenzuela and C. Westermeyer. Low-formaldehyde emission, fast pressing, pine and pecan tannin adhesives for exterior particleboard. *Holz Roh Werkst.* 52, 311–315 (1994).

54. F. Pichelin, A. Pizzi, A. Frühwald and P. Triboulot. Exterior OSB preparation technology at high moisture content. Part 1: Transfer mechanisms and pressing parameters. *Holz Roh Werkst.* 59, 256–265 (2001).

55. A. Pizzi. Pine tannin adhesives for particleboard. *Holz Roh Werkst.* 40, 293–303 (1982).

56. A. Pizzi, E. P. Von Leyser, J. Valenzuela and J. G. Clark. The chemistry and development of pine tannin adhesives for exterior particleboard. *Holzforschung* 47, 164–172 (1993).

57. A. Pizzi. Exterior wood adhesives by MDI cross-linking of polyflavonoid tannin B-ring. *J. Appl. Polym. Sci.* 25, 2123–2127 (1980).

58. A. Pizzi and T. Walton. Non-emulsifiable, water-based diisocyanate adhesives for exterior plywood. Part 1: Novel reaction mechanisms and their chemical evidence. *Holzforschung* 46, 541–547 (1992).

59. A. Trosa and A. Pizzi. A no-aldehyde emission hardener for tannin-based wood adhesives. *Holz Roh Werkst.* 59, 266–271 (2001).

60. A. Trosa. Développement et application industrielle de résines thermodurcissables a base de produits naturels de déchet et leur produits de copolymérisation avec des résines synthétiques pour application aux panneaux composites de bois. PhD thesis, University Henri Poincaré, Nancy 1, Nancy, France (1999).

61. F. Pichelin, C. Kamoun and A. Pizzi. Hexamine hardener behaviour: Effects on wood glueing, tannin and other wood adhesives. *Holz Roh Werkst.* 57, 305–317 (1999).

62. A. Ballerini, A. Despres and A. Pizzi. Non-toxic, zero-emission tannin-glyoxal adhesives for wood panels. *Holz Roh Werkst.* 63, 477–478 (2005).

63. A. E. McKenzie and Y. P. Yuritta. Starch corrugating adhesives. *Appita* 26, 30–34 (1974).

64. P. A. J. L. Custers, R. Rushbrook, A. Pizzi, and C. J. Knauff. Industrial applications of wattle-tannin/urea-formaldehyde fortified starch adhesives for damp-proof corrugated cardboard. *Holzforschung Holzverwertung* 31(6), 131–132 (1979).

65. A. Pizzi and D. G. Roux. The chemistry and development of tannin-based weather- and boil-proof cold-setting and fast-setting adhesives for wood. *J. Appl. Polym. Sci.* 22, 1945–1954 (1978).

66. A. Pizzi and D. G. Roux. Resorcinol/wattle flavonoid condensates for cold setting adhesives. *J. Appl. Polym. Sci.* 22, 2717–2718 (1978).

67. A. Pizzi. The chemistry and development of tannin-urea-formaldehyde condensates for exterior wood adhesives. *J. Appl. Polym. Sci.* 23, 2777–2797 (1979).

68. Specification for synthetic adhesive resins for wood, British Standard BS 1204, Parts 1 and 2, 1965.

69. A. Pizzi, D. du T. Rossouw, W. Knuffel and M. Singmin. "Honeymoon" phenolic and tannin-based fast setting adhesive systems for exterior grade fingerjoints. *Holzforschung Holzverwertung* 32(6), 140–151 (1980).

70. A. Pizzi and F. A. Cameron. Fast-set adhesives for glulam. *Forest Prod. J.* 34(9), 61–65 (1984).

71. K. H. Chung and Q. R. Hamed. Adhesives containing pine bark tannin for bonding nylon cord to rubber. In: *Chemistry and Significance of Condensed Tannins*, R. W. Hemingway and J. J. Karchesy (Eds.), pp. 479–492, Plenum Press, New York (1989).

72. A. Pizzi, B. Mtsweni and W. Parsons. Wood-induced catalytic activation of PF adhesives autopolymerization vs. PF/wood covalent bonding. *J. Appl. Polym. Sci.* 52, 1847–1856 (1994).

73. A. Nicollin, X. Li, P. Girods, A. Pizzi and Y. Rogaume. Fast pressing composites using tannin-furfuryl alcohol resin and vegetal fibers reinforcement. *J. Renew. Mater.* 1, 311–316 (2013).

74. U. H. B. Abdullah and A. Pizzi. Tannin-furfuryl alcohol wood panel adhesives without formaldehyde. *Eur. J. Wood Prod.* 71, 131–132 (2013).

75. N. Meikleham, A. Pizzi and A. Stephanou. Induced accelerated autocondensation of polyflavonoid tannins for phenolic polycondensates, Part 1: ^{13}C NMR, ^{29}Si NMR, X-ray and polarimetry studies and mechanism. *J. Appl. Polym. Sci.* 54, 1827–1845 (1994).

76. A. Pizzi and A. Stephanou. Comparative and differential behaviour of pine vs. pecan nut tannin adhesives for particleboard. *Holzforschung Holzverwertung* 45(2), 30–33 (1993).

77. A. Pizzi and N. Meikleham. Induced accelerated autocondensation of polyflavonoid tannins for phenolic polycondensates: Part III: CP-MAS ^{13}C NMR of different tannins and models. *J. Appl. Polym. Sci.* 55, 1265–1269 (1995).

78. A. Pizzi, N. Meikleham and A. Stephanou. Induced accelerated autocondensation of polyflavonoid tannins for phenolic polycondensates: Part II: Cellulose effect and application. *J. Appl. Polym. Sci.* 55, 929–933 (1995).

79. A. Pizzi, N. Meikleham, B. Dombo and W. Roll. Autocondensation-based, zero-emission, tannin adhesives for particleboard. *Holz Roh Werkstoff* 53, 201–204 (1995).

80. R. Garcia and A. Pizzi. Polycondensation and autocondensation networks in polyflavonoid tannins, Part 1: Final networks. *J. Appl. Polym. Sci.* 70, 1083–1092 (1998).

81. R. Garcia and A. Pizzi. Polycondensation and autocondensation networks in polyflavonoid tannins, Part 2: Polycondensation vs. autocondensation. *J. Appl. Polym. Sci.* 70(6), 1093–1110 (1998).

82. A. Pizzi. Tannin-based polyurethane adhesives. *J. Appl. Polym. Sci.* 23, 1889–1890 (1979).

83. D. Garcia, W. Glasser, A. Pizzi, A. Osorio-Madrazo and M.-P. Laborie. Hydroxypropyl tannin derivatives from *Pinus pinaster* (Ait.) bark. *Ind. Crops Prod.* 49, 730–739 (2013).

84. D. E. Garcia, W. G. Glasser, A. Pizzi, S. Paczkowski and M.-P. Laborie. Substitution pattern elucidation of hydroxypropyl *Pinus pinaster* (Ait.) bark polyflavonoids derivatives by ESI(-)-MS/MS. *J. Mass Spectrom.* 49, 1050–1058 (2014).

85. D. E. Garcia, W. G. Glasser, A. Pizzi, S. Paczkowski and M.-P. Laborie. Hydroxypropyl tannin from *Pinus pinaster* bark as polyol source in urethane chemistry. *European Polym. J.* 67, 152–165 (2015).

86. D. E. Garcia, W. G. Glasser, A. Pizzi, S. Paczkowski and M.-P. Laborie. Modification of condensed tannins: From polyphenols chemistry to materials engineering. *New J. Chem.* 40, 36–49 (2016).

87. W. G. Glasser, C. A. Barnett, T. G. Rials and S. Vasudev. Engineering plastics from lignin II. Characterization of hydroxyalkyl lignin derivatives. *J. Appl. Polym. Sci.* 29, 1815–1830 (1984).

88. W. G. Glasser, and R. H. Leitheiser. Engineering plastics from lignin. XI. Hydroxypropyl lignins as componenets of fire resistant foams. *Polymer Bulletin* 12, 1–5 (1984).

89. W. G. Glasser, S. S. Kelley, T. G. Rials, and S. L. Ciemniecky. Structure-property relationships of engineering plastics from lignin. *Proceedings of the 1986 TAPPI Research and Development Conference*, New Orleans, LA 157–161 (1986).

90. J. Valenzuela, E. von Leyser, A. Pizzi, C. Westermeyer and B. Gorrini. Industrial production of pine tannin-bonded particleboard and MDF. *Eur. J.Wood Prod.* 70, 735–740 (2012).

91. M. C. Basso, A. Pizzi, C. Lacoste, L. Delmotte, F. A. Al-Marzouki, S. Abdalla and A. Celzard, Tannin-furanic-polyurethane foams for industrial continuous plant lines. *Polymers* 6, 2985–3004 (2014).

92. G. Rokicki and A. Piotrowska. A new route to polyurethanes from ethylene carbonate, diamines and diols. *Polymer* 43, 2927–2935 (2002).

93. B. Nohra, L. Candy, J.-F. Blanco, C. Guerin, Y. Raoul and Z. Mouloungui. From petrochemical polyurethanes to biobased polyhydroxyurethanes, a review. *Macromolecules* 46, 3771–3792 (2013).

94. H. Blattmann, M. Fleischer, M. Bähr and R. Mülhaupt. Isocyanate- and phosgene-free routes to polyfunctional cyclic carbonates and green polyurethanes by fixation of carbon dioxide. *Macromol. Rapid Commun.* 35, 1238–1254 (2014).

95. F. Camara, S. Benyahya, V. Besse, G. Boutevin, R. Auvergne, B. Boutevin and S. Caillol. Reactivity of secondary amines for the synthesis of nonisocyanate polyurethanes. *Eur. Polym. J.* 55, 17–26 (2014).

96. M. Thebault, A. Pizzi, S. Dumarcay, P. Gerardin, E. Fredon and L. Delmotte. Polyurethanes from hydrolysable tannins obtained without using isocyanates. *Ind. Crops Prod.* 59, 329–336 (2014).

97. M. Thebault, A. Pizzi, H. Essawy, A. Baroum and G. Van Assche. Isocyanate free condensed tannin-based polyurethanes. *Eur. Polym. J.* 67, 513–523 (2015).

98. F. Braghiroli, V. Fierro, A. Celzard, A. Pizzi, K. Rode, W. Radke, L. Delmotte and J. Parmentier. Condensation reaction of flavonoid tannins with ammonia. *Ind. Crops Prod.* 44, 330–335 (2013).

99. M. Thebault, A. Pizzi, F.M. Al-Marzouki, S. Abdalla and S. Al-Ameer. Isocyanate-free polyurethanes by coreaction of condensed tannins with aminated tannins. *J. Renewable Mater.* 5, 21–29 (2017).

100. F. J. Santiago-Medina, G. Foyer, A. Pizzi, S. Calliol and L. Delmotte. Lignin-derived non-toxic aldehydes for ecofriendly tannin adhesives for wood panels. *Int. J. Adhesion Adhesives* 70, 239–248 (2016).

101. M. C. Basso, A. Pizzi, J. Polesel-Maris, L. Delmotte, B. Colin and Y. Rogaume. MALDI-TOF and ^{13}C NMR analysis of the cross-linking reaction of condensed tannins by triethyl phosphate. *Ind. Crops Prod.* (2017) DOI:10.1016/j.indcrop.2016.11.031

102. L. R. Calvé. Fast cure and pre-cure resistant cross-linked phenol-formaldehyde adhesives and methods of making same. Can. Pat. 2042476 (1999).

103. D. Gardner and T. Sellers, Jr. Formulation of a lignin-based plywood. *Forest Prod. J.* 36, 61–67 (1986).

104. N.-E. El-Mansouri, A. Pizzi and J. Salvadó. Lignin-based wood panel adhesives without formaldehyde. *Holz Roh Werkst.* 65, 65–70 (2007).

105. N.-E. El-Mansouri, A. Pizzi and J. Salvado. Lignin-based polycondensation resins for wood adhesives. *J. Appl. Polym. Sci.* 103, 1690–1699 (2007).

106. P. Navarrete, A. Pizzi, S. Tapin-Lingua, B. Benjelloun-Mlayah, H. Pasch, K. Rode, L. Delmotte and S. Rigolet. Low formaldehyde emitting biobased wood adhesives manufactured from mixtures of tannin and glyoxalated lignin. *J. Adhesion Sci. Technol.* 26, 1667–1684 (2012).

107. P. Navarrete, A. Pizzi, H. Pasch and L. Delmotte. Study on lignin-glyoxal reaction by MALDI-TOF and CP-MAS ^{13}C NMR. *J. Adhesion Sci. Technol* 26, 1069–1082 (2012).

108. P. Navarrete, A. Pizzi, K. Rode, M. Vignali and H. Pasch. MALDI-TOF study of oligomers distribution for stability-durable spray-dried glyoxalated lignin for wood adhesives. *J. Adhesion Sci. Technol.* 27, 586–597 (2013).

109. H. H. Nimz. Lignin-based adhesives. In: *Wood Adhesives: Chemistry and Technology*, Vol. 1, A. Pizzi (Ed.), pp. 247–288, Marcel Dekker, New York (1983).

110. H. H. Nimz and G. Hitze. The application of spent sulfite liquor as an adhesive for particleboards. *Cellulose Chem. Technol.* 14, 371–382 (1980).

111. D. P. C. Fung, K. C. Shen and L. R. Calvé. Spent sulphite liquor-sulphuric acid binder: Its preparation and some chemical properties, *Report OPX 180 E*, Eastern Forest Products Laboratory, Ottawa (1977).

112. K. C. Shen. Spent sulpite liquor binder for exterior wafer-board. *Forest Prod. J.* 27, 32–38 (1977).

113. K. C. Shen. Adhesive composition. Patent Convention Treaty patent WO 1998037148A3 (1998).

114. A. Kharazipour, A. Haars, M. Shekholeslami and A. Hüttermann. Enzymgebundene Holzwerkstoffe auf der Basis von Lignin und Phenoloxidase. *Adhäsion* 35(5), 30–36 (1991).

115. A. Kharazipour, C. Mai and A. Hüttermann. Polyphenols for compounded materials. *Polym. Degrad. Stabil.* 59, 237–243 (1998).

116. C. Felby, L. S. Pedersen and B. R. Nielsen. Enhanced auto adhesion of wood fibers using phenol oxidases. *Holzforschung.* 51, 281–286 (1997).

117. W. H. Newman and W. G. Glasser. Engineering plastics from lignin, XII: Synthesis and performance of lignin adhesives with isocyanate and melamine. *Holzforschung.* 39, 345–353 (1985).

118. A. Pizzi and A. Stephanou. Rapid curing lignins-based exterior wood adhesives, Part 1: Diisocyanates reaction mechanisms and application to panel products. *Holzforschung.* 47, 439–445 (1993).

119. A. Pizzi and A. Stephanou. Rapid curing lignins-based exterior wood adhesives, Part 2: Acceleration mechanisms and application to panel products. *Holzforschung.* 47, 501–506 (1993).

120. M. C. Basso, A. Pizzi, L. Delmotte and S. Abdalla. MALDI-TOF and ^{13}C NMR analysis of the Cross-linking Reaction of Lignin by Triethyl Phosphate, *Polymers*, 9, 206–221 (2017).

121. S. S. Kusumah, K. Umemura, K. Y. Oshioka, H. Miyafuji and K. Kanajama. Utilisation of sweet sorghum bagasse and citric acid for manufacturing particleboard I: Effects of predrying treatment and citric acid content on the board properties. *Ind. Crops Prod.* 84, 34–42 (2016).

122. S. Amirou, A. Pizzi and L. Delmotte. Citric acid as waterproofing additive in butt joints linear wood welding. *Eur. J. Wood Prod.* 75, 651–654 (2017).

Antonio Pizzi

CONTENTS

10.1 INTRODUCTION

The urea–formaldehydes (UFs), melamine–formaldehydes (MFs), and melamine–urea–formaldehydes (MUFs) are the most important and most used class of aminoresin adhesives. Aminoresins are polymeric condensation products of the reaction of aldehydes with compounds carrying amine or amide groups. Formaldehyde is by far the most used. The advantages of UF adhesives are their (1) initial water solubility (this renders them eminently suitable for bulk and relatively inexpensive production), (2) hardness, (3) nonflammability, (4) good thermal properties, (5) absence of color in cured polymers, and (6) easy adaptability to a variety of curing conditions [1,2].

Thermosetting aminoresins produced from urea or melamine are built up by condensation polymerization. Urea is reacted with formaldehyde, which results in the formation of addition products such as methylol compounds. Further reaction and the concurrent elimination of water leads to the formation of low molecular weight condensates, which are still soluble. Higher molecular weight products, which are insoluble and infusible, are obtained by further condensing the low molecular weight condensates. The greatest disadvantage of the aminoresins is their bond deterioration, caused by water and moisture. This is due to the hydrolysis of their aminomethylenic bond. Therefore, pure UF adhesives are used only for interior applications.

10.2 UREA–FORMALDEHYDE ADHESIVES

10.2.1 CHEMISTRY OF UF RESINS: UREA–FORMALDEHYDE CONDENSATION

The reaction between urea and formaldehyde is complex. The combination of these two chemical compounds results in both linear and branched polymers, as well as tridimensional networks, in the cured resin. This is due to a functionality of 4 in urea (due to the presence of four replaceable hydrogen atoms) (in reality, urea is only trifunctional, as tetramethylol urea has never been isolated, except in the formation of substituted urons [2]) and a functionality of 2 in formaldehyde. The most important factors determining the properties of the reaction products are (1) the relative molar proportion of urea and formaldehyde, (2) the reaction temperature, and (3) the pH value at which condensation takes place. These factors influence the rate of increase of the molecular weight of the resin. Therefore, the characteristics of the reaction products differ considerably when lower and higher condensation stages are compared, especially solubility, viscosity, water retention, and rate of curing of the adhesive. These all depend, to a large extent, on molecular weight.

The reaction between urea and formaldehyde is divided into two stages. First, the alkaline condensation to form mono-, di-, and trimethylolureas (tetramethylolurea has never been isolated). The second stage is the acid condensation of the methylolureas, first to soluble and then to insoluble cross-linked resins. On the alkaline side, the reaction of urea and formaldehyde at room temperature leads to the formation of methylolureas. When condensed, they form methylene-ether bridges between the urea molecules. The products from the reaction of urea and formaldehyde, and from the mono- and dimethylolureas, are as follows:

$$
\begin{array}{c}
O{=}C\!\!<\!\!{}^{NH_2}_{NH_2} + OH^- \longrightarrow O{=}C\!\!<\!\!{}^{NH_2\,H}_{NH\,H}\overset{+\delta\ -\delta}{C}{=}O \rightleftharpoons O{=}C\!\!<\!\!{}^{NH_2}_{NH{-}CH_2{-}OH} \\[2mm]
\Big\updownarrow OH^- \\[2mm]
O{=}C\!\!<\!\!{}^{NH{-}CH_2OH}_{NH{-}CH_2OH} \xrightleftharpoons{\ \overset{H}{\underset{H}{C}}\overset{+\delta\ -\delta}{=}O\ } O{=}C\!\!<\!\!{}^{\overline{N}H}_{NH{-}CH_2OH} \\[2mm]
\Big\updownarrow OH^- \\[2mm]
O{=}C\!\!<\!\!{}^{NH{-}CH_2OH}_{NH{-}CH_2OH} \xrightleftharpoons{\ \overset{H}{\underset{H}{C}}\overset{+\delta\ -\delta}{=}O\ } O{=}C\!\!<\!\!{}^{NH{-}CH_2OH}_{N{<}{}^{CH_2OH}_{CH_2OH}} \\[2mm]
\Big\downarrow \\[2mm]
O{=}C\!\!<\!\!{}^{NH_2}_{NHCH_2{-}O{-}H_2CHN}\!\!>\!\!C{=}O \qquad\qquad O{=}C\!\!<\!\!{}^{NHCH_2OH}_{NHCH_2{-}O{-}CH_2NH}\!\!>\!\!C{=}O \\[2mm]
O{=}C\!\!<\!\!{}^{NHCH_2{-}O{-}H_2CHN}_{\underset{CH_2OH}{NH}\ HOH_2CHN}\!\!>\!\!C{=}O \qquad O{=}C\!\!<\!\!{}^{NHCH_2{-}O{-}CH_2NH}_{\underset{CH_2OH}{N{-}CH_2OH}\ \underset{CH_2OH}{HN}}\!\!>\!\!C{=}O
\end{array}
\tag{10.1}
$$

The reaction also produces cyclic derivatives: uron, monomethyloluron, and dimethyloluron.

On the acid side, the products precipitated from aqueous solutions of urea and formaldehyde, or from methylolureas, are low molecular weight methyleneureas [3]:

$$H_2NCONH(CH_2NHCONH)_n H$$

These contain methylol end groups in some cases, through which it is possible to continue the reaction to harden the resin.

The monomethylolureas formed copolymerize by acid catalysis and produce polymers, and then highly branched and cured networks:

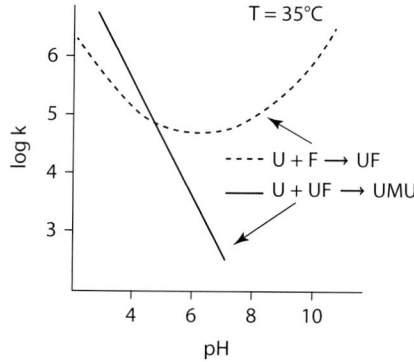

(10.2)

The kinetics of the formation and condensation of mono- and dimethylolureas and of simple urea–formaldehyde condensation products have been studied extensively. The formation of monomethylolurea in weak acid or alkaline aqueous solutions is characterized by an initial fast phase, followed by a slow bimolecular reaction [4,5]. The first reaction is reversible and is an equilibrium that proceeds to methylolureas, which then undergo the second reaction. The rate of reaction varies according to the pH, with a minimum rate of reaction in the pH range 5–8 for a urea:formaldehyde molar ratio of 1:1 and a pH of 6.5 for a 1:2 molar ratio [6] (Figure 10.1). The 1:2 urea–formaldehyde reaction has been proved to be three times slower than the 1:1 molar ratio reaction [7].

The rapid initial addition reaction of urea and formaldehyde is followed by a slower condensation that results in the formation of polymers [7]. The rate of condensation of urea with monomethylolurea to form methylenebisurea (or UF dimers) is also pH dependent. It decreases exponentially from a pH of 2–3 to a neutral pH value. No condensation occurs at alkaline pH values.

The initial addition of formaldehyde to urea is reversible and is subject to general acid and base catalysis. Different energies of activation are reported for the forward methylolation and backward demethylolation reaction. The forward bimolecular reaction is reported to have an activation energy of 13 kcal/mol, while the reverse unimolecular reaction has an activation energy of 19 kcal/mol [5].

FIGURE 10.1 Influence of pH on the addition and condensation reactions of urea and formaldehyde. U: urea; F: HCHO; M: $-CH_2-$.

Other sources report values of 17.5 and 17.1 kcal/mol for the same reactions, respectively [8]. If one considers that the monomethylation reaction of urea at pH 7 is of the order of 1×10^{-4} L/(mol.s) for each site [8] and of the order of 3×10^{-4} L/(mol.s) at rather alkaline pH, it is possible to deduce that at alkaline pH, urea reacts with formaldehyde to form methylolated ureas. However, the inverse reaction of decomposition of the methylol urea will limit somewhat the proportion of methylolated ureas obtained. The reaction goes to completion only as methylolated ureas react to form dimers and higher oligomers when the pH is lowered in the condensation phase. If the condensation phase is not effected, the calculation of the degree of advancement of the reaction of methylolation of urea under alkaline conditions can be carried out by using Equation 10.1 [9].

$$p/\left[2(1-p)\right] = \exp\left[\left(-\Delta G^{\theta}\right)/(2RT)\right]$$ (10.1)

where:

p is the degree of conversion at equilibrium of the methylolation and demethylolation reactions

ΔG^{θ} is the standard Gibbs energy variation

R is a constant (1.987 cal/g.mol K)

T is the temperature in kelvins

When introducing the reported activation energies of the urea forward methylolation reaction (17.5 kcal/mol) [8] and of the methylol urea demethylolation reaction (17.1 kcal/mol) [8], one obtains a degree of advancement ($p = 0.60$); hence, at equilibrium under the conditions used, 60% of the urea is present as methylol ureas. This compares well with the degree of conversion of 65% at equilibrium of the more reactive melamine extrapolated by reported kinetic values of the addition of formaldehyde [10], under the same conditions as used here. The advancement of the reaction may eventually proceed to even higher degrees of conversion, even in alkaline environments, as a consequence of the subsequent formation of methylene-ether-linked oligomers.

The rates of introduction into the urea molecule of one, two, and three methylol groups have been estimated to have the ratio 9:3:1, respectively. The formation of N,N'-dimethylolurea from monomethylolurea is three times that of monomethylolurea from urea.

Methylenebisurea and higher oligomers undergo further condensation with formaldehyde [11] and monomethylolurea [12], behaving like urea. The capability of methylenebisurea to hydrolyze to urea and methylolurea in weak acid solutions (pH 3–5) indicates the reversibility of the amidomethylene link and its lability under weak acid and moist conditions. This explains the slow release of formaldehyde over a long time in particleboard and other wood products manufactured with UF resins.

10.2.2 General Principles of Manufacture and Application

It is very important in the commercial production of UF resins to be able to control the size of the molecules by the condensation reaction, since their properties change continuously as they grow larger. The most perceptible change is the increase in viscosity. Low-viscosity syrups are formed first. These change into high-viscosity syrups, which are clear to turbid. Molecular weight may vary from a few hundred to a few thousand, with a wide range of molecular size. These molecules are built up by splitting off water at random between reactive groups of neighboring molecules, thereby increasing their size. Once their solubility, viscosity, pH, concentration, and so on, have been determined, they constitute the resins available commercially. The most important factors influencing the final properties of aminoresins in industrial manufacture are the purity of the reagents, the molar proportions of the materials used, the preparation process used, and the pH variation and control.

The most common method of preparation for commercial UF resin adhesives is the addition of more urea during the preparation reaction. This consists in reacting urea and formaldehyde in higher than equivalent proportions. Generally, an initial urea:formaldehyde molar ratio of 1:2.0 to 2.2 is used. Methylolation can, in this case, be carried out in a much shorter time, by using temperatures of 90°C–95°C. The mixture is then maintained under reflux. When the exotherm subsides (usually after 10–30 min), the methylol compounds are formed, and the reaction is completed under reflux by adding a trace of an acid to decrease the pH to the UF polymer-building stage (pH 5.0–5.3). As soon as the correct viscosity is reached, the pH is increased to stop polymers building and the resin solution is cooled to about 25°C–30°C. More urea (called *second urea*) is added to consume the excess of formaldehyde, until the molar ratio of urea to formaldehyde is in the range 1:1.1–1:1.7. After this addition of urea, the resin is left to react at 25°C–30°C for as long as 24 h. The excess water is eliminated by vacuum distillation until a resin solids concentration of 64%–65% is reached, and the pH adjusted to achieve suitable shelf life or storage life.

The final addition of urea can be carried out in one operation, or the urea may be added at suitable intervals in smaller lots. Second or further ureas can be added at a temperature slightly higher than ambient, or can be added at higher temperatures (60°C–90°C), depending on the type of final resin wanted [13–16]. Increasing second or further urea additions tends to improve bond quality, especially at low formaldehyde:urea molar ratios [13–16]. Higher molar ratio resins tend to exhibit an overall better initial bond quality [14], but present an exponentially increased formaldehyde emission problem [16], most often disqualifying them from many, or most, modern uses. Some UF resins used for joinery are also produced without a final, or second, urea addition. The pH used during the condensation reaction (not the methylolation) is generally in the range 4.8–5.3.

Control of the average molecular size of the finished resin is essential for proper flow in plywood and particleboard applications while in the hot press prior to curing. Too low a level of condensation (i.e., low molecular weight resins) may impart too much flow; the resin runs away from the wood or sinks into it rapidly under pressure, leaving starved gluelines. This can be corrected by lowering the pH by adding an acid or acid-producing substance, usually a curing agent, hardening catalyst, or simply a hardener. If a resin of too high a condensation stage (i.e., high molecular weight resins) is on hand, its flow under normal pressure and temperature may be too low to produce good results. This can usually be corrected by adding flow agents to it. It is generally an advantage to produce resins with ample flow in the factory. Their storage life is longer, and final adjusting of the resin characteristics can be done at any time, at short notice, to specification, particularly by adjusting the flow and speed of cure.

Resins that have lost part of their flow during manufacture or storage must be corrected by the addition of a flow agent. The simplest means is often the addition of water sprayed on the compound and mixed in well. If a resin is still capable of flowing, this procedure produces a resin with properties that are still acceptable. In cases where moisture content control is critical, it may be necessary to allow a little more time for heating to let the added moisture escape. However, if the flow is very low, and large quantities of water must be used to bring the flow back to normal, this method is not recommended. The large amount of water would cause longer breathing times to be necessary, due to excessive volatile components, and excessive shrinkage may take place, causing excessive stress on the gluelines. It must be kept in mind that excessive water addition causes UF resin precipitation. The best way to adjust flow in these cases is to mix the resin with large amounts of an equal resin of the same quality that has a higher flow. Any proportion may be used to bring the flow back to normal. If increased flow is desired, 0.5%–2.0% of spray-dried UF or MF resin can also be added to function as a flow agent. Methylol compounds, such as dimethylolurea, also increase flow, but they increase more the water released during reaction than do spray-dried resins. Lubricating agents such as calcium stearate are also able to provide a fair degree of flow increase.

Many substances have been suggested as curing agents. These include the following acid products: (1) boric acid, (2) phosphoric acid, (3) acid sulfates, (4) hydrochlorides, (5) ammonium salts of phosphoric or polyphosphoric acid, (6) sodium or barium ethyl sulfate, (7) acid salts of

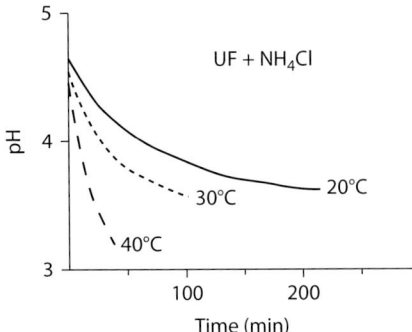

FIGURE 10.2 Change of pH of UF resins with ammonium chloride hardener as a function of temperature and time. (From Pizzi, A., *Wood Adhesives Chemistry and Technology*, Marcel Dekker, New York, 1983.)

hexamethylenetetramine, (8) phthalic anhydride, (9) phthalic acid, (10) acid resins such as poly(basic acid)–poly(hydric alcohol), (11) oxalic acid or its ammonium salts, and many others. However, the most widely used curing agents in the wood products industry are still ammonium chloride or ammonium sulfate. Their effect can be altered by retarding the reaction of the resin. This is done by the simultaneous addition of small amounts of ammonia solution (that are eliminated during hot curing) to lengthen the pot-life of the glue-mix. Latent catalysts that produce acid only on heating may also be used, such as dimethyl oxalate and other easily hydrolyzable esters, or halogenated substances such as 0.1%–0.2% of bromohydrocinnamic acid and others (Figure 10.2).

The driving force in the use of these salts as hardeners is their capacity to release acid, which decreases the pH of the resin and thereby accelerates curing. The speed of the reaction between the ammonium salt and formaldehyde (or ammonia and formaldehyde when this is present) also determines, together with the amount of heat supplied, the rate of acid release and therefore the rate of curing:

$$4NH_4Cl+6HCHO \rightarrow 4HCl+(CH_2)_8 N_4+6H2O$$

Hexamethylenetetramine

Ammonium chloride is a better hardener than hydrochloric acid, as the latter produces weaker joints. The effect of a fixed amount of ammonium chloride on the pH change and on the rate of resin curing as a function of time and temperature is shown in Figure 10.2.

Often, particularly in cold-setting UF resins for joinery, hardeners consisting of mixtures of a salt such as ammonium chloride or ammonium sulfate with an acid such as phosphoric acid, citric acid, or others are used to regulate pot-life and rate of curing. Both pot-life and rate of curing of the resin can thus be regulated by (1) varying the concentration of the hardener in the resin, (2) by changing the relative proportions of acid and salt, and (3) by changing the type of acid and/or salt composing the hardener. Acting on these three principles, setting times of between a few minutes and several hours can easily be obtained.

10.2.2.1 Plywood Adhesives

The UF adhesives for plywood generally contain less than 2 mol of formaldehyde per mole of urea, and most of them are condensed to a slightly viscous, hydrophilic stage and are quite soluble in water. The degree of polymerization, and hence the viscosity under comparable conditions, of UF resins for plywood is generally higher than those of UF resins for particleboard.

The application of UF resins for gluing purposes is based on the excellent control that can be exercised on the condensation reaction by varying the pH, a procedure easily applicable on a production scale. A small amount of an acid as hardener is added at ambient temperature. This produces no visible change at first, or possibly for a few hours; finally, thickening sets in, and the resin hardens into an insoluble material. While the adhesive is still in liquid form, it can be spread on the wood surfaces, which have to be glued and joined under pressure. These have to be cured, either at room temperature for a few days or at elevated temperature for a few minutes. Solutions of ammonium salts—generally, ammonium chloride or ammonium sulfate, or mixtures of ammonium chloride with urea—are used as resin hardeners. Often, ammonia solution is added to lengthen the usable life of the glue-mix. Hydraulic presses with multiple openings are generally used for the production of plywood or flat veneer. They can operate at pressures of 10–16 kg/cm², but usually operate in the range 12–14 kg/cm², depending on the wood species, to avoid destruction of the porous structure of the wood.

The temperature is usually in the range 120°C–160°C, depending on the type and moisture content of the veneers. It is chosen according to its capacity for the shortest press time and its ability to produce a good joint without blisters. Different pressing conditions are used in different countries, and the resin must be manufactured keeping the differences in the conditions of application in mind. There is quite a difference, for example, between a UF adhesive and glue-mix that is capable of giving good joints at 5%–8% moisture content of the veneer and a press temperature of 120°C, and a UF resin and glue-mix usable at a veneer moisture content of 0%–1% and press temperatures of 140°C–160°C. The former needs better flow characteristics and faster curing under standard conditions than does the latter if optimum press times and production schedules are to be maintained. Lower temperatures lengthen the curing time of the resins considerably, but have the advantage that when the cured plywood sheets are taken out of the hot press, they tend to warp less on cooling or drying.

The use of fillers with plywood UF adhesives has significant economic consequences and is necessary for technical reasons, because the fillers produce body in the glue solution and, therefore, prevent joint starvation in porous wood. Without filler, it would be difficult to prevent part of the adhesive from flowing away or flowing into the open pores of the wood; or, in the gluing of medium to thin veneers, from flowing through them to the other side, thereby causing undesirable resin patches on the outer veneer surfaces. As a rule, 20%–50% filler is used for joinery and up to 100% for plywood. The most common fillers are wheat flour, corn flour, rye flour, very fine hardwood flour, and gypsum. If gypsum is used, it must be free of calcium hydroxide, because this interferes with the acid curing agent.

10.2.2.2 Particleboard Adhesives

A very important application for UF adhesives is in the manufacture of particleboard. The glue-mix is generally composed of a liquid resin to which water has been added to decrease viscosity and to facilitate spraying, plus small amounts of ammonium chloride or sulfate and small amounts of ammonia solution. Small quantities of insecticides, wax emulsion, and fire-retarding agents (such as ammonium phosphates) are added before spraying the adhesive onto the wood chips. Press temperatures and maximum pressures used in the cycle are in the ranges of 150°C–200°C and 2–35 kg/cm², respectively.

The moisture content of glued wood chips is 7%–8% for the board core and 10%–12% for the surface. The resin contents used (i.e., solids) are 6%–8% for board core and 10%–11% for board surfaces, but such proportions may be higher for the weaker low emission adhesives used today and depending on the application envisaged (i.e., particleboard or medium density fiberboard [MDF]).

It must be realized that on curing, the viscosity of UF resins changes, not only at different rates but also in different manners, depending on the temperature. The viscosity gradually increases with temperature up to +50°C. Above 60°C, the viscosity quite rapidly reaches a maximum, and then decreases. This indicates that the resin tends to degrade under prolonged heating at high temperatures (Figure 10.3). To avoid this problem, UF-bonded particleboard must never be pressed for too long, and must never have a *hot-stack* or *postcure* after pressing. It must preferably be cooled after manufacture, to avoid deterioration in strength and quality. The cured UF resins degrade rapidly at any temperature at pH below 2. The viscosity for a good particleboard resin is of the order of

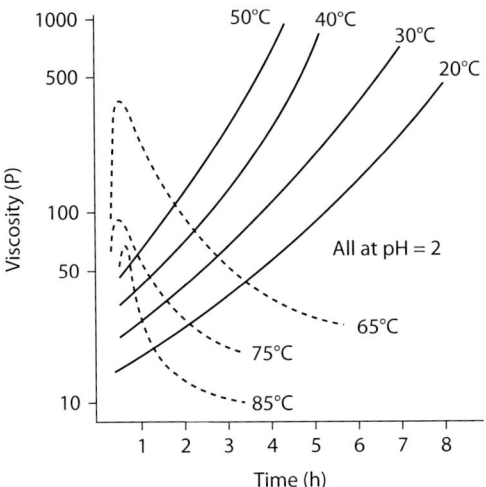

FIGURE 10.3 Viscosity of a UF resin as a function of time at different temperatures. Traditional resin of high F:U molar ratio.

100–450 cP (at 20°C) [17]. While this rule is true, developments of UF resins undergone in the last 15–20 years to decrease drastically the levels of formaldehyde emission have led to new formulations that have very different characteristics and behavior. In some respects, at least partially, some of the old rules are no longer completely valid. This is the case with the rule about trying to avoid hot-stacking of UF-bonded boards [2,18].

Thus, when a panel is taken out of the press, it gives off a considerable amount of moisture and its temperature is quite high. If a board in such a condition is immediately placed in an oven, the temperature of which is higher than 75°C, some degradation with consequent loss of performance will occur, this being shown to be due mainly to some progressive degradation of the UF adhesive hardened network [2,8]. Conversely, if the board is just cooled down, there will not be any further curing of the resin. The predominance of the effect derived from the first of these two considerations has led to the need to limit the heat conservation of UF particleboard after pressing and, hence, to today's widespread practice of cooling the board after pressing [2,8]. As a consequence, decrease of board performance by resin degradation is mainly avoided, but an unexploited reservoir of further potential strength of the resin achievable by further curing is wasted. It has been reported many times that the mechanical performance of aminoresin-bonded particleboards cannot be improved by hot posttreatment, with the exception of physical properties such as stabilizing the board moisture content throughout and stress reduction, thus improving the board dimensional stability [19].

Results obtained by a series of techniques for the curing of several resin systems [18,20–24] have indicated, however, that posttreatment and hot-stacking (postcuring) conditions capable of improving the mechanical performance of aminoplastic resin-bonded particleboards without any degradation should instead exist. This is of some importance, firstly because performance of UF- and MUF-bonded particleboard could be improved with very little process change from present industrial conditions to yield better board performance (or the same performance at lower adhesive content levels), and secondly because at parity of board performance, such an approach may well lead to the use of even shorter industrial press cycles than today, even for aminoresins.

From the experimental results obtained [18] it is evident that: (1) Postcuring (e.g., by hot-stacking in simpler cases, or by an oven or other heat treatment in more sophisticated cases) can be used in principle and under well-defined conditions to improve the performance of UF-bonded joints and panels without any further joint and hardened adhesive degradation, as the strength value reached after postcuring is always consistently higher than the value at which the strength stabilizes after complete curing during the press cycle [18]. (2) Postcuring could also be used, in principle, and for the same reasons, to further shorten the press time of UF-bonded joints and panels when appropriate postcuring conditions are used [18]. (3) There is clear indication that even when adhesive degradation starts, the application of posttreatment reestablishes the value of the joint strength to a value higher than its maximum value obtained after curing [18]. The molecular level reasons for this behavior can also be deduced from bonded wood-panel internal bond (IB) strength behavior. The IB performance improvements, for instance, are caused by the series of reactions pertaining to internal methylene-ether bridge rearrangements to a tighter methylene bridge network, which have already been observed and extensively discussed in terms of thermomechanical analysis (TMA) of aminoplastic and phenolic resins [20–24]. These are able to counterbalance well the degradative trend to which an aminoresin is subjected. Furthermore, in modern resins of lower F:U molar ratio, the amount of methylene-ether bridges formed in curing is much lower. Thus, disruption by postcuring of the already formed resin network by internal resin rearrangements will be milder, if at all present, and will definitely not cause the marked degradation nor the collapse of the structure of the network that characterizes older resins of much higher molar ratios when postcured under the same conditions [2,8]. In short, notwithstanding the internal rearrangement, the hardened resin network will be strong; no, or hardly any, decrease of IB strength will be noticeable. For modern, lower molar ratio aminoplastic adhesives, since the resin network does not noticeably degrade or collapse with postcuring, only the tightening of the network derived by further bridge formation by reaction within the network of a few formaldehyde molecules released by the now mild internal rearrangement will be noticeable; the IB strength value will thus improve with postcuring, within certain limits, in boards bonded with modern, lower formaldehyde aminoplastic adhesives [18].

A model has been developed to describe the decrease in temperature under different conditions of a particleboard after hot pressing, and this model is shown to correlate well with experimental results of board temperature variation after pressing, both on cooling and during postcuring under different conditions [18]. From this, conditions of temperature and time favorable to improve panel performance by postcuring treatments were also determined [18]. The validity of improvements forecast under such conditions was then confirmed at molecular level for UF adhesive–wood joints by TMA, and finally confirmed by testing the mechanical performance of laboratory boards prepared under postcuring treatment conditions [18]. The panel performance improvements observed were explained on the basis of already described [24] and well-known molecular level rearrangements of the cured adhesive network, and on the change in their relative importance in modern, lower formaldehyde content UF adhesives. The conclusion was that such adhesives can benefit considerably as regards board performance from short-period hot postcuring at temperatures in the 60°C–100°C range, a trend in clear contrast with the degradation and loss of performance this practice was known to induce [2,16] in the older, very much higher formaldehyde content aminoplastic resins of the past. Economic and technical consequences flow from this, as the findings also imply lower adhesive consumption and possibly even faster press cycles at parity with present resin performance, if simple postcuring procedures such as hot-stacking after pressing (rather than board cooling as at present) are implemented for UF-bonded particle and other types of boards [18].

Figure 10.4 [18,20] shows that the slower the heating rate, the more evident the entanglement plateau and the higher the value of the modulus due to the initial physical chain entanglement. This confirms that linear growth of the polycondensate can be maximized by decreasing the temperature at which polycondensation is carried out (this is likely to be valid in the reactor during preparation of the UF resin as well as in the resin curing stages on the substrate). Figure 10.4 indicates that this effect becomes more marked with slower rate of heating. It is implied that polycondensates grow mostly linearly to a higher degree of polymerization, before tridimensional cross-linking starts, with the slower rate of heating. This might depend on the respective reactivities of urea sites with formaldehyde that are in an approximate ratio of 9:3:1, respectively, for the first-reacted, second-reacted, and third-reacted urea sites [2]. The slower heating rates used decrease molecular movement and hence further decrease the chance of the third urea site reacting, hence favoring more linear growth of the polycondensate. Tridimensional covalent networking will still occur, and a tridimensional

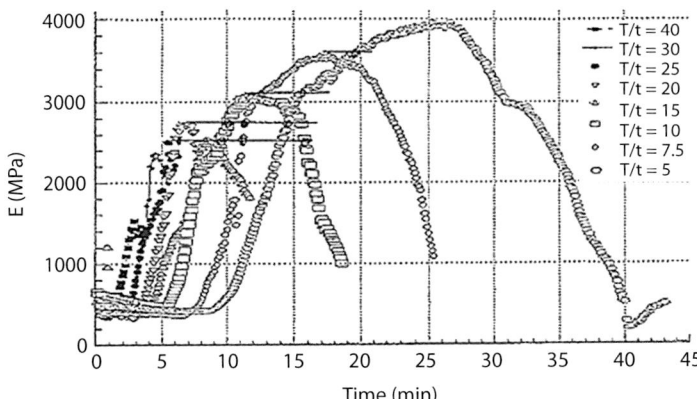

FIGURE 10.4 Increase of modulus of elasticity (MOE) at heating rates in the 5°C/min to 40°/min range as a function of time for a beech wood joint bonded with a UF resin. Increase of MOE corresponds to resin hardening.

cross-linked network will still be the final product of the reaction, but will occur later when the polymer has grown to greater lengths. The most important observation from Figure 10.4, however, is the considerably higher value of the modulus at slower heating rates, which must also be viewed in the same context as mentioned; it relates to the polymer having time to adjust by better utilization of empty volume spaces, the same reason that gives a lower value of the glass transition temperature T_g with a slower rate of heating. The extent of the effect observed is considerable: The maximum value of the modulus, once the resin is tridimensionally cross-linked for the 40°C/min case, is lower than the value of the modulus of just the entanglement network observed for the 15°C/min and slower heating rate curves, due to early tridimensional immobilization of the resin in a less tight tridimensional covalent network.

It is important here to point out the concept widespread in wood-panel manufacture that a resin capable of shorter pressing and curing times (e.g., a faster curing resin) gives better panel strength depending on the exact definition of the concept of time of curing (and of pressing) in Figure 10.4. Thus, in Figure 10.4, a fast resin, as fast as the 40 s/mm curve (which is in line with today's rate of curing for wood particleboard panels) will only be able to give to the joint the strength of 1.5 GPa modulus, while a slower resin, which is capable in principle of yielding a modulus three times stronger (4.5 GPa) (reproducing the 5 s/mm curve) has no strength (less than 0.2 GPa) at the same curing time as used to maximize the strength result of the faster resin. Thus, it is preposterous to define a resin as better than another one unless the concept of time is also defined and the two resins are seen in this time context. This insight leads also to two consequences: (1) It contributes to explaining why, in modern UF resins, one can improve strength by hot-stacking after pressing; this is the equivalent of transferring from the faster curve in Figure 10.4 to one of the slower ones after hot pressing, one that allows the system to reach a higher strength value, as shown in the figure. (2) A faster curing resin needs to be engineered not just to give a shorter curing time, as this will only yield an ultimately lower strength due to the looser and hence weaker network produced, but also to be able to concomitantly obtain a higher degree of cross-linking of the network to counterbalance the weakening caused by the faster curing rate of the resin; this needs to be introduced by varying resin parameters and other techniques. To obtain a good, yet faster, adhesive, both effects must be taken into account.

As important as viscosity is also resin flow, which reflects viscosity under hot-pressing conditions. Resin flow is a determining factor in manufacturing good particleboards. Excessive flow causes the resin to soak into the wood particles and causes glueline starvation; insufficient flow causes insufficient contact surface. The gel time generally used at 100°C for glue-mixes of UF-bonded particleboards is 3–12 min, with 30 s–3 min for board faces and cores, respectively. The actual gel time in the press depends on the press temperature and is considerably shorter.

10.2.2.3 UF Adhesives for Low-Formaldehyde-Emission Particleboard

UF resins in their cured state are nontoxic. Urea itself is also harmless. However, free formaldehyde and formaldehyde generated by slow hydrolysis of the aminoplastic bond are highly reactive and combine easily with proteins in the human body. This may cause a painful inflammation of the mucous membranes of the eyes, nose, and mouth [25]. Even a low concentration of formaldehyde vapor in the air can cause disagreeable irritation of the nose and eyes. However, such irritations usually disappear in a short time without permanent damage. Occasionally, allergic or anaphylactic reactions develop and complete removal from exposure to formaldehyde is necessary.

High temperatures and high relative humidity can result in odor problems in a room containing particleboard manufactured with UF resins [25]. The release of formaldehyde from UF particleboard is caused by two factors. It can be due to free formaldehyde present in the board that has not reacted, or to formaldehyde formed by hydrolysis of the aminoplastic bond as a result of the temperature and relative humidity [2,25]. While the first type of release lasts only a short time after manufacture of the particleboard, the second type of release can continue throughout the entire working life of the board. A considerable number of variables influence the emission of formaldehyde from a UF-bonded particleboard. The main ones are the molar ratio of urea to formaldehyde (which influences both types of releases), the press temperature, and in service, the ambient temperature and relative humidity.

UF resins for particleboard with urea:formaldehyde molar ratios of 1:1.45, 1:1.32, and 1:1.25 have free formaldehyde contents of 0.8%, 0.3%, and less than 0.2%, respectively [15]. While the current tendency internationally is to use UF resins that have a urea:formaldehyde molar ratio much lower than 1:1.3, which releases much less formaldehyde, these resins perform less well in the production of UF-bonded particleboard [15,17].

In particular, these resins do not allow as much flexibility in particleboard production as do those with higher formaldehyde:urea molar ratios. This fact stresses the need for greater control and supervision of the production at particleboard plants where UF resins of low molar ratio are used. An example of the variation in properties between particleboards manufactured with different molar ratio resins is given in Table 10.1.

It is also necessary to use more hardener when working with UF resins of a lower formaldehyde:urea molar ratio, as the gel time of the resin is slower. Up to 5% urea can sometimes be added to the glue-mix to decrease the amount of formaldehyde released during pressing and to decrease the initial amount of free formaldehyde present in the finished board immediately after manufacture. Strict norms have been established in many countries with regard to limits of formaldehyde emission from particleboards bonded with UF resins [25–27]. Recent work indicated that good E1-type UF resins of urea:formaldehyde molar ratio lower than 1:1.1 can be prepared in a variety of ways

TABLE 10.1

Comparison of Particleboards Prepared with UF Resins of Various Molar Ratios

U/F Molar Ratio	Approximate Density (g/cm³)	Internal Bond Strength (MPa)	Percentage of Water Swelling (2 h)	Percentage of HCHO Released, Perforator Method (mg HCHO/100 g Board)
1:1.4–1.5	0.680	0.7–0.8	4	50–70
1:1.3–1.35	0.680	0.6–0.7	4–5	25–30
1:1.1–1.25	0.680	0.45–0.55	5	5–20

Source: Pizzi, A., *Wood Adhesives Chemistry and Technology*, Marcel Dekker, New York, 1983.

[16]. Although the theoretical basis of this finding has been discussed in part elsewhere [16], it is of interest to apply these findings to the formulation and preparation of UF resins of low formaldehyde emission, first in the laboratory, and then at the industrial level. First, these resins can be divided into two broad classes: (1) those resins based on addition of melamine or MF resins to the UF resin, and (2) those UF resins in which very low formaldehyde emission capability is obtained exclusively by manipulation of their manufacturing parameters. The former class is nothing but a subset of the second.

The underlying principle for a low formaldehyde emission UF resin is that a certain amount of free urea needs to be present to (1) mop up a large amount of the free formaldehyde that may be present at the end of the preparation, and (2) to mop up most of the free formaldehyde that may be generated during hot curing of the resin. A third possible requirement would be that some free monomeric urea species should still be left to mop up, over a long period of time, some of the formaldehyde that may be liberated during the service life of the board.

Such requirements of a UF resin are fundamentally quite divergent and extreme. This means that the addition of large amounts of urea is needed, possibly at the end of the reaction; such urea will react with the free HCHO present or generated during hot curing, but will also react with the active methylol groups present on the urea resin itself, severely limiting the possibility of cross-linking of the resin and ultimately affecting adversely, and diminishing, its cured strength. These two sets of contradictory requirements indicate that, in general, a low formaldehyde emission UF formulation must be a compromise between strength and emission requirements. Once this basic conflict of requirements is understood, it can be overcome to attain formulations that give both good strength and low formaldehyde emission. A UF resin is a mixture of molecular species, namely, methylolurea, UF polymers, and methylolated UF polymers. It has already been shown, both theoretically [8,19] and by chemical analysis [14,28–30] that while monomeric and polymeric methylolated species contribute more to the adhesion of the resin to the wood substrate, it is the polymeric fraction (methylolated and nonmethylolated) that contributes most to the cohesion of the resin. Thus, a resin to which large amounts of final urea are added will have a proportionally high amount of urea and monomeric methylolated species, giving both good adhesion and low formaldehyde emission, and proportionally a lower amount of prebuilt polymeric species, giving poor cohesion, hence lower strength. Conversely, a resin of final higher urea:formaldehyde molar ratio such as the classic UF resins used for the last few decades will have a large number of polymeric species, and will still be heavily methylolated; however, most of the methylolated species will be polymeric and will still have a considerable amount of free and potentially free formaldehyde. These resins will have good cohesion and good adhesion, hence good strength, but very high HCHO emission.

The logical approach to avoid the conflicting requirements of the two properties required is then to prebuild, in some easy and convenient manner, the particular mixture of species that will give the correct balance of strength and emission for the applications required. Thus, although UF resins of very low urea:formaldehyde ratios [16] can be prepared by adding large amounts of second and third urea, the high predominance of urea and other monomeric species in relation to polymer proportions will give boards of poor strength, albeit with very low HCHO emission. The required balance of chemical species and properties can then be achieved more easily by preparing two or more UF resins and/or preresins that are mixed in various amounts to yield the desired balance of acceptable strength and low emission [28]. One aspect that has not been given sufficient attention in the past is the effect that the colloidal behavior of aminoresins has not only on their shelf life, due to aggregation on aging, [31], but also on their formulation.

10.2.2.4 Other UF Adhesive Applications

Although particleboard and plywood are the major users of UF adhesives, two other applications, although consuming much lower proportions of these resins, are also worthy of note.

The first is in the furniture and joinery industry, including the manufacture of hollow-core doors. While, in the latter application, thermosetting resins with characteristics and glue-mixes similar to those for plywood are used, often (but not always) cured by radio-frequency, the former can be simpler resins of higher urea:formaldehyde molar ratio to which cold-setting capability and different pot-lives are given by a variety of hardener types; in these, hardeners formed by an acid plus a salt are the norm. The second application of note is in foundries as sand core binders. In this application, UF resins compete with phenolic and furanic resins. In general, however, the resins used for the hot-box process are UF resins modified with 20–50% furfuryl alcohol to obtain UF–furanic resin copolymers and phenol–formaldehyde resins modified with urea. Small amounts of paraffin wax and corn flour are often added to facilitate mixing of the resin with the sand (generally between 1% and 2.5% resin based on sand).

10.2.3 FORMALDEHYDE ANALYSIS

Methods of formaldehyde analysis include the iodometric, sulfite [32], and mercurimetric [33,34] methods. The sulfite method measures only the formaldehyde present, whereas the iodometric method can also estimate the methylol groups. Another method is based on the partition of formaldehyde between water and isoamyl alcohol [35]. Estimation of the formaldehyde in the alcohol phase of a mixture of an aqueous solution of the resin and isoamyl alcohol allows deduction of the amount of free formaldehyde. This procedure has the advantage that no risk of reaction arises between free formaldehyde and the resin components.

Kappelmeier [36] has suggested the use of aniline, benzylamine, and phenyl–ethylamine as reagents for the identification and analysis of urea in UF resins. He has provided evidence that the methylene-ether groups form a bridge between urea residues in urea–formaldehyde resins. The use of benzylamine, in particular (which yields dibenzylurea from urea derivatives), has been developed as a method of analysis. In determining the ratio of urea to formaldehyde in UF resins, the benzylamine method has been coupled with a process of formaldehyde estimation that involves depolymerization with phosphoric acid, followed by distillation into alkaline potassium cyanide solution [37].

Chow and Steiner [14] advocate the use of bromination in CCl_4 and subsequent x-ray analysis to determine available reactive methylol groups. High-resolution nuclear magnetic resonance (NMR) has also been used to analyze UF resins and trace their kinetic behavior [38]. Particularly useful is ^{13}C NMR analysis of liquid UF resins, where clear identification of monomeric species—methylolated or not—methylol ureas, methylol groups on the polymer, methylene-ether linkages, methylene bridges, sites of branching, urons, free formaldehyde, and other features can be achieved easily and rapidly [39,40]. For example, this technique makes it possible to easily estimate the probable bonding ability and approximate emission class to which boards bonded with aminoresins are likely to belong [24].

10.2.4 URONS AND DERIVED RESINS

The potential introduction of an intermediate reaction step at very acid pH inducing the formation of some uron in the preparation of UF resins of lower formaldehyde emission has generated some industrial interest [40] and, nowadays, industrial UF resins manufactured with the introduction of a rapid, very acid step (pH 1–2) during preparation are available. The only published research work that can be found in the worldwide literature on this subject deals with the introduction of just such a strongly acid condensation step in the preparation of UF resins [41–43]. This work came to the conclusion that introduction of such an acid step can lead to UF resins of improved bonding strength [40–42] and also of lower postcure formaldehyde emission [2,3]. One of the effects of the introduction at lower reaction temperatures of the additional strongly acid condensation step was the formation of some considerable quantity of uron [41–43], thus of the well-known structure of a cyclic intramolecular urea methylene ether [8,44]. Urons have been found by ^{13}C NMR to be present as methylol urons, methylene urons, or methylene-ether urons, hence as structures [40] as follows:

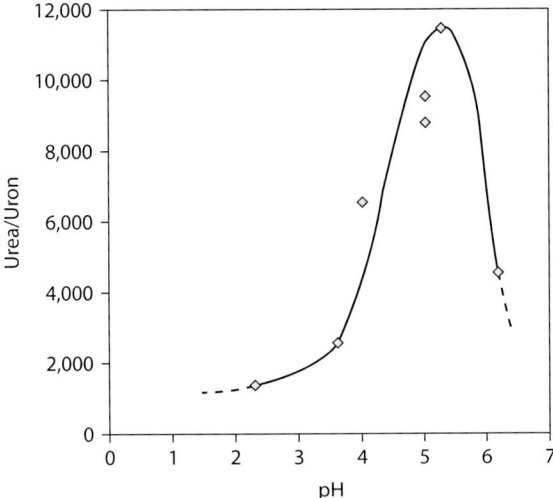

$$(10.3)$$

Even more interestingly, some completely substituted urons appear to exist in reaction mixture, this being the only case in which the existence of some form of tetrasubstituted urea has been noted [41].

The favored pH ranges for the formation of urons in UF resins preparation were determined, these being pH levels higher than 6 and lower than 4 at which the equilibrium urons ↔ N,N' dimethylol ureas is shifted in favor of the cyclic uron species [41] (Figure 10.5). Shifting the pH slowly during the preparation from one favorable range to the other causes a shift in the equilibrium and formation of a majority of methylol urea species, while a rapid change in pH does not cause this to any great extent [41]. UF resins in which uron constituted as much as 60% of the resin were prepared and the procedure to maximize the proportion of uron present at the end of the reaction described [41]. Uron has been found to be present in these resins also as linked by methylene bridges to urea and other urons, and also as methylol urons. The reactivity of the methylol group of this latter group has been shown to be much lower than that of the same group in methylol ureas. TMA and tests on wood particleboard prepared with uron resins to which relatively small proportions of urea were added at the end of the reaction showed that these were capable of gelling and yielding bonds of considerable strength [41]. Equally, mixing a uron-rich resin with a low F:U molar ratio UF resin yielded resins of greater strength than a simple UF of corresponding molar ratio, indicating that UF resins of lower formaldehyde emission with still acceptable strength could be prepared with these resins [41]. As the acid-step industrial resins are not prepared under conditions as extreme as the research on the potential for uron introduction

FIGURE 10.5 Variation of the urea:uron ratio of the ^{13}C NMR carbonyl peak areas as a function of pH during the total reaction.

has shown to be possible, it is clear that in this direction there is some room for further improvement for UF adhesives.

The reopened structures re-form the cyclic intramolecular uron methylene ether as the pH reaches the acid range in which the cyclic structure is again stable. Thus, the uron structure is in equilibrium with the open dimethylol ureas form and the pH range determines the direction toward which of the two forms such an equilibrium is more or less shifted.

(10.4)

The rate of opening or closure of the uron cycle structure is not very rapid, as, at the end of the reaction, when the pH is rapidly adjusted from the very acid to the alkaline range, the proportion of cyclic structures present does not alter much, while slow decrease of the pH during the reaction, passing through the middle pH range, causes disappearance of a greater proportion of the uron cyclic structures, which then form again once the acid pH range is reached.

10.3 MELAMINE–FORMALDEHYDE ADHESIVES

MF and MUF resins are among the most used adhesives for exterior and semiexterior wood panels and for the preparation and bonding of both low- and high-pressure paper laminates and overlays. Their much higher resistance to water attack is their main distinguishing characteristic from UF resins. MF adhesives are expensive. For this reason, MUF resins, which have been made less expensive by addition of a greater or lesser amount of urea, are also often used. Notwithstanding their widespread use and economical importance, the literature on melamine resins is only a small fraction of that dedicated to UF resins. Often, MFs and MUFs are described in the literature as part of UF aminoresins. This is not really the case, as they have peculiar characteristics and properties of their own, which in certain respects are very different from those of UF adhesives.

10.3.1 Uses for MF Resins

MF resins are used as adhesives for exterior- and semiexterior-grade plywood and particleboard. In this application, their handling is very similar to that of UF resins for the same use, with the added advantage of their excellent water and weather resistance. MF resins are also used for impregnation of paper sheets in the production of self-adhesive overlays for the surface of wood-based panel products and self-adhesive laminates. In this application, the impregnation substrate, a-cellulose paper, is thoroughly impregnated by immersing it in the resin solution, squeezing it between rollers, and drying without curing it to proper flow by passing it through an air-draft tunnel oven at 70°C–120°C at ±10 m/s. The dry MF-impregnated sheets can then be bonded by one of two main processes:

1. The sheets of MF-impregnated paper, consisting of one surface layer or a few surface layers, are bonded together with a substrate of paper sheets impregnated with phenolic resins to form laminates of variable thickness. In the impregnated papers, the MF resin is dry but still active, functioning as the adhesive for the MF-impregnated sheet to both MF-impregnated sheets and at the interface between MF-impregnated and phenol–formaldehyde (PF)impregnated layers. These laminates are high-pressure laminates.

2. The MF in an impregnated paper sheet is not completely cured, but still has a certain amount of residual activity and is applied directly in a hot press, in a single sheet, onto a wood-based panel, to which it bonds by completing the MF adhesive curing process.

Press platens are made from stainless steel or chromium-plated brass and copper. The chromium layer preserves surface quality longer than ordinary steel. The MF laminates exhibit a remarkable set of characteristics. Because of their unusual chemical inertness, nonporosity, and nonabsorbance, they resist most substances, such as mild alkalis and acids, alcohols, solvents such as benzene, mineral spirits, natural oils, and greases. No stains are produced on MF surfaces by these substances. In addition to almost unlimited coloring and decorating possibilities, this remarkable resistance has resulted in an extensive use of MF laminated wood-based panel products for tabletops, sales counters, laboratory benches, heavy-duty work areas in factories and homes, wall paneling, and so on.

10.3.2 CHEMISTRY

10.3.2.1 Condensation Reactions

The condensation reaction of melamine [1] with formaldehyde (Figure 10.6) is similar to but different from the reaction of formaldehyde with urea. As for urea, formaldehyde first attacks the amino groups of melamine, forming methylol compounds.

However, formaldehyde addition to melamine occurs more easily and completely than does addition to urea. The amino group in melamine easily accepts up to two molecules of formaldehyde. Thus, complete methylolation of melamine is possible, which is not the case with urea [2]. Up to six molecules of formaldehyde are attached to a molecule of melamine. The methylolation step leads to a series of methylol compounds with two to six methylol groups. Because melamine is less soluble than urea in water, the hydrophilic stage during the resin preparation is shorter in MF resins. Therefore, hydrophobic intermediates of the MF condensation appear early in the reaction. Another

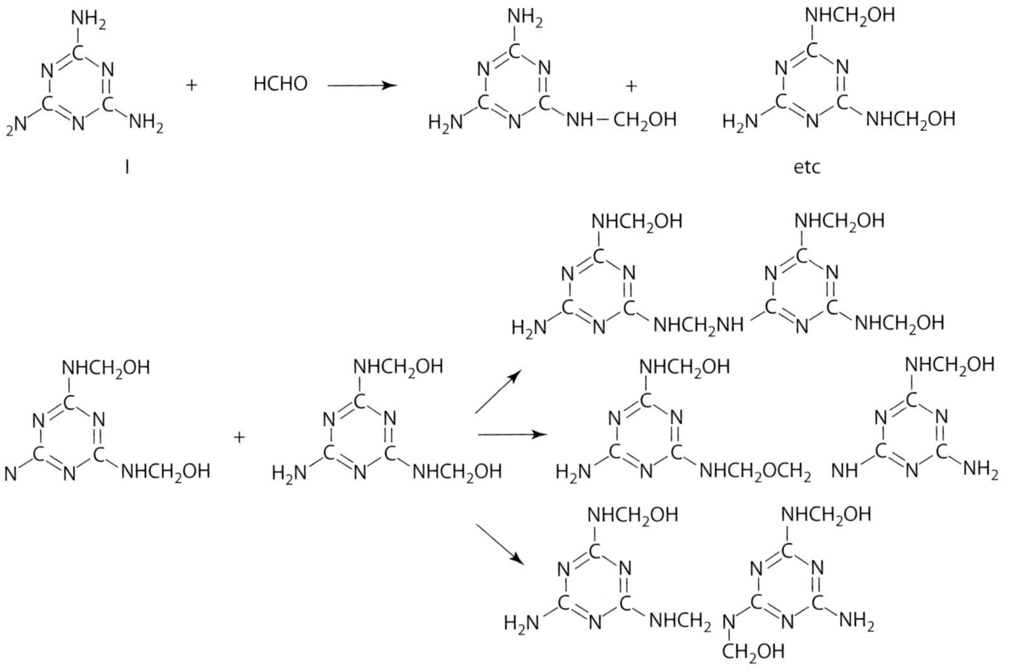

FIGURE 10.6 Methylolation (hydroxymethylation) and subsequent condensation reactions to form MF adhesive systems.

important difference is that MF condensation to give resins, and their curing, can occur not only under acid conditions, but also under neutral or even slightly alkaline conditions. The mechanism of further reaction of methylol melamines to form hydrophobic intermediates is the same as for UF resins, with splitting off of water and formaldehyde. Methylene and ether bridges are formed and the molecular size of the resin increases rapidly. These intermediate condensation products constitute the large bulk of the commercial MF resins. The final curing process transforms the intermediates to the desired MF insoluble and infusible resins through the reaction of amino and methylol groups that are still available for reaction.

A simplified schematic formula of cured MF resins has been given by Koehler [45] and Frey [46]. They emphasize the presence of many ether bridges besides unreacted methylol groups and methylene bridges. This is because, in curing MF resins at temperatures up to 100°C, no substantial amounts of formaldehyde are liberated. Only small quantities are liberated during curing to 150°C. However, UF resins curing under the same conditions liberate a great deal of formaldehyde.

At the condensation stage, attention must be paid to the formation of hydrolysis products of the melamine before preparation starts. The hydrolysis products of melamine are obtained when the amino groups of melamine are gradually replaced by hydroxyl groups. Complete hydrolysis produces cyanuric acid.

Melamine → Ammeline → Ammelide → Cyanuric acid (10.5)

Ammeline and ammelide can be regarded as partial amides of cyanuric acid. They are acidic, and have no use in resin production. They are very undesirable by-products of the manufacture of melamine, because of their catalytic effect in subsequent MF resin production, due to their acidic nature. If present, both must be removed from crude melamine by an alkali wash and/or crystallization of the crude melamine.

The mechanism of the initial stages of the reaction of melamine with formaldehyde, leading to the formation of methylolmelamines, is very similar to that of urea [8,47–50].

10.3.2.2 Mixed Melamine Resins

With regard to MUF, copolymers can be prepared that are generally used to lower the cost of MF resins, but that also show some worsening of properties. Copolymerization was confirmed from model compounds and polycondensates [51]. MUF resins obtained by copolymerization during the resin preparation stage are superior in performance to MUF resins prepared by mixing preformed UF and MF resins, especially because processing of such mixtures is quite difficult [52]. The relative mass proportions of melamine to urea used in these MUF resins is generally in the melamine:urea range 50:50 to 30:70. [53]. Melamine–phenol–formaldehyde resins, which in some respects show better properties than those of their corresponding MF and PF resins, have also been prepared [54–59]. Analysis of the molecular structure of these resins in both their uncured and cured states appeared to show that no cocondensates of phenol and melamine form and that two separate resins coexist.

This is due to the difference in reactivity of the phenolic and melamine methylol groups as a function of pH. Also, in their cured state, an interpenetrating network of separate PF and MF resins, as a polymer blend, is formed, not a copolymer of the two [57,60]. Today, MUF resins are produced in greater amount than MF resins in the field of adhesives, due to the relatively high cost of melamine; their formulations have progressed to such a level that often no difference in performance exists between a good MUF resin and a pure melamine MF resin. MF resins are still extensively used in the paper impregnation/laminates fields, although MUF copolymers, as well as separate double applications of UF (paper core) and MF (paper surfaces) resins to the same paper sheet, are making considerable inroads in this area. MUF resins, instead, totally dominate today the wood adhesives field. Paper laminates and wood adhesives are the two main application areas of these resins. The reaction of condensation of MUF resins has been followed in detail during its different stages by ^{12}C NMR, indicating the sequence of species forming and disappearing during the series of reactions involved in resin preparation [61–63].

A type of resin also used today are the so-called *PMUF* (or *MUPF* according to which author is writing) adhesives. These are fundamentally MUF resins in which a minor proportion of phenol (between 3% and 10%, P:M:U by weight of 10:30:60, just as an example) has hopefully been coreacted to further upgrade weather resistance of the bonded joint. Unfortunately, the alleged superior performance of such resins is often only wishful thinking, as the phenol has frequently not been properly reacted with the other materials and, consequently, the PMUF resin will perform worse than a comparable top of the range MUF resin. This was confirmed by the demonstration that it depends exclusively on the resin manufacturing parameters and materials reaction order used whether the phenol coreacts or not with many PMUF adhesives, showing that the phenol often remains as a useless pendent group in the hardened aminoplastic (MUF) network without contributing at all to its performance [64,65] (Figure 10.7).

The best reaction order necessary to obtain PMUF resins in which phenol also contributes to the performance of the hardened network has also been defined [64]. PMUF resins are still used, and some good resins of this type are indeed used in the unrealistic expectation that they outperform equivalent MUF resins, when it has clearly been shown that they perform at best as a MUF adhesive presenting the same number of moles of melamine for the total moles of phenol plus melamine of the PMUF itself. The idea that the addition of small percentages of phenol to a MUF resin yields resins of better exterior durability is, then, an incorrect myth traditionally perpetuated today in the wood-panel industry. Newer formulations of MUF resins always outperform the corresponding PMUF. PMUFs are not bad resins, they are simply resins in which one of the materials—phenol—is often wasted for no purpose.

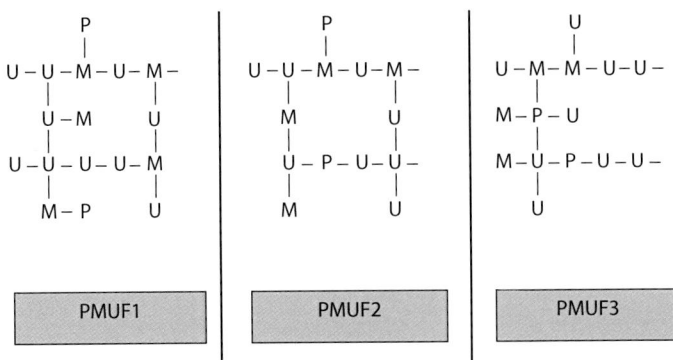

FIGURE 10.7 Schematic representation of the dependence on the type of formulation used for phenol in PMUF resins. (1) Phenol only present as unlinked free phenol/phenol derivatives, but mainly as a pendent group, neither participating in resin cross-linking nor contributing to resin performance and water resistance. (2) Intermediate case. (3) Case in which phenol is cocondensed and participating in the cross-linked network.

10.3.3 Resin Preparation, Glue Mixing, and Hardening

Due to their characteristic rigidity and brittleness in their cured state, when MF resins are used for impregnated paper overlays, small amounts of modifying compounds, typically 3%–5%, are often copolymerized with the MF resin during its preparation to impart better flexibility to the finished product and better viscoelastic dissipation of stress in the joint. The most commonly modifying compounds used are acetoguanamine, εcaprolactam, and p-toluenesulfonamide.

$$
\begin{array}{ccc}
\text{Acetoguonamine} & \varepsilon\text{-caprolactam} & \text{p-toluene sulfonamide}
\end{array}
\tag{10.6}
$$

The effect of these is to decrease cross-link density in the cured resin due to the lower number of amidic or aminic groups in their molecules. Thus, in resin segments where they are included, only linear segments are possible, decreasing the rigidity and brittleness of the resin. Acetoguanamine is most used for modification of resins for high-pressure paper laminates, while caprolactam, which in water is subject to the following equilibrium,

$$
\rightleftarrows \quad H_2N-(CH_2)_5-COOH
\tag{10.7}
$$

is used primarily for low-pressure overlays for particleboard. Small amounts of non-copolymerized plasticizers such as diethylene glycol can also be used for the same purpose. Due to the peculiar structure of the wood product itself, MF adhesives for particleboard generally do not need the addition of these modifiers. Often, a small amount of dimethylformamide, a good solvent for melamine, is added at the beginning of the reaction to ensure that all the melamine is dissolved and is available for reaction. Sugar is often added to lessen the cost of the resin. The aldehyde groups of sugars have been shown to be able to condense with the amine groups of melamine and, hence, to copolymerize in the resin. Their quantity in MF resins must be limited to very low percentages, and if possible, sugars should not be used at all, as, with aging, they tend to cause yellowing, crazing, and cracking of cured MF paper laminates and to have a bad effect on adhesive long-term water resistance in both plywood and particleboard.

MF adhesive resins for plywood and particleboard must be prepared to quite different characteristics than those for paper impregnation. The latter must have lower viscosity, but still high resin solids content, because they need to penetrate the paper substrate to a high resin load, to be dried without losing adhesive capability, and only later to be able to bond strongly to a substrate. Instead, MF adhesive resins for plywood and particleboard are generally more condensed, to obtain lower penetrability of the wood substrate (otherwise, some of the adhesive is lost by overpenetration into the substrate). The reverse applies for paper substrates, where the contrasting characteristics are desired—good paper penetration and fast curing—which can be obtained in several ways during resin preparation. These characteristics can be achieved by producing, for example, a resin with a lower degree of condensation and high methylol group content. Typically, a MF resin of a lower level of condensation with a melamine:formaldehyde molar ratio of 1:1.8–1:2 will give the desired characteristics. Its high methylol content and somewhat lower degree of polymerization will give low viscosity at a high resin solids content, favoring rapid wetting and impregnation of the paper substrate, while the high proportion of methylol groups will give it fast cross-linking and curing capabilities.

A second, equally successful approach is to produce a MF resin of lower methylol group content and higher degree of condensation, to which a small second proportion of melamine (typically, 3%–5% total melamine) is added toward the end of resin preparation. The shift to lower viscosity and higher solids content given by a second addition of melamine, shifting to lower values the average of the resin molecular mass distribution, yields a resin of rapid impregnation characteristics. Conversely, the higher degree of polymerization of the major part of the resin gives fast cross-linking and curing, due to the lower number of reaction steps needed to reach gel point. Typical total M/F molar ratios used in this system are 1:1.5–1:1.7.

Figures 10.8 and 10.9 show typical temperature and pH diagrams for the industrial manufacture of MF and MUF resins for adhesives and other applications. The important control

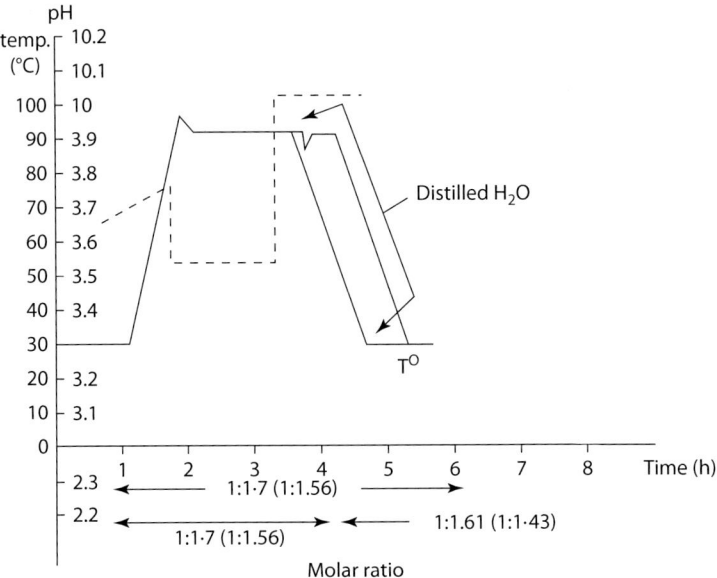

FIGURE 10.8 Typical temperature and pH diagram for the industrial manufacture of MF resins.

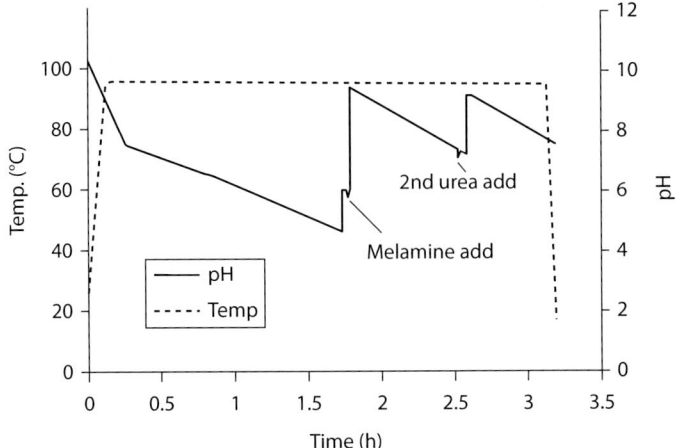

FIGURE 10.9 Typical manufacturing diagram for 40:60 to 50:50 melamine:urea weight ratio MUF resins.

parameters to consider during manufacture are the turbidity point (the point during resin preparation at which the addition of a drop of MF reaction mixture to a test tube of cold water gives
slight turbidity) and the water tolerance or hydrophobicity point, which marks the end of the
reaction. The latter is a direct measure of the extent of condensation of the resin and indicates
the percentage of water or mass of liquid on the reaction mixture that the MF resin can tolerate
before precipitating out. It is typically set for resins of higher formaldehyde:melamine ratios and
lower condensation levels at around 170%–190%, but for resins of lower formaldehyde:melamine
molar ratios and higher condensation levels it is set at around 120%. As can be seen from the
diagrams in Figure 10.8, once maximum reaction temperature is reached, pH is lowered to
9–9.5 to accelerate formation of the polymer. Once the turbidity point is reached, pH is again
increased to 9.7–10.0, to slow down and more finely control the end point, determined by reaching the required value of the water tolerance point. Industrial MF resins are generally manufactured to a 53%–55% resin solids content with a final pH of 9.9–10.4 (but lower pH values
are also used for low-condensation resins). To achieve acceptable rates of curing if higher pH
values are used, higher quantities of hardener need to be used, which is clearly uneconomi-
cal. For typical MF resins for low-pressure particleboard, self-adhesive overlay press times of
between 30 and 60 s at 170°C–190°C press temperature are required, depending on the type
of resin used. Press conditions for particleboard and plywood adhesives are identical to those
used for UF resins.

Glue-mix preparation is different depending on the final use of the MF resin. Hardeners are
either acids or materials that will liberate acids on addition to the resin or on heating. In MF and
MUF adhesives for bonding particleboard and plywood, the use of small percentages of ammonium
salts, such as ammonium chloride or ammonium sulfate, is well established and is indeed identical
to standard practice in UF resins. In MF adhesives for low- and high-pressure self-adhesive overlays
and laminates, the situation is quite different. Ammonium salts cannot be used for the latter application for three reasons. First, evolution of ammonia gas during drying and subsequent hot curing of
the MF-impregnated paper would cause high porosity of the cured MF overlay. Second, the instability of ammonium salts, in particular of ammonium chloride, might cause the MF-impregnated paper
to cure and deactivate at ambient temperature after a short time in storage, causing the resin to lose
its adhesive capability by the time it is needed in hot curing. Third, the elimination of ammonia during drying and curing would leave the cured, finished paper laminate essentially very acidic, due to
the residual acid of the hardener left in the system. This badly affects the resistance to water attack
of the cured MF surface, defeating the primary advantage for which such surfaces have become so
popular. Thus, a stable, self-neutralizing, non-gas-releasing hardener is needed for such an application. Several hardeners have been used. One of the most commonly used is the readily formed
complex between morpholine and p-toluenesulfonic acid. Morpholine and p-toluenesulfonic acid
readily react exothermically to form a neutral pH complex stable up to well above 65°C.

$$\qquad\qquad\qquad\qquad\qquad\qquad\qquad\qquad\qquad\qquad\qquad\qquad (10.8)$$

During heat curing of the MF paper overlay in the press, the complex decomposes, the MF
resin is hardened by the acid that is liberated, the morpholine is not vaporized but remains in
the system, and on cooling the complex is re-formed, leaving the cured glueline essentially
neutral.

In MF glue mixing for overlays and laminates, small amounts of release agents to facilitate
release from the hot press of the cured bonded overlay are added. Small amounts of defoamers
and wetting agents to further facilitate wetting and penetration of the resin in the paper are always
added. A typical glue-mix is shown in Table 10.2.

TABLE 10.2
Typical Paper Impregnation Glue-Mix for Self-Adhesive
Low-Pressure MF Overlays

Ingredient	Parts by Mass
MF resin, 53% solids content	99.1
Release agent	0.08
Wetting agent	0.16
Hardener (morpholine/p-toluenesulfonic acid complex)	0.64
Defoamer	0.02

Two strong trends have appeared recently regarding the preparation of melamine-impregnated paper laminates. First, impregnating machines capable of producing papers in which much cheaper UF resin substitutes—as little as 50% of the more expensive MF resin—have now been in operation for several years. This equipment is based on a double impregnating bath application: The paper passes through the first bath where it absorbs the UF resin first, the excess on the surfaces being scraped off in-line, and then passes through a second bath where it absorbs the MF resin. The concept is to limit the UF resin to the inside of the paper with the MF resin coating the outside of the paper; the hardened surface after final curing will then have all the waterproof characteristics of MF paper laminates but at a lower price. Good results are obtained, and many machines using this type of process are in industrial operation today. A more recent trend has been to develop MUF copolymers for use with the less costly single impregnating bath machines. A few cases of this route to cope with the high cost of melamine are on record.

10.3.4 MUF Adhesive Resins of Upgraded Performance

Several effective techniques to consistently and markedly decrease the melamine content in MUF wood adhesives without any loss of performance have also been recently developed. Some of these formulation systems and techniques are already in the early stages of industrialization. Melamine–acid salts, such as melamine acetate, function both as efficient hardeners of UF resins for plywood as well as upgrading a simple UF resin to the performance of a MUF resin. Grafting 10% mass melamine by this technique yields comparable strength durability of premanufactured MUF resins of 30%–40% melamine mass content; hence, of resins of much higher mass content of melamine. In short, a MUF resin of M:U weight ratio 10:90 will perform in certain applications such as exterior plywood as a premanufactured MUF resin of M:U between 30:70 and 40:60 [66–69]. The system works (1) by simple addition of the melamine salt in the UF glue-mix, eliminating the need to premanufacture a MUF resin. The effectiveness of melamine grafting in the glue-mix and during hot pressing has been found to depend on the relative solubility of the melamine salt, which depends on the acid strength of the acid, as well as the number of acid functions in the salt. (2) By adding a salt in which the excess acid has been eliminated from the salt—hence, melamine monoacetate with no loose acid residue—to a UF resin in the resin factory, and the mix is sold as a MUF resin with as long a pot-life. This latter resin needs the addition of a classical hardener for aminoplastic resins, such as sodium sulfate or chloride, for hardening. The solubility of the salts used increases with temperature. The reasons why traditional, premanufactured MUF resins waste two-thirds or more of the melamine used in them, and why such a melamine salt addition system is so much more effective by not wasting melamine, were presented in the same study [66].

$$H_2N \underset{N}{\overset{N}{\bigtriangleup}} NH_3{}^{+-}OOC{-}CH_3$$

$$CH_3COO^{-+}H_3N \underset{N}{\overset{N}{\bigtriangleup}} NH_3{}^{+-}OOC{-}CH_3$$

$$NH_2$$

$$NH_3{}^{+-}OOC{-}CH_3$$

$$H_2N \underset{N}{\overset{N}{\bigtriangleup}} NH_3{}^{+-}OOC{-}CH_3$$

$$NH_3{}^{+-}OOC{-}CH_3$$

(10.9

)

How is it possible that addition of a melamine salt in a UF glue-mix to a M:U mass ratio of 10:90 yields plywood of comparable water resistance as a prereacted MUF resin of M:U mass ratio in the range 30:70–40:60? As a consequence of what is presented here, it is now possible to answer such a question. In the preparation of pre-copolymerized MUF resins, hence of today's normal, commercial MUF resins, during the high-temperature preparation reaction the melamine also reacts with formaldehyde to form short MF chains that are then bound to the more abundant UF chains. Hardening of MUF resins has been shown to occur almost exclusively by cross-linking through $-CH_2-$ bridges connecting two melamines [70,71] as, due to its much lower reactivity, urea is not greatly involved. The use of melamine salts at ambient temperature in the glue-mix instead ensures that only single melamine molecules are singly and separately grafted on the UF resin chain

$$-U-CH_2-M-CH_2-M-CH_2-M \text{ against } -U-CH_2-M$$

to yield rather different cross-linked networks than those of a standard MUF reactor-made resin [66–71].

To cross-link the system, only a very small amount of melamine molecules for each UF chain is needed to achieve the same effect. To have several chains of MF, as in standard MUF resins, does not improve the bond strength because (1) only one of the melamines in the chain needs to react, as the others do not need to participate at all in final cross-linking; (2) the bonding strength will also not be improved by having even all the melamines of the MF chain react; on the contrary, the highly localized concentration on vicinal sites in the network of a high density of cross-links might well render the resin far too rigid and brittle (which is indeed the case for most melamine-based resins). It is clear, then, that at least two-thirds of the melamine presently used in MUF resins is actually wasted and does not contribute much to the final results other than in a damaging manner, this being unavoidable as a consequence of the system of preparation used. The new system presented greatly improves the present situation, not only because of ease of handling (only a UF resin and a melamine salt as a hardener are needed rather than a more sophisticated MUF resin), but also the amount of melamine needed is very small (just approximately one-third of present consumption for equal exterior-grade bonding performance) with potentially considerable economic advantages, as melamine is generally expensive.

The results of 2 years' field-weathering tests in Europe have confirmed that a UF resin to which 15% melamine acetate salt has been added at the glue-mix stage, to obtain a melamine:urea mass ratio of 10:90 solids on solids, imparts a better durability and better exterior performance to a plywood glueline than traditionally reactor-coreacted MUF resins where a melamine:urea mass ratio of 33:66 is most commonly used, and even better than commercial, prereacted PMUF resin, where the relative mass proportions of P:M:U in the resin are 10:30:60 [68].

Postcuring of aminoplastic-bonded wood joints has always been avoided, due to the evident degradation induced by heat and humidity of the aminoplastic resin-hardened network. This is a known fact, and it is for this reason that boards bonded with UF, MUF, and MF resins are traditionally cooled as rapidly as possible after manufacture. However, tightening of formaldehyde emission and

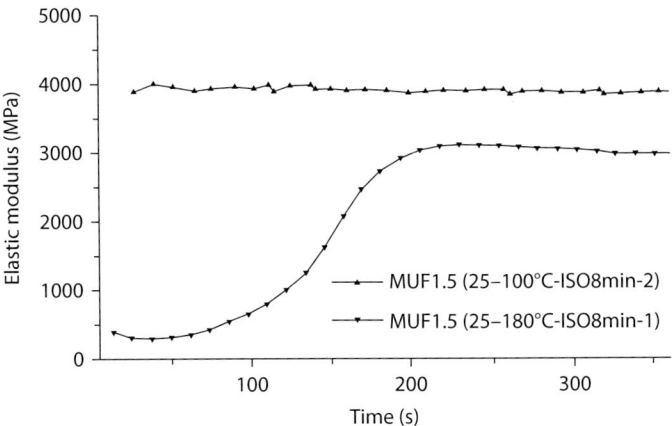

FIGURE 10.10 TMA of a joint glued with a modern MUF adhesive showing the advantage of hot-post-stacking for modern, low molar ratio aminoplastic adhesive bonded panels. Note the maximum modulus achieved during isothermal heating (180°C for 8 min) (lower curve) and maximum modulus achieved after cooling and reheating at 100°C for 8 min (upper curve); the difference in modulus is the potential gain due to hot-poststacking.

regulations has caused considerable progress in aminoplastic formulations, especially much smaller molar ratios, and hence today's aminoplastic adhesives are indeed very different materials than those of 10–20 years ago. Two studies [18,72] have shown that the postulation of avoiding postcuring of aminoplastic resin-bonded joints is, under many conditions, no longer valid (Figure 10.10). Thus, (1) postcuring (e.g., by hot-stacking in simpler cases, by an oven or other heat treatment in more sophisticated cases) can be used in principle and under well-defined conditions to improve the performance of UF- and MUF-bonded joints and panels without any further joint and hardened adhesive degradation, as the value of the modulus reached during postcuring is always consistently higher than the value at which the modulus stabilizes after complete curing during the press cycle. (2) Postcuring could also be used in principle, and for the same reasons, to further shorten the press time of UF-bonded joints and panels when well-defined postcuring conditions are used or to decrease the proportion of adhesive used at parity of performance [18,72]. (3) There is clear indication that under certain conditions, even when adhesive degradation starts, the application of posttreatment reestablishes the value of the joint strength to a value higher than its maximum value obtained during curing. Some of the best posttreatment schedules have also been presented [18].

The performance improvements in the IB strength of bonded wood panels are introduced by the series of reactions pertaining to internal methylene-ether bridge rearrangements to a tighter methylene bridge network, which have already been observed and extensively discussed for aminoplastic and phenolic resins [18,20,21,24]. These are able to counterbalance well the degradative trend to which the aminoplastic resins are subjected. Furthermore, in modern resins of lower F:(U + M) molar ratio, the amount of methylene-ether bridges formed in curing is much lower. Thus, disruption by postcuring of the already formed resin network by internal resin rearrangements will be milder, if at all present, and will definitely not result in the marked degradation and even collapse of the structure of the network that characterizes older resins of much higher molar ratios when postcured under the same conditions [2,8]. In short, notwithstanding the internal rearrangement, the network will remain quite strong; no, or hardly any, decrease in IB strength will be noticeable. For modern, lower molar ratio aminoplastic adhesives, the resin network does not noticeably degrade or collapse with postcuring. Thus, only the tightening of the network derived by further bridge formation by the few formaldehyde molecules released by the now mild internal rearrangement will be noticeable. The IB strength will then improve with postcuring in boards bonded with modern, lower formaldehyde aminoplastic adhesives.

There are important differences in the behavior of MUF resins prepared in different manners, and hence in the level of their performance as binders of wood panels, due both to their differences at the level of the resin structure and to the type and distribution of the molecular species formed before hardening, as well as to the differences in the structures of the final hardened networks. Examples of three types of MUF resins examined can illustrate this point: (1) A sequential MUF in which the UF was prepared first and then melamine coreacted afterward once the UF polymer had been formed [8]; a last small urea addition was also carried out for a final (M+U):F molar ratio of 1:1.5 and M:U weight ratio of 47:53. (2) An MUF resin in which the great majority of the urea and of the melamine were premixed and then reacted simultaneously to form the resin, followed by addition of small amounts of both the last melamine and last urea, for an (M+U):F molar ratio of 1:1.5 and M:U weight ratio of 47:53. (3) A UF resin of molar ratio 1:1.5, to which was added in the glue-mix 15% by weight on resin solids of monoacetate of melamine, to obtain a final (M+U):F molar ratio of 1:1.39 and M:U weight ratio of 14:86.

The proportion and type of chemical species formed that can be calculated from the molar proportions of the reagents, the manner in which these are combined during the reaction under different conditions, and the reaction rate constants of urea and melamine with formaldehyde lead to the conclusion, confirmed by ^{13}C NMR, that the distribution of species for resins (1), (2), and (3) are as follows:

Case (1) presents the following predominant chemical species

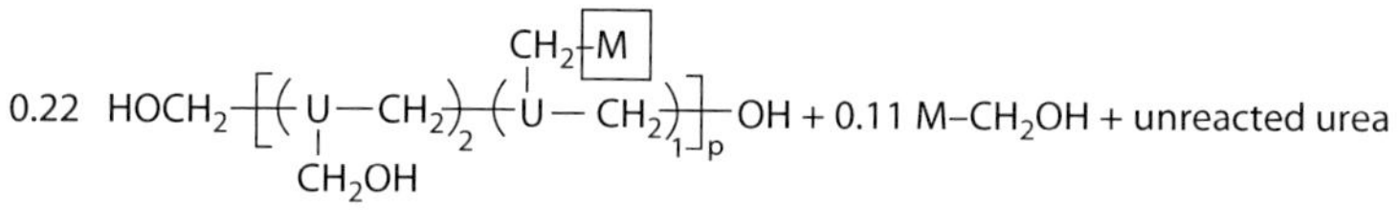

$$0.22 \ HOCH_2 \left[(U-CH_2)_2 (U-CH_2)_{1-p} \right] OH + 0.11 \ M-CH_2OH + unreacted\ urea \tag{10.10}$$

where the M attached to the UF polymer is in the form of both a single melamine as well as in the form of an MF short oligomer.

Case (3) presents instead just UF oligomers and melamine salts

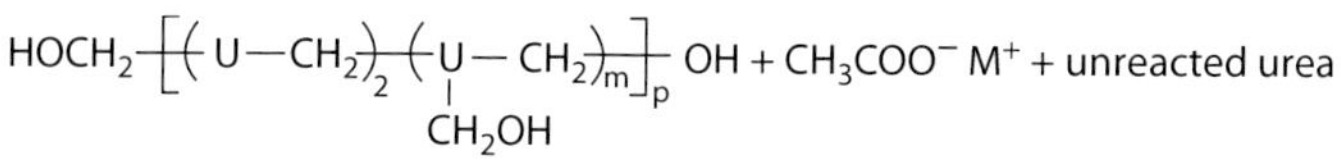

$$HOCH_2 \left[(U-CH_2)_2 (U-CH_2)_m \right]_p OH + CH_3COO^- M^+ + unreacted\ urea \tag{10.11}$$

where the M is always in the form of a single melamine molecule.

Case (2) presents the following predominant chemical species

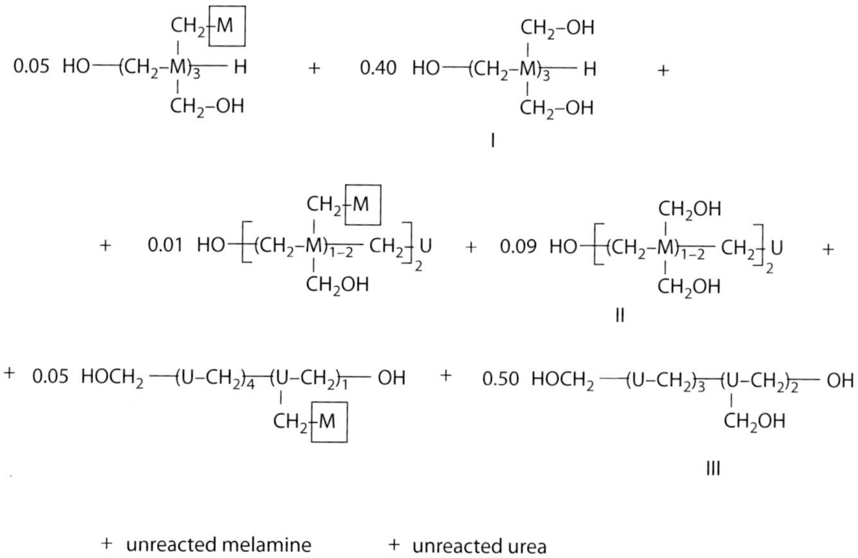

$$0.05 \ HO-(CH_2-M)_3-H \quad + \quad 0.40 \ HO-(CH_2-M)_3-H \quad +$$

I

$$+ \quad 0.01 \ HO-\left[(CH_2-M)_{1-2}-CH_2\right]_2 U \quad + \quad 0.09 \ HO-\left[(CH_2-M)_{1-2}-CH_2\right]_2 U \quad + \tag{10.12}$$

II

$$+ \quad 0.05 \ HOCH_2-(U-CH_2)_4(U-CH_2)_1-OH \quad + \quad 0.50 \ HOCH_2-(U-CH_2)_3(U-CH_2)_2-OH$$

III

$$+ \ unreacted\ melamine \quad + \ unreacted\ urea$$

where the M attached to the UF polymer is in the form of both a single melamine (M and M framed) as well as in the form of a melamine–formaldehyde short oligomer (M framed).

The structure of the three resins when still in liquid form is responsible for the appearance of their structure after hardening. Thus, hardened MUF resins of formulation type (2) will present structures as shown in Formula 10.13 and, thus, will waste the benefit of a considerable proportion of the melamine used. Hardened MUF resins of type (1) will present structures intermediate between those shown in Formulas 10.12 and 10.13 (but tending more to the type of Formula 10.13) and thus, while also wasting a considerable proportion of the melamine used, this will be less than for formulations of type (2); the strength and water-resistance results of MUFs of type (3) will then be noticeably better at parity of all other conditions than is obtainable with resins of type (2), as has indeed been shown to be the case. MUF resin formulations of type (3)—hence, the melamine acetate ones—will give hardened structures according to Formula 10.11, without wasting much melamine and, hence, giving the best performance, with the limitation of proportion that has already been mentioned and explained. This can be seen by comparing the strength results obtained by constant heating-rate TMA [71]. A MUF formulation of type (3) containing 20% melamine acetate performs almost as well as a good formulation of type (1) that contains two and a half times more melamine. They both perform much better than a formulation of type (2) [71].

Another recent approach that has shown considerable promise in markedly decreasing the percentage of adhesive solids on a board and, hence, in markedly decreasing melamine content also, has been found almost by chance. It is based on the addition to the MUF resin of certain additives. Additives capable of decreasing melamine content in MUF resins at parity of performance have been found, as well as additives capable of decreasing the percentage of a MUF resin needed for bonding while still conserving the same adhesive and joint performance. This second class of additives works for UF adhesives also, but less well, while it gives acceptable results for PF resins, but is at its best in the case of MUF resins. This second class of additives are the acetals [73,74], which do not release formaldehyde at pH values higher than 1, methylal and ethylal being the two most apt due to their cost:performance ratio. According to results reported by the Environmental Protection Agency (EPA), methylal has an LD_{50} value of 10,000 against a value of 100 in the case of formaldehyde, and is thus classed as completely nontoxic. The addition of these materials to the glue-mix of a formaldehyde-based resin improves its mechanical resistance and the performance of bonded joints considerably.

This is, in general, valid for MUF, UF, and PF resins, but the effect is particularly evident and particularly marked for the MUF resins [73]. Decreases in MUF resin solids content of as much as 33%, while conserving the same performance, are reported in the case of wood particleboard. In Figure 10.6 are shown the continuous heating-rate TMA curves of modulus as a function of temperature for two MUF resins of 1:1.2 and 1:1.5 (M+U):F molar ratios, respectively. Similar but much less extreme trends are also obtained for UF and PF resins. In the case of MUF resins, the addition of 10% additive on resin solids yields laboratory particleboard in which one can decrease the percentage of resin solids on the board between 20% and 25% without any loss of performance. Equally, at equal resin solids, the strength of a particleboard is 33% higher when 10% additive on resin solids is added to the glue-mix. Addition of 20% methylal on the board yields, in the case of the same resin, the same strength with 31% less adhesive (and hence less melamine) [73].

What is the mechanism of action of methylal, ethylal, and some other acetals to achieve such a feat? Three mechanisms have been found to be operative.

They have an excellent solvent action on melamine. The cases mentioned in the earlier part of this section referring to melamine salts and the loss of effectiveness due to wastage of melamine are applicable in this case also. When added to a reacting mixture during resin manufacture, melamine is not really soluble. It reacts, then, in heterogeneous phase with the other components of the resin, some of it being in a transient state in equilibrium between being in solution and out of solution and, thus, its efficacity is partially, but noticeably reduced. The introduction of an excellent solvent, such as an acetal, brings the totality of the reaction into homogeneous phase. The consequence is a noticeable improvement in both the effectiveness of the reaction and of melamine utilization.

As regards the class of additives capable of decreasing melamine content in MUF resins, at parity of performance, these are different from those just presented. They are based on the addition in the glue-mix of 1%–5% additive and allow the preparation of MUF copolymers, premanufactured in a traditional manner, in which either (1) the proportion of melamine is lower—for example, a 20:80 by weight M:U resin to which the additive has been added performs as well as an M:U 50:50 resin—or (2) alternatively, a top of the range M:U 50:50 MUF adhesive is upgraded to an exterior performance comparable and even superior to that of PF resins [75–77]. Several different types of additives are capable of achieving this, but they are all based on the preparation and acid stabilization of imines and iminomethylene bases (Figure 10.11) [77] and of their addition to the MUF resin. Thus, if the iminomethylene bases, which are acid anion-stabilized, are prepared by coreaction of ammonia and formaldehyde [77], or by acid anion-stabilized decomposition of hexamethylenetetramine [76], the effect is still the same, with the formation of stable amino di- and trimethylene bridges occurring in the hardening of the network. The structure of the imines and of the iminomethylene bases yielding this effect are very similar indeed to the structure of the acetal additives presented earlier in this

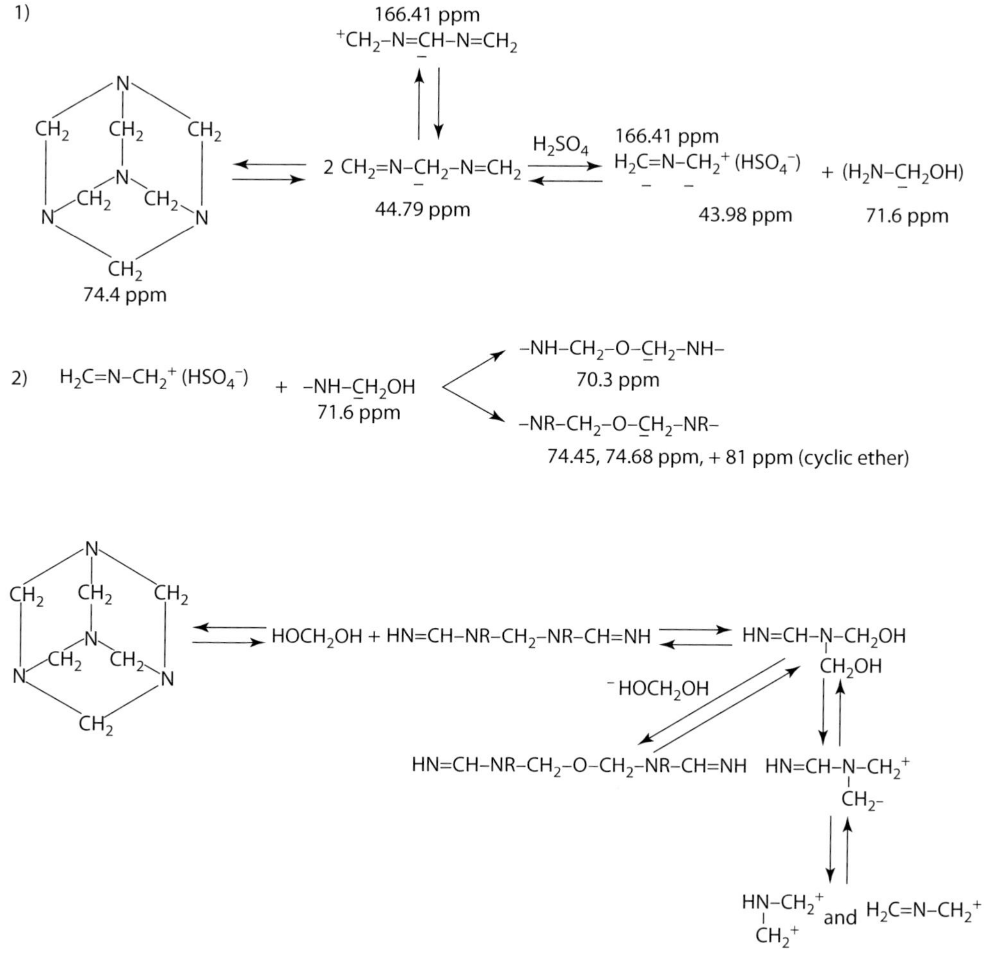

FIGURE 10.11 Mechanism of hexamethyenetetramine decomposition leading to the formation of anion-stabilized reactive iminomethylene bases. The same bases can be formed by reaction of ammonium salts such as ammonium sulfate and formaldehyde and constitute a metastable intermediate between hexamine and final decomposition products, and vice versa. (From Pizzi, A., *Proceedings Wood Adhesives 2000*, Forest Products Society, Madison, WI, 2001; Zanetti, M. et al., *Holz Roh Werkst.*, 61, 55–65, 2003; Kamoun, C. et al., *J. Appl. Polym. Sci.*, 90, 203–214, 2003. Permission given.)

section, the –NH– bridge of the imines having the same function as the –O– bridge of the acetals [78]. The imines/iminomethylene bases have the added dimension, however, that the nitrogen can function well as a site of tridimensional cross-linking itself that the oxygen bridge cannot obviously do; the amount of nitrogen-based additive that can be used is limited by its higher sensitivity to water in the hardened network. This is not the case with possibly less effective oxygen-based additives, which can be used in greater amount; one favorable property counterbalancing the disadvantage outlined. The oxygen bridge, conversely, presents perhaps a better longer-term thermal stability than the nitrogen-based bridges. These are only very relative, rather subjective advantages. What is instead significant is that the similarity of structures indicates that in the main (but not completely), the mode of action of all these additives may appear to be the same, but often different effects are also at work; first of all, a considerable improvement on the viscoelastic dissipation of energy of the glueline and bonded joint without a drop in cross-link density. The differences between the different additives is due to additional, although rather significant effects such as (1) the solvent effect of the acetals in the MUF resins, and (2) the increase in reaction rate [70] and buffer effect of the iminomethylene bases, as well as yet others.

It is on the basis of this similarity of structure and of effect that additives have been placed on a scale (Formula 10.13) according to their similarity of effects but at different levels. Thus,

$$CH_3OCH_2OCH_3 \quad CH_3NH_2 \quad CH_3OCH_2NHC=NCNHR \quad CH_3OCHOCH_3 \quad CH_3OCHOCH_3 \quad CH_3OCHOCH_3$$

Methoxymethyl melamines

TMA strength increase:

100%	50%	31%	25%	15%	8%
Accelarates (Structurally, not kinetically)	Slows down		No slowing No accelerating	No slowing No accelerating	Slows down

It must be pointed out that a TMA strength improvement of 100% on a MUF resin with 20% methylal addition corresponds to an increase of 33% in wood particleboard IB strength. This means that of all the compounds shown, only the acetals themselves, such as methylal and ethylal, as well as the similarly structured imine/iminomethylene bases discussed (for which the effect on strength is more marked), are capable of marked improvements in IB strength at the actual wood-panel level. A further significant cause of this effect is the strong buffering action of the additives on the resin, the buffer maintaining the resin while hardening within at the most appropriate pH range for optimal results [79–81]. A further point of considerable interest that must be kept in mind for MUF, UF, and MF resins is their aggregation into colloidal particles as a function of age, and how to counteract such a tendency, to improve their shelf life [31].

These developments are of use for MUF resins not just in the field of wood adhesives, or of other binders in general, but also to improve and upgrade the performance of such resins in other applications such as melamine-based impregnated paper laminates, where they have been shown to improve considerably the storage stability of paper impregnation resins [78].

Moreover, MUF resins can be applied in different ways; this also sometimes has a bearing on some types of additives used. Thus, to a MUF plus its hardener, one can add a formaldehyde depressant such as a low condensation MF, UF, or MUF precondensate or one of their mixes. This is sometimes carried out in combination with an accelerator, based on the same principle. Such an approach is more used in other resins, but it has been shown and reported as being feasible also for MUF resins [82].

An interesting development derived from application to UF resins is the use of small proportions of hyperbranched dendrimers as an upgrading additive for MUF resins [83–86]. Hyperbranched poly(amidoamine)s (Formula 10.14) exhibiting various levels of hydrophobicity have been used as modifiers for MUF adhesives. Cocondensability of these additives with the MUF resin was shown to occur. The use of these additives as MUF modifiers resulted in manifold advantages. Their addition, either immediately before the resin final use or at the last stage of resin preparation, yielded considerable upgrading of the dry IB strength of the particleboards bonded with the modified MUF resin.

10.3.5   COLD-SETTING MUF ADHESIVES

MUF resins can be used as cold-setting wood laminating adhesives for glulam and fingerjointing, using adequate acid hardeners. In all semiexterior and protected exterior structural applications where a clear/invisible glueline is preferred for esthetic reasons, a MUF adhesive is preferred to the classical phenol–resorcinol–formaldehyde (PRF) adhesives used for this purpose. MUF resins have, thus, taken a considerable hold today in Europe (in contrast to North America, where PRFs are by far preferred) and confidence in them for this application has been steadily growing.

PRF *honeymoon* fast-set, separate application adhesives for exterior-grade structural glulam and fingerjointing have now been used industrially for about 20 years [2,23] in several relevant variations developed over the years. MUF resins are now taking the same honeymoon direction: The use of an adhesive system based on a melamine resin and a separate resorcinol component [87] has been reported and tried industrially. However, for all the improvements to commercial MUF resins of this type in all their different variations, the adhesive systems are still based on resorcinol or a resorcinol-aided component. Thus, using a (MUF) resin of high melamine content as one component and resorcinol as a second component is simply unusual in its use of a MUF rather than a PRF resin, which is a very acceptable resin concept but for the fact that it is coupled with a phenol such as resorcinol. The coupling of an acid-setting MUF adhesive and of resorcinol might well present no advantages or even some potentially serious disadvantages. It has been shown, for example, that thermosetting PMUF resins do not present better performance than equivalent MUF resins and that, often, depending on their sequence of manufacture, present instead a much worse performance. There are very well-defined technical and chemical reasons for this [64,65] that boil down to the relevant differences in the reactivities of the two materials, namely the phenol (here resorcinol) and melamine. The reactivity of melamine and even urea at the acid-setting pHs that they need is much greater than that of any phenol, even resorcinol, as this pH range is the one of the lowest reactivity of any phenol. Thus, even resorcinol runs the risk of being scarcely linked to the MUF matrix, especially in a fast-setting system such as a honeymoon one, and at best it will not be able to fulfill fully the function for which it has been added.

More recently, an exclusively MUF-based honeymoon adhesive for glulam and fingerjoints has been developed and reported, in which one component is a high-performance MUF resin, while the second separate application component is based on just slightly acidified water thickened to the same viscosity as the first component by the addition of 1.5% carboxymethyl cellulose (CMC) [88,89] (Figure 10.12). The system has also been tried successfully in industry for fast production of fingerjointing (Figure 10.13) and glulam, but also for the fast production of ambient temperature-pressed plywood [88,89]. MUF-based, honeymoon-type, fast-setting, separate application adhesive systems that do not need any resorcinol are, then, capable of performing as adhesives for structural exterior-grade joints and glulam and of satisfying all the requirements of the relevant adhesive specifications for such an application. The parameter that was shown to be determining is mainly the performance of the MUF resin, if and once an excellent resin formulation is available. The ratios of melamine to urea and the resin molar ratio have a lesser effect. The performance of the resin system only starts to drop from the requirements of relevant standards from M:U weight ratios well below 20:80 and in the order of 10:90. Addition of resorcinol at these failing levels, while improving the performance slightly, did not solve the problem, and resorcinol addition does not allow the satisfaction of specification requirements. At higher M:U ratios, such as M:U = 47:53, but at even lower melamine content, addition of resorcinol does not improve the results at all. The reasons for such a behavior are those already presented and explained. The MUF honeymoons present all the other usual advantages associated with honeymoon adhesives, namely high curing rate, long pot-life, tolerance to higher moisture content of the substrate, and tolerance to even quite severe imbalances in viscosity and proportions between the two components.

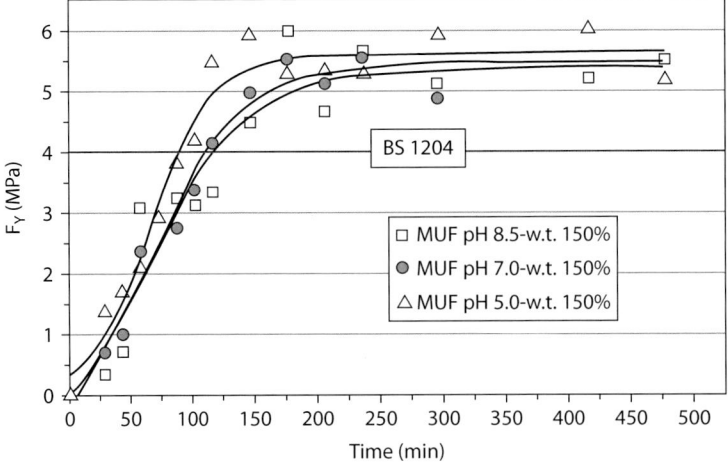

FIGURE 10.12 Tensile strength increase as a function of time of beech joints (BS 1204, Part 1) bonded with MUF-based honeymoon adhesive systems: effect of the variation of the initial application pH of the resin (Component A).

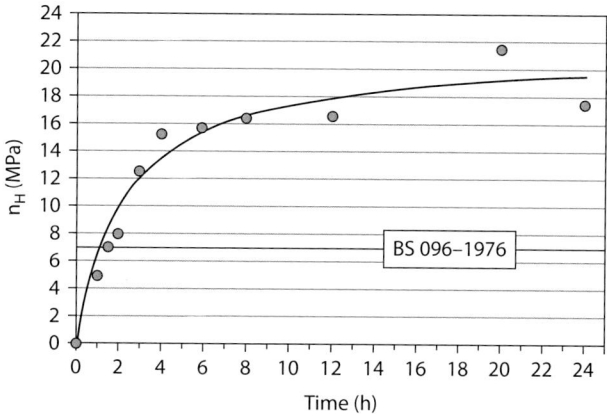

FIGURE 10.13 Four-point bending strength increase as a function of time of pine (*Pinus sylvestris*) finger-joints bonded with MUF-based honeymoon adhesive system.

Considerable effort has been made in the past on the chemical analysis of melamine resins. A specialized literature on the subject exists, ready to be consulted [34,90–96].

10.4 BIOBASED UREA AND MELAMINE RESINS

Urea is a natural raw material. It is also obtained industrially in enormous quantities by catalytic reaction of the oxygen and nitrogen of the air on glowing coals or other glowing carbon materials, even charcoal or wood. The material to substitute is formaldehyde, although even this can be of natural origin. Formaldehyde is now classified as toxic and oncogenic. To substitute it, the purely commercial need to maintain the resulting resin as white must be respected, a fact that greatly complicates formaldehyde substitution. Seeing the volumes involved, it pays to start concentrating on developing urea-based adhesives using aldehydes that are not toxic nor volatile, but still maintain the clear or white appearance of UF resins, as such adhesives can be classified as natural also. While many approaches can be taken to develop urea-based adhesives in this manner, a first important success in the bonding of plywood with this approach was recently achieved [97]. Resins based on urea–glyoxal for textiles are well known [98], but these are low-condensation resins that are not adaptable for wood. Hybrid resins of UF-type plus another aldehyde have been the initial target of several researchers, with good results for plywood [99–101]. The problem with these is that formaldehyde, although in much lower proportion, is still present. An old technology based on urea–furfural resins [102] works, but is not a good substitute for UF, not only because of the lower reactivity of furfural but mainly due to the dark color of the resin imparted by the condensation of furfural. Equally, the use of urea–furfuryl alcohol or urea–hydroxymethylfurfural, while yielding better cross-linking, also presents some of the drawbacks outlined with urea–furfural resins. Thus, the first truly urea adhesive for plywood without any formaldehyde has only recently been developed, opening a new chapter for natural environmentally friendly adhesives [97].

The nonvolatile and nontoxic aldehyde glyoxal (G) was used to substitute formaldehyde to react with urea (U) to synthesize a urea–glyoxal (UG) resin under weak acid conditions (pH = 4–5). The strength of the bonded plywoods was tested, and the curing process of the UG resin was studied by dynamic mechanical analysis (DMA). Some initial acid–catalytic reactions involved in the synthesis of the UG resin were theoretically investigated by quantum chemistry, using density functional theory. Furthermore, the UG resin was characterized by matrix-assisted laser desorption ionization time of flight mass spectrometry (MALDI-TOF-MS). The results showed that the bonded plywood, with dry shear strength of 0.98 MPa, could be directly used as interior decoration and furniture material without formaldehyde emission in dry conditions [97]. The results of

DMA analysis indicate that the cured system has the best mechanical properties when cured at 138.4°C–182.4°C. This is a relatively high temperature for plywood preparation, and a drawback for such an adhesive. The addition reaction of glyoxal with urea goes through a mechanism represented by a four-member ring transition state with a notable barrier (above 130 kJ/mol). On the other hand, the reactions of urea with different protonated forms of glyoxal follow two main pathways to form two reactive intermediates with lower energy barriers of ~30–40 kJ/mol. The problem of these resins is that while they are suitable for plywood application, they are unsuitable for the faster-pressing particleboard application due to cross-linking difficulty. Their cured behavior is mainly that of a physically entangled network formed by very long linear chains, rather than of a chemically cross-linked network. However, this considerable drawback was recently solved by using an ionic liquid as the hardener of the UG adhesive resin [103,104]. Thus, a UG resin was hardened with 1%–3% N-methyl-2-pyrrolidone hydrogen sulfate and produced particleboards with excellent IB strength results, marked lowering of the resin energy of activation, and resin gelling and hardening times comparable to UF resins at equivalent press temperatures, well in line with today's industrial practice [93,104]. This constitutes a breakthrough, as it renders feasible urea resins that can be classed as from renewable resources, using a nontoxic and nonvolatile aldehyde with no formaldehyde emission, simply because no formaldehyde is used.

Melamine is synthesized starting from urea according to the reaction $(NH_2)_2CO \rightarrow C_3H_6N_6 + 6 NH_3 + 3CO_2$. Thus, it can also be considered a biosourced chemical compound. Melamine–glyoxal (MG) resins also were also synthesized with different M:G molar ratios, and their properties were tested [105]. These MG resins were characterized by Fourier transform infrared spectroscopy (FT-IR), ^{13}C NMR and ^1H NMR, and MALDI-TOF-MS. The results show that the synthesized MG resins remain stable for at least 10 days after preparation at ambient temperature. Conjugated structures and large amounts of –OH, –NH–, and C–N groups with different substitution levels exist in the MG resins prepared. Again, long linear chains are preferentially formed. For these resins, however, chemical cross-linking was achieved by using chromium nitrate as a cross-linking catalyst. Gel times of around 3 min were obtained, but the temperature needed to obtain these was 150°C rather than the traditional 100°C. This means that the lower reactivity of the aldehyde involved causes a much higher energy of activation barrier to advance to curing for these adhesive resins [105].

Aminoresin precursors prepared by the addition of a new, colorless, nonvolatile, and nontoxic aldehyde, dimethoxyethanal (DME), a product derived from glyoxal, to melamine or urea gave resins for boards that were able to harden [106]. However, they underperformed because of the lower reactivity of DME in relation to formaldehyde. Melamine and urea react with one or two (or up to three in the case of melamine) molecules of DME to form M-DME and U-DME, but the subsequent cross-linking reaction to form bridges does not occur unless the reaction is catalyzed during resin preparation by the addition of glyoxylic acid. Such bridges between two melamine molecules are created only up to the formation of dimers and no further. The use of glyoxylic acid during the reaction allowed the formation of different oligomers by both aldol condensation and condensation of melamine and glyoxylic acid with two molecules of melamine to form dimers. These were observed by ^{13}C NMR and MALDI-TOF-MS. However, the addition of 20% isocyanate (pMDI) was necessary to satisfy the relevant mechanical strength standards for panels prepared with these resins. The pMDI contributed to cross-linking of M-DME and U-DME by its reaction to form urethane bridges according to reactions already described [107–109]. The adhesive resins so formed had excellent performance and were colorless, and they produced boards that satisfied well the requirements of the relevant standards for interior panels. Formaldehyde emission was down to what would be expected by just heating the wood chips in the absence of adhesives. The emission from the panels was sufficiently low to satisfy even the most strict relevant level (F****) of the JIS A 5908 and JIS A 1460 Japanese standards [110,111]. These adhesives are colorless as MUF and UF resins.

A further biosourced system that is under development to prepare urea–aldehyde totally biosourced resins is the one based on the reaction of urea with hydroxymethylfurfural (HMF). HMF is biosourced and can be obtained, as furfural, by acid treatment of waste lignocellulosic materials derived from wood or from agricultural waste. HMF presents an aldehyde function and a reactive hydroxymethyl group linked to its furanic ring, and presents, as such, a reactive group as reactive as furfuryl alcohol and a second group as reactive as furfural. The work on this system is only at its beginning [112].

10.5 FORMULATIONS

10.5.1 UF Adhesive Formulation

An introduction to the typical synthesis of a UF resin used as an adhesive for wood products and in industrial applications is given here. It constitutes a handy formulation for those who want to work in this field. It is not a low formaldehyde emission formulation. To 1000 parts by mass of 42% formaldehyde solution (methanol <1%) are added 22% NaOH solution of pH 8.3–8.5, 497 parts by mass of 99% urea, and the temperature is raised in approximately 50 min from ambient to 90°C, while maintaining the pH in the range 7.3–7.6 by small additions of 22% NaOH. The temperature is maintained at 90°C–91°C until the turbidity point is reached (generally another 15–20 min). The pH is then corrected to 4.8–5.1 by addition of 30% formic acid, and the temperature is raised to 98°C. The resin water tolerance point is reached in approximately 18 min and the pH is then adjusted to 8.7. Vacuum distillation of the reaction water with concomitant cooling is then initiated. After distillation of the required amount of water to reach a resin content of 60%–65%, the resin is cooled to 40°C, 169 parts by mass of second urea are added, the pH is adjusted to 8.5–8.7, and the resin is allowed to mature at 30°C for 24–48 h. The resin characteristics are: solids content, 60%; density, 1.268 g/cm^3; free HCHO, 0.4%; viscosity 200 cP; pH 8.

10.5.2 Formulation for Low-Pressure MF Paper-Impregnated Overlays

Following the same procedure as in Section 10.5.1, at the end of the water vacuum distillation, 1.7–1.9 parts by mass of the second melamine are added and the reaction mixture is heated to 95°C again, maintaining this temperature for 5–6 min. Then, the mixture is cooled rapidly.

10.5.3 MUF Formulation for Exterior Particleboard

This sequential MUF formulation can be successfully prepared at different (M+U):F molar ratios according to exactly the same procedure, and not only for the proportions as indicated in the example that follows. Thus, the same formulation gives excellent results, for example at (M+U):F molar ratios of 1:1.5 and 1:1.7, but also at other molar ratios. For more MUF formulations, see references [8,11].

To 113 parts by weight of formurea (a formaldehyde concentrate stabilized by urea, of mass content 57% formaldehyde and 23% urea [NB: there are also other concentrations]) are added 13 parts urea and 30 parts water. The pH is set at 10–10.4, and the temperature brought to 92°C–93°C under continuous mechanical stirring. The pH is then lowered to 7.8 and the reaction continued at the same temperature, allowing the pH to drop by itself over a period of 1 h 30 min–1 h 35 min to a pH of 5.2 (one should strictly prevent the pH falling under 5 to avoid uncontrollable reactions taking hold, as well a as decrease in the finished adhesive performance later). To bring the pH back to 9.5 or higher, 22% NaOH water solution is added, followed by 41 parts by weight of melamine premixed with 19 parts of water. One part of dimethylformamide and two parts of diethylene glycol are then added to the reaction mixture, maintaining a temperature of 93°C. The water tolerance is checked every 10 min while the pH is allowed to drop by itself. When the resin water tolerance reached is 180%–200% (this is often reached after 35–40 min, and the pH attained is 7.2), 6.5 parts by weight of second urea are added and the pH is again brought up to 9.5. The reaction is continued until the water tolerance reached is lower than 150% (the pH has generally reached 7.7 at this stage). The pH is then increased to 9.5 again and the reaction mixture cooled and stored. Resins produced using this procedure have solids contents of 58%–65%, a density of 1.260–1.280 at 20°C, a viscosity of 70–150 cP, free formaldehyde of approximately 0.32%, and gel times with 3% NH_4Cl of 51–57 s at 100°C. Increasing and lowering of the pH where the pH levels indicated are not reached simply via the reaction time can be carried out by addition of 22%–33% NaOH water solutions (increases in pH) and by addition of formic or acetic acid (decreases in pH).

10.6   SUMMARY

Aminoresin adhesives have dominated and still dominate, by volume, the field of adhesives. Approximately 11 million tn. of UF resin solids are used worldwide every year. Considering that adhesives for wood constitute more than 60% of all adhesives used in the world, one can understand their economic significance. They are mainly, but not only, used for wood. Their success is due to their relatively low cost and ease of manufacture. However, although they are considered a commodity, a considerable amount of technology has been developed to allow them to achieve the excellent performances they present today. They have been able to respond and adapt to several serious challenges in almost one century of their industrialization. For example, they adapted and survived well to the formaldehyde emission controversy thanks to intensive research and through novel technological developments. The new challenge now facing them is to adapt to the biosourced/biobased adhesives trend that is now receiving increasing attention. The chapter presents the most significant and key parameters in chemistry, manufacturing, and applications of these versatile adhesives, as well as their new lines of development.

REFERENCES

1. A. Pizzi. Urea-formaldehyde adhesives In: *Handbook of Adhesive Technology*, 2nd edn, A. Pizzi and K. L. Mittal (Eds.), pp. 635–652, Marcel Dekker, New York (2003).
2. A. Pizzi. Aminoresin wood adhesives. In: *Wood Adhesives Chemistry and Technology*, Vol. 1, A. Pizzi. (Ed.), pp. 59–104, Marcel Dekker, New York (1983).
3. G. Zigeuner. Abbauversuche an Harnstoff-Formaldehyd-Kondensaten 1. *Fette Seifen Anstrichmittel* 56, 973–978 (1954); Abbauversuche an Harnstoff-Formaldehyd-Kondensaten 2. *Fette Seifen Anstrichmittel* 57, 14, 100 (1955).
4. L. E. Smythe. A kinetic study of the urea-formaldehyde reaction. *J. Phys. Colloid Chem.* 51, 369–378 (1947); Urea formaldehyde kinetic studies. III. polarographic studies in dilute solution. *J. Am. Chem. Soc.* 75, 574–576 (1953).
5. G. A. Growe and C. C. Lynch. Urea-formaldehyde kinetic studies. *J. Am. Chem. Soc.* 70, 3795–3797 (1948); Polarographic urea-formaldehyde kinetic studies. *J. Am. Chem. Soc.* 71, 3731–3733 (1949).
6. L. Bettelheim and J. Cedwall. The slow condensation to polymers of methylol ureas. *Svenska Kem. Tidskr.* 60, 208–212 (1948).
7. A. Pizzi. *Advanced Wood Adhesives Technology*, Marcel Dekker, New York (1994).
8. B. Meyer. *Urea-Formaldehyde Resins*, Addison-Wesley, Reading, MA (1979).
9. S. Logan. *Fundamentals of Chemical Kinetics*, Longman, London (1996).
10. M. Gordon, A. Halliwell and T. Wilson. Kinetics of the addition stage in the melamine–formaldehyde reaction. *J. Appl. Polym. Sci.* 10, 1153–1170 (1966).
11. J. I. de Jong and J. de Jonge. Kinetics of the reaction between mono-methylolurea and methylene diurea. *Recl. Trav. Chim. Pays-Bas* 72, 207–212 (1953).
12. J. I. de Jong, and J. de Jonge. The reaction of methylene diurea with formaldehyde, *Recl. Trav. Chim. Pays-Bas* 72, 213–217 (1953).
13. K. Horioka, M. Noguchi, K. Moriya and A. Oguro. The crosslinked formaldehyde-urea copolymer effect. *Bull. Gov. For. Exp. Sta. Tokyo* 113, 20–25 (1959).
14. S. Chow and P. R. Steiner. The resistance to cyclic expansion at extreme low temperatures of urea resin bonds. *For. Prod. J.* 23 (12), 32–36 (1973).
15. B. Sundin. Formaldehyde release and indoor formaldehyde levels. In: *Proceedings of FESYP International Particleboard Symposium*, pp. 112–120, Hamburg, Germany (1978).

16. A. Pizzi, L. Lipschitz and J. Valenzuela. Theory and practice of the preparation of low formaldehyde emission UF adhesives for particleboard. *Holzforschung* 48, 254–261 (1994).

17. R. Marutzky and L. Ranta. Die Eigenschaften formaldehydarmer HF-Leimharze und daraus hergestellter Holzspanplatten. *Holz Roh Werkst.* 37, 389–393 (1979).

18. X. Lu and A. Pizzi. Curing conditions effects on the characteristics of thermosetting adhesives-bonded wood joints: Part 1: Substrate influence on TTT and CHT curing diagrams of wood adhesives. *Holz Roh Werkst.* 56, 339–346, 401 (1998).

19. F. P. F. Kollman and W. A. Côté. *Principles of Wood Science and Technology*, Vol. 2, Springer-Verlag, Berlin (1968).

20. A. Pizzi, X. Lu and R. Garcia. Lignocellulosic substrates influence on TTT and CHT curing diagrams of polycondensation resins. *J. Appl. Polym. Sci.* 71, 915–925 (1999).

21. C. Kamoun, A. Pizzi and R. Garcia. The effect of humidity on cross-linked and entanglement networking of formaldehyde-based wood adhesives. *Holz Roh Werkst.* 56, 235–243 (1998).

22. A. Pizzi. On the correlation of some theoretical and experimental parameters in polycondensation cross-linked networks. *J. Appl. Polym. Sci.* 63, 603–617 (1997).

23. A. Pizzi. *Advanced Wood Adhesives Technology.* Marcel Dekker, New York (1994).

24. R. Garcia and A. Pizzi. Cross-linked and entanglement networks in thermomechanical analysis of polycondensation resins. *J. Appl. Polym. Sci.* 70, 1111–1116 (1998).

25. R. Marutzky. Release of formaldehyde by wood products. In: *Wood Adhesives Chemistry and Technology*, Vol. 2, A. Pizzi (Ed.), pp. 307–388, Marcel Dekker, New York (1989).

26. H.-J. Deppe. Emission von organischen Substanzen aus Spanplatten. In: *Luftqualitat in Hinenraumen*, K. Auran, B Seifert, and J. Wegner (Eds.), pp. 91–128, Fischer Verlag, Stuttgart, Germany (1982).

27. H.-J. Deppe. Holzindustrielle Fertigung und Formaldehyd-emission. *Holz-Kunststoffverarb.* 20 (7/8), 12–15 (1985).

28. D. Levendis, A. Pizzi and E. Ferg. The correlation of strength and formaldehyde emission with the crystalline/amorphous structure of UF resins. *Holzforschung* 46, 260–267 (1992).

29. A. Pizzi. A molecular mechanics approach to the adhesion of urea-formaldehyde resins on cellulose. Part 1: Crystalline Cellulose I. *J. Adhes. Sci. Technol.* 4, 573–588 (1990).

30. A. Pizzi. UF resins adhesion to wood: A quantification method for adhesive formulators. *Holzforsch. Holzverwert.* 43, 63–67 (1991).

31. A. Despres and A. Pizzi. Colloidal aggregation of aminoplastic polycondensation resins: UF vs. MF and MUF resins. *J. Appl. Polym. Sci.* 100, 1406–1412 (2006).

32. J. I. de Jong and J. de Jonge. The reaction of urea and formaldehyde in concentrated solutions. *Recl. Trav. Chim. Pays-Bas* 71, 890–898 (1952).

33. J. I. de Jong. A determination of methylol groups in condensates of urea and formaldehyde. *Recl. Trav. Chim. Pays-Bas* 72, 653–654 (1953).

34. A. Petz and M. Cherubim. Die Bestimmung des freien Formaldehyds bei wässrigen Lösungen von Harnstoffharzen. *Holz Roh Werkst.* 13, 70–75 (1955).

35. G. Widmer. Quantitative Bestimmung von Harnstoff und Melamin in in Gemischen von Harnstoff-Melamin-Formaldehyd-Kondensationsprodukten. *Kunststoffe* 46, 359–362 (1956).

36. C. P. A. Kappelmeier. *Chemical Analysis of Resin-Based Coating Materials*, Wiley, New York (1951).

37. P. P. Grad and R. J. Dunn. Determination of mole ratio of urea to formaldehyde. *Anal. Chem.* 25, 1211–214 (1953).

38. B. Tomita and Y. Hirose. Urea-formaldehyde resins: NMR study on base-catalyzed reaction of formaldehyde with urea in deuterium oxide. *J. Polym. Sci* 14, 387–401 (1976).

39. J. R. Ebdon and P. E. Heaton. Characterization of urea-formaldehyde adducts and resins by ^{13}C-NMR spectroscopy. *Polymer* 18, 971–974 (1977).

40. R. M. Rammon, W. E. Johns, J. Magnuson and A. K. Dunker. The chemical structure of UF resins. *J. Adhes.* 19, 115–135 (1986).

41. C. Soulard, C. Kamoun and A. Pizzi. Uron and uron-urea-formaldehyde resins. *J. Appl. Polym. Sci.* 72, 277–289 (1999)

42. J. Gu, M. Higuchi, M. Morita and C. Y. Hse. Synthetic conditions and chemical structures of urea-formaldehyde resins. Part 1: Properties of the resin synthesized by three different procedures. *Mokuzai Gakkaishi* 41, 1115–1121 (1995)

43. J. Gu, M. Higuchi, M. Morita and C.-Y. Hse. Synthetic conditions and chemical structures of urea-formaldehyde resins. 2: Synthetic procedures involving a condensation step under strongly acidic conditions and the properties of the resin obtained. *Mokuzai Gakkaishi* 42, 149–156 (1996).

44. J. F. Walker. Formaldehyde. *Am. Chem. Soc. Monogr. Ser.* 159, 1–701 (1964).

45. A. Pizzi. Melamine-formaldehyde adhesives. In: *Handbook of Adhesive Technology*, 2nd edn, A. Pizzi and K. L. Mittal (Eds.), pp. 653–680, Marcel Dekker, New York (2003).

46. R. Frey. Kondensationsreaktionen von Anilin mit Formaldehyd und deren Bedeutung für die Herstellung plastischer Massen. *Helv. Chim. Acta* 18, 491–513 (1935).

47. K. Sato and T. Naito. Studies on melamine resin: Kinetics of acid-catalysed condensation of di and trimethylolmelamine, *Polym. J.* 5, 144–157 (1973).

48. M. Akano and Y. Ogata. Kinetics of the condensation of melamine with formaldehyde. *J. Am. Chem. Soc.* 74, 5728–5732 (1952).

49. K. Sato and S. Ouchi. Studies on formaldehyde resin. XVI. Hydroxymethylation of melamine with formaldehyde in weakly acidic region and without additional catalyst. *Polym. J.* 10, 1–11 (1978).

50. D. Braun, M. De Lourdes Abrao and H.-J. Ritzert. Gemeinsame Kondensation von Harnstoff, Melamin und Formaldehyd. I. Einfluß der Reaktionsparameter. *Angew. Makromol. Chem.* 135, 193–210 (1985).

51. D. Braun and H.-J. Ritzert. Gemeinsame Kondensation von Harnstoff, Melamin und Formaldehyd. II. Modellreaktionen. *Angew. Makromol. Chem.* 156, 1–20 (1988).

52. D. Braun and H.-J. Ritzert. Eigenschaften und Verarbeitbarkeit von Harnstoff-Melamin-Formaldehyd-Harzen. *Kunststoffe* 77, 1264–1267 (1987).

53. T. A. Mercer and A. Pizzi. Consideration on the principles of preparation of melamine-urea-formaldehyde adhesive resins for particleboard. *Holzforschung Holzverwertung* 46, 51–54 (1994).

54. A. Bachmann and T. Bertz. *Aminoplaste*, p. 81, VEB Verlag für Grundstoffindustrie, Leipzig (1967).

55. A. Knop and W. Scheib. *Chemistry and Application of Phenolic Resins*, p. 134, Springer-Verlag, Berlin (1979).

56. R. Vieweg and E. Becker. *Kunststoffhandbuch*, Vol. 10, p. 352, Hanser, Munich (1968).

57. D. Braun and W. Krausse. Gemeinsame Kondensation von Phenol, Melamin und Formaldehyd. I. Modellreaktionen. *Angew. Makromol. Chem.* 108, 141–159 (1982).

58. D. Braun and W. Krausse. Gemeinsame Kondensation von Phenol, Melamin und Formaldehyd. II. Chromatographische und spektroskopische Untersuchungen, *Angew. Makromol. Chem.* 118, 165–182 (1983).

59. D. Braun and H.-J. Ritzert. Gemeinsame Kondensation von Phenol, Melamin und Formaldehyd. III. Vorkondensation. *Angew. Makromol. Chem.* 125, 9–26 (1984).

60. D. Braun and H.-J. Ritzert. Gemeinsame Kondensation von Phenol, Melamin und Formaldehyd. IV. Härtung. *Angew. Makromol. Chem.* 125, 27–36 (1984).

61. A. Despres, A. Pizzi, H. Pasch and A. Kandelbauer. Comparative ^{13}C NMR and MALDI-TOF of species variation and structure maintenance during MUF resins preparation. *J. Appl. Polym. Sci.* 106, 1106–1128 (2007).

62. A. Kandelbauer, A. Despres, A. Pizzi and I. Taudes. Testing by FT-IR species variation during MUF resins preparation. *J. Appl. Polym. Sci.* 106, 2192–2197 (2007).

63. A. Kandelbauer, P. Petek, S. Medved, A. Pizzi and A. Teischinger. On the performance of a melamine-urea-formaldehyde resin for decorative paper coatings. *Eur. J. Wood Prod.* 68, 63–75 (2010).

64. C. Cremonini, A. Pizzi, and P. Tekely. Influence of PMUF resins preparation method on their molecular structure and performance as adhesives for plywood. *Holz Roh Werkst.* 54, 85–88 (1996).

65. M. Higuchi, J.-K. Roh, S. Tajima, H. Irita, T. Honda, and I. Sakata. Polymeric structures of melamine-based composite adhesives. In: *Proceedings No. 4735 of the Adhesives and Bonded Wood Symposium*, Forest Products Society, Madison, WI, pp. 429–449 (1994).

66. M. Prestifilippo, A. Pizzi, H. Norback, and P. Lavisci. Low addition of melamine salts for improved UF adhesives water resistance. *Holz Roh Werkst.* 54, 393–398 (1996).

67. C. Cremonini and A. Pizzi. Improved waterproofing of UF plywood adhesives by melamine salts as glue-mix hardeners. *Holzforschung Holzverwertung* 49, 11–15 (1997).

68. C. Cremonini and A. Pizzi. Field weathering of plywood panels bonded with UF adhesives and low proportion of melamine salts. *Holz Roh Werkst.* 57, 318 (1999).

69. C. Kamoun and A. Pizzi. Performance effectiveness of addition to UF of melamine salts vs. melamine alone in MUF adhesives for plywood. *Holz Roh Werkst.* 56, 86 (1998).

70. A. Pizzi and L. A. Panamgama. Diffusion hindrance vs. wood-induced catalytic activation of MUF adhesives polycondensation. *J. Appl. Polym. Sci.* 58, 109–115 (1995)

71. A. Pizzi. High performance MUF resins of low melamine content by a number of novel techniques. In: *Proceedings Wood Adhesives 2000*, Forest Products Society, Madison, WI, pp. 219–240 (2001).

72. C. Zhao and A. Pizzi. Hot postcuring improvement of MUF-bonded particleboards and its temperature forecasting model. *Holz Roh Werkst.* 58, 307–308 (2000).

73. A. Pizzi, M. Beaujean, C. Zhao, M. Properzi and Z. Huang. Acetals-induced strength increases and lower resin contents in MUF and other polycondensation adhesives. *J. Appl. Polym. Sci.* 84, 2561–2571 (2002).

74. M. Zanetti, A. Pizzi, M. Beaujean, H. Pasch, K. Rode and P. Dalet. Acetals induced strength increase of MUF polycondensation adhesives, Part 2: Solubility and colloidal state disruption. *J. Appl. Polym. Sci.* 86, 1855–1862 (2002).

75. A. Pizzi, P. Tekely and L.A. Panamgama. A different approach to low formaldehyde emission aminoplastic wood adhesives. *Holzforschung.* 50, 481–485 (1996).

76. P. Mouratidis, E. Dessipri and A. Pizzi. New adhesive system for improved exterior-grade wood panels. European Union Final Contract Report, FAIR TC 96–01604, European Commission, Brussels, Belgium (2000).

77. F. Pichelin, C. Kamoun and A. Pizzi. Hexamine hardener behavior: Effects on wood glueing, tannin and other wood adhesives. *Holz Roh Werkst.* 57, 305–317 (1999).

78. G. Lambiotte and A. Pizzi. Aminoplastic or phenoplastic adhesive with improved mechanical strength. US Patent application 2004/0034185A1 (2004); European Patent application EP1174480A1 (2000).

79. M. Zanetti, A. Pizzi and C. Kamoun. Upgrading of MUF particleboard adhesives and decrease of melamine content by buffer and additives. *Holz Roh Werkst.* 61, 55–65 (2003).

80. C. Kamoun, A. Pizzi and M. Zanetti. Upgrading of MUF resins by buffering additives: Part 1: Hexamine sulphate effect and its limits. *J. Appl. Polym. Sci.* 90, 203–214 (2003).

81. M. Zanetti and A. Pizzi. Upgrading of MUF resins by buffering additives: Part 2: Hexamine sulphate mechanisms and alternate buffers. *J. Appl. Polym. Sci.* 90, 215–226 (2003).

82. L. A. Panamgama and A. Pizzi. ^{13}C NMR analysis method for MUF and MF resins strength and formaldehyde emission. *J. Appl. Polym. Sci.* 59, 2055–2068 (1996).

83. X. Zhou, H. Essawy, A. Pizzi, X. Li, K. Rode and G. Du. Upgrading of MUF wood adhesives for particleboard by highly-branched polymers. *J. Adh. Sci. Technol.* 27, 1058–1068 (2013).

84. X. Zhou, H. Essawy, A. Pizzi, X. Li, H. Pasch, N. Pretorius and G. Du. Poly(amidoamine)s dendrimers of different generations as components of melamine urea formaldehyde (MUF) adhesives used for particleboards production: What are the positive implications? *J. Polym. Res.* 20, 267–280 (2013).

85. X. Zhou, H. A. Essawy, A. Pizzi, J. Zhang, H. Pasch and G. Du. Second generation of dendritic poly(amidoaldehyde) as modifying components of melamine-urea-formaldehyde (MUF) adhesives: Subsequent use in particleboard production. *J. Polym. Res.* 21, 379–393 (2014).

86. S. Amirou, J. Zhang, A. Pizzi, H. Essawy, A. Zerizer, J. Li and L. Delmotte. Utilization of hydrophilic/hydrophobic hyperbranched poly(amidoamine)s as additives for melamine urea formaldehyde (MUF) adhesives. *Polym. Composites* 36, 2255–2264 (2015).

87. J. R. Parker, J. B. Taylor, D. V. Plackett and R. E. Lomax. Method of joining wood. US Patent 5674338 (1991).

88. M. Properzi, A. Pizzi and L. Uzielli. Honeymoon MUF adhesives for exterior grade glulam. *Holz Roh Werkst.* 59, 413–421 (2001).

89. M. Properzi, A. Pizzi and L. Uzielli. Performance limits of pure MUF honeymoon adhesives for exterior grade glulam and fingerjoints. *Holzforschung Holzverwertung*, 53, 114–117 (2001).

90. C. Hirt, F. T. King and R. G. Schmitt. Detection and estimation of melamine in wet-strength paper by ultraviolet spectrophotometry. *Anal. Chem.* 26, 1273–1274 (1954).

91. R. W. Stafford. Identification of melamine and urea resins in wet strength paper. *Paper Trade J.* 120, 51–56 (1945).

92. H. Schindlebauer and J. Anderer, Eine erste Charakterisierung von Melamin-Formaldehyd-Kondensaten mit Hilfe der ^{13}C-NMR-Spektroskopie. *Angew. Makromol. Chem.* 79, 157–162 (1979).

93. B. Tomita and H. Ono. Melamine–formaldehyde resins: Constitutional characterization by Fourier transform ^{13}C-NMR spectroscopy. *J. Polym. Sci. Chem. Ed.* 17, 3205–3215 (1979).

94. L. A. Panamgama and A. Pizzi. ^{13}C NMR analysis method for phenol-formaldehyde resins strength and formaldehyde emission. *J. Appl. Polym. Sci.* 55, 1007–1015 (1995).

95. T. A. Mercer and A. Pizzi. ^{13}C NMR analysis method for MF and MUF resins strength and formaldehyde emission from wood particleboard: Part 1: MUF resins. *J. Appl. Polym. Sci.* 61, 1687–1696 (1996).

96. T. A. Mercer and A. Pizzi. ^{13}C NMR analysis method for MF and MUF resins strength and formaldehyde emission from wood particleboard: Part 2: MF resins. *J. Appl. Polym. Sci.* 61, 1697–1702 (1996).

97. S. Deng, G. Du, X. Li and A. Pizzi. Performance and reaction mechanism of zero formaldehyde-emission urea-glyoxal (UG) resin. *J. Taiwan Inst. Chem. Eng.* 45, 2029–2038 (2014).

98. H. Petersen. Process for the production of formaldehyde-free finishing agents for cellulosic textiles and the use of such agents. *Textilveredlung* 2, 51–62 (1968).

99. S. Deng, A. Pizzi, G. Du, J. Zhang and J. Zhang. Synthesis and chemical structure of a novel glyoxal-urea-formaldehyde (GUF) co-condensed resins with different MMU/G molar ratios by ^{13}CNMR and MALDI-TOF-MS. *J. Appl. Polym. Sci.* 131, 21, 41009, (2014).

100. J. Zhang, H. Chen, A. Pizzi, Y. Li, Q.Gao and J. Li. Characterisation and application of urea-formaldehyde-furfural co-condensed resins as wood adhesives. *BioResources* 9, 6267–6276 (2014).

101. Y. F. Zhang, X. R. Zeng and B. Y. Ren. Synthesis and structural characterization of urea-isobutyraldehyde-formaldehyde resins. *J. Coatings Technol. Res.* 6, 337–344 (2009).

102. E. E. Novotny and W. W. Johnson. Furfural-urea resin and process of making the same. U.S. Patent 1827824 (1931).

103. H. Younesi-Kordkheili and A. Pizzi. Acid ionic liquids as a new hardener in urea- glyoxal adhesive resins. *Polymers* 8 (3), 57–71 (2016).

104. H. Younesi-Kordkheili and A. Pizzi. Ionic liquids as enhancers of urea-glyoxal panel adhesives as substitutes of urea-formaldehyde resins. *Eur. J. Wood Prod.* 75, 481–483 (2017).

105. S. Deng, A. Pizzi, G. Du, M. C. Lagel, L. Delmotte and S. Abdalla. Synthesis, structure characterization and application of melamine-glyoxal adhesive resins. *Eur. J. Wood Prod.* 1–14, DOI: 10.1007/s00107-017-1184-9 (2017).

106. A. Despres, A. Pizzi, C. Vu and H. Pasch. Formaldehyde-free aminoresin wood adhesives based on dimethoxyethanal. *J. Appl. Polym. Sci.* 110, 3908–3916 (2008).

107. A. Pizzi and T. Walton. Non-emulsifiable, water-based diisocyanate adhesives for exterior plywood, Part 1: Novel reaction mechanisms and their chemical evidence. *Holzforschung* 46, 541–547 (1992).

108. A. Pizzi, J. Valenzuela and C. Westermeyer. Non-emulsifiable, water-based diisocyanate adhesives for exterior plywood, Part 2: Industrial application. *Holzforschung* 47, 69–72 (1993).

109. A. Despres, A. Pizzi and L. Delmotte. ^{13}C NMR investigation of the reaction in water of UF resins with blocked emulsified isocyanates. *J. Appl. Polym. Sci.* 99, 589–596 (2006).

110. Japanese Standard JIS A 5908: Particleboards. Testing method (2003).

111. Japanese Standard JIS A 1460: Building boards determination of formaldehyde emission (2001).

112. M. C. Lagel, F. J. Santiago-Medina, A. Pizzi and L. Delmotte. Development and characterization of formadehyde-free biobased adhesive for particle boards production: urea and hydroxymethylfurfural resin. *Int. J. Adhes. Adhes.* submitted (2017).

11 Polyurethane Adhesives

*Dennis G. Lay, Paul Cranley, and Antonio Pizzi**

CONTENTS

11.1 INTRODUCTION

The development of polyurethane adhesives can be traced back more than 60 years to the pioneering efforts of Otto Bayer and coworkers. Bayer extended the chemistry of polyurethanes initiated in 1937 [1] into the realm of adhesives in about 1940 [2] by combining polyester polyols with di- and polyisocyanates. He found that these products made excellent adhesives for bonding elastomers to fibers and metals. Early commercial applications included life rafts, vests, airplanes, tires, and tanks [3]. These early developments were soon eclipsed by a multitude of new applications, new technologies, and patents at an exponential rate.

The uses of polyurethane adhesives have expanded to include bonding of numerous substrates, such as glass, wood, plastics, and ceramics. Urethane prepolymers were first used in the early 1950s [4] to bond leather, wood, fabric, and rubber composites. A few years [5] later, one of the first two-component urethane adhesives was disclosed for use as a metal-to-metal adhesive. In 1957 [6], the first thermoplastic polyurethane used as a hot-melt adhesive (adhesive strips) was patented for use in bonding sheet metal containers. This technology was based on linear, hydroxy-terminated polyesters and diisocyanates. Additional thermoplastic polyurethane adhesives began appearing in the 1958–1959 period [7,8]. During this period, the first metal-to-plastic urethane adhesives were developed.

---

* Indicates corresponding author

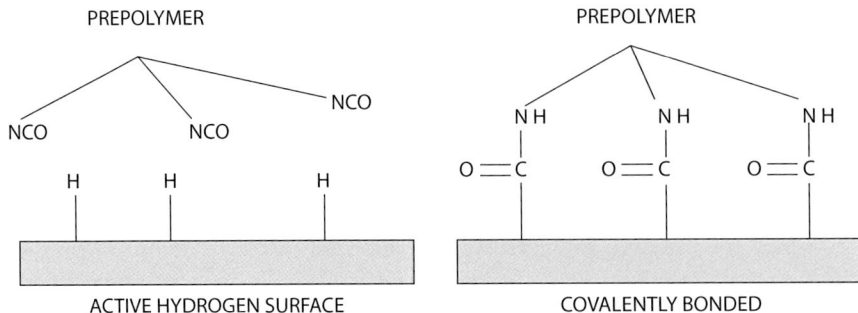

FIGURE 11.1   Typical mechanism for a urethane adhesive bonding covalently to a polar surface.

Waterborne polyurethanes were also being developed, with a polyurethane latex claimed to be useful as an adhesive disclosed in 1961 by du Pont [9]. A commercial urethane latex was available by 1963 (Wyandotte Chemicals Corporation) [10]. The adhesive properties of urethane latexes were explored further by W.R. Grace in 1965 [11]. In the early 1960s, B.F. Goodrich developed thermoplastic polyester polyurethanes that could be used to bond leather and vinyl [12]. In 1968, Goodyear introduced the first structural adhesive for fiberglass reinforced plastic (FRP), used for truck hoods [13].

Polyurethane pressure-sensitive adhesives began appearing in the early 1970s [14]. By 1978, advanced two-component automotive structural adhesives (Goodyear) were commercially available. Waterborne polyurethane adhesives received additional attention during this period [15]. In 1984, Bostik developed reactive hot-melt adhesives. Polyurethane adhesives are sold in an ever-widening array of markets and products, where they are known for their excellent adhesion, flexibility, low-temperature performance, high cohesive strength, and high cure speeds that can readily be tailored to the manufacturer's demands [16].

Urethanes make good adhesives for a number of reasons: (1) They effectively wet the surface of most substrates (the energy level of very low energy surfaces such as polyethylene or polypropylene must be raised before good wetting occurs) [17], (2) they readily form hydrogen bonds to the substrate, (3) their small molecular size allows them to permeate porous substrates, and (4) they form covalent bonds with substrates that have active hydrogens. Figure 11.1 shows the typical mechanism for a urethane adhesive bonding covalently to a polar surface.

Polyurethane adhesive consumption has been estimated at 828 million lb. in 2003, having a value of approximately US$2.3 billion [18]. While the packaging market is the fourth-largest market in terms of pounds of urethane adhesives sold, it is substantially larger than the forest products market and the foundry core binder market in terms of dollars. Specific market segments such as automotive and recreational vehicles have easily surpassed the gross national product (GNP) growth rate. In the next few years, a number of specific market segments are expected to grow at about 5% per year. These would include vehicle assembly (automotive and recreational vehicles), electronics, furniture, and curtain wall manufacture.

11.2   APPLICATIONS OVERVIEW

The textile market has traditionally been the largest consumer of polyurethane adhesives. There are a number of high-volume applications, including textile lamination, integral carpet manufacture, and rebonded foam. Textile lamination occurs through either a solution coating process or flame bonding. Flame bonding textile lamination is accomplished by melting a polyurethane foam by flame and then nipping the foam between two textile rolls while it is still tacky. Integral carpet manufacture describes carpeting that is manufactured by attaching either nylon, wool, or polypropylene tufts that are woven through a polypropylene scrim with a urethane adhesive to a polyurethane foam cushion in a continuous process. Rebonded foam is made using scrap polyurethane foam bonded together with a urethane prepolymer and is used primarily as carpet underlay. Durability, flexibility, and fast-curing speeds are all critical parameters for these applications.

Foundry core binders are isocyanate-cured alkyd or phenolic adhesives used as binders for sand used to produce foundry sand molds. These sand molds are used to cast iron and steel parts. A fast, economical cure of the sand mold is required under ambient conditions.

Packaging adhesives are adhesives used to laminate film to film, film to foil, and film to paper in a variety of packaging constructions. A broad variety of products are sold to this market, with solvent-based, high solid content, 100% solid content, and waterborne adhesives all being used. Polyurethane adhesives are considered one of the high-performance products offered to this industry because of their excellent adhesive properties, heat resistance, chemical resistance, and fast-curing properties. Polyurethane adhesives can also be designed to meet U.S. Food and Drug Administration approval, a requirement for food packaging applications.

Solvent-borne adhesives represent the majority of the volume in the packaging market, with both one- and two-component systems being used. Waterborne polyurethane adhesives are a much smaller segment that has been driven by environmental considerations. Growth has slowed in recent years, because of generally inferior performance compared with solvent-based adhesives and because most of the major converters have already made capital investments in solvent recovery systems.

Isocyanates are used in the forest products industry to adhesively bond wood chips, which are then pressed to form particleboard and oriented strandboard. Urethanes are also used to fill knot-holes and surface defects in finished plywood boards (*plywood patch*). These filled systems must cure rapidly and be sanded easily.

The transportation market has used polyurethane adhesives for such diverse applications as bonding FRP and sheet molding composite (SMC) panels in truck and car applications, polycarbonate headlamp assemblies, door panels, and weather strip flocking.

The construction market for polyurethane adhesives consists of a variety of applications, such as laminating thermal sandwich panels, bonding gypsum board to wood ceiling joists in modular and mobile homes, and gluing plywood floors. Early green strength, low shrinkage, and high bond strength are critical properties.

The furniture industry uses polyurethane adhesives to bond veneers of various compositions onto board and metal substrates. Both waterborne and solvent-based adhesives are used.

Footwear is a sizable niche for polyurethane adhesives, which are used to attach the soles. Polyurethane adhesives compete primarily with neoprene-based adhesives and have replaced much of the neoprene volume due to improved performance. However, the overall market has declined, as U.S. manufacturers have moved production overseas.

11.3 BASIC URETHANE CHEMISTRY

Isocyanates react with active hydrogens, as depicted in Figure 11.2. This addition reaction occurs with the active hydrogen adding to the nitrogen atom and the electron-rich nucleophile (Lewis base) reacting with the carbonyl group. Generally, the stronger the base, the more readily it reacts with the isocyanate. Table 11.1 shows typical reaction rates of some active hydrogen-containing compounds.

As expected, the aliphatic amines and aromatic amines (the strongest bases in the table) react the fastest. The urethanes industry has taken advantage of this reactivity in two-component commercial processes, demanding fast cure by using specially designed metering equipment and spray heads.

Alcohols and water react readily at room temperature. Most urethane adhesives depend on the –NCO group reacting with either water or alcohols. Primary OH groups are two to three times as fast reacting as sterically hindered secondary OH groups under equivalent conditions. The reaction rates shown in Table 11.2 reflect uncatalyzed reaction rates, and should be used as an indication of relative reaction rates. Actual rates are dependent on solvent, temperature, and the presence of catalysts. Catalysts can significantly accelerate these reactions and can, in some cases, alter the order of reactivity [19].

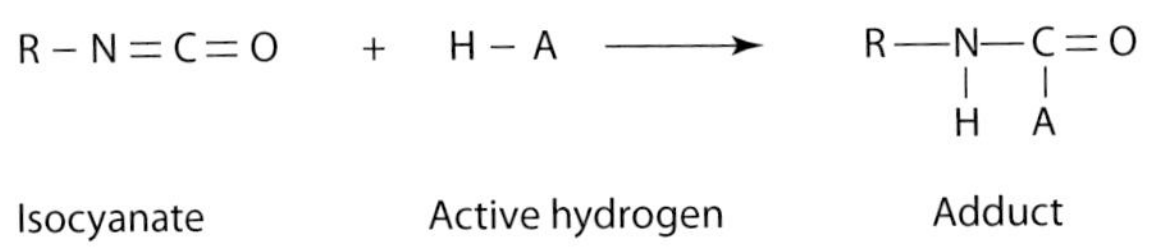

$$R-N=C=O \quad + \quad H-A \quad \longrightarrow \quad R-\underset{\underset{H}{|}}{N}-\underset{\underset{A}{|}}{C}=O$$

Isocyanate   Active hydrogen   Adduct

FIGURE 11.2   Reaction of isocyanate with active hydrogen.

11.3.1 Branching Reactions

There are a number of complex reactions that can occur besides the desired reaction of the polyol hydroxyl group with the isocyanate group to form a urethane, as shown in Figure 11.3.

Isocyanates can continue to react with undesirable consequences under conditions of high heat or strong bases. Basic impurities and excess heat catalyze branching reactions, leading to variations in prepolymer viscosity, gelation, and exotherms. Most basic impurities arise from the polyol, since polyols are typically produced under basic conditions. As such, the net acidity of the overall system (contribution of acidic or basic components from the reactants) plays a critical role in determining the final viscosity achieved [20,21].

The presence of water will lead to the formation of ureas and evolve CO_2, as shown in Figure 11.4. This mechanism is thought to proceed through the formation of an unstable intermediate, carbamic acid, which then decomposes to give CO_2 and an aromatic amine. The amine will then react further with another isocyanate to give a urea linkage. All common moisture-cured urethanes give off CO_2 on curing, which can pose problems if not properly controlled. Urea groups are known to cause high prepolymer viscosity, because of increased hydrogen bonding and because of their ability to react further with excess isocyanate groups to form a biuret, as shown in Figure 11.5.

At room temperature, the biuret reaction proceeds very slowly; however, elevated temperatures and even traces of a base will catalyze the biuret reaction as well as other branching reactions. These would include the formation of allophanate groups, as shown in Figure 11.6 (due to the reaction of urethane groups with excess isocyanate groups), or trimerization of the terminal NCO group (to form an isocyanurate), as shown in Figure 11.7. Biurets and allophanates are not as stable thermally or hydrolytically as branch points achieved through multifunctional polyols and isocyanates. The allophanates shown in Figure 11.6 can continue to react with excess isocyanates to form isocyanurates (as shown in Figure 11.7), a trimerization reaction that will liberate considerable heat. In most cases, the desired reaction product is the simple unbranched urethane or a urea formed by direct

---

TABLE 11.1
Typical Reaction Rates for Selected Hydrogen-Containing Compounds

Active Hydrogen Compound	Typical Structure	Relative* Reaction Rate
Aliphatic amine	$R-NH_2$	100,000
Secondary aliphatic amine	R_2-NH	20,000–50,000
Primary aromatic amine	$Ar-NH_2$	200–300
Primary hydroxyl	$R-CH_2OH$	100
Water	$H-O-H$	100
Carboxylic acid	$R-CO_2H$	40
Secondary hydroxyl	R_2CH-OH	30
Urea proton	$R-NH-CO-NH-R$	15
Tertiary hydroxyl	R_3C-OH	0.5
Urethane proton	$R-NH-CO-OR$	0.3
Amide	$R-CO-NH_2$	0.1

* Uncatalyzed reaction rate, 80°C [1].

TABLE 11.2

Gelation Times in Minutes at 70°C for Different Catalyzed Isocyanates

Catalyst	TDI	m-Xylene Diisocyanate	Hexamethylene Diisocyanate
None	>240	>240	>240
Triethylamine	120	>240	>240
Triethylenediamine	4	80	>240
Stannous octoate	4	3	4
Dibutyltin di(ethylhexoate)	6	3	3
Bismuth nitrate	1	0.5	0.5
Zinc naphthenate	60	6	10
Ferric chloride	6	0.5	0.5
Ferric 2-ethylhexoate	16	5	4
Cobalt 2-ethylhexoate	12	4	4

Source: D.G. Lay and P. Cranley, Polyurethane Adhesives, in *Handbook of Adhesive Technology*, Second edition (A. Pizzi and K.L. Mittal Eds), Taylor & Francis (2003).

Note: TDI: toluene diisocyanate.

reaction of an isocyanate with an amine. Ureas are an important class, because they typically have better heat resistance, higher strength, and better adhesion. By controlling the reaction temperature (typically less than 80°C) and stoichiometry, and using a weakly basic catalyst (or none at all), the reaction will stop at the urethane or urea product. Increasing the functionality of the polyol or the isocyanate will achieve branching or cross-linking in a more controlled fashion.

11.3.2 CATALYSTS

As noted previously, strong or weak bases that are sometimes present in the polyols will catalyze the urethane reaction. The effect of catalysts on the isocyanate reaction is well documented. Indeed, the first reported examples occur in the literature well before urethanes became a commercially significant class of compounds. The first use of a catalyst with an isocyanate was reported by Leuckart and Schmidt in 1885 [22]. Other early reports were from French and Wirtel (1926), who used triethylamine to catalyze the reaction of phenols with 1-naphthyl isocyanate [23]. Baker and Holdsworth (1947) detailed the mechanism of the urethane reaction [24].

Commercial catalysts consist of two main classes: organometallics and tertiary amines. Both classes have features in common, in that the catalytic activity can be described as a combination of electronic and steric effects. Electronic effects arise as the result of the molecule's ability to donate or accept electrons. For example, in the tertiary amines, the stronger the Lewis base, generally the stronger the polyurethane catalyst. Empty electronic orbitals in transition metals allow reactants to

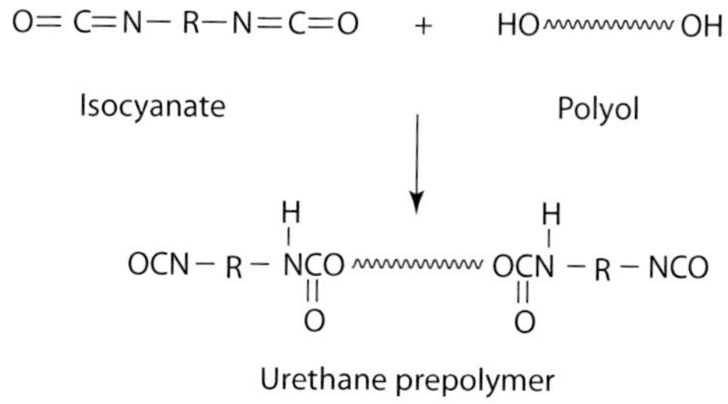

FIGURE 11.3   Reaction of polyol hydroxyl group with isocyanate group to form a urethane.

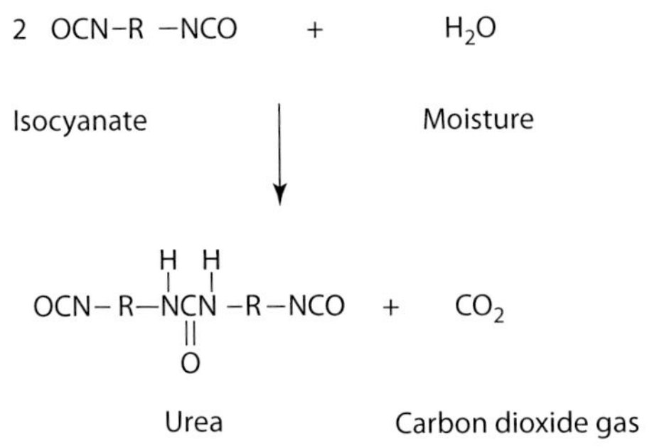

$$2 \ OCN-R-NCO \quad + \quad H_2O$$

Isocyanate   Moisture

$$OCN-R-\overset{\overset{H}{|}}{N}\overset{\overset{H}{|}}{C}N-R-NCO \quad + \quad CO_2$$
$$\overset{||}{O}$$

Urea   Carbon dioxide gas

FIGURE 11.4   Reaction of isocyanate with water.

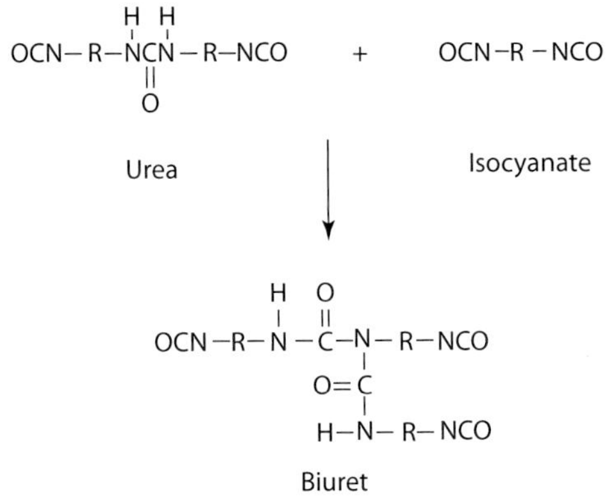

$$OCN-R-\overset{\overset{H}{|}}{N}C\overset{\overset{H}{|}}{N}-R-NCO \quad + \quad OCN-R-NCO$$
$$\overset{||}{O}$$

Urea   Isocyanate

$$OCN-R-\overset{\overset{H}{|}}{N}-\overset{\overset{O}{||}}{C}-N-R-NCO$$
$$O=\overset{|}{C}$$
$$H-N-R-NCO$$

Biuret

FIGURE 11.5   Reaction of urea with isocyanate.

coordinate to the metal center, activating bonds and placing the reactants in close proximity to one another.

Steric effects arise from structural interactions between substituents on the catalyst and the reactants that will influence their interaction. The importance of steric effects can be seen by comparing the activity for triethylenediamine with that of triethylamine. The structure of triethylenediamine (see Figure 11.8) forces the nitrogens to direct their lone electron pairs outward in a less shielded position than is true of triethylamine. This results in a rate constant for triethylenediamine that is four times that of triethylamine at 23°C [1].

Organometallic complexes of Sn, Bi, Hg, Zn, Fe, and Co are all potent urethane catalysts, with Sn carboxylates being the most common. Hg catalysts have long induction periods that allow long open times. Hg catalysts also promote the isocyanate–hydroxyl reaction much more strongly than the isocyanate–water reaction. This allows their use in casting applications where pot-life and bubble-free parts are critical. Bismuth catalysts are replacing mercury salts in numerous applications, as the mercury complexes have come under environmental pressure.

Catalysts will not only accelerate reaction rates, but may also change the order of reactivity. Table 11.2 illustrates this behavior. These data indicate that amines do not affect the relative reactivities of different isocyanates, and show that Zn, Fe, and Co complexes actually raise the reactivity of aliphatic isocyanates above aromatic isocyanates.

11.4   URETHANE POLYMER MORPHOLOGY

One of the advantages that a formulator has using a polyurethane adhesive is the ability to tailor the adhesive properties to match the substrate. Flexible substrates, such as rubber or plastic, are obvious

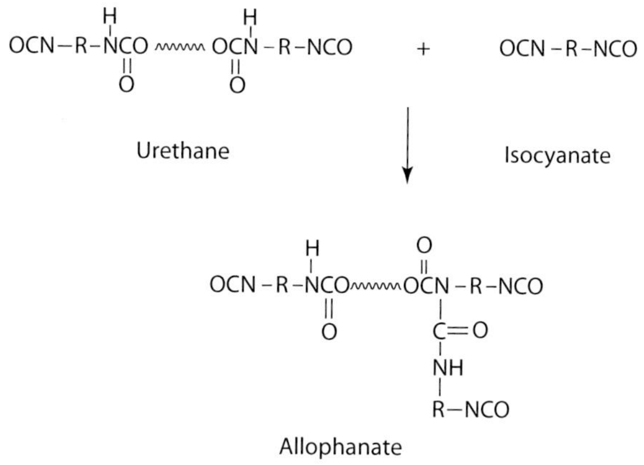

FIGURE 11.6 Reaction of urethane with isocyanate.

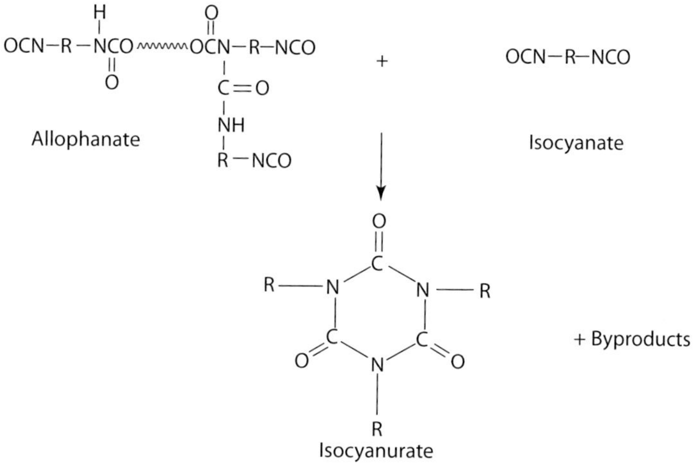

FIGURE 11.7 Reaction of allophanate with isocyanate.

matches for polyurethane adhesives because a tough elastomeric product can easily be produced. Polyurethanes derive much of their toughness from their morphology.

Polyurethanes are made up of long polyol chains that are tied together by shorter hard segments formed by the diisocyanate and chain extenders if present. This is depicted schematically in Figure 11.9. The polyol chains (typically referred to as *soft segments*) impart low-temperature flexibility and room-temperature elastomeric properties. Typically, the lower molecular weight polyols give the best adhesive properties, with most adhesives being based on products of molecular weight of less than 2000. Generally, the higher the soft-segment concentration, the lower the modulus, tensile strength, hardness, and tear strength, while elongation will increase. Varying degrees of chemical resistance and heat resistance can be designed by proper choice of the polyol.

Short-chain diols or diamines are typically used as chain extenders. These molecules allow several diisocyanate molecules to link, forming longer-segment hard chains with higher glass transition temperatures. The longer-segment hard chains will aggregate together because of similarities in polarity and hydrogen bonding to form a pseudo-cross-linked network structure. These hard domains affect modulus, hardness, and tear strength and also serve to increase resistance to compression and extension. The hard segments will yield under high shear forces or temperature and, in fact, determine the upper use temperature of the product. Once the temperature or shear stress is reduced, the domains will re-form.

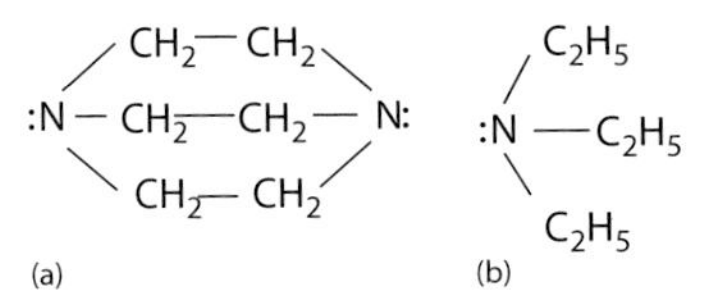

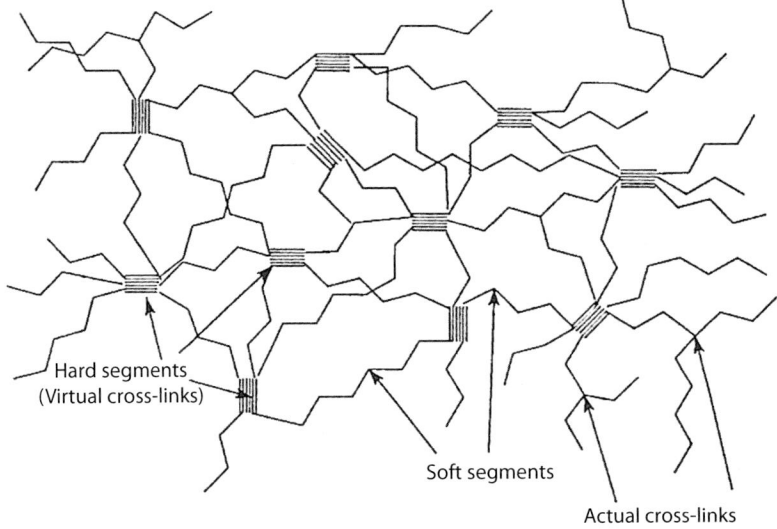

FIGURE 11.8   Structure of (a) triethylenediamine and (b) triethylamine.

FIGURE 11.9   Polyol-chain structure of polyurethane.

The presence of both hard-segment and soft-segment domains for polyurethanes gives rise to several glass transition temperatures: one below −30°C that is usually associated with the soft segment, transitions in the range 80°C–150°C, and transitions above 150°C. Transitions in the range 80°C–150°C are associated with the breakup of urethane hydrogen bonds in either the soft segment or the hard segment. Transitions higher than 150°C are associated with the destruction of hard-segment crystallites or aggregates. Linear polyurethane segmented prepolymers can act as thermoplastic adhesives that are heat activated. A typical use for this type of product is in the footwear industry.

By proper choice of either the isocyanate or the polyol, actual chemical cross-links can be introduced in either the hard or soft segments that may be beneficial to some properties. The effectiveness of these cross-links is offset by a disruption of the hydrogen bonding between polymer chains. Highly cross-linked polyurethanes are essentially amorphous in character, exhibiting high modulus, high hardness, and low elasticity. Many adhesives fall into this category.

11.5   PREPOLYMER FORMATION

Most urethane adhesives are based on urethane prepolymers. A prepolymer is made by reacting an excess of diisocyanate with a polyol to yield an isocyanate-terminated urethane, as shown in Figure 11.10. Prepolymers may have excess isocyanate present (*quasi-prepolymers*) or they may be made in a 2:1 stoichiometric ratio to minimize the amount of free isocyanate monomer present. Most moisture-cured prepolymers are based on 2:1 stoichiometric ratios. Two-component adhesives are generally based on quasi-prepolymers, which use the excess isocyanate to react with either the chain extenders present in the other component or with the substrate surface.

Prepolymers are isocyanates, and react similarly to isocyanates with several important differences. Prepolymers are, typically, much higher in molecular weight, higher in viscosity, lower in

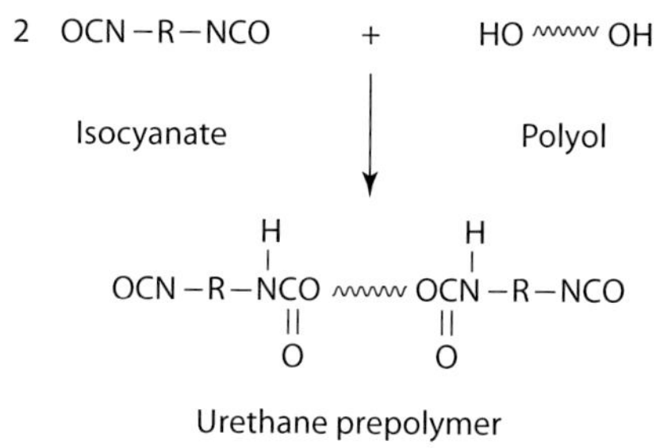

$$2 \quad OCN-R-NCO \quad + \quad HO \sim OH$$

Isocyanate   Polyol

$$OCN-R-\overset{H}{\underset{\underset{O}{||}}{N}}CO \sim OCN-R-\overset{H}{\underset{\underset{O}{||}}{N}}CO$$

Urethane prepolymer

FIGURE 11.10   Reaction of isocyanate with polyol.

isocyanate content by weight percentage, and have lower vapor pressures. Prepolymers are important to adhesives for a number of reasons. The desired polymeric structure of the adhesive can be built into the prepolymer, giving a more consistent structure with more reproducible physical properties. In addition, since part of the reaction has been completed, reduced exotherms and reduced shrinkage are normally present. For two-component systems, better mixing of components usually occurs, since the viscosities of the two components more closely match. In addition, the ratios of the two components match more closely. Side reactions such as allophanate, biuret, and trimer are lessened. Finally, prepolymers typically react more slowly than does the original diisocyanate, allowing longer pot-lives.

11.6   ADHESIVE RAW MATERIALS

Polyols for adhesive applications can be generally divided into three main categories: (1) polyether polyols, (2) polyester polyols, and (3) polyols based on polybutadiene. Polyether polyols are the most widely used polyols in urethane adhesives, because of their combination of performance and economics. They are typically made from the ring-opening polymerization of ethylene, propylene, and butylene oxides, with active proton initiators in the presence of a strong base, as shown in Figure 11.11.

Polyether polyols are available in a variety of functionalities, molecular weights, and hydrophobicities, depending on the initiator, the amount of oxide fed, and the type of oxide. Capped products are commercially available, as well as mixed-oxide feed polyols, as shown in Figure 11.12. Polyether polyols typically have glass transitions in the $-60°C$ range, reflecting the ease of rotation about the backbone and only little chain interaction. As one would expect from such low glass transition temperatures, they impart very good low-temperature performance. The polyether backbone is resistant to alkaline hydrolysis, which makes them useful for adhesives used on alkaline substrates such as

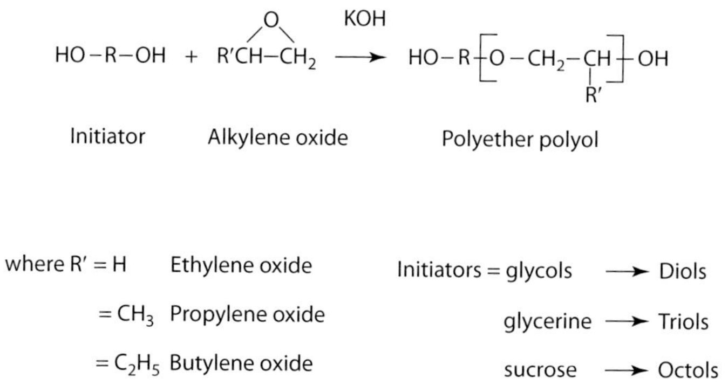

$$HO-R-OH \quad + \quad R'CH-CH_2 \quad \xrightarrow{KOH} \quad HO-R\left[O-CH_2-\underset{R'}{\overset{}{CH}}\right]OH$$

Initiator   Alkylene oxide   Polyether polyol

where R′ = H   Ethylene oxide   Initiators = glycols $\longrightarrow$ Diols

= CH₃ Propylene oxide   glycerine $\longrightarrow$ Triols

= C₂H₅ Butylene oxide   sucrose $\longrightarrow$ Octols

FIGURE 11.11   Ring-opening polymerization to form polyether polyols.

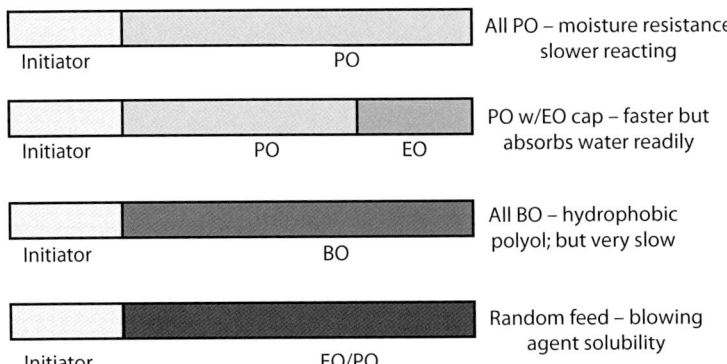

FIGURE 11.12   Various commercially available capped products and mixed-oxide feed polyols. BO: butylene oxide; EO: ethylene oxide; PO: propylene oxide.

$$HO-(-CH_2-CH_2-CH_2-CH_2-O-)_n-H$$

FIGURE 11.13   Structure of polytetramethylene oxide.

concrete. They are typically very low in viscosity and exhibit excellent substrate wetting. In addition, their low cost and ready availability from a number of suppliers add to their attractiveness.

The more commonly used polyether polyols range in molecular weight from 500 to 2000 for diols and 250 to 3000 for triols. Lower molecular weight, higher-functionality polyols are traditionally used in rigid-foam applications, but have also been used as cross-linkers for two-component, fast-curing urethane adhesives. Polytetramethylene oxides (PTMOs; see Figure 11.13) can be considered a subset of polyether polyols. They offer improved physical properties compared with polyethers based on ethylene oxide, propylene oxide, or butylene oxide, combining high tensile strength (due to stress crystallization) with excellent tear resistance. They are also noted for their excellent resistance to hydrolysis. They are typically priced at a premium to other polyols.

Polyester polyols are used widely in urethane adhesives, because of their excellent adhesive and cohesive properties. Compared with polyether-based polyols, polyester-based polyol adhesives have higher tensile strengths and improved heat resistance. These benefits come at the expense of hydrolytic resistance, low-temperature performance, and low chemical resistance. One of the more important application areas for these products is in the solvent-borne thermoplastic adhesives used in shoe sole binding. These products are typically made from adipic acid and various glycols (see Figure 11.14).

Some glycerine or trimethylolpropane may be used to introduce branching structures within the polyester backbone. Phthalic anhydride may also be used to increase hardness and water resistance. Inexpensive terephthalic acid-based polyesters from recycled poly(ethylene terephthalate) (PET) resins have become popular more recently.

Polycaprolactones (see Figure 11.15), another type of polyester polyol, offer improvements in hydrolysis resistance and in tensile strength (can stress crystallize) over adipic acid-based polyester polyols. They are typically higher in viscosity and higher in cost than polyether polyols of comparable molecular weight. When moisture resistance is critical, urethane adhesives incorporating polybutadiene polyols are used. These products are hydroxy-terminated, liquid polybutadiene resins. The hydrocarbon backbone greatly decreases water absorption, imparting excellent hydrolytic stability. Polybutadiene compounds also have exceptional low-temperature properties, with glass transition temperatures being reported below −70°C [24]. These products are priced at a 40%–50% premium over comparable polyether polyols. The structure of polybutadiene polyol is shown in Figure 11.16.

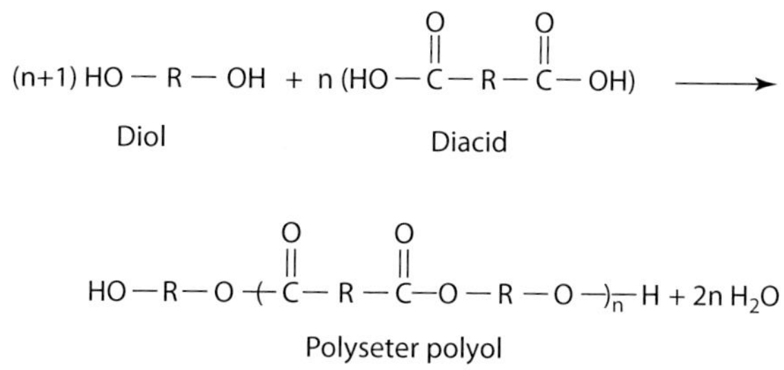

$$(n+1)\ HO-R-OH\ +\ n\ (HO-\underset{\substack{\parallel\\O}}{C}-R-\underset{\substack{\parallel\\O}}{C}-OH) \longrightarrow$$

Diol   Diacid

$$HO-R-O\left(\underset{\substack{\parallel\\O}}{C}-R-\underset{\substack{\parallel\\O}}{C}-O-R-O\right)_{\!\overline{n}}H + 2n\ H_2O$$

Polyseter polyol

FIGURE 11.14 Reaction of diol with diacid to form polyester polyol.

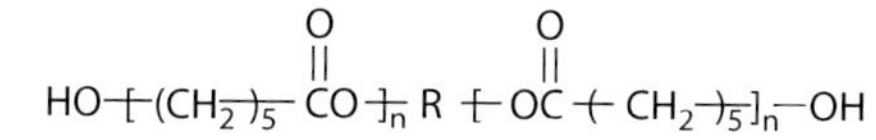

$$HO\left((CH_2)_{\overline{5}}\ \underset{\substack{\parallel\\O}}{C}O\right)_{\!\overline{n}}R\left(O\underset{\substack{\parallel\\O}}{C}(CH_2)_{\overline{5}}\right)_{\!n}-OH$$

FIGURE 11.15 Structure of polycaprolactone diol.

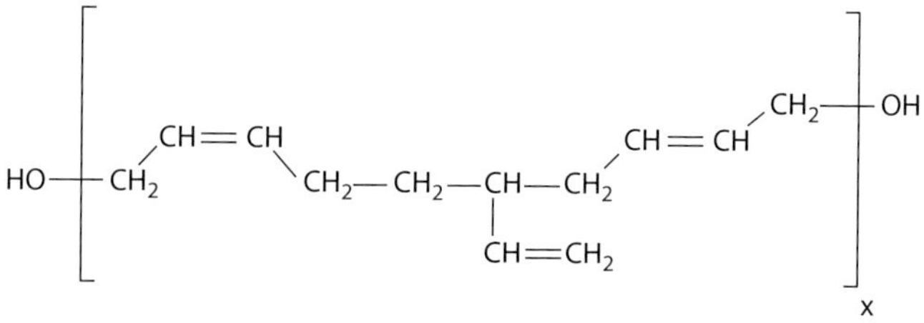

FIGURE 11.16 Structure of polybutadiene polyol.

11.6.1 ISOCYANATES FOR ADHESIVE APPLICATIONS

Toluene diisocyanate (TDI) is a colorless, volatile, low-viscosity liquid commonly used in the adhesives area to manufacture low-viscosity prepolymers for flexible substrates. The structure of TDI is shown in Figure 11.17. TDI is typically supplied as an 80:20 mixture of the 2,4 and 2,6 isomers, respectively, with two grades of acidity available. Type I TDI is low in acidity (10–40 ppm); Type II TDI is higher (80–120 ppm). Type II TDI is generally used for prepolymer applications, because the additional acidity is available to neutralize trace bases found in polyether polyols. These trace bases can cause branching reactions during prepolymer preparation reactions, resulting in high viscosities and even gelation if not properly controlled (see Section 11.5). The extra acidity present also serves to stabilize the prepolymer, extending the shelf stability. In addition, since TDI is predominately the 2,4 isomer, a reactivity difference is noted for the isocyanate groups. Since the less hindered site reacts first, the sterically hindered site is left when prepolymers are formed, leading to prepolymers that are more shelf stable. TDI prepolymers are used in adhesives for the textile and food packaging laminates industries, where a fit is found for their low viscosity and low cost. The volatility of TDI and additional handling precautions that must be taken when using TDI have limited its growth in adhesive applications.

Methylene diphenyl diisocyanate (MDI) is used where high tensile strength, toughness, and heat resistance are required. MDI is less volatile than TDI, making it less of an inhalation hazard. The acidity levels in MDI are very low, typically in the order of 0–10 ppm, so the trace base levels in the polyols are much more critical in prepolymer production than with TDI. The structure of MDI is shown in Figure 11.18. There are several commercial suppliers of MDI that typically supply grades with 98% or better 4,4′ isomer. MDI is a solid at room temperature (melting point 38°C, 100°F),

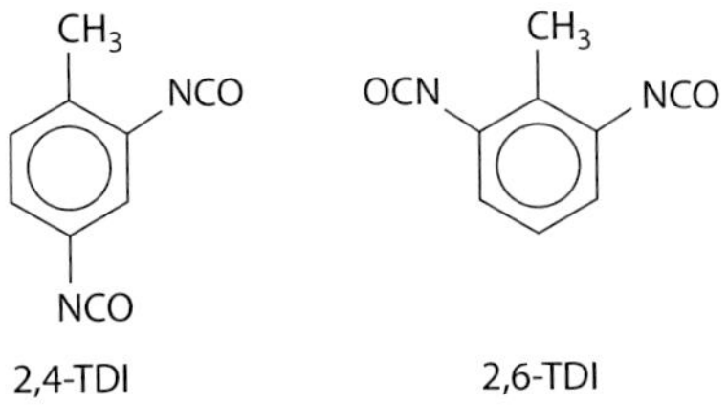

FIGURE 11.17   Structure of the 2,4 and 2,6 isomers of toluene diisocyanate.

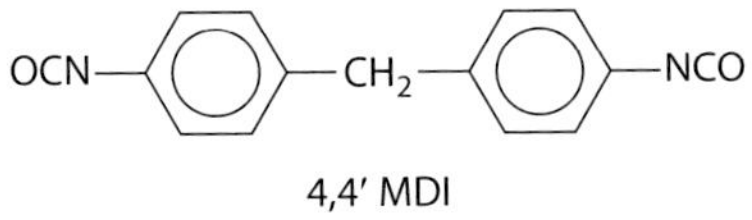

FIGURE 11.18   Structure of methylene diphenyl diisocyanate.

requiring handling procedures different from those for TDI. MDI should be stored as a liquid at 46°C or frozen as a solid at −28°C to minimize dimer growth rate. MDI reacts faster than TDI, and because the NCO groups in MDI are equivalent, they have the same reactivity, in contrast to TDI. MDI is used in packaging adhesives, structural adhesives, shoe sole adhesives, and construction adhesives.

Several MDI products have been introduced that address the inconvenience of handling a solid. They have seen increased usage in the adhesives industry and are expected to experience a higher growth rate. Most MDI producers offer a uretonimine-modified form of MDI that is a liquid at room temperature. The uretonimine structure is shown in Figure 11.19. In addition, several producers have introduced MDIs containing elevated levels of the 2,4′ isomer, as shown in Figure 11.20. At approximately 35% 2,4′ isomer level, the product becomes a liquid at room temperature, greatly increasing the handling ease. A number of advantages are seen: slower reactivity, longer pot-life, lower-viscosity prepolymers, prepolymers with lower residual monomeric MDI, and improved shelf stability.

Polymeric MDIs are made during the manufacturing of monomeric MDI. These products result as higher molecular weight oligomers of the reaction of phosgene on the reaction products of aniline and formaldehyde. A typical structure for these products is shown in Figure 11.21. These oligomers average 2.3–3.1 in functionality and contain 30%–32% NCO. Much of the hydrolyzable chlorides and other side-reaction compounds produced in the manufacturing process of MDI are left behind in these products. The acidity levels can be 10–50 times the level found in pure MDI, and the products are dark brown in color. The higher acidity level decreases reactivity; however, this decrease is offset somewhat by the higher functionality.

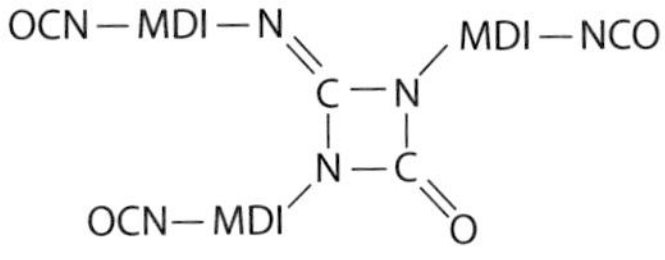

FIGURE 11.19   Structure of uretonimine.

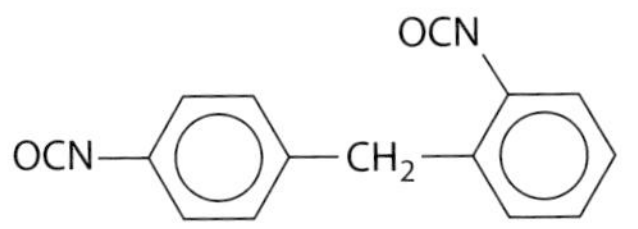

FIGURE 11.20   Structure of the 2,4′ isomer of MDI.

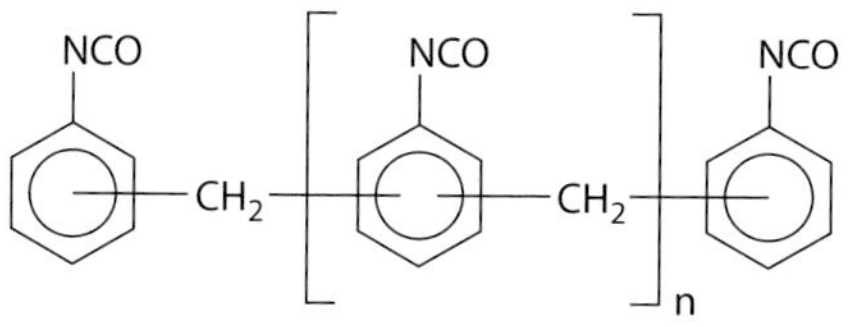

FIGURE 11.21   Structure of pMDI.

Polymeric MDIs are typically lower in cost than pure MDI, and because of the increased asymmetry, they have a lower freezing point (liquid at room temperature). They are less prone to dimerization and, as a consequence, are more storage stable than are pure MDI and its derivatives. Polymeric MDIs are used whenever the color of the finished adhesive is not a concern. They are generally not used for prepolymers, because high-viscosity branched structures are typically formed. They are widely used as adhesives in the foundry core binder area, in oriented strandboard or particleboard, and between rubber products and fabric or cord. It is interesting to note that the polymeric isocyanates used commercially today are structurally very similar to the Desmodur R (Bayer) products used over 50 years ago [2].

Aliphatic isocyanates are used whenever resistance to ultraviolet light is a critical concern. Examples of aliphatic isocyanates are hexamethylene diisocyanate, hydrogenated MDI, isophorone diisocyanate, and tetramethylxylene diisocyanate. Structures for these molecules are shown in Figure 11.22. The aliphatic isocyanates are usually more expensive than aromatic isocyanates and find limited use in adhesive applications. Resistance to ultraviolet light is usually not a critical concern in adhesives, because the substrate shields the adhesive from sunlight.

Blocked isocyanates are also used in urethane adhesives. Blocking or *masking* of the isocyanates refers to reacting the isocyanate groups with a material that will prevent the isocyanate from reacting with active hydrogen-containing species at room temperature, but will allow that reaction to occur at elevated temperatures. Blocked isocyanates are easily prepared, and their chemistry has been developed extensively since their inception by Bayer and coworkers during the early 1940s [25–28]. As an example, the preparation of a methyl ethyl ketoxime blocked isocyanate is shown in Figure 11.23.

Blocked isocyanates offer a number of advantages over unblocked isocyanates. The traditional concern for moisture sensitivity can be addressed by blocking the isocyanate. Heat activation is then required, but most commercial adhesive applications can meet this requirement. Water-based dispersions and dispersions of the isocyanate in the polyol or other reactive media become possible using blocked isocyanates. There are a number of blocked isocyanates commercially available that could be used in adhesive applications. Miles (Bayer) produces a series of aromatic and aliphatic blocked isocyanates marketed for primers, epoxy flexibilizers, wire coatings, and automotive topcoat applications. Blocked isocyanates are widely patented for fabric laminating adhesives, fabric coating adhesives [29,30], and tire cord adhesives [31].

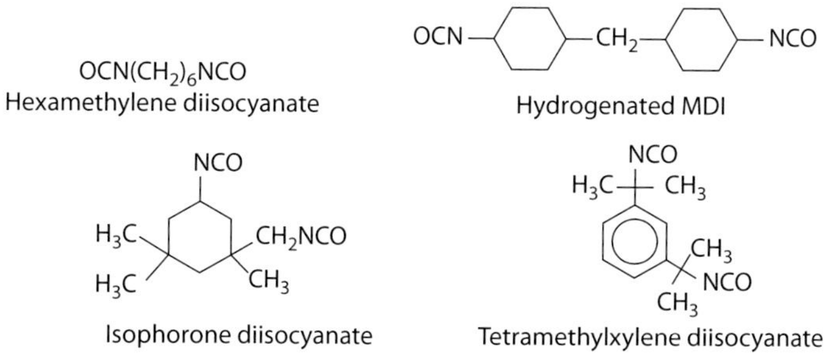

OCN(CH$_2$)$_6$NCO
Hexamethylene diisocyanate

Hydrogenated MDI

Isophorone diisocyanate

Tetramethylxylene diisocyanate

FIGURE 11.22   Structures of various aliphatic isocyanates.

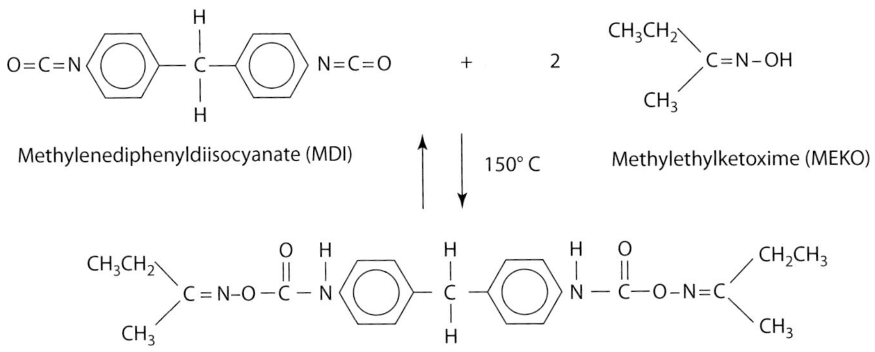

FIGURE 11.23   Preparation of a methylethylketoxime blocked isocyanate.

11.6.2 TOXICOLOGY

Polyether polyols are generally considered to be low in toxicity with respect to eye and skin irritation; however, amine-initiated polyether polyols have been found to be more irritating to the skin and eyes. The manufacturer's material safety data sheet (MSDS) should always be consulted before use. Oral toxicity is generally a secondary concern in an industrial environment. The vapor pressure of polyols is generally negligible; thus, vapor inhalation is not usually a concern [32]. Low molecular weight glycols (chain extenders) are considered more problematic than polyether polyols. Generally, while the vapor pressure of these products is low, there are processes that could potentially result in vapor concentrations close to the exposure limits [32]. The exposure guidelines for chain extenders may be written to differentiate between aerosols and vapors. For more specific handling information, the manufacturer should be consulted.

The toxicology of isocyanates is a primary concern when developing or using polyurethane adhesives. Respiratory effects are the primary toxicological manifestation of repeated overexposure to diisocyanates [33–37]. In addition, most of the monomeric isocyanates are eye and skin irritants. Precautions should be taken in the workplace to prevent exposure. The risk of overexposure is primarily (but not limited to) allergic sensitization with asthma-type symptoms. Manufacturers' guidelines (the MSDS) should be consulted for the most current information and legal requirements.

11.6.3 FILLERS AND ADDITIVES

Fillers are used in adhesives to improve physical properties, control rheology, and lower cost. The most common polyurethane fillers are calcium carbonate, talc, silica, clay, and carbon black. A more rigorous treatment of this subject can be found in Katz and Milewski [38]. Fumed silicas and carbon blacks are used primarily as thixotropes in application areas that require a nonsagging bead. Calcium carbonates, clays, and talcs are used to improve the economics of an adhesive formulation. A major concern using fillers with urethane prepolymers is the moisture content associated with the fillers. Fillers must typically be dried prior to use with urethane prepolymers or isocyanates. Hygroscopic fillers should be avoided, as moisture introduced by the filler can lead to poor shelf stability of the finished product.

Pigments are sometimes used in polyurethane adhesive systems, but since most adhesives are generally hidden from view, pigments do not play major roles. Pigments may be used to color the adhesive to match the substrate. Pigments are more typically used to color one side of a two-component system, to help the user distinguish between the isocyanate and the polyol. They are also sometimes used as an aid to judge mix ratios. Carbon black and titanium dioxide are two commonly used pigments.

Plasticizers can also be used in polyurethane adhesives to lower viscosity, improve filler loadings, improve low-temperature performance, and plasticize the polyurethane adhesive. Phthalate esters, benzoate esters, phosphates, and aromatic oils are common examples. Plasticizers should be used sparingly, as adhesion will generally decrease as levels increase.

11.7 SURFACE PREPARATION AND PRIMERS

Proper surface preparation is the key to obtaining good adhesive bonds having a long service life. Substrate surfaces may have dirt, grease, mold-release agents, processing additives, plasticizers, protective oils, oxide scales, and other contaminants that will form a weak boundary layer. When the adhesive bond fails, it is usually in the weak boundary layer, giving a low-strength bond. Some form of surface treatment is necessary to obtain optimum bond strength. The primary goal of surface treatment is to remove any weak surface boundary layer on the substrate [39]. A large number of surface treatments have been developed, with many targeted toward specific substrates. These include mechanical abrasion, etching, solvent cleaning, detergent washing, laser and flame treatments, chemical treatments, and plasma and corona discharges [17,40–48].

Primers are also used in conjunction with a surface treatment, either to improve adhesive performance or to increase production flexibility in a bonding operation. Isocyanates have been used for over 50 years as primers on substrates such as rubber, plastic, fibers, and wood [49]. Isocyanates will react with polar groups on the surface and promote bonding.

Silane coupling agents are commonly used as primers for glass, fiber composites, mineral-filled plastics, and cementitious surfaces. The silane coupling agents have been found to be especially effective with glass substrates. One end of the coupling agent is an alkoxysilane that condenses with the silanol groups on the glass surface. The other end of the coupling agent is an amino, mercapto, or epoxy functionality that will react with the isocyanate group in the adhesive. Epoxy silanes have also been used as additives to adhesives to improve water resistance [49–51]. Other organometallic primers are based on organotitanates, organozirconates, and some chromium complexes [39].

11.8 COMMON POLYURETHANE ADHESIVE TYPES

11.8.1 One-Component Adhesives

The oldest types of one-component polyurethane adhesives were based on di- or triisocyanates that cured by reacting with active hydrogens on the surface of the substrate or moisture present in the air or on the substrate. The moisture reacts with the isocyanate groups to form urea and biuret linkages, building molecular weight, strength, and adhesive properties. Prepolymers are also used either as 100% solids or solvent-borne one-component adhesives. Moisture-cured adhesives are used today in rebonded foam, tire cord, furniture, and recreational vehicle applications.

A second type of one-component urethane adhesive comprises hydroxypolyurethane polymers based on the reaction products of MDI with linear polyester polyols and chain extenders. There are several commercial suppliers of these types of thermoplastic polyurethanes. The polymers are produced by maintaining the NCO:OH ratio at slightly less than 1:1 to limit molecular weight to the range 50,000–200,000, with a slight hydroxy content (approximately 0.05%–0.1%). These are typically formulated in solvents for applications to shoe soles or other substrates. After solvent evaporation, heat is used to melt the polymer (typically 50°C–70°C; at these temperatures, the polymers reach a soft, rubbery, amorphous state), so the shoe upper can be press fitted to the sole. On cooling, the adhesive recrystallizes to give a strong, flexible bond [52]. More recently, polyisocyanates have been added to these to increase adhesion and other physical properties on moisture curing (see Section 11.9.2 for more detail).

The use of waterborne polyurethane adhesives has grown in recent years, as they have replaced solvent-based adhesives in a number of application areas. There are a number of papers and patents covering the use of waterborne polyurethanes in shoe soles, packaging laminates, textile laminates, and as an adhesive binder for the particleboard industry [53,54]. Because waterborne polyurethane adhesives have no volatile organic content (VOC) emissions and are nonflammable, they are more environmentally friendly. Typically, they can be blended with other dispersions without problems and exhibit good mechanical strength. Water-based systems are fully reacted linear polymers that are emulsified or dispersed in water. This is accomplished by building hydrophilicity into the polymer backbone with either cationic or anionic groups or long hydrophilic polyol segments or, less frequently, through the use of external emulsifiers. Figure 11.24 illustrates the more common functional groups that can be built into the urethane molecule that will confer hydrophilicity.

A typical example of how these groups are built into the polymer backbone is shown in Figure 11.25. A urethane prepolymer is reacted with chain extenders containing either carboxylates or sulfonates in a water-miscible solvent (e.g., acetone). The reaction product is an isocyanate-terminated polyurethane or polyurea with pendant carboxylate or sulfonate groups. These groups can easily be converted to salts, which, as water is added to the prepolymer–solvent solution, allows the prepolymer to be dispersed in water. The solvent is then stripped, leaving the dispersed product. There are variations on this theme that allow lower solvent volumes to be used [55]. Long hydrophilic polyol segments can also be introduced. Chain extenders with hydrophilic ethylene oxide groups pendent to the backbone are reacted with the prepolymer to form a nonionic self-emulsifying polyurethane. This reaction is also carried out in a water-miscible solvent that can later be stripped from the solvent–water solution.

Blocked isocyanates can also be considered one-component adhesives. The use of a blocking agent allows the isocyanate to be used in a reactive medium that can be heat activated. Thus, one-component adhesives based on blocked isocyanates are not amenable to room-temperature curing applications. The chemistry of these products is covered in more detail in Section 11.6.

11.8.2 Two-Component Adhesives

The second major classification of common polyurethane adhesives is the two-component system. Two-component polyurethane adhesives are widely used where fast cure speeds are critical, as on original equipment manufacturers (OEM) assembly lines that require quick fixture of parts, especially at ambient or low bake temperatures. Two-component urethanes are required in laminating applications where no substrate moisture is available or where moisture cannot penetrate through to the adhesive bond. Two-component urethanes are also useful where CO_2 (generated by a one-component moisture cure) or a volatile blocking agent would interfere with the adhesive properties.

Two-component adhesives typically consist of low equivalent weight isocyanate or prepolymer that is cured with a low equivalent weight polyol or polyamine. They may be 100% solids or solvent borne. Since the two components will cure rapidly when mixed, they must be kept separate until just before application. Application is followed quickly by mating of the two substrates to be bonded.

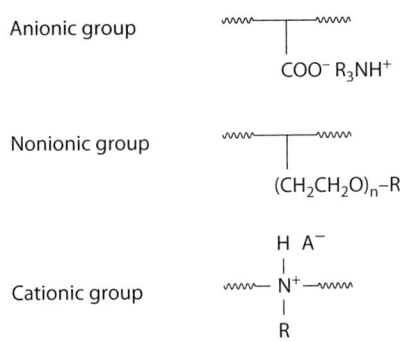

FIGURE 11.24 Common functional groups that confer hydrophilicity in the urethane molecule.

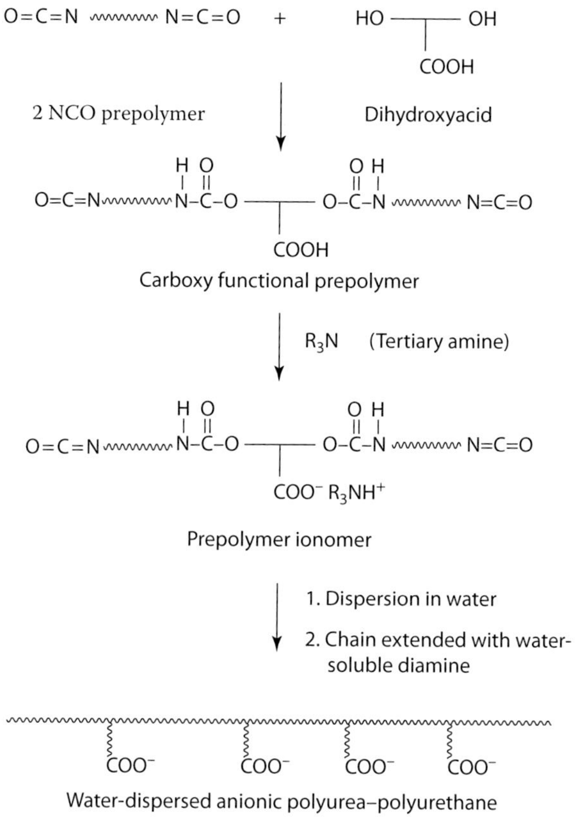

FIGURE 11.25 How functional groups are built into the polymer backbone of urethane.

Efficient mixing of the two components is essential for complete reaction and full development of designed adhesive properties. In-line mixing tubes are adequate for low-volume adhesive systems. For larger-volume demands, sophisticated meter mix machines are required that will mix both components just prior to application. Commercial systems for delivering two-component adhesives are segmented, based on the viscosity ranges of the components. The ranges can be divided into low, middle, and high viscosity.

In present-day high-speed assembly line operations, adhesives are applied robotically. The adhesive bead is applied quickly and evenly to parts on a conveyor line just prior to being fitted. These operations, especially the need to handle the adhered substrates soon after assembly, demand fast-curing adhesive systems. Two-component adhesives are used to bond metals to plastics in automobiles, to laminate panels in the construction industry, to laminate foams to textiles, to laminate plastic films together, and to bond poly(vinylidene chloride) films to wood for furniture. A commercial waterborne two-component adhesive is sold by Ashland under the trademark ISOSET. This system is used for exterior sandwich panels by recreational vehicle manufacturers, and is composed of a water-emulsifiable isocyanate and a hydroxy-functionalized emulsion latex.

11.9 RECENT DEVELOPMENTS

11.9.1 HYBRID ADHESIVES

Over the last four decades, there have been a number of attempts to combine the unique benefits of polyurethane adhesives with those of other adhesive systems. These attempts have led to a variety of urethane hybrids. Early work focused on simple blends; for example, in 1964 Union Carbide blended organic isocyanates with ethylene–vinyl acetate copolymers [56]. These blends were used as an adhesive interlayer in glass laminations, particularly safety glass laminates. Similarly, polylurethane–epoxy blends for safety glass laminates have been reported since 1970 [57].

More recent efforts have focused on developments that create true hybrids. For example, blocked isocyanate prepolymers have been mixed with epoxy resins and cured with amines [58,59]. These blocked prepolymers will react initially with amines to form amine-terminated prepolymers that cross-link the epoxy resin. Several blocked isocyanates are commercially available. The DESMOCAP (Bayer) 11A and 12A products are isocyanates (believed to be blocked with nonylphenol) used as flexibilizing agents for epoxy resins. ANCAREZ (Pacific Anchor) 2150 is a blocked isocyanate–epoxy blend used as an adhesion promoter for vinyl plastisols. A one-package, heat-cured hybrid adhesive has been reported, consisting of isophorone diisocyanate, epoxy resin, and a dispersed solid curative based on the salt of ethylenediamine and bisphenol A [60]. Urethane amines are offered commercially that can be used with epoxy resins to develop hybrid adhesive systems.

Urethane acrylic hybrids have been reported based on several approaches. Pacific Anchor has developed a urethane acrylate that is commercially available (ANCAREZ 300A). Acrylic polyols have been synthesized in the presence of polyether polyols by Saunders for use in two-component structural adhesives with improved tensile and impact strengths [61]. Pressure-sensitive acrylic prepolymers with hydroxyl groups have been formulated with isocyanate prepolymers to obtain adhesives with improved peel strength [62,63]. Aqueous-based vinyl-to-fiberboard adhesives were reported by Chao, using water-dispersible MDI with a functionalized acrylic latex and an aqueous dispersions polyurethane to give improved shear and hot-peel strengths [64]. Acrylonitrile dispersion graft polyether polyols have also been used in two-component structural metal composite (SMC) adhesives [65].

Urethanes have also been used to toughen vinyl-terminated acrylic adhesives for improved impact resistance. Thus, rubber-toughened urethane acrylates [66,67], water-dispersible urethane acrylates [68], and high-temperature performance urethane–acrylate structural adhesives have been reported [69]. Polyurethanes terminated with acrylic functionality are also used for anaerobic or radiation-cured adhesives with improved toughness [70].

11.9.2 Reactive Hot-Melts

Polyurethane reactive hot-melts are 100% solid, hot-melt thermoplastic prepolymers that moisture cure slowly after application. Conventional hot-melts are known for their quick setting, excellent green strength, ease of application, and low toxicity. Their primary limitation is low heat resistance (at elevated temperatures, the adhesive will soften and flow) and poor adhesion to some substrates, due to insufficient wetting. The use of a polyurethane prepolymer with low levels of free isocyanates as a hot-melt offers distinct advantages: High initial green strength is still achieved and, in addition, the isocyanate will moisture cure slowly, converting the thermoplastic adhesive to a thermoset. There are a number of patents on reactive hot-melts [71–73]. The tensile strength of the adhesive increases, heat resistance is improved, and the final cured adhesive will not flow at elevated temperatures. A limitation of this technology is the need for porous substrates or bond designs that will allow the diffusion of moisture into the adhesive, so that moisture curing will occur. The adhesive itself must be protected from moisture prior to use. This technology should be applicable to assembly line operations that require an adhesive that provides high initial green strength.

11.9.3 Pressure-Sensitive Adhesives

The use of polyurethanes in the pressure-sensitive adhesives market has been relatively small. Polyurethanes have been somewhat limited to being used as additives to pressure-sensitive adhesives to improve their cohesive strength. Recent developments in the institutional carpet backing or automotive carpet floor mat markets suggest that pressure-sensitive urethanes can succeed commercially.

11.9.3.1 Biobased Polyurethane Adhesives

The use of biobased vegetable oils for the preparation of polyurethanes by their reaction with polyisocyanates is a long-known practice. Dedicated and extensive reviews on this aspect of polyurethane synthesis already exist [74,75] and this chapter does not pretend to scan the whole literature on such reactions. However, in this abundant literature, mostly dedicated to the preparation of foams and lubricants, the references strictly aimed at the preparation and application of polyurethane adhesives are not too numerous [76–78]. Equally, it must be made clear that most of the developments made in this field for applications other than for adhesives can also be used for adhesives. The reaction is the same as for synthetic polyols (Figure 11.26), with the difference that HO–R–OH is a natural, biosourced polyol that substitutes a synthetic one partially or totally.

The use of polyols of natural origin, such as castor oil, is a well-established practice. This is due to the presence of hydroxyl groups on these materials, groups that are the preferential sites for the reaction with the isocyanate groups to form urethanes. Apart from castor oil and its derivatives, which are relatively expensive, the use of other vegetable oils to react with isocyanates to prepare polyurethanes is increasing due to the demand for environmentally friendly green products. The field of adhesives is no exception to this trend. The adhesive field where these materials have made important inroads is, first of all, that of pressure-sensitive adhesives. For example, epoxidized soybean oil has been cross-linked with dicarboxylic acids such as sebacic, adipic, and other acids, whereas chromium(III) organometallic derivative has been successfully used as the cross-linker [77,78]. Such a pressure-sensitive adhesive had good peel and shear strengths and good resistance to aging (Figure 11.27). The only problem that somehow disqualifies this as fully biosourced was the use of epichloridrin for the preparation of the epoxidized vegetable oil.

The second adhesive application is wood adhesives. Polyester polyols synthesized from potato starch and natural vegetable oils by transesterification were used in the preparation of polyurethane wood adhesives by reaction with toluene diisocyanate. Their performance was reported as being

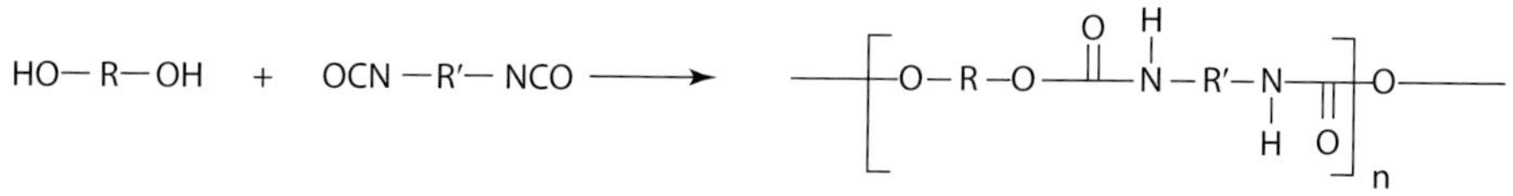

FIGURE 11.26 Schematic reaction of formation of a polyurethane from a polyisocyanate and a polyol.

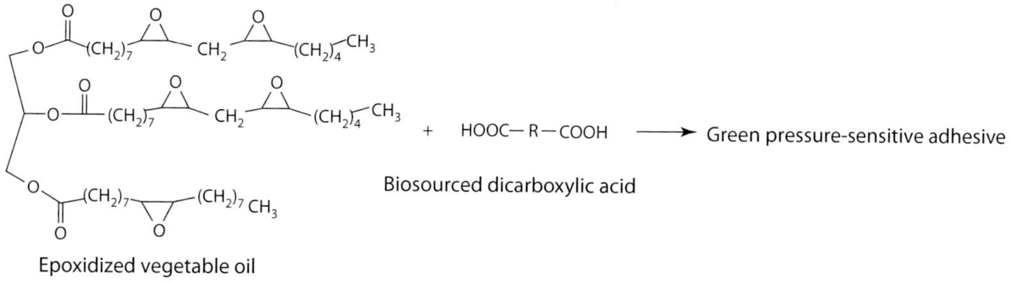

Epoxidized vegetable oil

FIGURE 11.27 Green pressure-sensitive polyurethane adhesive prepared from glycerol and a dicarboxylic acid. Note that epichloridrin is still used to epoxidize the glycerol.

comparable or even superior to commercial, fully synthetic polyurethane adhesives for the same application [79].

As regards the preparation of polyurethane adhesives starting from condensed or hydrolyzable tannins, two approaches have been taken: (i) Modification of the flavonoid tannin to render easier the reaction with isocyanates. This is due to the difficulty in reacting the flavonoid hydroxyl groups directly with isocyanates. In this type of application, the tannins are in direct competition with more suitable natural polyols, of which the literature abounds, to prepare semi-biosourced polyurethanes. (ii) The use of a total no-isocyanate approach to improve the environment-friendly character of such adhesives.

For the first approach, benzoylation to reduce the number of hydroxyl groups of the tannin before reaction with diisocyanate was tried a long time ago [79]. A more studied approach has been to use lignin and lignosulfonate hydroxypropylated to prepare urethanes [80–82], although this approach has been more directed toward coatings than adhesives. The same and more studied approaches have been to prepare novel thermosetting tannin-based polyurethane adhesive resins, using the hydroxypropyl and hydroxybutyl derivatives of purified condensed tannins from *Pinus pinaster* bark and other condensed tannin species by reaction with diisocyanates [83–86]. Hydroxypropyl tannins with a degree of substitution rising stepwise from 1 to 4 were cross-linked with either an aromatic (polymeric methylene diphenyl diisocyanate [pMDI]) or an aliphatic isocyanate (hexamethylene diisocyanate [HDI]) with good results. Hydroxypropylation and hydroxybutylation are one of the approaches used to react polyphenolic materials with isocyanates to obtain polyurethane adhesives and resins [83–86] (Figures 11.28 and 11.29). This also renders easier the reaction of flavonoid tannins with isocyanates, due to the introduction of much more approachable hydroxyl groups into the tannin structure, thus increasing reaction yield.

The same approach, thus to introduce into the tannin structure more available –OH groups but through a reaction totally different from hydroxypropylation, consisted of reacting an aldehyde with the tannin and then using the hydroxyl groups formed by its addition onto the flavonoid structure, before their further condensation with other flavonoids, to react with an isocyanate [87–89] (Figure 11.30). Such a system has been used and is used industrially for wood adhesives [89].

More recently, the system has been adapted by eliminating formaldehyde and substituting it with glyoxal with good results [90], rendering the approach even more environmentally interesting. The reactions involved are shown in the case of glyoxal in Figure 11.31; the first reaction step, namely the reaction of the tannin with the aldehyde, being easier than hydroxyalkylation.

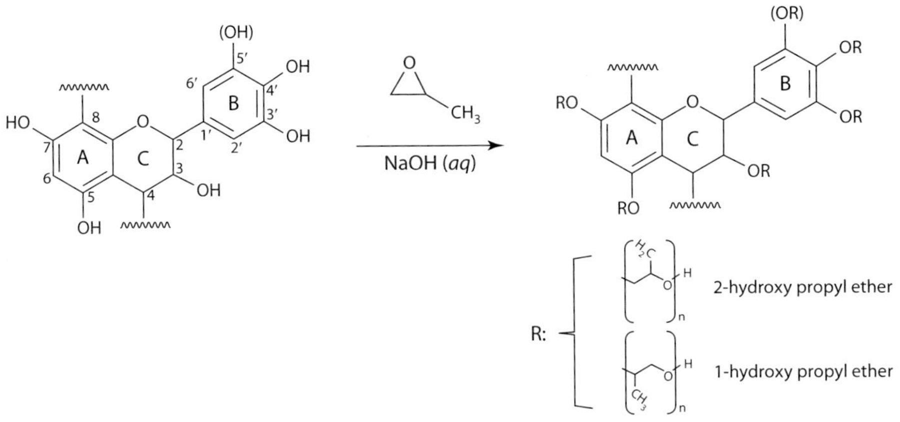

FIGURE 11.28   Reaction of *P. pinaster* bark tannin with propylene oxide to produce hydroxypropyl ethers tannin derivatives (considering full hydroxyl propylation [HP]).

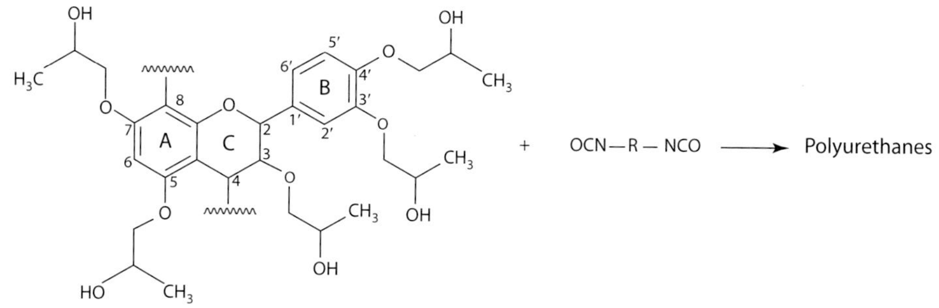

FIGURE 11.29 Formation of polyurethane adhesive and resins by reaction of a hydroxypropylated polyphenolic tannin with an isocyanate.

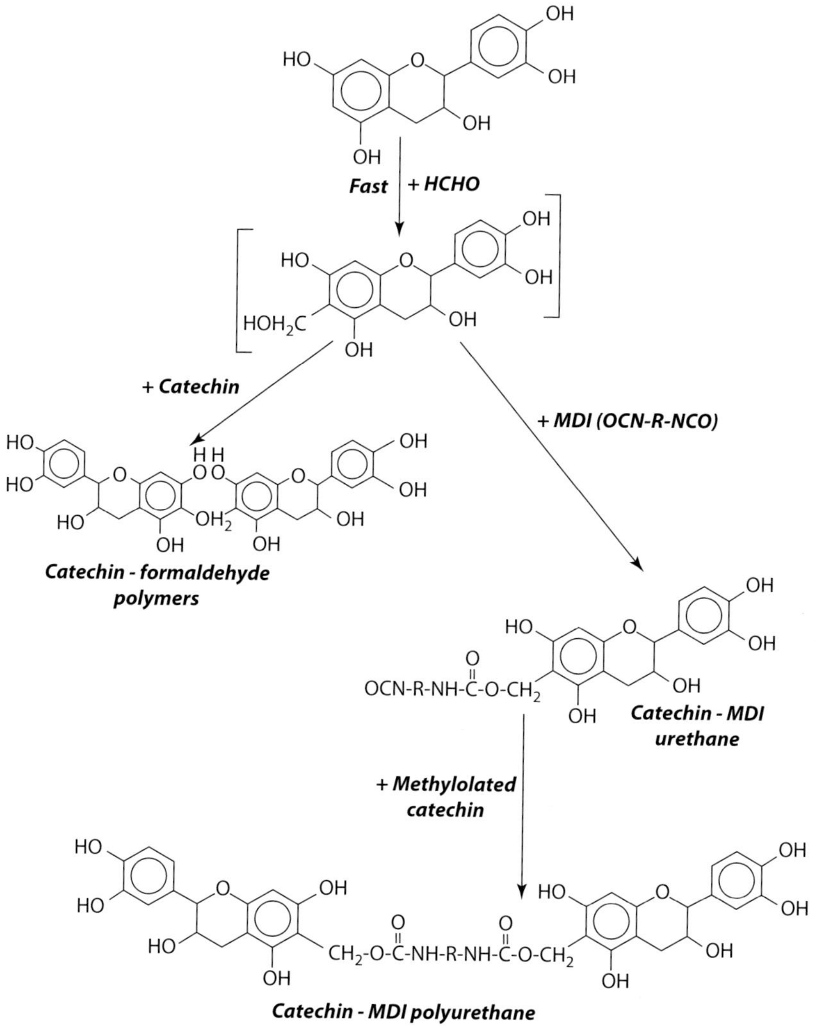

FIGURE 11.30 Reaction of the flavonoid tannin/formaldehyde system with isocyanates to form polyurethanes. Note that the reaction also occurs in water.

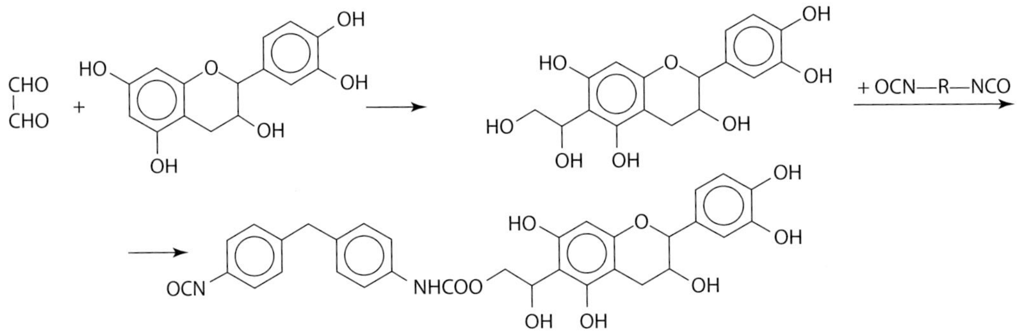

FIGURE 11.31 Formation of polyurethane by reaction of a glyoxalated flavonoid tannin with a polyisocyanate.

However, isocyanates are harmful to human health. Thus, the preparation of polyurethane adhesives of a high level of biosourced and environment-friendly material involves, by necessity, the synthesis of nonisocyanate polyurethanes. This approach, which avoids the use of any isocyanate, is attracting increasing interest. It is based on the polycondensation of diamines with dicyclocarbonates to lead to polyhydroxyurethanes. This reaction has been studied by a few research groups [91–106] and leads to polyhydroxyurethanes of relatively low glass transition temperatures and Mn lower than 30,000 g/mol. All the work on nonisocyanate polyurethanes has concentrated on synthetic materials. Thus, while isocyanates are definitely not used, synthetic diamines and synthetic dicyclocarbonates, both of nonnatural origin, are used, yielding polyurethane resins that are not biosourced. While use of glycerol has been reported [106], the percentage of biosourced material is still lower than in the approach of using an isocyanate reacting with a natural polyol.

This second approach—for example, the more radical elimination of the isocyanate in the preparation of polyurethane resins—is based on the reaction of a double cyclic carbonate with a diamine that has been proposed based exclusively on synthetic, not biosourced materials [91–106] (Figure 11.32).

Only very recent approaches to nonisocyanate polyurethane adhesives have aimed at not only eliminating the toxic isocyanate, but also at improving the percentage of biosourced material composing the green polyurethane adhesive. The same reaction used for synthetic materials has been used with both a hydrolyzable tannin and condensed flavonoid tannins—natural renewable materials—by reacting them first with dimethyl carbonate—thus, a noncyclic carbonate—followed by reaction with hexamethylenediamine [107,108] to obtain nonisocyanate urethanes of the type shown in Figure 11.33.

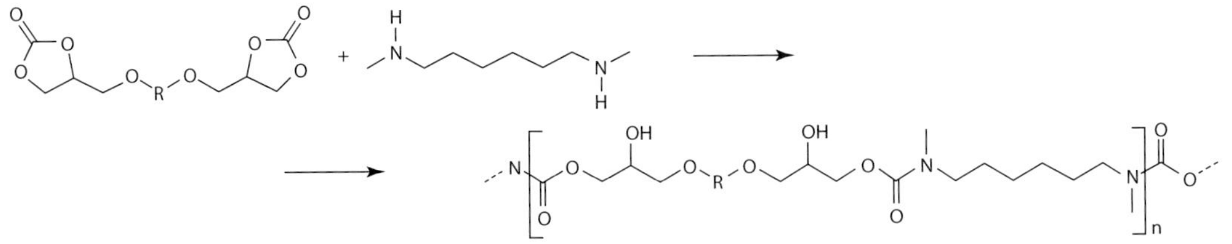

FIGURE 11.32 Nonisocyanate polyurethane formation by reaction of a dicyclic organic carbonate with a diamine.

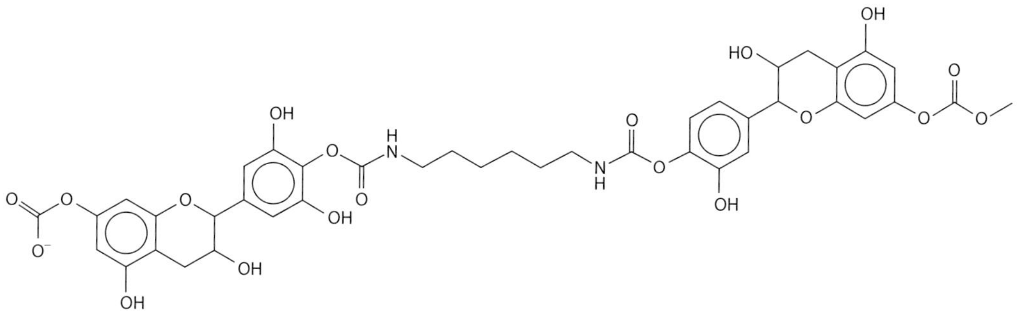

FIGURE 11.33 Nonisocyanate diurethane obtained by reacting a precarbonated flavonoid tannin with a diamine. (From Thebault, M. et al., *Eur. Polym. J.*, 67, 513–523, 2015. With permission.)

This was the first approach to obtain a biosourced polyurethane without using isocyanates, although here, too, as in approach (i), the percentage level of biosourced material was only 45%–50%, the diamine and the carbonate being still synthetic materials. As the amination of condensed tannins—thus, the conversion of their hydroxyl groups into amino groups—is an easy reaction [109], the aminated tannin was used to substitute the hexamethylenediamine, further improving to more than 70% of biosourced material in the final polyurethane, and this without using any isocyanate [110] to obtain flavonoid oligomers linked by urethane bridges, as shown in Figure 11.34.

What was of further interest in this approach was that it was not necessary to use a purified tannin, as the 10%–12% carbohydrate fraction of commercial, industrial tannin extracts was also found to undergo the same two reactions leading to urethane bridges [107,108]. Thus, species in which either flavonoid tannin oligomers were linked by urethane bridges to a carbohydrate monomer when an aminated tannin was used as diamine (Figure 11.35) or where the urethane bridges were formed between the carbonated carbohydrate monomer and a synthetic diamine were also identified, such as shown **in** Figure 11.36.

Such a finding opens new possibilities for the preparation of nonisocyanate urethane adhesives of higher biosourced content.

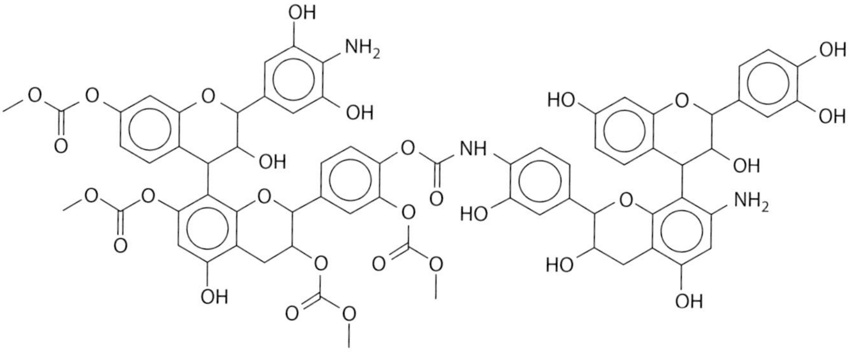

FIGURE 11.34 Nonisocyanate urethane bridge linking a precarbonated flavonoid tannin dimer with an aminated tannin dimer. Note that only the dimethyl carbonate is of synthetic origin, all the rest being renewable biosourced materials. (From Thebault, M. et al., *J. Renewable Mat.*, 51, 2–29, 2017. With permission.)

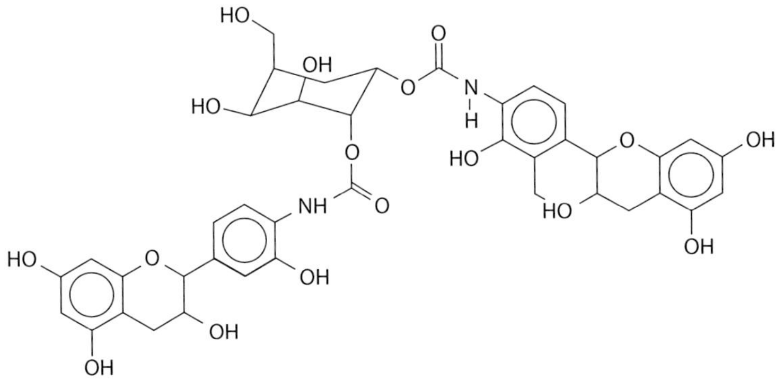

FIGURE 11.35 Nonisocyanate diurethane obtained by reaction of a carbonated carbohydrate monomer with two aminated flavonoid tannin oligomers. (From Thebault, M. et al., *J. Renewable Mat.,* 51, 2–29, 2017. With permission.)

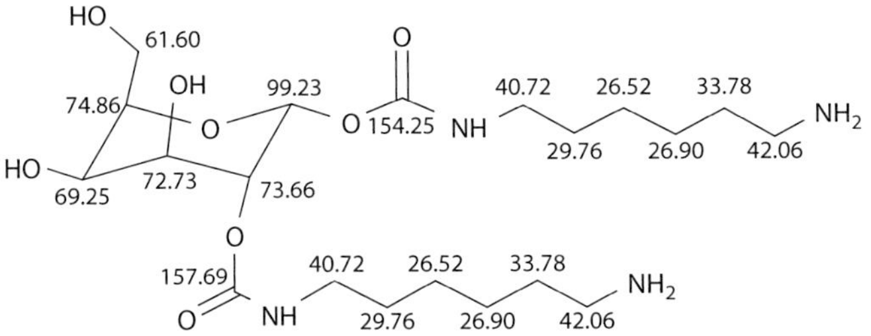

FIGURE 11.36 Nonisocyanate diurethane obtained by reaction of a carbonated carbohydrate monomer with a diamine. The numbers indicate the ^{13}C NMR shifts. (From *Eur. Polym. J.,* 67, 513–523, 2015. With permission.)

11.10 SUMMARY

Polyurethane adhesives, as a class, can no longer be perceived as new raw materials. From a base of 217 million lb., double-digit growth can no longer be expected. Even so, significant growth will continue. Formulators are taking advantage of the tremendous flexibility of urethane chemistry in designing new adhesive products. Specialty niches such as waterbornes and reactive hot-melts, for example, will continue to emerge and fuel growth. Exciting times lie ahead for innovative formulators of polyurethane adhesives!

REFERENCES

1. J. Saunders and K. Frisch. *Polyurethanes: Chemistry and Technology,* Pt. 1, Interscience, New York (1963).
2. J. M. DeBell, W. C. Goggin and W. E. Gloor. *German Plastic Practice,* DeBell and Richardson, Cambridge, MA (1946).
3. B. A. Dombrow. *Polyurethanes,* Reinhold, New York (1957).
4. W. Erwin. Adhesive composition comprising a diisocyanate modified polyester. U.S. patent 2,650,212, assigned to Farbenfabriken-Bayer (1953).
5. W. Y. Clayton. Isocyanate-castor oil metal cement. U.S. patent 2,769,826, assigned to Stoner-Mudge Co. (1956).
6. J. F. Anderson and L. F. Fiedler. Container bonded with a polyesterurethane. U.S. patent 2,801,648, assigned to B.F. Goodrich (1957).

7. C. S. Schollenberger, H. Scott, and G. R. Moore. Polyurethane VC: A virtually cross-linked elastomer. *Rubber World* 13, 549 (1958).

8. C. S. Schollenberger. Simulated vulcanizates of polyurethane elastomers. U.S. patent 2,871,218, assigned to B.F. Goodrich (1959).

9. J. E. Mallone. Stable polyurethane latex and process of making same. U.S. patent 2,968,575, assigned to E.I. du Pont de Nemours and Co. (1961).

10. Experimental urethane latex E-204. Bulletin, Wyandotte Chemicals Corporation (1963).

11. S. P. Suskind. Polyurethane latex. *J. Appl. Polym. Sci.* 9, 2451–2458 (1965).

12. C. S. Schollenberger. Thermoplastics polyesterurethanes. U.S. patent 3,015,650, assigned to B.F. Goodrich (1962).

13. M. E. Kimball. Polyurethane adhesives: Properties and bonding procedures. *Adhes. Age* 24(6), 21–24 (1981).

14. R. Dahl. New polyurethane pressure-sensitive adhesive products and processes. U.S. patent 3,802,988, assigned to Continental Tapes (1974).

15. D. Dieterich and J. N. Rieck. Aqueous polyurethane systems and their possible uses. *Adhes. Age* 21(2), 24–27 (1978).

16. P. E. Cranley. Polyurethane adhesives. In: *Reaction Polymers*, W. F. Gum, W. Riese and H. Ulrich (Eds.), p. 692, Oxford University Press, Hanser Publishers, New York (1992).

17. R. A. Bragole. *Urethanes in Elastomers and Coatings*, p. 136, Technomic, Westport, CT (1973).

18. U. Meyer-Westhuis. *Polyurethanes: Coatings, Adhesives and Sealants*, pp. 20–22, Vincentz Narwork, Hanover, Germany (2003).

19. J. W. Britain and P. G. Gemeinhardt. Catalysis of the isocyanate-hydroxyl reaction. *J. Appl. Polym. Sci.* 4, 207–211 (1960).

20. H. G. Scholten, J. G. Schumann and R. E. TenHoor. Urethane polyether prepolymers and foams. Influence of chemical and physical variables on reaction behavior. *J. Chem. Eng. Data* 5, 395–400 (1960).

21. Preparation of prepolymers. Form 109-008-95-290-SAI, Dow Chemical Co., Freeport, TX (1990).

22. R. Leuckart and M. Schmidt. Ueber die Einwirkung von Phenylcyanat auf Phenole und Phenoläther. *Ber. Dtsch. Chem. Ges.* 18, 2338–2341 (1885).

23. H. E. French and A. F. Wirtel. Alpha-naphthylisocyanate as a reagent for phenols and aliphatic amines. *J. Am. Chem. Soc.* 48, 1736–1739 (1926).

24. J. W. Baker and J. B. Holdsworth. The mechanism of aromatic side-chain reactions with special reference to the polar effects of substituents. Part XIII. Kinetic examination of the reaction of aryl *iso*cyanates with methyl alcohol. *J. Chem. Soc.* 713–726 (1947).

25. Hydroxyl terminated poly BD resins: Functional liquid polymers. Electrical applications brochure, Atochem North America (1990).

26. Z. W. Wicks. New developments in the field of blocked isocyanates. *Progr. Org. Coatings* 3, 73–99 (1975).

27. Z. W. Wicks. New developments in the field of blocked isocyanates. *Progr. Org. Coatings* 9, 3–28 (1981).

28. S. Petersen. Lower molecular reaction products of aliphthalic diisocyanates. *Liebigs Ann. Chem.* 562, 205–207 (1949).

29. J. F. Levy and J. Kucsan. Splitting blocked isocyanates with a carboxylic acid salt of calcium, strontium, magnesium, or barium. U.S. patent 3,705,119; C.A. 78: 59887, assigned to Rhom and Haas (1973).

30. H. L. Elkin. Hydrophilic urethane compositions and process for preparation of moisture containing breathable fabrics. U.S. patent 3,384,506, assigned to Thiokol Chemical Corp. (1968).

31. R. Miller, J. L. Witt and M. L. Tidmore. Adhesion of rayon textile to rubber with aqueous dispersion of blocked isocyanate or isocyanate dimer, rubber latex and resorcinol-formaldehyde resin. U.S. patent 3,707,178, assigned to Uniroyal Inc. (1972).

32. V. K. Rowe and M. A. Wolf. In: *Patty's Industrial Hygiene and Technology,* Engineering control and personal protection. Vol. 2C, F. D. Clayton and F. E. Clayton (Eds.), Wiley, New York (1982).

33. T. D. Landry and C. A. Steffens. In: *Reaction Polymers*, Toxicology overview of isocyanates. W. F. Gum, W. Riese and H. Ulrich (Eds.), 747, Oxford University Press, Hanser Publishers, New York (1992).

34. R. J. Davies. Respiratory hypersensitivity to diisocyanates. *Clin. Immunol. Allerg.* 4(1), 103–124 (1984).

35. M. H. Karol. Respiratory effects of inhaled isocyanates. *CRC Crit. Rev. Toxicol.* 16, 349–379 (1986).

36. D. E. Banks, B. T. Butcher and J. E. Salvaggio. Isocyanate-induced respiratory disease. *Ann. Allergy* 57, 389–396 (1986).

37. A. W. Musk, J. M. Peters and D. H. Wegman, Isocyanates and respiratory disease: Current status. *Am. J. Ind. Med.* 13, 331–349 (1988).

38. H. S. Katz and J. V. Milewski. *Handbook of Fillers and Reinforcements for Plastics*, Van Nostrand Reinhold, New York (1978).

39. A. J. Kinloch. *Adhesion and Adhesives*, pp. 101–170, Chapman & Hall, London (1987).

40. J. J. Martin. In: *Adhesion and Adhesives*, Vol. 2, R. Houwink and G. Salomon (Eds.), Elsevier, New York (1967).

41. R. C. Snogren. *Handbook of Surface Preparation*, Palmerton, New York (1975).

42. Recommended Practice for the Preparation of Surfaces of Plastics Prior to Adhesive Bonding, *ASTM D 2093-69*, ASTM, Philadelphia (1980).

43. D. M. Brewis. *Surface Analysis and Pretreatment of Plastics and Metals*, Applied Science Publishers, London (1982).

44. J. Shields. *Adhesives Handbook*, Butterworth, London (1985).

45. M. Strobel, C. S. Lyons and K. L. Mittal (Eds.), *Plasma Surface Modification: Relevance to Adhesion*, CRC Press, Boca Raton, FL (1994).

46. M. Thomas and K. L. Mittal (Eds.). *Atmospheric Pressure Plasma Treatment of Polymers: Relevance to Adhesion*, Wiley-Scrivener, Beverly, MA (2013).

47. K. L. Mittal and T. Bahners (Eds.). *Laser Surface Modification and Adhesion*, Wiley-Scrivener, Beverly, MA (2015).

48. D. G. Lay and P. Cranley. Polyurethane adhesives. In: *Handbook of Adhesive Technology*, A. Pizzi and K. L. Mittal (Eds.), pp. 405–429, CRC Press, Boca Raton, FL (1994).

49. L. D. George. Composition. U.S. patent 2,277,083, assigned to Du Pont (1942).

50. A. F. Lewis, L. M. Zaccardo and A. M. Schiller. Polyurethane based adhesive systems and laminates prepared therewith. U.S. patent 3,391,054, assigned to American Cyanamid Company (1968).

51. M. Huang and E. R. Pohl. Organofunctional silanes for sealants. In: *Handbook of Sealant Technology*, K. L. Mittal and A. Pizzi (Eds.), pp. 27–49, CRC Press, Boca Raton, FL (2009).

52. G. Oertel. *Polyurethane Handbook*, p. 554, Carl Hanser, Munich (1985).

53. L. Maempel, Wässrige PUR-Dispersionen. *Adhäsion* 32(5), 14–18 (1988).

54. J. M. Gaul, T. Nguyen and J. S. Babiec, Jr. Novel isocyanate binder system for composite wood panels. *J. Elastomers Plastics* 16, 206–210 (1984).

55. D. Dieterich. Aqueous emulsions, dispersions and solutions of polyurethanes: Synthesis and properties. *Progr. Org. Coatings* 9, 281–340 (1981).

56. B. O. Baum. Safety glass. U.S. patent 3,157,563, assigned to Union Carbide Corporation (1964).

57. V. E. Hamilton and L. M. Roseland. Adhesive and glass laminate bonded therewith. U.S. patent 3,546,064, assigned to McDonnell Douglas Corporation (1970).

58. R. Grieves and K. G. M. Pratley. Two-component adhesive or sealing composition. U.S. patent 4,623,702, assigned to Pratley Invest. Ltd. (1986).

59. V. C. Markevka, S. J. Mackey and W. L. Bunnelle. Thermally stable hot melt moisture-cure polyurethane adhesive composition. U.S. patent 4,775,719, assigned to H.B. Fuller Co. (1988).

60. A. B. Goel. Single component, latent curing epoxy resin composition. U.S. patent 4,737,565, assigned to Ashland Oil (1988).

61. F. Saunders. 4,4′-Biphenylene diisocyanate and other diisocyanates. U.S. patent 4,731,416, assigned to Dow Chemical Co. (1988).

62. W. E. DeVry, R. S. Drake and R. T. Morissey, Polyurethane pressure-sensitive adhesive and laminates therefrom. U.S. patent 4,145,514, assigned to B.F. Goodrich (1979).

63. Y.-S. Lee. Pressure-sensitive adhesive. U.S. patent 4,214,061, assigned to B.F. Goodrich (1980).

64. Y.-Y. H. Chao. Process for preparing modified polymer emulsion adhesives. U.S. patent 4,636,546, assigned to Rohm and Haas (1987).

65. T. E. Gismond and D. J. Damico. Structural adhesive compositions. U.S. patent 4,742,113, assigned to Lord Corp. (1988).

66. R. F. Schappert, J. M. Makhlouf and M. M. Chau. Polyurea-polyurethane acrylate dispersions. U.S. patent 4,721,751, assigned to PPG Industries (1988).

67. J. W. Saracsan. Peroxide cured urethanes for application to rim and adhesives. U.S. patent 4,452,964, assigned to Ashland Co. (1984).

68. A. Trovati. Aqueous dispersions of polyurethanes from oligo urethanes having unsaturated terminal groups. U.S. patent 4,497,932, assigned to Resem (1985).

69. T. Dawdy. Structural adhesive formulation. U.S. patent 4,452,944, assigned to Lord (1984).

70. L. J. Baccei. Unsaturated curable poly(alkylene)ether polyol-based resins having improved properties. U.S. patent 4,309,526, assigned to Loctite Co. (1982).

71. F. Reischle, J. Windhoff and W. Bernig, Fusion adhesive which can comprise an isocyanate prepolymer, a thermoplastic polymer and/or a lower molecular weight ketone resin. U.S. patent 4,585,819, assigned to H.B. Fuller Co. (1986).

72. V. C. Markevaka, J. M. Zimmel, E. R. Messman and W. L. Bunnelle. Thermally stable reactive hot melt urethane adhesive composition having a thermoplastic polymer, a compatible, curing urethane poly-alkylene polyol prepolymer and a tackifying agent. U.S. patent 4,820,368, assigned to H.B. Fuller Co. (1989).

73. H. Gilch, H. von Voithenberg and K.-H. Albert. Adhesive compositions. U.S. patent 4,618,651, assigned to USM Co. (1986).

74. M. Desroches, M. Escouvois, R. Auvergne, S. Caillol and B. Boutevin. From vegetable oils to polyure-thanes: Synthetic routes to polyols and main industrial products. *Polymer Rev.* 52(1), 38–134 (2012).

75. B. Nohra, L. Candy, J.-F. Blanco, C. Guerin, Y. Raoul and Z. Mouloungui. From petrochemical polyure-thanes to biobased polyhydroxyurethanes, a review. *Macromolecules* 46, 3771–3792 (2013).

76. S. D. Desai, J. V. Patel and K. Sinha. Polyurethane adhesive system from biomaterial-based polyol for bonding wood. *Int. J. Adhesion Adhesives* 23, 393–399 (2003).

77. A. Li and K. Li. Pressure-sensitive adhesives based on epoxidized soybean oil and dicarboxylic acids. *ACS Sustain Chem. Eng.* 2, 2090–2096 (2014).

78. B. Kolbe Ahn, S. Kraft, D. Wang and S. Sun. Thermally stable, transparent, pressure-sensitive adhesives from epoxidized and dihydroxyl soybean oil. *Biomacromolecules* 12, 1839–1843 (2011).

79. A. Pizzi, Tannin-based polyurethane adhesives. *J. Appl. Polym. Sci.* 23, 1889–1890 (1979).

80. W. G. Glasser, C. A. Barnett, T. G. Rials and S. Vasudev. Engineering plastics from lignin II. Characterization of hydroxyalkyl lignin derivatives. *J. Appl. Polym. Sci.* 29, 1815–1830 (1984).

81. W. G. Glasser and R. H. Leitheiser. Engineering plastics from lignin XI. Hydroxypropyl lignins as components of fire resistant foams. *Polym. Bull.* 12, 1–5 (1984).

82. W. G. Glasser, S. S. Kelley, T. G. Rials and S. L. Ciemniecky. Structure-property relationships of engi-neering plastics from lignin. Proc. 1986 TAPPI Research and Development Conference, 157–161 (1986).

83. D. Garcia, W. Glasser, A. Pizzi, A. Osorio-Madrazo and M.-P. Laborie. Hydroxypropyl tannin deriva-tives from *Pinus pinaster* (Ait.) bark. *Ind. Crops Prod.* 49, 730–739 (2013).

84. D. E. Garcia, W. G. Glasser, A. Pizzi, S. Paczkowski and M.-P. Laborie. Substitution pattern elucida-tion of hydroxypropyl *Pinus pinaster* (Ait.) bark polyflavonoids derivatives by ESI(-)-MS/MS. *J. Mass Spectrom.* 49, 1050–1058 (2014).

85. D. E. Garcia, W. G. Glasser, A. Pizzi, S. Paczkowski and M.-P. Laborie, Hydroxypropyl tannin from *Pinus pinaster* bark as polyol source in urethane chemistry. *Eur. Polym. J.* 67, 152–165 (2015).

86. D. E. Garcia, W. G. Glasser, A. Pizzi, S. Paczkowski and M.-P. Laborie, Modification of condensed tannins: From polyphenols chemistry to materials engineering. *New J. Chem.* in press, DOI:10.1039/C5NJ02131F 40, 36–49 (2016).

87. A. Pizzi and T. Walton. Non-emulsifiable, water-based diisocyanate adhesives for exterior plywood, Part 1: Novel reaction mechanisms and their chemical evidence. *Holzforschung* 46, 541–547 (1992).

88. A. Pizzi, E. P. von Leyser, J. Valenzuela and J. G. Clark. The chemistry and development of pine tannin adhesives for exterior particleboard. *Holzforschung* 47, 164–172 (1993).

89. J. Valenzuela, E. von Leyser, A. Pizzi, C. Westermeyer and B. Gorrini. Industrial production of pine tannin-bonded particleboard and MDF. *Eur. J. Wood Prod.* 70, 735–740 (2012).

90. M. C. Basso, A. Pizzi, C. Lacoste, L. Delmotte, F. A. Al-Marzouki, S. Abdalla and A. Celzard, Tannin-furanic-polyurethane foams for industrial continuous plant lines. *Polymers* 6, 2985–3004 (2014).

91. G. Rokicki and A. Piotrowska. A new route to polyurethanes from ethylene carbonate, diamines and diols. *Polymer* 43, 2927–2935 (2002).

92. J. M. Whelan, Jr., M. Hill and R.J. Cotter. Multiple cyclic carbonate polymers. U.S. patent 3,072,613 (1963).

93. N. Kihara and T. Endo. Synthesis and properties of poly(hydroxyurethane)s. *J. Polym. Sci., Part A: Polym. Chem.* 31, 2765–2773 (1993).

94. N. Kihara, Y. Kushida and T. Endo. Optically active poly(hydroxyurethane)s derived from cyclic car-bonate and L-lysine derivatives. *J. Polym. Sci., Part A: Polym. Chem.* 34, 2173–2179 (1996).

95. H. Tomita, F. Sanda and T. Endo. Structural analysis of polyhydroxyurethane obtained by polyaddition of bifunctional five-membered cyclic carbonate and diamine based on the model reaction. *J. Polym. Sci., Part A: Polym. Chem.* 39, 851–859 (2001).

96. H. Tomita, F. Sanda and T. Endo. Polyaddition behavior of bis(five- and six-membered cyclic carbonate) s with diamines. *J. Polym. Sci., Part A: Polym. Chem.* 39, 860–867 (2001).

97. H. Tomita, F. Sanda and T. Endo. Model reaction for the synthesis of polyhydroxyurethanes from cyclic carbonates with amines: Substituent effect on the reactivity and selectivity of ring-opening direction in the reaction of five-membered cyclic carbonates with amine. *J. Polym. Sci., Part A: Polym. Chem.* 39, 3678–3685 (2001).

98. O. Birukov, R. Potashnikova, A. Leykin, O. Figovsky and L. Shapovalov. Advantages in chemistry and technology of non-isocyanate polyurethane. *J. Sci. Israel-Technol. Adv.* 11, 160–167 (2009).

99. O. Figovsky and L. Shapovalov. Features of reaction amino-cyclocarbonate for production of new type polyurethanes. *Macromol. Symp.* 187, 325–332 (2002).

100. F. Camara, S. Benyahya, V. Besse, G. Boutevin, R. Auvergne, B. Boutevin and S. Caillol. Reactivity of secondary amines for the synthesis of nonisocyanate polyurethanes. *Eur. Polym. J.* 55, 17–26 (2014).

101. H. Blattmann, M. Fleischer, M. Bähr and R. Mülhaupt. Isocyanate- and phosgene-free routes to polyfunctional cyclic carbonates and green polyurethanes by fixation of carbon dioxide. *Macromol. Rapid Commun.* 35, 1238–1254 (2014).

102. A. Boyer, E. Cloutet, T. Tassaing, B. Gadenne, C. Alfos and H. Cramail. Solubility in CO_2 and carbonation studies of epoxidized fatty acid diesters: Towards novel precursors for polyurethane synthesis. *Green Chem.* 12, 2205–2213 (2010).

103. M.-R. Kim, H.-S. Kim, C.-S. Ha, D.-W. Park and J.-K. Lee. Syntheses and thermal properties of poly(hydroxy)urethanes by polyaddition reaction of bis(cyclic carbonate) and diamines. *J. Appl. Polym. Sci.* 81, 2735–2743 (2001).

104. B. Ochiai, S. Inoue and T. Endo. Salt effect on polyaddition of bifunctional cyclic carbonate and diamine. *J. Polym. Sci., Part A: Polym. Chem.* 43, 6282–6286 (2005).

105. L. Ubaghs, N. Fricke, H. Keul and H. Höcker. Polyurethanes with pendant hydroxyl groups: Synthesis and characterization. *Macromol. Rapid Commun.* 25, 517–521 (2004).

106. M. Fleischer, H. Blattmann and R. Mülhaupt. Glycerol-, pentaerythritol- and trimethylolpropane-based polyurethanes and their cellulose carbonate composites prepared *via* the non-isocyanate route with catalytic carbon dioxide fixation. *Green Chem.* 15, 934–942 (2013).

107. M. Thebault, A. Pizzi, S. Dumarcay, P. Gerardin, E. Fredon and L. Delmotte. Polyurethanes from hydrolysable tannins obtained without using isocyanates. *Ind. Crops Prod.* 59, 329–336 (2014).

108. M. Thebault, A. Pizzi, H. Essawy, A. Baroum and G. Van Assche. Isocyanate free condensed tannin-based polyurethanes. *Eur. Polym. J.* 67, 513–523 (2015).

109. F. Braghiroli, V. Fierro, A. Celzard, A. Pizzi, K. Rode, W. Radke, L. Delmotte and J. Parmentier. Condensation reaction of flavonoid tannins with ammonia. *Ind. Crops Prod.* 44, 330–335 (2013).

110. M. Thebault, A. Pizzi, F.M. Al-Marzouki, S. Abdalla and S. Al-Ameer. Isocyanate-free polyurethanes by coreaction of condensed tannins with aminated tannins. *J. Renewable Mat.* 51, 2–29 (2017).

12 Reactive Acrylic Adhesives

Emmanuel Pitia* and John Hill

CONTENTS

* Indicates corresponding author

12.1 INTRODUCTION

Reactive acrylic adhesives are high-strength, flexible, and environmentally durable structural adhesives. They are also known as *toughened acrylics* or *second-generation acrylics*. They cure rapidly via radical polymerization at room temperature to create a highly cross-linked structure. In general, they are less sensitive to surface preparation and bond to a variety of materials.

Reactive acrylic adhesives are different from anaerobics, cyanoacrylates, acrylic solutions, and emulsion adhesives. These related chemistries use different formulating materials, cure via different curing mechanisms, and often have inferior durability when exposed to aggressive environments [1–7]. Reactive acrylics use proprietary initiating systems to initiate the radical curing of the adhesive.

12.1.1 History

Although a significant amount of the most recent work in developing high-performing acrylic adhesives has been performed in the United States, they were first developed in Germany in the late 1960s [1,4,6,8–10]. The initial formulations found use in bonding aluminum for assembling windows and doors. The earliest generation of acrylic adhesives consisted of low molecular weight poly(methyl methacrylate) dissolved in methyl methacrylate (MMA) monomer. These polymer solutions were then toughened with elastomers, and the adhesive systems were typically cured with peroxide. Since those early days, significant advances have been made in a number of areas: unprepared metal adhesion, environmental durability, high- and low-temperature strength, fatigue resistance, impact (crash) toughness, low odor, low flammability, and high elongation for flexible applications.

12.1.2 Applications

Reactive acrylic adhesives' unique ability to rapidly bond to a variety of unprepared substrates has led to their acceptance as a significant component in assembly of electronics, automotive, boat, signs, and wind and solar energy equipment. In automotive original equipment manufacturing (OEM), nonflammable acrylic adhesives bond to unprepared and oily metals without heat cure, which is a significant advantage. They shorten cycle times, reduce energy consumption, improve environmental resistance, and typically eliminate the need for any welding. Eliminating welding further improves corrosion performance and the cosmetic appearance of finished panels.

In addition, unlike welding, acrylic adhesives are able to bond dissimilar metals, facilitating the increasing use of aluminum, magnesium, and other lightweight alloys in vehicle manufacturing.

In aftermarket vehicle repair, where the heat cure available to OEM manufacturers is not typically an option, metal bonding acrylic adhesives are again widely used in repair assembly of trucks, buses, trains, and trailers. Specifically, these are used in applications where metals are welded, riveted, or screwed together.

In more general product assembly, reactive acrylic adhesives are formulated for high-speed microelectronics assembly, bonding of painted metals and plastics panels, architectural cladding, and in the assembly of large composite structures such as boats and wind turbine blades that require long open times and an adhesive that can fill large gaps between surfaces.

Some areas that are not ideally suited for acrylic adhesives are wood bonding, as certain types of woods contain compounds that can inhibit cure, and there can be issues with wicking of monomers into the wood; and certain polymers such as acrylic sheets need to be evaluated for crazing that may weaken them or be detrimental to their cosmetic appearance. Finally, as with most types of adhesives, primary load-bearing structures are not typically bonded without very careful design considerations that minimize forces on the adhesive bondline [11,12].

TABLE 12.1

Performance and Processing Features of Epoxies, Urethanes, and Reactive Acrylic Structural Adhesives [11–13]

	Epoxies	Urethanes	Reactive Acrylics
	Performance Features		
Benefits	Wide formulation range, high strength	Excellent flexibility, toughness	Ease of processing, good impact resistance
Limitations	Mixing required	Sensitive to moisture	Odor/flammable
Temperature range	−40°C to 204°C	−30°C to 121°C	−40°C to 204°C
Fluid resistance	Excellent	Good	Very good
Adhesion to metals	Excellent	Good	Excellent
Adhesion to plastics	Fair	Very good	Excellent
Adhesion to glass	Excellent	Good	Very good
Adhesion to rubber	Fair	Good	Poor
Tensile strength	High	Medium	High
Elongation/flexibility	Low	High	Medium
Cost	Medium	Low	High
	Processing Features		
Number of components	1 or 2	2	1 or 2
Cure temperature	Room temperature/heat	Room temperature/heat	Room temperature/heat
Work time[a] (min)	5–180	4–120	1–60
Handling time	2–12 h	0.5–24 h	2–60 min
Max gap fill (mm)	3	3	12.7
Dispensing/mixing equipment required	Yes	Yes	Yes, with exception of UV cured systems

[a] Time available to reposition the substrates without affecting the final bond strength.

12.1.3 Comparison with Epoxy and Urethane Adhesives

Reactive acrylic adhesives compete with conventional structural adhesives such as epoxies and urethanes. Over the years, reactive acrylic adhesives have been gaining acceptance and share of the structural adhesive market due to their unique application and process advantages, combined with excellent performance [1,9,13].

One main difference between reactive acrylics and epoxy or urethane adhesives is the ability to bond to various metals with minimum surface preparation at ambient conditions, including the ability to bond through most cutting oils. This can provide a significant advantage in minimizing adhesive processing time and complexity. Another key difference is the curing mechanism, where rapid radical curing is unique to reactive acrylics. This curing mechanism allows for a wide formulation range, a wide range of processing variables, and excellent performance. Table 12.1 shows the generalized performance and processing features of reactive acrylic, epoxy, and urethane structural adhesives.

12.2 CURE SYSTEMS AND FORMULATION

12.2.1 Cure Kinetics (Free Radical Polymerization)

Reactive acrylic adhesives generally consist of two parts. The *A-side* contains unsaturated monomers and polymers such as MMA monomers and poly(MMA), while the other part, the *B-side*, contains a radical generating material such as benzoyl peroxide. Once in contact, the system can

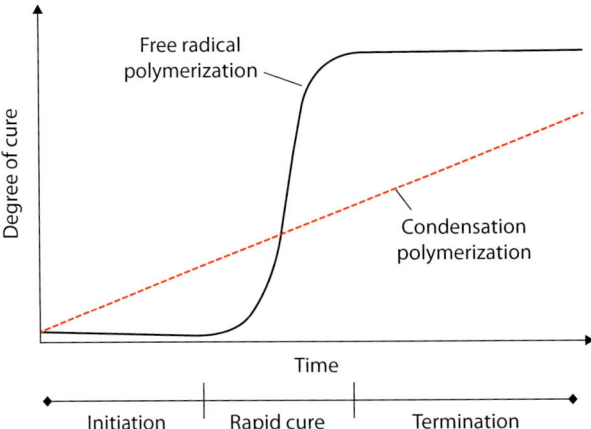

FIGURE 12.1 Cure profile of free radical versus condensation polymerization. Reactive acrylic adhesives cure via free radical polymerization, while urethane and epoxy adhesives cure via condensation polymerization. The free radical polymerization consists of initiation, rapid cure, and termination stage.

cure rapidly via free radical polymerization of the unsaturated mixture of monomers and polymers. This cure mechanism is significantly more rapid than condensation polymerization found in epoxy and urethane adhesives, as shown in Figure 12.1 [14].

Immediately after the two-part adhesive is mixed, radicals are generated during the initiation stage, where little polymerization takes place. The presence of inhibitors can also delay propagation of radicals. The delay in the early stage of the reaction allows the end user to position and reposition parts for longer periods after the adhesive is mixed. This can be an advantage for some end users who are concerned with assembly of critical components [1,3,10].

In the propagation stage, sufficient radicals are present, and thus the adhesive rapidly polymerizes to near full cure, characterized by a high-strength network structure. The rapid cure allows end users to readily control the assembly process, predicting consistently when parts are securely bonded and finished goods are suitable for shipment. Toward the end of the reaction, radicals are terminated in a slow process until full cure is achieved.

12.2.2 Cure Initiating Chemistry

As mentioned in Section 12.2.1, free radical generation is a critical step in the process of curing reactive acrylic adhesives. A typical route for generating free radicals is shown in Figure 12.2. This method of initiating the cure of acrylic adhesives is known as a redox reaction [14,15]. This redox reaction for acrylics involves the reaction between an aromatic amine and a peroxide. The amine is present in the component of the adhesive containing the unsaturated monomer–polymer mixture (A-side), while the peroxide is present in the B-side. The B-side is typically formulated with fumed silica as a rheological aid to help in mixing with the A-side. The reaction of amine and peroxide results in the formation of free radicals that initiate the curing of reactive acrylic adhesives (Figure 12.2). The peroxide is formally a free radical initiator, and the amine accelerates decomposition of the peroxide and, therefore, also accelerates the adhesive cure. However, in common nomenclature, the peroxide contained in the B-side is usually referred to as the *accelerator* and the amine as an *initiator*.

The curing reaction will generally continue as long as there are unreacted monomer molecules present and there are no chemical or other factors present that could interfere with the reaction's completion. With some formulations, certain surfaces and contaminants can interfere with the reaction, resulting in incomplete cure of the adhesive and low bond strength. Surfaces and contaminants to avoid are usually mentioned in the manufacturer's literature. Without unwanted interfering reactions, the curing reaction will ultimately lead to the formation of a high molecular weight polymer useful as an adhesive.

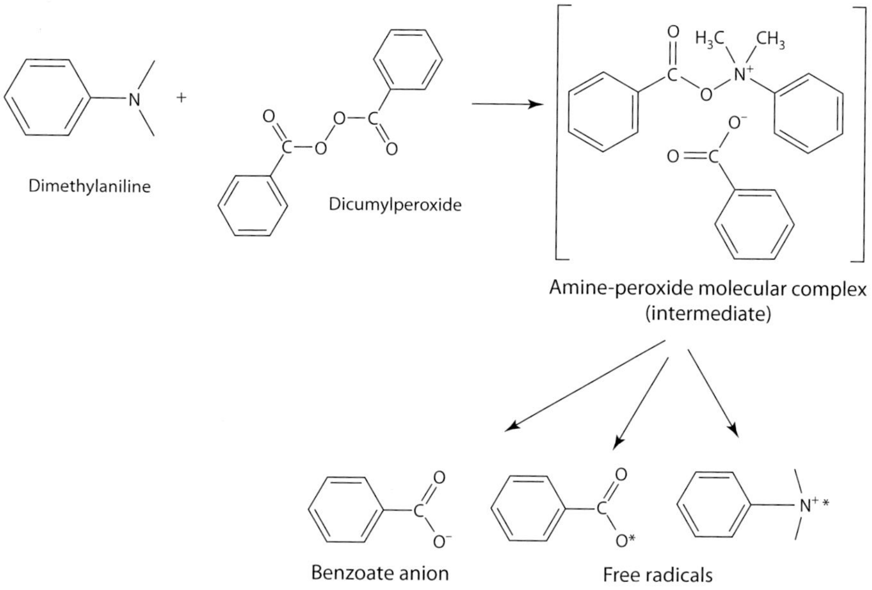

FIGURE 12.2   Redox reactions for reactive acrylics [14,15].

It should be noted that in addition to the use of amines and peroxides, there have been other initiating reactions reported as methods to generate free radicals and cure acrylic adhesives. One example is shown in Figure 12.3 [16]. Another important example, used in relatively new commercialized acrylic adhesives that bond particularly well to unprepared polyolefin substrates, is shown in Figure 12.4. This latter initiator system utilizes an aminoborane compound [17–19]. The organoborane–amine complex is stable at ambient conditions. However, once mixed with a decomplexer such as a carboxylic acid, the aminoborane decouples and reacts quickly in the presence of oxygen to generate radicals.

As seen in Figures 12.2 through 12.4, organic compounds used to initiate the reaction can vary considerably; however, they all result in the formation of critically important free radicals. The complex details of properly initiating and appropriately controlling the chain reactions in reactive acrylic adhesives are highly proprietary and are kept secret by adhesive suppliers. However, when accomplished skillfully, the result is a well-controlled and predictable curing mechanism that leads to a tough, high-strength, and useful adhesive.

In addition to the mentioned examples, a literature search will reveal numerous other mechanisms to initiate a free radical cure [14]. The availability of a large number of formulating tools and choice of initiators allows formulators of reactive acrylic adhesives to develop systems with a wide spectrum of cure rates. This provides manufacturing and design engineers the freedom and flexibility to choose systems that best fit their assembly process and end-use conditions.

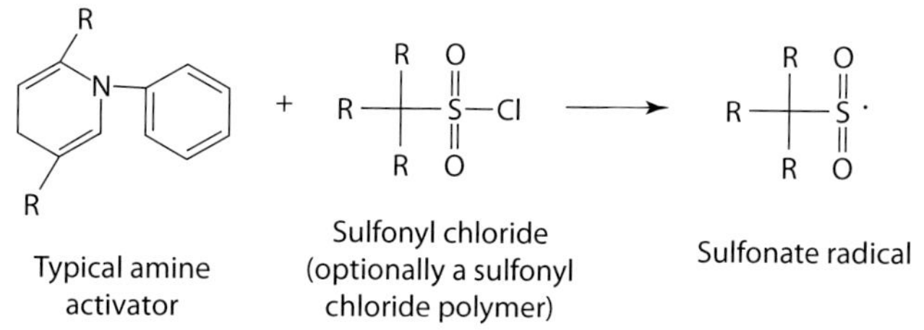

FIGURE 12.3   Amine activator and sulfonyl chloride initiation reaction.

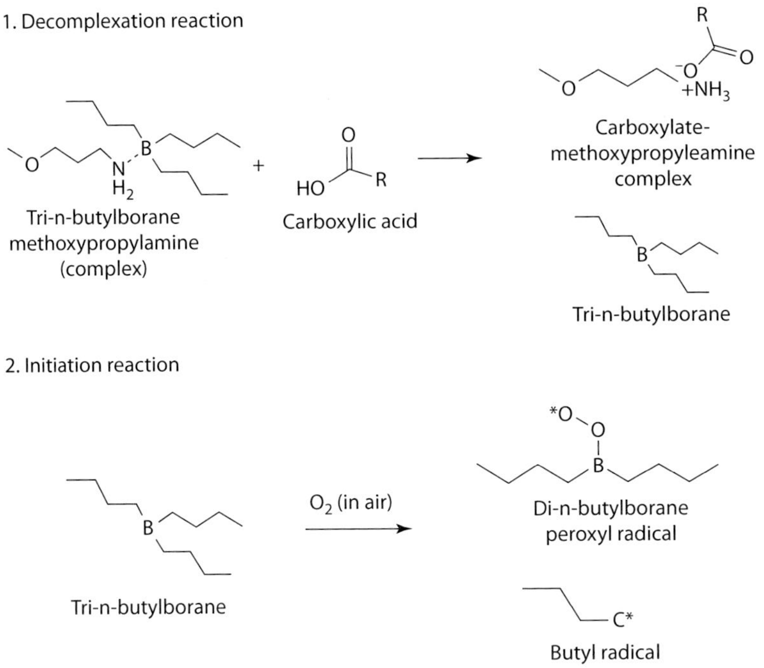

1. Decomplexation reaction

Tri-n-butylborane
methoxypropylamine
(complex)

Carboxylic acid

Carboxylate-
methoxypropyleamine
complex

Tri-n-butylborane

2. Initiation reaction

Tri-n-butylborane

O_2 (in air)

Di-n-butylborane
peroxyl radical

Butyl radical

FIGURE 12.4 Organoborane–amine complex initiation reactions. The organoborane–amine complex decouples in the presence of acid followed by reaction of tributlyborane with oxygen to generate radicals.

12.2.3 COMPONENTS AND FUNCTIONS

Reactive acrylic adhesives consist primarily of acrylic monomers and toughening polymers. They also contain initiators, accelerators, and various other organic and inorganic components designed to achieve desired handling properties, flexibility, specific adhesion, and shelf stability. A summary of the materials in reactive acrylic adhesives and their main functions are shown in Table 12.2.

12.2.3.1 Monomers

As mentioned in Section 12.1.1, MMA is the first and currently most prevalent type of monomer used in reactive acrylic adhesives. The main function of MMA is to compatibilize components and provide a cured network structure. MMA is, in general, an excellent solvent, having the ability to dissolve many polymers and other organic components in the formulation. This solvating ability also gives reactive acrylic adhesives a greater ability to bond plastics and unprepared metal substrates [1,4,9].

Although MMA offers excellent performance in acrylic adhesives at a minimum cost, its odor and flammability can be a limitation to some end users [1,9,20]. Figure 12.5 shows the chemical structures of MMA and other related monomers that are typically used in reactive acrylic adhesives.

These monomers vary in volatility, with the lower molecular weight members of this family having high odor and low flash point, while higher molecular weight members have lower odor and higher flash point, and some can impart greater flexibility in the finished adhesives. However, the higher molecular weight members are, without exception, more expensive. They also tend to exhibit reduced adhesion to polar substrates and reduced bond durability at higher temperatures, depending on molecular structure. As an example, tetrahydrofurfuryl methacrylate (THFMA) is one common alternative to MMA. Adhesive formulators have incorporated THFMA into commercially successful products with acceptable odor and flammability for use in the automotive industry [10,20–22].

TABLE 12.2

Components of Reactive Acrylic Adhesives and Their Effects on Performance and Handling Properties

Component	Main Functions	Possible Disadvantages
Acrylic monomers/ cross-linkers	Dissolve polymer and substrate contaminants	Odor
	Cross-link to high-strength network	Flammable
	Provide fluid and temperature resistance	Brittle
		Reduced shelf life
Polymer	Provide toughness and flexibility	Increased viscosity (poor application on surface)
		Reduced fluid and temperature resistance
Filler	Reduce cost	Reduced toughness
		Lower adhesion
Initiator system	Control cure rate and profile (work time, cure time, handling time)	Reduced shelf life
Stabilizers	Improve shelf life	Reduced cure rate and delayed profile
Adhesion promoters	Improve adhesion to various substrates	Shortened work and handling times
		Reduced shelf life
Pigments/colorants	Improve quality (help to distinguish A-side from B-side)	
Rheology modifiers	Improve processing characteristics (e.g., spray coverage, settling, viscosity)	Lower adhesion
		Reduced toughness

12.2.3.2 Polymers

Polymers are incorporated into reactive acrylic adhesives to improve their low-temperature toughness, flexibility, and adhesion strength. This is particularly important in systems based primarily on MMA. Polymerized methyl methacrylate (PMMA) is a hard, brittle material with glass transition ~110°C. Hence, PMMA by itself is not very useful as an adhesive. As a result, adhesive formulators develop unique, proprietary ways to add flexible polymers to acrylic monomers. Polymers used in

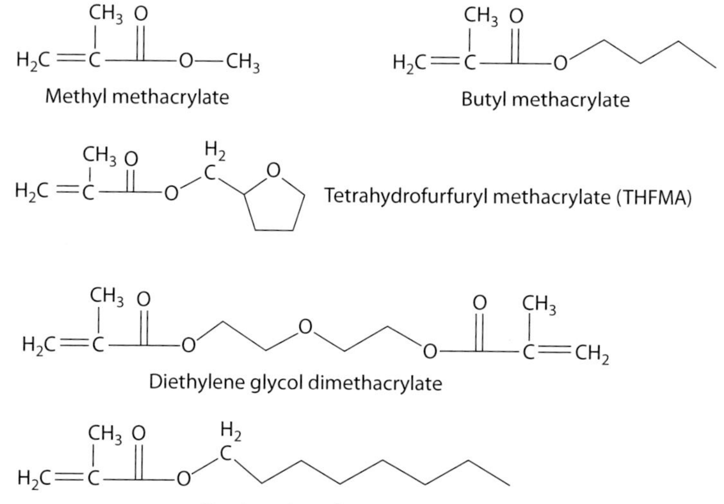

FIGURE 12.5 Reactive acrylic monomers used in reactive acrylic adhesives.

reactive acrylic adhesives include polyacrylonitrile, poly(styrene–butadiene), polychloroprene, and chlorosulfonated polyethylene. In recent years, acrylic formulations have been developed with combinations of multiple monomers and rubber tougheners that can deliver cured adhesives with tensile elongations as high as 100%, or even higher, without a substantial sacrifice in tensile strength.

12.2.3.3 Stabilizers and Other Additives

Stabilizers are used in reactive acrylic adhesives to stop unwanted side reactions and assure good shelf life. This is critical, because of the tendency of acrylic monomers to polymerize spontaneously and quickly. Stabilizers also play a role in delaying the propagation of free radicals in the cure initiation mechanism, described in Section 2.2, and can thus prolong the acrylic adhesives' cure times.

In addition to monomers, polymers, stabilizers, and initiators, many reactive acrylics also contain fillers, pigments, colorants, thixotropes, and other additives [15,18,21,22]. These additional components can be incorporated to reduce cost, improve handling properties, adjust cure rates, and improve adhesion. Table 12.2 summarizes the major components of acrylic adhesives and their effect on performance and handling properties.

12.3 FEATURES, BENEFITS, AND DISADVANTAGES

12.3.1 Features and Benefits

12.3.1.1 Fast Cure

Reactive acrylic adhesives have much a faster cure rate in comparison with epoxy and urethane adhesives, as shown in Figure 12.1. For example, reactive acrylics can reach structural strength in <15 min at room temperature, while epoxies and urethanes with the same *work time* (time available to reposition the substrates without affecting the final bond strength) can require hours to reach the same level of strength. This unique ability to adjust the work time and maintain rapid cure provides valuable processing flexibility for end users. The fast cure allows end users to improve productivity.

12.3.1.2 No-Mix Cure

The polymerization of reactive acrylic adhesives via free radical polymerization (chain reaction) allows them to display some unique handling and curing characteristics. These characteristics follow from the fact that the free radicals formed during initiation stage will continue to propagate throughout the curing adhesive. This means that cure, even if it begins in a localized region, can proceed throughout the adhesive bondline without additional mixing. As a result, a *no-mix cure* was developed. In this approach, the A-side of the adhesive is applied to one surface, while the B-side of the adhesive is applied to the other. Once the two surfaces are joined, a rapid cure of adhesive takes place as shown in Figure 12.6. This behavior is in contrast to epoxy and urethane types, where intimate mixing is critical for curing. However, the no-mix cure method is limited to low adhesive thickness (<0.5 mm) due to limited transport of active agents as the adhesive cures. The no-mix cure approach is attractive to some end users, because it uses inexpensive dispensing equipment.

12.3.1.3 Gap Filling

Two-component reactive acrylics have much higher gap filling ability compared with epoxies and urethanes, because of their robust and rapid cure. The cure mechanism of advanced reactive acrylic adhesives resists oxygen inhibition and propagates without mixing. They fill gaps of up to 12.7 mm and still maintain performance. As a result, they find application in bonding large components in the assembly of boats and windmills [1,3,9].

12.3.1.4 Substrate Versatility: Plastics and Contaminants

Acrylic monomers such as MMA have an exceptional ability to dissolve organic materials. As a result, reactive acrylic adhesives can bond a wide variety of substrates with minimum surface

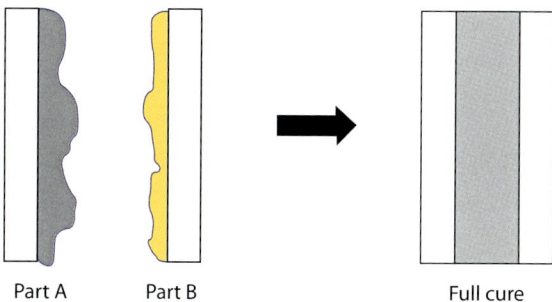

Part A   Part B   Full cure

FIGURE 12.6   In a no-mix cure method for reactive acrylic adhesives, the Part A and Part B of the adhesive are coated onto different substrates that are then joined together to initiate the mixing and curing of Part A and Part B at the substrate surface.

preparation. For example, in the case of bonding to plastics, the MMA monomer is able to wet and penetrate deeper into the substrate, thus increasing the quality of interaction at the interface. In addition, the acrylic monomers can dissolve organic contaminants such as processing oils and release agents that typically interfere with bonding to metal substrates for epoxy and urethane adhesives. Eliminating or reducing the steps required for substrate preparation improves productivity and reduces cost for end users.

12.3.1.5   Balanced Strength and Flexibility: Impact Resistance

Toughened acrylic adhesives have excellent balance of strength and flexibility. Acrylic adhesives are replacing urethane adhesives in some applications, due to their combined fast cure, high strength, and flexibility. Urethanes are inherently flexible materials, while epoxies are brittle and high-strength materials. Formulating epoxies and urethanes for improved balanced toughness and strength typically results in limited performance and handling characteristics when compared with reactive acrylics. The balance of strength and flexibility for acrylic adhesives provides end users with the potential to reduce cost by converting to a single acrylic adhesive system in their plant, instead of using epoxies and urethanes for different applications.

12.3.2   DISADVANTAGES

12.3.2.1   Odor and Flammability

Typical concern with reactive acrylic adhesives, specifically those containing MMA as a component, is the sharp odor of the MMA monomer. Most people can smell MMA when the level in the air is even considerably below the level that is hazardous to health (~100 ppm) [23]. This odor may be objectionable to some people. MMA is also a flammable liquid, and this can limit its use in some large-scale applications. Although odor and flammability are characteristics of some reactive acrylic adhesives, there are many commercial formulations with lower and acceptable odor and flammability. These formulations offer performance characteristics that approach MMA-containing products, although at a higher price.

12.3.2.2   Poor Polar Solvent Resistance

Reactive acrylic adhesives are inherently polar materials, and as such they are susceptible to swelling and degradation in polar solvents such as aromatics, alcohols, and ketones. Depending on the product and conditions, the performance may drop by as much as 50%–80% in polar solvents. However, reactive acrylics have excellent performance and are durable in nonpolar solvents such as diesel fuel, lubricants, and gasoline [20].

12.3.2.3 Poor Temperature Resistance

In general, reactive acrylics are limited to a lower operating temperature when compared with epoxies. Typical operating temperature for reactive acrylics is ~ −40°C to 140°C, while epoxies can go up as high as 150°C–175°C. The operating temperature is limited by the glass transition of the cured poly(MMA) (Tg ~ 110°C). Cross-linking additives can increase the glass transition of the system; however, negative effects on cure rate and adhesion strength limit this approach. Compared with urethane adhesives, reactive acrylics do not suffer from the tendency to decompose above 150°C, and thus they can have bake resistance (short-term exposure) to temperatures as high as 204°C or even higher.

12.4 PHYSICAL FORM AND HANDLING CHARACTERISTICS

Most acrylic adhesives are two-part, high-viscosity liquids or pastes. Examples of typical physical properties of select acrylic adhesives are given in Table 12.3. Most of these types of products can be purchased in ready-to-use cartridges. The mix ratios for commercial acrylic adhesive cartridge systems are most commonly 10:1, 4:1, and 2:1 for benzoyl peroxide-based systems, and 1:1 for systems that utilize the amine plus sulfonyl chloride initiation system. For large, continuous operations, various automated types of meter-mix-and-dispense (MMD) equipment are marketed by a number of engineering companies. In theory, MMD equipment can deliver adhesive in an infinitely variable mix ratio.

Some acrylic adhesives can also be cured by a no-mix accelerator lacquer method. Accelerator lacquers are low-viscosity liquids that are brush or spray applied to one or both of the substrates to be bonded. Following accelerator application, the unmixed adhesive is applied to the primed substrate. This method eliminates the need to premix the adhesive and accelerator, which eliminates pot-life concerns.

An advantage of acrylic adhesives compared with other adhesive systems is that the cure time can be adjusted over a wide margin, ranging from minutes to hours. There are a number of common practical terms used to describe the application features of reactive acrylic adhesives. *Open time* is usually understood as the time allowed before mating the substrates together that will still deliver ideal strength and failure mode on cure. This is in contrast to *work time*, which is the continued time after the substrates are mated, but can still be repositioned relative to each other. Open time can be affected by the rate of free radical initiation, rate of polymerization, evaporation of monomers, or migration of formulation components to the surface, all of which are strongly influenced by the application temperature. Open time mostly depends on the thickness of the applied adhesive bead.

TABLE 12.3

Typical Properties of Reactive Acrylic Structural Adhesive for Mixture of Adhesive A or B with Accelerator A

Properties	Adhesive A*	Adhesive B**	Accelerator A
Appearance	Off-white liquid	Off-white tan paste	Gray paste
Viscosity (cP)	8,000–32,000	80,000–180,000	200,000–500,000
Density (kg/m³)	1150–1240	1007–1066	1438–1546
Work time (min)	1–2	18–24	—
Handling time (min)	2–4	48–72	—
Cure time (min)	2–24	24	—
Mix ratio by volume	—	—	2:1 or 4:1

*Adhesive A: Low-viscosity acrylic adhesive with quick cure package. Typically used for bonding small surfaces.

**Adhesive B: High-viscosity acrylic adhesive with slow cure package. Typically used for bonding large surfaces and filing large gaps between surfaces.

Work time mostly depends on the adhesive cure rate, bondline thickness, application temperature, and thermal conductivity of the substrates. Work time and open time are interchangeable terms. In some cases, work time is understood to include open time.

Purge time is the time at which fresh adhesive must be transferred to the substrate through the mix tip of a cartridge static mixer before it becomes too hard to dispense. The purge time strongly depends on the polymerization rate of the formulation and temperature, and is closely related to work time. *Handling time* is the time required before the adhesive gains enough bond strength for the assembled part to be handled without it shifting or coming apart (this occurs after work time).

Finally, *cure time* is the time required for the adhesive to gain at least 90% of its ultimate strength. Open time, work time, purge time, and cure time are common practical terms that relate to cure profile, but each can individually be of greater or lesser importance depending on the requirements of specific adhesive bonding applications.

In two-component *mix-in accelerator* adhesive systems, it is common to incorporate different colors into the two sides of the system to assist in indicating proper mixing. Many acrylic adhesive products can also go through a notable color change as they cure, which is unique to these materials. This color change can be used as a rough indicator of cure state, and in many manufacturing situations it is an effective means of verifying the presence of both adhesive components in a failure analysis.

Because acrylic adhesives can cure rapidly with a significant exotherm, manufacturers generally recommend that quantities greater than 0.5 L (or 1 pt.) should not be mixed at one time. With very fast-curing systems, literal boiling of the MMA monomers can sometimes be observed even at much smaller volumes, creating the potential to cause voids in thick and wide bondlines and thus compromise the bond strength [11–13].

12.5 STORAGE

Like most adhesive types, acrylic adhesives do not have an indefinite shelf life after manufacture. Typically, suppliers of acrylic adhesives list a shelf life of at least 6 months, with some more than 1 year. One difference with acrylic adhesives is that they often contain large amounts of highly reactive substances (monomers) that are used to achieve some of the unique properties associated with this family of products. Methacrylate monomers have a tendency to autopolymerize (cure without the use of externally added accelerators or hardeners); consequently, shelf stability with some systems can be limited even at relatively low temperatures. Even at 35°C, most acrylic formulations will begin to thicken and gel in as little as a few weeks if maintained at that temperature. Larger containers above 20 L can worsen this problem.

The accelerator or B-side, which is commonly a formulation containing benzoyl peroxide, is the common limiting component of shelf stability. Benzoyl peroxide is inherently unstable, and the rate of decomposition is strongly temperature dependent. For a benzoyl peroxide-containing accelerator, the recommended storage temperature is typically below 15°C. Two-component acrylic adhesive cartridges with benzoyl peroxide curative must typically be stored below 25°C to obtain the listed shelf life. Users of these products should consult with suppliers to confirm the recommended storage conditions.

12.6 TYPES OF REACTIVE ACRYLIC ADHESIVES

Acrylic adhesives, like other adhesive types on the market today, have had a series of technical improvements often referred to as *generations* or *families*. These so-called generations of acrylic adhesives did originally represent fundamental changes in the formulations and significant improvement options to potential users. In recent years, these distinctions have been blurred considerably. Most products on the market today are advanced formulations offering performance, handling, low odor, and safety features well beyond products introduced decades earlier. Nevertheless, some of the early nomenclature is still used, and is certainly common in older literature. Consequently, it is reviewed here for historical purposes. In the following sections, each of the important acrylic types is covered separately in an attempt to clarify its key attributes and differences [9].

12.6.1 First-Generation Acrylic Adhesives

First-generation, also known as *conventional* acrylic adhesives, are the earliest examples of this technology. They contain methacrylate monomers, a variety of polymers, an initiator/stabilizer package, and methacrylic acid. Such formulations, a typical example of which is shown in Table 12.4 [9], emerged in the early 1970s and found considerable utility in bonding thermoplastics.

Products classified as first-generation are still on the market and are sold primarily for bonding thermoplastics. They provide excellent gap filling characteristics, rapid cure and adhesion, and are generally stronger than the substrates themselves. Typical bond performance of adhesives of the first-generation type is shown in Table 12.5 [9]. The adhesives have good adhesion to thermoplastics such as polystyrene, acrylonitrile–butadiene–styrene (ABS), and poly(vinyl chloride) and to clean steel and aluminum. Bonding to galvanized steel was a problem with first-generation acrylic adhesives, but this has been overcome with more recent embodiments of this technology.

It should be noted that if an adhesive of the type shown in Table 12.4 is applied and cured using an accelerator lacquer, bond strengths equivalent to those achieved with mix-in peroxide pastes are obtained. In bonding ABS to itself, for example, the literature reports cohesive failure of the substrate with force in the range of 5.9 MPa when either a mix-in accelerator or an accelerator lacquer are used.

Data have also been reported in company literature claiming that first-generation acrylic adhesives will resist aggressive environments such as exposure to alcohols and hot-water immersion. The patent literature also reports that when using mix-in accelerators, good bond strengths are retained for at least 35 days in aggressive environments (e.g., in condensing humidity cabinets) [9].

12.6.2 Dexter Hysol Acrylic Adhesives

The so-called *Dexter Hysol (DH)* acrylic adhesives are an example of one of the later generation products and are, loosely, an extension of the original technology. DH and other generations are often lumped together under a general term, *second-generation* acrylic adhesives. These adhesives differ from the original offerings in the polymers used, the monomers used, and an increasing use of novel cure chemistry and specialty adhesion promoters. In some cases, the term DH is used interchangeably with second-generation, although adhesive suppliers would argue that this is not accurate and the two terms represent quite different types of products.

Nevertheless, one obvious difference between these materials and earlier types is that the accelerator lacquers used with these newer systems were often oily, and peroxides other than benzoyl peroxide were often used. The oily accelerator lacquers that are available for curing DH acrylics are difficult to apply, and once applied, care must be taken to ensure the oily accelerator is not wiped away or contaminated.

TABLE 12.4
Typical Formulation of First-Generation Reactive Acrylic Adhesive

Component	Parts by Weight
Styrene-methyl methacrylate copolymer syrup	40
Methacrylic acid	9
Poly(methyl methacrylate) syrup	49
Accelerator-stabilizer package	2

TABLE 12.5
Bond Performance of First-Generation Acrylic Adhesive

Substrate	Lap Shear Strength (MPa)[a]
Clean steel	26.2
Clean aluminum	21.4
Rigid poly(vinyl chloride)	17.9
Steel to nylon	20.5
Steel to polystyrene	16.7
Steel to ABS[b]	17.1
Galvanized steel	2.9

[a] ASTM D1002, 6.9 MPa–1000 psi.
[b] Acrylonitrile–butadiene–styrene.

Second-generation and DH acrylic adhesives were notably different from earlier systems, in that they were the first type of acrylic adhesives that bonded well to aluminum and steel surfaces that were contaminated with a variety of oils and organic lubricants. This ability marked a significant advancement in acrylic adhesives and established them as a unique family of adhesive materials for bonding oily and unprepared metals.

Second-generation and DH acrylic adhesives enjoyed considerable popularity when they were first introduced. More recently, however, they have been replaced by other, more "user-friendly" adhesive formulations that do not use oily accelerators. In addition, some of the key formulation raw materials had limited availability, interfering with manufacturers' ability to supply product consistently.

12.6.3 HIGH-PERFORMANCE ACRYLIC ADHESIVES

The emergence of the DH acrylic adhesives occurred concurrently with the introduction of the so-called *high-performance (HP)* acrylic adhesives. This type of system, as with the DH types, offered users the ability to bond through oily metals, but did not require the use of oily primers. They were also based on formulations that contained specialty adhesion promoters, often contained monomers that were less flammable than first-generation products, and were often also considerably lower in odor.

Adhesives of HP type have been shown to bond bronze, lead, nickel, magnesium, copper, aluminum, steel, and stainless steel, in addition to most of the other substrates that earlier offerings were capable of bonding. They did, however, continue to show weaknesses when zinc surfaces were bonded. So, these adhesives may not be well suited for certain applications in the automotive industry where galvanized steel is being bonded. Note that later generations covered in this chapter do not necessarily have this shortcoming.

Table 12.6 shows the lap shear bond strength of typical substrates bonded with HP acrylic adhesives. It can be seen from these data that bonds to oily metals are at least as strong, and often stronger, than bonds obtained with clean surfaces. When HP acrylics were used to bond oily metals, and the bonded parts were subjected to aggressive environments (e.g., condensing humidity, salt spray, water immersion, and gasoline immersion) for up to 1000 h, only little or no reduction in overall bond strength was reported.

12.6.4 HIGH-IMPACT AND CURRENT FAMILIES OF ACRYLIC ADHESIVES

Another advancement in acrylic adhesives has been in the area of acrylics with improved low-temperature properties, sometimes called *high-impact (HI)* acrylics. These variants of the technology not only offer outstanding low-temperature properties, but also significant heat resistance

TABLE 12.6
Bond Performance of High-Performance Acrylic Adhesives on Oily Metals

Substrate	Lap Shear Strength (MPa)[a]
SAE 1010 steel (oily)	44.1
SAE 1010 steel (solvent wiped)	42.1
SAE 1010 steel (grit blasted)	41.0
6061-T6 aluminum (oily)	35.1
6061-T6 aluminum (solvent wiped)	33.6
6061-T6 aluminum (grit blasted)	34.7
6061-T3 aluminum (oily)	35.1
6061-T3 aluminum (solvent wiped)	32.5

[a] ASTM D1002, 6.9 MPa–1000 psi.

that makes them ideal for applications where product performance over a wide temperature range is required. Formulators have also further advanced the art in the use of ingredients that are nonflammable and have little or no odor.

These products represent additional advances made by polymer scientists and formulating chemists. They use unique molecular building blocks and polymers, sometimes of different families than strictly acrylic, leading to numerous "hybrids" involving other emerging polymer types. In addition to the properties noted, HI products retain virtually all of the performance features of earlier variants.

12.6.5 EQUAL-MIX AND NO-MIX ACRYLIC ADHESIVES

A disadvantage of many acrylic adhesive products on the market today is that those that require mixing may require a mix ratio that is difficult to automate. The first-generation products, for instance, generally required a mix of 20 parts adhesive to one part curing agent, making highly automated production very difficult to achieve effectively. As the various generations of acrylics have emerged, more convenient mix combinations have also been introduced: first 10:1, then 4:1, then 2:1 variants, and even certain products with 1:1. Recently, more equal-mix products have emerged, and even products that do not require significant mixing at all have come to the market. The no-mix technology is unique, in that it has found extensive use in the electronics industry, and is a very useful technique in bonding magnets for electrical motors, where the fast cure and easy application technique have found extensive utility.

12.6.6 PHOTOCURABLE (ONE-PART NO-MIX) ACRYLIC ADHESIVES

A separate class of acrylic adhesives, not always listed with this family of materials, are the systems that will cure via exposure to light of various wavelengths. Of most interest are ultraviolet (UV) radiation (below 300 nm wavelength) and, to some extent, the visible range (400–700 nm wavelength). These types of adhesives are obviously of utility in bonding substrates that are transparent in the UV and visible regions, which is typical of many clear plastics and glass.

UV cure requires a specific type of initiator called a photoinitiator. These materials are added to the formulations and result in the formation of free radicals when exposed to UV or visible light. Representative compounds that are sensitive to UV light, particularly of 200–300 nm wavelength, are the benzoin ethers, benzophenones, and similar compounds. These and other compounds that are activated in the visible region are shown in Figure 12.7. When these photoinitiators are exposed to the correct wavelength of light, they dissociate into smaller segments to form free radical-containing molecular fragments that ultimately result in a cure much like other acrylic adhesives [3,9,24].

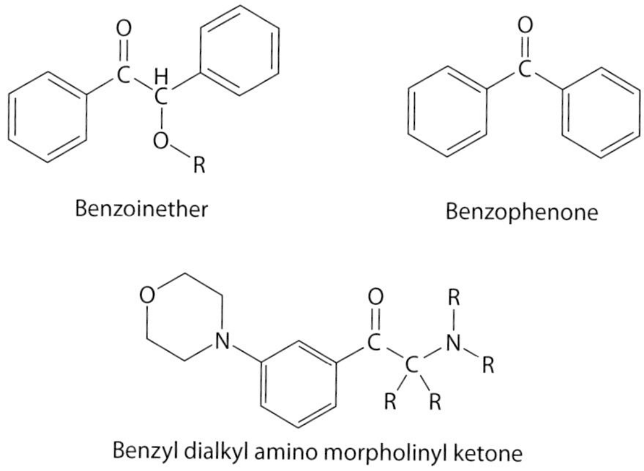

Benzoinether

Benzophenone

Benzyl dialkyl amino morpholinyl ketone

FIGURE 12.7 Typical photoinitiators used in reactive acrylic adhesives.

Compositionally, photoinitiated reactive acrylics are much like the formulations already reported in this chapter, but with special photoinitiators in place of redox and other mechanisms of generating free radicals. The only significant difference is found in the common use of acrylate, rather than methacrylate, as the carrier monomer. The acrylates are much more prone to UV initiation and propagation than the methacrylates and are, consequently, the better choice for formulators.

Company literature from several adhesive suppliers is currently available that lists UV curable and visible light curing acrylic adhesives as part of their product package. This type of acrylic adhesive is being used successfully in polycarbonate lens bonding.

12.6.7 POLYOLEFIN BONDING ACRYLIC ADHESIVES

Low surface energy polymers, such as polyolefin substrates, are difficult to bond adhesively because they are unreactive, saturated polymers. The most common practice for bonding polyolefins such as polypropylene and polyethylene is to pretreat the surface with a high-energy process such as plasma [25] or flame treatment [26]. Such pretreatments create functional groups (primarily R–OH and R–COOH) on the polyolefin surface with which the adhesive can then react.

A relatively new class of acrylic adhesives has been developed that can bond very strongly to unprepared polyolefin substrates. The key to improved performance of these MMA-based adhesives is the aminoborane accelerator system shown in Figure 12.4. With this accelerator, the bond strength of an acrylic adhesive can improve from less than 1 MPa with 100% interfacial failure to greater than 10 MPa with substrate cohesive failure [18,19,27]. The aminoborane accelerator system displays some disadvantages in handling characteristics relative to the benzoyl peroxide cured acrylics. While benzoyl peroxide cured acrylics can be formulated with both short and long open times, aminoborane cured systems are mainly limited to short open time. In addition, aminoborane-based acrylics are typically more expensive. However, the trade-off to higher cost is an excellent ability to bond low surface energy substrates without surface treatment, and continual improvements are being made in an attempt to address the known handling deficiencies.

12.6.8 FLEXIBLE ACRYLIC ADHESIVES

The latest advancement in acrylic adhesive technology is the commercialization of *flexible* acrylic adhesives [28]. Common methacrylates used in acrylic adhesives homopolymerize to create a brittle polymer matrix; as a result, formulated acrylic adhesives tended to have high tensile strength

(>20 MPa) and low elongation (<10%). Advancements in rubber toughening of adhesive formulations led to increased elongation, in the range of 30%, but recent years have seen the introduction of acrylic adhesives with elongation in the range of 100% or higher.

Whereas higher tensile strength of an adhesive tends to deliver higher lap shear bonding performance, it is usually at the expense of peel performance. Higher flexibility typically leads to improved peel strength. Flexibility in an adhesive formulation can be gained simply by softening the polymer matrix, and thus creating an adhesive with low tensile strength (and low lap shear bond strength). Silicone adhesives are an example of this; they are soft materials that deliver low shear strength but good peel strength. However, for high-performance structural adhesives, it is desirable to have a combination of high elongation and high tensile strength. This is characteristic of the newest generation of acrylic adhesives, which can have elongations in the range of 100% in combination with tensile strengths still around 20 MPa. Such adhesives are especially useful for bonding composites, metals, and dissimilar substrates where both high shear and high peel performances are desirable.

12.7 SUMMARY

Reactive acrylic adhesives are gaining market share in the structural adhesives market by replacing adhesives based on polyurethane and epoxies in niche areas, due to their high performance plus excellent handling and processing features. Since their introduction in the late 1960s, significant advances have occurred in this technology, making them tougher, fast curing, more resistant to aggressive environments, low odor, and easier to handle.

ACKNOWLEDGMENTS

This chapter is an update of the previous version written by Dennis Damico. Published in *Handbook of Adhesive Technology*, second edition, A. Pizzi and K. L. Mittal (Eds), CRC Press, Boca Raton, FL (2003). The authors acknowledge his significant contribution. The authors also acknowledge the willing and valuable contribution by colleagues at Lord Corporation: Ian Graham and Mark W. Pressley.

APPENDIX

Supplier	Product	Attributes
Lord Corporation	Lord, Versilok, Fusor	General purpose, automotive, and aftermarket application
ITW Plexus	Plexus, Devcon	Composites and general purpose
Henkel	Loctite, Hysol	General purpose and metal/magnet bonding
IPS	Sci-Grip	Composites and general purpose
Engineered Bonding	Solutions Acralock	Marine and general purpose
3M	Scotch-Weld	Plastics and general purpose
Sika	SikaFast	Construction application
Permabond	Permabond	"Toughened" acrylics
Dymax	Dymax	UV curable
Beacon Adhesives	Magna-cryl	Laminating, plastics
Saf-T-Lok	Saf-T-Lok	Syrup, bonds porous surfaces and fills large gaps in between substrates
Denka	Hardloc	Syrup, magnet, and golf ball applications

REFERENCES

1. A. G. Bachmann. Aerobic acrylic adhesives. *Adhesives Age* August, 19–23 (1982).
2. R. D. Rich. Anaerobic adhesives. In*: Handbook of Adhesive Technology*, 2nd edn, A. Pizzi and K. L. Mittal (Eds.), pp. 750–762, Marcel Dekker, New York (2003).
3. A. G. Bachmann. Aerobic acrylics: Increasing quality and productivity with customization and adhesive/process integration. In: *Handbook of Adhesive Technology*, 2nd edn, A. Pizzi and K. L. Mittal (Eds.), pp. 763–787, Marcel Dekker, New York (2003).
4. J. M. Rooney and B. M. Malofsky. Anaerobic adhesives. In: *Handbook of Adhesives*, 3rd edn, I. Skeist (Ed.), pp. 451–462, Chapman and Hall, New York (1990).
5. H. W. Coover, D. W. Dreifus and J. T. O'Connor. Cyanoacrylate adhesives. In: *Handbook of Adhesives*, 3rd edn, I. Skeist (Ed.), pp. 463–477, Chapman and Hall, New York (1990).
6. P. Klemarczyk and J. Guthrie. Advances in anaerobic and cyanoacrylate adhesives. In: *Advances in Structural Adhesive Bonding*, D. Dillard (Ed.), pp. 96–127, CRC Press, Boca Raton, FL (2010).
7. A. Mori, K. Tashiro, K. Makita, M. Takatani and T. Okamoto. Development of room-temperature curing aqueous emulsion-type acrylic adhesive 1: Effect of monomer composition on the initial adhesive strength. *J. Wood Sci.* 51, 33–37 (2005).
8. W. H. Brendley. Fundamentals of acrylics. *Paint Varnish Prod.* 63, 19–27 (1973).
9. D. J. Damico. Reactive acrylic adhesives. In: *Handbook of Adhesive Technology*, 2nd edn, A. Pizzi and K. L. Mittal (Eds.), pp. 747–759, Marcel Dekker, New York (2003).
10. P. C. Briggs and G. L. Jialanella. Advances in acrylic structural adhesives. In: *Advances in Structural Adhesive Bonding*, D. A. Dillard (Ed.), pp. 132–150, CRC Press, Boca Raton, FL (2010).
11. Lord Corporation. Lord Maxlok™ acrylic adhesives. http://www.lord.com/Documents/Product%20 Brochures/PB3083_Maxlok.pdf (2015).
12. Henkel Corporation. Structural adhesives and NVH selector guide. http://www.henkelna.com/us/content_data/330380_LT4809_10458_Structural_Bro_LR_v11_1.pdf (2015).
13. Lord Corporation. Lord structural adhesives selector guide. http://www.lord.com/Documents/ Selector%20Guides/SG1018_LORDStructuralAdhesives.pdf (2015).
14. G. Odian. *Principles of Polymerization*, 4th edn, pp. 199–238, John Wiley (2004).
15. W. Owston. Fast curing polychloroprene acrylic adhesive. US Patent 3725504, assigned to Lord Corporation, Hoboken, NJ (1973).
16. P. C. Briggs, Jr. and L. C. Muschiatti. Novel adhesive composition. US Patent 3890407A, assigned to Du Pont (1973).
17. M. F. Sonnenschein, S. P. Webb, P. E. Kastl, D. J. Arriola, D. R. Harrington and N. G. Rondan. Mechanism of trialkylborane promoted adhesion to low surface energy plastics. *Macromolecules*, 37, 7974–7978 (2004).
18. D. M. Moren. Organoborane amine complex initiator systems and polymerizable compositions therewith. US Patent 6384165 B1, assigned to 3M Innovative Properties Co. (2001).
19. C. Ollivier and P. Renaud. Organoboranes as a source of radicals. *Chem. Rev.* 101, 3415–3434 (2001).
20. 3M Corporation. Acrylic structural adhesives: Features and recent advancements. http://multimedia.3m. com/mws/media/1054052O/acrylic-adhesives-recent-advancements-white-paper.pdf (2015).
21. J. Huang, R. F. Righettini and F. G. Dennis. Adhesive formulations. US Patent 6225408 B1, assigned to Lord Corporation (1999).
22. S. Tsuno and R. D. Cooman. (Meth)acrylic adhesive with low odor and high impact resistance. US Patent 20060155045 A1, assigned to Sika Technology AG (2005).
23. H. J. Cloft, D. N. Easton, M. E. Jensen, D. F. Kallmes and J. E. Dion. Exposure of medical personnel to methyl methacrylate vapor during percutaneous vertebroplasty. *Am. J. Neuroradiol.* 20, 352–353 (1999).
24. B. Goss. Bonding glass and other substrates with UV curing adhesives. *Intl. J. Adhesion Adhesives* 22, 405–408 (2002).
25. M. Strobel, C. S. Lyons and K. L. Mittal (Eds.). *Plasma Surface Modification of Polymers: Relevance to Adhesion*, CRC Press, Boca Raton, FL (1994).
26. L. Mazzola and A. Cusma. Flame treatment of polymeric materials—relevance to adhesion: A critical review. *Rev. Adhesion Adhesives* 2, 433–466 (2014).
27. J. V. Zharov and J. N. Krasnov. Polymerizable compositions made with polymerization initiator systems based on organoborane amine complexes. US Patent 5539070 A, assigned to 3M Company (1995).
28. R. D. Cooman, M. Werner and R. Flury. Elastic meth(acrylic) adhesive compositions. US Patent 20050004304 A1, assigned to R. D. Cooman, M. Werner and R. Flury (2002).

13 Anaerobic Adhesives

David Birkett and David Condron*

CONTENTS

13.1 INTRODUCTION

Anaerobic adhesives were first developed in the 1950s in the United States. Further information on their history can be found in the chapter by Richard D. Rich in the second edition of this handbook [1]. The formulations are complex, as are the reaction mechanisms. Therefore, we devote the

---

* Indicates corresponding author

first two sections of this chapter to a discussion of the typical ingredients used in anaerobic adhesive formulations and an outline of the current understanding of the reaction mechanism. Later sections go into greater detail regarding the chemistries of the different components and the differing requirements of the most common applications.

13.1.1 WHAT IS AN ANAEROBIC ADHESIVE?

Anaerobic adhesives are generally single-component systems that cure by a redox-initiated free radical polymerization. They are designed for the bonding of metals. The basic units that polymerize are usually methacrylate monomers, but acrylate monomers and oligomers with methacrylate or acrylate functionality are also commonly employed.

The cure system normally contains free radical initiators such as hydroperoxides or peroxides, reducing agents such as amines or hydrazines, and acid salts such as saccharin. The metal substrate provides the metal ions critical to the cure of the adhesive. The important characteristics and typical properties of anaerobic adhesives are outlined in Tables 13.1 and 13.2, respectively.

13.1.2 WHY USE ANAEROBIC ADHESIVES?

Anaerobic adhesives have been specially developed for key engineering applications such as threadlocking, thread sealing, retaining, flange sealing, and porosity sealing. Details of these applications

TABLE 13.1
Characteristics of Anaerobic Adhesives

Characteristics	Comments
Single-part adhesives; liquid or paste form	Easy to apply to parts of different sizes
Cure at room temperature	Cure in the absence of oxygen (anaerobic) and in the presence of redox active metals
Stable in the presence of oxygen	Stored in small breathable packs such as low-density polyethylene (LDPE) that allow transmission of oxygen
Suitable for metal bonding applications with small gaps	Gaps less than 0.25 mm (250 µm)
Main applications	Threadlocking, thread sealing, retaining, flange sealing, porosity sealing
Nonflammable, reactive adhesives	Solvent free

TABLE 13.2
Properties of Anaerobic Adhesives

Properties	Typical	Comments
Viscosity	10–1,000,000 mPa.s	Low-viscosity liquids to high-viscosity pastes
Fixture time: active substrates, e.g., brass, steel	3–30 min	Sensitive to gap and temperature
Fixture time: inactive substrate, e.g., stainless steel	5–360 min	Sensitive to gap and temperature
Strength (breakaway torque); M10 nut and bolt	1–60 N.m	Strength can be adjusted
Service temperature range	−55°C to 150°C	Specialty grades −55°C to 230°C
Chemical resistance	Good	To fluids such as motor oil, acetone, ethylene glycol
Flexibility (elongation at break)	Rigid <10%	Specialty grades up to 200%

and the advantages provided by anaerobic adhesives are given in Sections 8.1 through 8.5. The key advantage of anaerobic adhesives over other adhesive technologies is that they provide a single-part, room-temperature curing solution that meets the needs of these applications. The cured adhesive exhibits excellent thermal and chemical resistance, and properties such as strength, cure speed, and ease of disassembly with hand tools can be optimized.

13.2 FORMULATION

The section in Richard Rich's chapter in the second edition of this handbook is still fully valid, so the following is just an overview.

13.2.1 MONOMERS

The most common monomers in anaerobic adhesives are methacrylate esters. Acrylate esters, acrylic acid and other unsaturated acids, and other vinyl polymerizable monomers are also sometimes used for particular performance characteristics.

13.2.2 INITIATOR

The most commonly used initiator in anaerobic formulations is cumene hydroperoxide (Structure 13.I), although other hydroperoxides and related compounds have occasionally found use.

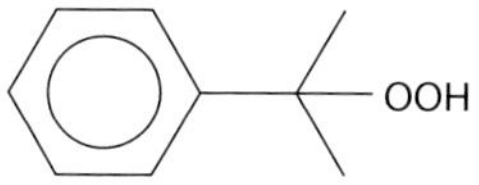

I

13.2.3 ACCELERATORS

Saccharin (benzoic sulfimide) (Structure 13.II) is the most commonly used accelerator. It is usually found in combination with aromatic amines such as toluidines, acetylphenylhydrazine, and tetrahydroquinoline (Structures 13.III, 13.IV, and 13.V, respectively).

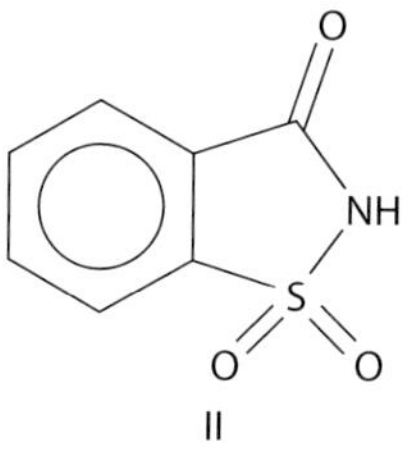

II

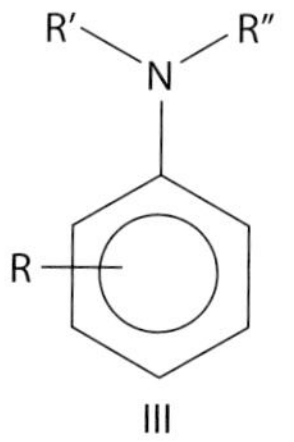

III

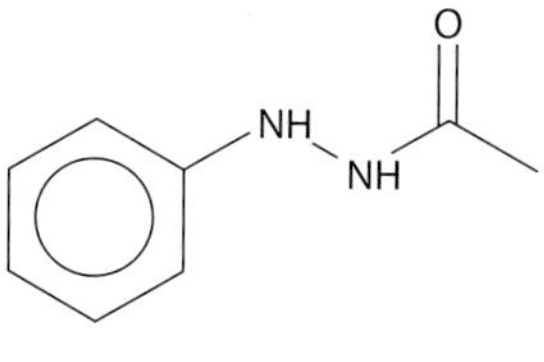

IV

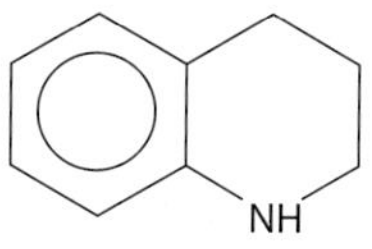

V

13.2.4 STABILIZERS

The methacrylate monomers generally contain traces of free radical stabilizers, but it is usually necessary to employ additional free radical stabilizers such as naphthoquinone (Structure 13.VI).

In addition, to prevent trace quantities of metal ions causing premature polymerization, metal scavengers such as salts of ethylenediaminetetraacetic acid (Structure 13.VII) are frequently included in the formulation.

VI   VII

13.2.5 MODIFIERS

A wide range of modifiers are found that do not typically affect the cure mechanism. These include coloring and fluorescent agents, rheological modifiers, and plasticizers, to name but a few.

13.2.6 SURFACE ACTIVATORS

Powerful accelerators (including metal salts and amine reducing agents) that cannot be included in one-part formulations without premature polymerization can find application either in true two-part formulations, or more commonly when applied to the surface of the substrate in solution form.

13.3 ANAEROBIC ADHESIVE CURE MECHANISM

Anaerobic adhesives cure by a free radical-initiated addition polymerization:

$$A^{\bullet} + nCH_2 = CXY \rightarrow A(CH_2 - CXY)_n^{\bullet}$$

where $A^{\bullet}$ is the initiating free radical. (For a methacrylate monomer, X would be $-CH_3$ and Y would be $-CO_2R'$). Eventually, the growing chain will terminate through radical recombination or disproportionation.

However, there are two important features that distinguish the anaerobic adhesive cure mechanism. The first is the pair of redox reactions between a transition metal ion and a hydroperoxide [2]:

$$Fe^{2+} + ROOH \rightarrow Fe^{3+} + RO^{\bullet} + OH^-$$

$$Fe^{3+} + ROOH \rightarrow Fe^{2+} + ROO^{\bullet} + H^+$$

Similar reactions apply to other transition metals that exist in different oxidation states such as Cu^+ and Cu^{2+}. The peroxy radical generated by the higher oxidation state metal ion is a much less powerful initiator for polymerization than the alkoxy radical generated by the lower oxidation state metal ion. In the presence of oxygen, the higher oxidation states will predominate.

The second feature is that even if polymerization is initiated by an alkoxy radical, the oligomeric chain will eventually encounter an oxygen molecule if oxygen is present. This will convert an active tertiary alkyl radical into a much less active peroxy radical:

$$A(CH_2 - CXY)_n^{\bullet} + O_2 \rightarrow A(CH_2 - CXY)_n OO^{\bullet}$$

Hence, the polymerization is doubly inhibited by the presence of oxygen.

Although the combination of hydroperoxide, transition metal substrate, and absence of oxygen will on its own lead to a bonded assembly, the majority of all modern commercial anaerobic adhesives also contain saccharin (benzoic sulfimide) and an amine (generally, an aromatic amine). Both function as accelerators in their own right. Unfortunately, most of the work elucidating the detailed mechanisms is unpublished industrial work, but there is sufficient information in the open literature to get some idea of what is going on. Saccharin probably plays several roles. As a weak acid, it may help transport metal ions from the surface of the substrate into the bulk of the adhesive, and it may catalyze the decomposition of hydroperoxides [3]. The amine, likewise, probably plays several roles. However, it is clear that the combination of amine and saccharin is much more effective, suggesting a degree of synergy, and it is probable that some sort of complex is involved. Both a charge transfer complex [4] (Structure 13.VIII) and an aminal (Structure 13.IX) have been proposed [5]. However, when the aminal was synthesized, it was found to reduce Cu^{2+} but not Fe^{3+} [6].

VIII   IX

Indeed, formulations containing both amine and saccharin can even cure in the absence of a hydroperoxide [4]. Conceivably, this could result from autoxidation of hydrogens alpha to the nitrogen atom, but this has not been conclusively shown to be the case in the literature.

The amine most studied in the literature is dimethyl-p-toluidine, which was used in the 1980s in commercial products. However, it has now largely been replaced by a number of alternatives: a blend of dimethyl-o-toluidine and diethyl-p-toluidine (which probably follows a similar mechanism), 1,2,3,4-tetrahydroquinoline (THQ; Structure 13.V), and acetylphenylhydrazine (APH; Structure 13.IV); slightly different mechanisms have been proposed for these last two [7].

Finally, both maleic acid [7] and acrylic acid have accelerating effects, and a mechanism for maleic acid has been proposed. Both are used in commercial formulations.

13.4 MONOMERS

Anaerobic adhesives contain monomers with unsaturated groups capable of polymerization under anaerobic conditions. Most anaerobic adhesives contain methacrylate ester monomers, although acrylate esters are also sometimes employed. In principle, any methacrylate monomer can be used, but in practice considerations such as odor, volatility, health and safety (H&S), and formulation compatibility determine the selection. Anaerobic adhesives traditionally contain methacrylates of higher functionality such as dimethacrylates, which promote cross-linking. For convenience, the following discussion groups the most popular monomers under their (meth) acrylate functionality.

13.4.1 MONO(METH)ACRYLATES

13.4.1.1 Hydroxyl and Carboxyl Functionalized Monomers

Acrylic and methacrylic acids (Structure 13.X) are widely used in anaerobic adhesives. They promote adhesion to metal surfaces [8]. However, their H&S aspects have recently come under scrutiny, and there is a trend to find safer alternatives.

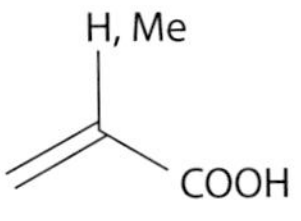

X

2-hydroxyethyl methacrylate (Structure 13.XI) and 2-hydroxypropyl methacrylate (Structure 13.XII) are examples of hydroxyl methacrylates used in anaerobic formulations [9]. In practice, the commercial grade of hydroxypropyl methacrylate is a mixture of two isomers. These monomers have also been used to improve adhesion to metal surfaces because of their polar nature.

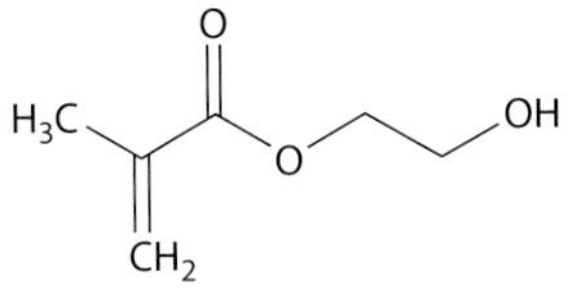

XI 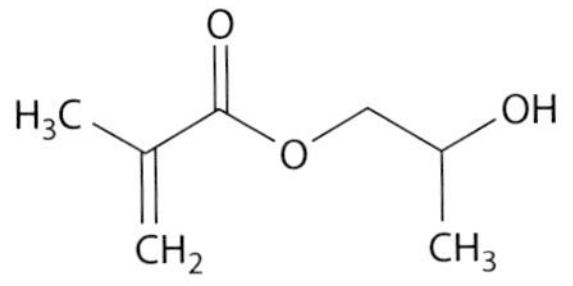   XII

A number of related materials have also been employed as metal adhesion promoters. Examples include methacryloyloxyethyl succinate (Structure 13.XIII), methacryloyloxyethyl maleate (Structure 13.XIV) and bis[2-(methacryloyloxy)ethyl] phosphate (Structure 13.XV).

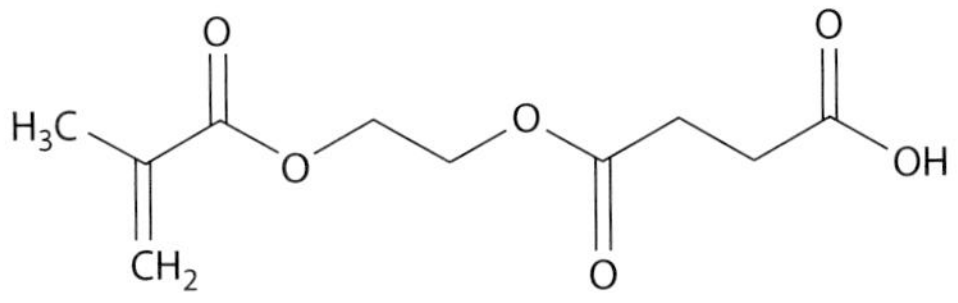

XIII

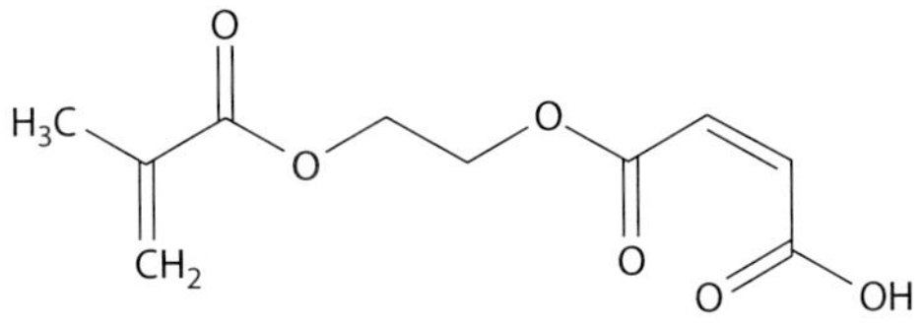

XIV

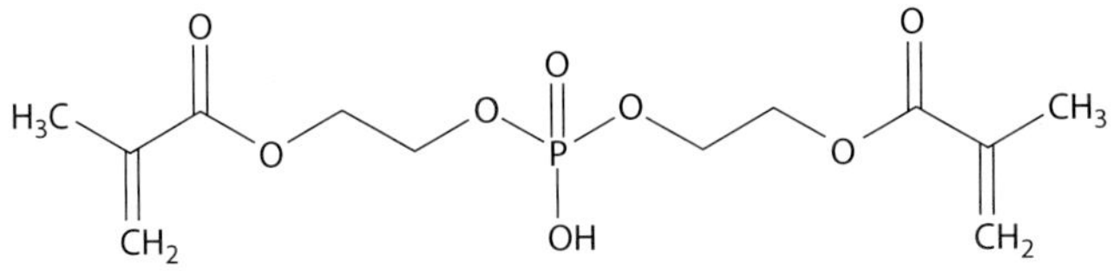

XV

13.4.1.2 Alkyl and Aryl Methacrylates

Alkyl and aryl methacrylates are also widely used in anaerobic adhesives. They impart a more hydrophobic character to the polymerized adhesive and are frequently used as flexibilizers.

Alkyl methacrylates are widely commercially available and longer chain molecules have been employed in anaerobic adhesives. Examples include isodecyl methacrylate having a C10 chain (Structure 13.XVI), lauryl methacrylate (C12), and longer chains up to C17–C18.

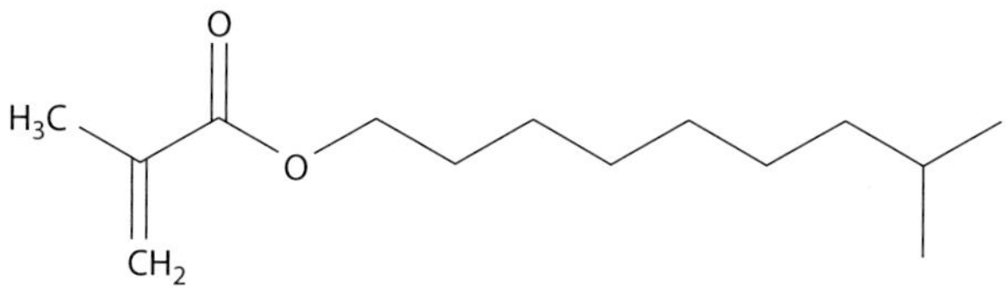

XVI

Cycloaliphatic methacrylates include isobornyl methacrylate (Structure 13.XVII) and various derivatives of cyclohexyl methacrylate such as 3,3,5-trimethylcyclohexyl methacrylate (Structure 13.XVIII) and 4-tert-butylcylcohexyl methacrylate (Structure 13.XIX). Examples of aryl methacrylates include benzyl methacrylate and phenoxyethyl methacrylate (Structure 13.XX)

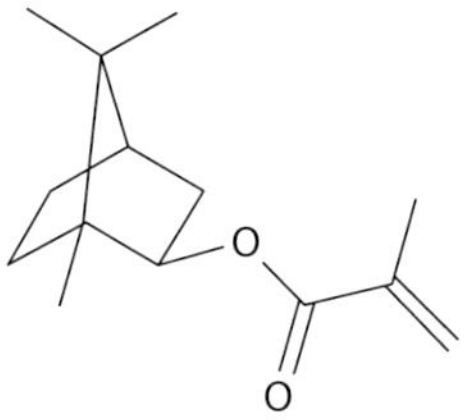

XVII

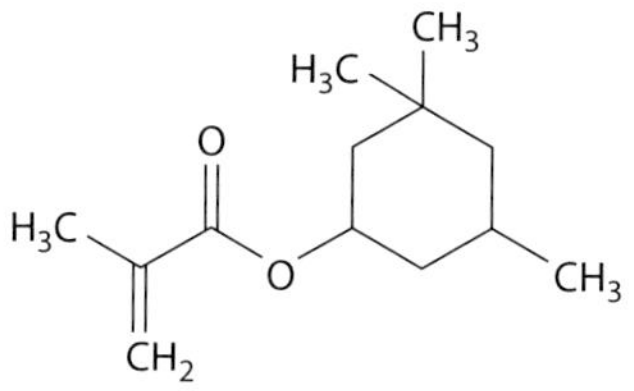

XVIII

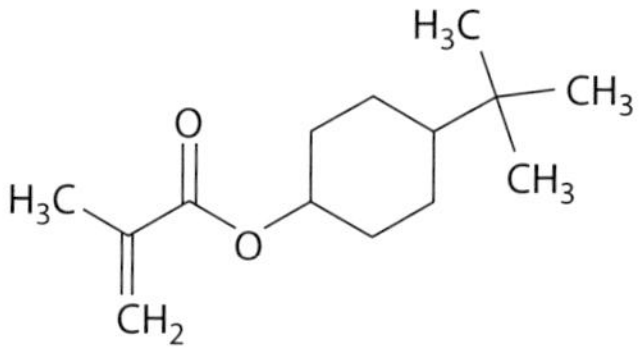

XIX

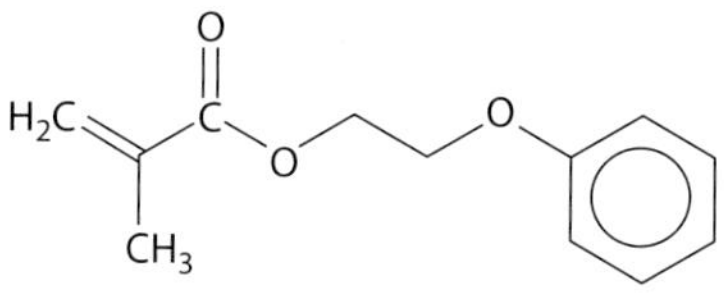

XX

13.4.2 DIMETHACRYLATES

Difunctional methacrylates are most widely used in anaerobic adhesives. They have the ability to act as cross-linkers, forming a branched polymer network. This promotes desirable features such as chemical and thermal resistance.

Ethylene glycol dimethacrylates have been widely used in anaerobic adhesives. Examples include diethylene glycol dimethacrylate, triethylene glycol dimethacrylate (Structure 13.XXI) and polyethylene glycol dimethacrylate.

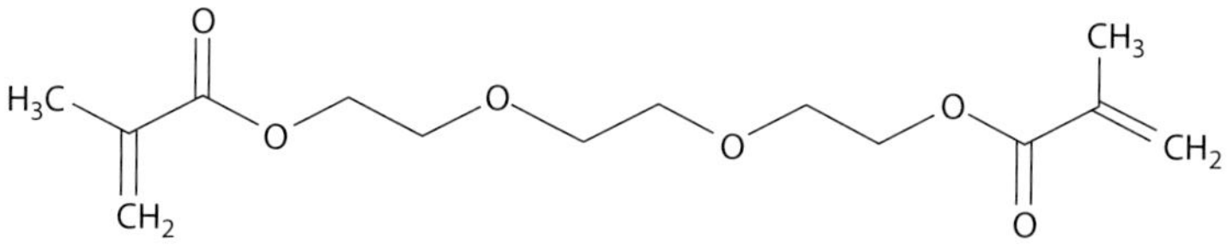

XXI

Other commonly used dimethacrylate monomers include 1,4-butanediol dimethacrylate (Structure 13.XXII), 1,6-hexanediol dimethacrylate, and 1,12-dodecanediol dimethacrylate. Characteristics such as volatility, hydrophobicity, and flexibility can be varied depending on the chain length.

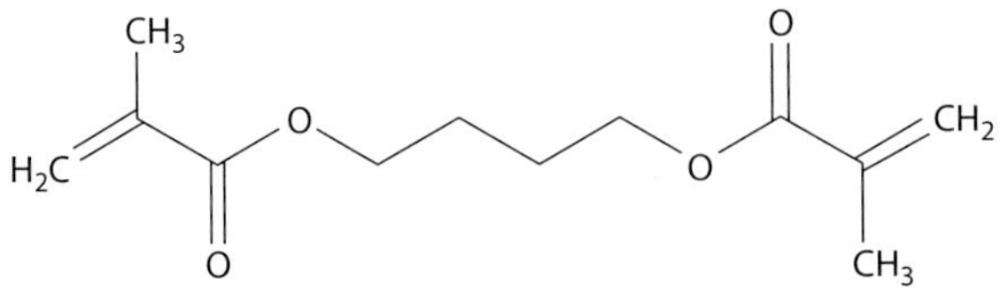

XXII

Another important family of dimethacrylate monomers is those based on bisphenol A. Examples include ethoxylated (2) bisphenol A dimethacrylate (Structure 13.XXIII), ethoxylated (3) bisphenol A dimethacrylate, and ethoxylated (10) bisphenol A dimethacrylate. The lower molecular weight monomers provide an adhesive with good thermal resistance properties.

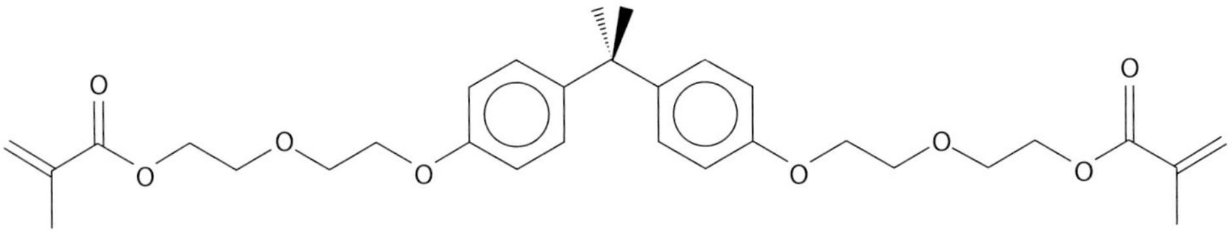

XXIII

Tricyclodecane dimethanol dimethacrylate (Structure 13.XXIV) has also been employed in anaerobic formulations. It has a high Tg and exhibits low volume shrinkage when polymerized.

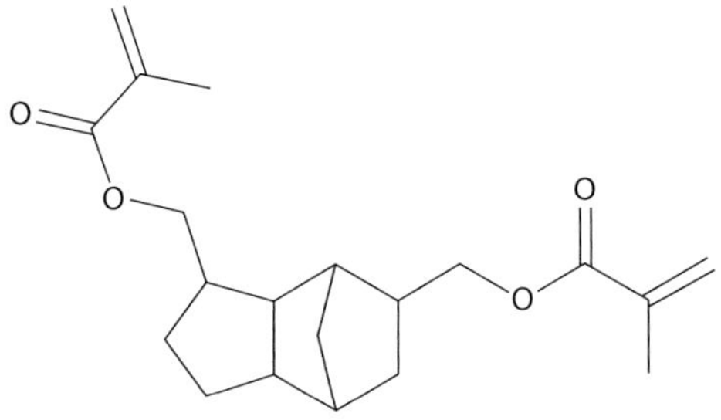

XXIV

13.4.3 Trifunctional and Higher Functionality Methacrylates

Trifunctional monomers are sometimes used to increase cure speed and increase the degree of cross-linking. The most widely used example is trimethylolpropane trimethacrylate (TMPTMA).

Alkoxylated pentaerythritol tetramethacrylate is a high functional methacrylate monomer developed for use as a cross-linker for a wide range of applications. A recent Evonik patent [10] mentions the preparation of higher functional methacrylates such as pentaerythritol tetramethacrylate (Structure 13.XXV).

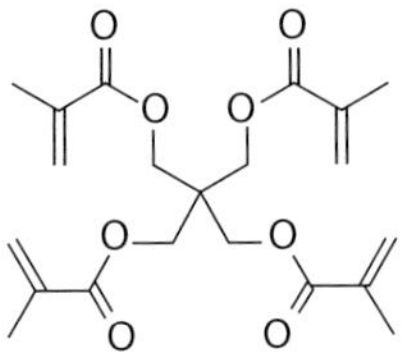

XXV

13.4.4 OTHER MONOMERS

A variety of other nonmethacrylate monomers have been used in anaerobic formulations for various purposes, including cross-linking at high temperature. This increases the functional operating temperature of standard anaerobic adhesives from 150°C up to 230°C.

Examples include 1,1'-(methylenedi-4,1-phenylene) bismaleimide (Structure 13.XXVI) and N,N'-1,3-phenylenedimaleimide (Structure 13.XXVII) as described in a Loctite patent [11].

XXVI   XXVII

Recently, there has been a trend toward monomers made from renewable raw materials in an effort to reduce the carbon footprint of formulated products. Arkema (France) now promotes a range of monomers with biorenewable carbon content. Examples include isobornyl methacrylate and lauryl methacrylate.

13.5 CURATIVES AND STABILIZERS

13.5.1 INITIATORS

The most common initiator used in commercial products is cumene hydroperoxide. This is because it is relatively inexpensive and because it has a high 10 h half-life temperature of 158°C. However, the patent literature (e.g., [12]) discloses a wide range of hydroperoxides, and some such as t-butyl hydroperoxide, p-menthane hydroperoxide (Structure 13.XXVIII), and tetramethylbutyl hydroperoxide (Structure 13.XXIX) can be found in commercial formulations or have found use in the past.

XXVIII   XXIX

In addition, a recent Henkel patent [13] discloses a number of complexes of hydrogen peroxide with urea, substituted ureas (Structure 13.XXX), and other amides, such as *diacetone acrylamide* (Structure 13.XXXI):

XXX   XXXI

It is also sometimes possible to obtain slow cure without added hydroperoxide, due to traces of native hydroperoxides that may be present.

13.5.2 ACCELERATORS

Although there is a Loctite patent from the 1990s [14] on substituted saccharin derivatives for use in anaerobic formulations, we are not aware of any commercial examples that use any such molecule or any other molecules that play the same role as saccharin.

On the other hand, there is much more variety in the amines that have been used. One of the earliest was tributylamine [15], but aromatic amines are more commonly used. Beyond the commonly used toluidines, THQ and APH, discussed in Section 13.3, other hydrazine derivatives such as p-toluenesulfonylhydrazine (Structure 13.XXXII), as referenced in a Three Bond patent [16], and APH analogues such as succinylphenylhydrazine (Structure 13.XXXIII) [17] have found application.

XXXII   XXXIII

Henkel has recently been active in this area, and has had patents granted for (inter alia) conjugated (but nonaromatic) enamines such as Structure 13.XXXIV [18], benzoxazines such as Structure 13.XXXV [19] and THQ derivatives such as those formed by reaction with glycidol (Structure 13.XXXVI), benzomorpholine (Structure 13.XXXVII), and julolidine (Structure 13.XXXVIII) [20].

XXXIV   XXXV   XXXVI

XXXVII   XXXVIII

Accelerators can also be bound into urethane methacrylate resins [21,22] or onto maleic anhydride copolymers [23].

13.5.3 STABILIZERS

It is important to have a significant concentration of oxygen in the adhesive to ensure that it does not cure prematurely. To accomplish this, anaerobics are traditionally stored in bottles made of low-density polyethylene (which has a relatively high oxygen permeability), and the bottle is not completely filled, so that there is a head space containing air above the surface of the liquid.

However, to maintain a storage stable one-component product, it helps not only to have oxygen present, but also small quantities of materials to sequester stray free radicals or metal ions. In a bonded assembly, these are swamped by the flux of free radicals and metal ions, and cure can proceed. By far the most common metal scavenger used is the tetrasodium salt of ethylenediamine tetraacetic acid, Na_4EDTA [24]. Free radical scavengers frequently found include quinones (hydroquinone, benzoquinone, naphthoquinone, anthraquinone), butylated hydroxytoluene (BHT), and p-methoxyphenol [25].

13.6 PRIMERS AND ACTIVATORS

One approach commonly used to improve the performance of anaerobic adhesives on certain lower activity metals (such as stainless steel) and/or to increase the depth of cure is to treat the substrate with a solution of a transition metal compound (e.g., copper naphthenate or ferrocene) and/or a powerful accelerator in a volatile solvent. The organic activator can be any of the accelerators discussed, but is often a material that would be too active to include in one-component systems; for example, dihydropyridines, as in Structure 13.XXXIX, or mercaptobenzothiazole (Structure 13.XL) [26] or thiourea derivatives (Structure 13.XLI) [27,28].

XXXIX   XL   XLI

Furthermore, transition metal compounds and/or accelerators can be used in the so-called *two-component anaerobics* where one part is a conventional anaerobic adhesive and the other contains the activating material.

13.7 MODIFIERS

Anaerobic adhesive chemistry is versatile and can accommodate the inclusion of a wide variety of materials designed to modify the properties of the adhesive. A number of additives have been employed to modify the viscosity and rheology. Examples of polymeric thickening agents that have been employed include polymethacrylates, polyesters, poly(vinyl acetate), polyacrylonitrile butadiene, and poly(vinyl chloride) [29]. The flow characteristics of anaerobic formulations can also be controlled by the addition of fumed silica, modified castor oil derivatives, and polyamides. These materials impart *thixotropic* properties.

Strength modification is also an important consideration for anaerobic adhesives and sealants. Reduced strength is a desirable feature for components that require disassembly for repair and replacement. Many liquid plasticizers have been used for this purpose, including adipates, benzoates, citrates, phthalates, phosphates, and trimellitates. The toughness properties of anaerobic adhesives can be enhanced by the addition of a reactive elastomer [30].

Dyes, fluorescent agents, and pigments are commonly added to anaerobic adhesives and sealants. They assist in product identification and inspection. Automated inspection is made possible with dyes that fluoresce under ultraviolet light.

Comments made by Rich [1] with respect to solid fillers are still valid today. Mica, talc, and other mineral fillers can help to impart instant seal capability to anaerobic pipe sealants. Powdered graphite, polytetrafluoroethylene, and polyethylene can function as lubricants in pipe sealants and threadlocking compounds. This lubrication can prevent galling in close-fitting pipe threads [31].

Lubricating additives in threadlocking sealants can provide control of the clamping force exerted by a fastener at a given tightening torque [32]. Materials added with the intention to improve temperature resistance, such as bismaleimides, are discussed in the section on monomers (Section 13.4), as these materials polymerize at higher temperature.

13.8 APPLICATIONS

The relevant section in Richard Rich's chapter in the second edition of this handbook is still fully valid. Applications of anaerobic adhesives nearly all involve metal-to-metal bonding through a limited gap (although the gap limitation can be circumvented using two-component products or surface activators).

13.8.1 THREADLOCKING

Threadlocking products prevent self-loosening and secure threaded fasteners against vibration and shock loads.

This was the earliest application for anaerobic products. If two plates are fastened together with a nut and bolt, then the plates are under compression and the bolt is under tension. All that prevents the assembly from coming apart under vibration is the friction between the nut and the bolt. This, in turn, comes from the microscopic roughness of their surfaces—there are only a few points of true contact. Threadlockers fill in the empty spaces with a solid material that dramatically increases friction. There are many different anaerobic threadlockers, with varying strengths, speeds, and rheologies.

It is also possible to coat threaded parts with threadlockers (and with thread sealants, for that matter) in the form of dry-to-the-touch films containing microcapsules. The *preapplied* products only polymerize on assembly, whereupon the capsules rupture and release a quick curing resin.

The effectiveness of Loctite threadlockers versus mechanical locking devices is demonstrated in Figure 13.1. This graphically represents the clamp load retention of a nut and bolt assembly subjected to vibration in a Junkers machine (transverse vibration test). It can be seen that the clamp load of the nut and bolt assembly is best maintained when a threadlocker is applied to the nut and bolt. In contrast, a significant drop in bolt tension is observed when mechanical methods are employed in the absence of a threadlocking adhesive.

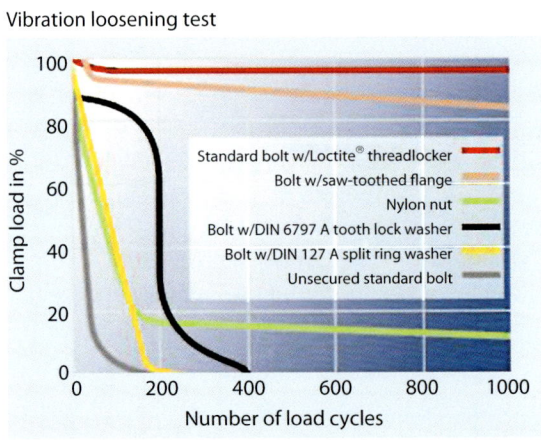

FIGURE 13.1 Vibrational resistance comparison: threadlocker versus other mechanical methods.

13.8.2 Thread Sealing

Anaerobic thread sealants, available in liquid or paste form, prevent leakage of gases and liquids from connections involving threaded pipe assemblies. Designed for low-to-high-pressure applications, they fill the space between threaded parts and provide an instant, low-pressure seal. When fully cured, they seal to the burst strength of most pipe systems.

EN 751-1:1996 is a standardized test to determine the sealing capability of curing compounds on pipe assemblies. Table 13.3 summarizes seal testing results for Loctite 577, an anaerobic thread sealant, tested according to the methodology specified by EN 751-1:1996. Three stages of the test were investigated in succession on the same assembled parts: soundness, hot-water resistance, and temperature cycling at 150°C. The results show that Loctite 577 thread sealant seals successfully under these conditions with no leaks observed.

13.8.3 Retaining

Anaerobic retaining compounds are used to secure bearings, bushings, and cylindrical parts into housings or onto shafts. They are designed to achieve high load transmission and uniform stress distribution, and eliminate fretting and corrosion. Applied as a liquid, retaining compounds fill the inner space between the components to be bonded and cure to form a strong precision assembly (Figure 13.2).

Anaerobic retaining compounds offer advantages over mechanical retainers. Because there is 100% surface-to-surface contact, load and stress are distributed evenly over the joint. Corrosion and

TABLE 13.3
Seal Test Results for Loctite 577: Anaerobic Thread Sealant
EN 751-1 Seal Tests

	Loctite 577		
	Test 1	Test 2	Test 3
Soundness test	Pass	Pass	Pass
Hot-water resistance test	Pass	Pass	Pass
Temperature cycle up to 150°C	Pass	Pass	Pass

FIGURE 13.2   Application of a liquid-retaining compound to a gear wheel.

fretting of the joint are prevented by the presence of the adhesive, which fills any voids. Anaerobic retaining compounds are often used in combination with interference fits, augmenting the strength of the assembly and facilitating the use of relaxed tolerances. A more detailed treatment of the retaining application is provided by Haviland [33].

13.8.4 Gasketing

Gaskets are used to prevent leakage of liquids and gases by forming impervious barriers. The seal should remain intact and leak free over a long period of time. The gasket should be resistant to liquids and/or gases and withstand the operating temperatures and pressures to which it is subjected. Anaerobic gasketing products are self-forming gaskets that provide an excellent seal between components, with maximum face-to-face contact. They are also useful in eliminating flange face corrosion. A low-pressure seal is formed immediately on assembly. On full curing, the cured adhesive provides a joint that will not shrink, crack, or relax.

Anaerobic adhesives provide several benefits over traditional precut gaskets, and overcome many of the issues associated with compression gaskets such as low surface contact, relaxation, extrusion, and bolt-hole distortion (Figure 13.3). The adhesives can be applied automatically or by hand (Figure 13.4). The adhesives are applied as liquids or pastes to the flange and subsequently cure to a thermoset plastic. Hence, anaerobic flange sealants enable reduction of inventory compared with precut gaskets.

13.8.5 Impregnation

Sintered metal parts and porous castings can be sealed against leakage of liquids or gases using anaerobic adhesive formulations. The adhesive cures inside the pores where anaerobic conditions prevail. In the 1960s, Loctite Corporation developed a vacuum impregnation process to achieve greater penetration of parts with anaerobic sealant [34]. The benefits of impregnating porous metal parts include strengthening of the part, improved sealing, easier plating and bonding of the part, and improved machining.

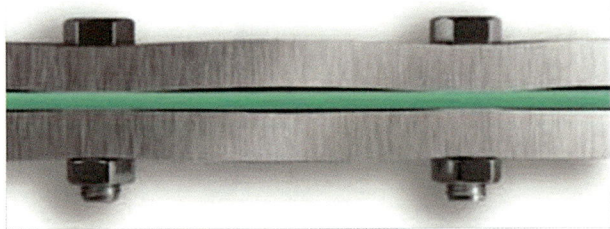

FIGURE 13.3   An illustration of flange bowing of a compression gasket due to excessive bolt tension.

FIGURE 13.4   Application of an anaerobic adhesive to a flange.

13.9   STANDARDS AND SPECIFICATIONS

Testing standards and performance specifications for anaerobic adhesives and sealants have been established by government agencies and industrial organizations in several countries. In the United States, military specifications (MIL-Specs) have been established for different classes of anaerobic products. In the United Kingdom, the Ministry of Defence established specification DTD 5628-5633 for a range of products in five strength and four viscosity bands.

MIL-S-22473, April 12, 1983, "Sealing, Locking and Retaining Compounds: (Single Component)" covers the earliest "letter grade" products. MIL-S-46163A, July 12, 1983, "Sealing, Lubricating and Wicking Compounds: Thread-Locking, Anaerobic, Single-Component" covers products for sealing (Type I), Lubricating (Type II), and Wicking (Type III). MIL-R-46082B, June 10, 1983, "Retaining Compounds Single Component, Anaerobic" (Amendment 6, January 9, 1990) covers three types of retaining compounds. ASTM D5363-16 is intended to replace these MIL specifications and covers single-component adhesives suitable for locking, sealing, and retaining threaded or cylindrical assemblies. The specification also aims to be a means of classifying anaerobic adhesives.

European Standard EN 751-1:1996 specifies requirements and test methods for anaerobic jointing compounds to be used on threaded metallic joints in contact with first, second, and third family gases and hot water. This is the relevant standard for anaerobic adhesives, and has been widely adopted by a number of European countries as national standards such as BS EN 751-1 (United Kingdom) and DIN EN 751-1 (Germany). Deutscher Verein des Gas und Wasserfaches (DVGW) CERT GmbH is the largest certification body in Europe and DVGW certification has gained wide acceptance in the industry. The Australian Gas Association (AGA) certifies jointing compounds according to Standard AS 4623-2008. AGA has certified anaerobic sealants for use at pressures up to 20 bar (2000 kPa). The Bundesanstalt für Materialforschung und-prüfung (BAM) institute in Germany has certified some anaerobic thread sealants for use in oxygen cylinders under defined temperature and pressure limits. Underwriters Laboratories Inc. (UL) also certifies thread sealants for use under defined pressure limits in contact with certain liquids.

Anaerobic adhesives have been approved for potable water sealing applications by a number of certification bodies. Different assessment procedures apply, but the common objective is to ensure that drinking water that comes into contact with adhesives is safe to use. NSF International (originally the National Sanitation Foundation, United States) has gained broad global acceptance and developed the NSF/AINSI 61 standard for materials in contact with drinking water. The Water Regulations Advisory Scheme (WRAS, United Kingdom) is a conformance mark that demonstrates that a material meets the requirements for contact with potable water in the United Kingdom. An interesting feature of this approval is the flavor and odor (organoleptic) evaluation of water in contact with the bonded test piece. In Germany, plastic materials that come into contact with drinking water must meet the KTW guidelines issued by the Umwelt Bundesamt (UBA). UBA has recently reviewed the requirements for anaerobic adhesives and, at the time of writing, new proposals are in the process of implementation. Under these proposals, German drinking water approval will be based on a set of guidelines relating to formulation requirements and proper use of the adhesives. Inspection Institute for Water Supply Articles (KIWA - the Netherlands) also operates a set of guidelines for materials in contact with drinking water, while another popular standard in Australia and New Zealand is AS/NZS 4020.

13.10   SUSTAINABILITY

It is now possible to make anaerobic adhesives having excellent H&S labeling. For example, Henkel manufactures a range of Loctite anaerobic products that are label free under Classification, Labelling and Packaging Regulation (CE) 1272/2008 (CLP) regulations [35]. However, as H&S legislation continues to evolve with differing interpretations from region to region, it is expected that there will be continued pressure on formulators to maintain good H&S labeling by removing raw materials that acquire adverse H&S labeling and replacing them with safer alternatives.

13.11 SUMMARY

Anaerobic adhesives are an important and fascinating segment of the adhesives industry in general and the engineering adhesives industry in particular, driven by sophisticated engineering customers requiring a plethora of customized products.

After a quiet decade from about 1995 in the anaerobic adhesive patent literature worldwide, there has been a resurgence of technological development, focusing on both sustainability and improved performance. This drive for improved performance is likely to result in anaerobic technology merging with other technologies not only to bond new substrates in the traditional anaerobic application areas, but also to expand into new structural bonding- and sealing-type applications. The future is bright for anaerobic adhesives. We certainly find it a rewarding sector of the adhesives industry to work in!

REFERENCES

1. R. D. Rich. Anaerobic adhesives. In: *Handbook of Adhesive Technology: Revised and Expanded*, 2nd edn, A. Pizzi and K. L. Mittal (Eds.), Chapter 39, CRC Press, Boca Raton, FL, 761–774 (2003).
2. D. J. Stamper. Curing characteristics of anaerobic sealants and adhesives *Br. Polym. J.* 11, 34–39 (1983).
3. V. Krieble. Accelerated anaerobic curing compositions. US Patent 3046262, assigned to American Sealants Co. (July 24, 1962).
4. Y. Okamoto. Anaerobic adhesive cure mechanism-I. *J. Adhes.* 32, 227–235 (1990).
5. S. Wellmann and H. Brockmann. New aspects of the curing mechanism of anaerobic adhesives. *Int. J. Adhes. Adhes.* 14, 47–55 (1994).
6. D. Raftery, M. Smyth, R. Leonard and D. Heatley. Effect of copper(II) and iron(III) ions on reactions undergone by the accelerator 1-acetyl-2-phenylhydrazine commonly used in anaerobic adhesives. *Int. J. Adhes. Adhes.* 17, 151–153 (1997).
7. D. Raftery, P. O'Dea, M. R. Smyth, R. G. Leonard, and M. Brennan. An electroanalytical investigation of reactions undergone by elemental iron and copper in the presence of anaerobic adhesive based cure components. *Int. J. Adhes. Adhes.* 17, 9–16 (1997).
8. J. Gorman and B. Nordlander. Anaerobic adhesive composition and method of producing same. US Patent 3300547, assigned to Loctite Corporation (January 24, 1967).
9. B. Nordlander. Anaerobic sealant composition containing monoacrylate esters. US Patent 3435012, assigned to Loctite Corporation (March 25, 1969).
10. B. Schmidt, J. Knebel and G. Graeff. Process for the preparation of (meth)acrylates of tetra- or polyhydric alcohols. US patent 7521578, assigned to Evonik Roehm GmbH (April 21, 2009).
11. B. Malofsky. Anaerobic adhesive composition having improved strength at elevated temperature consisting of unsaturated diacrylate monomer and maleimide additive. US Patent 3988299, assigned to Loctite Corporation (October 26, 1976).
12. K. Azuma, H. Kato, K. Kondo, A. Motegi, O. Suzuki, H. Tatemichi and I. Tsuji. Anaerobic curable compositions. US patent 3925322, assigned to Toa Gosei Chem Ind (December 9, 1975).
13. P. Klemarczyk, D. Birkett, D. Farrell, P. Wrobel, C. McArdle and G. Clarke. Hydrogen peroxide complexes and their use in the cure system of anaerobic adhesives. US Patent 8524034, assigned to Henkel IP and Holding (September 3, 2013).
14. A. Jacobine and D. Glazer. Substituted saccharin compounds and curable compositions containing same. European patent 0232364, assigned to Loctite Corporation (February 23, 1994).
15. V. Krieble. Anaerobic curing compositions containing acrylic acid diesters. US patent 3041322 (June 26, 1962).
16. O. Hara. Anaerobic adhesive composition. European patent 0499483 A1, applicant: Three Bond Co. Ltd. (published August 19, 1992).
17. P. Klemarczyk and K. Brantl. Cure accelerators for anaerobic curable compositions. US patent 6835762, assigned to Henkel Corporation (December 28, 2004).
18. D. Birkett and M. Wyer. Cure accelerators for anaerobic adhesive compositions. US patent 7951884, assigned to Loctite R&D Ltd (May 31, 2011).
19. S. Attarwala, Q. Zhu, D. Birkett, M. Wyer, D. Mullen and L. McGarry. Anaerobically curable compositions. European patent 2488580, assigned to Henkel IP and Holding GmbH (September 9, 2015).

20. D. Birkett, M. Wyer, A. Messana, D. Dworak and A. Jacobine. Cure accelerators for anaerobic curable compositions. US patent 8481659, assigned to Henkel Ireland Ltd and Henkel Corporation (January 29, 2013).

21. A. Jacobine, A. Messana, D. Glaser and S. Nakos. Cure accelerators for anaerobic curable compositions. US patent 8106141, assigned to Henkel Corporation (January 31, 2012).

22. D. Birkett, M. Wyer, A. Messana, D. Dworak and A. Jacobine. Cure accelerators for anaerobic curable compositions. US patent 8362112, assigned to Henkel Ireland Ltd and Henkel Corporation (January 29, 2013).

23. A. Messana and A. Jacobine. Adducts useful as cure components for anaerobic curable compositions. US patent 8598279, assigned to Henkel IP US LLC (December 3, 2013).

24. E. Frauenglass and G. Werber. Highly stable anaerobic compositions. US patent 4038475, assigned to Loctite Corporation (July 26, 1977).

25. R. Kriebel. Anaerobic curing sealant composition having extended shelf stability. US patent 3043820 (July 10, 1962).

26. A. Toback and J. O'Connor. Polymerizable acrylate composition and curing accelerator therefor. US patent 3591438, assigned to Loctite Corporation (July 6, 1971).

27. D. Birkett, A. Jacobine, A. Messana, J. Schall, D. Mullen, M. Wyer, L. Hurlburt, J. Ouyang and S. Shah. Accelerators for two step adhesive systems. US patent 9371473, assigned to Henkel IP and Holding (June 21, 2016).

28. D. Birkett, D. Mullen and M. Wyer. Primer composition for anaerobic adhesive. British patent 2502554, assigned to Henkel Ireland Ltd. (February 3, 2016).

29. G. Piestert and H. Gilch. Adhesive composition and method. US patent 4285755, assigned to USM Corporation (August 25, 1981).

30. T. Baldwin, D. Bennett and W. Lees. Anaerobic curing compositions. US patent 4138449 (February 6, 1979).

31. C. Fairey, E. Frauenglass and L. Vincent. Petroleum equipment tubular connection, US patent 4813714, assigned to Loctite Corporation (March 21, 1989).

32. L. O'Connor. Controlling the turn of the screw. *Mech. Eng.* 113 (9), 52–57 (1991).

33. G. S. Haviland. *Machinery Adhesives for Locking, Retaining, and Sealing*, 1st edn, pp. 220–272, Marcel Dekker, New York (1986).

34. G. S. Haviland. *Machinery Adhesives for Locking, Retaining, and Sealing*, 1st edn, pp. 313–315, Marcel Dekker, New York (1986).

35. Loctite Sustainable Solutions. Sustainable Products and Solutions http://www.loctite.com.au/sustainable-products-and-solutions-6802.htm (2016).

14 Aerobic Acrylic Adhesives

Nigel Sweeney

CONTENTS

14.1 INTRODUCTION

Acrylic adhesives based on methacrylate monomer systems can be classed as reactive, anaerobic, or aerobic. These adhesives are 100% reactive ingredient systems and do not contain carrier solvents. Reactive and anaerobic acrylics cures are based on redox chemistry. Typically, a hydroperoxide is decomposed by a metal catalyst in the presence of accelerators to form free radicals. These free radicals initiate a free radical polymerization of the methacrylate monomers. In reactive systems, the cure occurs when two parts of the adhesive are mixed. For anaerobic systems, the cure occurs in the absence of oxygen, such as when parts are assembled in a typical threadlocking application.

Oxygen acts as an inhibitor for the redox cure by forming peroxy radicals, which are poor free radical initiators. An initiator (I) forms a free radical that adds to a (meth)acrylate monomer (M). This free radical monomer then adds to another monomer, starting a chain propagation process. Termination can occur when two propagating chains meet. Oxygen can also form a peroxy radical on the chain, inhibiting further growth:

$$I \rightarrow I^{\bullet}\,(\text{Initiation})$$

$$I^{\bullet} + M \rightarrow I - M^{\bullet}$$

$$I - M^{\bullet} + nM \rightarrow I - M_n - M^{\bullet}\,(\text{Propagation})$$

$$2I - M_n - M^{\bullet} \rightarrow I - M_n - M - M - M_n - I\,(\text{Chain termination})$$

$$I - M_n - M^{\bullet} + O_2 \rightarrow I - M_n - M - OO^{\bullet}\,(\text{Chain termination})$$

This is very often seen in bonding applications where the bond fillet or *squeeze out* remains tacky to touch when it is exposed to air outside the bondline in anaerobic and reactive two-component (2K) acrylic adhesives.

Aerobic acrylic adhesives are classed as acrylic adhesives that can overcome this oxygen inhibition or can be initiated by oxygen. They are called aerobic acrylic adhesives due to their tolerance of oxygen during curing, in contrast to reactive and anaerobic acrylic adhesives.

Examples include heat curable two-step acrylics, radiation-curable acrylics, and 2-component (2K) organoborane initiated acrylics. The examples shown in this chapter will highlight the main advantages due to their tolerance to oxygen as well as other properties that offer benefits to users.

14.2 TWO-STEP AEROBIC ACRYLICS

Two-step aerobic acrylics consist of an adhesive resin containing methacrylate monomers, oligomers, and toughening agents alongside other fillers and a peroxide. The surfaces to be bonded are coated (usually by brush or aerosol spray) with a solvent-based activator. The activator commonly contains an amine–aldehyde condensate that accelerates hydroperoxide decomposition by metal catalysts. This reaction overcomes the inhibition of free radicals by atmospheric oxygen by accelerating the rate of redox free radical formation.

These adhesives provide bonding that is structural and environmentally resistant. Properties of the two-step systems are compared with other typical structural adhesives in Table 14.1. The main advantages of two-step acrylics are their high-temperature resistance and short times to reach handling strengths for bonded parts. Their weakness lies in their relatively poor performance at larger bond gaps in comparison with other adhesives.

A well-established use for two-step acrylics is for magnet bonding applications [1]. These applications typically involve the bonding of ferrite, aluminum–nickel–cobalt, and neodymium–iron–boron permanent magnets (Figure 14.1). These magnets are used in electric motors, engines, speakers, and transformers. The quick times to achieve handling strengths at room temperature means that two-step acrylics can optimize production line assembly of these parts in comparison with 2K room-temperature curable epoxies or one-component (1K) heat curable epoxies.

TABLE 14.1
Comparison of Properties for Structural Adhesives

	Two-Part Epoxies	Urethanes	Two-Part Acrylics	Two-Step Acrylics
Mixing required	Yes	Yes	Yes	No
Typical temperature range for use	−54°C–82°C	−54°C–121°C	−54°C–121°C	−54°C–149°C
Adhesion				
Metals	Excellent	Good	Excellent	Excellent
Plastics	Fair	Very good	Excellent	Fair
Glass	Excellent	Good	Good	Excellent
Rubber	Fair	Good	Poor	Poor
Wood	Very good	Fair	Good	Good
Material Properties				
Peel strength	Medium	Medium	High	High
Tensile strength	High	Medium	High	High
Gap Fill				
Optimal range	1–1.5 mm	1–1.5 mm	2–10 mm	0.01–0.5 mm
Maximum	30 mm	30 mm	125 mm	1 mm

FIGURE 14.1   Magnet bonding using two-step aerobic acrylics.

14.3   LIGHT OPTICALLY CLEAR ADHESIVES (LOCAs)

Radiation-curable acrylics use radiation at selected wavelength intensities to irradiate a molecule that provides a source of free radicals that cure the acrylic adhesive (Figures 14.2 and 14.3). The irradiation can occur over a wide range of wavelengths (mercury arc lamps) or a precise range (light emitting diodes [LEDs]). Customers now frequently use LEDs with very precise wavelength emissions as the radiation source [2]. This represents a move away from mercury arc lamps. Although these lamps have a broad range from the ultraviolet (UV) to the visible regions of the electromagnetic spectrum, the technology is not as environmentally friendly or as energy efficient as LEDs. LEDs are also safer to operate with respect to exposure to UV radiation, which is limited to UV-A and radiation at longer wavelengths.

LOCAs constitute a commonly used UV curable acrylic technology. They are used in the bonding of display devices, as shown in Figure 14.4. The market for these adhesives has increased substantially in the last few years, due to the use of smartphones, and will be expected to grow even further with the advent of smart device technology and its widespread use.

For display bonding applications, LOCAs must have low levels of yellowing and reduced shrinkage of the cured adhesive. These challenges are overcome by the use of oligomeric diacrylate monomers that contain polyurethane or polybutadiene functionality [3].

Figure 14.4 shows that LOCAs can be used in several different layers between the direct bonding surface, the optically conductive indium tin oxide (ITO) glass layers, and the cover lens. The cured

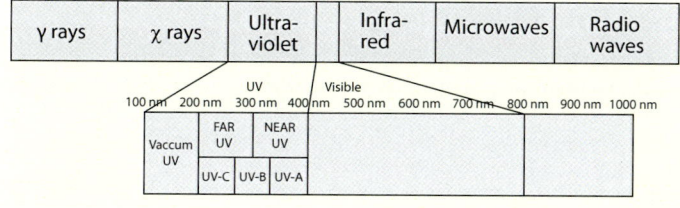

FIGURE 14.2   The electromagnetic spectrum. For photocuring of acrylics, the light used comes from the UV region.

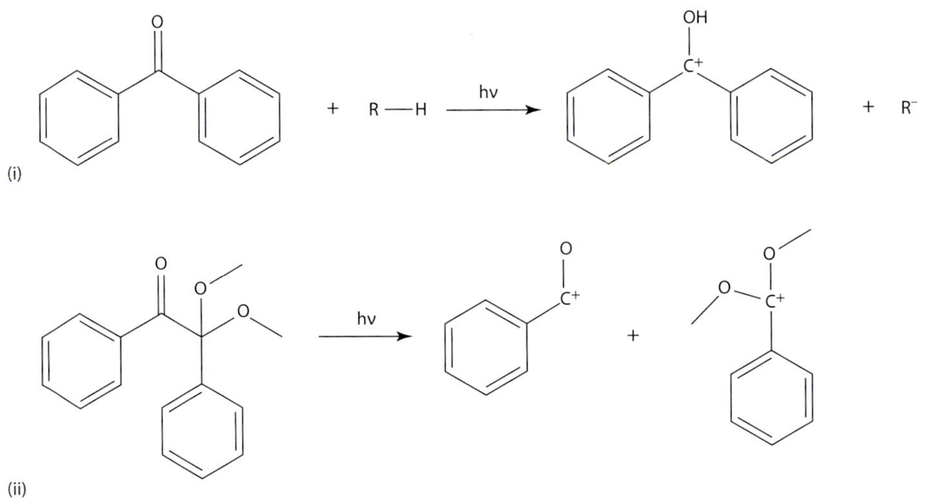

FIGURE 14.3 Examples of photoinitiators used to cure UV acrylic adhesives. The radical generation is based on hydrogen abstraction from a donor (i) or direct cleavage of the molecule (ii).

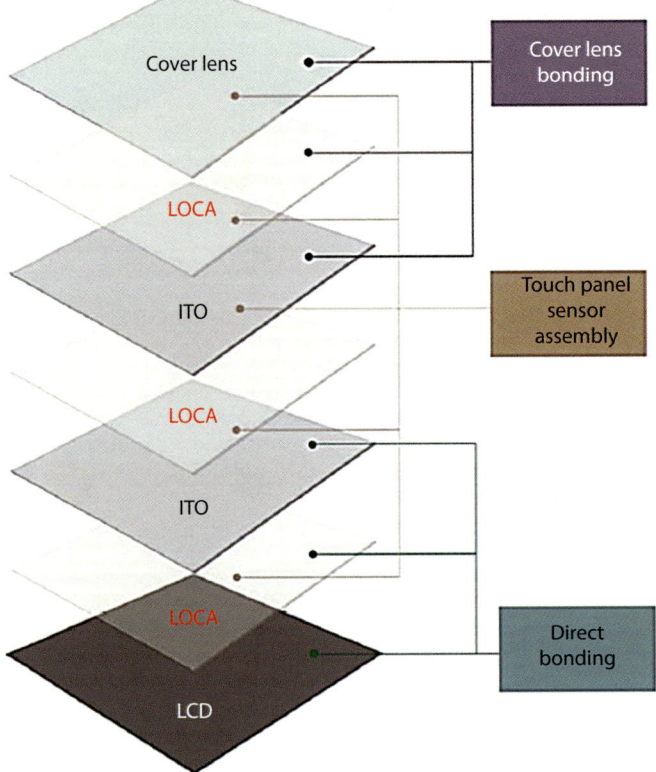

FIGURE 14.4 Examples where LOCAs are used in display devices.

LOCA helps to optimize the performance of the display devices by bonding these layers and reducing air gaps. Properties that are optimized by replacing the air gap layer include:

• Increased contrast ratio by using adhesives with a refractive index matching the glass layer's refractive index. Air gaps cause losses of light due to reflectance. Cured LOCAs prevent this from occurring.

TABLE 14.2

Comparison of LOCA and OCA Tape Adhesives for Display Bonding

	LOCA	OCA Tape
Adhesion/reliability	High	Low
Die-cutting required	No	Yes
Reworkable	Yes at precuring stage	No
Large panel	Suitable	Unsuitable
Gap filling (at edges)	Good	Poor
Equipment	Application, precuring, and curing	Roller lamination and autoclaving
Surface treatment	Fewer requirements	More often required
Assembly line	Full automation possible	Only semiautomation possible

- Increased impact resistance.
- Increased display life.
- Reduced power consumption (due to reduced loss of light).

LOCAs based on UV–LED curable acrylics offer several advantages when compared to tape-based optically clear adhesives (OCAs). Table 14.2 shows a comparison between the two systems.

Although equipment requirements for LOCAs versus OCA tapes are higher and more expensive, the benefits are automation of the assembly process and elimination of die-cutting. The application of tape adhesives over a large panel is a less robust and reproducible procedure compared to automated LOCA dispensing and curing.

The process uses LED curing systems to provide curing at up to three different stages of the assembly process (Figure 14.5):

- Spot curing using UV–LED single arrays. A partially cured adhesive layer is formed between the bonded layers. It is possible to debond and rework the bonded layers if required.
- Flood curing using UV–LED multiple arrays to ensure the display area is covered.

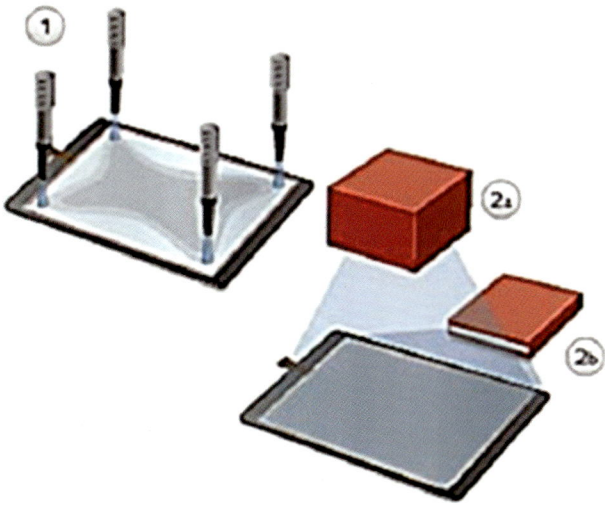

FIGURE 14.5 UV–LED curing of LOCAs. Spot curing (1), bulk curing with a flood cure (2a), and curing of the exposed sides of the display with a line cure (2b).

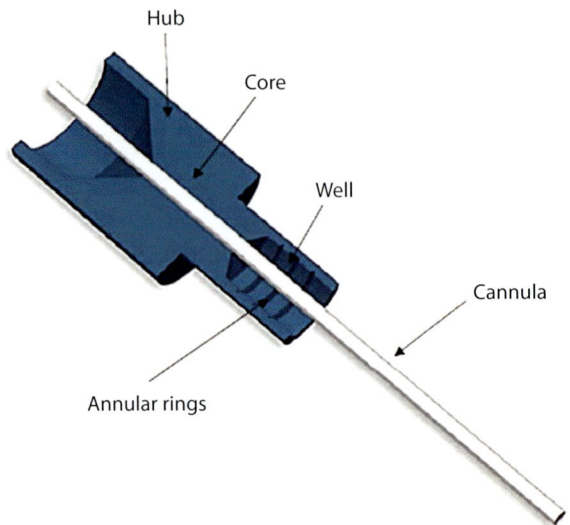

FIGURE 14.6   Structure of needle (cannula) bonded in a plastic hub.

• Line curing of exposed sides not reached by the flood curing process.

14.4   UV ACRYLIC NEEDLE BONDERS

Another application area for UV curable acrylics is in needle bonding. Typically, in this application, a stainless steel cannula is bonded within a plastic hub. The plastics used range from polyethylene (PE), polypropylene (PP), polycarbonate (PC), acrylics, polyurethanes (PUs), acrylonitrile–butadiene–styrene (ABS) and polysulfones (PSs). For adhesion to polyolefins, plasma surface treatment is required to make the surfaces bondable [4]. Figure 14.6 shows a cross-section of a needle in its hub. The hub, needle, and the adhesive must meet certain requirements so that the needle is securely bonded.

• Cannula gauge size can influence the mode of failure of an adhesive. The adhesive forms a bond within the well to the cannula. As the cannula gauge increases, the circumference of the cannula becomes smaller. This can cause interfacial failure on the cannula during a pull-out test.
• The well depth can be optimized to increase the amount of adhesive that flows into it, forming a good bond to the hub. Annular rings can be included to form a stronger bond on more difficult-to-bond plastic substrates.
• The core is where the cannula forms an engagement point with the hub. Adhesive (depending on viscosity) can also flow into this area between the needle and the hub. For pull-out testing, if the engagement length is optimized, the adhesive joint strength will be greater than the hub material.

Table 14.3 lists the advantages and disadvantages of UV curable acrylics versus other types of adhesives used in this application area: UV curable cyanoacrylates and 1-component heat curable epoxies.

The biggest advantage of UV curable acrylics is their combination of rapid cure and the ability of formulators to meet the requirements of the needle manufacturers [5]. These requirements include:

• Fluorescence as an in-line method for detection of adhesive for quality control purposes.

TABLE 14.3

Advantages and Disadvantages of the Three Classes of Adhesives Used for Needle Bonding

	Advantages	Disadvantages
UV curable acrylics	Cure in 6–20 s	Lack of shadow curing
	Broad range of T_gs and other physical properties possible	High intensity light is required to overcome oxygen inhibition of cure (tackiness)
	Fast flow, low-viscosity formulations	
	Rigid resins with good thermal and chemical resistance	
UV curable cyanoacrylates	Cure in 3–6 s	Highest volumetric cost
	<5 s cure with low intensity light	Blooming can occur in shadow curing areas
	Shadow curing possible	Lowest fluorescence
		Lowest thermal and chemical resistance
1K heat curable epoxies	Lowest volumetric cost	Slowest cure
	Highest fluorescence	Viscosities of less than 5000 cP not possible
	Shadow curing possible	
	Best chemical and thermal resistance	

- Viscosity to match needle manufacturers' varying designs. Needle designs can differ in several ways. For example, the cannula can protrude from the core rather than sitting in it perfectly or in a recessed way. The viscosity of the adhesive must be matched to the needle design so that its flow does not block the protruding cannula.
- Resistance to methods of sterilization such as ethylene oxide, gamma irradiation, and autoclaving.
- Heat and humidity resistance.

14.5 ORGANOBORANE INITIATED ACRYLIC ADHESIVES

As described in the Introduction section of this chapter, oxygen is commonly an inhibitor of 2K acrylic redox cure systems. However, there is a notable exception to this oxygen inhibition. Organoborane initiated acrylics contain a cure system that is initiated by atmospheric oxygen, that is, an autoxidation. These adhesives have found a unique field of applications for acrylic adhesives; for instance, adhesion to polyolefins without the need for surface treatments.

Typically, the organoborane is a trialkylborane in the form of an amine complex (a so-called *borane:amine complex*). These compounds are relatively stable toward autoxidation. In the presence of acids or isocyanates that react with the amine, the trialkylborane is decomplexed and undergoes autoxidation. In a 2K system, the borane:amine complex is present in one part and the decomplexing agent is present in the other. The reaction occurs on mixing of the two parts. 3M and Dow have patented the use of borane:amine complexes as initiators for acrylic adhesives [6,7]. Henkel has patented the use of trialkylborohydride salts as initiators, which are analogous to the borane:amine complexes in terms of reactivity [8].

Sonnenschein and coworkers proposed a mechanism for the generation of free radicals from a trialkylborane by autoxidation [9]. This is shown in Figure 14.7. Step (1) shows that oxygen forms a borate ester, which is a source of alkoxy and alkyl radicals, as shown in Steps (2), (3), and (4). Step (5) shows that the alkyl radicals can form methacrylate monomers for a propagating radical species. Steps (6) and (7) show how these free radicals not only initiate polymerization of methacrylates but allow them to graft onto hydrocarbon surfaces (denoted by R') leading to adhesion to the surface.

$$R_3B + O_2 \longrightarrow R_2BOOR \qquad (1)$$

$$R_2BOOR \longrightarrow R_2BO^{\cdot} + RO^{\cdot} \qquad (2)$$

$$RO^{\cdot} + R'\text{-}H \longrightarrow ROH + R'' \qquad (3)$$

$$3R_2BO^{\cdot} \longrightarrow R_3B_3O_3 + 3R^{\cdot} \qquad (4)$$

$$R^{\cdot} + nM \longrightarrow RM_n^{\cdot} \qquad (5)$$

$$R'' + nM \longrightarrow R'M_n^{\cdot} \qquad (6) \text{ Adhesion}$$

$$R'' + RMn^{\cdot} \longrightarrow R'M_nR \qquad (7) \text{ Adhesion}$$

FIGURE 14.7 Autoxidation of trialkylboranes to initiate free radical polymerization. (Reprinted from Sonnenschein, M. F. et al., *Macromolecules*, 39, 2507–2513, 2006. With permission. Copyright 2006 American Chemical Society.)

Henkel, Dow, and 3M have 2K acrylic products based on this technology. These products combine the structural bonding and environmental resistance associated with acrylics alongside adhesion to polypropylene and polyethylene, without the need for plasma and corona discharge surface treatments.

These adhesives allow for the use of polyolefin structures in general industry or automotive applications. This can lead to cost savings related to weight reduction and the use of polyolefins in comparison with more expensive engineering thermoplastics [10].

14.6   PROSPECTS

Aerobic acrylic adhesives are a well-established technology addressing issues of concern to manufacturers: speed, automation, and reliability. Advantages shown in the previous sections of this chapter (e.g., rapid cure, OCAs, polyolefin adhesion) are ones that will also be of importance for the future. For LOCA applications, development will be required to provide solutions for bonding the layers in, for example, curved displays or touch screen displays of wearable devices [11]. Areas of potential for LOCAs also include encapsulation of organic light emitting diodes (OLEDs) and quantum (Q) dots [12,13].

The curing of UV curable acrylics is also an area of continuous development. Although LED lamps address concerns about mercury arc lamps, there is a desire to move to even more sustainable processes. As an example, there is ongoing research to develop photosensitizers that efficiently use visible light to initiate free radical curing [14].

An example of a breakthrough technology for UV curable acrylics is the potential for reversible adhesion on demand [15]. This shows that the prospects for aerobic acrylics as a technology to address not just current needs, but more demanding new ones, are indeed promising.

14.7   SUMMARY

Aerobic acrylic adhesives are capable of rapid curing due to their cure systems, which overcome oxygen inhibition. The adhesives are based on (meth)acrylate oligomers and monomers and can be used to bond a wide variety of substrates. Adhesion to polyolefins is possible with certain 2K formulations.

The applications that these adhesives are used in (e.g., display bonding and needle bonding) are highly demanding, both for optimizing the manufacturers' assembly processes and meeting the end users' performance expectations.

REFERENCES

1. Sticky business–how to glue neodymium magnets. https://www.kjmagnetics.com/blog.asp?p=sticky-business-how-to-glue-neodymum-magnets (2016).
2. UV LED curing products. http://www.phoseon.com/products/uvled-vs-mercury-arc-lamps (2016).
3. Y. Yuan and D. Lu. Liquid optically clear photocurable adhesive for display application. US Patent Appl. 20150166860, assigned to Henkel (China) Co. Ltd. and Henkel IP & Holding GmbH (2015).
4. M. Strobel, C. S. Lyons and K. L. Mittal (Eds.). *Plasma Surface Modification of Polymers: Relevance to Adhesion*, CRC Press, Boca Raton, FL (1994).
5. S. M. Tavakoli, D. A. Pullen and S. B. Dukkerton. A review of adhesive bonding techniques for joining medical devices. *Assembly Automation*, 25, 10–15 (2005).
6. J. V. Zharov and J. N. Krasnov. Polymerizable compositions made with polymerization initiator systems based on organoborane amine complexes. US Patent 5539070, assigned to 3M (1996).
7. M. F. Sonnenschein, S. P. Webb and N. G. Rondan. Amine organoborane complex polymerization initiators and polymerizable compositions. US Patent 6706831 B2, assigned to Dow Global Technologies Inc. (2004).
8. B. J. Kneafsey and G. Coughlan. Metal alkyl borohydride polymerisation initiators, polymerisable compositions and uses thereof. US Patent 6844080 B2, assigned to Loctite (R&D) Limited (2005).
9. M. F. Sonnenschein, S. P. Webb, O. D. Redwine, B. L. Wendt and N. G. Rondan. Physical and chemical probes of the bond strength between trialkylboranes and amines and their utility as stabilized free radical polymerization catalysts. *Macromolecules*, 39, 2507–2513 (2006).
10. Simpler adhesive bonding for low surface energy polyolefin plastics. http://www.designworldonline.com/simpler-adhesive-bonding-for-low-surface-energy-polyolefin-plastics/ (2013).
11. 10 big trends in display technology. http://www.samsungvillage.com/blog/2015/06/23/10-trends-display-technology-2015/ (2015)
12. J. Cao and D. Herr. Radiation-curable rubber adhesive/sealant. US Patent Appl. 2014/0190736 A1, assigned to Henkel Corporation (2014).
13. 3M quantum dot enhancement film. http://multimedia.3m.com/mws/media/985375O/3mtm-quantum-dot-enhancement-film-qdef-white-paper.pdf?fn=Quantum%20Dot%20QDEF%20Whitepaper.pdf (2013).
14. W. Wu, H. Guo, W. Wu, S. Ji and J. Zhao. Organic triplet sensitizer library derived from a single chromophore (BODIPY) with long-lived triplet excited state for triplet–triplet annihilation based upconversion. *J. Org. Chem*, 76, 7056–7064 (2011).
15. C. Heinzmann, S. Coulibaly, A. Roulin, G. L. Fiore, and C. Weder. Light-induced bonding and debonding with supramolecular adhesives. *ACS Appl. Mater. Interfaces*, 6, 4713–4719 (2014).

15 Biobased Acrylic Adhesives

Nigel Sweeney

CONTENTS

15.1 INTRODUCTION

The use of biobased materials as adhesives contains many examples from ancient to modern times. These adhesives came from natural sources such as animals, fish, casein, and starch [1]. In the twentieth century, the use of petroleum feedstocks provided the building blocks for the materials used in contemporary adhesives such as acrylics, epoxies, silicones, and polyurethanes. In addition to petroleum, natural gas also acts as a feedstock.

The use of petroleum feedstocks for synthetic chemicals is of great concern for the impact on the global environment, the finite nature of petroleum and natural gas, and the potential for increased costs associated with more costly and demanding extraction methods. The adhesives industry recognizes the need to move away from petroleum-based feedstocks to more sustainable and environmentally friendly alternatives, with many examples of companies committed to sustainability [2]. These alternatives can include the use of biorenewable feedstocks from biological sources and the use of processes to make chemicals using green technology (e.g., enzyme catalysis). The overall objective is to reduce CO_2 and other greenhouse gases or environmentally harmful emissions from the manufacturing processes.

The amount of biobased carbon content in a chemical can be determined using ASTM D6866. The percentage of biobased carbon content is a measure of the carbon obtained from biomass source out of the total organic carbon content [3].

Acrylic adhesives are an important class of adhesives. This class of adhesives consists of solvent- and aqueous solution-based pressure-sensitive adhesives (PSAs), methacrylate structural adhesives, anaerobic adhesives, and radiation-curable acrylics. The basis of these adhesives are acrylate or methacrylate monomers, which are either polymerized to form high molecular weight polyacrylates for PSAs or which undergo a free radical-initiated polymerization to initiate a cure in methacrylates, anaerobics, and radiation-curable acrylics.

The focus of this chapter is on biobased materials for use in these types of adhesives, notably the potential for biobased (meth)acrylate monomers.

15.2 BIOBASED ACRYLATE MONOMERS

15.2.1 BIOBASED ACRYLIC ACID

Acrylic acid itself, or acrylates derived from it by reactions with alcohols, are fundamental ingredients in the making of acrylic adhesives. The process currently used widely in industry is the double oxidation of propene (Figure 15.1).

Acrylic acid and its derivatives are used in many applications outside of adhesives such as coatings, superabsorbent polymers, paints, construction materials, and textiles. The demand is driven by growth in the Asia-Pacific market. In 2011, global consumption of acrylic acid and its derivatives had reached almost 10.1 million metric tons [4]. There is a growing demand for acrylic acid and its derivatives that has encouraged several of the major producers to develop processes to produce biobased acrylic acid.

15.2.1.1 3-Hydroxypropionic Acid Process

The 3-hydroxypropionic acid (3-HP) process is the fermentation of sugars to yield 3-HP, which is then dehydrated to give acrylic acid (Figure 15.2) [4].

BASF was formerly in partnership with Novozymes and Cargill to develop the 3-HP process. In 2013, they showed that 3-HP could be made using the process on the pilot scale, with conversion to glacial acrylic acid in 2014 [5]. Dow and OPX Biotechnologies also collaborated to develop the 3-HP process. In 2015, Cargill acquired OPX Biotechnologies, with the objective to scale up the process [6].

15.2.1.2 Glycerol Process

Glycerol can be obtained as a by-product from the process of making biodiesel. Dehydration of glycerol forms acrolein, which can then be oxidized into acrylic acid (Figure 15.3) [4].

Arkema has developed this process on the pilot scale, but to date has not scaled it up, due to the potential increased cost of obtaining the glycerol from biodiesel versus the conventional propene process [7].

FIGURE 15.1 Double oxidation of propene to form acrylic acid. This method is the most commonly used in the chemical industry.

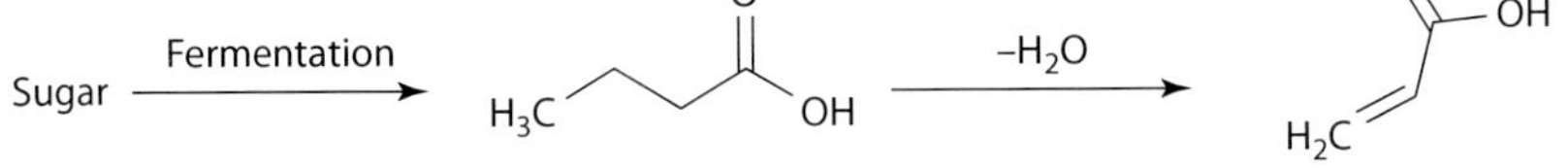

FIGURE 15.2 3-HP approach to biobased acrylic acid.

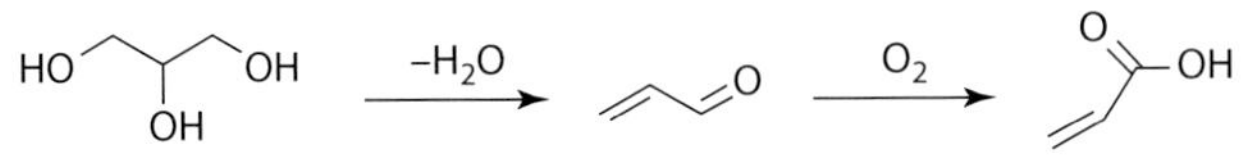

FIGURE 15.3 Glycerol (sourced from biodiesel) is converted to acrylic acid via acrolein.

15.2.1.3 Other Biobased Acrylic Acid Processes

Other processes that have been developed include the dehydration of lactic acid obtained from sugars (Myriant, based in Quincy, Massachusetts), and using bioengineered poly(3-hydroxypropionate) as a source of acrylic acid [8,9].

15.2.2 BIOBASED METHYL METHACRYLATE

Lucite and Evonik have developed more sustainable processes for the production of methyl methacrylate (MMA). Lucite uses its Alpha process based on ethane, carbon monoxide, and ethanol feedstocks [10]. Evonik uses the AVENEER process, which is an acetone-cyanohydrin sulfur-free process [11]. Biobased MMA is a target of development by use of biomass and fermentation technologies to produce either the feedstocks (Lucite) or MMA directly (Lucite and Evonik), which can then potentially be used in the Alpha and AVENEER® processes.

15.2.3 OTHER BIOBASED (METH)ACRYLATE MONOMERS

Biobased acrylic acid and MMA can be reacted with biobased alcohols to give reactive monomers suitable for use in acrylic adhesive formulations. For example, n-butanol can be derived from the fermentation of sugarcane. In combination with biobased acrylic acid, this could lead to the formation of biobased butyl acrylate, a key ingredient of PSAs [12]. Further primary alcohols of interest include medium hydrocarbon chain fatty alcohols (Figure 15.4). These can be obtained from the hydrogenation of naturally occurring fatty acids or their esters [13].

Multifunctional acrylates can be obtained from the naturally occurring soybean oil, by reacting acrylic acid with the hydroxylated soybean oil [14].

As shown in this chapter, it is possible to obtain biobased building blocks for a variety of (meth) acrylates that formulators require for their acrylic adhesives. Arkema and Evonik have produced and made available in commercial quantities (meth)acrylates with varying degrees of biobased carbon content.

Arkema, through its Sartomer range of (meth)acrylates, has a product line based on biorenewable raw materials called Sarbio. Sarbio products have obtained U.S. Department of Agriculture certification of their biobased content [15].

Evonik, through their VISIOMER® *Terra* range, have three methacrylate monomers available with various amounts of biobased carbon [16]. These are isobornyl methacrylate (IBOMA) and the so-called C13 and C17.4 MAs, which are medium hydrocarbon chain methacrylate esters (Figure 15.5).

FIGURE 15.4 Lauryl acrylate, an acrylic ester with a linear C12 chain. Monomers with linear hydrocarbon chains have potential use in PSAs, as they can increase cohesive strength and adhesion to low surface energy substrates.

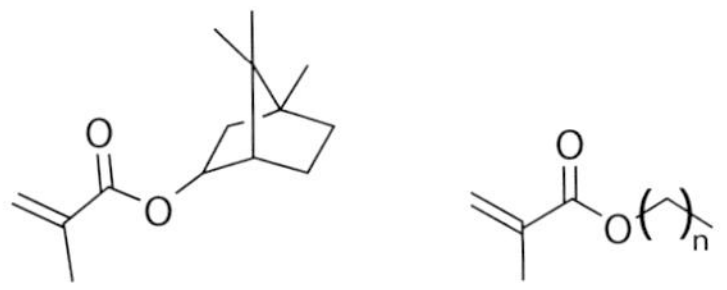

FIGURE 15.5 Evonik has produced isobornyl methacrylate using camphene derived from pine tree resin. Two other products containing medium length hydrocarbon chain methacrylate monomers have also been produced using raw materials obtained from natural oils.

15.2.4 Biobased (Meth)acrylate Macromonomers

The use of polysaccharides is a good example of the successful use of biobased materials in commercialized acrylic PSAs. Ecosynthetix (based in Burlington, Ontario, Canada) has successfully developed PSAs that are based on the copolymerization of alkyl polyglycoside maleic acid esters with acrylic monomers [17]. These products are available as their Eco-Stix® Bio PSA range.

Another sugar-based technology has been developed by Severtson and coworkers. Lactide (derived from corn) and caprolactone can be reacted with 2-hydroxylethyl methacrylate to form a ring-opening macromonomer product (Figure 15.6). This macromonomer can be copolymerized with other acrylates to form PSAs with high biobased carbon content [18].

15.2.5 Emerging Biobased Acrylic Technology

Many of the examples shown in the previous sections of this chapter are direct replacements or equivalents of existing monomers as currently used in acrylic adhesives. There are examples of recent findings in academia that are attempting to bring enhanced properties using biobased materials.

For example, Long and coworkers have made polyacrylates containing nucleoside-base pairs that provide a platform to produce PSAs with excellent cohesive strength and the ability to tune adhesion based on H-bonding interactions of the base pairs [19].

Other examples include acrylate monomers functionalized with groups that are derived from natural sources. Wool and coworkers have synthesized a vanillin methacrylate monomer (Figure 15.7) [20]. The vanillin group is derived from lignin, a natural material with structural reinforcement properties.

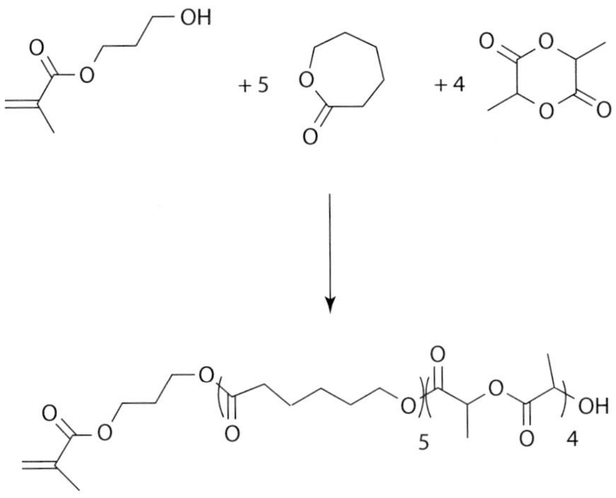

FIGURE 15.6 Macromonomer formed by ring opening of caprolactone and lactide in the presence of 2-hydroxypropyl methacrylate. This monomer can be used in PSAs, resulting in a high biobased carbon content.

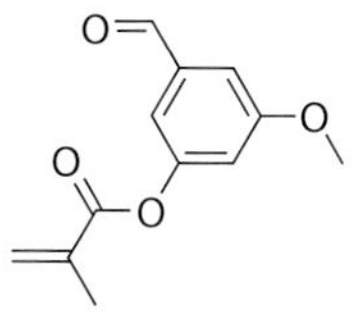

FIGURE 15.7 Vanillin methacrylate monomer derived from biobased lignin.

Holmberg and coworkers have synthesized a series of lignin-based methacrylate polymers. These polymers have tunable flow behavior and thermal properties based on the various lignin derivatives used to make methacrylate monomers. The methacrylates are polymerized by means of reversible addition–fragmentation chain transfer (RAFT) [21]. The polymers have significantly reduced shear viscosities than poly(methyl methacrylate) (PMMA). This means they can be processed from a molten state at lower temperatures than PMMA. The polymers also have potential to replace PMMA-based high T_g blocks in block copolymer polyacrylate-based PSAs.

Another recent example of biobased thermoplastic polymers related to polyacrylates is the work done by Satoh and coworkers [22]. Itaconic acid is derived from fermentation of starch. Ester and imide derivatives of itaconic acid were polymerized using RAFT to form block copolymers with soft poly(itaconate) and hard poly(itaconimide) blocks (Figure 15.8). Block copolymers are often used as additives to toughen acrylic adhesives or as ingredients in hot-melt PSAs.

Itaconic acid can also be used to form biobased unsaturated polyester resins. Dai and coworkers have prepared waterborne biobased ultraviolet (UV) curable coatings from itaconic acid and biobased diols [23].

Sadler and coworkers have synthesized isosorbide methacrylate (Figure 15.9) using the biobased isosorbide [24]. Isosorbide is obtained from glucose. When the monomer is free-radically polymerized, a thermoset material with a T_g of 240°C and a decomposition temperature greater than 400°C is obtained. These physical properties are comparable to high-performance epoxies and outside of the range of free-radically curable vinyl esters.

Cardanol (Figure 15.10) is a phenolic lipid that is derived from cashew nutshell liquid. There are several examples of its use in the synthesis of free-radically polymerizable acrylates. John and Pillai

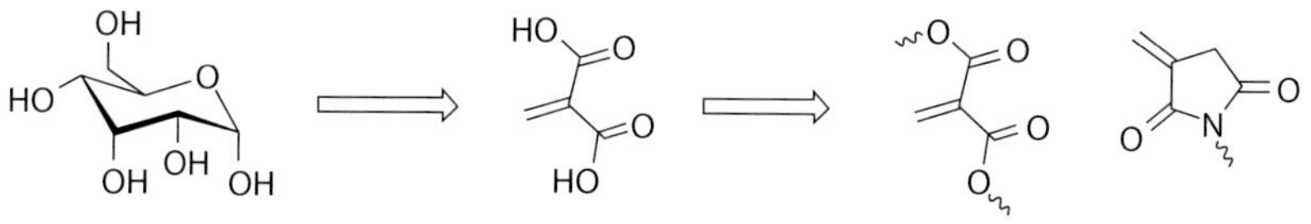

FIGURE 15.8 Itaconic acid can be derived from sugar. It can subsequently be reacted to form itaconate esters and itaconimides.

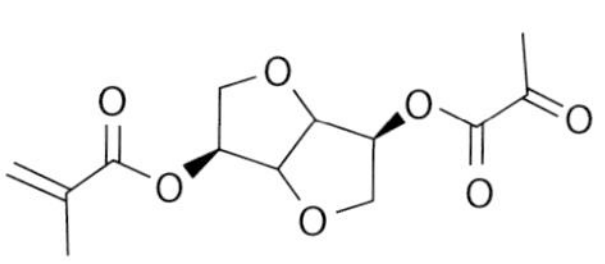

FIGURE 15.9 Isosorbide methacrylate. The fused bicyclic ring structure gives the resultant thermoset polymer material excellent high-temperature performance.

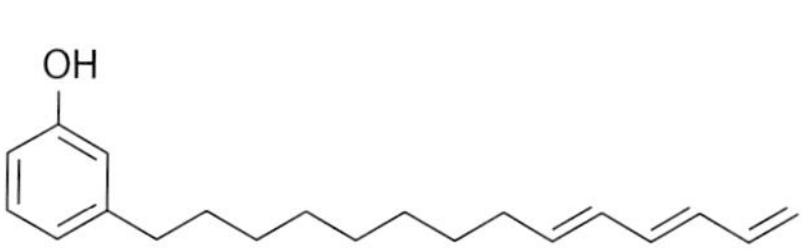

FIGURE 15.10 Cardanol, a chemical derived from cashew nutshell oil.

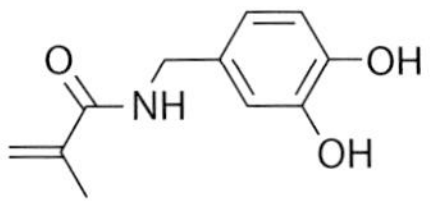

FIGURE 15.11 Dopamine methacrylamide monomer, which has biomimetic properties. The dopamine functionality improves adhesion to wet surfaces.

synthesized cardanol acrylate. The resultant polyacrylate polymer showed the ability to cross-link in the presence of ambient oxygen due to the autoxidation of the allylic groups on the lipid side chain [25]. This technology can be used as a drying agent for acrylic-based adhesives. Liu and coworkers synthesized multifunctional acrylate oligomers using cardanol and biodegradable diols as starting materials [26]. These oligomers can be UV cured, and the polymers formed give better hardness, adhesion strength, and acid resistance when compared with commercially available polyester acrylates.

Dopamine methacrylamide (Figure 15.11) contains the L-dopamine group, which is believed to be the active ingredient in mussel secretions that gives them the ability to adhere to wet surfaces. Glass and coworkers have shown that a polyacrylate containing dopamine methacrylamide as one of the comonomers enhances adhesion in submerged wet environments [27].

Another biobased strategy is the incorporation of biological adhesion phenomena into synthetic adhesives. Crosby and coworkers used the principle of the fibrillar structure of the gecko's foot to form a poly-n-butyl acrylate adhesive film with increased surface wrinkling [28]. The effect of this increased wrinkling is seen in Figure 15.12. It is proposed that the highly wrinkled structure has an increased surface contact perimeter versus the smooth film. This leads to greater adhesion strength.

15.3 APPLICATIONS OF BIOBASED ACRYLIC ADHESIVES

Companies are addressing customer needs for sustainable adhesives with improved health and safety. For example, Henkel has a range of anaerobic adhesives that are nonhazardous [29]. These products are considered *label free*, as they do not possess any pictograms that denote hazardous contents. Further developments in the area of sustainable acrylic adhesives are highly likely to make use of monomers containing biorenewable content. This strategy will involve the replacement of acrylic monomers with biobased equivalents (with the same chemical structures or with similar structures and functionalities).

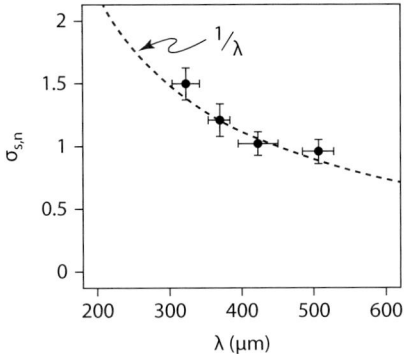

FIGURE 15.12 Separation forces comparison of the replicated wrinkled adhesives versus the smooth, non-wrinkled analogue as a function of λ (the wrinkle wavelength). The results are summarized as a normalized separation strength ($\sigma_{s,n}$) that is determined by the ratio of the σ_s for the wrinkled adhesive versus the σ_s for the smooth adhesive. The dashed curve represents the empirical fit, where $\sigma_{s,n}$ is proportional to $1/\lambda$. (Reprinted from Chan, E. P. et al., *Adv. Mater.*, 20, 711–716, 2008. With permission. Copyright 2008 WILEY-VCH Verlag GmbH & Co. KGaA, Weinheim, Germany.)

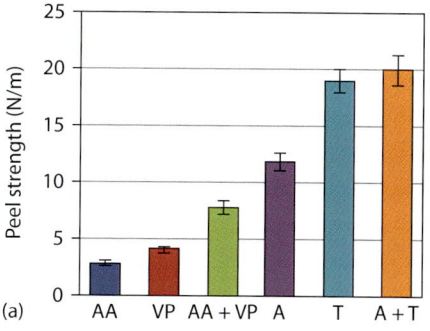 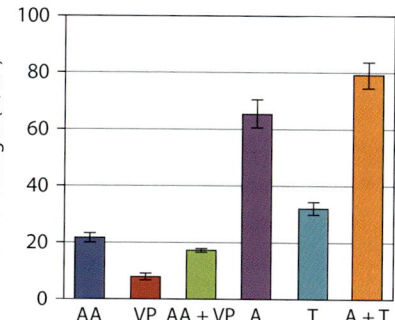

FIGURE 15.13 Peel strength and shear strength of poly(n-butyl acrylate-co-acrylic adenine) (A) and poly(n-butyl acrylate-co-acrylic thymine) (T) in comparison with acrylic acid (AA)- and 4-vinyl pyridine (VP)-based complementary hydrogen-bonding polymer analogues. Tests were carried out according to ASTM standard D-3300. (Reprinted from Cheng, S. et al., *Macromolecules*, 45, 805–812, 2012. With permission. Copyright 2012 American Chemical Society.)

Ecosynthetix has combined the use of biobased acrylic technology for both improved sustainability and performance. In the area of label removal, their PSAs, which are based on sugars, can be removed more easily. The conventional process of label removal requires soaking the labeled bottles in an alkaline solution. Using the biobased label adhesive, water is sufficient enough to remove the label. The removed labels and PSAs are then able to undergo repulping. The recycling process is simplified as a result of this.

BASF has patented the use of biobased (meth)acrylate monomers for making copolymers containing butadiene and vinyl aromatic compounds as comonomers. These copolymers are used as binders or as coatings [30].

As mentioned in Section 15.2.5, emerging biomimetic technology shows promise for improved performance. An example of this is shown for PSAs in Figure 15.13. The nucleoside-base pair containing polyacrylates synthesized by Long and coworkers show improved peel and shear strengths versus their non-biobased analogues [19]. This is based on the stronger hydrogen-bonding interactions between the nucleoside-base pairs. This interaction leads to a combination of high peel and shear strengths, due to greater internal cohesion and interfacial adhesion of the polymer.

Other potential areas of use for biobased methacrylates include the replacement of styrene as a reactive diluent for unsaturated polyester resins [31]. These UV curable systems can be used as coatings and as adhesives for bonding composite structures. They can also be used as resins for the formation of glass-filled composite structures.

15.4 PROSPECTS

The technology for biobased acrylic adhesives has developed considerably within the last 10 years, with both large and small companies in the adhesive industry investing considerable resources in developments.

As this chapter was being written, none of the world's major producers of acrylic acid had begun large-scale production of acrylic acid. In the first decade of the twenty-first century, biodiesel production from glycerol seemed to be financially favorable in comparison with petrochemicals. This is no longer the case. End users (i.e., companies using acrylic acid and its derivatives to formulate products) will not tolerate large price increases in comparison with conventional acrylic acid [4].

Pricing of biobased processes for acrylic acid and MMA may become more competitive with petroleum- or gas-based processes in the future [32]. There is a move from petroleum feedstocks to shale gas feedstocks. Shale gas feedstocks are a commercially less favorable source for extracting higher C-content petrochemicals in comparison with crude oil. This has the potential to make the biobased processes economically more favorable.

If biobased acrylic acid and MMA become more readily available and at an appropriate price, then this would significantly increase their usage in acrylic adhesives of all types.

To tolerate higher costs of biobased adhesives, consumers must see advantages in performance. For biobased acrylic adhesive technology to expand even further, new resins, not just replacements of existing ones, will need to be found. Biomimetic solutions to obtain better adhesive mechanical properties, adhesion in demanding environments, and debonding on demand are potential areas for future development by investigating how nature addresses these issues and how to incorporate them into adhesive products.

15.5 SUMMARY

Although, at present, there are not many examples of biobased acrylic adhesives available for consumers, Arkema and Evonik have commercialized monomers that have biobased carbon content. Ecosynthetix has commercialized biobased PSAs.

Technology development is ongoing, with companies developing processes to produce biobased acrylic acid and MMA, which would act as a platform for a range of biobased monomers. A longer-term objective is a move to biomimetic technology.

REFERENCES

1. F. A. Kiemel. Historical development of adhesives and adhesive bonding. In: *Handbook of Adhesive Technology*, 2nd edn, A. Pizzi and K. L. Mittal (Eds.), Chapter 1, CRC Press, Boca Raton, FL (2003).
2. i) Henkel listed in companies leading sustainability. http://the-mea.co.uk/news/henkel-listed-companies-leading-sustainability (2014). (ii) 3M Sustainability Report 2015. http://www.3m.com/3M/en_US/sustainability-report/ (2016) (iii) BASF Annual Report 2014. https://www.basf.com/documents/it/publications/BASF_Report_2014.pdf (2015).
3. ASTM D6866-11: Standard test methods for determining the biobased content of solid, liquid, and gaseous samples using radiocarbon analysis.
4. A. H. Tullo. Hunting for bio-based acrylic acid. *Chem. Eng. News*, 91, 18–19 (November 18, 2013). http://cen.acs.org/articles/91/i46/Hunting-Biobased-Acrylic-Acid.html?h=1016647717
5. Novozymes and Cargill continue bio-acrylic acid partnership as BASF exits. http://novozymes.com/en/news/news-archive/Pages/Novozymes-Cargill-continue-bio-acrylic-acid-partnership-BASF-exits.aspx (2015).
6. Cargill acquires OPX Biotechnologies. http://www.biofuelsdigest.com/bdigest/2015/04/29/cargil-lacquires-opxbio-technologies/ (2015).
7. Arkema is fully engaged in the bio-based chemical processes. http://www.arkema.com/en/innovation/renewable-raw-materials/bio-based-factories/index.html (2014).
8. Biobased product pipeline. http://www.myriant.com/products/product-pipeline.cfm (2011).
9. Metabolix Awarded Two U.S. Patents for Technology to Produce Biobased Polymers and Industrial Chemicals. http://ir.yield10bio.com/releasedetail.cfm?releaseid=667211 (2012).
10. Acrylics for the future. http://www.ingenia.org.uk/Content/ingenia/issues/issue45/harris.pdf (2010).
11. Evonik Paves The Way For Construction of the First MMA Production Plant Using the AVENEER® Process. http://corporate.evonik.com/en/investor-relations/news-reports/investor-relations-news/Pages/news-details.aspx?newsid=30165 (2012).
12. R. Vendamme, N. Schüwer and W. Eevers. Recent synthetic approaches and emerging bio-inspired strategies for the development of sustainable pressure-sensitive adhesives derived from renewable building blocks. *J. Appl. Polym. Sci*, 131, 17–37 (2014).
13. Y. Y. Lu and K. S. Anderson. Microsphere pressure sensitive adhesive composition. US Patent Appl. 2010/167614 A1, assigned to 3M Innovative Properties Company (2010).
14. R. Wool, S. Kusefoglu, G. Palmese, S. Khot and R. Zhao. High modulus polymers and composites from plant oils. US Patent 6121398 A, assigned to University of Delaware (1998).
15. Sartomer earns USDA biobased product certification for Sarbio® product line. http://news.thomasnet.com/companystory/sartomere-arns-usda-biobased-product-certification-for-sarbio-productline-20024446 (2014).

16. VISIOMER® *Terra*: Methacrylate monomers based on biorenewable raw materials. http://methyl-methacrylate-monomersevonik.com/product/visiomer/en/sustainability/visiomer-terra/Pages/default.aspx

17. S. Bloembergen, I. J. McLennan and R. Narayan. Environmentally friendly sugarbased vinyl monomers useful in repulpable adhesives and other applications. US Patent 6242593 B1, assigned to Ecosynthetix Inc. (1998).

18. J. Zhang, S. J. Severtson and M. R. Lander. Pressure-sensitive adhesives having high bio-based content and macromonomers for preparing same. US Patent Appl. 2015/0210907 A1, assigned to Regents of University of Minnesota, Minnesota (2010).

19. S. Cheng, M. Zhang, N. Dixit, R. B. Moore and T. E. Long. Nucleobase self-assembly in supramolecular adhesives. *Macromolecules* 45, 805–812 (2012).

20. J. F. Stanzione III, J. M. Sadler, J. J. La Scala, K. H. Reno and R. P. Wool. Vanillin-based resin for use in composite application. *Green Chemistry* 14, 2346–2352 (2012).

21. A. L. Holmberg, N. A. Nguyen, M. G. Karavolias, K. H. Reno, R. P. Wool and T. H. Epps III. Softwood lignin-based methacrylate polymers with tunable thermal and viscoelastic properties. *Macromolecules*, 49, 1286–1295 (2016).

22. K. Satoh, D. H. Lee, K. Nagai and M. Kamigaito. Precision synthesis of bio-based acrylic thermoplastic elastomer by RAFT polymerization of itaconic acid derivatives. *Macromol. Rapid. Comm.* 35, 161–167 (2014).

23. J. Dai, S. Ma, X. Liu, L. Han, Y. Wu, X. Dai and J. Zhu. Synthesis of bio-based unsaturated polyester resins and their application in waterborne UV-curable coatings. *Prog. Org. Coat.* 78, 49–54 (2015).

24. J. M. Sadler, A.P. T. Nguyen. F. R. Toulan, J. P. Szabo, G. R. Palmese, C. Scheck, S. Lutgen and J. J. La Scala. Isosorbidemethacrylate as a biobased low viscosity resin for high performance thermosetting applications. *J. Mater. Chem.* A, 1, 12579–12586 (2013).

25. G. John and C. K. S. Pillai. Synthesis and characterization of a self-crosslinkable polymer from cardanol: Autooxidation of poly(cardanyl acrylate) to crosslinked film. *J. Polym. Sci. A Polym. Chem.* 39, 1069–1073 (1993).

26. J. Liu, R. Liu, X. Zhang, H. Tong and X. Liu. Preparation and properties of UV-curable multi-arms cardanol-based acrylates. *Prog. Org. Coat.* 90, 126–131 (2016).

27. P. Glass, H. Chung, N. R. Washburn and M. Sitti. Enhanced reversible adhesion of dopamine methacrylamide-coated elastomer microfibrillar structures under wet conditions. *Langmuir* 25, 6607–6612 (2009).

28. E. P. Chan, E. J. Smith, R. C. Hayward and A. Crosby. Surface wrinkles for smart adhesion. *Adv. Mater.* 20, 711–716 (2008).

29. Henkel releases anaerobic adhesives designed to improve health and safety in the workplace. https://www.theengineer.co.uk/supplier-network/product/henkel-releases-anaerobic-adhesives-designed-to-improve-health-and-safety-in-the-workplace/ (2015).

30. P. C. Hayes, Copolymers including biobased monomers and methods of making and using same. US Patent 8889783 B2, assigned to BASF SE (2014).

31. S. Cousinet, A. Ghadban, E. Fleury, J.P. Pascault and D. Portinha. Toward replacement of styrene by bio-based methacrylates in unsaturated polyester resins. *Eur. Polym. J.* 67, 539–550 (2015).

32. K. Wagemann. Production of basic chemicals on the basis of renewable resources as an alternative to petrochemistry? *ChemBioEng Reviews* 2, 315–334 (2015).

16 Silicone Adhesives and Sealants

Jerome M. Klosowski

CONTENTS

16.1 INTRODUCTION

Silicone adhesives and sealants were introduced approximately 60 years ago and many of the silicones used in the early days are still performing. Products are available in a variety of forms, from paste-like materials to flowable adhesives. Both single- and multicomponent versions are available, with several different cure chemistries. Most of the silicones of commerce are based on polydimethylsiloxane (PDMS) polymers. Other siloxane polymers may be used when resistance to ultrahigh temperature, ultralow temperature, solvents, or radiation is required.

Applications are extremely broad. A partial list includes construction, highway, automotive, appliance assembly, original equipment manufacture, maintenance, electronics, aerospace, and consumer uses. Depending on the industry, silicones compete with other materials, such as polyurethanes, polysulfides, epoxies, acrylics, hybrids, and others. Silicones are often chosen for long-term durability in a given application. Silicones are often chosen for their excellent resistance to weathering and temperature extremes, their adhesion, and their ability to accommodate substrate movement. When silicone sealants and adhesives are mentioned, the thought of excellent durability comes to most readers' minds. Silicones (named for the similarity of the $(CH_3)_2SiO$ polymer repeat unit to the analogous organic ketones, $R_2C=O$) occupy a unique position between inorganic and organic materials. The saturated inorganic Si–O–Si polymer backbone provides flexibility and stability to sunlight, while the pendent methyl groups ensure low intermolecular forces. Some of the key attributes of silicones that are responsible for their unique properties and durability are [1]:

- Low surface tension
- High water repellence
- Partial ionic backbone
- Large free volume
- Low apparent energy of activation for viscous flow
- Low glass transition temperature
- Freedom of rotation around bonds
- Small temperature variations of physical constants
- High gas permeability
- High thermal and oxidative resistance

- Low reactivity
- Insolubility in water
- High silicon–oxygen bond energy

Selected properties of PDMS are as follows:

Critical surface tension of wetting	24 mN/m
Water contact angle	110°
Glass transition temperature	150 K
Energy of rotation	0 kJ/mol
Activation energy for viscous flow	14.7 kJ/mol
Si–O bond energy	445 kJ/mol
Percentage polar contribution	41%

The saturated backbone and high Si–O bond energy result in products that perform very well in applications involving exposure to sunlight. Since the silicone polymer does not absorb energy in the ultraviolet (UV) region of the light spectrum, one must be cautious with the use of clear silicones. The silicones need no UV absorbers to be stable (and typically contain none); thus, the UV light from the sun can pass through clear silicones to the surface below the sealant. If the surface is sensitive to UV light, deterioration of the substrate may occur. Except for light-protected areas and nonsensitive substrates (such as glass), the most judicious choice is a pigmented silicone. The pigment acts as a UV blocker and protects the substrate beneath the silicone.

Because of their stability to UV radiation and elastomeric nature, silicones are the sealants of choice for wet glazing, and they are the only generic class of sealants allowed for structural glazing (the adhering of glass and other building materials to structures with no attachment other than the silicone). Structural glazing with silicone adhesive/sealant is used in all-glass buildings, for glass ribbons on buildings, on skyscrapers, and in many other important structures. Silicone structurally glazed buildings have significantly less damage than buildings with more conventional facades in major earthquakes, because of the elastomeric nature of the structural adhesive.

Other types of sealants often contain large amounts of filler and UV stabilizers, to afford some degree of longevity in sunlight. This makes the nonsilicones satisfactory for some applications, but not in applications in which the sun shines directly on the bondline. This application is reserved for silicones. A specialty application for silicones that further illustrates their UV light durability is in the sealing of accelerated UV weathering test machines. The excellent stability to UV light is true only for pure silicones and is not true of *siliconized organics* or *modified silicones*. These contain very little silicone and thus have durability characteristics determined primarily by their base non-silicone polymer systems.

Silicones have low intermolecular forces that result in relatively flat physical property response with temperature change. An example of this flat response is shown in Figure 16.1, in which the viscosity of silicone polymers and a hydrocarbon oil are plotted as a function of temperature [2,3]. The relatively low response of silicone properties to temperature is important during sealant application (e.g., no heating needed in cold weather and no excessive flow in hot weather). Even more important, however, is the fact that the performance of the cured sealant or adhesive will be less temperature dependent than most organic-based products. This has practical implications: in building joints, for example. In cold weather, the building components shrink, and joint sealants must maintain elasticity to accommodate this movement. This is also fundamental to their use as a structural glazing sealant/adhesive. The sides of all-glass buildings can get very warm in the summer sun, and the silicone must not lose strength at these temperatures. While this rather constant performance is critical in some construction applications, it is also important in many industrial and appliance applications, such as steam irons, where the sealant simultaneously prevents water leakage and acts as an assembly adhesive.

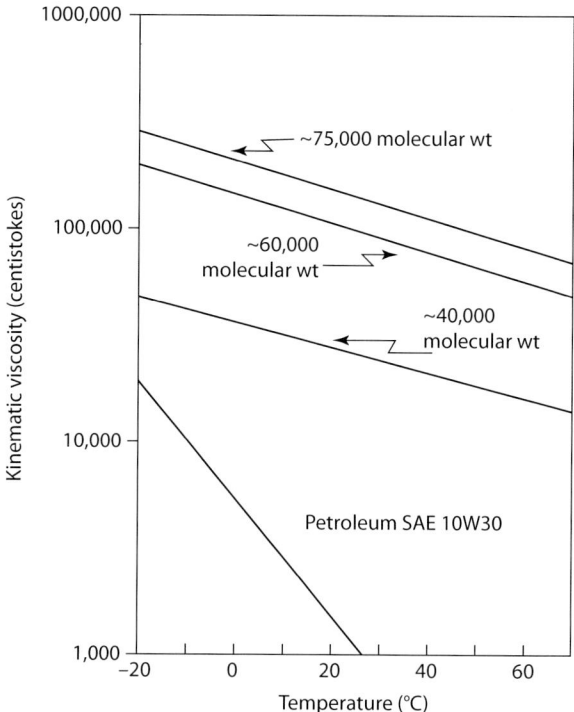

FIGURE 16.1 Viscosity versus temperature.

Silicone sealants are rated for their movement capability, with classes at ±12.5%, ±25%, ±50%, and even higher joint movement capability. This, too, is quite unique, since high-movement nonsilicone sealants rarely perform for long periods of time above ±25% joint movement. To demonstrate this, the sealants being compared have to have stresses and strains imposed when weathering. This applies to either accelerated weathering or actual outdoor weathering. A further note of importance is that when comparing sealants with stresses and strains during weathering, the acceleration factor for most sealants, for most U.S. climates, is between 4 and 8. This teaches that important applications that include weathering have to be tested for months and maybe years in an accelerated weathering machine to obtain significant data. Weathering of sealants, when expansion and contraction are not imposed alternately during the weathering, simply should not be done, since the data will not be relevant to most sealant applications.

The high-temperature capability of many silicone sealants (200°C or even 250°C) make them the sealant of choice for many automotive and electrical applications. Sealing light emitting diode (LED) lighting, sealing headlights, some ovens, and many more applications that require stability of an elastomer at high temperature are sealing applications for silicones.

Combine the high-temperature capability with the water repellency and insulating characteristics, and more applications may come to mind. However, if one wants conductivity, a silicone sealant can be made with electrical-conductive or heat-conductive fillers to open up new applications.

16.2 CURE CHEMISTRY

Silicones are available in one- and multicomponent forms. The one-component condensation curing types are commercially the most important and will be the focus of most of this discussion. These products, which generally cure by reaction with atmospheric moisture, are called *room-temperature vulcanizing (RTV)* sealants or adhesives. The surface cure rate of these products is a function of the cure system (cross-linker reactivity, catalyst activity), but the rate of cure in depth depends on the ability to transmit water vapor through the mass of sealant and the concentration of reactive components in the bulk of the sealants. Silicones are highly permeable to moisture vapor, and generally the one-component types cure at a rate of about 0.3 cm to 0.7 cm/day. Due to the high vapor permeability, the one-component silicones typically cure faster than do their one-component nonsilicone counterparts.

The multicomponent products generally do not rely on moisture penetration from outside of the sealant for cure. Their chief attribute is fast cure in very deep sections. Thus, many industrial production lines that demand fast cure use a two-component sealant (including the use of silicone encapsulants for electrical components). Cure of these two-part systems can be accelerated further by additional catalyst or exposure to elevated temperatures.

One of the more common two-part cure chemistries is based on the addition reaction of Si–H cross-links with vinyl functional polymers using platinum catalysts. This chemistry is shown here. One advantage of this addition chemistry is that it produces no cure by-products. Another common two-part chemistry involves condensation cure with alkoxysilane cross-linkers using Sn(IV) catalysts.

$$-Si(Me)_2 \, OSi(Me)_2 \, CH = CH_2 + HSi \rightarrow -Si(Me)_2 \, OSi(Me)_2 \, CH_2 - CH_2 Si -$$

A simplified cure mechanism for the one-component silicone RTV sealants or adhesives is shown here.

Reaction of cross-linker with polymer ends:

$$2RSiX_3 + HOSi(Me)_2 \, O\Big[(Me)_2 \, SiO\Big]_n Si(Me)_2 \, OH \rightarrow$$

$$\rightarrow X_2(R)SiO\Big[(Me)_2 \, SiO\Big]_n Si(R)X_2 + 2HX$$

Reaction of cross-linker with polymer ends:

$$X_2(R)SiO\Big[(Me)_2 \, SiO\Big]_n Si(R)X_2 + H_2O \rightarrow HO(X)(R)SiO\Big[(Me)_2 \, SiO\Big]_n Si(R)X_2 + HX$$

Reaction of resultant polymer end with another polymer:

$$A + B \rightarrow X_2(R)SiO\Big[(Me)_2 \, SiO\Big]_n Si(R)X-O-Si(X)(R)O\Big[(Me)_2 \, SiO\Big]_n Si(R)X_2 + HX$$

As indicated, these X groups are hydrolyzable. Repeated hydrolysis and reaction of resultant polymer end groups leads to full cure, with elimination of HX as the leaving group. Examples of leaving groups, cross-linkers, and the common cure system names are given in Table 16.1.

TABLE 16.1
Examples of Leaving Groups, Cross-Linkers, and Cure Systems

Leaving Group (HX)	Cross-Linker	Cure System
$HOC(O)CH_3$	$CH_3Si[OC(O)CH_3]_3$	Acetic acid
$HOCH_3$	$CH_3Si(OCH_3)_3$	Alcohol
$HONC(CH_3)(C_2H_5)$	$CH_3Si[ONC(CH_3)C_2H_5]_3$	Oxime
$CH_3C(O)CH_3$	$CH_3Si[OC(CH_2)CH_3]_3$	Acetone
$HN(CH_3)C(O)C_6H_5$	$CH_3Si[N(CH_3)C(O)C_6H_5]_3$	Benzamide

Numerous other cross-linkers may be used. For the trifunctional cross-linkers, the R group may be methyl, ethyl, vinyl, and several other groups, with methyl the most common. In some cases, tetra-functional and higher-functionality cross-linkers or polymeric cross-linkers may also be employed. The acetic acid cure system should be avoided where substrates are subject to acid corrosion.

Two other classes of silicones deserve mention. These are the water-based silicones that are used in sealant and coating applications and the silicone pressure-sensitive adhesives. Water-based silicones can be prepared by anionic polymerization of siloxanes in water using a surface-active catalyst such as dodecylbenzenesulfonic acid [4]. The resulting emulsion can then be cross-linked in several ways, including the use of alkoxysilane copolymerization or tin catalysts in conjunction with colloidal silica. The result is essentially an emulsion of cured PDMS in water. Various fillers and other components are added, resulting in a sealant (coating) composition. On evaporation of water, the system coalesces to form the elastomer.

These sealants have the advantages of low odor, ease of installation, and easy cleanup. Their properties are rather close to those of their conventional silicone counterparts. Their disadvantages are relative to the surfactant that complicates the adhesion.

The components of the silicone pressure-sensitive adhesives (PSAs) are analogous to their organic counterparts [5]. Generally, a silicate resin and a silicone polymer or gum are dissolved in solvent. Both the resin and the polymer typically contain silanol (Si–OH) groups that are reacted during processing of the PSA, leading to a cross-linked network. Additional reactions can be accomplished through the use of free radical catalysts. The extent to which these cross-linking reactions occur and the resin:polymer ratio, as well as the respective molecular weights of these components, are important in setting the properties of the PSA.

Silicone PSA products are used in a number of medical and industrial applications, ranging from a variety of PSA tapes and transfer films to automotive bonding. Advantages for the silicone PSA products include resistance to temperature extremes, chemical resistance, conformity to irregular surfaces, and electrical properties. They are also unique to most PSAs in their ability to adhere to difficult low-energy substrates, such as polytetrafluoroethylene and other silicones.

16.3 PROCESSING CONSIDERATIONS

Silicone adhesives and sealants typically contain polymer, fillers, cross-linkers, catalysts, and other additives. The most common fillers are the reinforcing fumed silicas and/or a variety of calcium carbonates. Other fillers and pigments, such as carbon black and titanium dioxide, are also used. Silicones are typically made in high-shear, vertical change-can mixers, but continuous processing equipment may also be employed. Processing details are generally held proprietary, but some general guidelines are in order. Since most silicones cure through reaction with water, it is important that the moisture content of fillers and other additives be controlled. The moisture content of fumed silica, for example, can vary from 0.2% to over 2%, depending on the humidity conditions during storage. It is also critical that introduction of moist air be kept to a minimum during mixing. Generally, air incorporated during processing must be removed to reduce the tendency toward cured gels and related appearance problems in the final product.

The dispersion of the filler particles is also important to final sealant appearance. With increasing costs of fillers, it is also important to optimize dispersion to maximize the rheological and reinforcement benefits provided by the fillers. Manufacturers must balance the mixing time and energy required for complete dispersion with resultant product appearance and physical properties.

16.4 PROPERTY DETERMINATIONS

Since most silicone adhesives and sealants are elastomeric in nature, their physical property testing often parallels classical rubber testing approaches. Common tests include durometer, tensile strength, elongation, and modulus. Several methods are available for the measurement of rubber properties, but the most commonly used are the American Society for Testing and Materials (ASTM) D-412, "Test Method for Rubber Properties in Tension," and ASTM C-661, "Standard Test for Indentation Hardness of Elastomeric-Type Sealants by Means of a Durometer." These properties vary widely with the product and its intended application. Durometer measurements can range from a Shore A of less than 20 to over 50. Tensile strength ranges from less than 0.2 to greater than 5 MPa, and elongation varies from about 100% to 2000%.

Cure time testing tends to be somewhat subjective, but again there are methods available, such as ASTM C-679, "Tack-Free Time of Elastomeric Sealants." Tack-free time is the curing time required for the product to develop a skin that is not damaged when subjected to application and removal of a plastic film. It is important in all cases to determine the cure characteristics of the product in actual working conditions. Since most silicones cure by reaction with moisture in the air, the sensitivity of cure time to humidity should be determined. Surface cure rates can usually be tailored to meet application requirements.

The rheological properties of adhesives and sealants are important in many applications. When these products must be pumped or applied through automated equipment, the flow characteristics at pertinent shear rates are critical. Sophisticated rheological measurements can be performed to predict performance. The rheology of silicone adhesives and sealants can be tailored through adjustment of polymer viscosity, filler loading, and incorporation of various additives.

Often, only the extrusion rates of adhesives and sealants are measured, which is accomplished by subjecting the product to a given pressure and measuring its flow rate through a nozzle of known diameter (see ASTM C-1183, "Extrusion Rate of Elastomeric Sealants"). For many sealant applications, the sealant must not flow under its own weight in conditions of low shear rate. In this case, some measurement of *slump* is generally made. Several methods are available for measuring slump [see ASTM D-2202, "Standard Test for Slump of Sealants," and ASTM C-639, "Standard Test for Rheological (Flow) Properties of Elastomeric Sealants"]. Again, it is important to determine the rheological performance of the product in the actual application.

Adhesion testing is a matter of some controversy. There is, however, a growing trend among manufacturers, specifiers, and standards organizations to move toward tests that better predict performance in application. The 180° peel adhesion test is often used as an internal quality control tool by manufacturers. This test allows for measurement of lot-to-lot consistency of products. The methodology for this test can be found in ASTM C-794, "Test Method for Adhesion In-Peel of Elastomeric Joint Sealants." In its typical form, this test involves placing a bead of the product onto the substrate, with a flexible member embedded in the product. The product is allowed to cure, and the member is then pulled away from the substrate. The force required for peel is noted, along with the mode of failure. The advantages of this test are that it is fast and lends itself well to subjecting the adhesive bond to environmental stresses such as hot-water exposure. This peel test is not to be used to compare different sealants, since the modulus of elasticity of the sealant must be considered when evaluating the sufficiency of the strength of the bond, and that is not considered in this test method.

A series of tests that are better predictors of performance of sealants involves the preparation of tensile–adhesion joints or H-pieces. In this case, the sealant bead is placed between blocks of the two substrates of interest and allowed to cure. This joint can then be pulled to destruction, with measurement of strength and failure mode. In addition, various environmental stresses may be applied, such as UV light exposure (weatherometry), water exposure, and cyclic movement. Testing in this way allows for improved prediction of movement capability and long-term performance. The methodology involved in this testing is included in the following methods: ASTM C-1135, "Determining Tensile Adhesion Properties of Structural Sealants," and ASTM C-719, "Test for Adhesion and Cohesion of Elastomeric Joint Sealants Under Cyclic Movement" and the standard practice ASTM C-1589 and C-1589C for outdoor weathering with joint movement during weathering.

Any sealant or adhesive that is expected to perform in outdoor applications should be tested after exposure to light, heat, and water. Even better is if the sealant is put into alternating strain and compression while being exposed to the light, heat, and water. For most forms of weatherometry, 500–1200 h is considered approximately the equivalent of 1 year outside in most climates (United States). If a minimum of 5 years of service is expected from the sealant, no less than 2500 h and probably 5000 h in a UV fluorescent accelerated weathering machine or a xenon lamp device should be used (as the conditioning cycle for rubber property testing). This is true for silicones and all other sealants that are expected to perform in such applications. This is a startling contrast to the 250–500 h used in most present standards (see references [6,7]). When picking a specification or test method, the hours of exposure to accelerated weathering need to be considered relative to the application and its location in the world. Sometimes, 20,000 h in the accelerated weathering machines are needed. Obviously, outdoor weathering is not an acceleration, but it is the best indicator of suitability to an outdoor application. The outdoor tests also indicate dirt pickup and fungus growth and the influence of other environmental factors on sealant behavior.

16.5 BASIC FORMULATIONS

As mentioned in Section 16.3, silicone sealants and adhesives generally contain PDMS polymers, cross-linkers, fillers, catalysts, and other additives. These additives may be pigments, plasticizers (often unreactive PDMS polymers), and adhesion additives (such as silane coupling agents). Given here are simple formulations and properties (Table 16.2) for oxime-cured silicone sealants [8]. In these examples, the use of a nonreactive silicone plasticizer and a nonreinforcing carbonate filler results in substantial modulus reduction. This approach can also be used to modify the physical properties of silicones based on other cure chemistries. Low-modulus sealants are often used in sealant applications requiring high movement capability. High-modulus sealants are used more in structural and adhesive applications.

	Percentage by Weight
A. High-Modulus Oxime Sealant	
Hydroxyl-ended PDMS polymer	80–85
Fumed silica	5–10
Oxime cross-linker	5–7
Sn(IV) catalyst	0.05–0.10
B. Medium-Modulus Oxime Sealant	
Hydroxyl-ended PDMS polymer	60–80
Silicone plasticizer	5–20
Fumed silica	2–6
Calcium carbonate	20–30
Oxime cross-linker	5–7
Sn(IV) catalyst	0.05–0.10

TABLE 16.2
Properties of Oxime-Cured Silicone Sealants

	Sealant	
	High Modulus	**Medium Modulus**
Tack-free time (min)	20–30	30–60
Durometer (Shore A)	25–35	20–30
Tensile strength (MPa)	1.2–2.1	0.9–1.4
Elongation (%)	200–400	400–700
100% modulus (MPa)	0.5–0.9	0.35–0.5

16.6 SUBSTRATE BONDING

Applications for silicones in bonding are numerous. Generally, one sealant will not bond to all substrates, and it is common practice to develop new formulations to meet the ever-increasing list of requirements. In some instances, primers are used for certain substrates, but some silicones are self-priming. This self-priming feature is important from the standpoints of reducing installation costs and reducing dependency on high-solvent primers, which are sometimes subject to environmental regulations. The surface characteristics for a given type of substrate can vary considerably between substrate manufacturers. For this reason, it is always advisable to check adhesion before specifying a particular sealant. In addition, the importance of proper substrate cleaning and preparation should not be overlooked. Most adhesive and sealant producers will recommend the proper procedures for surface preparation. Some of the more common substrates and related applications for silicones are given in Table 16.3.

The last of the most fundamental considerations of silicone adhesives and sealants is application precautions. Except for PSAs, most silicone adhesives and sealants need to be cured for some

TABLE 16.3
Applications of Substrates

Substrate	Examples	Typical Applications
Masonry	Concrete	Construction
	Mortar	Highway
	Brick	Consumer
Natural stone	Marble	Construction
	Granite	
	Sandstone	
Wood	Unpainted	Construction
	Painted	Glazing
		Consumer
		Maintenance
Glass	Float	Construction
	Reflective	Glazing
		Maintenance
		Original equipment manufacturing
		Consumer
Metals	Aluminum	Construction
	Steel	Glazing
	Copper	Electronics
	Stainless steel	Maintenance
	Galvanized steel	Original equipment manufacturing
		Consumer
Coated metals	Paints	Construction
	Fluorocarbon	Glazing
	Polyester	Maintenance
		Original equipment manufacturing
Plastics	PVC	Construction
	PMMA	Glazing
	Polyester	Maintenance
	Engineering plastics	Automotive
		Sanitary
		Original equipment manufacturing
		Consumer

time before adhesion is developed. This does not seem to be a problem at first glance, but if there is movement or stress while the cure is taking place, most often this will reduce the strength of the bond. Thus, for many silicone sealant/adhesive applications, the most common recommendation is to restrict movement or stress on the bond until the cure is complete or mostly complete [9].

For applications that need to have movement quickly after application, the use of faster curing one-part products is a help; two-part products for faster cure are often used and, for some applications, the use of a silicone-curing PSA is warranted. Each application will have special considerations, and while general notes, such as those given here, are interesting and useful, the specific conditions of each application need to be carefully considered, and then products and application procedures can be specified.

REFERENCES

1. M. J. Owen and J. M. Klosowski. Adhesives, sealants and coatings for space and harsh environment In: *Adhesives, Sealants and Coatings for Space and Harsh Environments*, L. H. Lee (Ed.), pp. 283, Plenum Press, New York (1988).
2. E. G. Rochow and H. G. LeClair. On the molecular structure of methyl silicone. *J. Inorg. Nucl. Chem.* 1, 92 (1955).
3. J. M. Klosowski and G. A. L. Gant. *Plastic Mortars, Sealants, and Caulking Compounds*, ACS Series, Vol. 113, R. B. Seymour (Ed.), p. 117, American Chemical Society, Washington, DC (1979).
4. D. T. Liles and N. E. Shephard. Silicone rubber latex sealants. In: *Science and Technology of Building Seals, Sealants, Glazing and Waterproofing*, Vol. 2, ASTM STP 1142, J. M. Klosowski (Ed.), American Society for Testing and Materials, Philadelphia, PA (1992).
5. L. A. Sobieski and T. J. Tangney. Silicone pressure sensitive adhesives. In: *Handbook of Pressure Sensitive Adhesive Technology*, D. Satas (Ed.), pp. 508–517, Van Nostrand Reinhold, New York (1989).
6. L. B. Sandberg. Comparison of silicone and urethane sealant durabilities. *J. Mater. Civil Eng.* 3, 278–291 (1991).
7. G. R. Fedor. Usefulness of accelerated test methods for sealant weathering. Second Symposium on Science and Technology of Building Seals, Sealants, Glazing and Waterproofing, Ft. Lauderdale, Fla., ASTM C-24 FT. American Society for Testing and Materials, Philadelphia, PA (1992).
8. J. M. Klosowski. *Sealants in Construction*. Marcel Dekker, New York, pp. 269–270 (1989).
9. J. M. Klosowski and A. T. Wolf. *Sealants in Construction*, 2nd edn. Taylor & Francis, New York (2015).

17 Epoxy Adhesives

Anna Rudawska

CONTENTS

17.1   INTRODUCTION

Epoxy resins are a kind of thermosetting polymer, based on the reaction between epoxide precursor molecules themselves or on the reaction between the epoxy group and other kinds of reactive molecules called *hardeners*, with or without the help of a catalyst [1,2].

An epoxy resin is a polymer or oligomer containing reactive epoxy groups. Epoxy resins have been known for decades, and the earliest documented technologically viable method of obtaining epoxies dates back to 1860 [3]. The method exploited the ammonia and epichlorohydrin reaction. It was not, however, until 1946 that the first company, Ciba, introduced the first commercial type of epoxy resin, produced from diane and epichlorohydrin. This initiated a rapid development of epoxy resin technology. New epoxy diane resins appeared and, in particular, advanced hardeners and compositions. At present, with 85%, diane- and epichlorohydrin-based resins maintain the highest share in resin production [4].

Specific properties of epoxy resins have resulted in their application as adhesives. A great variety of both resins and hardeners, together with their excellent capability for modification, enables preparation of a composition most suitable for a given application; for example, joining particular substrates and particular conditions of service [5].

The polar character of epoxy resins causes their high adhesiveness, and thus of epoxy adhesives, to different materials. Low chemical contraction occurring during curing enables the production of casts precisely copying the shape of the mold, in a process that, furthermore, does not require the application of any additional pressure. Low chemical contraction generates low stresses, which can be further eliminated by the addition of plasticizers [3].

Proper wetting of a substrate surface with a liquid adhesive is governed by Equation 17.1:

$$\gamma_{LV} \leq \gamma_c \tag{17.1}$$

where γ_{LV} is the surface tension of the liquid adhesive and γ_c is the critical surface tension of the substrate.

The surface tension of a liquid epoxy resin amounts to 47–49 mN/m; whereas, with the addition of a curing agent, it lies in the range of 30–40 mN/m. The surface tension of the mix of epoxy resin and diethylaminopropylamine is 33 mN/m, while with the addition of diethylenetriamine it is 44 mN/m at room temperature [3]. A temperature increase or the addition of surface-active substances leads to a lowering of the surface tension that, in turn, contributes to a higher strength of the adhesive joint [6,7].

The presence of polar hydroxyl and ester groups results in a good adhesive property of epoxy resins to metals and other materials of polar character. Hydroxyl groups found in the resin or generated during curing, which create hydrogen bonds, cause an increase in adhesion strength.

Epoxy resins are gaining popularity in various fields of engineering, such as the electrical industry, for structural applications in both the commercial and the military aircraft industries [2,8,9]. Epoxy resins are widely used for protective coatings, due to their strong adhesion and corrosion protection of metals [10,11,12].

Epoxy adhesives find a broad spectrum of applications in joining various materials. In addition, adhesive technology offers numerous advantages over other joining technologies [13]. The effectiveness of adhesive joining depends on several factors, that is to say, properly selected adhesive composition and proper preparation of the adherend surface. Adhesive bonding is applied in a growing number of industries, including aircraft, aerospace, building, automotive, and shipbuilding [13].

17.2   EPOXY ADHESIVES

17.2.1   INTRODUCTION

Due to their characteristic properties, epoxy resins are regarded as one of the best adhesives and can be applied in various branches of industry. One interesting instance of epoxy resin application is a footbridge in Scotland over the River Tay, where the resin was used to join the polymer structural elements of the construction [4]. The bridge offers load capacity of 1 t/m² and has been assembled with no metal screws or bolts.

Epoxy adhesives demonstrate far better properties than most adhesives, predominantly due to their superior adhesion and chemical resistance [5].

Wide varieties of resins and hardeners, as well as virtually endless possibilities of modification, allow adjustments to the epoxy adhesive compositions for particular materials and conditions of use. Notably, epoxy adhesives enable joining of materials of dissimilar chemical and mechanical properties and, furthermore, can be applied alongside traditional joining methods, such as riveting or welding [13].

The use of epoxy adhesives is facilitated by the fact that the curing process does not require additional pressure. No curing by-products are released in the process. In addition, epoxy adhesives are characterized by low chemical contraction, and concern about excessive thermal contraction, which depends on the coefficient of thermal expansion and temperature, may be solved by the addition of suitable fillers [4,10].

One notable feature of epoxy adhesives is that their properties can be modified to a great extent by selecting a particular type of resin and the type and amount of hardener, as well as by the introduction of additional substances such as nanomaterials, which could, *inter alia,* increase bond elasticity or thixotropy and other properties [5,14,15]. Yoon et al. [16] have investigated hydrogenated bisphenol A (HBA) epoxy adhesives, containing organic–inorganic hybrid nanomaterials, and noticed that the tensile strength increased with the addition of nanomaterials containing a glycidyl epoxy group. Mimura and Ito [17] have presented the results of curing epoxy resin with a curing agent formed from radical copolymerization of vinyl monomers during the cure process of epoxy resin to improve its heat resistance, together with a reduction of the viscosity of the resin composition.

Elastic epoxy adhesives find application where there is a difference between the coefficients of expansion of the cured resin and the substrates. The coefficient of thermal expansion can be reduced by the addition of mineral fillers or powdered metals.

Furthermore, the high resistance of epoxy resins to ionizing radiation makes them suitable for application in nuclear equipment.

A great diversity of compositions, which can be cured over a wide range of temperatures, and considerable freedom in designing their properties to obtain the required and specific features are just a few advantages of epoxy adhesives over other adhesives; this indicates that the range of their applications will continue to expand [15,17].

17.2.2 Epoxy Resins

Epoxide molecules in the pure state at room temperature normally do not react with each other, and can stay unchanged for years in a dry container without mutual reaction. The types of chemicals added to the epoxide to effect network formation fall into two categories: curing agents and catalysts. Curing agents, sometimes called *hardeners*, are added in significant amounts to the epoxide and react with it to become a part of the cross-linked network [18].

Epoxy resins are compounds cured by chemical catalysts [1,19,20,21]—one of the types of chemicals added to the epoxide. Their molecules contain at least two epoxy groups capable of polyreaction, that is, curing, as a result of which the resin becomes a cross-linked, insoluble and infusible product [3,4]. They are typically of low molecular mass, that is, lower than 4000. Catalysts, on the other hand, are added in extremely small amounts to cause the epoxide molecules to react directly with each other, that is, to homopolymerize. Sometimes the chemicals used as catalysts for homopolymerization can be used for another purpose and, when added in small amounts to epoxide–curing agent mixtures, they will accelerate the curing reaction. In this role, they are called *accelerators* rather than catalysts [18]. The discovery of amine catalysis in the curing of epoxy resins with dicyandiamide led to an increasing use of such adhesives in the 1960s [19].

Epoxy resins are processed as compounds, that is, mixtures of resin with such components as diluents, plasticizers, solvents, fillers, curing agents, and so forth. The basic epoxy composition consists of resin and curing agent, and the latter can be introduced into the resin immediately prior to processing or can be a component of a thermosetting commercial product.

Synthesis of such epoxy resins usually involves the use of epichlorohydrin and di- and polyhydroxy phenols. Epoxy resins usually contain polyphenols, or less commonly, polyglycols, and epichlorohydrin or oligomers with epoxy pendent groups.

Epoxy resins are also hardened by curing agents (the second type of chemical added to the epoxide), such as amines, amides and polyamides, and anhydride hardeners, as well as certain Lewis acids and bases and others [1,18,19,20]. Hardened resins become insoluble and infusible, highly adhesive to virtually all types of materials (including metals, glass, ceramics, or wood), chemically resistant, and possess excellent dielectric properties (high electrical resistivity).

There are three methods used for obtaining epoxy resins in industrial practice [3,4]:

1. Base-catalyzed reactions of epichlorohydrin and compounds containing reactive hydrogen (usually diphenols or, less frequently, diamines or polyglycols), followed by oligomerization of remaining compounds
2. Direct oxidation with organic acids of unsaturated aliphatic or cycloaliphatic hydrocarbons of particular epoxy compounds
3. Addition of hypochlorous acid (HClO) to unsaturated compounds and base-induced dehydrohalogenation

17.2.2.1 Epoxy Resins Based on Epichlorohydrin

Among a variety of epoxy compounds available commercially, the most important group of epoxy resins is derived from bisphenol A [15,19,20]. Epoxy resins are produced from bisphenol A in a reaction with epichlorohydrin to generate diglycidyl ether of bisphenol A, a liquid resin widely used in adhesive manufacture (Figure 17.1).

Epoxy resins are derived from direct polycondensation of 1-chloro-2,3-epoxypropane (epichlorohydrin) with dihydroxy phenols or polyglycols in alkaline conditions (Figure 17.2).

Synthesis of diane epoxy resins is conducted in a multiphase system in a highly alkaline environment. There are two types of processes for manufacturing epoxy resins: continuous or batch. Component A, that is, the bisphenol mass, referred to as the *primary product*, can be combined with amines, which act as hardeners and are often referred to as *Component B*.

Epoxy resins produced from the process of polycondensation demonstrate different degrees of polymerization, ranging between $n=0$ and $n=12$, where n denotes the number of polymerized subunits and is typically in the range from 0 to 25 (Figure 17.3). The degree of polymerization (DP) is usually defined as the number of monomeric units in a macromolecule or polymer or oligomer molecule [23].

Depending on the degree of polymerization, epoxy resins can be divided into three groups [3,4]:

- Low molecular mass resins: Number average degree of polycondensation $n < 1$, epoxide number between 0.35 and 0.58, and molecular weight less than 450.

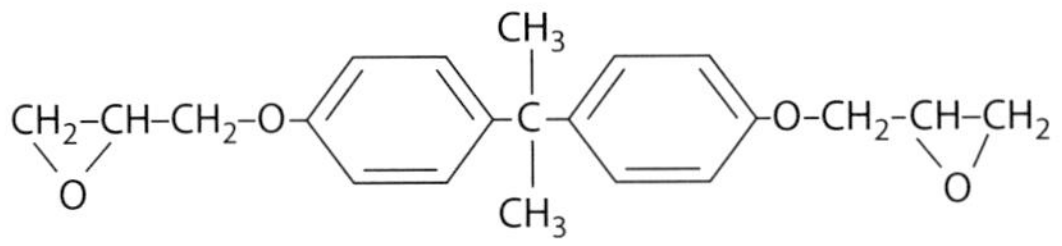

FIGURE 17.1 Diglycidyl ether of bisphenol A. (From E.M. Pertie, *Epoxy Adhesive Formulation*, pp. 3–19, 26–41, 43–53, 71–82, 85–98, 207–225, 229–236, McGraw-Hill, New York (2006).

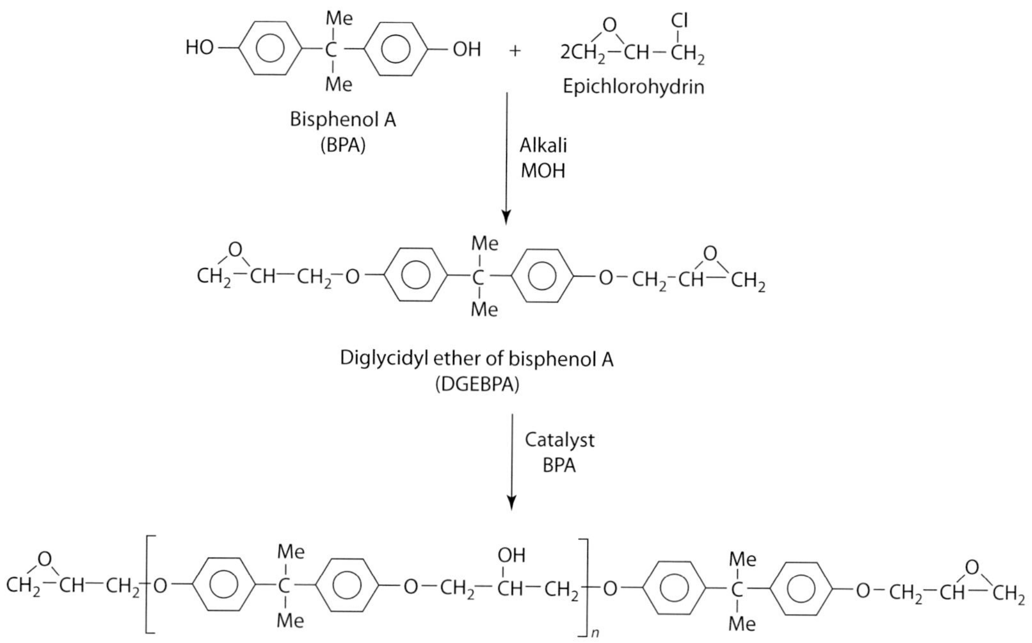

FIGURE 17.2 Reaction scheme for the formation of epoxy resin. BPA: Bisphenol A; MOH: sodium or potassium hydroxide. (B. Ellis (Ed.), *Chemistry and Technology of Epoxy Resins*, first edition, pp. 1–35, 37–68, 72–99, 121–126, 206–239. Springer Science+Business Media, Dordrecht (1993).)

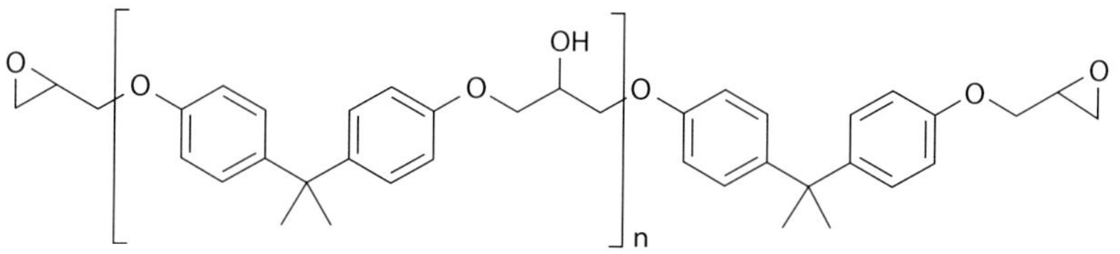

FIGURE 17.3 Structure of bisphenol A diglycidyl ether epoxy resin. (From https://en.wikipedia.org/wiki/Epoxy, 2016.)

- Average molecular mass resins: Low softening point (below 100°C), epoxide number between 0.15 and 0.30, and molecular weight ranging between 500 and 1000.
- Hard high-molecular mass resins: Characterized by high softening point (100°C–150°C), low epoxide number (0.1–0.02), and molecular weight ranging between 900 and 4000.

Epoxy resins are composed of long, chain-like molecules similar to that of vinyl ester, with reactive parts at both ends (Figure 17.3). The difference is that the reactive parts are composed of epoxy, not ester, groups. The absence of ester groups means that epoxies are characterized by very high water resistance. A large part of the epoxy resin backbone contains aromatic rings, which provide a high degree of heat and chemical resistance [4].

17.2.2.2 Epoxy Resins Based on Unsaturated Compounds: Cycloaliphatic Resins

Cycloaliphatic resins (diepoxides) are produced by direct epoxidation (oxidation) of cycloaliphatic diunsaturated compounds, which have a unique system of double bonds. These intermediate products are normally synthesized by coupling cycloaliphatic compounds with one unsaturated bond and one functional group capable of coupling [3,4,19].

Epoxy resins obtained by oxidation of olefins can be divided into:

- Aliphatic resins, of linear structure, where epoxy groups are at the ends or in the aliphatic chain
- Cycloaliphatic resins, where epoxy groups are incorporated in the cycloaliphatic ring

The coupling method determines the reactivity of epoxy groups. Groups incorporated into the cycloaliphatic ring are particularly reactive with acids and acid anhydrides, particularly in the presence of catalysts; however, their reaction with amines requires higher temperatures and is quite slow. In the case of the glycidyl group, an exact opposite relationship is observed: reaction with aliphatic amines at low temperatures and with anhydrides at higher temperatures.

There exist over a thousand compounds of cycloaliphatic resins, which differ in terms of structure and properties, but share certain characteristics and are different from classical diane resins [4,19,20]. Epoxy groups in cycloaliphatic resins demonstrate higher nucleophilicity than glycidyl ethers and, what is more, compared with diane resins, these are more reactive with anhydride hardeners than with amine ones.

Compared with diane resins, cycloaliphatic resins that are heat-cured with anhydrides display high resistance to ultraviolet (UV) radiation, leakage currents, and electric arc due to the lack of aromatic backbone. Certain cycloaliphatic resins of dense structure, once cured, exhibit high heat resistance (much higher than that of diane resins) on the one hand, but exhibit inferior mechanical properties on the other.

17.2.2.3 Types of Epoxy Resins

There exist hundreds of epoxy compounds, described in numerous works [4,10,18,19,20]. Epoxy resins can be individual chemical compounds (crystalline solids: e.g., dicyclopentadiene dioxide [DCPDO]), active liquids (e.g., vinylcyclohexene dioxide [VCHDO]), or can be of polymer structure (e.g., diane epoxy resins, novolac epoxies) [4]. The resins differ in terms of molecular size and number of reactive epoxy and hydroxyl groups. One of the distinct features of epoxy resins is their *epoxide number*, which denotes the number of epoxy groups in 100 g of resin, and has fundamental meaning to determine the correct amount of curing agent [3,24].

Epoxy resins may contain two, three, or more epoxy groups in a molecule, and the resins that contain two groups, called *diepoxides*, constitute their major proportion. According to their chemical structure, commercial epoxy resins can be divided into the following categories [3,4,10,11,18,19]:

- Diane epoxy resins: Depending on the molecular mass, these can be further divided into: low molecular weight resins (liquids), average molecular weight resins (high-viscosity liquids with low softening point) and high molecular weight resins (solids with high softening point).
- Epoxy resins of low viscosity, obtained from a combination of bis(4-hydroxyphenyl)methane (bisphenol F) and epichlorohydrin.
- Self-extinguishing epoxy resins (particularly with 4-bromodione and epichlorohydrin).
- Novolac epoxy resins.
- Cycloaliphatic epoxy resins.
- Epoxy resins with special properties.

17.2.3 Epoxy Adhesives

There exist several criteria for the classification of epoxy adhesives. Depending on the number of components, these can be distinguished [5,25,26,27] as follows:

- One-component
- Two-component
- Multicomponent

With regard to the curing method, adhesives can be divided into those setting at [5,19,25,28,29]:

- Room temperature
- Average elevated temperature
- Highly elevated temperature

To obtain particular parameters, resins are transformed into insoluble and infusible products in a cross-linking reaction, taking place between the functional groups of the hardener.

The process of curing an epoxy resin is a reaction of the epoxy and hydroxyl groups of the resin molecule with reactive groups of the hardeners. The higher the molecular mass of the resin, the higher the number of epoxy groups it contains at the ends. The content of hydroxyl (alcohol) groups, found with regular interspaces along the whole chain of the macromolecule, is small in low molecular weight resins. With the increase in molecular weight, the content of the hydroxyl groups rises rapidly at the beginning, and then gradually becomes slower.

17.2.3.1 Two-Component Epoxy Adhesives

Epoxy adhesives represent an important group of reactive adhesives. Owing to varied formulation possibilities, such organic compounds provide tailor-made adhesives for a range of applications.

A characteristic structure of monomers, that is, a specific arrangement of carbon and oxygen as part of the molecule, allows for high reactivity with other monomers. In the epoxy group, two carbon atoms and one oxygen atom are combined into a *triangle*, which is shown in Figure 17.4 [25].

The curing reaction is initiated by adding Component B (hardener) to the resin, which opens the *epoxy triangle* (Figure 17.5). One representative of the group of hardeners is amine (Figure 17.6) [25].

The polyaddition reaction is the most commonly used type of reaction for the cure of epoxy resins. The curing agents used in this type of reaction have a hydrogen compound, and these include amines, amides, and mercaptans. For this reaction mechanism, the most important curing agents for adhesives are primary and secondary amines containing at least three active hydrogen atoms and various di- or polyfunctional carboxylic acids and their anhydrides [4]. The schematic of a

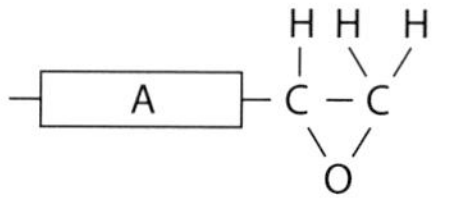

FIGURE 17.4   Resin component A with epoxy group. (From Habenicht, G., *Applied Adhesive Bonding. A Practical Guide for Flawless Results,* pp. 23–26, Wiley-VCH, Weinheim, Germany, 2009.)

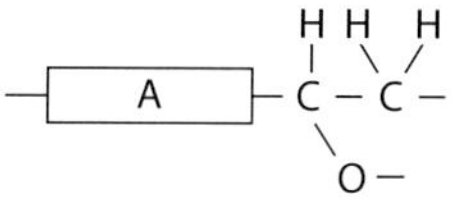

FIGURE 17.5   Resin component A with open epoxy group. (From Habenicht, G., *Applied Adhesive Bonding. A Practical Guide for Flawless Results,* pp. 23–26, Wiley-VCH, Weinheim, Germany, 2009.)

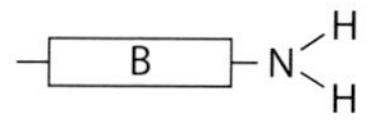

FIGURE 17.6   Hardener component B with amine group. (From Habenicht, G., *Applied Adhesive Bonding. A Practical Guide for Flawless Results,* pp. 23–26, Wiley-VCH, Weinheim, Germany, 2009.)

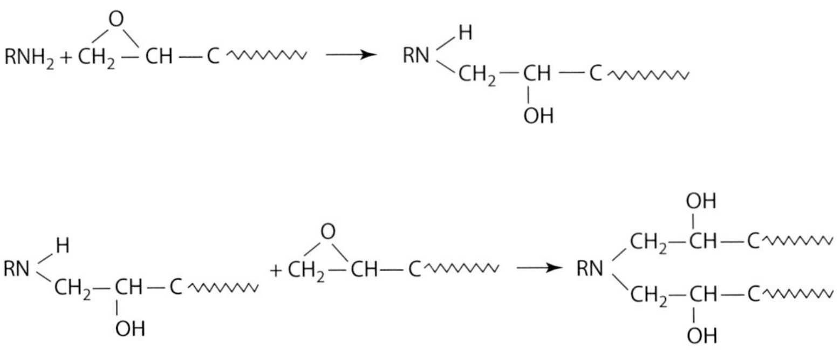

FIGURE 17.7 Schematic of polyaddition reaction of an epoxy resin with an active hydrogen compound. (From Pertie, E.M., *Epoxy Adhesive Formulation*, pp. 3–19, 26–41, 43–53, 71–82, 85–98, 207–225, 229–236, McGraw-Hill, New York, 2006.)

polyaddition reaction of an active hydrogen compound and an epoxy molecule is presented in Figure 17.7.

Hydroxyl groups, especially phenolic hydroxyls, and tertiary amines catalyze this reaction. The bulk and nature of substituent groups on the epoxy and curing agent molecules also play a major role in the reaction rate. Due to the various possibilities regarding the chemical structures of Components A (resin with epoxy group) and B (curing agent), curing behavior can also be influenced.

Therefore, the following differences are distinguished [4,5,19,25,26]:

- Curing at room temperature (cold curing)
- Curing at elevated temperature (heat curing up to approximately 120°C and hot curing up to approximately 250°C)
- Applications with short or long pot-lives (minutes, hours, days)

17.2.3.2 One-Component Epoxy Adhesives

One-component epoxy adhesives are found almost exclusively in industrial applications (aircraft construction, car manufacture, and electronics). In these types of epoxy adhesives, the resin and the hardener are already mixed; however, they are prevented from reacting with each other at room temperature and thus from curing. As a result of this reaction, the adhesive cures more quickly. The curing time of reactive adhesives can be reduced by the heat supply. This impact of temperature also enables the production of reactive adhesives. For this purpose, monomers are chosen that are not inclined to react with each other at room temperature or below, due to their chemical inertness.

In a mixed state, they are nonreactive and can be applied without pot-life limitation. The required temperature depends on the monomer structure. Hot-curing adhesives set in the range of temperatures of approximately 60°C–150°C; adhesives setting beyond these temperatures are referred to as *heat-curing adhesives* [25].

17.2.3.3 Hot-Curing Adhesives

Hot-curing adhesives are compositions of molten epoxy resin with a softening point of about 70°C and hardener. For heat-curing adhesives, dicyandiamide is usually used. This is a solid whose melting point is 209°C and it reacts at sufficient speed at temperatures above 130°C.

The curing temperature depends on the type of adherend and whether it is possible to heat the joint at temperatures ranging between 120°C and 200°C. Cure time is generally temperature dependent, and curing for 1–2 h at 180°C is most commonly regarded as optimal conditions [4].

Hot-curing adhesives with dicyandiamide are produced in the form of blocks, rods, or powder. At a temperature of 100°C–120°C, the substance is liquid, below 100°C it is thermoplastic, and at 40°C–50°C it is of pasty texture. Adhesives in the form of powder may be applied on a hot surface by spraying. These adhesives may also occur in the form of solutions or emulsions, and can similarly be applied by spraying.

These adhesives are used to join steel, cast iron, aluminum and its alloys, and nonferrous metals, as well as some polymeric materials that can withstand temperatures above 120°C. Adhesives of this type are not used to join materials of different coefficients of thermal expansion [3,4].

17.2.3.4 Cold-Curing Adhesives

Cold-curing adhesives are obtained from a liquid diane resin of average molecular weight. Aliphatic polyamides (diethylenetriamine, triethylenetetramine, dimethylaminopropylamine), adducts of amines, or polyaminoamides are most commonly used as hardening agents. The pot-life of resin and amine mixtures is short; for example, in the case of diethylenetriamine, it is 30–60 min. The amount of amine should be carefully calculated according to the content of epoxy groups in the resin because, with an excess of unreacted amine, the bond strength is reduced. Curing at room temperature most frequently lasts for 1–3 days; nevertheless, the adhesive joint acquires optimal strength in no sooner than approximately 7 days [3]. Gladkikh et al. [30] investigated the properties of fast-curing adhesive materials based on modified cold-curing epoxy resins.

The mechanical strength and chemical resistance of cold-cured adhesives can be increased when the joint is subjected to a secondary cure at elevated temperatures. Adhesives cured at room temperature are resistant to petrol and oils up to 60°C. Water, alcohols, and aromatic hydrocarbons decrease their bond strength, whereas acetone and chlorinated hydrocarbons cause adhesional (interfacial) failure of the bond. Depending on the application, adhesives can be modified to decrease their viscosity by the addition of active or inactive diluents, or plasticizers (e.g., dibutyl phthalate), which simultaneously increases the elasticity of the bond.

Adhesives cured at room temperature are applied when adherends could be damaged at elevated temperature or when heat treatment is impossible due to, for instance, the large size of the adherends. These adhesives may be used to join glass to metal, as their low curing temperature does not cause the glass substrate to crack.

17.2.4 Properties of Epoxy Adhesives

Adhesives cured at room temperature are characterized by lower shear strength (10–20 MPa on average) than adhesives cured at elevated temperatures (above 100°C). Adhesives cured at elevated temperatures are characterized by high shear strength (25–45 MPa), and exhibit resistance to the impact of elevated temperature. It should be noted that cure temperature depends, to a large extent, on the type of hardening agent [4].

Solvent-free adhesives, which consist of resin and hardener, and which may optionally contain diluents or fillers, do not require predrying. Although solvent-based adhesives obtained from resins are of high molecular weight and have low viscosity, they do require predrying prior to the actual curing process. A certain amount of solvent that may remain in the joint could have an adverse effect on its strength.

At room temperature, epoxy resins cured with hardeners (e.g., aliphatic and aromatic amines or acid anhydrides) are usually rigid and brittle materials with high flexural, tensile, and compressive strengths and a high modulus of elasticity. Nevertheless, these adhesives have a very low elongation-at-break and low impact strength. At room temperature, cured epoxy resins are in a glassy state, and only above the glass transition temperature do they become highly plastic. Depending on the type of resin and hardener, the glass transition temperature may vary between 70°C and 280°C; the higher the cross-link density of the cured resin, the higher its glass transition temperature [3,4].

17.2.5 Curing of Epoxy Adhesives

Curing of an adhesive occurs as a result of monomer cross-linking to form a solid, insoluble adhesive as a result of a hardener, catalyst, or elevated temperature impact. Following the curing of the adhesive, the hardener becomes a constituent of the cross-linked structure of the adhesive [19,29]. Simultaneously, the catalyst accelerates the cure, yet is not a part of the structure of the adhesive. Raising the curing temperature only accelerates this process. Stewart et al. [18] investigated the cohesive properties of an epoxy adhesive through analysis of mechanical behavior as a function of curing conditions. Various aspects of curing epoxy adhesives and epoxy resins are presented in numerous studies [7,19,29–37].

Curing conditions are of crucial importance. Various parameters can influence rheological properties, such as composition and amounts of hardener and diluents, curing temperature, and holding time [3,4,8].

When selecting an appropriate amount of hardening agent, certain guidelines must be followed. Hardeners react with epoxy resins at stoichiometric ratios (excluding hardener-initiated ionic polymerization) and, therefore, they are used in amounts according to stoichiometric calculations, based on the content of epoxy groups in the resin and the content of functional groups with active hydrogen atoms in the hardener [20,21].

During the cure, the physical properties of the resin–hardener complex change. Curing reactions are exothermic, and the heat emitted increases the temperature of the system and accelerates the reaction between the resin and hardener.

After mixing resin and hardener (and possibly heating), an increase in the viscosity of the system is observed. During this period, primarily small molecules join to form branched macromolecules that are still not cross-linked. After a certain amount of time, viscosity begins to rise rapidly, and the resin reaches the gel point, which is when gelling occurs. At this point, spatially cross-linked structures develop, and the resin with hardener becomes infusible and only partially soluble in organic solvents. With a further increase in cross-link density, the material turns into a glassy solid of increasing mechanical strength and hardness.

As a result of the curing reaction, there is a decrease (contraction) in the volume of the mixture, that is, chemical shrinkage [4]. The chemical structure of the hardener has a significant effect on the strength properties of the cured epoxy resin and on the dependence of the resin properties on the temperature. Hardening agents with a large number of functional groups, simultaneously containing aromatic or cycloaliphatic rings, allow the acquisition of cured products with high cross-link density and rigid structure; that is, with high heat resistance and good mechanical strength, as well as good dielectric properties at elevated temperatures. The use of hardeners, with which the functional groups are interconnected with short aromatic or cycloaliphatic chains, contributes to obtaining products with lower heat resistance and poorer properties at elevated temperatures. The cured products are rigid and brittle at room temperature.

Moreover, the choice of hardener depends on the required curing conditions, as well as on the required properties of the epoxy resin [4,25,26]. An important criterion is the curing temperature. For curing at room temperature (cold curing), aliphatic polyamines are used. A disadvantage of compositions cured in such a way is the relatively low heat resistance. In curing at temperatures below 80°C–100°C, tertiary amines and aromatic primary amines are used. In the case of primary aromatic amines, additional curing (i.e., post- or secondary curing) at higher temperature imparts high heat resistance. The use of acid anhydrides requires curing at higher temperatures (heat curing), in the range 120°C–180°C, thereby obtaining high heat resistance and excellent dielectric properties [26,27].

The hardener, often an amine, cures the epoxy by an attachment (addition) reaction, in which both substances react with each other (Figure 17.8). Normally, it proceeds in such a way that each amine group is combined with two epoxy active sites to form a three-dimensional molecular structure.

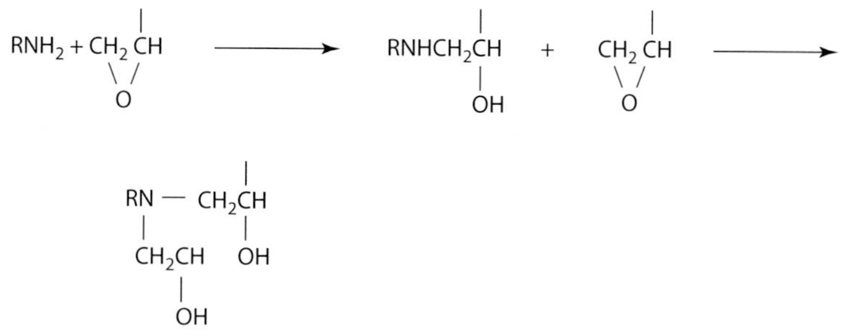

$$RNH_2 + CH_2\overset{|}{C}H \xrightarrow{\quad\quad} RNHCH_2\overset{|}{C}H \;+\; CH_2\overset{|}{C}H \xrightarrow{\quad\quad}$$

FIGURE 17.8 Addition of primary or secondary amine to epoxy ring. (From Pertie, E.M., *Epoxy Adhesive Formulation*, pp. 3–19, 26–41, 43–53, 71–82, 85–98, 207–225, 229–236, McGraw-Hill, New York, 2006.)

Since the molecules of amines react with the epoxy molecules in a fixed ratio, the resin and hardener should be mixed in suitable proportions to ensure completion of the reaction. If the proportions are incorrect, the unreacted resin or hardener will remain in the matrix, which will, furthermore, affect the properties obtained following curing. To facilitate mixing, manufacturers offer products in which mix proportions are simple and easy to measure out by volume or weight.

Curing adhesive joints may be carried out in two stages [3,4]. In the first stage, curing is carried out at lower temperature and, subsequently, for a short time at elevated temperature (hot or heat curing). The second phase of the two-stage curing is referred to as *secondary curing* or *postcuring* of the adhesive. In the first stage, the joint is solidified at ambient temperature (15°C–20°C) under pressure suitable for the particular adhesive. By the end of this phase, the solid adhesive acquires approximately 60%–70% of its final strength. In the second step, the adhesive joint is finally cured without the use of curing pressure. Most commonly, it consists in leaving the joint to cure for several hours at elevated temperature (approx. 50°C–100°C), following which the joint obtains full strength and full hardening. Due to two-stage curing, the total time of cure is shortened several times compared with curing at ambient temperature.

For cold-curing adhesives, apart from the two-stage cure, a single-step curing cycle is also carried out at elevated temperatures. The process of heat curing is carried out at temperatures of 60°C–120°C, under suitable curing pressure. In relation to the two-stage cure, it shortens the cure time several times.

Hot-cure of adhesives occurs in three stages. The first phase consists in bringing the adhesive to curing temperature. The time of this step depends on the mass of adherends and the type of adhesive joint. During the second phase, the joint curing temperature is maintained, which often requires the use of autoclaves. The third step of adhesive heat curing is the slow cooling of the joint to room temperature; here, the important factor is the cooling time. The time is most frequently determined experimentally. It should be long enough so as not to induce shrinkage stress, and short enough not to prolong the production cycle. Most often, it amounts to 10–20 min.

The mechanism of curing epoxy resins and adhesives depends, *inter alia*, on the type of hardening agent. The following hardeners are commonly added to epoxy adhesives [3,4,5,14,19,20,36,38–41]:

- Anhydrides of organic acids, for example:
  - Maleic anhydride
  - Succinic anhydride
  - Phthalic anhydride
  - Endomethylenetetrahydrophthalic anhydride
  - Pyromellitic anhydride
  - Benzophenone tetracarboxylic anhydride
  - Others

- Primary or secondary amines, for example:
  - Diamine and aliphatic polyamides
  - Aromatic and cycloaliphatic diamines
  - Adducts of amines with ethylene oxide or propylene
  - Polyaminoamides
  - Others
- Tertiary amines (Lewis bases), for example:
  - Aliphatic amines
  - Cycloaliphatic amines
  - Aliphatic amines from aromatic rings
  - Amine salts
- Adducts of boron fluoride
- Phenolic compounds (e.g., phenol–formaldehyde resins)
- Proton-deficient acids (Lewis acids)
- Others

Epoxy resins are generally cured by polyaddition with hardeners, which is also true in the case of acid anhydrides and primary and secondary amines. Amines contain at least two active hydrogen atoms in the molecule. Tertiary amines (also known as *Lewis bases*) cure epoxy resins by anionic polymerization. In contrast, proton-deficient acids (*Lewis acids*) cause cationic polymerization of epoxy resins [19].

Acid anhydrides were the earliest introduced and are currently the most widely used hardeners for epoxy resins. While full curing with an acid anhydride requires heating at elevated temperature for a long time, cross-linking reactions are characterized by a small exothermic heat effect and low volumetric shrinkage. Resins cross-linked with anhydrides are devoid of internal stress and associated mechanical defects, and they are characterized by high thermal stability.

Acid anhydrides contain no active hydrogen atoms capable of reacting with epoxy groups, but these atoms are formed in the reaction of anhydride groups with hydroxyl groups of epoxy resins. Anhydride hardeners for epoxy resins may be added by the manufacturer of the adhesives. These hardeners remain reactive for at least 2 years; it is necessary, though, to increase the temperature for the curing reaction to proceed. Curing temperatures are in the range 100°C–260°C, and increasing the temperature accelerates the reaction.

Amine hardeners are cold-curing substances [4,19]. These are liquids that are usually added to epoxide resin immediately before the application of adhesive in suitable stoichiometric quantity. The reaction of resin curing is exothermic and takes place at room temperature in 8–48 h. An amine-cured epoxy adhesive must, therefore, be used almost immediately after mixing the resin and hardener. Its pot-life, depending on the type of resin and hardener used, is between 30 min and several hours. If the process of epoxy resins curing with amines is conducted at a temperature from −10°C to +20°C, then 3% diphenyl phosphite can be used as the accelerator. The curing process with aliphatic polyamines takes place at room temperature or slightly above, while the use of aromatic amines requires a higher curing temperature. Epoxy resins may also be cured with phenolic resins, and such compositions are characterized by particularly good thermal stability.

The curing process of epoxy resins can also be performed by [2,32,33,40–46]:

- UV radiation with wavelength in the range of 200–300 nm, which is the dominant type of radiation used in the process.
- Electron beam.
- X- or γ-radiation.
- Microwave radiation.

Industrial application of curing with electron beam and x- or γ-rays is becoming increasingly widespread [2,41,42]. These technologies are gradually replacing the conventional cationic polymerization initiators in cross-linking commercial types of epoxy resins (based on bisphenol A and F), novolac epoxy resins, and cycloaliphatic resins [44]. Radiation-cured resins demonstrate higher glass transition temperature, lower water absorption, higher thermal stability, and higher mechanical strength as compared with hot-cured resins. Taking only a few seconds, the radiation-curing process is extremely time efficient, generates lower volumetric shrinkage, and, also, does not require the addition of any toxic or carcinogenic hardening agents. Moreover, cross-linked layers may have higher thickness, from 50 nm (electron beam) to 200 nm (x-ray or γ). Alessi et al. [32] studied γ-curing at various controlled temperatures of a system consisting of difunctional epoxy resin in the onium salt as cationic initiator.

Another method consists in curing epoxy resins with the application of conventional hardeners, which is, however, conducted at elevated temperature by microwave irradiation heating. This method accelerates the curing process, and resins cross-linked under microwave irradiation manifest higher glass transition temperature than their conventionally hot-cured counterparts [45].

17.2.6 Modification of Adhesive Compositions

Epoxy adhesives are modified to develop particular properties, such as to increase shear and tensile strengths, tensile modulus, flexibility, and heat resistance.

Increases in the flexibility of the adhesive and the peel strength can be achieved in a variety of ways [2,8,15,20,27,28,40]:

- Use of high molecular weight resin to reduce cross-linking density
- Introduction of aliphatic resin or hardener comprising long aliphatic chains
- Modification with thermoplastic polymers, polysulfides, and so forth
- Introduction of liquid oligomeric butadiene–acrylonitrile rubbers

The increase in flexibility often leads to deterioration of other properties of epoxy adhesives, such as shear strength; and, moreover, the heat distortion temperature is lowered [5,26].

Resins containing multiple aromatic rings demonstrate higher rigidity (in comparison with diane epoxy resins). Improved heat resistance is a characteristic of resins containing more functional groups, due to higher cross-link density. Adhesives with high heat resistance should have high cross-link density, but this causes high rigidity of joints formed and low compression and peel strengths.

Good mechanical properties of adhesives at elevated temperatures can be maintained through the use of aromatic amines or certain acid anhydrides as hardeners. Moreover, epoxy and phenolic resin compositions hardened with, for example, hexamethylenediamine are capable of retaining excellent mechanical properties in extreme thermal conditions, that is, at temperatures amounting to 300°C. Finally, addition of antioxidants imparts further improvement in properties: in particular, resistance to aging [4]. Good heat resistance is displayed by adhesives obtained from epoxy–novolac precondensates that have been acquired in preliminary condensation of novolacs with a low molecular weight epoxy resin.

A wide range of modifications of epoxy adhesive properties, through the proper choice of resin, hardener, fillers, and other ingredients, has led to the development of conductive adhesives, which find application in electronics [47]. Conductive adhesives contain fillers in the form of powdered gold or silver flakes [46].

17.2.7 Preparation of Modified Epoxy Adhesives

The choice of a suitable method for formulation of modified adhesives is dictated by the type of adhesive and filler [16,26,48]. Epoxy adhesives containing filler particles are obtained either by direct mixing or mixing in solution [45,46]. Direct mixing is based on the introduction of filler directly into liquid adhesive and mixing the ingredients mechanically or ultrasonically.

Mechanical mixing can be carried out using a rolling mill, a high-speed mixer, an extruder (typically a twin-screw extruder), or a calender. In this way, it is possible to add fillers in amounts up to 20% by weight. After the introduction of a small amount of filler, the viscosity of certain adhesives may rapidly increase, which means that their processing is difficult or virtually impossible. Wettability of nanofillers can be improved by modifying their surface with surfactants. Such a method is used to produce polymer nanocomposites with ceramic nanoparticles, metallic nanotubes, nanosilica, and layered silica.

The intensity of mixing—thus, breaking of agglomerates—can be increased by the application of ultrasonic methods. Stirring by ultrasonics consists in using alternating pressure caused by the acoustic waves in the region above the threshold of cavitation (cavities) production in solutions. This leads to the creation of numerous cavities oscillating at the frequency of the applied pressure (typically 20 kHz), while the gas inside remains unchanged. Cavities that are not entirely filled with gas collapse as a result of compressive stress caused by sonic waves; thus, the induced energy turns into a pulsating pressure. This results in an intensification of many physicochemical processes occurring in the mixture, such as wetting, diffusion, dissolution, emulsification, or dispersion.

Mixing in solution, otherwise known as the *solvent method*, is another widely used alternative to direct mixing. It is a two-step process. The first phase involves mixing nanofillers with a polymer solution obtained by dissolving the polymer in a suitable solvent (e.g., acetone, toluene, isopropanol, or dimethylformamide). Afterward, the solvent is removed by evaporation. This is a widely used method for production of, *inter alia,* polymer nanocomposites with carbon nanotubes and layered silica [39].

17.3 STRENGTH OF EPOXY ADHESIVE JOINTS OF SELECTED STRUCTURAL MATERIALS

17.3.1 STRENGTH OF ADHESIVE JOINTS CURED AT ELEVATED TEMPERATURES

17.3.1.1 Characteristics of Joints and Adhesive

Tests were conducted on DX 51D+Z275MA zinc-coated sheet, with zinc coating thickness of approximately 18 μm, which formed single-lap adhesive joints. The dimensions and shape of the joint are shown in Figure 17.9.

Joints were formed with a medium viscosity, two-component epoxy adhesive, Loctite EA 9497, whose selective properties are shown in Table 17.1 (for the uncured material) and Table 17.2 (cured material).

The preparation of the adhesive consisted in mixing the proper amounts of resin and hardener (in a 2:1 ratio). The mixing process was performed manually for approximately 15 s, until a uniform light-gray color of the composition was obtained. During manual mixing, the total weight of the adhesive mass was 4 g. The adhesive was applied on one of the adherends immediately after mixing.

The test plan devised was based on the characteristics of the adhesive, shown in Figure 17.10, where cure temperature was the variable.

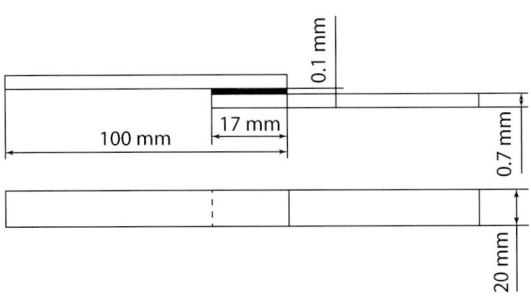

FIGURE 17.9   Adhesive joint specimen used in static strength tests.

TABLE 17.1

Selective Properties of Loctite EA 9497 (Uncured)

Property	
Chemical type of resin	Epoxy resin
Specific gravity @ 25°C	2.50–2.13
Casson viscosity, cone and plate rheometer (Pa·s); temperature 25°C	5–16
Chemical type of hardener	Epoxy
Mix ratio by volume (resin:hardener)	2:1
Mix ratio by weight (resin:hardener)	100:50
Pot-life @ 25°C (min); 267 g resin, 133 g hardener	165–255

Source: www.loctite-kleje.pl (2015).

TABLE 17.2

Selective Properties of Loctite EA 9497 (cured)

Property	
Tensile strength (MPa)	52.6
Tensile modulus (MPa)	2420
Hardness, ISO 868 (Shore D)	83
Elongation, ISO 37 (%)	2.9

Source: www.loctite-kleje.pl (2015).

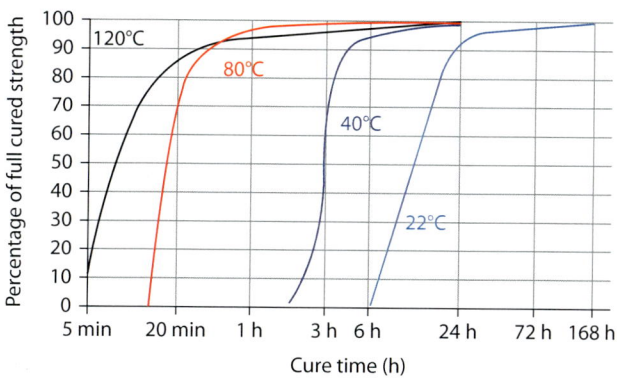

FIGURE 17.10 Strength of Loctite EA 9497 as functions of cure time and cure temperature. (From www. loctite-kleje.pl, 2015.)

17.3.1.2 Formation of Adhesive Joints

The surfaces of substrates were mechanically pretreated with a P500 abrasive tool (from Matador, Poland [50]). This increased the surface roughness, which enabled the formation of a stronger adhesive bond as a result of adhesive anchorage to the substrate. The surfaces were then profusely sprayed with Loctite 7036 degreasing agent and dried with a clean cloth to remove contamination.

The procedure was repeated and followed by another application of the degreasing agent, which was eventually left to evaporate to obtain a dry surface [51].

The adhesive was spread on prepared surfaces, and a load of 0.03 MPa was applied. The joint formation conditions were: temperature $20 \pm 2°C$, relative humidity $25 \pm 2\%$. The joint was cured in two variants. Variant I consisted of a 24 h single-step cure cycle at a temperature of $20 \pm 2°C$ and humidity $25 \pm 2\%$. The constant load applied during curing was 0.03 MPa.

Specimens cured in Variant II followed a two-step cure cycle: The first stage was identical to Variant I, whereas during the second phase the specimens were cured at the following temperatures: 60°C, 70°C, 80°C, 90°C, 100°C, 110°C. The heat treatment time was 60 min, with humidity 50%. The additional heat curing of the joint was performed in an SH 66-1 climatic chamber (from ESPEC, Poland [52]) (Figure 17.11).

The prepared specimens were then subjected to destructive testing, to determine the relationship between static strength and different heat treatment temperatures. The tests were conducted on a Zwick/Roell Z150 materials testing machine, according to the DIN EN 1465 standard [53].

17.3.1.3 Strength Test Results

The results of the strength tests conducted on shear loaded adhesive joints are shown in Figure 17.12. The results presented (Figure 17.12) indicate that the highest joint strength was recorded in the case of specimens subjected to cure at 70°C, and the lowest strength was demonstrated at a temperature of 20°C. Moreover, the results obtained for these specimens were characterized by high consistency. Comparably high strength of joints was measured in the case of secondary cure conducted at 70°C; nevertheless, this batch of specimens was characterized by a considerable scatter in results. The strength of adhesive joints following secondary cure at elevated temperatures is distinctly higher than the joints subjected to a single-stage cure at ambient temperature. In the latter case, for joints cured at $20 \pm 2°C$, the strength was the lowest of all the adhesive joints measured and amounted to

(a)   (b)

FIGURE 17.11 SH 66-1 temperature and humidity chamber: (a) general view; (b) inside view.

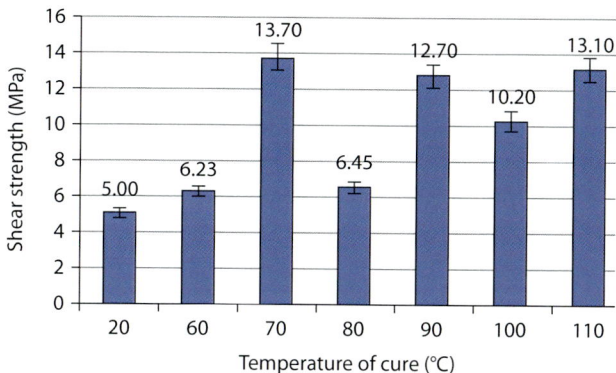

FIGURE 17.12   Relationship between adhesive joint shear strength and temperature of cure.

5.05 MPa, while the maximum load carried by the joint was equal to 1752 N. Furthermore, it was noted that the repeatability of results was considerably high in this case.

The test results shown in Figure 17.12 were analyzed statistically. Maximum shear strength values were compared, which showed that in almost all cases studied the values were different.

17.3.1.4   Summary

The analysis of results obtained in strength tests on zinc-coated sheet adhesive joints bonded with Loctite EA 9497 indicates a beneficial effect of secondary cure at elevated temperatures on the maximum load capacity of the joint. The conclusions are consistent with test results presented in the literature [31,54]. However, it should be noted that test results are influenced by numerous factors, such as the type of adhesive and substrates, not to mention the method of joint formation or secondary cure. The highest joint strength was measured in specimens subjected to secondary cure at 110°C. A similarly high strength of joints was obtained for the adhesives cured at 70°C and 90°C; nevertheless, these batches of specimens show a considerable scatter in results. This requires further testing, because it is suspected that the range of temperatures from 70°C to 100°C has a substantial impact on adhesive properties and thermal stability. This issue will be studied in the future. The results presented were confirmed by the statistical analysis.

Therefore, it seems justified to claim that the decision whether or not to conduct a two-stage cure process should not be dictated by workshop capabilities alone, but be predominantly based on its impact on the properties of materials.

17.3.2   Strength of Adhesive Joints Modified with Fillers

17.3.2.1   Characteristics of Adhesive Joints

Tests were carried out on 1.5 mm C45 structural steel single-lap joint specimens. The diagram of the analyzed joint is shown in Figure 17.13.

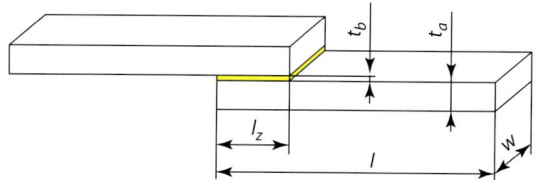

FIGURE 17.13   Adhesive joint specimen used in comparative analysis.

The dimensions of the joint were as follows: specimen length: $l = 100$ mm, adherend thickness: $t_a = 1.5$ mm, bondline thickness: $t_b = 0.1$ mm, specimen width: $w = 20$ mm, overlap length: $l_z = 23$ mm.

17.3.2.2   Formation of Adhesive Joints

The factors analyzed in the tests were the types of adhesive compounds, modified by different adhesive:nanofiller ratios and adhesive:curing agent ratios.

Tests were conducted in the following conditions:

1. Adherend surface preparation: mechanical treatment with P400 abrasive disks, degreasing with Loctite 7063 degreasing agent
2. Joint formation conditions: temperature: $20 \pm 2°C$, humidity: $40 \pm 2\%$, load: 0.02 MPa, conditioning time: 144 h
3. Adhesive compounds preparation conditions: dispersion speed: 128 m/min, mixing time: 2 min, gas bubble removal time after mixing: 30 min

The adhesive joints were formed with epoxy resin (Epidian 57 [Organika–Sarzyna Chemical Works, Poland] [24]), selected hardeners (PAC, Z-1, PF [Organika–Sarzyna Chemical Works, Poland] [24]) and modified with nanofillers: NanoBent ZR2 and ZS1 (organophilized montmorillonite [ZG-M "Zębiec" SA, Poland] [55]).

The following epoxy adhesive compounds were tested (Table 17.3):

- Epidian 57/PAC/1:1 epoxy resin modified with 1%, 2%, and 3% NanoBent ZR2 and 1%, 2%, and 3% NanoBent ZS1
- Epidian 57/Z-1/10:1 epoxy resin modified with 1%, 2%, and 3% NanoBent ZR2 and 1%, 2%, and 3% NanoBent ZS1
- Epidian 57/PF/10:5 epoxy resin modified with 1%, 2%, and 3% NanoBent ZR2 and 1%, 2%, and 3% NanoBent ZS1

For the purpose of comparison, the strength tests were also carried out on joints formed with Epidian 57-based and PAC, Z-1, and PF hardeners epoxy adhesive compounds. The hardeners were mixed with the resin at the correct ratios (Table 17.3) without the addition of modifiers.

The compositions were prepared in a mixer equipped with a special attachment providing uniform dispersion of the nanofiller in the entire volume of epoxy resin. The process of dispersion was conducted at a mixing rate of 128 m/min for 2 min. Next, after 30 min (in that time, the gas bubbles naturally created during mixing partially disappeared from the mix) the hardener was added to the modified resin in a given weight ratio.

Identical conditions for the preparation of adhesive joints and curing of all the compositions with three different hardening agents were used, to ensure repeatability of results. At 144 h after forming the joint, the strength tests were conducted to determine the breaking loads.

TABLE 17.3
Types of Epoxy Adhesive Compounds Used

	Type of Nanofiller and Weight Proportion Used	
	NanoBent ZR2	NanoBent ZS1
Epidian 57/PAC/1:1	1%, 2%, and 3%	1%, 2%, and 3%
Epidian 57/Z-1/10:1	1%, 2%, and 3%	1%, 2%, and 3%
Epidian 57/PF/10:5	1%, 2%, and 3%	1%, 2%, and 3%

17.3.2.3 Strength Test Results

Figures 17.14 through 17.16 collate the test results for the sake of comparative analysis.

The analysis of the results, shown in Figures 17.14 through 17.16, indicates that at ambient temperature ZS1 and ZR2 fillers increase the strength of joints more effectively in the presence of Z-1 hardening agent (Figure 17.14). In the case of PF hardening agent, it seems justified to claim that the analyzed fillers demonstrate negligible impact on the strength of adhesive joints (Figure 17.15). At the same time, modifying the compounds with ZS1 filler in the presence of PAC hardening agent produced the best results (Figure 17.16). For ZR2 filler and PAC hardener, lowering of the joint strength was observed at ambient temperature.

The results on the strength of the adhesive joints formed with adhesive compounds used with the addition of selected hardening agents, modified by 2% ZR2 and ZS1 fillers, are collated in Figure 17.17.

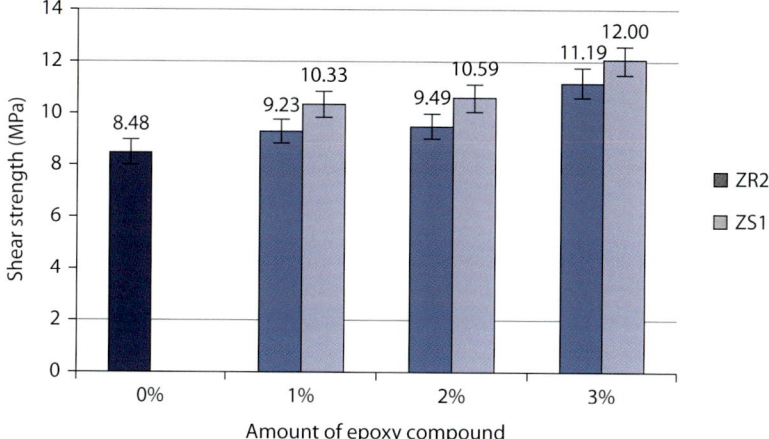

FIGURE 17.14 Shear strengths of steel sheet adhesive joints formed with adhesive compounds of Epidian 57/Z-1/10:1 resin: 0% unmodified; modified with ZR2 and ZS1 fillers in the amounts of 1%, 2%, 3%.

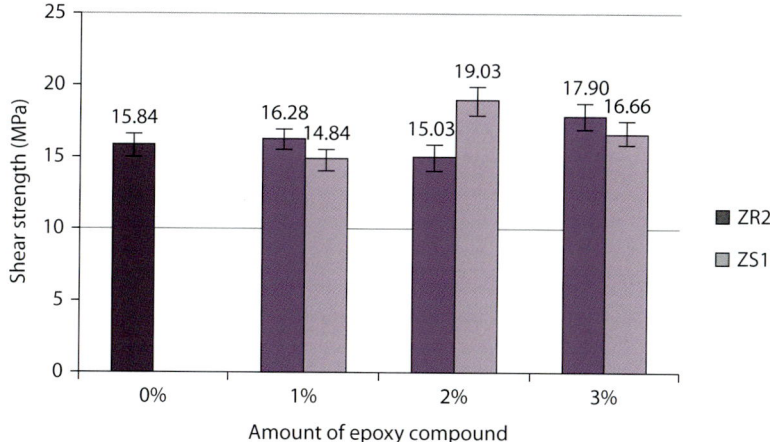

FIGURE 17.15 Shear strengths of steel sheet adhesive joints formed with adhesive compounds of Epidian 57/PF/10:5 resin: 0% unmodified; modified with ZR2 and ZS1 fillers in the amounts of 1%, 2%, 3%.

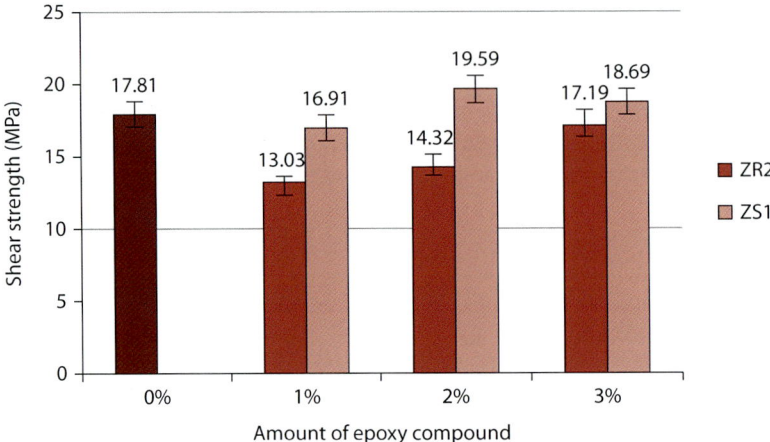

FIGURE 17.16 Shear strengths of steel sheet adhesive joints formed with adhesive compounds of Epidian 57/PAC/1:1 resin: 0% unmodified; modified with ZR2 and ZS1 fillers in the amounts of 1%, 2%, 3%.

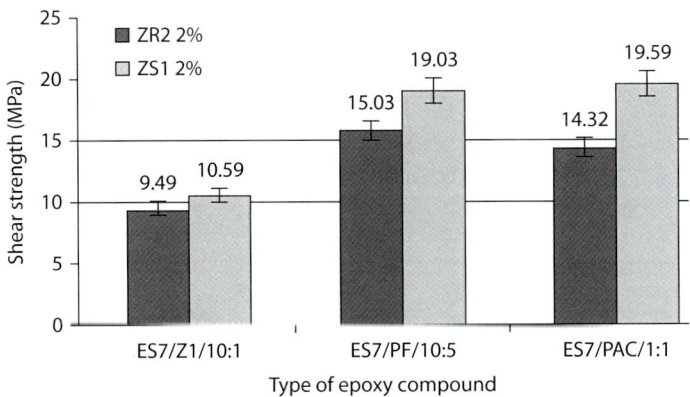

FIGURE 17.17 Shear strengths of steel sheet adhesive joints formed with adhesive compositions of Epidian 57/PAC/1:1 resin modified with 2% ZR2 and ZS1 fillers.

A comparison of test results (Figure 17.17) reveals that the strength of steel sheet adhesive joints was highest when the adhesive compound was cured with hardeners modified with ZS1 and ZR2 fillers (in the same filler:resin ratio). Depending on the type of hardening agent, different strength values were recorded; nevertheless, regardless of the type of hardening agent used, modifying the compounds with ZS1 filler tended to produce higher strength values than when using ZR2 filler.

17.3.2.4 Summary

The analysis of test results allowed the following general conclusions to be drawn:

1. The efficiency of epoxy material modification with low-particle size fillers depends on the type of hardener used.
2. The application of ZS1 and ZR2 fillers does not lower the strength of adhesive joints at ambient temperature, compared with unmodified adhesives.
3. It appears that when using PAC hardening agent and ZR2 filler, the viability of modification should be assessed, as lower strength parameters were recorded as compared with unmodified compositions.

4. Compared with ZR2 filler, ZS1 increases the strength of adhesive joints regardless of the type of hardening agent applied.

5. The technology of producing such epoxy compounds requires further development, which has already been undertaken by us in the tests that include an ultrasonic mixer. The objective of this research was, first, to analyze the impact of hardeners and, secondly, to prove that the fillers analyzed have no negative impact on the strength of adhesive joints.

17.3.3 Strength of Adhesive Joints Produced with Different Epoxy Adhesives Using Different Hardening Agents

17.3.3.1 Characteristics of Epoxy Adhesives

Adhesive joints of stainless steel sheets were produced with several types of epoxy adhesives, formed by mixing the epoxy resin with a hardening agent. The resin constituents of the adhesive compositions were Epidian 53 and Epidian 57, and hardening agents comprised PAC, Z-1, and PF. Adding a hardener to the resin initiates the bond cure process, which consists in transformation of the adhesive from liquid to solid state.

Epidian 57 is a transparent resin of characteristic yellow or dark-brown hue. It serves as a base for an all-purpose room-temperature adhesive ($20°C \pm 2°C$), suitable for metals, glass, ceramics, thermoset polymers, and leather. Adding PAC, Z-1, or PF hardening agents changes Epidian 57 into a two-component adhesive, and at that point the pot-life of the mixture is very important. The adhesive should be prepared in small batches, immediately before application, because it loses its properties with time. The pot-life of Epidian 57/Z-1 amounts to approximately 90 min, whereas that of Epidian 57/PAC is several hours [24,33].

Similarly to Epidian 57, Epidian 53 is a resin that serves as a primary component for a room-temperature adhesive, suitable for joining structural elements. In addition, since its bond does not possess good elastic properties, this adhesive finds application in rigid structures. The bond using Epidian 53 provides good electrical insulation, which is only highly desirable in electronics.

Supplementing the resin with PAC hardening agent produces bonds of greater elasticity than in the case of Z-1 hardener. Such adhesive also provides higher shear strength. PAC hardening agent is frequently added to resins that join elements subject to strain, such as thin sheets or rubber and metal substrates.

Z-1 hardener, consisting of triethylenetetramine, is a light-yellow liquid with a characteristic sharp and unpleasant odor. Addition of Z-1 hardener to Epidian 57 resin produces a higher stiffness bond compared with PAC hardener [24].

PF hardening agent is applied whenever low viscosity and relatively short gelation time are required. When added to resin, PF hardener produces bonds of lower elasticity than in the case of PAC; however, it shows higher compressive strength, heat, and chemical resistance [24].

The data regarding the type of resin and hardener, resin:hardener ratio, and designation of certain adhesive compositions are shown in Table 17.4.

TABLE 17.4
Description of Adhesive Compositions Used in Tests

Resin	Resin: Hardener Ratio	Hardener	Adhesive Designation
Epidian 57	1:1	PAC	Epidian 57/PAC/1:1
Epidian 57	10:1	Z-1	Epidian 57/Z-1/10:1
Epidian 57	10:5	PF	Epidian 57/PF/10:5
Epidian 53	10:5	PAC	Epidian 53/PAC/10:5
Epidian 53	10:1	Z-1	Epidian 53/Z-1/10:1
Epidian 53	10:4	PF	Epidian 53/PF/10:4

Resin and hardener were mixed in the correct ratio to produce compositions with the required properties. Any imbalance—for example, excessive or insufficient amounts of hardening agent—could decrease the mechanical strength as well as the thermal and chemical resistance of the joint. It is, therefore, necessary to determine the correct stoichiometric resin:hardener ratio, depending on the type of resin and hardening agent.

17.3.3.2 Formation of Adhesive Joints

A graphic representation of single-lap adhesive joints formed with different types of epoxy adhesives is given in Figure 17.18. The dimensions of the analyzed joint are presented in Table 17.5.

The substrate surface preparation consisted in mechanical treatment with an abrasive grinding tool, performed with a P320 abrasive paper. The treatment consists of 30 repetitions of circular movements on each of the adherends' surface. The abraded surfaces were subsequently cleaned three times with a Loctite 7061 degreasing agent, prior to application of the adhesive.

Adhesive joints were formed at ambient temperature $24°C \pm 2°C$ and humidity in the range of 30%–32%. The pressure applied during cure was 0.03 MPa and the curing time was 168 h.

The strength tests were carried out on a Zwick Roell Z150 materials testing machine, and were in accordance with the standard DIN EN 1465 [53]. Test specimens were clamped with the wedge-screw grips of the machine and subjected to a tensile shear test.

17.3.3.3 Test Results

The tensile shear test results on the joints made with the adhesive compositions used are compared in Figure 17.19.

The results of the strength tests (Figure 17.19) show that the highest tensile shear strength was shown by joint specimens formed with Epidian 57/PAC/1:1 adhesive. Here, the recorded joint strength was higher than in the case of all other compositions. Furthermore, it became evident that joints bonded with Epidian 57 are stronger than those bonded with Epidian 53 resin-based adhesive; for example, 53/PAC/10:5 produced a joint strength that amounted to 55% of the joints made with Epidian 57/PAC/1:1.

The greatest difference in strength was noted for PAC hardening agent. The smallest impact on the strength was noted for adhesive compositions with Z-1 hardening agent. This claim is confirmed by the results of Epidian 53/Z-1/10-bonded joints, which correspond to 77%

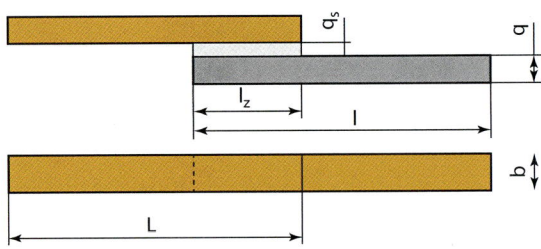

FIGURE 17.18   Adhesive joint specimen used in tests.

TABLE 17.5
Dimensions of Single-Lap Adhesive Joint Specimen

Overlap Length, l_z (mm)	Specimen Thickness, g (mm)	Specimen Width, b (mm)	Length of Joint Specimen, L (mm)	Adhesive Layer Thickness, g_s (mm)	Specimen Length, l (mm)
16 ± 1.96	0.98 ± 0.01	20 ± 0.09	183.37 ± 2.54	0.07 ± 0.04	100 ± 0.98

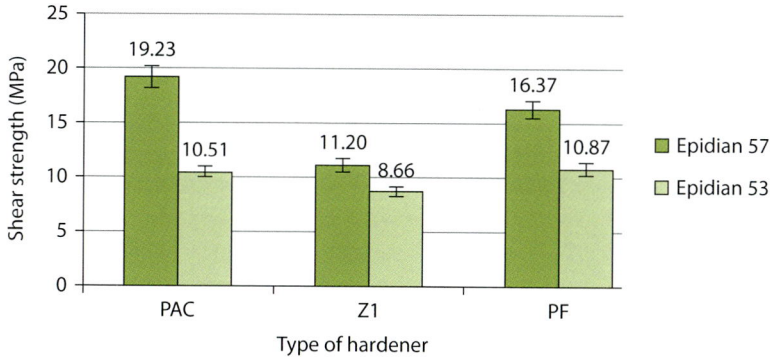

FIGURE 17.19 Tensile shear strength of adhesive joints formed with the epoxy adhesives used.

strength of the adhesive joints formed with Epidian 57/Z-1/1:1. In conclusion, it is apparent that apart from the type of resin, it is the type of hardening agent that affects the adhesive joint strength.

A comparison of maximum load and elongation-at-break of adhesive joints formed with PAC hardening agent compositions with different epoxy resins is shown in Figure 17.20.

The results in Figure 17.20 show that higher values of elongation-at-break were obtained when the composition consisted of PAC hardener and Epidian 57 resin. The elongation was 1.95 mm, and it was nearly twice as long as for Epidian 53/PAC/10:5; that is, the value of elongation-at-break observed for Epidian 53/PAC/10:5 amounted to 46% of that obtained for Epidian 57/PAC/1:1.

The maximum load for all the joint variants tested was exhibited in the case of Epidian 57/PAC/1:1, and amounted to 2398 N, whereas the highest maximum load for Epidian 53/PAC/10:5 was 57% of the maximum (failure) load applied to the joint formed with Epidian 57/PAC/1:1.

A comparison of the maximum load and elongation-at-break of adhesive joints formed with Z-1 hardening agent and different epoxy resins is shown in Figure 17.21, where the two adhesives used were Epidian 57/Z-1/10:1 and Epidian 53/Z-1/10:1.

The collated test data for Z-1 hardener and two different resins (Figure 17.21) indicate that a higher elongation-at-break is obtained for Epidian 57/Z-1/10:1, amounting to 0.83 mm at a maximum load of 1398 N. There was a slight difference observed in the elongation-at-break values, depending on the epoxy resin component of the adhesive. This is confirmed by comparison of the

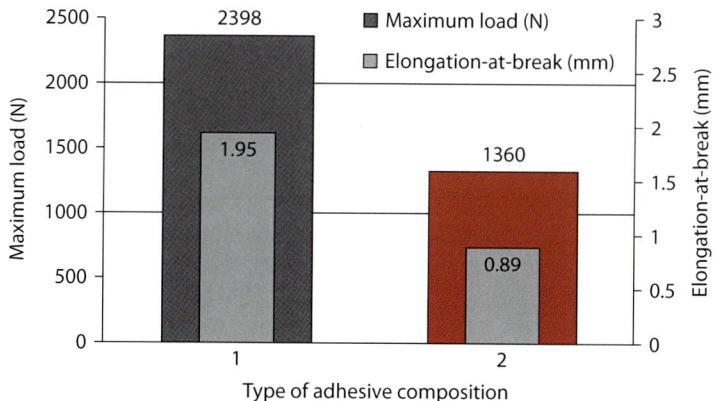

FIGURE 17.20 Maximum load and elongation-at-break of adhesive joints formed with epoxy adhesives with PAC hardening agent: 1. Epidian 57/PAC/1:1 and 2. Epidian 53/PAC/10:5.

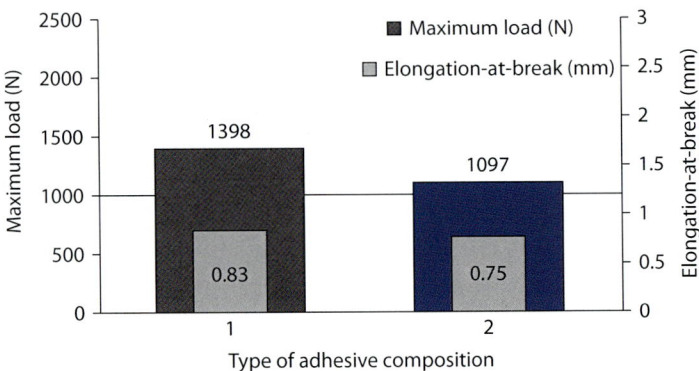

FIGURE 17.21 Maximum load and elongation-at-break of adhesive joints formed with epoxy adhesives with Z-1 hardening agent: 1. Epidian 57/Z-1/10:1 and 2. Epidian 53/Z-1/10:1.

strengths of joints formed with Epidian 53/Z-1/10:1 and Epidian 57/Z-1/10:1. The strengths of joints formed with Epidian 53/Z-1/10:1 correspond to 90% and 78% of the adhesive joint strength formed with 57/Z-1/10:1.

A comparison of the maximum load and elongation-at-break on the failure of joints bonded with PF hardening agent and different adhesives are presented in Figure 17.22. This shows that Epidian 57 provides higher joint strength. Here, the elongation-at-break amounted to 1.36 mm, and the maximum load exceeded 2100 N. Epidian 53/PF/10:4 produced elongation-at-break that amounted to 68% of the elongation of the specimen formed with Epidian 57/PF/10:5; however, the maximum (breaking) load values for both adhesives were comparable.

An analysis of diagrams led to the observation that an increase of load value applied to determine the tensile shear strength of adhesive joints increases the elongation-at-break of the specimen. Its highest value was recorded in specimens joined with Epidian 57/PAC/1:1. This adhesive, with elongation-at-break of 1.95 mm, significantly outperformed the other adhesive compositions analyzed. Simultaneously, the lowest value of elongation was obtained when Epidian 53/Z-1/10:1 was applied. The value of elongation-at-break in this instance was merely 0.75 mm, which was measured at the adhesive joint failure at a load of 1097 N. The difference between the highest and the lowest elongations-at-break was over 1 mm. Another composition worth mentioning is Epidian 57/PF/10:5, which resulted in an elongation-at-break of 1.36 mm. The performance of other

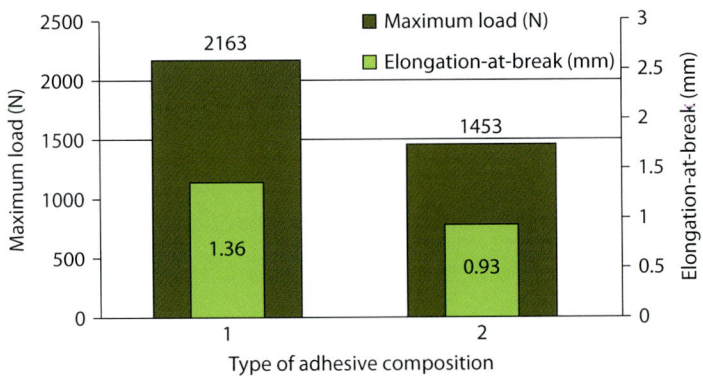

FIGURE 17.22 Maximum load and elongation-at-break of adhesive joints formed with epoxy adhesives with PF hardening agent: 1. Epidian 57/PF/10:5 and 2. Epidian 53/PF/10:4.

adhesives was intermediate. Epoxy adhesives with PAC hardening agent manifest higher elasticity than Z-1 hardener-based ones, and the results for PF hardener are intermediate.

17.3.3.4 Summary

Having analyzed the results obtained in strength tests of adhesive joints, we observed a direct correlation between the type of adhesive and the quality and strength of adhesive joints. The strength of joints depends on the failure load (maximum load) that causes joint failure and elongation-at-break of the specimen. In terms of strength properties, it was the adhesive composition based on Epidian 57 resin and PAC hardening agent that produced, by far, the best results among all the investigated compositions. With elongation-at-break equal to 1.95 mm and maximum breaking load of 2398 N, the joints formed with Epidian 57/PAC/1:1 exhibited tensile shear strength of 19.23 MPa. The joints formed with Epidian 53/Z-1/10:1 were the weakest, as the values recorded for this composition were 0.75 mm, 1097 N, and 8.66 MPa for elongation-at-break, breaking load, and tensile shear strength, respectively.

17.4 SUMMARY

Advantageous properties obtained with relatively simple modification of epoxy resins contribute to their growing popularity in the industry. Due to their polar character, epoxy resins display high adhesion to various substrates. Low chemical contraction during curing enables the production of casts precisely copying the shape of the mold, which, moreover, does not require application of additional pressure. Furthermore, low chemical contraction generates low stresses, which can be eliminated by the addition of plasticizers.

The characteristic properties of epoxy resins have found them a place among the best adhesive materials. Epoxy adhesives produced from epoxy resins demonstrate far better properties than other major adhesive materials, predominantly owing to their excellent adhesion to a variety of structural materials and superior chemical resistance, to name but a few.

A large variety of compositions that can be cured in a range of temperatures and an array of modification methods, allowing the attainment of desired specific parameters, as well as other marked benefits of epoxies over other adhesives, indicate that the increasing use of such materials is inevitable.

The curing process of epoxy adhesives does not require additional pressure to ensure close and consistent contact between the substrates, which, together with the point that no by-products of curing are released in the process, facilitates their practical use. Moreover, epoxy adhesives are characterized by low chemical contraction, whereas the concern of excessive thermal contraction, which depends on the coefficient of thermal expansion and temperature, can be solved by the addition of suitable fillers.

REFERENCES

1. O. Martin and J. Maslega. The use of 9-anthroic acid and new amide derivatives to monitor curing of epoxy resins. *J Mater. Process. Technol* 143–144, 851–855 (2003).
2. K. D. V. P. Yarlagadda and S.-H. Hsu. Experimental studies on comparison of microwave curing and thermal curing of epoxy resin used for alternative mould materials. *J. Mater. Process. Technol* 155–156, 1532–1538 (2004).
3. B. Ellis (Ed.). *Chemistry and Technology of Epoxy Resins*, 1st edn, pp. 1–35, 37–68, 72–99, 121–126, 206–239. Springer Science+Business Media, Dordrecht (1993).
4. E. M. Pertie. *Epoxy Adhesive Formulation*, pp. 3–19, 26–41, 43–53, 71–82, 85–98, 207–225, 229–236, McGraw-Hill, New York (2006).
5. I. Hamerton. *Recent Developments in Epoxy Resins*, Rapra Review Reports, 2nd *edn*, pp. 7–8, 21–34, 49–55. Rapra Technology, Shawbury, UK (1997).

6. S. Ebnesajjad. *Adhesives Technology Handbook*, 2nd edn, pp. 80–84, William Andrew Publishing, Norwich, NY (2008).

7. K. L. Mittal. The role of the interface in adhesion phenomena. *Polym. Eng. Sci.* 17, 467–473 (1977).

8. S. G. Prolongo, G. del Rosario and A. Ureña. Comparative study on the adhesive properties of different epoxy resins. *Int. J. Adhesion Adhesives* 26, 125–132 (2006).

9. L. Merad, M. Cochez, S. Margueron, F. Jauchem, M. Ferriol, B. Benyoucef and P. Bourson. In-situ monitoring of the curing of epoxy resins by Raman spectroscopy. *Polymer Testing* 28, 42–45 (2009).

10. M. Sadeghinia, K. M. B. Jansen and L. J. Ernst. Characterization and modeling the thermo-mechanical cure-dependent properties of epoxy molding compound. *Int. J. Adhesion Adhesives* 32, 82–88 (2012).

11. C. A. May. *Epoxy Resins, Chemistry and Technology*, 2nd edn, pp. 9–20, 465–485, 653–670. Marcel Dekker, New York (1988).

12. H. L. Lee and H. Neville. *Handbook of Epoxy Resins*, McGraw-Hill, New York (1988).

13. K. Tsuchida and J. P. Bell. A new epoxy/episulfide resin system for coating applications: Curing mechanism and properties. *Int. J. Adhesion Adhesives* 20, 449–456 (2000).

14. F. Lapique and K. Redford. Curing effects on viscosity and mechanical properties of a commercial epoxy resin adhesive. *Int. J. Adhesion Adhesives* 22, 337–346 (2002).

15. H. Brockmann, M. Haufe and J. O. Schulenburg. Mechanism of the curing reaction of model epoxy compounds with monuron. *Int. J. Adhesion Adhesives* 20, 333–340 (2000).

16. I.-N. Yoon, Y. Lee, D. Kang, J. Min, J. Won, M. Kim, Y. S. Kang, S.-H. Kim and J.-J. Kim. Modification of hydrogenated Bisphenol A epoxy adhesives using nanomaterials. *Int. J. Adhesion Adhesives* 31, 119–125 (2011).

17. K. Mimura and H. Ito. Characteristics of epoxy resin cured with in situ polymerized curing agent. *Polymer* 43, 7556–7559 (2002).

18. I. Stewart, A. Chambers and T. Gordon. The cohesive mechanical properties of toughened epoxy adhesive as a function of cure level. *Int. J. Adhesion Adhesives*, 27, 277–287 (2007).

19. L. S. Penn and H. Wang. Epoxy resins. In: *Handbook of Composites*, Chapter 3, S. T. Peters (Ed.), pp. 48–74. Chapman and Hall, London (1998).

20. W. Brockmann, P. L. Geiß, J. Klingen and B. Schröder. *Adhesive Bonding. Materials, Applications and Technology,* pp. 58–65, Wiley-VCH, Weinheim, Germany (2009).

21. C. Ramirez, M. R. A. Torres, L. Barral, J. Lopez and B. Montero. Epoxy/POSS organic–inorganic hybrids: ATR-FTIR and DSC studies. *Eur. Polym. J.* 44, 3035–3045 (2008).

22. J.-P. Pascault and R. J. J. Williams (Eds.). *Epoxy Polymers. New Materials and Innovations*, pp. 1–11. Wiley VCH, , Weinheim Germany (2010)

23. https://en.wikipedia.org/wiki/Epoxy (2016).

24. http://www.zch.sarzyna.pl (2015).

25. G. Habenicht. *Applied Adhesive Bonding. A Practical Guide for Flawless Results*, pp. 23–26. Wiley-VCH, Weinheim, Germany (2009).

26. R. D. Adams (Ed.). *Adhesive Bonding. Science, Technology and Applications*, pp. 221–222. Woodhead Publishing, Cambridge, UK (2010).

27. R. D. Adams, J. Comyn and W. C. Wake. *Structural Adhesive Joints in Engineering*, 2nd edn, pp. 177–180, 190–195. Springer, UK (1997).

28. O. Moussa, A. P. Vassilopoulos and T. Keller. Effects of low-temperature curing on physical behaviour of cold-curing epoxy adhesives in bridge construction. *Int. J. Adhesion Adhesives* 32, 15–22 (2012).

29. H. Preu and M. Mengel. Experimental and theoretical study of a fast curing adhesive. *Int. J. Adhesion Adhesives* 27, 330–337 (2007).

30. S. Gladkikh, V. Kolobova and L. Kuznetsova. Fast-curing adhesive compositions based on modified epoxy resins. *Polymer Sci. Series C*, 49, 193–194 (2007).

31. S. Du, Z.-S. Guo, B. Zhang and Z. Wu. Cure kinetics of epoxy resin used for advanced composites. *Polymer Int.* 53, 1343–1347 (2004).

32. S. Alessi, A. Parlato, C. Dispenza, M. De Maria and G. Spadaro. The influence of the processing temperature on gamma curing of epoxy resin for the production of advanced composites. *Radiation Phys. Chem.* 76, 1347–1350 (2007).

33. A. Rudawska and M. Czarnota. Selected aspects of epoxy adhesive compositions curing process. *J. Adhesion Sci. Technol.*, 27, 1933–1950 (2013).

34. C. Czaderski, E. Martinelli, J. Michels and M. Motavalli. Effect of curing conditions on strength development in an epoxy resin for structural strengthening. *Composites: Part B* 43, 398–410 (2012).

35. P. Musto, E. Martusceli, G. Ragosta and L. Mascia. Cure kinetics and ultimate properties of a tetrafunctional epoxy resin toughened by a perfluoro-ether oligomer. *Polymer* 42, 5189–5198 (2001).

36. K. P. Unnikrishnan and E. T. Thachil. Toughening of epoxy resins. *Designed Monomers Polymers* 9, 129–152 (2006).
37. T. H. Chiang and T.-E. Hsieh. A study of monomer's effect on adhesion strength of UV-curable resins. *Int. J. Adhesion Adhesives* 26, 520–531 (2006).
38. H. Okuhira, T. Kii, M. Ochi and H. Takeyama. Characterization of epoxy resin hardening with ketimine latent hardeners. *J. Adhesion Sci. Technol.* 18, 205–211 (2004).
39. J. Lee and L. T. Drzal. Surface characterization and adhesion of carbon fibers to epoxy and polycarbonate. *Int. J. Adhesion Adhesives* 25, 389–394 (2005).
40. D. A. Nishitsuji, G. Marinucci, M. C. Evora and L. G. de Andrade e Silva Study of electron beam curing process using epoxy resin system. *Nucl. Instrum. Methods Phys. Res. B* 265, 135–138 (2007).
41. Y. C. Nho, P. H. Kang and J. S. Park. The characteristic of epoxy resin cured by γ-ray and E-beam. *Radiation Phys. Chem.* 71, 241–244 (2004).
42. M. Bajpai, V. Shukla and A. Kumar. Film performance and UV curing of epoxy acrylate resins. *Prog Org. Coatings* 44, 271–278 (2002).
43. J. M. Morancho, A. Cadenato, X. Ramis, X. Fernández-Francos and J. M. Salla. Thermal curing and photocuring of an epoxy resin modified with a hyperbranched polymer. *Thermochimica Acta* 510, 1–8 (2010).
44. V. J. Lopata, C. B. Saunders, A. Singh, C. J. Janke, G. E. Wrenn and S. Havens. Electron-beam-curable epoxy resins for the manufacture of high-performance composites. *Radiation Phys. Chem.* 56, 405–415 (1999).
45. J. Wei, M. C. Hawley, J. D. Delong and M. Demeuse. Comparison of microwave and thermal cure of epoxy resins. *Polym. Eng. Sci.* 33, 1132–1140 (1993).
46. Y. Guan, X. Chen, F. Li and H. Gao. Study on the curing process and shearing tests of die attachment by Ag-epoxy electrically conductive adhesive. *Int. J. Adhesion Adhesives* 30, 80–88 (2010).
47. R. Gomatam and K. L. Mittal (Eds.). *Electrically Conductive Adhesives*, CRC Press, Boca Raton, FL (2008).
48. P. Czub and I. Franek. Epoxy resin modified with palm oil derivatives: Preparation and properties. *Polimery* 58, 135–139 (2013).
49. www.loctite-kleje.pl (2015).
50. http://www.klakiernik.pl (2017).
51. A. Rudawska. Selected aspects of the effect of mechanical treatment on surface roughness and adhesive joint strength of steel sheets. *Int. J. Adhesion Adhesives* 50, 235–243 (2014).
52. http://www.klimatest.eu (2017)
53. DIN EN 1465. Adhesives: Determination of tensile lap-shear strength of bonded assemblies.
54. T. Chihani, P. Bergmark, P. Flodin and T. Hjertberg. Surface properties of an anhydride-epoxy resin cured against different mould surfaces. *J. Adhesion Sci. Technol.* 7, 569–582 (1993).
55. http://www.zebiec.pl/pl/oferta/produkcja-przerobcza/bentonity (2016).

18 Bio-Sourced Epoxy Monomers and Polymers

Sylvain Caillol*, Bernard Boutevin, and Jean-Pierre Pascault

CONTENTS

18.1 INTRODUCTION

Thermoset materials represent less than 20% of total plastic production. A large variety of families of thermosetting polymers are used in industry. Typical examples are phenolic- and urea–formaldehyde resins, unsaturated polyesters, and epoxy resins. Globally, epoxy resins account for approximately 70% of the market for thermosetting polymers. Epoxy resins are usually low molar mass prepolymers that normally contain at least two epoxide groups. The epoxide group is also referred to as a *glycidyl* or *oxirane* group. Epoxy monomers may be reacted either with themselves through anionic or cationic homopolymerization, or with a wide range of coreactants including polyfunctional amines, acids, anhydrides, phenols, alcohols, and thiols. These coreactants are often referred to as *hardeners*. Epoxy formulations also contain some additives and fillers. The aim of blending is to achieve the desired processing and/or final properties, or just to reduce cost. Since the beginning of their commercial production in the late 1940s, epoxy polymers have evolved dramatically to now find acceptance in diverse industrial applications requiring superior strength, excellent adhesion, good chemical resistance, and excellent performance at elevated temperatures. Due to their properties, they are used in coatings, electrical electronic laminates, adhesives, flooring and paving applications, and high-performance composites [1].

The global epoxy thermosetting polymers production was 2 million tn. in 2010 and is projected to reach 3 million tn. by the year 2017. Their global market was roughly US$18 billion in 2012 and was forecast to reach US$21.5 billion by 2016 [2]. More than 60% of the global production is used in the coatings industry [3–6]. Furthermore, epoxies are probably the most versatile family of engineering/structural adhesives, because they are compatible with many substrates and can easily be modified to achieve widely varying properties. Epoxy adhesives are nowadays mainly used in automobile assembly, concrete, sandwich panels, aircraft and aerospace, DIY (Do-It-Yourself), and so forth. New markets for structural composites in automotive parts or wind turbine blades are new challenges for epoxies. Due to the need to reduce the weight of transportation vehicles (airplanes, trains, automobiles, etc.) to reduce fuel consumption, structural composites and adhesives will be the key market drivers for epoxy resins over the next few years.

* Indicates corresponding author

Uncertainty in terms of price and availability of petroleum and global political tendencies toward the principle of sustainable development have urged the chemical industry to develop sustainable chemistry and, particularly, the use of renewable resources to develop biobased chemicals and products. A biobased product is a product obtained from renewable resources (vegetal, animal, or fungal), but this does not mean that it is a biodegradable material. Hence, partially or fully biobased epoxy cross-linked polymers are an important target nowadays, and also a real challenge from both academic and industrial points of view. The use of renewable resources for epoxy monomers synthesis results in reduction of environmental impacts, such as consumption of nonrenewable resources. However, chemical hazards should also be taken into account to reduce human toxicity and ecotoxicity impacts.

The most popular epoxy monomers are those derived from the reaction of bis(4-hydroxy phenylene)-2,2 propane (called *bisphenol A* or *BPA*) and 1-chloroprene 2-oxide (called *epichlorohydrin* or *ECH*) in the presence of sodium hydroxide. The structure of the major product BPA diglycidyl ether (DGEBA or BADGE) and its condensed forms is dependent on the stoichiometry of the reactants (Schemes 1 and 2a). The aromatic ring of BPA is particularly suitable, since it confers good thermal resistance to epoxy networks, and more than 90% of epoxy cross-linked polymers use BPA as reactant. BPA is not only classified as reprotoxic R2, but is also a compound that was initially synthesized as a chemical estrogen [7]. Indeed, this endocrine disruptor can mimic the body's own hormones and may lead to several negative health effects [8–10], including alterations in both immune and reproductive systems, along with a modification in brain chemistry [11]. The negative impact of BPA on human health and environment necessitates focusing research on a substitution for BPA, especially since some countries have recently banned the use of BPA in food contact materials. Recently, BPA was banned from epoxy cross-linked polymers used for feeding bottles or printing inks, and also for can coatings and pipeline coatings for drinking water. So far, several works have been carried out from both the academic and industrial sectors to create epoxy cross-linked polymers by using safer reagents [12]. Therefore, there is an increasing interest within the chemical industry in nonharmful aromatic compounds allowing synthesis of epoxy thermosets without BPA.

Renewable molecules and derivatives of biomass can be sources for alternatives to epoxies from BPA. The present chapter summarizes the investigations to obtain, either partially or fully, biobased epoxide materials. In the recent literature, there are a lot of papers and reviews on the use of biomass for the creation of polymers and materials. Most of them provide an overview on biobased plastics [13] or focus on thermoplastic polymers [14]. Some interesting reviews have already been published concerning the synthesis of biobased epoxy materials, such as the contribution of Gandini [15]. However, some authors [16–20] have only reviewed epoxies from vegetable oils or have not taken into account some interesting natural aromatic resources such as tannins, and generally, no review reports the synthesis of biobased curing agents or material properties to discuss the applications envisaged. Therefore, this chapter complements the previous reviews. Biorefinery processes and retreatment of biomass will be out of scope, and in this chapter, we will be limiting ourselves to the use of natural polymers and molecular biomass for the preparation of epoxy formulations and networks.

In Section 18.2, some background on epoxy chemistry is given, including the structures and reactivity of epoxy monomers and the description of reactions leading to epoxy networks. The third section will focus on the synthesis of biobased building blocks, including epoxy precursors from all sorts of biomass: carbohydrates, tannins, or vegetable oil sources. In the fourth section, some examples of biobased networks are discussed. Finally, we give our opinion on what could be the future of these new materials in different applications.

18.2 BACKGROUND

A full description of both the chemistry and the properties of conventional epoxy cross-linked polymers can already be found in several books and reviews [1,2,21]. The mechanism of the condensation reaction between phenols and ECH can be described as in Figure 18.1. In all procedures described, ECH is used in excess and reaction is carried out in an aqueous solution of sodium

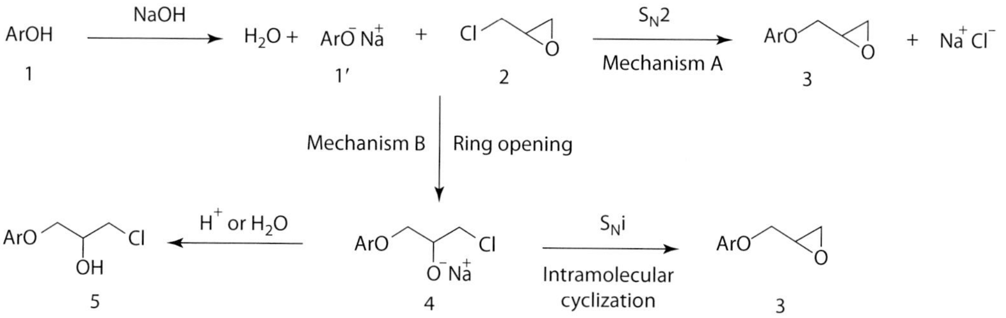

FIGURE 18.1 Mechanism of epoxidation of phenolic compounds.

hydroxide (NaOH). The literature [22,23] clearly shows that the reaction between the phenate ion (ArO⁻) **1′** and ECH **2** corresponds to two competitive mechanisms: one-step nucleophilic substitution (mechanism S_N2) with cleavage of the C–Cl bond; and two-step mechanism based on ring opening of ECH **2** with ArO⁻ **1′** followed by intramolecular cyclization (S_Ni) of corresponding alcoholate **4**, containing one atom of chlorine in the β-position.

Depending on the substituent position or nature of phenol **1**, it takes 6–20 h on reflux or 24–26 h at room temperature to complete the reaction. It is important to note that the glycidylation of phenol first requires the synthesis of the phenate group with NaOH, which can entail some inconvenience, such as only moderate isolated yields, moderate purity of products, and poor reaction selectivity. Indeed, 1-chloro-3-aryloxypropan-2-ol **5** is a by-product formed in a nonnegligible amount. Another by-product is the abnormal addition of the phenolic hydroxyl group to ECH.

The most popular epoxy monomers (Figure 18.2) are those derived from the reaction of bis(4-hydroxy phenylene)-2,2 propane (BPA) and ECH in the presence of NaOH. The structure of the

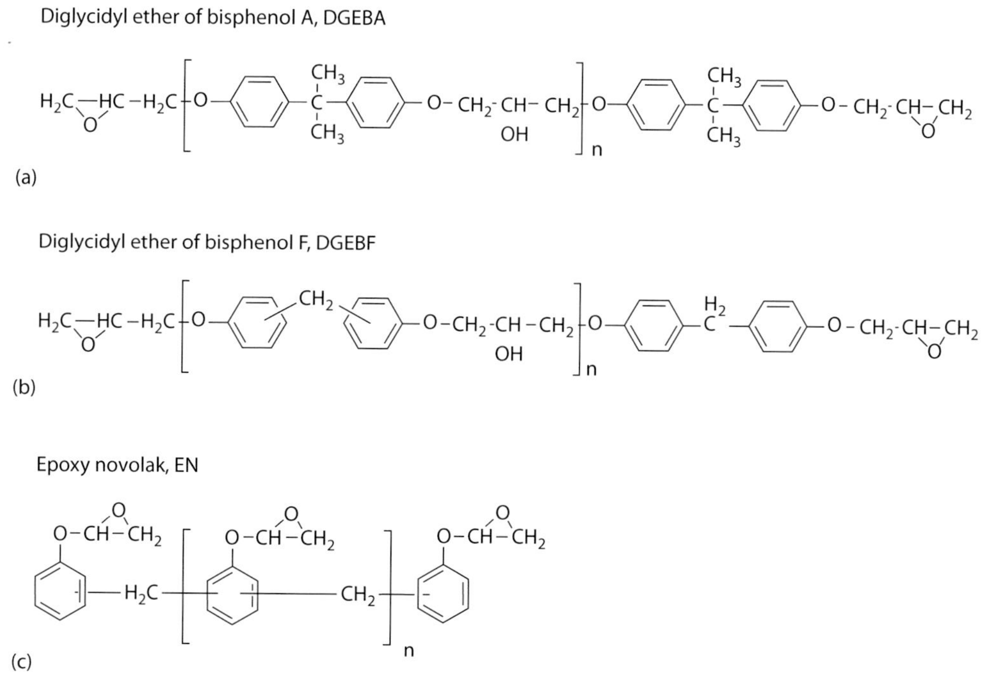

FIGURE 18.2 Main commercial epoxy monomers from phenolics.

major product, BPA diglycidyl ether (DGEBA) and its condensed forms (Scheme 2), is dependent on the stoichiometry of the reactants. Typically, monomers are marketed with n in the range 0.03–10. At room temperature, these monomers are crystalline solids for values of n close to zero, liquids up to $n \sim 0.5$ and vitreous ($T_g \sim 40°C–80°C$) for higher values of n.

Another major group of epoxy monomers derived from ECH is that comprising monomers synthesized with an aromatic amine, such as triglycidyl methylenedianiline (TGMDA) with methylenedianiline or triglycidyl para-aminophenol (TGpAP) with para-aminophenol. For these monomers, the side reaction consists of intramolecular cyclization between two adjacent oxiranes.

The reaction of ECH with an alcohol is more difficult, with many side reactions, since this reaction generates new alcohol groups with similar pKa that are able to react with the epoxy group of ECH, thus leading to its homopolymerization, and the final product is often a mixture of chlorohydrin and epoxy monomers [24]. This particular point, that is, reaction of ECH with biobased aliphatic polyols, will be discussed in Section 18.3.3. Liquid monomers based on butanediol, neopentylglycol, trimethylolpropane (TMP), and OH-terminated polypropylene oxide oligomers are the most common. Due to their low viscosity compared with DGEBA, they are often used as reactive diluents.

Another approach to the formation of oxirane groups is the peroxidation of a carbon–carbon double bond. The reaction conditions differ according to the type of double bonds. When the double bonds belong to the aliphatic chains, their oxidation is rather simple and only requires the use of hydrogen peroxide (H_2O_2). But, when the double bonds are of glycidyl type, the oxygen in the β-position may alter the reactivity of the double bonds. Stronger oxidative reagents are then required, such as m-chloroperoxybenzoic acid. Purification of the resulting epoxy synthons becomes rather complex, which limits their industrial development [25–29].

The epoxy–oxirane group is characterized by its reactivity toward both nucleophilic and electrophilic species, and it is thus receptive to a wide range of reagents. Epoxy monomers polymerize through step-growth and chain-growth processes. Linear or cross-linked epoxy polymers are obtained by reaction of the epoxy monomers with comonomers (hardeners) and/or initiators. The most typical example of a step-growth polymerization of epoxy monomers is the reaction with amines, which are the most common curing agents/hardeners used to build up epoxy networks. One epoxy ring reacts with one amino proton (Figure 18.3).

The reactivity of the amine increases with its nucleophilic character [30]: aliphatic > cycloaliphatic > aromatic. While, for aliphatic and cycloaliphatic amines, primary and secondary amine hydrogens exhibit similar reactivities, for aromatic amines, the reactivity of the secondary amine is 2–5 times less than the reactivity of the primary amine hydrogen. This means that once the primary amine reacts, the generated secondary amine exhibits a lower reactivity, a fact that is called a *substitution effect*. Hydroxyl groups catalyze the reaction through the formation of a trimolecular complex that favors the nucleophilic attack of the amino group. Apart from species containing OH groups that may be added as catalysts, the epoxy–amine reaction generates OH groups. Therefore, the reaction is self-catalyzed by reaction products. In most cases, when stoichiometric amounts of epoxy and amine comonomers are used, no side reaction takes place. When there is an excess of epoxy groups, the reaction shown in Figure 18.4 occurs after most of the amine hydrogens have reacted.

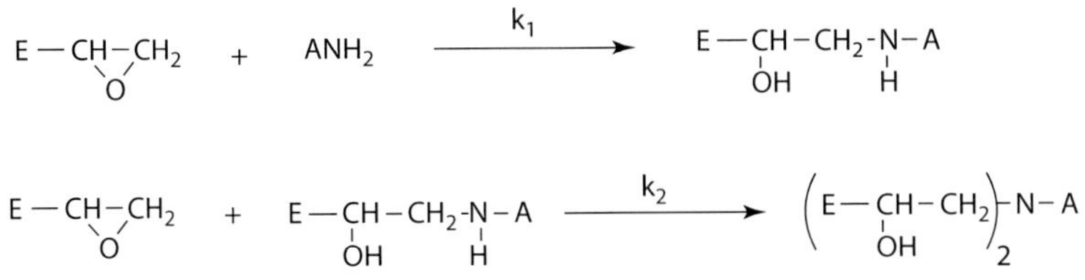

FIGURE 18.3 Reactions between epoxy groups and primary and secondary amines.

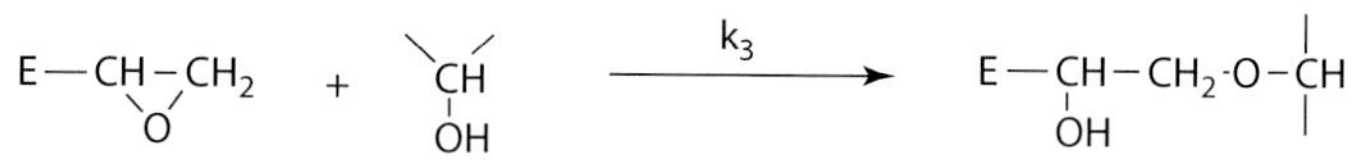

FIGURE 18.4 Reaction between epoxy and hydroxyl groups.

The epoxyhydroxy reaction (also called *etherification*) modifies the initial stoichiometric ratio based on epoxy to amino hydrogen groups. Other factors may also influence the path of the curing reaction, such as the presence of a catalyst or an initiator.

The synthesis of amines should be mentioned; this is not trivial, since they generally show a strong reactivity, which may lead to by-products. It is widely known that amines are obtained from nitrile hydrogenation, and the nitrile compounds can be obtained from two pathways: either from carboxylic acids, or from alcohols (addition reaction on acetonitrile). Nitrile compounds can thus be hydrogenated, according to different methods, such as hydrogenation based on $LiAlH_4$. Poly(oxyalkylene) amines showing terminal $-CH_2-CH(CH_3)-NH_2$ or $-CH_2-CH(C_2H_5)-NH_2$ are obtained by reaction of ammonia with corresponding secondary alcohol compounds. Nevertheless, alkyl groups situated in the α-position of the amine strongly decrease amine reactivity.

The polyaddition reaction between the difunctional DGEBA and a tetrafunctional diamine (containing four active amine hydrogens in the structure) leads to a polymer network (gelation occurs at a conversion where percolation of a giant molecule takes place throughout the system). At this critical conversion, the system consists of a large number of finite molecules (the *sol fraction*) and one cross-linked molecule (the *gel fraction*). After gelation, the mass fraction of the insoluble cross-linked structure increases continuously, as does the elastic modulus of the sample. As mentioned in Section 18.1, the main characteristic of thermosetting polymers is that polymerization and final material shaping are performed in the same process. Reactions with phenols (Ar–OH), isocyanates (R–NCO), or acids (RCOOH) are not as clear as expected, because catalysts that can also initiate a chain polymerization are practically always used. With acids, other reactions such as esterification and transesterification are also possible.

18.3 BIOBASED BUILDING BLOCKS

Biobased compounds have recently gained strong interest and have allowed new developments for epoxy cross-linked polymers, especially from both academic and industrial points of view. As already mentioned, ECH is conventionally used to obtain epoxy monomers by reaction with hydroxyl compounds. Nevertheless, ECH is classified as *CMR*, that is, a Category 1B carcinogen. Usually, ECH was obtained from chlorhydrination of allyl chloride, which was synthesized by reaction of chlorine with propylene [31]. Interestingly, ECH is now industrially produced from biobased glycerol (Figure 18.5) [32], and such process affords about 100 kt per year [33]. It has been found that conversion of glycerol to ECH is economically attractive [34]. The biobased weight content of DGEBA is about 25% if the synthesis is carried out from this biobased ECH.

As mentioned in Section 18.2, epoxy monomers can also be commonly obtained from double-bond oxidation. Nevertheless, two different strategies are generally employed, according to the type of double bond in the natural products: when the natural products already contain double bonds in the aliphatic chains—for example, vegetable oils—the epoxidation reaction is carried out in the presence of H_2O_2. Hydrogen peroxide is probably the best oxidative agent with O_2 from both economic

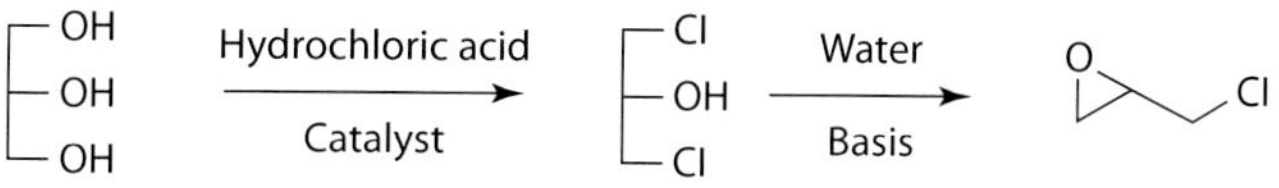

FIGURE 18.5 Synthesis of biobased ECH.

and environmental considerations. Lane and Burgess [35] reviewed the use of hydrogen peroxide as an oxidative agent and especially focused on the use of a metal catalyst. They also showed that oxidation with H_2O_2 was highly efficient only with electron-rich alkenes. In some cases, the double bonds can also be of the allyl type. Oxygen atoms are often in the β-position of the double bond, that is, of the glycidyl type, and may deactivate the double bond; thus, oxidative reagents stronger than H_2O_2 must be used to provide epoxy groups. Some examples of natural product oxidation from the two distinct strategies to provide biobased epoxy monomers are given in the next sections. We show here the recent trends concerning both the epoxy monomers and the cross-linking agents.

18.3.1 FROM POLYPHENOLS, TANNINS, AND CARDANOL

Tannins are nonhazardous compounds with antiallergenic, antiatherogenic, and antimicrobial properties [36–40]. These positive attributes toward human health can be associated with the antioxidative property of these compounds [41–43]. Tannins are natural poly(phenol) compounds [44] and can be found as hydrolyzed tannins (ellagitannins or gallotannins), condensed tannins, complex tannins, and phlorotannins [45]. Both hydrolyzed and condensed tannins are employed for the preparation of epoxy monomers. Gallotannins are composed of gallic acid groups grafted onto polyols (i.e., D-glucose), whereas ellagitannins result from the oxidative coupling of at least two gallic units of gallotannins. Condensed tannins are obtained from flavan-3-ols polymerization (Figure 18.6). Depending on the substituents R1 to R4, different flavan-3-ols exist and can be polymerized [46].

It should be noted that catechin and gallic acid are among the most studied tannin derivatives. The epoxy monomers from such compounds are given in Scheme 7.

Catechin was epoxidized either by reaction with ECH or by alkylation with an unsaturated halogenated compound, followed by its oxidation [47]. A full characterization of the obtained compounds shows the presence of a by-product with a benzodioxane group that then decreases the average epoxy functionality [48]. These by-products result from an internal cyclization reaction between phenolic alcohol in the ortho position and ECH after addition by the S_N2 mechanism (Figure 18.7).

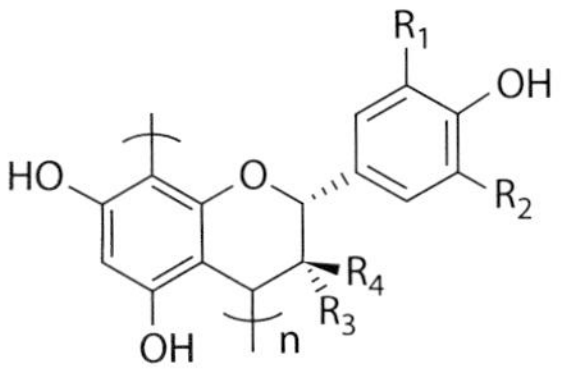

FIGURE 18.6 Structures of flavan-3-ol monomers: (+)-afzelechin ($R_1 = R_2 = R_3 = H$; $R_4 = OH$); (−)-epiafzelechin ($R_1 = R_2 = R_4 = H$; $R_3 = OH$); (+)-catechin ($R_1 = R_3 = H$; $R_2 = R_4 = OH$); (−)-epicatechin ($R_1 = R_4 = H$; $R_2 = R_3 = OH$); (+)-gallocatechin ($R_3 = H$; $R_1 = R_2 = R_4 = OH$); (−)-epigallocatechin ($R_4 = H$; $R_1 = R_2 = R_3 = OH$).

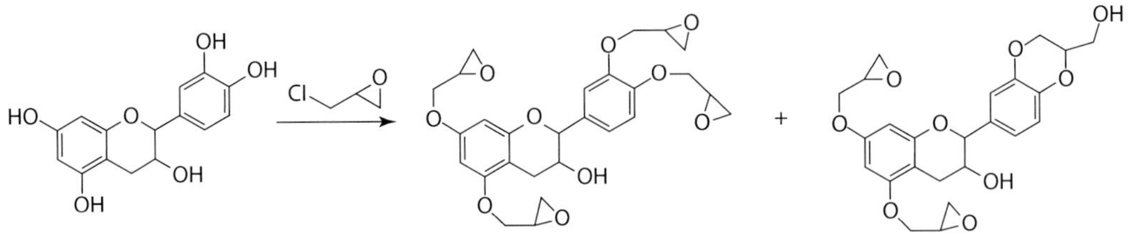

FIGURE 18.7 Epoxy monomer and by-product from catechin epoxidation.

More recently, two studies by Jahanshahi et al., one on the formation and testing of polyflavonoid tannins and epoxy resins, were also published, the second of which was of particular interest, referring to an epoxidized tannin acrylic that allowed rapid curing of the epoxy tannin without having to use a hardener [49,50]. The tannin-based epoxy acrylate resin was prepared by reaction between glycidyl ether tannin (GET) and acrylic acid in the presence of a catalyst and hydroquinone [50] and tested for block shear strength.

Gallic acid is found in most astringent vegetables, and especially in gallic nuts. Gallic acid may be combined with a tannin. Gallic acid epoxidation with epichlorohydrin was first reported by Tomita and Yonezawa [51]. The addition occurs on both the carboxylic acid group and also on at least one phenol group in the presence of an ammonium-type phase transfer catalyst in an anhydrous medium. Thus, Tomita and Yanezawa report epoxy equivalent weight (EEW) values ranging from 137 to 160, corresponding to epoxy functionality ranging from 1 to 4. The epoxidized compound was then cross-linked with a conventional cross-linking agent (i.e., polyamine or acid anhydride) [51]. However, theoretical calculations show that epoxy functionalities of 1, 2, 3, and 4 correspond, respectively, to EEWs of 226, 141, 113, and 99. This demonstrates that Tomita and Yonezawa obtained an average epoxy functionality of 2, despite there being four epihalohydrin units per phenol group.

In a recent publication, Aouf et al. [52] reported conditions allowing the synthesis of tetra epoxy gallic acid. Previous studies only reported functionalization with two to three epoxy groups. Moreover, they investigated the mechanism of O-glycidylation for several model phenolic compounds to establish a relationship between the chemical structures of phenolic compounds and their behavior toward ECH. Indeed, an understanding at molecular level of the reactivity of phenolic monomers for glycidylation represents a crucial step in the development of biobased epoxy monomers based on polymeric tannins as phenolic sources [52].

Nouailhas et al. [48] also recently proposed another synthesis pathway leading to epoxy prepolymers from gallic acid allylation by the reaction of gallic acid with allyl bromide, followed by double-bond oxidation using m-chloroperoxybenzoic acid. This method allows the attainment of epoxy prepolymers with an epoxy functionality of up to 3.

Following a former study based on model molecules of tannins, Benyahya et al. [53] proposed direct use of condensed tannins naturally rich in polyphenols for the synthesis of aromatic biobased epoxy oligomers. These natural and inexpensive green tea tannins are directly used after extraction from renewable resources and open a route of a new aromatic renewable resource usable for polymer synthesis.

Some other polyphenols were investigated to provide new biobased epoxy monomers, such as curcuma and resveratrol. The double bonds can be epoxidized by reaction with m-chlorobenzoic acid, resulting in glycidyl ether compounds, thus enabling cross-linking to provide epoxy cross-linked polymers [54].

Resorcinol is produced by mixing resins (e.g., galbanum, asafoetida) with potassium hydroxide, and demonstrates an aromatic structure [55]. Prior to its use for the synthesis of epoxy monomers, resorcinol is reacted with mesityl oxide [56]. Then, the polyphenol compound is able to react with ECH in the presence of $(CH_3)_4NBr$. Different patents show the synthesis of epoxy cross-linked polymers from epoxy monomers of resorcinol (or phloroglucinol), previously epoxidized by reaction with ECH [57–59].

Cashew nutshell liquid (CNSL) is a renewable natural resource obtained from the cashew (*Anacardium occidentale*) nut as a by-product during the process of removing the cashew kernel from the nut. The total production of CNSL approaches 1 million tn. annually [60], and CNSL is one of the few major and economic sources of naturally occurring phenols. CNSL can be regarded as a versatile and valuable raw material for polymer production and represents a good natural alternative to petrochemically derived phenols [61]. CNSL constitutes nearly 25% of the total mass of the nut and is composed of anacardic acid (3-n-pentadecylsalicylic acid) and smaller amounts of cardanol (3-n-pentadecylphenol), cardol (5-n-pentadecylresorcinol), and methylcardol

(2-methyl-5-n-pentadecylresorcinol); the long aliphatic side chain is saturated—monoolefinic (C8), diolefinic (C8, C11), and triolefinic (C8, C11, C14)—with an average value of two double bonds per molecule. The thermal treatment of cashew nuts and CNSL induces a partial decarboxylation of anacardic acid that is completed by subsequent purification by distillation. The result is an industrial-grade cardanol, in the form of a yellow oil containing mainly cardanol (about 90%), with a smaller percentage of cardol and methylcardol [62]. Cardanol itself is a mixture of four meta-alkyl phenols differing in the unsaturation degree of the aliphatic chain: 3% of 3-(pentadecyl)-phenol, 42% of 3-(8-pentadecenyl)-phenol, 17% of 3-(8,11-pentadecadienyl)-phenol and 38% of 3-(8,11,14-pentadecatrienyl)-phenol, with mainly cis conformation [63,64].

Due to both its aromatic and aliphatic structures, cardanol seems to be a promising candidate for the substitution of petroleum-based phenol derivatives. Furthermore, cardanol enhances coating properties such as both chemical and mechanical resistances, anticorrosion, and flexibility [65]. This will be discussed in more detail in Section 18.4. A patent claims the use of commercial epoxidized cardanol from Cardolite Company [66]. PPG Industries also suggest the use of cardanol to provide biobased epoxy cross-linked polymers. Cardanol was, in fact, mixed with BPA [67]. Tan [68] reported the epoxidation of phenolated lignin by cardanol. Cardanol was added to lignin between 0.3 and 0.9 mol/lignin. Moreover, a study [69] was performed on the enzymatic epoxidation of cardanol. More interestingly, epoxy monomers were synthesized from cardanol [70–73] by double-bond epoxidation in the presence of enzymes (lipase) and acetic acid in toluene. A continuous addition of H_2O_2 is required for 6 h to quantitatively epoxidize the reactive double bonds. The epoxidized compound is then polymerized with H_2O_2 in the presence of 2-propanol at ambient temperature. After 24 h, the crude product is concentrated under vacuum. The cross-linking reaction is allowed with phenylamine at 150°C [69].

The synthesis of monoepoxidized cardanol is actually carried out from direct reaction with ECH, which is used as a reactive diluting compound. The synthesis of diepoxy cardanol is carried out via a two-step reaction: cardanol phenolation followed by reaction with ECH of the resulting diphenol. It should be noted that the phenolation mechanism of unsaturated aliphatic chains is well known and occurs in the presence of strong acids such as HBF_4 [74]. The idealized structure of diepoxy cardanol is given in Figure 18.8

It is also possible to obtain a cardanol–novolak prepolymer [75] from the conventional reaction of cardanol with formaldehyde catalyzed with p-toluene sulfonic acid at 120°C for 7 h. Epoxidation is performed with ECH in a basic medium.

18.3.2 FROM WOODY BIOMASS AND LIGNIN

Only a few publications deal with the modification of lignin for use as an epoxy monomer [76,77–80]. Mostly, three processes have been described in the literature [20] enabling the synthesis of epoxidized lignin. The first process principally aims at increasing the content of phenolic groups onto lignin [81–85]. The phenols used are basically phenol and BPA, but acetone is also used to create

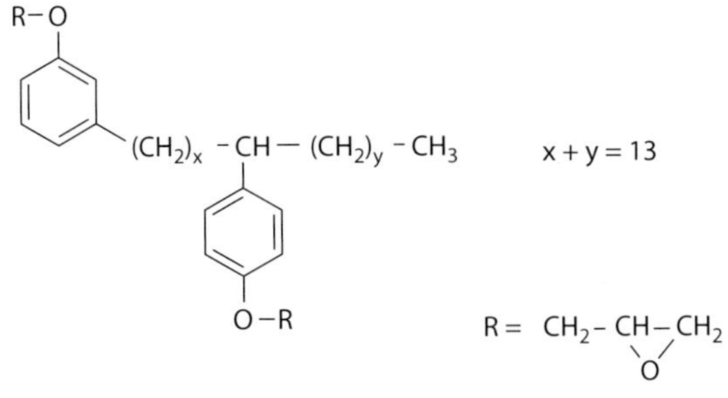

FIGURE 18.8 Structure of diepoxy cardanol.

interphenol bridges from lignin *in situ*. The phenolated lignin is then reacted with ECH, leading to epoxidized lignin, which can be cross-linked with a polyamine curing agent. It was particularly shown that the lignin–epoxy precursor provided better water-resistant adhesion strength after curing at 140°C with triethylenetetramine (TETA). Recently, Hitachi Company claimed the use of lignin to provide epoxy cross-linked polymers based on this process. Hitachi also claimed, through two patents, that their performances as thermal insulating material are similar to that of conventional epoxy cross-linked polymers (based on DGEBA) [86]. It was also mentioned that high molar mass lignins were removed to reach biobased lignins with a number average molar mass, M_n, ranging from 300 to 10,000 g/mol. Below 300 g/mol, the prepolymer will not possess enough reactive groups (hydroxyl, carbonyl, or acids), and above 10,000 g/mol, the material will lose its solubility in organic solvents. Such compound is primarily epoxidized with ECH in a basic medium. Then, the epoxy monomer, based on lignin, is cross-linked with a very low molar mass lignin compound. The glass transition temperature, T_g, of the resulting epoxy network, based on lignin, is above 180°C [87]. Araco Corporation, a Japanese company, also claimed the use of lignin as an epoxy monomer [88]. They actually use a phenolated lignin mixed with a phenol compound to increase the amount of aromatic groups and, thus, to increase both heat resistance and hydrophobicity of the final material.

In a second process, lignin reacts with an aliphatic diepoxide compound in DiMethyl Formamide (DMF) at 80°C to afford epoxide-containing lignin [89,90]. Hirose et al. also proposed the synthesis of epoxy monomers from reaction between kraft lignin and diglycidyl ether of poly(ethylene) glycol. The cross-linking occurred by reaction with poly(azealic anhydride), leading to an epoxy network of very low T_g, that is, below 0°C, irrespective of the epoxide:acid ratio used for the cross-linking reaction [91].

The third process is based on oxypropylation of lignin to afford a lignin decorated with hydroxyl groups. These hydroxyl groups are then converted into epoxide groups by reaction with ECH, and cross-linking occurs in the presence of diamines. The authors [92] mentioned that the cross-linking rate was rather low as compared with a conventional system, due to the poor mobility of epoxide groups. With aromatic diamines, the T_g ranged from 80°C to 200°C, depending on both the type of lignin and the degree of oxypropylation [92].

One route for biobased building blocks involves chemical degradation and transformation of natural polymers. Considering their aromatic, polymeric, and renewable properties and vast supply, lignin sources of lignocellulosic biomass undoubtedly represent a significant sustainable feedstock for chemicals [93]. However, the depolymerization of lignin into chemically useful fragments is not a mature technology. For example, different processes used in biorefineries lead only to primary synthons such as vanillin (VN), 2-pyrone-4,6-dicarboxylic acid, and p-coumarylic, coniferic, sinapylic, and muconic acid [34].

Epoxy polymers are widely employed thermosets, especially in applications with demanding thermomechanical conditions such as in the aeronautics or electronics industry. Indeed, epoxy monomers usually have an aromatic structure that brings stability to the network. Thus, VN and its derivatives are excellent candidates for biosourcing of epoxy polymers. A few teams have worked on this topic and have used various strategies. In a review article [94], Koike described an interesting diepoxy monomer based on a coupling strategy. The first step is performed by acetalization of two VN molecules with pentaerythritol. The second step is the glycidylation of the two phenolic hydroxyl groups to obtain a difunctional epoxy monomer. This compound is interesting, as the reactants involved in its synthesis are biobased and widely available, and the reactions are well known and easy to implement. The major drawback of this monomer, however, is the fragility of the acetal group. Indeed, acetals are usually employed as protective groups for carbonyl compounds, as the deprotection is conducted at room temperature in an aqueous acidic medium. Nevertheless, Koike reports an epoxide equivalent weight of 270 g/eq and a cross-linking reaction with diaminodiphenylmethane, as well as good impact strength, tensile strength, and elongation of the final material.

Aouf et al. also used a coupling strategy to prepare a diepoxy monomer from vanillic acid [95]. This diepoxy monomer is based on vanillic acid, which is easily obtained from VN and is also an available synthon from lignin depolymerization. Ultimately, the epoxidation step is a chemoenzymatic process, a very interesting reaction in terms of sustainability, due to the use of a catalytic system (enzyme) and mild conditions (40°C). It led to a mixture of diepoxidized product (66%) and monoepoxidized product. However, the use of DMF as a solvent (Category 1B reproductive toxicity) in the allylation step is less ideal in terms of safety, and an alternative solvent could be a possible improvement for this reaction. Also, the presence of ester linkages might limit the thermostability of the final material. The same team also used another strategy by directly functionalizing vanillic acid in a two-step procedure to prepare another diepoxy monomer. Once again, the diepoxydation was only partial (70%) and the same concerns about the use of allyl bromide and DMF can be raised, even if this enzymatically catalyzed epoxidation reaction is an interesting alternative to the use of ECH. These compounds were envisaged as epoxy monomers, but polymers were not prepared. Our group also prepared diepoxy monomers from VN derivatives [96]. Figure 18.9 shows the three diepoxy monomers prepared from the same direct functionalization strategy.

The three VN derivatives chosen were in different oxidation states, either oxidized or reduced. It is worth noting that methoxyhydroquinone was prepared through a green VN oxidation process leading to the loss of one carbon atom. Briefly, VN was subjected to a Dakin oxidation with sodium percarbonate as an environmentally friendly oxidant in a THF–water mixture at room temperature. Sodium percarbonate is a chemical commonly used in laundry detergents. It is a good alternative to hydrogen peroxide, the storage and transportation of which is difficult for safety reasons. In contrast, sodium percarbonate is an environmentally friendly solid compound composed of sodium carbonate and hydrogen peroxide. It is a powerful oxidant, while being cheap, nontoxic, stable in storage, and easily handled.

The syntheses of both vanillic acid and vanillyl alcohol are conventional reactions, well described in the literature. ECH was used to perform a one-step glycidylation of these compounds and obtain

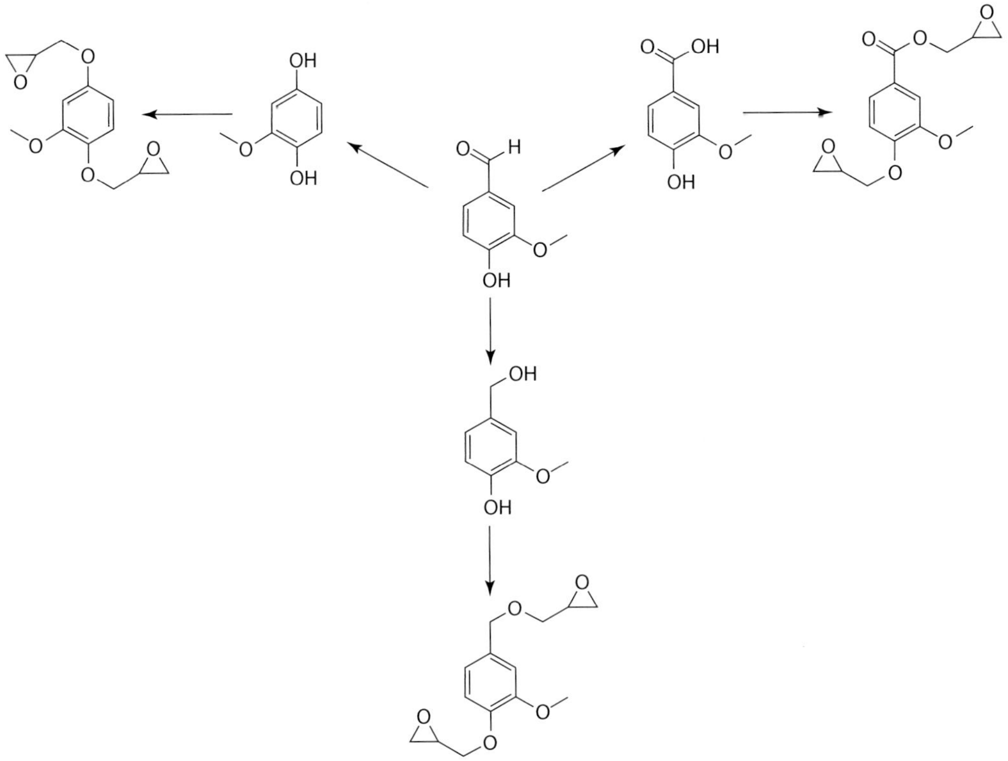

FIGURE 18.9 Diepoxy monomers prepared by direct functionalization of vanillin derivatives.

the three diepoxy monomers shown. It is worth noting that acid- and amine-functionalized VN derivatives were also prepared in the same work. These could potentially be employed as hardeners in epoxy thermosets.

These diepoxy monomers were used to prepare biobased epoxy thermosets [97]. The polymers, all hardened with the same diamine, displayed excellent thermomechanical properties, matching the ones of the current industrial reference, DGEBA. For instance, a T_g of 152°C was found for the thermoset prepared from the diglycidyl ether of vanillic acid. Moreover, properties could be tuned depending on the starting monomer. They are candidates for the substitution of BPA in epoxy polymers. Work is underway to prepare oligomers from these diepoxy monomers, using the strategy currently used for the preparation of industrial epoxy resins.

In another study, epoxy polymers were prepared using VN in an unexpected way. Indeed, Shimasaki et al. [98] first prepared a calixarene by condensing pyrogallol (PG; also biobased) and VN. The next step in their work was to prepare a composite with a sorbitol-based epoxy matrix hardened with the calixarene that they synthesized from VN. They obtained a polymer with a T_α of 148°C, which is high for biobased epoxy polymers. The next step in their work was to use wood flour as a filler in this epoxy matrix, leading to a fully biobased epoxy composite. One of the major highlights of this work is the preparation of the calixarene used as hardener. This compound was formed by the condensation reaction used to prepare phenolic resins, but in their case, the oligomer formed had a well-defined, polyphenolic cyclic structure. The performance of this compound in other fields might be worth investigating; for instance, in catalysis or supramolecular chemistry for its potential host–guest interactions.

18.3.3 FROM STARCH AND SUGAR

As glucose is produced from starch, but also from cellulose, or from woody biomass by chemical transformation, biomass carbohydrates are the most abundant renewable resource.

Numbers of poly(epoxide) prepolymers obtained from biobased polyols such as glycerol and sorbitol are commercially available [99,100] and are mainly aliphatic glycidyl ether types. Some studies report the synthesis of poly(epoxide) prepolymers from glycerin [101,102] (Figure 18.10).

As mentioned in Section 18.2, the reaction of ECH with natural aliphatic polyols, bearing both primary alcohol groups at the end of the chain and secondary alcohol groups within the chain, is not trivial. Indeed, this reaction generates new alcohol groups, being also reactive toward ECH [24]. Unlike the reaction between phenol groups and ECH, the hydroxyl groups resulting from the reaction between the polyols and ECH show reactivity fairly similar to that of the starting polyols. This thus implies multiple addition of ECH on the same alcohol, which is then not able to undergo intracyclization and could also lead to the presence of chlorine atoms. Furthermore, sorbitol is not soluble in water, that is, the reaction medium, and so an excess of ECH is required. This, in fact, reveals that the abovementioned structures provided by the chemical companies are idealized structures,

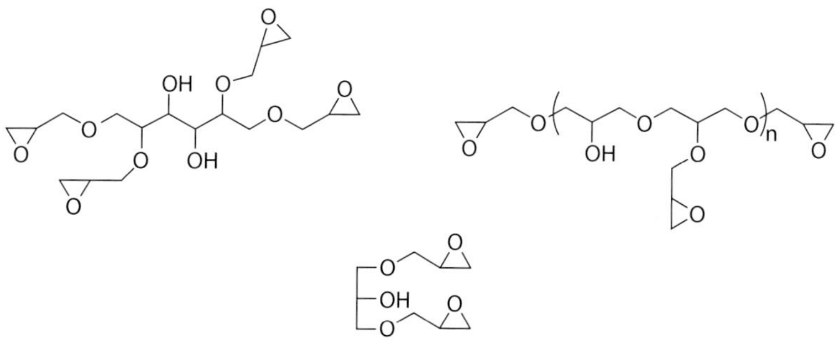

FIGURE 18.10 Epoxy monomers based on glycerin and sorbitol.

since other by-products containing chlorine atoms are also obtained. It is important to mention that most of these commercial products contain about 10%–20 wt% of Cl. This actually has two consequences: First, the epoxy formulations become harder, and secondly, the resulting epoxy networks may undergo HCl formation. Sorbitol and maltitol have also been converted to multifunctional epoxy monomers through oxidation of allyl [103] or crotonic [104] double bonds, and in this case, there was no remaining chlorine atom. Interest in the production of 1,4:3,6-dianhydrohexitols, especially isosorbide, has been generated by potential industrial applications such as the synthesis of polymers. Isosorbide is obtained from dehydration (~2 mol) of sorbitol. As already shown by Fenouillot et al. [105], isosorbide is a nontoxic and chiral molecule that also impacts stiffness to polymer chains such as polyesters, polycarbonates, or polyurethanes. Different processes [106–111] allow synthesis of diglycidyl ethers of isosorbide as well as oligoglycidyl ethers of isosorbide. Furthermore, two methods lead to either monomers or oligomers of epoxidized isosorbide by reaction with ECH. The first one, fairly similar to the synthesis of DGEBA, is based on the reaction between diol, ECH, and a strong base in an aqueous medium (Figure 18.11) [104,111,112]. This method often generates epoxy oligomers of isosorbide as well. In this case, the chlorine content remains slightly higher than that of polyphenols but much lower than that of sorbitol. This behavior can be ascribed to a better solubility of isosorbide in the reaction medium.

The second method is a two-step reaction in an anhydrous medium between ECH and isosorbide. The first step consists of generating the alcoholate of isosorbide in the presence of NaH, for instance, which leads to chlorohydrin ECH ring opening. The second step corresponds to the formation of the oxirane ring in the presence of a strong base (Figure 18.11) [113]. This method is selective, since it only leads to diglycidyl ether of isosorbide.

Epoxidized isosorbide was also obtained by applying a method based on the use of allyl bromide via a two-step reaction [110–112]. First, diallyl of isosorbide was prepared from the reaction between isosorbide and allyl bromide in the presence of potassium hydroxide. Then, modified isosorbide was oxidized using meta-chloroperoxybenzoic acid (MCPBA) in dichloromethane at room temperature to epoxidize the double bonds. It is, nevertheless, important to mention that this synthesis was not industrially developed due to the high content of MCPBA required.

Common acids including lactic acid, succinic acid, itaconic acid (IA), and levulinic acid can be extracted from starch and sugar. Lactic acid, used as lactide and glycidol, may provide either linear or hyperbranched epoxide prepolymer, depending on the experimental conditions [114].

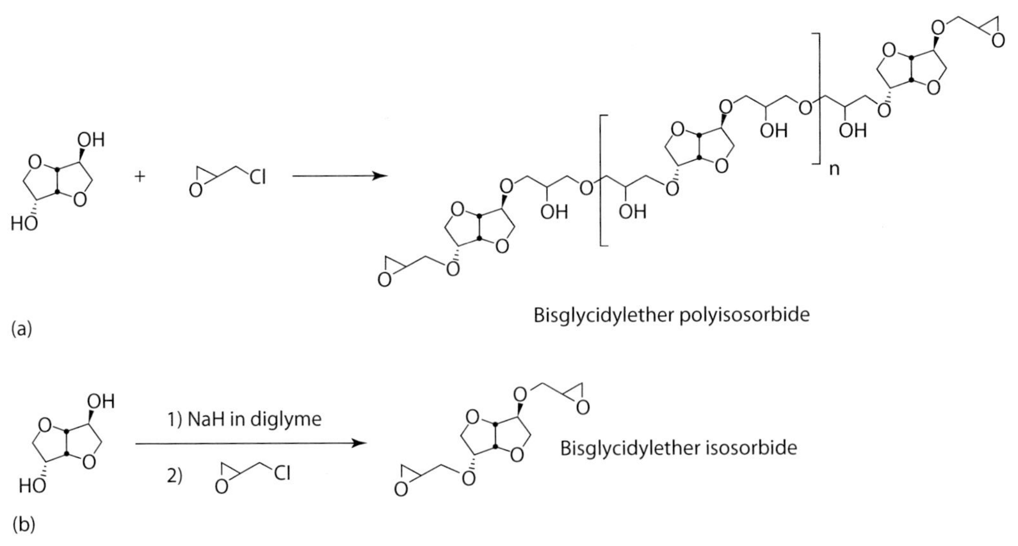

(a)

Bisglycidylether polyisosorbide

(b)

1) NaH in diglyme

2)

Bisglycidylether isosorbide

FIGURE 18.11   Synthesis of diglycidyl ether of isosorbide and structure of diisosorbide.

Epoxy monomers from sugars are usually obtained by double-bond oxidation, which requires the use of strong oxidative agents such as MCPBA in large excess. Sachinvala et al. [104] performed oxidation of sucrose bearing both allyl and crotonic groups. The use of aqueous peracetic acid (32% solution in the presence of sodium acetate) allows the performance of an efficient epoxidation reaction. The authors were able to evaluate the content of epoxide groups from both fast atom-bombardment mass spectroscopy and ^{13}C NMR, and showed that from octa-allyl sucrose, 3.7 epoxide groups were obtained, whereas 7.2 epoxide groups could be reached from crotonyl sucrose due to the electron-donor character of the crotonyl. They also specified that magnesium peroxy phthalate, m-chloroperoxybenzoic acid, oligomers of phosphotungstic acid with H_2O_2, and molybdenum hexacarbonyl with t-butyl hydroperoxide were unable to allow satisfactory oxidation yields.

IA is an unsaturated dicarboxylic acid produced via fermentation of carbohydrates. It was used to synthesize copolymers from radical copolymerization with vinyl monomers such as styrene followed by functionalization with ECH [115].

Epoxy monomers can also be synthesized by the esterification reaction between an acid and ECH. Recently, an IA-based epoxy monomer with curable double bonds was prepared. Different characterization techniques indicate the presence of oligomers. If ECH always reacted with the carboxy groups of IA, linear oligomers would be formed. When the reaction was carried out between ECH and the secondary hydroxyl groups, which were formed during the esterification reaction between IA and ECH, branched oligomers would be found [116].

Glycidyloxystyrene (Figure 18.12) is obtained from the reaction of p-hydroxystyrene (p-HS or 4-vinyl phenol) with ECH. Microbial production of p-HS from glucose is carried out in three steps: (1) Glucose is first converted to the aromatic amino acid L-tyrosine, (2) which is deaminated by an enzyme to yield p-hydroxycinnamic acid (or coumaric acid), (3) and subsequent decarboxylation of p-HCA, which gives p-HS [117–124].

18.3.4 FROM VEGETABLE OILS

Among the natural molecular biomass, vegetable oils are derived from plants and belong to the hydrocarbon-rich biomass. Vegetable oils are liquid materials at room temperature; they are currently the most employed renewable resources in the chemical industry [18,125]. The main components of this important class of abundant natural resource are triglycerides. A triglyceride usually has three ester bonds that can be hydrolyzed to form glycerol and three fatty acids. The fatty acids take up about 95% of the total weight of triglycerides. Vegetable oils have various compositions of fatty acids, but most of the fatty acids have natural functional groups such as double bonds, hydroxyl, and also, sometimes, epoxy groups. Natural carbon–carbon double bonds with low reactivity can be replaced with new functional groups, which are then readily polymerized.

Vegetable epoxidized oils are generally used as either plasticizers or stabilizers for poly(vinyl chloride) (PVC) synthesis or for paints and coatings [126]. They are also used to replace conventional poly(epoxide)s such as DGEBA to synthesize epoxy cross-linked polymers. Arkema is one of the main industrial producers of epoxidized vegetable oils, with the development of VIKOFLEX based on epoxidized soybean oil. Interestingly, Vernonia oil is a natural vegetable oil having an epoxide group in the aliphatic chain (Figure 18.13) [127].

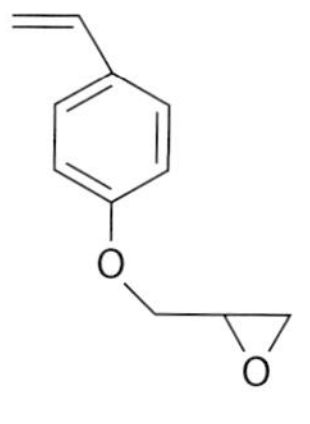

FIGURE 18.12 Structure of glycidyloxystyrene.

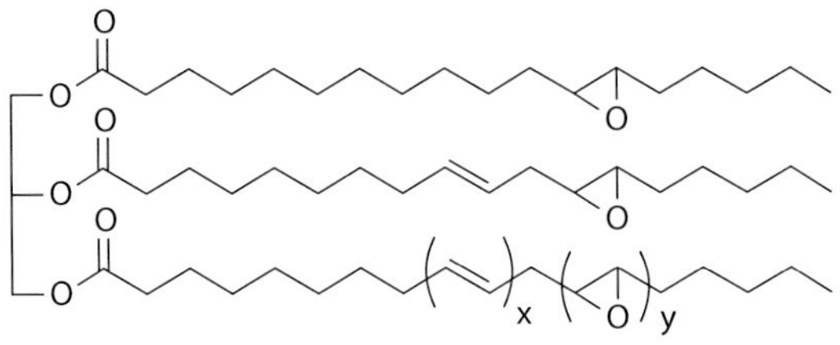

FIGURE 18.13 Chemical structure of vernonia oil.

18.3.5 From Terpenes, Terpenoids, and Resin Acids

Terpenes, terpenoids, and resin acids are also a class of hydrocarbon-rich natural molecular biomass. Many plants and trees (particularly pine and conifer trees) produce this class of molecular biomass.

Limonene is a six-membered ring terpene [128], present in agricultural wastes derived from citrus peels. Its world production is estimated to be between 110 and 165 million lb. per year. Limonene is a relatively stable terpene that has two double bonds—one vinylene group allocated in the ring and one vinylidene side group—and can be distilled without decomposition. It is used as a fragrance as well as a solvent for coatings. Limonene oxidation allows the acquisition of limonene mono- and diepoxide, which are commercially used as reactive diluents in epoxy applications (Figure 18.14).

Another study [129] shows the possibility of using limonene to build steric epoxy monomers by reaction with naphthol. Sellers [130] describes another synthesis path, based on the reaction of ECH on a polyhydric phenol, previously obtained from reaction between terpene and phenol. Finally, Fenn et al. [131] performed radical copolymerization of terpene with monomers carrying glycidyl groups.

Rosin is the solid nonvolatile fraction of resin. Depending on its specific origin (gum, wood, and tall oil), rosin consists primarily of abietic- and pimaric-type resin acids (or rosin acids) with characteristic hydrophenanthrene structures [132]. The intrinsic acidity, rigidity, and renowned hydrophobicity, coupled with other chemical properties, enable rosin acids to be converted to a large number of derivatives.

According to their structures, resinic acids behave like aromatic or cycloaliphatic compounds in terms of rigidity. Thus, they are solid compounds, and solvents have to be used for chemical modifications or processing. They can be good candidates for the preparation of rigid epoxy cross-linked polymers [133]. Some works [134] were especially conducted to provide first imidodicarboxylic resins of maleic anhydride. This compound was then epoxidized with ECH yielding the corresponding glycidyl ether. Wang et al. also suggested the synthesis of epoxide synthon from abietic acid. The reaction with epibromohydrin leads to monoepoxy monomers (Figure 18.15) [135].

Liu et al. performed the synthesis of a rosin-based epoxy monomer by modification of rosin acid, leading first to maleopimaric acid. The corresponding triglycidyl ester of maleopimaric acid was obtained by reaction with ECH and aqueous sodium hydroxide with tetrabutyl ammonium bromide as the phase transfer catalyst [136]. The Diels–Alder reaction of maleic anhydride on levopimaric

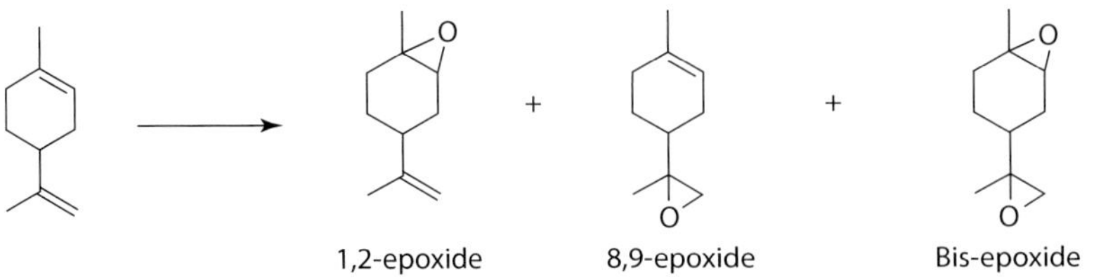

| 1,2-epoxide | 8,9-epoxide | Bis-epoxide |

FIGURE 18.14 Limonene and epoxy derivatives.

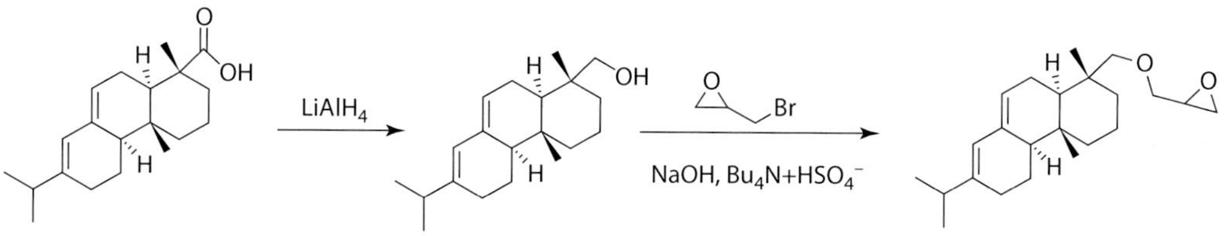

FIGURE 18.15 Synthesis of epoxy monomer from abietic acid.

acid leads to maleopimaric anhydride and its corresponding diacid, from which Atta et al. [137] synthesized a diepoxide by reaction with ECH. Resin dimer adducts with maleic anhydride and acrylic acid were also used to prepare epoxy networks [138–140]. The epoxy precursors were prepared by the reaction of the Diels–Alder adducts with diethanolamine, followed by treatment with ECH under alkaline conditions. We should, nevertheless, mention that the extraction of such acid compounds is very complicated and not environmentally friendly, so the availability of these renewable resources is rather poor. Thus, the development of epoxy cross-linked polymers from resinic acid remains low so far.

18.4 BIOBASED FORMULATIONS AND NETWORKS

In Section 18.3, we tried to list most of the biobased molecules found in the literature that are able to be used in epoxy formulations. If monomers, oligomers, and hardeners are the main components that define a formulation, initiators, catalysts, rubbers, fillers, short or continuous fibers, pigments, and so on can also be very important. This means that for all these reasons, synthesis of biobased monomers is just the first step before preparation of fully or partially biobased materials for well-defined applications.

Owing to their superior mechanical properties, ease of processing, and excellent chemical resistance, epoxy-based polymers have been widely used as structural, coating, and adhesive materials in many demanding application fields such as automotive, aeronautics, electronics, and electrical engineering. Each application has its unique performance requirements, and so formulations are a complex compromise between performance, processing, but also cost aspects. However, some special considerations are common to many applications, and knowledge of them is essential in determining solutions for biobased formulations.

Formulations have to be adapted to the process conditions. An overview of the processing techniques for epoxy thermosets is out of scope of this chapter. But, four different steps that depend on temperature, pressure, reaction rate, and the evolution of viscoelastic properties should be taken into account: (1) the pot-life of the reactive system at the storage temperature; (2) flow inside the processing machine and into the mold; (3) reactions in the mold or on a substrate (for coatings and adhesives); and (4) demolding and possible postreaction [141].

While some epoxy monomer–hardener combinations will cure at ambient temperature, many require heat, with temperatures up to 150°C being quite common, and up to 200°C for some particular systems. On the other hand, a reactive system can be a one-pack or two-pack system. The end user-desired solution is a one-component system with a long pot-life at room temperature and a high reactivity in the mold at a temperature as low as possible, which is clearly in contradiction to the Arrhenius equation, which can, however, be bypassed by the use of latent/blocked reactants or by reactions activated by radiation.

Another example of complexity of epoxy formulations is the incorporation of a reactive diluent that is used primarily to reduce viscosity; it also permits higher filler loading and gives better wetting and impregnation of fibers (for composites applications); but it also very often decreases the thermal properties of the final epoxy network. In some cases, additives (or modifiers) are also used to improve thermal shock resistance, increase elongation, and to obtain higher impact strength and flexibility; the problem, however, is that they also very often increase viscosity.

All these examples mean that it is not easy to optimize the different requirements for a given formulation and that development of fully biobased epoxy formulations is a long way off. Most of the authors, after having prepared biobased monomers, only give some information concerning reactivity and thermal properties. Of course, these parameters are important to be known, but from our point of view this is only the first step. So, instead of giving a list of T_g and thermogravimetric analysis results, our aim is to focus on some potential formulated systems that illustrate some important points for the future development of epoxy formulations based on renewable resources. Lignin- and rosin-based epoxies have largely been discussed in Section 18.3 and will not be discussed again in this section.

There have been a few reports on the combination of both biobased epoxy monomers and hardeners. For example, the thermal and mechanical properties of epoxidized soybean oil cured with maleinated soybean oil [142] or terpene-derived acid anhydride [143] were reported. Recently, networks from epoxidized linseed oil cross-linked with a vegetable oil polyamine cross-linker (prepared by thiol-ene chemistry) [144], with elastomeric properties and low T_g ($T_\alpha \sim -40°C$) were prepared. Shibata and coworkers [101,102] have focused their work on commercial glycerol polyglycidyl ether (GPE, epoxy functionality = 2) and polyglycerol polyglycidyl ether (PGPE, epoxy functionality = 4.1) reacted with two hardeners.

First, biobased nanocomposites have been prepared by curing GPE or PGPE with ε-poly(L-lysine) (PL) at 110°C in the presence of nonmodified montmorillonite (MMT) [100]. Interesting mechanical properties (modulus, tensile strength, etc.) have been reported, but always with a T_g of less than 60°C. Aerobic biodegradability of the PGPE–PL networks in an aqueous medium was about 4% after 90 days and decreased with MMT content. The second hardener studied was tannic acid (TA) in the presence of cellulose nanofibers [101] cured at 160°C. With this hardener, the T_g could reach 100°C, again with high mechanical properties (in both cases, the use of water as a mixing solvent for the monomers and hydrophilic nanofillers is interesting). But, as explained in Section 18.3, it can be said that the main drawback of these commercial epoxy monomers is that they contain a large number of chlorine atoms, which may also explain the water resistance of the final networks.

Pressure-sensitive adhesives (PSAs) constitute a different topic. The synthesis of renewable PSAs via cationic ultraviolet-initiated polymerization of epoxidized vegetable oils or other epoxidized vegetable oils has been disclosed in a patent application [145]. Vendamme and Eevers [146] have synthesized functional PSAs derived from carboxylic acid-terminated polyesters. The polyesters were prepared from bulk polycondensation of dimerized fatty acids with several diols such as dimer fatty diol, butanediol, or isosorbide. The resulting polymers were then cured with two commercially available epoxidized plant oils with different oxirane functionalities (epoxidized linseed oil and epoxidized soybean oil) to form viscoelastic bioelastomers with a tunable degree of stickiness. The authors highlighted how the viscoelastic and adhesion properties of the polymers could be tailored by incorporation of isosorbide. Interestingly, these fully renewable coatings combined the intrinsic flexibility of lipids with the polarity of sugars, and demonstrated interesting performances.

As there are only a few biobased hardeners, most of the literature was focused on partially biobased epoxies, and mainly on how to enhance the initial poor properties (low T_g) of networks based just on vegetable oils, as the epoxidized oils react more rapidly and with better properties with anhydrides than with amine hardeners. This explains why the use of aromatic or cycloaliphatic anhydride has been preferred [147–151]. The hardener concentration is certainly also one explanation of the better properties.

As an example, epoxidized linseed oil (ELO) has been cross-linked using different cyclic acid anhydride hardeners and tertiary amine or imidazole as initiators [148]. Both the T_g and flexural modulus of the resulting networks depend on the anhydride content. Results comparable to classical epoxies, $T_g \sim 110°C$ and flexural modulus of 30°C $\sim$ 2100 MPa, were obtained with tetrahydrophthalic anhydride (THPA).

One way to enhance the properties of networks based on epoxidized vegetable oils is to increase the functionality of the epoxy precursor. This has been done by preparing epoxidized sucrose esters of fatty acids (ESEFAs) [152] (Figure 18.16). These new monomers were cross-linked with a liquid cycloaliphatic anhydride, 4-Methyl-1,2-cyclohexanedicarboxylic anhydride (MHHPA), and 1,8-diazabicyclo[5.4.0]undec-7-ene (DBU) was used as initiator. The properties of coatings on steel substrates were studied to determine coating hardness, adhesion, solvent resistance, and mechanical durability. Compared with a commercial Epoxidized Soybean Oil (ESO), the anhydride-cured ESEFAs have high modulus and are hard and ductile. Also, these are high-performance thermoset materials that maintain a high biobased content (~75%). The exceptional performances are attributed to the unique structure of these precursors, that is, well-defined compact structures with high epoxide functionality. These biobased thermosets may have potential uses in applications such as composites, adhesives, and coatings.

On the other hand, a large number of thermosetting polymers were synthesized just by partial replacement of the classical epoxy prepolymers, such as DGEBA, with epoxidized vegetable oils. Very often, the modified natural oils are used as reactive diluents and/or flexibilizers [153]. As an example, blends of DGEBA with increasing amounts of ESO and with methyl tetrahydrophthalic anhydride (MTHPA) as hardener and 1-methyl imidazole (1-MI) as an initiator have been studied. As expected, the T_gs of the resulting epoxy cross-linked polymers decreased with the amount of epoxidized vegetable oil from 108°C to 57°C (without DGEBA) [149]. For a composition that is DGEBA–epoxidized oil 40%:60% (wt), a good compromise of mechanical properties was obtained but the biobased content was still rather low. Similar results have been obtained with mixtures of 50% DGEBF with both soybean and linseed oils [154] reacted with MTHPA or a triamine [poly(oxypropylene) triamine].

In their studies on biobased nanocomposites with microfibrillated cellulose and TA as hardener, Shibata and Nakai [101] also used industrially available sorbitol polyglycidyl ether (SPE; 172 g/eq

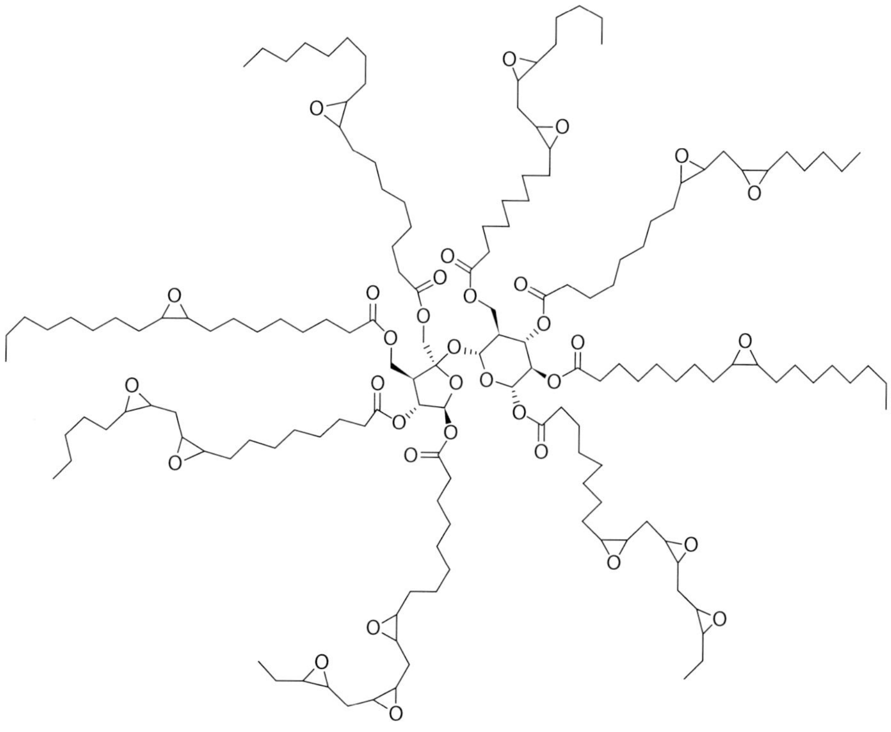

FIGURE 18.16   Epoxidized sucrose ester of fatty acids (ESEFAs).

for an average number of epoxy groups of 3.6). Similar results as with GPE were obtained. Expecting an increase of thermal resistance, Shibata and Nakai tested other hardeners having a lower hydroxyl value and a higher aromatic content than TA. Quercetin (3,3',4',5,7-pentahydroxyflavone [QC]) is one of the most abundant flavonoids found in plants such as onion, capers, and tea.

Monomers were first dissolved in tetrahydrofuran to obtain a homogeneous solution. After solvent vaporization at 40°C for 24 h, the resulting viscous liquid was prepolymerized at 150°C for 0.5 h, and then compression-molded at 170°C for 3 h. Thermomechanical measurements showed that the SPE–QC network had a higher tan δ peak temperature than SPE–phenol novolak (PN). A similar comparison was obtained with liquid DGEBA. These results indicate that QC is a superior epoxy hardener for generating epoxy networks with high heat resistance. In their research for biobased phenols with high aromatic content for a low hydroxyl value, Shibata and coworkers [155] also prepared a PG–VN calixarene (PGVNC). This phenolic hardener was obtained by reacting PG (prepared by decarboxylation of gallic acid) and VN in the presence of p-toluenesulfonic acid. The PGVNC was blended with SPE using tetrahydrofuran as a solvent. After a similar curing schedule, measurements indicate that the tan δ peak temperature of the SPE–PGVNC was 148°C, which was much higher than that of the SPE–PN.

Based on these monomers, biocomposites were also prepared with different percentages of wood flour with the aim of increasing the tensile modulus and strength at room temperature, and also the storage modulus at the rubbery plateau. These phenols seem to be interesting hardeners, but as the chlorine content of the SPE monomer used in these three studies was 9.6%, it is worth testing them with a free chlorine epoxy monomer. Another point that is questionable concerns the toxicity and carcinogenicity of the quercetin molecule [156].

As mentioned in Section 18.3.3.1, sorbitol and maltitol were converted to multifunctional epoxy monomers through oxidation of allyl double bonds and were used in combination with diethylene triamine (DETA) to produce biobased networks with T_g in the range 70°C–80°C [103].

Epoxy allyl and epoxy crotyl sucroses (EAS and ECS) were synthesized in one step from octa-O-allyl and octa-O-crotyl sucroses by epoxidation. The average number of epoxy groups (functionality, f) per sucrose was controlled by varying the concentration of peracetic acid in the reaction mixture [104].

Although if it is possible to find old patents [106,107], a number of studies have been devoted in the last 5 years to isosorbide-based epoxies as safe and renewable alternatives to petroleum-based BPA epoxies. Isosorbide is also known to impact stiffness to the polymer chain of linear polyesters or polycarbonates [105]. Depending on the chemistry used, different types of diglycidyl ethers of isosorbide (DGEDAS) have been prepared. The cross-linking reaction was performed with aliphatic triamine (Jeffamine T403) and cycloaliphatic diamine (isophorone diamine [IPDA]) with different curing schedules, but similar to those used for DGEBA with the same hardeners. The T_gs of networks synthesized from DGEDAS were always lower than the corresponding DGEBA-based network: about 40°C with Jeffamine T403. This result was a slight disappointment. However, they have higher rubbery modulus, due either to the lower molar mass or to the higher functionality of the prepolymer.

These biobased prepolymers must be stored in dry conditions, because these compounds are hygroscopic. But, the main problem to solve is the water uptake of the networks that can be very high, which deteriorates the properties of the networks. To optimize the mechanical properties and decrease the water uptake (less than 2%) of the isosorbide-derived epoxies for industrial applications such as can coating, some solutions have been proposed. Feng et al. [157] have added hydrophobic functional groups (such as 4-allyloxybenzoyl chloride) into the backbone of the isosorbide epoxy (leading to isosorbide bis-glycidyl benzoate). It was also possible to add into the formulation hydrophobic epoxidized cardanol (ECD) [158]. But a more efficient method that is under testing is to adjust the amount and type of cross-linker, such as terpene diphenol (TPD, from terpene compounds and phenol as raw materials, melting point, $T_m = 85°C$) [159], 4,4-*bis*-(4'-hydroxyphenyl) pentanoic acid, or diphenolic acid (DPA, from levulinic acid and phenol, $T_m = 171°C$), biobased phenols with high aromatic content for a low hydroxyl value (such as the ones used with SPE monomers), or other phenols extracted from lignin [20].

The epoxy networks with high T_gs are required to retain the dimensional stability and rigidity at a high temperature in electrical and structural materials. However, we have seen in previous sections that the T_gs of the biobased epoxy networks and their composites were still lower than those of cured materials based on the popular DGEBA, or other epoxy precursors with higher functionality such as TGMDA. The development of heat-resistant biobased epoxy monomer–hardener having a comparable T_g to conventional epoxy curing systems is a real challenge. Strategies using epoxidized vegetable oils and complex chemistries leading to biobased epoxy cross-linked polymers that can be used as comonomers or be directly polymerized have demonstrated their limitations.

Compared with vegetable oils, phenols have a higher potential. They can be used as hardeners, but also as epoxy precursors (after an epoxidation reaction).

Furthermore, gallic acid epoxidation leads to monomers with molar mass values ranging from 137 to 160 g/mol, which corresponds to epoxy functionality from 1 to 3. The epoxidized compounds were then cross-linked with a conventional cross-linking agent [47–48].

Epoxidized tannins have also been prepared. An epoxy monomer was formed from commercialized hydrolyzed tannin. This compound was pretreated with alcohol or was ether extracted to obtain polyphenol oligomers (molar masses from 500 to 5000 g/mol) soluble in organic solvents [160]. Then, phenol groups were epoxidized by using ECH, utilizing a phase transfer catalyst to avoid reaction with ester groups. Interestingly, in this study, the cross-linking reaction was performed using phenolic groups of hydrolyzed tannins. Furthermore, Nouailhas et al. enhanced the tannin reactivity for the synthesis of epoxy monomers [48]. An epoxy monomer was indeed obtained from the glycidylation of tannin units, that is, (−)-epicatechin and (+)-catechin. This monomer was mixed with DGEBA (25%:75% and 50%:50 wt%), and the mixture was cross-linked with a commercially available cycloaliphatic amine at 60°C for 24 h. It can be said that tannin imparts softness to the epoxy network both in the glassy and elastic regions, as compared with the neat network obtained from DGEBA only.

The resulting epoxy monomer of resorcinol was cross-linked with diaminodiphenylsulfone (DDS, which is known as *nontoxic aromatic amine*). The thermal analysis Differential Scanning Calorimetry (DSC) of the epoxy network did not reveal any glass transition in the 50°C–250°C region. Furthermore, the thermal stability of the polymer was clearly improved compared with conventional DGEBA [161]. A different approach to accelerate curing of epoxy tannins based on commercial tannin oligomers was taken by Jahanshahi et al., by preparing an epoxidized tannin acrylic system that yielded good bonding results [50].

Recently, Benyahya et al. [53] have prepared epoxy monomers based on catechin- and green tea-extracted phenols. These monomers were reacted with isophorone diamine (IPDA). The DSC results showed a high reactivity for these epoxy monomers. The T_gs were in the range 140°C–190°C, depending on the epoxy functionality, and a high elastic modulus was obtained at the rubbery plateau ($T_g + 30°C$).

Another example is the cross-linking reaction of epoxy obtained from resinic acid with 1,2-cyclohexanedicarboxylic anhydride (CHDB) in the presence of 2-ethyl-4-methylimidazole as initiator at 100°C for 2 h and at 180°C for 2 h. The epoxy network exhibits $T_g = 153°C$, 10°C higher than the corresponding network based on DGEBA as epoxy monomer [162].

Synthons extracted from lignin certainly have a high potential for structural epoxies [20]. Nevertheless, they are solid, whereas at room temperature a liquid allows easier mixing and processing without the help of solvents. As an example, the difunctional epoxy monomer based on VN [163] is a white solid with a high melting point, $T_m \sim 175°C$. This epoxy monomer was reacted with methylenedianiline. It was reported that, due to several β-relaxations, the impact strength, tensile strength, and elongation were improved [163,164].

Typically, DGEBA and other di- or multifunctional monomers containing the aromatic ring structure will cure to hard, rigid compositions having rather low impact and elongation characteristics.

There are many approaches to improving these properties. Among these are modifications with vegetable oils, polyamide or polysulfide curing agents, or long-chain polyglycols that increase network flexibility, but always with a decrease of T_g and modulus. To overcome this drawback, particular modifiers such as a rubber or a thermoplastic that are miscible before reaction of the epoxy formulation but become phase separated in the course of polymerization (reaction-induced phase separation) are employed to produce different appropriate morphologies and increase toughness in the final products [165].

As an example, natural rubber, that is, poly(1,4-isoprene), was first exploited to perform as a biobased epoxy monomer. Epoxidation by using a peracid compound was performed on the unsaturations, leading to epoxidized natural rubber (ENR). Gandini [15], in his review, described the process leading to ENRs, which was operated on an industrial scale. ENRs have been cured in the presence of diamines, which also acted as radical scavengers able to delay ENR degradation [167]. Further, Hong and Chan [168] incorporated ENR in a conventional epoxy formulation, that is, DGEBA, dicyandiamide, and 2-methylimidazole as initiator. The authors pointed out the (macro) heterogeneous nature of the resulting material, probably due to the high molar mass of the modifier. Thus, the degradation of ENR, which leads to liquid ENR with low molar mass, may be more suitable for incorporation in conventional epoxy formulations [15,169] and can increase the toughness of the thermosetting polymer.

A totally different example is the use of a synthesized cardanol-based epoxy curing agent (NX54) [170]. This partially biobased hardener results from the Mannich reaction between cardanol, formaldehyde, and DETA. NX54 is less reactive with DGEBA than common phenalkamines; because of a decrease in solubility of NX54 during reaction, the authors observed via scanning electron microscopy a morphology of the cured sample that consisted of cavities dispersed within a continuous epoxy matrix. These cavities markedly improved the lap shear strength and impact strength of the network.

18.5 SUMMARY AND FUTURE TRENDS

Even if biorefinery processes and retreatment of biomass were out of scope for this chapter, it is clear that the future of biobased polymers, including building blocks for thermosetting epoxies, is strongly dependent on the future of biorefineries [171]. The largest waste source for carbohydrates and lignin is from lignocellulosic biomass residues, which are estimated to exceed 2×10^{11} t/year worldwide [172]. Indeed, lignin is the only large-volume renewable feedstock that comprises aromatics. Up till now, despite extensive research, there are very few reports of efficient ways of recovering such aromatic products. The only noticeable commercial process has been the production of VN from lignosulfonates with a yield less than 10% by weight, and of levulinic acid, which is an interesting biobased chemical [173].

Food residues and by-products are produced in very significant quantities by both the food industry and the agricultural sector [174]. As a first conclusion, there are plenty of opportunities to develop molecular biomass and monomers for the preparation of renewable epoxy formulations and materials. But, the biggest changes for thermosetting epoxy formulations in the near future will certainly come from regulation changes such as the environmental directives for the "Registration, Evaluation and Authorisation of Chemicals" (REACH). Biobased compounds will not be the only solution for these policy pressures, but will be an opportunity.

Developing highly efficient, safe, low-waste, low-toxicity, and *atom-economy* processes are the key words for green chemistry. Chemistry has to be simple, practical, and operational, and catalysts are expected to play an important role. ECH is the preferred way to prepare epoxy monomers; but even if ECH can be biobased, it remains a toxic molecule that has to be handled in a safe environment. Epoxidation without the use of ECH is a key challenge. We have seen that allylation or crotonization of alcohols can be a first step for epoxidation, but the second step, that is, double-bond oxidation, requires the use of expensive and hazardous catalysts.

Some industrialized countries have declared BPA to be a toxic substance that causes risks to human health as well as to the environment; thus, it has to be banned from all food contact applications in the coming years, and in the immediate future the pressure is on finding a replacement. BPA is industrially produced from condensation of acetone with phenols. Furthermore, the wastes produced from the forest industry can be distilled to lead to phenol compounds. One option is the synthesis of biobased BPA, which has been tested to produce biobased epoxy oligomers (EPOBIOX from AMROY Cie). Nevertheless, whether from petroleum or renewable resources, BPA remains classified as reprotoxic, and this route is certainly not the solution. Other biobased phenols discussed in this chapter can be proposed to replace BPA. The oldest one is diphenol acid, synthesized from levulinic acid. Based on the same hydroxyalkylation synthesis, TPD or diphenol based on VN could also be excellent choices. Natural phenols extracted from tannins and derivatives are also good candidates. Because of its long aliphatic chain, the diphenol of cardanol leads to materials that are too flexible to replace BPA. But the main problem remaining with phenols is their toxicity. On the other hand, isosorbide is known to be nontoxic and is also rigid, but the problem to solve in this case is the low water resistance of the resulting epoxy networks. Therefore, biobased aromatic/rigid epoxy monomers are still needed to obtain a good compromise between processing and properties, and should be able to replace BPA.

An additional remark concerns biodegradability, which is a special functionality conferred on a material (there are few results in the literature [102], which means that except for some biomedical applications, it is not a required property for thermosetting epoxies). In conclusion, one idea to maximize the utilization of natural resources and to improve material performance is to combine natural polymers and natural molecular biomass. This route merits further development.

REFERENCES

1. I. Hamerton. High performance polymers. *Polym. Int*. 41, 101–102 (1996).
2. E. M. Petrie. *Epoxy Adhesive Formulation*, McGraw-Hill, New York (2006).
3. D. Bogdal, J. Pielichowski, P. Penczek, J. Gorczyk and G. Kowalski. Synthesis of elevated-molecular-weight epoxy resins with aid of microwaves. *Polimery* 47, 842–845 (2002).
4. D. P. Nick. The market: Epoxy resins. *TRFA-Thermosetting Resin Formulators Association Annual Meeting*, Philadelphia, PA (2003).
5. P. Michaud. Epoxy resins in composites. *JEC-Composites Conference*, Paris, France (2004).
6. J. E. White. Thermoplastic epoxy polymers. In: *Epoxy Polymers: New Materials and Innovations*, J. P. Pascault and R. J. J. Williams (Eds.), Wiley-VCH, Weinheim, Germany (2010).
7. E. D. Dodds and W. Lawson. Synthetic estrogenic agents without the phenanthrene nucleus. *Nature* 137, 996 (1936).
8. J. C. O'Connor and R. E. Chapin. Critical evaluation of observed adverse effects of endocrine active substances on reproduction and development, the immune system, and the nervous system. *Pure Appl. Chem*. 75, 2099–2123 (2003).
9. A. Matsushima, X. Liu, H. Okada, M. Shimohigashi and Y. Shimohigashi. Bisphenol AF is a full agonist for the estrogen receptor ERα but a highly specific antagonist for ERβ. *Environ. Health Perspect*. 118, 1267–1272 (2008).
10. F. S. Vom Saal and J. P. Myers. Bisphenol A and risk of metabolic disorders. *JAMA-(J. Am. Med. Assoc.)* 300, 1353–1355 (2008).
11. F. S. Vom Saal and C. Hughes. An extensive new literature concerning low-dose effects of bisphenol A shows the need for a new risk assessment. *Environ. Health Perspect*. 113, 926–933 (2005).
12. J.-P. Pascault, and R. J. J. Williams. Background in *Epoxy Polymers: New Materials and Innovations*, J.-P. Pascault and R. J. J. Williams (Eds.), Wiley-VCH, Weinheim, Germany (2010).
13. R. Muelhaupt. Green polymer chemistry and bio-based plastics: Dreams and reality. *Macromol. Chem. Phys*. 214, 159–174 (2013).
14. R. T. Mathers. How well can renewable resources mimic commodity monomers and polymers? *J. Polym. Sci., Part A Polym. Chem*. 50, 1–15 (2012).
15. A. Gandini. Epoxy polymers based on renewable resources. In: *Epoxy Polymers: New Materials and Innovations*, J.-P. Pascault and R. J. J. Williams (Eds.), Wiley-VCH, Weinheim, Germany (2010).

16. J. M. Raquez, M. Deleglise, M. F. Lacrampe and P. Krawczak. Thermosetting (bio) materials derived from renewable resources: A critical review. *Prog. Polym. Sci.* 35, 487–509 (2010).

17. S. G. Tan and W. S. Chow. Biobased epoxidized vegetable oils and its greener epoxy blends: A review. *Polym. Plastics Technol Eng.* 49, 1581–1590 (2010).

18. R. Wang and T. P. Schuman. Vegetable oil-derived epoxy monomers and polymer blends: A comparative study with review. *Express Polym. Lett.* 7, 272–292 (2013).

19. Y. Xia, R. L. Quirino and R. C. Larock. Bio-based thermosetting polymers from vegetable oils. *J. Renewable Mat.* 1, 3–15 (2013).

20. T. Koike. Progress in development of epoxy resin systems based on wood biomass in Japan. *Polym. Eng. Sci.* 52, 701–717 (2012).

21. S. Zheng, J.-P. Pascault and R. J. J. Williams. Nanostructured epoxies by the use of block copolymers. In: *Epoxy Polymers: New Materials and Innovations*, J.-P. Pascault and R. J. J. Williams (Eds.), Wiley-VCH, Weinheim, Germany (2010).

22. W. Bradley, J. Forrest and O. Stephenson. The catalysed transfer of hydrogen chloride from chlorohydrins to epoxides. A new method of preparing glycidol and some of its derivatives. *J. Chem. Soc.* 359, 1589–1597 (1951).

23. Y. M. Beasley, V. Petrow and O. Stephenson. Analgesics. Part 1; Some aryloxypropanolamines. *J. Pharm. Pharmacol.* 10, 47–59 (1958).

24. A. Lopez Quintela, M. Pazos Pellin and S. Paz Abuin. Epoxidation reaction of trimethylolpropane with epichlorohydrin: Kinetic study of chlorohydrin formation. *Polym. Eng. Sci.* 36, 568–573 (1996).

25. E. E. Royals and L. L. Harrell. Oxygenated derivatives of d-α-pinene and d-limonene. Preparation and use of monoperphthalic acid1. *J. Am. Chem. Soc.* 77, 3405–3408 (1955).

26. C. L. Stevens and J. Tazuma. Epoxyethers. V. The reaction of perbenzoic acid with vinyl ethers. *J. Am. Chem. Soc.* 76, 715–717 (1954).

27. Y. Fort, A. Olszewski-Orter and P. Craubee. Synthesis of epoxy (meth) acrylic esters by selective epoxidation of unsaturated (meth) acrylic esters using the system H 2 O 2-Na 2 WO 4 under phase transfer. *Tetrahedron* 48, 5099–5110 (1992).

28. B. E. Rossiter, T. R. Verhoeven and K. B. Sharpless. Stereoselective epoxidation of acyclic allylic alcohols. A correction of our previous work. *Tetrahedron* 49, 4733–4736 (1979).

29. C. Flosbach and R. Fugier. Epoxy functional acrylic polymers for high performance coating applications. In: *Epoxy Polymers: New Materials and Innovations*, J.-P. Pascault and R. J. J. Williams (Eds.), Wiley-VCH, Weinheim, Germany (2010).

30. J.-P. Pascault, H. Sautereau, J. Verdu and R. J. J. Williams, *Thermosetting Polymers*, CRC Press, New York (2002).

31. F. Grunchard. Process for the manufacture of epichlorohydrin. European Patent EP 561441, assigned to Solvay Belgium (1993).

32. P. Krafft, P. Gilbeau, B. Gosselin and S. Classens. Production of organic compounds comprising producing glycerol from renewable raw materials and producing organic compounds from the glycerol. French Patent FR 2862644, assigned to Solvay Belgium (2005).

33. L. Shen, E. Worrell and M. Patel. Biofuels: Present and future development in plastics from biomass. *Bioprod. Biorefin.* 4, 25–40 (2010).

34. C. O. Tuck, E. Perez, I. T. Horvath, R. A. Sheldon and M. Poliakoff. Valorization of biomass: Deriving more value from waste. *Science* 337, 695–699 (2012).

35. B. S. Lane and K. Burgess. Metal-catalyzed epoxidations of alkenes with hydrogen peroxide. *Chem. Rev.* 2003, 103, 2457–2474 (2003).

36. D. Ferreira, G. G. Gross, A. E. Hagerman, H. Kolodziej and T. Yoshida. Tannins and related polyphenols: Perspectives on their chemistry, biology, ecological effects, and human health protection. *Phytochemistry* 69, 3006–3008 (2008).

37. O. Benavente-Garcia, J. Castillo, F. R. Marin, A. Ortuno and J. A. Del Rio. Uses and properties of citrus flavonoids. *J. Agric. Food Chem.* 45, 4505–4515 (1997).

38. C. Manach, A. Mazur and A. Scalbert. Polyphenols and prevention of cardiovascular diseases. *Curr. Opin. Lipidol.* 16, 77–84 (2005).

39. E. Middleton and C. Kandaswami. Potential health-promoting properties of citrus flavonoids. *Food Technol.* 48, 115–119 (1994).

40. R. Puupponen-Pimia, L. Nohynek, C. Meier, M. Kahkonen, M. Heinonen, A. Hopia, and K. M. Oksman-Caldentey. Antimicrobial properties of phenolic compounds from berries. *J. Appl. Microbiol.* 90, 494–507 (2001).

41. M. G. Hertog, E. J. Feskens, P. C. Hollman, M. B. Katan and D. Kromhout. Dietary antioxidant flavonoids and risk of coronary heart disease: The Zutphen elderly study. *Lancet* 342, 1007–1011 (1993).

42. A. J. Parr and G. P. Bolwell. Phenols in the plant and in man. The potential for possible nutritional enhancement of the diet by modifying the phenols content or profile. *J. Sci. Food Agr.* 80, 985–1012 (2000).

43. K. E. Heim, A. R. Tagliaferro and D. J. Bobilya. Flavonoid antioxidants: Chemistry, metabolism and structure-activity relationships. *J. Nutr. Biochem.* 13, 572–584 (2002).

44. S. Haettenschwiler, A. E. Hagerman and P. M. Vitousek. Polyphenols in litter from tropical montane forests across a wide range in soil fertility. *Biogeochem.* 64, 129–148 (2003).

45. K. Khanbabaee and T. Van Ree. Tannins: Classification and definition. *Nat. Prod. Rep.* 18, 641–649 (2001).

46. P. M. Aron and J. A. Kennedy. Flavan-3-ols: Nature, occurrence and biological activity. *Mol. Nutr. Food Res.* 52, 79–104 (2008).

47. B. Boutevin, S. Caillol, C. Burguiere, S. Rapior, H. Fulcrand and H. Nouailhas. Novel method for producing thermosetting epoxy resins. WO 2010136725, assigned to Centre National de la Recherche Scientifique, France (2010).

48. H. Nouailhas, C. Aouf, C. Le Guerneve, S. Caillol, B. Boutevin and H. Fulcrand. Synthesis and properties of biobased epoxy resins. Part 1: Glycidylation of flavonoids by epichlorydrin., *J. Polym. Sci., Part A Polym. Chem.* 49, 2261–2270 (2011).

49. S. Jahanshahi, A. Pizzi, A. Abdolkhani, K. Doosthoseini, A. Shakeri, M. C. Lagel and L. Delmotte. MALDI-TOF and 13C-NMR and FT-MIR and strength characterization of glycidyl ether tannin epoxy resins. *Ind. Crops Prod.* 83, 177–185 (2016).

50. S. Jahanshahi, A. Pizzi, A. Abdolkhani and A. Shakeri. Analysis and testing of bisphenol-A-free bio-based tannin epoxy-acrylic adhesives. *Polymers* 8, 143 (2016).

51. H. Tomita and K. Yonezawa. Epoxy resin and process to prepare the same. European Patent EP 0095609, assigned to Kanegafuchi Chemical Ind. (1983).

52. C. Aouf, S. Le Guerneve, S. Caillol and H. Fulcrand. Study of the O-glycidylation of natural phenolic compounds. The relationship between the phenolic structure and the reaction mechanism. *Tetrahedron* 69, 1345–1353 (2013).

53. S. Benyahya, C. Aouf, S. Caillol, B. Boutevin and H. Fulcrand. Functionalized green tea tannins as phenolic prepolymers for bio-based epoxy resins. *Ind. Crops Prod.* 53, 296–307 (2014).

54. F. Dasgupta. Biocompatible and biodegradable polymers from renewable natural polyphenols. WO 2011041487 assigned to Innovotech LLC (2011).

55. M. A. V. Ribeiro da Silva and A. I. M. C. L. Ferreira. Experimental standard molar enthalpies of formation of some methylbenzenediol isomers. *J. Chem. Thermodyn.* 41, 1096–1103 (2009).

56. P. Livant, T. R. Webb and W. Xu. Reaction of resorcinol with acetone. *J. Org. Chem.* 62, 737–742 (1997).

57. M. Suzuki, H. Saito and T. Ikegawa. 5-Fluorouracil derivatives and antitumor agents. Japan Patent JP 60199888, assigned to Nippon Kasei Chemicals (1985).

58. J. Robins. Epoxy resin curing agent, process and composition. US Patent 4,503,211, assigned to Minnesota Mining and Manufacturing Co. (1985).

59. Y. Saito, A. Morii and H. Nakamura. Epoxy resin composition. European Patent EP 161576, assigned to Sumitomo Chemicals (1985).

60. P. Campaner, D. D'Amico, L. Longo, C. Stifani and A. Tarzia. Cardanol-based novolac resins as curing agents of epoxy resins. *J. Appl. Polym. Sci.* 114, 3585–3591 (2009).

61. C. K. S. Pillai, V. S. Prasad, J. D. Sudha, S. C. Bera and A. R. R. Menon. Polymeric resins from renewable resources. II. Synthesis and characterization of flame-retardant prepolymers from cardanol. *J. Appl. Polym. Sci.* 41, 2487–2501 (1990).

62. P. P. Kumar, R. Paramashivappa, P. J. Vithayathil, P. V. S. Rao and A. S. Rao. Process for isolation of cardanol from technical cashew (*Anacardium occidentale* L.) nut shell liquid. *J. Agric. Food Chem.* 50, 4705–4708 (2002).

63. M. Sultania, J. S. P. Rai and D. Srivastava. Process modeling, optimization and analysis of esterification reaction of cashew nut shell liquid (CNSL)-derived epoxy resin using response surface methodology. *J. Hazard. Mater.* 185, 1198–1204 (2011).

64. G. Scott. *Degradable Polymers: Principles and Applications*, 2nd edn, Springer, Berlin (2002).

65. K. P. Unnikrishnan and E. T. Thachil. Studies on the modification of commercial epoxy resin using cardanol-based phenolic resins. *J. Elastom. Plast.* 40, 271–286 (2008).

66. C. G. Weller, E. J. Siebert, Z. Yang, R. K. Agarwal, W. E. Fristad and B. D. Bammel. Autodeposition compositions for polymeric coatings of reduced gloss, good corrosion resistance and uniform appearance. WO 2003026888, assigned to Henkel DGGG, Germany (2003).

67. D. R. Fenn, G. R. Webster and G. J. McCollum. Modified epoxy resins comprising the reaction product of a biomass derived compound and an epoxy resin, and aqueous dispersions and coatings comprising such resins. WO 2009014842, assigned to PPG Ind Ohio Inc. (2009).

68. T. T. M. Tan. Cardanol-lignin-based epoxy resins: Synthesis and characterization. *J. Polym. Mater.* 13, 195–199 (1996).

69. Y. H. Kim, E. S. An, S. Y. Park and B. K. Song. Enzymatic epoxidation and polymerization of cardanol obtained from a renewable resource and curing of epoxide-containing polycardanol. *J. Mol. Catal. B: Enzymatic* 45, 39–44 (2007).

70. A. Devi and D. Srivastava. Enzymatic epoxidation and polymerization of cardanol obtained from a renewable resource and curing of epoxide-containing polycardanol. *J. Appl. Polym. Sci.* 102, 2730–2737 (2006).

71. K. P. Unnikrishnan and E. T. Thachil. Synthesis and characterization of cardanol-based epoxy systems. *Des. Monomers Polym.* 11, 593–607 (2008).

72. J. He, S. Xu and X. Xu. Study on the anacardol modified epoxy resin. *Tuliao Gongye* 29, 5–11 (1999).

73. N. Nieu, T. T. M. Tan and N. L. Huong. Epoxy–phenol–cardanol–formaldehyde systems: Thermogravimetry analysis and their carbon fiber composites. *J. Appl. Polym. Sci.* 61, 2259–2264 (1996).

74. M. Ionescu and Z. S. Petrovic. Phenolation of vegetable oils. *J. Serbian. Chem. Soc.* 76, 591–606 (2011).

75. M. Sultania, J. S. P. Rai and D. Srivastava. Studies on the synthesis and curing of epoxidized novolac vinyl ester resin from renewable resource material. *Eur. Polym. J.* 46, 2019–2032 (2010).

76. W. O. S. Doherty, P. Mousavioun and C. M. Fellows. Value-adding to cellulosic ethanol: Lignin polymers. *Ind. Crops Prod.* 33, 259–276 (2011).

77. K. Hofmann and W. G. Glasser. Engineering plastics from lignin. 21.1. Synthesis and properties of epoxidized lignin-poly (propylene oxide) copolymers. *J. Wood Chem. Technol.* 13, 73–95 (1993).

78. C. I. Simionescu, V. Rusan, M. M. Macoveanu, G. Cazacu, R. Lipsa, C. Vasile, A. Stoleriu and A. Ioanid. Lignin/epoxy composites. *Compos. Sci. Technol.* 48, 317–323 (1993).

79. M. V. Alonso, M. Oliet, J. Garcia, F. Rodriguez and J. Echeverria. Gelation and isoconversional kinetic analysis of lignin–phenol–formaldehyde resol resins cure. *Chem. Eng. J.* 122, 159–166 (2006).

80. D. Feldman, M. Lacasse and L. M. Beznaczuk. Lignin-polymer systems and some applications. *Prog. Polym. Sci.* 12, 271–276 (1987).

81. S. Tai, M. Nagata, J. Nakano and N. Migita. Studies on utilization of lignin. IV. Epoxidation of thiolignin. *Mokuzai Gakkaishi* 13, 102–107 (1967).

82. S. Tai, J. Nakano and N. Migita. Studies on utilization of lignin. V. Adhesive from lignin epoxide. *Mokuzai Gakkaishi* 13, 257–262 (1967).

83. S. Tai, J. Nakano and N. Migita. Studies on utilization of lignin. VI. Activation of thiolignin with phenol. *Mokuzai Gakkaishi* 14, 40–45 (1968).

84. H. Ito and N. Shiraishi. Epoxy-resin adhesives from thiolignin. *Mokuzai Gakkaishi* 33, 393–399 (1987).

85. N. Shiraishi. Recent progress in wood dissolution and adhesives from kraft lignin. *ACS Symp. Ser.* 397, 488–495 (1989).

86. Y. Okabe and H. Kagawa. Biomass-derived epoxy compound and manufacturing method thereof. US Patent 20100155122, assigned to Hitachi Ltd. (2010).

87. Y. Okabe and H. Kagawa. Biomass-derived epoxy resin compound and manufacturing method thereof. US Patent 8,232,365, assigned to Hitachi Ltd. (2012).

88. S. Inoue and H. Matsuzaki. Adhesive material and woody molded article containing adhesive material and method for producing woody molded article. Japan Patent JP 2004300229, assigned to Araco Corp. and Toagosei Co. Ltd. (2004).

89. B. Zhao, G. Chen, Y. Liu, K. Hu and R. Wu. Synthesis of lignin base epoxy resin and its characterization. *J. Mater. Sci. Lett.* 20, 859–862 (2001).

90. G. Sun, H. Sun, Y. Liu, B. Zhao, N. Zhu and K. Hu. Comparative study on the curing kinetics and mechanism of a lignin-based-epoxy/anhydride resin system. *Polymer* 48, 330–337 (2007).

91. S. Hirose, M. Kobayashi, H. Kimura and H. Hatakeyama. Synthesis and thermal properties of epoxy resins derived. In: *Recent Advances in Environmentally Compatible Polymers*, J. F. Kennedy, G. O. Phillips and P. A. Williams (Eds.), Woodhead Publishing, Cambridge, UK (2001).

92. A. Gandini and M. N. Belgacem. Partial or total oxypropylation of natural polymers and the use of the ensuing materials as composites or polyol macromonomers. In: *Monomers, Polymers and Composites from Renewable Resources*, M. N. Belgacem and A. Gandini (Eds.), Elsevier, Amsterdam (2008).

93. J. E. Holladay, J. J. Bozell, D. Johnson and J. F. White. Top value added chemicals from biomass: Vol 2: Results of screening for potential candidate from biorefinery lignin. U.S. Department of Energy, Technical Report Office of Scientific and Technical Information, N. R. E. L., Oak Ridge, TN (2007).

94. T. Koike. Progress in development of epoxy resin systems based on wood biomass in Japan. *Polym. Eng. Sci.* 52, 701–717 (2012).

95. C. Aouf, J. Lecomte, P. Villeneuve, E. Dubreucq and H. Fulcrand. Chemo-enzymatic functionalization of gallic and vanillic acids: Synthesis of bio-based epoxy resins prepolymers. *Green Chem.* 14, 2328–2336 (2012).

96. M. Fache, E. Darroman, V. Besse, R. Auvergne, S. Caillol and B. Boutevin. Vanillin, a promising bio-based building-block for monomer synthesis. *Green Chem.* 16, 1987–1998 (2014).

97. M. Fache, S. Caillol, R. Auvergne and B. Boutevin. New vanillin-derived diepoxy monomers for the synthesis of biobased thermosets. *Eur. Polym. J.* 67, 527–538 (2015).

98. T. Shimasaki, S. Yoshihara and M. Shibata. Preparation and properties of biocomposites composed of sorbitol-based epoxy resin, pyrogallol-vanillin calixarene, and wood flour. *Polym. Composites.* 33, 1840–1847 (2012).

99. J. Krivanek and E. Kolarova. Waterborne epoxy resins for coatings. Advances in Coatings Technology, ACT 98, 3rd International Conference, Katowice, Poland, October 20–23 (1998).

100. A. Poliscuk, J. Hyrsl and I. Slivkova. Preparation and properties of biocomposites composed of bio-based epoxy resin, tannic acid, and microfibrillated cellulose. Sbornik Prispevku: Mezinarodni Konference o Naterovych Hmotach, 36th, Sec, Czech Republic, May 23–25, 2005, p. 94 (2005).

101. M. Shibata and K. Nakai. Preparation and properties of biocomposites composed of bio-based epoxy resin, tannic acid, and microfibrillated cellulose. *J. Polym. Sci., Part B Polym. Phys.* 48, 425–433 (2010).

102. Y. Takada, K. Shinbo, Y. Someya and M. Shibata. Preparation and properties of bio-based epoxy montomorillonite nanocomposites derived from polyglycerol polyglycidyl ether and ε-polylysine. *J. Appl. Polym. Sci.* 113, 479–484 (2009).

103. D. Acierno, P. Russo and R. Savarese. Oxidation of maltitol and sorbitol as a way to bioepoxies, Polymer Processing Society, PPS-24, Salerno, Italy (2008).

104. N. D. Sachinvala, D. L. Winsor, R. K. Menescal, I. Ganjian, W. P. Niemczura and M. H. Litt. Sucrose-based epoxy monomers and their reactions with diethylenetriamine. *J. Polym. Sci., Part A, Polym. Chem.* 36, 2397–2413 (1998).

105. F. Fenouillot, A. Rousseau, G. Colomines, R. Saint-Loup, J.-P. Pascault. Polymers from renewable 1, 4: 3, 6-dianhydrohexitols (isosorbide, isomannide and isoidide): A review. *Prog. Polym. Sci.* 35, 578–622 (2010).

106. J. G. Morrison. Polyglycidyl ethers of cyclic anhydro hexitols and their aqueous solutions. US Patent 3,041,300, assigned to Martin Marietta Co. (1962).

107. J. D. Zech and J. W. L. Maistre. Diglycidyl ethers of isohexides. US Patent 3272845 assigned to Atlas Chemical Ind. (1966).

108. A. East, M. Jaffe, Y. Zhang and L. Catalani. Thermoset epoxy polymers from renewable resources such as anhydrosugars. WO 2008147473, assigned to New Jersey Techn. Inst. (2008).

109. D. Achet, M. Delmas and A. Gaset. Biomass as a source of chemicals. VI. Synthesis of new polyfunctional ethers of isosorbide in solid-liquid heterogeneous mixtures. *Biomass* 9, 247–254 (1987).

110. A. East, M. Jaffe, Y. Zhang and L. Catalani. Ethers of bisanhydrohexitols. US Patent 20080021209, assigned to New Jersey Techn. Inst. (2008).

111. X. Feng, A. J. East, W. Hammond and M. Jaffe. Sugar-based chemicals for environmentally sustainable applications. In: *Contemporary Science of Polymeric Materials*, L. Korugic-Karasz (Ed.), ACS Symp. Ser. 1061, pp. 3–27 (2010).

112. X. Feng, A. J. East, W. B. Hammond, Y. Zhang and M. Jaffe. Overview of advances in sugar-based polymers. *Polym. Adv. Technol.* 22, 139–150 (2011).

113. I. Wiesner, J. Kriz, V. Bruthans and J. Kolinsky. Glycidyl ethers of phenols and polyphenols for synthesis of epoxy resins. CS 113977 (1965).

114. L. M. Pitet, S. B. Hait, T. J. Lanyk and D. M. Knauss. Linear and branched architectures from the polymerization of lactide with glycidol. *Macromolecules* 40, 2327–2334 (2007).

115. G. Ahmetli, H. Deveci, U. Soydal, S. P. Gurler and A. Altun. Epoxy resin/polymer blends: Improvement of thermal and mechanical properties. *J. Appl. Polym. Sci.* 125, 38–45 (2012).

116. S. Ma, X. Liu, Y. Jiang, Z. Tang, C. Zhang and J. Zhu. Bio-based epoxy resin from itaconic acid and its thermosets cured with anhydride and comonomers. *Green Chem.* 15, 245–254 (2013).

117. A. Ben-Bassat and D. J. Lowe. A method for producing p-hydroxystyrene and other multifunctional aromatic compounds using two-phase extractive fermentation. WO 2004092392, assigned to Du Pont (2004).

118. S. L. Haynie, A. Ben-Bassat, D. J. Lowe and L. L. Huang. A method for preparing p-hydroxystyrene by biocatalytic decarboxylation of p-hydroxycinnamic acid in a biphasic reaction medium. WO 2004092344, assigned to Du Pont (2004).

119. K. Kunitsky, M. C. Shah, S. W. Shuey, B. M. Trost and M. E. Wagman. Method for preparing hydroxystyrenes and acetylated derivatives thereof by decarboxylation of phenolic compounds. US patent 7,586,013 assigned to Du Pont de Nemours (2009).

120. K. J. Kunitsky, M. T. Sheehan, J. R. Sounik and M. E. Wagman. Methods for preparing polymers from phenolic materials and compositions relating thereto. US Patent 7,728,082, assigned to Du Pont Electronic Polymers (2010).

121. K. Kunitsky, M. C. Shah, S. W. Shuey and M. E. Wagman. Method for preparing glycidyloxystyrene monomers from hydroxycinnamates and polymers thereof. US Patent 7,468,415, assigned to Du Pont de Nemours (2008).

122. A. Ben-Bassat, S. Breinig, G. A. Crum, L. Huang, A. L. B. Altenbaugh, N. Rizzo, R. J. Trotman, T. Vannelli, F. S. Sariaslani and S. L. Haynie. Preparation of 4-vinylphenol using pHCA decarboxylase in a two-solvent medium. *Org. Process Res. Dev*. 11, 278–285 (2007).

123. W. W. Qi, T. Vannelli, S. Breinig, A. Ben-Bassat, A. A. Gatenby, S. L. Haynie and F. S. Sariaslani. Functional expression of prokaryotic and eukaryotic genes in *Escherichia coli* for conversion of glucose to p-hydroxystyrene. *Metab. Eng.* 9, 268–275 (2007).

124. T. Vannelli, Z. Xue, S. Breinig, W. W. Qi and F. S. Sariaslani. Functional expression in Escherichia coli of the tyrosine-inducible tyrosine ammonia-lyase enzyme from yeast Trichosporon cutaneum for production of p-hydroxycinnammic acid. *Enzyme Microb. Techn.* 41, 413–422 (2007).

125. M. A. R. Meier, J. O. Metzger, U. S. Schubert. Plant oil renewable resources as green alternatives in polymer science. *U. S. Chem. Soc. Rev*. 36, 1788–1802 (2007).

126. J. O. Metzger. Fats and oils as renewable feedstock for chemistry. *Eur. J. Lipid Sci. Technol.* 111, 865–876 (2009).

127. F. S. Guener, Y. Yagci and A. T. Erciyes. Polymers from triglyceride oils. *Prog. Polym. Sci.* 31, 633–670 (2006).

128. R. K. Robinson. *Handbook of Citrus By-Products and Processing Technology*, 2nd edn, R. J. Braddock (Ed.), Florida Sci. Source, LLC, Boca-Raton, FL (2000).

129. K. Xu, M. Chen, K. Zhang and J. Hu. Synthesis and characterization of novel epoxy resin bearing naphthyl and limonene moieties, and its cured polymer. *Polymer* 45, 1133–1140 (2004).

130. R. F. Sellers. Epoxy resins from polyhydric phenol-terpene addition products. US Patent 3,378,525, assigned to Union Carbide (1968).

131. D. R. Fenn, S. Coca and J. O'Dwyer. Epoxy functional polymers comprising the reaction product of terpene and an epoxy functional monomer and coatings comprising terpene-modified resins. US Patent 7,776,960, assigned to PPG Industries Ohio (2008).

132. A. J. D. Silvestre and A. Gandini. Rosin: Major sources, properties and applications. In: *Monomers, Polymers and Composites from Renewable Resources*, M. N. Belgacem and A. Gandini, (Eds.), Elsevier, Amsterdam (2008).

133. H. Wang, X. Liu, B. Liu, J. Zhang and M. Xian. Synthesis of rosin-based flexible anhydride-type curing agents and properties of the cured epoxy. *Polym. Int.* 58, 1435–1441 (2009).

134. X. Liu, J. Zhu and Y. Jiang. Full-bio-based epoxy resin composition and condensate. China Patent CN 102206324, assigned to Ningbo Inst. Mat. Tech. and Eng., China (2011).

135. H. Wang, B. Liu, X. Liu, J. Zhang and M. Xian, Synthesis of biobased epoxy and curing agents using rosin and the study of cure reactions. *Green Chem*. 10, 1190–1196 (2008).

136. X. Liu, W. Xin and J. Zhang. Rosin-based acid anhydrides as alternatives to petrochemical curing agents. *Green Chem*. 11, 1018–1025 (2009).

137. A. M. Atta, S. M. El-Saeed and R. K. Farag. New vinyl ester resins based on rosin for coating applications. *React. Funct. Polym.* 66, 1596–1606 (2006).

138. A. M. Atta, R. Mansour, M. I. Abdou and A. M. Sayed. Epoxy resins from rosin acids: Synthesis and characterization. *Polym. Adv. Technol*. 15, 514–522 (2004).

139. I. Bicu and F. Mustata. Polymers from a levopimaric acid–acrylonitrile diels–alder adduct: Synthesis and characterization. *J. Polym. Sci., Part A Polym. Chem*. 43, 6308–6322 (2005).

140. A. M. Atta, R. Mansour, M. I. Abdou and A. M. Sayed. Synthesis and characterization of tetra-functional epoxy resins from rosin. *J. Polym. Res*. 12, 127–138 (2005).

141. J.-P. Pascault and R. J. J. Williams. Thermosetting polymers. In: *Handbook of Polymers, Synthesis, Characterisation and Processing*, E. Salvidar-Guerra and E. J. Vivaldo-Lima (Eds.), Wiley, New York (2013).

142. H. Warth, R. Mühlhaupt, B. Hoffmann and S. Lawson. Polyester networks based upon epoxidized and maleinated natural oils. *Angew. Makromol. Chem*. 249, 79–92 (1997).

143. T. Takahashi, K.-I. Hirayama, N. Teramoto and M. Shibata. Biocomposites composed of epoxidized soybean oil cured with terpene-based acid anhydride and cellulose fibers. *J. Appl. Polym. Sci.* 108, 1596–1602 (2008).

144. M. Stemmelen, F. Pessel, V. Lapinte, S. Caillol, J. P. Habas and J. J. Robin. A fully biobased epoxy resin from vegetable oils: From the synthesis of the precursors by thiol-ene reaction to the study of the final material. *J. Polym. Sci., Part A Polym. Chem.* 49, 2434–2444, (2011).

145. C. A. Koch. Pressure sensitive adhesives made from renewable resources and related methods. WO 2008144703, assigned to Avery Dennison Corp. (2008).

146. R. Vendamme and W. Eevers. Sweet solution for sticky problems: Chemoreological design of self-adhesive gel materials derived from lipid biofeedstocks and adhesion tailoring via incorporation of isosorbide. *Macromolecules*. 46, 3395–3405 (2013).

147. J. Roesch and R. Muelhaupt. Polymers from renewable resources: Polyester resins and blends based upon anhydride-cured epoxidized soybean oil. *Polym. Bull.* 31, 679–685 (1993).

148. A. P. Gupta, S. Ahmad and A. Dev. Development of novel bio-based soybean oil epoxy resins as a function of hardener stoichiometry. *Polym-Plast. Technol.* 49, 657–661 (2010).

149. F. I. Altuna, L. H. Esposito, R. A. Ruseckaite and P. M. Stefani. Thermal and mechanical properties of anhydride-cured epoxy resins with different contents of biobased epoxidized soybean oil. *J. Appl. Polym. Sci.* 120, 789–798 (2011).

150. N. Boquillon and C. Fringant. Polymer networks derived from curing of epoxidised linseed oil: Influence of different catalysts and anhydride hardeners. *Polymer* 41, 8603–8613 (2000).

151. S. G. Tan and W. S. Chow. Thermal properties of anhydride-cured bio-based epoxy blends. *J. Therm. Anal. Calorim.* 101, 1051–1058 (2010).

152. X. Pan, P. Sengupta and D. C. Webster. High biobased content epoxy–anhydride thermosets from epoxidized sucrose esters of fatty acids. *Biomacromol.* 12, 2416–2420 (2011).

153. P. Czub. Application of modified natural oils as reactive diluents for epoxy resins. *Macromol. Symp.* 242, 60–64 (2006).

154. H. Miyagawa, M. Misra, L. T. Drzal and A. K. Mohanty. Novel biobased nanocomposites from functionalized vegetable oil and organically-modified layered silicate clay. *Polymer* 46, 445–453 (2005).

155. T. Shimasaki, S. Yoshihara and M. Shibata. Preparation and properties of biocomposites composed of sorbitol-based epoxy resin, pyrogallol–vanillin calixarene, and wood flour. *Polym. Composite* 33, 1840–1847 (2012).

156. M. Harwood, B. Danielewska-Nikiel, J. F. Borzelleca, G. W. Flamm, G. M. Williams and T. C. Lines. A critical review of the data related to the safety of quercetin and lack of evidence of in vivo toxicity, including lack of genotoxic/carcinogenic properties. *Food Chem. Toxicol.* 45, 2179–2205 (2007).

157. X. Feng, A. East, W. Hammond, Z. Ophir, Y. Zhang and M. Jaffe. Thermal analysis characterization of isosorbide-containing thermosets. *J. Therm. Anal. Calorim.* 109, 1267–1275 (2012).

158. M. Chrysanthos, J. Galy and J.-P. Pascault. Influence of the bio-based epoxy prepolymer structure on network properties. *Macromol. Mater. Eng.* 298, 1209–1219, (2013).

159. H. Kimura, Y. Murata, A. Matsumoto, K. Hasegawa, K. Ohtsuka and A. Fukuda. New thermosetting resin from terpenediphenol-based benzoxazine and epoxy resin. *J. Appl. Polym. Sci.* 74, 2266–2273 (1999).

160. Y. Okabe and H. Kagawa. Epoxy resin composition. US Patent 20100255315, assigned to Okabe and Kagawa (2010).

161. J. Cheng, J. Chen and W. T. Yang. Synthesis and characterization of novel multifunctional epoxy resin. *Chinese Chem. Lett.* 18, 469–472 (2007).

162. X. Liu and J. Zhang. High-performance biobased epoxy derived from rosin. *Polym. Int.* 59, 607–609 (2010).

163. M. Ochi, T. Shiba, H. Takeuchi, M. Yoshizumi and M. Shimbo, Effect of the introduction of methoxy branches on low-temperature relaxations and fracture toughness of epoxide resins. *Polymer* 30, 1079–1084 (1989).

164. M. Ochi, M. Shimbo, M. Saga and N. Takashima. Mechanical and dielectric relaxations of epoxide resins containing spiro-ring structure. *J. Polym. Sci., Part B Polym. Phys.* 24, 2185–2195 (1986).

165. R. J. J. Williams, B. A. Rozenberg and J.-P. Pascaullt. Reaction-induced phase separation in modified thermosetting polymers. *Adv. Polym. Sci.* 128, 95–156 (1997).

166. M. C. S. Perera. Reaction of aromatic amines with epoxidized natural rubber latex. *J. Appl. Polym. Sci.* 39, 749–758 (1990).

167. S.-G. Hong and C.-K. Chan. The curing behaviors of the epoxy/dicyanamide system modified with epoxidized natural rubber. *Thermochim. Acta* 417, 99–106 (2004).

168. C. Nakason, A. Kaesaman, W. Sainamsai and S. Kiatkamjonwong. Rheological behavior of reactive blending of epoxidized natural rubber with cassava starch and epoxidized natural rubber with natural rubber and cassava starch. *J. Appl. Polym. Sci.* 91, 1752–1762 (2004).

169. K. Huang, Y. Zhang, M. Li, J. Lian, X. Yang and J. Xia. Preparation of a light color cardanol-based curing agent and epoxy resin composite: Cure-induced phase separation and its effect on properties. *Prog. Org. Coat.* 74, 240–247 (2012).

170. C. W. Reeb, L. A. Lucia and R. A. Venditti. Novel screening technique: Integrated combinatorial green chemistry and life cycle analysis (CGC-LCA). *Bio. Res.* 8, 1513–1524 (2013).

171. M. L. Zhang, Y. T. Fan, Y. Xing, C.-M. Pan, G.-S. Zhang and J. J. Lay. Enhanced biohydrogen production from cornstalk wastes with acidification pretreatment by mixed anaerobic cultures. *Biomass Bioenerg.* 31, 250–254 (2007).

172. Q. Song, F. Wang, J. Cai, Y. Wang, J. Zhang, W. Yu and J. Xu. Lignin depolymerization (LDP) in alcohol over nickel-based catalysts via a fragmentation–hydrogenolysis process. *Energ. Environ. Sci.* 6, 994–1007 (2013).

173. C. O. Tuck, E. Perez, L. T. Horvath, R. A. Sheldon and M. Poliakoff. Valorization of biomass: Deriving more value from waste. *Science* 337, 695–699 (2012).

19 Pressure-Sensitive Adhesives

István Benedek

CONTENTS

19.1 INTRODUCTION

This chapter presents the theoretical and practical state of the art of pressure-sensitive adhesives (PSAs) and products. The buildup, manufacturing, testing, and applications of PSAs and products are discussed. The basic principles of this technical domain are always given, and this chapter summarizes our present understanding of the construction and functioning of PSAs and products.

Pressure-sensitive tapes have been used for 170 years. Pressure-sensitive labels came on the market after 90 years. About 10 years later, pressure-sensitive protective films were manufactured also. The manufacture and application field of pressure-sensitive products has developed during the last three decades from a largely empirical body of accumulated practical knowledge to an increasingly sophisticated science, utilizing the most advanced techniques of physics, chemistry, and engineering. Pressure-sensitive labels, tapes, protective films, seals, business forms, and so on are used in medical and pharmaceutical applications, electronic circuits, assembly of machine parts, as well as in product promotion, coding, and packaging. In the next sections, the principle of functioning and a short history of pressure-sensitive products (PSPs) as permanently tacky, self-adhering materials, with or without adhesive, are given.

19.1.1 HISTORY

As discussed in [1], PSAs have been used since the last century for medical tapes and dressings. Natural adhesives mixed with natural resins, waxes, and fillers were applied as the first PSP in the form of medical plasters [2]. In 1845, Horace H. Day prepared and patented a plaster from a mixture of natural rubber and tackifier resin coated on a cloth [3]. According to [4,5], the first tape patented by Paul Beiersdorf was (zinc oxide–rubber)-based plaster. At the end of the last century, masking tapes and cellophane tapes were the early nonmedical PSPs. For such products, natural rubber has been preferred as the raw material. In this period, PSAs were used for plasters, labels, and tapes [6]. Industrial tapes were introduced on the market in the 1920s and 1930s, and self-adhesive labels in 1935–1936 [7,8]. Stanton Avery developed Kum-Kleen labels in the 1930s [9]. In 1955, the firm Sassions, of York, was licensed by Avery to produce PSA labels in Europe [10].

In view of the wide utility and consumer acceptance of these products, a high level of basic research and product development has evolved over the last years and is continuing. Pressure sensitivity, the main characteristic of these various products, possesses a common scientific basis: macromolecular chemistry and physics. However, the ways and means to achieve it differ, and include also the technology of adhesives and plastics.

PSAs represent only a part of the science and technology of pressure-sensitives. The whole domain includes them as well as other engineering solutions. Advances in macromolecular science and engineering have led to new PSPs and applications by using multiple ways to design them. Practice confirmed that the attempt to integrate the various PSPs—which may differ in their manufacturing technology, but do not differ in their use—in a sole category of self-adhesives was correct and necessary [11,12]. The author's major aim here is to bridge the gap between theory and practice, between the engineering of plastics and the technology of adhesives and their conversion, and to integrate the practical aspects with the engineering fundamentals.

Reference [11] was the first comprehensive book on bridging the gap between the fundamental concepts of pressure sensitivity and its application, and the first attempt to examine comparatively the various products with or without adhesive that show pressure-sensitive behavior during their application.

There is a need for a guide to the entire field of PSPs, with or without adhesives, discussing all the steps of engineering (of paper, plastics, adhesives, and other materials) required for the manufacture of PSPs. There are only a few books about PSAs and even fewer also discussing the competitive technologies.

The scope of this chapter is to highlight their end-use parameters, which must be generally valid. Although the composite structure and the mechanism of functioning of adhesive- and/or plastic-based PSP products can vary their competition in application practice is real. Therefore, a strong emphasis is placed on establishing a clear understanding of the complex interaction between the fundamentals of pressure-sensitive adhesion and the manufacture and application technology of self-adhesive products.

19.1.2 DEFINITIONS

PSPs, as PSA-coated webs, have been defined through the special nature of this adhesive, although their definition is not sufficiently clear [1]. The German technical term *Haftkleber* (i.e., adhesives that adhere) assumes the possibility of differentiating between adhesion and the buildup of an adhesive bond. In English, the term *pressure-sensitive adhesives* (i.e., adhesives that bond by pressure) admits pressure as a *conditio sine qua non* for their function. In reality, as known from loop-tack measurements and touch-blow labeling, almost no pressure is required for label application, but high pressures are needed for protective films in coil coating. The French term *autocollants* does not define the application conditions; it refers to the bonding behavior only. The common characteristic of these products is assured by a special viscoelastic behavior, manifested as permanent cold flow, where the chemical nature of the adhesive plays only a secondary role [13]. The development of PSPs without a coated PSA layer (in the classic sense, known from the conversion industry) makes the definition of this product group more difficult. The definition, construction, and physical and chemical basis of PSPs are described in detail in [1,14].

Adhesive-free PSPs were developed some decades ago. In this case, *adhesive-free* means that the self-adhesive component is not coated on the product surface. It is included in the carrier; that is, the carrier *per se* is pressure-sensitive. In some cases, an adhesive-free composition is used and adhesivity is provided by physical treatment of the carrier surface and/or application conditions (temperature, pressure). According to the definition given in [15]. PSAs are adhesives "which in dry form are aggressively and permanently tacky at room temperature … and adhere without the need of more than finger or hand pressure, require no activation by water, solvent or heat."

It is certain that most PSPs do not fulfill these requirements. However, they manifest self-adhesivity and (under well-defined conditions) can be applied similarly to a PSA-coated classic PSP, that is, similarly to an adhesive acting via viscous flow and debonding similarly to a viscoelastic compound. Such behavior is achieved by a complex buildup and reciprocal interaction of the product components, and in some cases by special application or removal conditions (Table 19.1). Obviously, a physically treated hot-laminating plastomer film applied under pressure or a warm-laminating film based on a partially viscoelastic olefin copolymer and also applied under pressure cannot have the same chemical basis as a PSA-coated product or a plastic carrier material that includes a PSA. As mentioned, their application conditions are quite different also.

However, all these products work as viscoelastic bonding elements and are used in the application domains of classic PSA-coated products. Therefore, in practice they can be considered as PSPs. Plastic processing specialists possess the know-how of the manufacture of plastic-based PSPs. The specialists in conversion/coating are skilled in the design and testing of PSPs. Therefore, from an economic point of view, both domains belong together.

A comparison of PSAs with thermoplasts and rubber, respectively, with other adhesives is given in [16].

19.2 BUILDUP AND CLASSES OF PSPs

A treatise about pressure-sensitives has to include the buildup (construction) and classification (classes) of PSPs, the physical basis of PSPs (rheology, mechanical properties, and other physical characteristics), the chemical basis of PSPs (the macromolecular basis and chemical composition of

TABLE 19.1

Application and Removal Conditions for Main PSPs

Pressure-Sensitive Product	Application Conditions	Removal Conditions
Label	Impact or nonimpact labeling	Mechanical, chemical, or thermal removal
	High-speed application	Repeated removal possible
	Room-temperature or low-temperature application	Partial removal possible
	Very low application pressure	
	Repeated application possible	
	Manual or automatic application	
	Conversion during application	
	Conversion after application	
Tape	Impact lamination	Mechanical removal
	Room-temperature, low- or high-temperature application	
	Low- or high-pressure application	
	High-stress application	
	Low- or high-speed application	
	Conversion during application	
	Hand-made or automatic application	
Protective film	Impact lamination	Mechanical removal
	High-temperature and high-pressure lamination	High stresses during removal
		Removal is more important than lamination
		High-temperature, high-pressure processability after lamination

carrier material, the adhesive, and the abhesive), adhesive properties of PSPs, manufacture of PSPs, conversion and end-use properties, and test methods. There are only a few books about PSAs, and most of them have focused on the adhesive only, which is quite normal. (According to the chronological development of this industry, we always spoke about PSA-coated products).

Plastic carrier materials were introduced two decades ago, and plastics displaying self-adhesive properties were only developed later.

Now, the situation is more complex. Actually, there is a broader range of products having pressure-sensitive properties, and some of them are not coated with PSAs; they have an inherent pressure sensitivity. Other products are adhesive-free. From the current point of view, PSPs include PSA-coated products, but the two product groups are not identical. PSPs are a bigger category than PSAs. Actually, there are some domains in PSP application where a technical solution may be given by engineering plastics or adhesives.

During its end use, no one asks about the construction of the product—whether it is pressure-sensitive and with or without adhesive.

19.2.1 BUILDUP OF PSPs

Pressure sensitivity is the result of special chemical and macromolecular buildup. Macroscopic product buildup may contribute to pressure sensitivity also. The range of potential product components useful for pressure sensitivity is very large. The number of technologies allowing the manufacture of PSPs is also large.

Pressure sensitivity—the permanence of adhesivity—is the result of special flow characteristics (rheology). PSPs have to display cold flow. Therefore, they have to possess a contact surface able to flow. Generally, the active, adhesive surface of PSPs belongs to a solid-state carrier material. Usually, this carrier material and the coating components (adhesive, abhesive, and printing components, etc.) build up the PSP, but coating-free constructions are possible too.

How, where, and why adhesive-free PSPs (plastics) can compete with adhesive-coated ones was described in previous works by this author [17–19]. Carrier components, coating components, product buildup (monoweb, multiweb, coated, uncoated, etc.), according to different product classes (labels, tapes, forms, etc.) were discussed [20]. Definition and construction of pressure-sensitive laminates are described in [21].

The classic manufacture of the finished product utilizes the buildup of a pressure-sensitive product from its components, that is, carrier material(s) and PSA. In practice, other solid-state product components, such as release liner, cover film, and so on, and liquid-state components—for example, primers, antistatic agents, and so forth—can be built in, in one or more layers, giving a laminate with a complex structure. Principally, monoweb, laminate, and multiweb constructions are known.

The mono- or multiweb construction of a PSP also depends on its adhesivity. Principally, PSPs coated with a contact adhesive (e.g., envelopes) with low tack that adhere only to themselves can also be considered as monowebs [22]. Only a few monoweb labels have been developed. Because of the balanced adhesive performances of labels (see Section 19.2.2.1), and their low-pressure, high-speed application technology (automatic labeling), it is very difficult to manufacture labels without a separate release liner, that is, labels with a monoweb structure. However, special monoweb labels (roll-labels) have also been developed. Linerless labels are supplied as a continuous tape-like monoweb material.

PSP manufacture may require the use of other supplemental nonadhesive-coated layers also; for example, primer, lacquer, printing ink, release agent, and so on. Thus, actually, the main part of so-called *monoweb PSPs* are multiwebs.

The pressure-sensitive laminate manufactured to protect the adhesive-coated surface of a tape or a label is a temporary construction. In such products, the release liner may protect the adhesive or other components incorporated in the adhesive layer. Another role of the separate release liner is that of a continuum passing through the coating, converting, and labeling machines carrying the discontinuous label. Decals or labels applied manually do not need this function.

Principally, PSPs with a soft multilayer and PSPs with a multilayer carrier are known. Postfinishing of such a web during its conversion (see Section 19.5.2.3) can lead to its coating with other supplemental soft layers; for example, printing ink, lacquer, and so on. Labels are variants of such constructions. For certain products, the carrier material is a laminate. In other cases, the product itself has a sophisticated multilayer structure, including various solid-state components.

Products with special adhesive buildup (filled or discrete adhesive) are also manufactured. Coating of the adhesive as a noncontinuous layer with discrete areas allows control of the bonding–debonding properties. Adhesive geometry (i.e., the form of the adhesive layer and its thickness) strongly influences the peel resistance. There are special coatings with an adhesive layer with discontinuous character, mostly to avoid the buildup of high peel resistance. On the other hand, the structure of the continuous adhesive layer also influences the peel resistance. The coating device affects the shape of the adhesive layer (see Section 19.5.2.1). For instance, cross-linked adhesives coated with a gravure cylinder give peel values that differ from those obtained with a Meyer bar, for the same weight of coating.

As special products, adhesiveless, carrierless, and linerless PSPs are also known.

PSPs may have a simple or sophisticated construction, depending on their end use. They can be classified according to their fundamental buildup; however, the main classes of PSPs (labels, tapes, and protective webs) cover a wide range of application fields, thus, they can have various specific constructions. On the other hand, more or less expensive products can be used in the same application field [22]. Monoweb constructions (uncoated and coated) and multiweb constructions have been described, uncoated and coated monoweb constructions have been investigated comparatively with coated ones, and various coated monoweb constructions have been examined comparatively.

In this section, only the principles of buildup of the main PSPs are presented. Their classification and a detailed discussion of their applications is given in [13]. Generally, PSPs are web- or sheet-like constructions that exhibit self-adhesion. In principle, such products include a component(s) that ensures the required mechanical properties and a component(s) that provides adhesivity. It should be emphasized that the development of the control of debonding resistance, that is, advances in removable products, were decisive for the development of adhesiveless self-adhesive products. Generally, removability requires peel reduction. Peel reduction is correlated to tack reduction. On the other hand, the application of PSPs needs minimum tack. Thus, certain removable low-peel and low-tack products must undergo a forced lamination (at RT or at elevated temperature) during their application with a higher application pressure; that is, they need the same application technology as adhesive-free plastic films. This means that for certain end uses, self-adhesive products without PSAs may also be suggested.

The discussion of the buildup of pressure-sensitive materials must include their composite structure also.

Principally, a composite includes a matrix and a reinforcing element. The matrix is the soft (plastic/elastic) *continuum* that contains a rigid (elastic/plastic) *discontinuum*. The flow of such a structure is ensured by the matrix, and its deformation stability is the result of the interaction of the reinforcing element with the matrix. Bonding needs flow; debonding needs stability against deformation. Such deformation may be plastic or elastic, permanent or temporary.

The best-known pressure-sensitive raw material (the model compound) is natural rubber. Its buildup illustrates that an adequate balance of the plasticity and elasticity of a macromolecular compound can be achieved by a network structure. Such network contains rigid, cross-linked parts and flexible, linear parts, where the cross-linked parts may suffer deformation and the non-cross-linked parts may undergo chain entanglement. According to a general concept, such a network is a composite.

Recent advances in PSPs have led to composites on the macromolecular level, on the macroscopic level, and on the product-construction level. The influence of the composite structure on PSAs was discussed in [23,24]. Flexural resistance and the multilayer structure of PSPs strongly affect their conversion properties.

Multilayer composites and their influence on peel resistance were discussed in detail in [25].

19.2.2 Classes of PSPs

A general description is given in [26] to understand PSPs, and to integrate them within a whole, with products having quite different buildup, or manufacture, but showing the same pressure sensitivity. This is now followed by a detailed discussion of the main product classes.

The classification of PSPs, according to different (functional, raw material-related, manufacture-related, etc.) criteria must include a detailed description of the main PSPs, their buildup, the principle of their functioning, and special characteristics and classes of the most important product groups (e.g., labels, tapes, protective films, etc.). Pressure-sensitive product classes can differ according to the PSAs' physical state, bond character, environmental behavior, and use.

19.2.2.1 Labels

Labels are self-adhesive, laminated carrier materials. Generally, they possess a continuous web-like character during their manufacture only. Labels produced as separate items are also known. In quite a different manner from tapes, labels are used as discontinuous items, having a well-defined geometry. Because of their adhesive and surface characteristics, the self-adhesive layer of labels must be protected with a supplemental solid-state abhesive material (release liner). The first release material was wax, as used by Stanton Avery [27]. Silicone release coatings have been in the market since the mid-1950s [28,29]. Labels conserve their laminate character until their application.

Because of their discontinuous character and limited contact surface (and time), high application speed, and low application pressure, labels have to exhibit well-balanced adhesive characteristics. Therefore, most of them are manufactured in the classic way, that is, coating a nonadhesive carrier material (face stock) with a pressure-sensitive adhesive.

19.2.2.2 Tapes

Tapes are continuous web-like PSPs applied in continuous form. The PSA tape can be defined as a pressure-sensitive, adhesive-coated substrate in roll form, wound on a core, used (mainly) as a continuous web. Generally, their role is to ensure the bonding, fastening, and/or assembling of adherend components due to their mechanical characteristics.

Generally, tapes are produced by coating a nonadhesive web with a pressure-sensitive adhesive, but tapes having a self-adhesive carrier material are also known. The pressure-sensitive layer is protected by the back of the carrier material. The permanent web-like character of tapes allows the use of higher forces in their application. The lower conversion degree of common tapes permits their design mainly for adhesive properties. In this case, the adhesive performances must not be balanced. Therefore, theoretically, tapes may be formulated as PSA-free products also.

19.2.2.3 Protective Films

Protective films are removable, self-adhesive webs, based on a carrier material that possesses built-in/on self-adhesive properties. The role of protective films is to protect a product adhering to it, by covering its surface with a mechanically resistant supplemental layer. This is a time-limited function; that is, the bond should be removable, allowing the separation of the protective sheet from the protected surface. The protective films are packaging materials—not in a legal sense, more a functional one. (This is different from classic packaging materials, where functionality concerns protection of the product during transport and storage, and esthetic, publicity-related design characteristics are determinant; protective films are technological components of a product, attached in many cases to the raw product, and going through the whole manufacturing process until the product is finished—that is, undergoing the working steps of manufacture.) Such products are applied by lamination/delamination of large surfaces. Therefore, the resultant bonding/debonding forces are much higher than for labels or tapes. On the other hand, protective films have to be removable. Therefore, the instantaneous adhesive performances of protective films play a secondary role only. It is evident that in this case, adhesive-free constructions may be equivalent to PSA-coated products. Table 19.2 summarizes the main protective webs and their buildup.

TABLE 19.2
Main Protective Webs

Application Domain	PSP Buildup	
	Adhesiveless	**Adhesive Coated**
Coil coating	Adhesiveless hot-laminating film	Adhesive-coated film
Plastic plate protection	Adhesiveless warm-laminating film	Adhesive-coated film
Automotive storage and transport	Adhesiveless warm-laminating film	Adhesive-coated film
Deep drawing	Adhesiveless film	Adhesive-coated film
Carpet protection	—	Adhesive-coated film
Product security	—	Adhesive-coated film
Furniture protection	Adhesiveless warm-laminating film	Adhesive-coated film
Building protection	—	Adhesive-coated film

19.3 SCIENTIFIC BACKGROUND OF PRESSURE SENSITIVITY

Unfortunately, the chemistry and science of PSAs developed more slowly than the industry using their products. Because of the rapid growth of the number of PSA-coated products and their application fields, there have been only a few trials to systematize the technical results on a scientific basis. What kind of science should be used? The science of adhesives, or that of the carrier materials? Paper or plastics?

In the last decades, advances in contact physics and mechanics have allowed the correlation of the macroscopic aspects of adhesive bonding and debonding to the macromolecular basis of the viscoelastomers. The most important aspects of this progress are described in [30].

19.3.1 PHYSICAL BACKGROUND

As stated in Section 19.1.1, pressure sensitivity, the main characteristic of these various products, possesses a common scientific basis: macromolecular chemistry and physics.

The physics of pressure sensitivity covers various aspects of adhesion. The surface phenomena on a solid–liquid interface and the rheology of pressure-sensitives were discussed in [31]; diffusion and adhesion were described in [32]. Transition zones were evaluated in [33]. Viscoelastic behavior, as the main characteristic of this rheology in the course of bonding and debonding processes, was discussed in [34]. Viscoelastic properties and windows of PSAs were discussed in [35]. Probe tack was described in [36]. The molecular fundamentals of pressure-sensitive adhesion were discussed by Feldstein in [37,38]. Pressure-sensitive adhesion as a material property and as a process was evaluated in [39]; durability of viscoelastic adhesive joints was discussed in [40]. The significance of relaxation for adhesion was explained in [41].

The application of PSAs requires a thorough knowledge of basic rheological and viscoelastic phenomena. Adhesive and polymer scientists, however, are not very often employed as industrial managers or machine operators. Therefore, the need arises to investigate and summarize the most important features of PSA technology and to explain the phenomenon scientifically. Viscoelastic properties depend on material characteristics and time/temperature.

How to characterize the degree of viscoelasticity and pressure sensitivity? Although the viscous or elastic character of a product is reflected in its practical (e.g., mechanical) properties, there is a need to correlate the macroscopically measurable material characteristics with the intrinsic buildup of a macromolecular product, that is, to the polymer characteristics. (The tools for such correlation are given by the glass transition temperature T_g and modulus of the material.)

19.3.1.1 Glass Transition Temperature

The glass transition temperature characterizes a second-order transition of molecular mobility. Above this temperature, molecular mobility is enhanced and the macromolecular compound may suffer viscoelastic deformation. Above the T_g, coalescence, tackiness, and rubber-like elasticity appear. The position of the glass transition temperature is an index of the molecular mobility and pressure sensitivity. RT PSAs should possess a T_g between −15°C and +50°C. Adjustment of T_g by formulation is described in [23].

Debonding energy can be described as the product of two terms: the thermodynamic work of adhesion and a dimensionless function. The thermodynamic work of adhesion depends on the surface tension. The dimensionless function depends on the debonding speed and temperature, and reflects the viscoelastic properties of the adhesive.

Viscoelastic deformability alone is not sufficient for adequate pressure sensitivity. The level of such deformability must be controlled also. The value of the modulus characterizes the magnitude of the deformability.

19.3.1.2 Modulus

The modulus, as the ratio between stress and deformation, is a common characteristic of the mechanical behavior of solid-state materials. Because of the different kinds of forces applied during end use (e.g., flexural, tensile, compressive stresses, etc.), different modulus values are employed. The modulus of elasticity was proposed as a characteristic of the deformability of macromolecular compounds used as raw materials for PSPs. Its reciprocal value is known as the *Dahlquist criterion* [42]. This states that the modulus value (plateau) should be lower than 3.3×10^6 Pa. PSA elasticity requires a modulus value of the order of 10^5 Pa. Generally, polymers used as PSAs have a dynamic modulus of 10^4–10^5 Pa. The Dahlquist criterion has been proven to be a practical guide for various PSA applications also; for instance, the creep compliance of PSAs for medical and surgical applications should be about 1.2×10^{-5} Pa, preferably 1.3×10^{-5} Pa. The significance of the Dahlquist criterion regarding tack at a molecular level is that the elasticity modulus is a measure of the ratio between the values of cohesion energy and free volume of adhesive material.

Low modulus (pronounced softness) of the material is required to allow its conformation to the substrate surface during bonding. High modulus is needed for adhesion of excellent resistance. The values of the T_g and modulus define the so-called *application window* of a product.

The nature of macromolecular compounds affects the T_g and the modulus. Polar functional groups, comonomers, side-chain length, cross-linking, molecular weight, and molecular weight distribution influence the rheology of the adhesive. To fully characterize a polymer, the macrostructure contributions as well as the microstructure contributions must be known. The macrostructure can be characterized by weight average molecular weight (M_w), number average molecular weight (M_n), molecular weight distribution (MWD), and degree of branching. Bonding and debonding are energy-related phenomena linked to molecular motions that absorb energy. Relaxations are molecular motions that occur after stress. Such molecular motions are strongly dependent on molecular structure.

The influence of macromolecular characteristics (molecular weight, comonomer content, molecular structure, sequence length and distribution, and compatibility) on T_g and modulus was discussed in detail in [43]. Adjustment of modulus by formulation is described in [23].

19.3.1.3 Time–Temperature Superposition Principle

The viscosity and modulus, measured as macroscopic characteristics during processing or application, are time and temperature dependent. On the macromolecular scale, the time–temperature dependence of the modulus is the result of hindered molecular motion (relaxation) after stress application. Time and temperature have the same effect. The temperature, as an index of the energetic state, favors molecular motion; the increase of the time allowed for relaxation acts in the same direction. The position of the glass transition temperature enables a prediction of the influence of the temperature on the modulus. The frequency of the applied stress determines the magnitude of the viscoelastic response of the material. The elastic modulus, as used in the rheology of macromolecular products, is not an absolute material characteristic; it depends on time and temperature. Because of their mutual interdependence, equations have been developed that allow correlation of the effect of time to the effect of temperature on the modulus. This is the time–temperature superposition principle. The applicability of such correlations strongly depends on the nature of the macromolecular compounds and on application temperature.

The viscoelastic behavior of amorphous polymers is characteristic of time–temperature dependence, dependence on the ratio between recoverable energy during a given deformation (i.e., the elastic part), and energy losses on the experimental conditions characterized by stress rate and temperature; that is to say, the validity of the time–temperature superposition principle. This principle states that viscoelastic properties at different temperatures can be superposed by a shift of the isotherm data along the logarithmic time–frequency scale. The shift factors are related to the temperature-dependent activation energy of flow, accessible via the slopes of the isochrone and isotherm data curves of viscoelastic functions.

19.3.1.4 Dynamic Modulus

The dynamic modulus is a frequency-dependent material function composed of the storage modulus (G') and the loss modulus (G''). Defining G' as the ratio of the in-phase stress to the in-phase strain (i.e., as the storage modulus) and G'' as the ratio of the off-phase stress to the off-phase strain (i.e., as the loss modulus), the following expressions are obtained for G' and G'':

$$G' = \tau_o \cos \delta / \gamma_o \tag{19.1}$$

$$G'' = \tau_o \sin \delta / \gamma_o \tag{19.2}$$

The phase angle δ, which reflects the time lag between the applied strain and stress, is defined by a ratio called the *dissipation factor* (tan δ):

$$\tan \delta = G'' / G' \tag{19.3}$$

where tan δ is a damping term and is a measure of energy dissipated as heat.

An index of the energy absorption is the value of the loss modulus. Cold flow should be at its maximum during bonding and at its minimum during debonding (except for removable formulations). This means that high loss modulus (peak) at the practical frequency of bonding ensures adequate bonding. On the other hand, at higher debonding frequency, the elastic behavior of the adhesive (storage modulus) is required. Formulation windows with boxes shifted to the left and upward (i.e., for lower frequencies and higher plateau modulus) are more difficult.

Advances in contact physics and mechanics of viscoelastic materials have allowed the correlation of macromolecular characteristics to the mechanical properties of the polymers and bonding/debonding mechanism. The basis for viscoelastic adhesion to solid surfaces has been set out in [44]. In the last decade, appreciable progress has been made in the quantitative description of the micromechanics of PSA debonding [45,46]. These works consider nucleation of cavities within pressure-sensitive polymers and the extension of fibrils as major factors leading to the dissipation of applied energy. The debonding of PSAs from the surface of a rigid substrate often occurs with the formation of a fibrillar structure bridging the surface and the adhesive. This fibrillar structure is responsible for the large (of the order of 1 kJ/m²) adhesion energy despite the very low level of applied stress. For a material obeying the simple kinetic theory of rubber elasticity, the critical pressure at which the cavity will grow without limit has been shown to be proportional to the initial elastic modulus of the material. On the other hand, the Dahlquist criterion correlates pressure sensitivity to the modulus values. Thus, taking into account the correlation of macromolecular characteristics (e.g., free volume and cohesion) to pressure sensitivity, a theoretical basis was given for Dahlquist's empirical criterion of a critical modulus value and for the industrial practice of pressure-sensitive bonding/debonding (measured by tack and peel resistance). Feldstein [37] discusses pressure-sensitive adhesion not only in its traditional aspect, as a specific property of viscoelastic materials under applied external compressive and tensile forces, but also as a continuous three-stage process that includes adhesive bond formation, adhesive relaxation, and debonding. The specific quality that unifies all three stages into a single process is the relaxation of adhesive material, accompanying the behaviors of the adhesive in the courses of bond formation under pressure and bond failure under tensile stress.

The rheology of uncoated adhesives—for example, PSA solutions and dispersions—was discussed in detail in [44]. (The wet-out of dispersions related to surface tension, contact angle, and dynamic wet-out was described.)

The rheology of pressure-sensitive laminates was discussed in [45]. The nature and elasticity of face stock influence the adhesive and conversion properties, as well as the coating technology (see also Section 19.4.2.2).

19.3.2 CHEMICAL BASIS

The chemical aspect of PSPs includes the macromolecular basis and chemical composition of (nonadhesive/adhesive) carrier material, the macromolecular basis and chemical composition of the adhesive (elastomeric, viscoelastic, and viscous components, plastomers, etc.), and the macromolecular basis and chemical composition of the (coated/built-in) abhesive.

Special raw materials allow the design, formulation, and manufacture of PSAs and PSPs. Their chemical basis was described in detail in our previous works [46–49]. The fundamentals of pressure sensitivity, based on this chemical and macromolecular basis, were discussed in [30].

The development of different product classes has been conditioned by the development of raw materials and coating technology and application technology. Pressure-sensitive labels, stickers, and other products have seen considerable development in the past years, with new materials and combinations as well as new processing technologies. At the end of the 1920s, acrylics were synthesized. They possess adequate die-cutting properties and do not manifest migration (they contain no low molecular products). Their introduction made possible the development of label manufacture. Acrylics display aging resistance and plasticizer resistance. These performances allowed their use on transparent carrier materials. Tapes and protective films on poly(vinyl chloride) and polyethylene carriers have been produced. Envelopes, wall coverings, forms, and so on have been manufactured. The nonirritant behavior of acrylics allows their use for medical tapes. Fixing, transfer, carpet, or electrical insulating tapes are made with acrylics also.

The development of thermoplastic elastomers has allowed the use of a less expensive coating equipment via hot-melts. Styrene–olefin block copolymers have been used since 1965 [50]. Hydrocarbon-based resins were introduced as tackifiers in the mid-1930s [51]. Poly(vinyl acetate) emulsions have been produced since 1940 [52]. Vinylacetate copolymer dispersions, together with water-based acrylics, allowed the development of water-based technology. In 1960, ethylene–vinylacetate copolymers were introduced on the market. The first class of raw materials appeared that were versatile for use as hot-melts, as nonadhesive as well as adhesive carriers. Due to this development, new ways have been opened up to manufacture products having pressure-sensitive properties without coatings and without PSAs. Some years previously, the plasticizing technology of poly(vinyl chloride) proved the manufacture of plastomers, elastomers, and adhesives on the basis of the same raw material. At the same time, advances have been made in the field of plastomers and elastomers. The development of cling films, tackified carrier films, sheet-like laminating adhesives, and so on confirmed the possibility of the manufacture of self-adhesive products without PSA. Non-PSA-related pressure-sensitive product technology gained an important market segment, especially in the field of protective films and tapes. On the other hand, curable prepolymer technology advanced from an experimental to an industrial domain (see Section 19.5.1.2). In this situation, pressure-sensitive product technology became a complex field of industrial procedures, from adhesive manufacture (synthesis and formulation) and coating to plastomer/elastomer manufacture and processing (see also Section 19.5). Some of the most important features of PSAs and their chemistry and technology have been described in [53]. The goal of this chapter is to discuss PSP technology as a whole, linking adhesive and adhesiveless manufacture based on their common background: macromolecular science.

19.3.2.1 Basic Products for PSAs

In their first stage of development, PSAs were formulated on the basis of natural macromolecular products. Natural rubber and natural resins were used. Some decades ago, synthetic elastomeric components (rubbers) and viscous raw materials (tackifiers, plasticizers, etc.) were developed and blended to produce (formulate) adhesives. Advances in macromolecular chemistry allowed synthesis of raw materials having built-in viscoelastic properties, that is, pressure sensitivity [e.g., acrylics, vinyl acetate copolymers, carboxylated rubber, poly(vinyl ether)s, polyurethanes, polyesters, etc.]. Some of the polymers used for PSAs are raw materials for other adhesives or plastics also. Their synthesis constitutes a special chemical–macromolecular technology.

The main basic products used for pressure-sensitive formulations, that is, the elastomers (random, alternative, and block copolymers) and the viscoelastomers (e.g., acrylics, vinyl acetate copolymers, and other vinyl polymers) are briefly described in [54–61]. The role of PSA formulation based on these raw materials is discussed in [62,63], the tackifiers and tackification are described in [64,65], and the manufacture of PSAs and PSPs in [66–68]. Raw materials for coated PSPs, (e.g., elastomer–viscous component, viscoelastomer, viscous component–solvent and viscoelastomer), are used for solvent-based PSPs. [69]. Water-based PSAs include viscoelastomer, or viscoelastomer–viscous component–technological additives - water or elastomer -viscous component. The category of 100% solids includes thermoplastic elastomer–viscous components and oligomer–viscous components [69]. The various raw materials used—for example, monomers, oligomers, natural rubber, polymers with segregated structures (styrene block copolymers, multiblock copolymers), synthetic rubber dispersions, polymeric alkene derivatives (polybutene and amorphous polyolefins) heteropolymers (silicones, polyurethanes), vinylacetate copolymers, pure acrylics, poly(vinyl ether), and tackifier resins (rosin derivatives, hydrocarbon resins, coumarone–indene resins, polyterpene resins, terpene–phenol resins, and phenolic resins and resin dispersions)—are described in [69]. The chemical additives (cross-linking agents, initiators, antioxidants, fillers, plasticizers, compatibilizers, detackifiers) and the technological additives (wetting agents, neutralizing agents, thickeners, viscosity reducers, humidification agents, antimigration agents, and cuttability additives) are also described in [69].

A brief description of the main raw materials is given in Section 19.5.1.3.

19.4 PERFORMANCE CHARACTERISTICS OF PSPs

The general performances of PSAs were described by Lim and Kim in [70]. Performance characteristics of PSPs include their conversion properties and end-use properties also. Adhesive properties are the most important end-use performance characteristics of PSAs. The main performance characteristics of PSPs also include adhesive and mechanical characteristics. Because of their supported character (laminate) during manufacturing and application, both the adhesive performance and the mechanical resistance of the carrier are important.

The definition, characterization, and control of adhesive properties and their interdependence on other performance characteristics are described in [70,71]. In a similar manner, the definition, characterization, and control of mechanical properties and their interdependence on other application properties is shown. The importance of the adhesion–cohesion balance and the influence of adhesive properties on other characteristics of PSAs (e.g., conversion properties and end-use properties) are discussed in [10,72].

19.4.1 ADHESIVE PERFORMANCE CHARACTERISTICS

Adhesion to substrate surfaces is a general characteristics of PSPs. It should be achieved with products having quite different buildup, under quite different application conditions for various end-use times, and different removal conditions and methods. Generally, adhesion is evaluated via adhesive properties. These properties are characterized by general adhesive performance and special product-dependent properties. Formulation-specific adhesive properties for solvent-based, water-based, and hot-melt adhesives are compared in [72]. The adhesive properties for various product classes and their end uses are presented in [72]. The adhesion–cohesion balance may differ for reel and sheet labels [73].

Adhesive properties can be controlled by the adhesive, the carrier, the manufacturing technology, and the product application technology.

The main adhesive performance characteristics of PSAs were described in detail in [74,75]. Special features of the adhesion of PSPs (labels, tapes, and protective films) were discussed in [76,77]. The main parameters that decisively influence adhesion are: properties of the adhesive; properties of the carrier material (face stock and liner); construction and geometry of the laminate; coating weight; coating technology of the PSA; product application technology; and the age of the laminate.

The basic adhesive characteristics of a PSA are defined by its tack, peel resistance, and shear resistance (see also Section 19.6). These characteristics are strongly influenced by the coating weight. The effect of the coating weight on adhesive properties was described in detail in [72]. General adhesive performances have been discussed in [73].

19.4.1.1 Tack

The instantaneous bonding ability of a PSP to a substrate is characterized by its tack. Although *standard tack* refers to the bonding ability tested under well-defined standard conditions, that is, on a dry, standardized surface at a given temperature, wet tack is known also. Wet tack is the performance characteristic of PSAs to adhere to humid surfaces. Tack depends on the nature of the PSA, the solid-state components of the laminate and joint, product buildup, and testing conditions. Tack also depends on chain mobility. Principally, tackification is the result of increased molecular mobility. Such increased molecular mobility is provided by diluting the macromolecular chains with low molecular substances (see also Section 19.5.1.3). Such substances are (solid or liquid) resins or solvent-like plasticizers, oils, and liquid polymers.

The combination of classic dry tackification with hygroscopicity-related wet tackification allows the formulation of adhesives that bond to humid surfaces covered with condensed water.

Tackification with resins or plasticizers allows increase of tack. Compatibility plays a decisive role in tackification. It depends on the chemical nature and molecular weight. The term *miscibility* or *compatibility* of polymers is used for their dispensing at molecular level. Many polymers having high molecular mass are immiscible.

The sensitivity of different tack-testing methods to changes in molecular mobility (reflected in the T_g) differs. Tack values obtained with different testing methods can differ strongly. Tack values obtained using different testing methods are presented in [63]. The dependence of tack on testing speed is discussed in [78]. For certain tack-testing methods, tack depends linearly on the loss modulus. As is known, the loss modulus is inversely proportional to the molecular weight.

The measurement of tack, peel, and shear resistance was discussed by Lim and Kim in [70].

PSAs possess adhesion for bonding and debonding, and cohesion is necessary against debonding. The special balance of these properties, the *adhesion–cohesion balance*, embodies the pressure-sensitive character of the adhesive. Related to their application technology (labeling) and end use, permanent labels require a balanced formulation, that is, high instantaneous adhesion (tack), high delamination resistance (peel resistance), and (related to their converting technology) an acceptable cohesion (shear resistance). The fundamentals of such characteristics were described in detail in [38,41]; the practice of characterization of adhesive properties was discussed in [70]. *Balanced formulation* refers to a formulation giving usable adhesive properties with a common coating weight of ca. 20 g/m^2. In this case, *usable* means an adhesive-coated product with a permanent bond by low pressure, instantaneous application, after a standard dwell time, and a common processability (convertibility) of the PSP.

Unbalanced formulations ensure higher adhesion-related or cohesion-related performance characteristics. For unbalanced formulations required for special labels and tapes, tack and peel resistance (e.g., special labels) or peel resistance and shear resistance (e.g., tapes) are increased comparatively with other adhesive characteristics.

19.4.1.2 Peel Resistance

Peel refers to the strength of the joint tested by debonding, using another stress direction than that of the joint. The peel resistance of a PSA characterizes the bond strength after a well-defined time, measured as the debonding force, by peeling as the delamination method. The peel resistance is influenced by the specific adhesion (i.e., electrostatic, van der Waals, mechanical, chemical, etc. interactions), the contact buildup ability of the adhesive at the substrate surface ($1/G'$), and the energy dissipated during deformation of the adhesive (G'').

The dependence of the peel resistance on the application conditions (temperature, substrate, dwell time, debonding angle, speed, etc.) is discussed in detail in [63,75,79].

Variations in the peel energy occur due to the deformation of the stripping members. Such deformation depends on the peel speed and on other peel test parameters [74].

Classic peel control is achieved by tackification with viscous components (e.g., tackifiers, resins, plasticizers, etc.) and by cross-linking. New possibilities include control (via synthesis) of the cross-linking degree of the base polymers. Special formulation components or accidental formulation components (impurities) can also modify the peel resistance.

Removability is a special case of peel control. Formulation for removability is discussed in [72]. Removable, nonpermanent PSAs should display peel values of 2.7–9.0 N/25 mm.

19.4.1.3 Shear Resistance

Shear characterizes the internal cohesion of the adhesive, tested using shear stresses parallel to the joint direction.

Shear measurement is a possibility to test the internal cohesion of the adhesive. Except in some special applications of a PSP where a shear resistance is *de facto* required, in most cases the shear resistance is taken into account as a component of the adhesion–cohesion balance only. The shear resistance depends on the PSA, on the solid-state components of the laminate/ joint, on the buildup of the laminate, and on the testing conditions.

The molecular weight and the buildup of a molecular network increase the cohesion and the shear resistance. The molecular structure (linear or branched) also affects the cohesion. The molecular weight between the entanglements (M_e) has a strong influence on the plateau modulus. A higher plateau modulus and a pronounced phase separation (network) provide more elastic film. Generally, tackification leads to decrease of the shear resistance. Shear resistance increases continuously as the testing speed increases, while tack does not increase continuously [76].

Tack, peel, and shear are interdependent. They are controlled by the adhesion–cohesion balance. This (theoretically) depends on the material. In practice, because of the non-Newtonian flow of the macromolecular compounds used as the chemical basis for PSA, this balance depends on time and temperature, that is, on the practical testing/application conditions also. During the end use of the product, it depends on the rate of application and removal. Moreover, because of the influence of the laminate components (substrate, carrier) on the rheology of PSA, that is, the reciprocal influence of laminate components, the adhesion–cohesion balance will depend on the buildup, that is, on the manufacturing parameters of the protective film, and also the protected laminate. In practice, depending on the end-use requirements, high application speed (e.g., labels), or the long-time high-strength joint (e.g., mounting tape), the balance of tack, peel, and shear may be different.

19.4.2 Conversion Properties

The adhesive and mechanical characteristics of PSPs allow their end use. Before their application, they have to be finished. In this case, *finishing* means the transformation of a continuous, web-like product having the optimal geometry to be manufactured in a product with optimal characteristics for use. This manufacturing step is called *conversion*. For some PSPs, coating is a part of conversion; for others, conversion includes cutting, laminating, or delaminating, etc. (called *confectioning*).

On the other hand, PSPs are used for different applications where the adhesive and mechanical performances are accompanied by other special properties. Therefore, there are a lot of properties called *conversion performances* that must be discussed separately. Table 19.3 summarizes the main conversion and end-use properties of PSPs.

Formulation also affects the conversion properties [77]. Convertibility depends on the physical state of the adhesive [80].

One of the most important conversion and end-use properties of web-like PSPs is their cuttability and die-cuttability; it is to be assumed that the cohesion of the adhesive—its shear resistance—allows the characterization of cuttability performances. Problems arise from the lack of correlation between the shear values measured at RT and cuttability. Cuttability and printability are discussed in detail in [81].

19.4.3 End-Use Characteristics

The end use of PSPs is described depending on their application field and application conditions. For the main product groups (labels, tapes, protective films, etc.), application fields are discussed for common and special uses. Application conditions are discussed depending on the application equipment and product end use.

The end-use properties, together with the adhesive properties, are the most important performance characteristics of PSPs [82–85,86]. The end-use properties of PSPs must fulfill different requirements for a number of application fields. Some of them are general requirements, related to the application technology of a product class (e.g., labels, tapes, protective webs, etc.), the substrate, or the processing of the labeled, taped, or protected products. Label application technology is discussed in [87]. There are products manufactured with different technologies that are proposed for the same application (e.g., certain adhesive-coated and adhesiveless protective films). Table 19.4 illustrates the application conditions for the main protective film grades. On the other hand, some classic products have the same buildup, but quite different applications, or vice versa [88]. The adhesive properties of the PSAs can be considered as end-use performance characteristics also. The influence of adhesive properties on end-use properties was discussed by Benedek in [89]. Convertibility of adhesives is also a function of adhesive properties [90]. Design and formulation of PSAs for various applications of the main adhesive classes (labels, tapes, and protective films) was described by Benedek in [91].

TABLE 19.3
Main Conversion and End-Use Properties of PSPs

Pressure-Sensitive Product	Conversion Properties	End-Use Properties
Label	Slitting ability	Printability
	Cutting ability	Labeling ability
	Die-cutting ability	De-labeling ability
	Printability	Medical characteristics
Tape	Slitting ability	Tear ability
	Die-cutting ability	Cuttability
	Printability	Mechanical characteristics
		Special electrical characteristics
		Special thermal characteristics
		Special dosage characteristics
Protective film	Slitting ability	Laminating ability
	Printability	Mechanical and thermal processibility
		Delaminating ability

TABLE 19.4
Application Conditions for Main Protective Film Grades

Application Domain	Protective Film		Application Conditions	
	Adhesiveless	Adhesive Coated	Temperature (°C)	Pressure
Metal coils	Hot-laminating	—	220–250	High
	—	AC or RR	RT	Low
Plastic plates	Warm-laminating	—	50–70	Medium
	—	AC or RR	RT- 70	Low
Finished plastic products	—	AC or RR	RT	Low
Carpet	—	AC or RR	RT	Low

End use-related formulation was discussed in [92], where formulation for environmental resistance (e.g., water resistance/solubility, temperature resistance, and biodegradability) was described. Special end-use characteristics- and application domains-related formulations were discussed in [92]. Table 19.5 illustrates the end-use requirements for medical PSPs.

In [93], the principles of buildup of the main PSPs were discussed, which allow their classification according to their construction. Construction of PSPs is decisive for their manufacturing technology and application domain (as labels, tapes, protective films, etc.). Product construction-related formulation (i.e., carrier-related, release liner-related, and PSP buildup-related) was discussed in [66]. PSPs can be divided into classes according to their end use. Some of these classes contain well-known products having a broad range of representatives (e.g., labels, tapes, etc.); others are special or one-of-a-kind products. Because of the continuous expansion of their application field, the number of special products is increasing.

A comparison of the requirements for various product classes [93] shows that labels and tapes have similarities with respect to the nature of the adhesive, and protective films and tapes exhibit common features concerning the nature of the carrier.

Formulation affects the dimensional stability (shrinkage and lay-flat) of PSPs [63].

TABLE 19.5
End-Use Requirements for Medical PSPs

Product	Requirement
Label, tape, electrode	Physiologically compatible
	Body fluid nondegradable
	No skin irritation
	Skin adhesion, 95%
	Low-adhesive transfer to skin (cohesion)
	Conformability
	Moisture insensitivity
	Cold water insoluble, hot-water (65°C) soluble adhesive
	Vapor transmittance (500 g/m^2, 24 h, 38°C)
	Light resistance
	Heat resistance (176°C)
Label, tape	Breathability (50 s/100 cm^3/in.^{2})
	Liquid transmittance (6–36 h)
	Gamma radiation resistance
Label	Detachable
	Ethylene oxide resistance
Bioelectrode	Electrical conductivity

19.5 MANUFACTURE OF PSPs

The manufacture of PSAs and products was described in detail in [11,66,67]. In reality, PSP technology has moved far beyond its original purpose to manufacture adhesive-coated paper labels or tapes by coating a web with a softened, plasticized elastomer. New special raw materials have been developed, having a built-in pressure-sensitive character, that is, PSAs. Instead of paper, almost all film-like materials can be used as carrier material, and different techniques have been tried to coat molten, dissolved, or dispersed adhesives. A conversion technology specializes in pressure-sensitive adhesive-coated webs, or laminates, and a special application technology has been developed for such products.

The manufacture of PSPs includes the manufacture of PSP components (PSA, carrier, etc.) and the manufacture of PSPs. The manufacture of PSA raw materials (synthesis) is followed by the processing of raw materials (mixing) to obtain pressure-sensitive compounds, and then the processing of PSA to obtain PSPs.

The adhesive layer can be applied using various (coating) techniques, such as those used in the converting (coating and/or printing) industry, or in plastics manufacture (extrusion and calendering). Because of the broad range of raw materials available for PSAs, and the sophisticated adhesive-coating technologies now in use, the adhesive properties of a PSP can be easily regulated by the coating of a solid-state carrier material with a low-viscosity PSA. Table 19.6 summarizes the manufacturing principles and technologies of PSPs.

19.5.1 MANUFACTURE OF PRODUCT COMPONENTS

The manufacture of PSPs includes the manufacture of the product components and of the finished product. The main raw materials for different pressure-sensitive product classes include acrylic dispersions (tackified and nontackified); rubber-resin solutions and acrylic solutions (for labels); rubber-resin solutions, acrylic dispersions, and thermoplastic elastomer-based hot-melts (for general tapes); special tackified elastomer solutions, special viscoelastomer solutions, tackified plastomers, and self-adhesive plastomers (for special tapes); and rubber-resin solutions, untackified acrylic solutions, acrylic dispersions, tackified plastomers, and self-adhesive plastomers for protective films [69].

TABLE 19.6
Principles and Technologies for Manufacture of PSPs

Type of PSP	Scope of Manufacturing Process	Manufacturing Technology	PSP Components	PSP
Without PSA	Web manufacturing	Extrusion	Carrier	Tape, sealant, protective film
		Coating	PSA	Tape, sealant
	Web manufacturing	Calendering	Carrier + PSA	Tape
		Extrusion	Carrier + PSA	Tape, protective film
With PSA		Coating	Carrier + PSA	Label, tape
	Web coating	Calendering	Carrier + PSA	Tape
		Extrusion	Carrier + PSA	Label, tape

Although the main synthesis of pressure-sensitive raw materials (elastomers, viscoelastomers, and viscous additives) has been the subject of many works, specializing in macromolecular chemistry and technology, advances in in-line manufacture technology of PSAs (especially in the development of radiation-curing and web-finishing technology) impose a basic discussion about the synthesis and manufacturing technology of pressure-sensitive raw materials. In the next sections, the polymerization technology, the technology based on polymer-analogous reactions, and the formulation of off-line manufactured PSA raw materials are described comparatively to the in-line synthesis.

The manufacture (synthesis, formulation, and technology) of the coating (adhesive, abhesive, etc.) components and the carrier materials (paper, films, others) was discussed in detail in [11]. The manufacture of finished products (coating and conversion) is described in [94].

The manufacture of PSAs can be carried out off-line or in-line. Both off-line and in-line manufacture can include the mixing of the adhesive components to be coated as ready-to-use PSAs, but generally, off-line synthesis supplies components of PSAs, that is, raw materials that may be or may not be pressure-sensitive, or are not pressure-sensitive enough; therefore, they have to be formulated, that is, mixed with other micro- or macromolecular compounds. This is the subject of formulation (see also Section 19.5.1.3).

In-line manufacture of adhesives consists of simultaneous coating and curing or postpolymerization of the adhesive raw materials. In this case, first, a special *ready-to-coat* mixture of polymerizing and cross-linking monomers, oligomers, or polymers (e.g., radiation-cured hot-melts) is applied on a temporary or definitive carrier material. Such a reaction mixture is transformed after coating to a *ready-to-use* adhesive or pressure-sensitive product. The postcoating synthesis of the PSA must be carried out by the converter. Off-line as well as in-line synthese were described in detail in 64,95,96; in the next sections, only a brief presentation of them will be given.

19.5.1.1 Off-Line Synthesis of Pressure-Sensitive Raw Materials

Off-line synthesis of PSAs is the common method of producing PSAs. Such products can be macromolecular compounds having ready-to-use or ready-to-formulate adhesive properties, or ready-to-postpolymerize reaction mixtures (monomer- or polymer based), manufactured and supplied by the chemical industry to be formulated and/or coated in-line, that is, to convert them into PSPs. The raw materials and the technology used to transform them into macromolecular compounds with ready-to-use or ready-to-formulate pressure sensitivity are the main parameters of off-line synthesis. It should be noted that due to the wide use of synthetic elastomers in domains other than PSA, and the existence of a sophisticated chemical technology, off-line synthesis, that is, the manufacture of macromolecular compounds by polymer specialists, remains a fundamental part of pressure-sensitive technology. The raw materials (monomers and additives) and the technology are the main parameters of off-line synthesis.

The raw materials for PSA synthesis due to the advances in macromolecular chemistry were discussed in detail in [97]. The glass transition temperature and modulus of elasticity as the main parameters of pressure-sensitive design and formulation, as a function of macromolecular characteristics (i.e., molecular weight and its distribution, comonomer content, molecular structure, sequence length and distribution, compatibility, cohesive strength/free volume balance, etc.) were described in [98].

Although polymerization and polymer-analogous reactions remain the main modalities for off-line synthesis, recent advances in macromolecular chemistry have allowed the synthesis of PSAs by simultaneous cross-linking and tackification, leading to hydrophilic, biocompatible polymers [99,100]. Monomers are the most important raw materials for polymerization and polymer modification. The choices of monomers, the polymerization procedure, and additives decisively influence the performance characteristics of the PSA [69].

Off-line synthesis was discussed in detail in [64]. The parameters of polymerization (choice of monomers, role of polarity and functionality, hydrophilicity, stabilizing effect), the choice of additives and of the polymerization procedure, and polymer-analogous reactions were discussed in detail. Built-in cross-linkable functional groups, hydrophilic monomers for water-soluble compositions, stabilizing hydrophilic comonomers, and so on were also discussed. The role of surfactants and the effect of particle size were described.

19.5.1.2 In-Line Synthesis of Pressure-Sensitive Raw Materials

Generally, for the PSA converter, the manufacture of adhesives consists of their formulation, that is, mixing of macromolecular compounds and additives. In certain cases, it would be desirable for the adhesive converter to carry out his own polymerization. Such necessity arises for low-volume special products, or for adhesives where the formulation is based on the mixing of low molecular products, oligomers, and/or prepolymers, and the conversion technology of this mixture must supply the PSA and the PSP simultaneously. There are also some special cases where the formulation is made to allow the modification of the coated adhesive by postpolymerization. Full or partial postmanufacturing of a PSA is the result of chemical development induced by the trend of solvent-free fabrication. Such formulations with 100% solids include hot-melts or radiation-curable reaction mixtures. In-line polymerization, which includes polymerization *in situ* or postpolymerization, uses special, multifunctional monomers or macromers and classical (chemical) or physicochemical polymerization techniques [101]. Polyaddition, polycondensation, or cross-linking are carried out. Postapplication cross-linking can be considered as a special case of off-line synthesis (finishing) of an adhesive. It is proposed to improve shear resistance, or to ensure delamination (detachment) after use. Such post-cross-linking is achieved using thermal, free radical, or photoinitiated reactions [102]. Radiation curing was developed as a general coating finishing technology. The cross-linking of hot-melts by radiation was developed to improve their temperature-related cohesion. For such adhesives, higher service temperatures can be obtained by supplemental cross-linking. In principle, cross-linking of styrene block copolymers (SBCs) can be carried out by UV light-induced or electron-beam (EB) curing, modification of polystyrene end blocks, and chemical reaction with functionalized block copolymers. SBCs include unsaturated segments; therefore, such products are potentially cross-linkable by radiation also.

The cross-linking agent, the radiation dosage, and the nature and level of the tackifiers are the main parameters in control of the properties of radiation-cured hot-melt PSAs (HMPSAs), but antioxidants may also influence it.

First, postpolymerization (growth of the macromolecular chain or its cross-linking) was practiced. In principle, the gelling of plasticized poly(vinyl chloride), vulcanization of rubbery articles, or curing of zinc oxide-filled, roll-pressed tapes can be considered as *in situ* polymerization. Such procedures are based on common, thermally initiated reactions. Later, radiation-induced polymerization was developed. The use of radiant energy to improve the performances of PSAs can be found as early as 1960. The first pressure-sensitive tapes were produced using such technology.

In situ reactions can be carried out on polymers, oligomers, or monomers or on mixtures of such compounds as polymer-analogous reactions or as polymerization.

Generally, in-line synthesis is based on polymer-analogous reactions (which include functionalization and cross-linking of polymers), polymerization, and macromerization [2,64,103]. Polymer-analogous reactions include functionalization of the carrier and/or modification (functionalization or cross-linking) [64]. Polymerization can also be limited to the surface of an existing base polymer. For instance, an increase of the reversible work of adhesion can be achieved by grafting polar groups onto the polymer. Surface treatment procedures also cause polymer-analogous reactions. Cross-linking required to increase molecular weight and to shift the adhesion balance can be carried out through classic, thermally initiated reactions, or by radiation-induced reactions.

In situ polymerization of monomers or oligomers was developed as a technology to manufacture special products with high productivity. Radiation-induced polymerization and curing offer new possibilities for such procedures [64]. Macromerization includes the synthesis of a relatively low molecular weight polymer and its co-polymacromerization or cross-linking [64].

In-line synthesis of raw materials for PSAs includes the raw materials, the monomers, oligomers, and macromers, and the technology for in-line synthesis. For both in-line or off-line syntheses of PSA formulation, skill is needed. Radiation curing and siliconizing through radiation are discussed as steps in the simultaneous manufacture of PSA and PSA laminates in [104].

The main coating component is the PSA. Depending on its adhesive characteristics, carrier nature, and product geometry, protection of the adhesive layer may be necessary. This protection is given by a release layer. The nature of this release layer (a coated layer or a separately coated, solid-state component) depends on the product construction (product class) and application conditions. For both, the release effect is provided by a coated abhesive component. Other coating components used to improve the anchorage of the adhesive, release, or printing inks may also be applied. Additives for some special performances (e.g., electrical conductivity) may be coated on the carrier material (see also [63]).

19.5.1.3 Formulation of Raw Materials

In principle, the purpose of formulation is to develop a recipe with end use-tailored properties (Figure 19.1).

The main criteria for end use are the adhesive characteristics. On the other hand, the formulation must allow the use of the adhesive for manufacture of the PSPs (coating and conversion properties) and the use of the final product to be applied (conversion and other end-use properties; see also Sections 19.4.3 and 19.4.4). Figure 19.1 illustrates the requirements and modalities for formulation. The design and formulation of PSAs constitutes the subject of [105–107]. Advances in the theory of pressure sensitivity have allowed the science-based formulation of PSAs to replace empiricism. Due to the development of macromolecular chemistry, pressure-sensitive formulation has been shifted

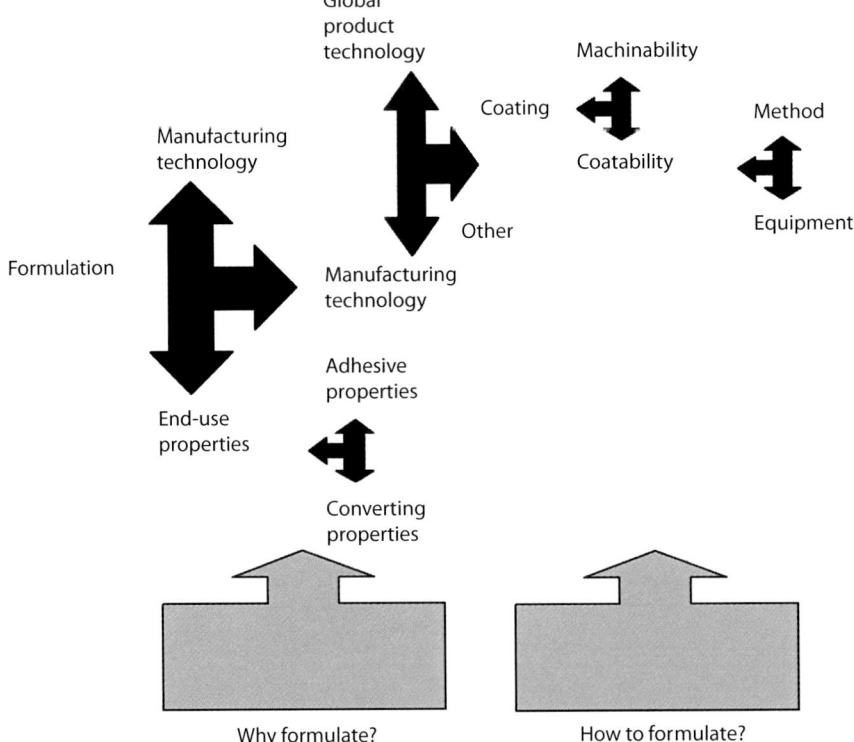

FIGURE 19.1 Influence of the formulation on the manufacture technology and properties of pressure-sensitive products (PSPs).

to pressure-sensitive design [65]). The parameters of pressure-sensitive design include the glass transition temperature, the modulus of elasticity, macromolecular characteristics, compatibility, the time–temperature superposition principle, fibrillation, the cohesive strength–free volume balance, macroscopic product buildup, and application conditions.

The first step in manufacturing technology involves formulation. In industrial practice, formulation is the blending of different components of a recipe to achieve adequate technological and end-use performance characteristics. In the domain of PSPs, formulation includes theoretical, experimental, and industrial know-how concerning the design and manufacture of a product component or the whole pressure-sensitive product. As discussed in [95,108], formulation is synthesis and end use-related. Formulation affects global product technology, adhesive-coating technology (productivity-related and environment-related considerations), adhesive manufacturing, and conversion technology (see Section 19.5.2.4). Details (the nature and viscosity of the dispersing medium, solids content, drying, running speed. and recycling) are given in [63].

The various components of a PSA have to be mixed; that is, the adhesive has to be formulated. Formulation cannot be separated from the engineering of the products and their end use. There is also a need to discuss their mutual influence.

Formulation by mixing includes tackification, cohesion control, detackification, polymer degradation, elastomer-based formulation and viscoelastomer-based formulation [103], and uses as components: high polymer (elastomer), oligomer (tackifier), and micromer (tackifier) [64].

19.5.1.3.1 Tackification

Tackification is the formulation made to improve tack and peel resistance. According to the effects of such formulation, tackification can be dry or wet, temporary or permanent. Figure 19.2 presents the scope and limitations of tackification. Concerning the site of tackification, it can act on the whole polymer (e.g., acrylics or rubber-resin formulations), or on preferred segments of the polymer [64]. Permanent tackification of hot-melts plays a temporary role in viscosity control during processing. This is called *tackification for processing*. Tackification for postcuring takes into account the radiation-induced behavior of tackifiers. For some polymers (e.g., chloroprene [CR], atactic polypropylene [APP]), tackification works only temporarily.

Tackification decreases the storage modulus and increases the loss modulus. The increase of the loss modulus is due to the T_g increase caused by the high T_g resin. A compatible tackifier works similarly to a solvent and an incompatible tackifier works similarly to a filler. It increases the modulus

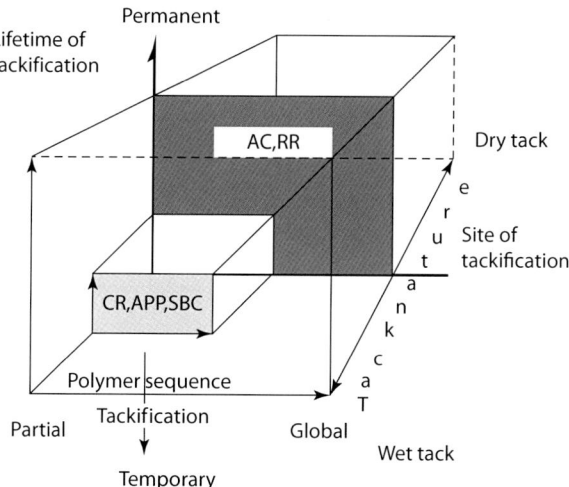

FIGURE 19.2 Scope and limits of tackification (dry or wet, permanent or temporary tack, partial or global tackification on a part of or on the whole polymer).

over the entire frequency range. Considering the tackified elastomer as a polymer solution in the resin, and the polymer solution as ideal (zero excluded volume effect), the plateau modulus (G_n') decreases as a function of the volume fraction of the polymer (V_p), the density of the blend (ρ), and the molecular weight between the entanglements (M_e).

There are various procedures for tackification. It is carried out with high polymers (elastomers, viscoelastomers, viscous polymers, and plastomers) with resins or with plasticizers. Tackification by mixing of high polymers is limited by the available range of pressure-sensitive polymers and by their compatibility. A significant reduction in the elasticity of rubbery-elastic products can be achieved by the use of viscous compounds (resins and plasticizers). Their application constitutes the classical domain of tackification. Tackifying resins generally contain more low molecular weight species and increase the T_g of the system to a greater degree than would be possible with polymeric tackifiers. It has been found that peel resistance is proportional to G''/G'. Higher T_g resins result in blends with higher G' in the bonding region and higher G'' in the debonding region. Viscoelastic responses are shifted toward lower frequencies as the resin T_g increases (i.e., the system is harder). Thus, debonding properties are improved but bonding properties decrease.

The scope of tackification and the effects of tackification of natural rubber (simultaneous tack and peel increase, tackification covering the whole polymer, reduction of shear resistance) and of thermoplastic elastomer (TPE) (no simultaneous tack and peel increase, tackification not covering the whole polymer, increase of shear resistance) are described in detail in [64]. Tackification and tackifiers are described in [86]. Criteria for the choice of tackifiers, the versatility of tackifiers, the polymer:tackifier ratio, reinforcement with resins, and tackification for processing and for postcuring are discussed in [64].

Tackification with plasticizers (the role and working mechanism of plasticizers, liquid resins as plasticizers, oils as plasticizers, liquid polymers as plasticizers, technological influence of plasticizers, plasticizer resistance) is described in [64]. Plasticizers influence the drying or cross-linking speed of adhesive formulations. As discussed, in HMPSA formulation the plasticizers play a primary role as processing aids.

19.5.1.3.2 Role of Formulation

The methods for cohesion enhancement (increase of molecular weight, cross-linking, reinforcing/filling) are described in [64]. Cohesion regulation by cross-linking, the scope of cross-linking, cross-linking modalities, the theory of cross-linking (role, parameters, cross-linking of ordered structures), the basis of cross-linking (functionality vs. chemical cross-linking agents, curing of SBCs, radiation curing, cross-linking of rubber-resin formulations, cross-linking of acrylics, cross-linking of segregated polymers), simultaneous cross-linking and plasticizing, and the limits of cross-linking are discussed in [109].

Similarly to the formulation for the synthesis of adhesive raw materials, high polymer-related formulation is carried out principally to regulate the adhesion–cohesion balance and to improve the end-use (adhesive) properties. Formulation for adhesive properties includes formulation for permanent and removable adhesives [69,110]. Special recipes for removable adhesives (the use of primers, flexibilizers, fillers, softening of the adhesive, contact reduction, etc.) are described in [111].

Formulation for special adhesive properties includes formulation for water resistance/solubility and also for low-temperature applications. Formulation for adhesive properties is influenced by the physical status of the formulating components. Except for hot-melts, other formulations need a dispersing medium or a diluting agent. The true solution character or emulsion character of a formulation affects its adhesive performance characteristics also.

Formulations for various product classes, for example, labels, tapes, and protective films, are discussed in [69,112]. Screening of formulations for such products is given in [64].

Formulations as a function of substrates and carriers are discussed in [113,114].

Formulation for other than adhesive properties includes formulation for removability, substrate-dependent formulation, carrier-dependent formulation, formulation for environmental stability, formulation for water resistance/solubility, product-related formulation, formulation of carrierless products, and formulation for conductivity and economic considerations. As discussed in [72,77], formulation also affects conversion properties. Dimensional stability (shrinkage and lay-flat) are affected by formulation.

Pressure-sensitive adhesion, according to its classic definition, assumes instantaneous adhesion characterized by tack. Self-adhesion includes pressure-sensitive adhesion, but certain self-adhesive products do not display instantaneous tack, although they are based on analogous chemical and macromolecular buildup. Such products mainly include removable PSPs, where low coating weight, low contact area, or sterical hindrance (i.e., trapped adhesives) do not allow the buildup of a strong joint, and the adhesive is cohesive enough to ensure failure at the contact surface. Such removable self-adhesive products can be manufactured by using PSAs or without them. Their bonding and debonding must be the same in industrial practice. Their application conditions are also the same. In some cases, high pressure, high temperature, or special physical treatment provide the conditions required to bond. Thus, there is a need to standardize such properties as application tack, low-temperature adhesion, wet or dry adhesion, and so on.

The design and formulation must take into account the practical requirements for PSPs, that is, their application conditions.

19.5.1.3.3 Limitations of Formulation

Formulation, as a technological process used to design and manufacture adequate PSAs or PSPs, possesses theoretical and practical limitations (product buildup, end-use, technological and economic considerations) [72]. In a first approximation, formulation assumes the additivity of the characteristics of the components used (according to a linear or nonlinear correlation). Correlations based on the additivity of the formulation components are largely used in engineering practice for the calculation of the parameters of multicomponent recipes (e.g., viscosity, temperature, pH, etc.). Such correlations are applied under well-defined (time–temperature) conditions for the evaluation of the parameters of molten macromolecular blends also. Among the main parameters influencing the rheology of polymer systems, the glass transition temperature is calculated using correlations based on the additivity rule. Solubility parameters are also computed based on such correlations.

The theoretical limitations of formulation depend on the base characteristics of the components intended to be used for a recipe, their miscibility, and the assumed additivity of the component characteristics. As discussed earlier, the base characteristics of the components determine the application window of a pressure-sensitive product. The glass transition temperature and the modulus, as the main rheological parameters, are both additive (under well-defined conditions). This means that the correct choice of the formulation components allows (at least theoretically) the control of both parameters. In practice, the theoretical basis necessary to clear the reason for the required modifications is not sufficiently advanced. Product buildup and end use can act as formulation limits. The technological limitations of the formulation are given by the processing parameters of the materials used and by the technological equipment available. The processing parameters include the factors related to time–temperature, that is, the processing temperature and processing speed (stress rate) of the materials. Economic considerations are important in the formulation also.

The formulation and the manufacturing technology are interdependent. The existence of a given manufacturing technology imposes a tailored formulation. A given formulation (required by the end use of the product) requires a special manufacturing procedure. In the case of a permanent carrier material (playing the role of a reinforcing web), the formulator primarily has to design the PSA composition to obtain the required adhesive characteristics.

19.5.1.3.4 Formulation Equipment

The formulation equipment depends on the processability of the components. The blendability of the components depends on their physical and chemical composition. Wet blending, dry blending, and blending in molten state are known [69].

19.5.2 Manufacture of Finished Product

According to the classical manufacturing procedure, PSPs are produced by the coating of a carrier web with a PSA. Depending on the product class and product application, other coating and carrier components may also be used.

Although paper and other special nonpaper, nonpolymeric materials may also be used for most PSPs, plastics are the main carrier material. In the first development step, packaging materials were applied as carrier films. Later, special carrier films were produced; most of them were manufactured by the converters [82]. This trend is accentuated by the introduction of coating-free and adhesive-free PSPs.

Generally, independently of their manufacture method, all PSPs undergo lamination during their processing.

19.5.2.1 Lamination

The coating of different carrier materials with adhesives was first used to bond them permanently, that is, to produce laminates. The same coating and laminating technology was used later for PSPs. Actually, depending on the laminate components, the nature of the adhesive, and the site of lamination, the adhesive-coated finished products may be supplied as permanent or temporary laminates (e.g., laminated films, labels) or as a monoweb (e.g., tapes, protective films). For the latter products, lamination is carried out during application.

The manufacture of permanent laminates is the field of classical adhesives bonding chemically. Surface coating with an active adhesive is the domain of PSAs. Because of the necessity to protect this adhesive layer (depending on its nature), PSA-coated products are manufactured as temporary laminates with another solid-state component (e.g., labels) or by themselves (e.g., tapes).

The main coating component is the PSA. Depending on its adhesive characteristics, carrier nature, and product geometry, protection of the adhesive layer may be necessary. This protection is given by a release layer (see also Section 19.2). The nature of this release component (coated layer or separate, solid-state component) depends on the product construction (product class) and application conditions. For both, the release effect is provided by a coated adhesive component.

19.5.2.2 Web Preprocessing Technology

Web preprocessing technology includes operations required to ensure coating of the web. Web pretreatment and priming are the main web preprocessing technologies. Such procedures use physical and chemical treatments or coatings to impart chemical affinity to the web, required for wetting-out of the adhesive and its anchorage. Surface treatment is used mainly for plastic carrier materials (but transfer-coated PSAs may be corona treated also); priming is suggested for both plastic and paper carrier materials. Although surface treatment is carried out during carrier manufacture [64], generally, refreshing of the surface treatment is required before coating the carrier.

19.5.2.3 Web Postprocessing Technology

Web postprocessing technology includes operations that finish the coated adhesive layer, that is, curing, foaming, and complete confectioning of the web. Lamination and overcoating of the web are carried out. Overcoating includes web printing and lacquering. Confectioning of the web includes slitting and cutting, waste stripping, cutting, die-cutting, perforating, embossing, and folding.

19.5.2.4 Manufacture by Coating

Generally, pressure-sensitive technology leads to a special web with pressure-sensitive properties. Such properties may be intrinsic characteristics of the web itself (that is, the carrier *per se* is pressure-sensitive) or may belong to a special pressure-sensitive coating on this web. Coating technology is described and coating machines are discussed in [86].

If someone is asked about the quintessence of this new industry, two characteristics are given: This is a coating industry, and the coated material is a pressure-sensitive adhesive. From the classic point of view, PSPs are pressure-sensitive adhesive-coated products.

Coatability depends on both components: carrier and adhesive. It is a surface phenomenon, which means that (at least theoretically) it can also be improved unilaterally (e.g., modifying the surface characteristics of the carrier only). Coatability of dispersed systems is related to the surface tension and viscosity of such systems. For water-based formulations, the surface tension is regulated mostly by surface-active agents; for solvent-based systems such regulation (if necessary) uses different solvents. For 100% solids systems, the viscosity can vary over very large ranges, because of the different possible buildups of such systems (as is known, such formulations are monomer-, oligomer-, or polymer based).

Coating and extrusion are the main PSP manufacturing procedures. Their choice as manufacturing technology is decisively controlled by the formulation. There are different criteria that impose the choice of coating as a manufacturing procedure. Coating allows regulation of product properties by the carrier material and the coated (adhesive and auxiliary) layers. The chemical composition and the geometry of the coated layers provide multiple possibilities to control the adhesive performance characteristics of a PSP. The chemistry of coatings is based on a very broad range of raw materials. Sophisticated coating images allow a fine regulation of the adhesive contact surface. Thus, in comparison with embedded adhesive or tacky carrier materials, formulation for coated PSAs constitutes a more perfect procedure to design PSPs.

Coating assumes the existence of a coatable adhesive and a (permanent or temporary) solid-state carrier component; extrusion may or may not need a supplemental solid-state carrier. Coating can use the carrier made by extrusion; thus, extrusion and coating can be combined. Coating can be employed for carrier manufacture also (e.g., radiation-curing technology). Coating is carried out continuously or discontinuously; extrusion is a continuous technology.

Concerning global product technology, the use of a given adhesive-coating technology with solvent-based, water-based, or solvent-free adhesive can influence the choice of the coating technology for other coating components; for example, primer, release, or ink. In-line siliconizing has been developed together with HMPSA. For certain silicone adhesives, a silicone release coating may be used as the primer. The primer coating must be cured prior to the application of the silicone PSA. The main product class-related coating procedures are listed as follows: coating for in-line manufacture (release–corona–primer–PSA–corona–primer–PSA); coating for very low coating weight; coating for medical products; coating of crêpe papers (without defolding effects); coating of high coating weights; coating of labels where the optical quality (image) of the adhesive layer may be determinant; and high solids coatings.

Generally, the manufacture technology of PSPs includes the following operations: manufacture of the adhesive and other materials to be coated; coating of the liquid product components; confectioning of the coated material and its transformation to the finished product; solvent recovery and testing of raw materials and semifinished and finished products. Taking into account the need to manufacture (formulate or synthesize) the adhesive, the mixing versatility of different raw materials can also influence the formulation. The mixing ability of solvent-based, water-based, or hot-melt formulations can differ strongly.

Adhesive-coating technology includes the machinability or processability of the adhesive. Machinability comprises versatility for storage, handling and transfer, and the runnability and recycling ability of the PSA. Runnability is related to the coatability of the adhesive. Coatability is affected by buildup of the wet adhesive layer and by its drying speed. Thus, the choice of the formulation depends on the existing coating and drying equipment. Drying is discussed in detail in [115].

19.5.2.4.1 Coating Equipment

The viscosity and solids content of the adhesive are the main rheological parameters for the coating device. The usability of various coating device constructions is limited, depending on the viscosity of the adhesive.

The usable range of adhesive viscosities and coating weights is imposed by the coating device. The drying equipment limits the maximum coating weight. The coating image (continuous or discontinuous, shaped, etc.) is influenced by the coating device. Theoretically, for a given piece of coating equipment, the formulator has the freedom of choice of adhesives with various raw material bases and different physical statuses. In practice, the choice is limited by the characteristics of the solidification unit.

The ability of various coating systems—for example, roll-coating, die-coating, wire-rod coating, and so on—for different adhesives (of various natures, viscosity, solids content, etc.) is discussed in detail in [82].

Assuming that the existing coating equipment allows the use of adhesives having different rheological characteristics and different physical statuses, productivity-related considerations can decisively influence the choice of the coating technology. In-line coating of multiple layers and laminating substantially improves productivity. Hot-melt PSAs exhibit the advantage that they can be laminated instantaneously (after coating at a high temperature) with the release liner; that is, in-line siliconizing and adhesive coating are possible. In-line hot-melt coaters allow coating of the material of choice with hot-melt before the release film or paper is applied. This removes the need for printers to buy precoated label stock, and enables them to specify face and release liner depending on customer specifications. In-line siliconizing is more economic than off-line siliconizing. The coatability of solvent-based, water-based, and HMPSA formulations is discussed in [116]. The influence of the formulation on the drying and on the running speed is described in [116]. Special coating methods (slot-die coating, screen printing, etc.) and contactless coating devices are also used.

The 1977 amendments to the Clean Air Act required cutbacks in the amounts of volatile compounds. Converters can meet the requirements of the Environmental Protection Agency Clean Air Act through a variety of methods: the use of incinerators, solvent recovery, solvent-free reactive systems, high solids coating, and water-based coatings. A comparative examination of solvent-free (hot-melts, radiation-cured systems), solvent-based, and water-based formulations is given in [72]. Recycling systems are discussed in detail in [116].

19.5.2.5 PSP Manufacture by Other Methods

Table 19.6 summarizes the main principles of PSP manufacture and the production technologies based on such principles. The main manufacturing technologies for PSPs include coating of the adhesive, extrusion of the (plastic) carrier or the (hot-melt) adhesive, and casting of the (plastic) carrier or of the adhesive (*in situ* manufacture) [69].

As seen from Table 19.6, the classic manufacturing technology of PSPs is based on conversion of a solid-state carrier material to a PSP by coating it with a PSA or by compounding it with a PSA. New technologies allow the manufacture of adhesive-free PSPs. Generally, PSA-related PSP production requires at least two components: a solid-state nonadhesive carrier material and a PSA. The manufacture of adhesive-free PSPs is one-component-based. It requires a tacky carrier material. Obviously, quite different raw materials and manufacturing technologies are necessary for adhesive-coated and adhesive-free PSPs.

19.5.2.5.1 Extrusion

The development of macromolecular chemistry allowed synthesis of rubber-like or plastomer-like products that can be processed as films, and (under well-defined conditions) exhibit self-adhesivity and pressure sensitivity. Such products are made (similarly to common plastic films) by extrusion. It should be mentioned that the manufacture and application of such self-adhesive products are possible by the use of special application conditions that may include elevated temperature and pressure.

Embedding of an adhesive in the carrier web depends on the carrier and adhesive material. Extrusion coating or extrusion are preferred for thermoplastic materials. Coating of a carrier material with an adhesive supposes coatability and bonding of the coated adhesive layer on the carrier. Blending of an adhesive with a carrier material supposes its incompatibility (rarely compatibility) with the carrier material. Compoundability is related to compatibility. It is a phenomenon influenced by the chemical characteristics of both components as well as by the diffusability (barrier properties) of the solid-state component.

19.5.2.5.2 Calendering

The so-called *press-rolling* technology allows impregnation of special webs with high-viscosity PSAs. Auxiliary equipment (winders, printing lines, electrostatic control, humidifiers, deionizing, web-cleaning, thickness control, and curing stations) was described in [66].

19.6 TESTING OF PSPs

The testing methods for PSPs include general and special testing methods for the manufacture of product components (carrier and adhesive) and of the finished products.

As in the case of many other finished products, a wide range of testing methods have been developed for PSA-based labels, tapes, and coatings. A number of organizations such as Federation Internationale des Fabricants Transformateurs, d'Adhesifs et Thermocollants sur Papiers et Autres Supports (FINAT), the Association des Fabricants Européens de Rubans Auto-Adhesifs (AFERA), the Pressure-Sensitive Tape Council (PSTC), and the Tag and Label Manufacturers Institute (TLMI) have established (standard) testing methods that are widely used in the industry, although there remain significant differences among the various methods used. These methods provide a good basis for the evaluation of adhesives, but some modifications or additional tests are required when testing materials for specific applications. Therefore, the principal PSA manufacturers and converters have developed their own methodology. It is not the aim of this section to discuss the standardized PSA testing methods in detail. Special methods were worked out for special end uses, or for specific PSA applications. It is evident that for a PSP manufacturer, the properties of the liquid adhesive, its coating behavior, and PSP performance characteristics are very important. From the point of view of methodology, these areas differ quite a bit.

The testing methods for PSAs and products serve to control the manufacturing process, the materials used in this process, the technological discipline, and the finished products. Some of them are equipment-related measurements that belong to process control; others are laboratory tests.

Testing methods for PSAs and PSPs are discussed in detail in [72,117]. The fundamental aspects of the tests for PSAs are described in [31–41].

For PSAs, the testing methods are used to control the formulation components and the finished adhesives. Testing for formulation components includes testing of raw materials for PSAs (elastomers, viscoelastomers, tackifier resins, plasticizers, etc.), testing of main pressure-sensitive formulation components, and of additives.

Specification-related tests of PSAs include testing of solid-state adhesives (mechanical characteristics, softening point, glass transition temperature) and PSA-specific tests (e.g., testing of cold flow, testing of environmental stability, testing for U.S. Food and Drug Administration [FDA] and European standard compliance [72]) Specification-related tests of liquid-state PSA cover processability (coatability, coating weight, solids content, viscosity, and density) also. The influence of the PSA on the carrier and on the abhesive must also be controlled. Specific tests for processability include hot-melts and dispersed systems. Tests for mechanical stability (shear stability), foaming, particle size, storage stability, and drying characteristics must be carried out.

Testing of PSPs must cover the common and product-specific confectioning properties of pressure-sensitive webs, e.g., testing of cuttability and die-cuttability, winding, telescoping, and laminate processability for labels, tapes, protective films, forms, and so on.

PSP tests must also include the solid-state laminate components. The main solid-state component of a pressure-sensitive laminate is the carrier material. Tests of carrier materials control the surface characteristics of the carrier and its bulk characteristics. Similar tests cover the release liner.

Testing of surface characteristics of carrier materials covers testing for coatability, wettability, printability, and smoothness. The tests for bulk characteristics of carrier materials include testing of dimensional stability, shrinkage, plasticizer resistance, elongation, lay-flat and curl, flagging, wrinkle buildup, stiffness, migration/bleeding and oozing/edge bleeding).

There are various testing methods for the end-use characteristics of a finished PSP. The most important methods measure the adhesive characteristics; for example, tack, peel resistance, and shear resistance.

19.6.1 Testing of Tack

Tack test methods include the standard ones [rolling ball tack (RBT), rolling cylinder, loop tack (LT)/quick stick, Polyken] and special tack test methods; for example, the toothed wheel method, the mechanooptical tack tester, and tack methods for special products. Tack is measured in terms of the coefficient of friction, cohesion, fracture energy, peel, plasticity, and so on [116].

Admitting that tack is the ability of a viscoelastic macromolecular compound to build up, by pronounced viscous flow, an instantaneous bond that behaves elastically during debonding (depending on the relaxation phenomenon in such polymer), the elastic debonding response may be different for the same polymer tested by different methods. This means that as a supplemental variable parameter, the chemical and macromolecular buildup of the adhesive has also to be taken into account. In practice, this means that for a given class of macromolecular compounds, the correspondence between the different tack measuring methods may also be different.

Generally, the testing procedure to measure tack consists of two steps: bond formation and bond separation. Some of the testing methods (quick stick or LT and Polyken tack) simulate the real bond formation conditions; others, such as RBT, measure the tack under quite different experimental conditions. All of these tests measure the ability of the adhesive to form a bond in a very short time with minimal contact force. None of these—RBT, LT, or probe-tack test—produce results that correlate well with the others.

The (axisymmetric) tack test (with an indenter) can be viewed as a constant deformation rate test for fibrillated materials. Therefore, this tack testing can be used to determine a phenomenological model for fibril deformation (see also Section 19.3.1.4). Flat punch tests are known as *probe-tack* or *Polyken tests*, and are also used for ASTM standards.

Over the last few years, appreciable progress has been made in the quantitative description of the micromechanics of PSA debonding. Attempts to correlate the hydrodynamics of the material flow and the fracture mechanics during debonding with the modulus of the material has allowed correlation of the macroscopic behavior (debonding of a strained polymer bond) with the macromolecular characteristics of the polymers. The microscopical phenomenon of flow, related to debonding, includes cavitation buildup, buildup of air fingers, and fibrillation. Nucleation of cavities within the PSA and the extension of fibrils are the major factors in the dissipation of applied energy. The debonding resistance in PSA layers appears to be dominated by the cavitation behavior of the adhesive.

19.6.2 Testing of Peel Resistance

Cavitation and fibrillation by debonding, observed on a microscopic scale, are mainly modulus dependent; however, experimental data demonstrate that they also depend on the adherend surface. It is evident that the other solid-state component of a pressure-sensitive joint (that is, the carrier material) also influences energy transfer in the adhesive through its surface and its own bulk deformation [89]. Such deformation may cause joint failure. In modeling the failure of an adhesive layer in a peel test, Lin and Hui [118] assume that plastic yielding of the cover sheet will not occur as long as it is sufficiently thick. Thus, the cover sheet is modeled as an elastica and the fibrils are modeled as elastic strings. The industrial problems that arise from the carrier deformation of thin plastic film-based PSPs were discussed in detail in [89] and the notion of *critical thickness* was introduced.

The influence of simultaneous deformation of the carrier and adhesive during debonding was evidenced by several authors [119].

Tests for peel resistance cover the angle-dependent peel test methods (90° peel adhesion, T-peel adhesion, and 180° peel resistance; see also [72,116]), debonding rate-dependent peel resistance tests, the dwell time-dependent test for peel resistance [72], the carrier-dependent test, and the substrate-dependent test.

Peel test methods for special products include zipp-peel, the drum peel test, the shaft-loaded blister test, and the peel test to evaluate release force.

19.6.3 TESTING OF SHEAR RESISTANCE

Testing of shear resistance includes standard methods for static testing of shear (testing of RT shear resistance, modified tests of RT shear resistance, testing of hot shear resistance, shear adhesion failure temperature [dead load hot strength test]), standard testing methods for the dynamic test of shear resistance, and special testing methods for shear resistance (20° hold test, automotive PSA shear test, thick adherend shear test, and creep test).

19.6.4 OTHER TESTING METHODS

There are also other special adhesion testing methods; for example, global evaluation of adhesive properties, product-related adhesion tests, and adhesion tests based on other principles (Williams plasticity and rheological evaluation of adhesive properties).

The adhesion-related end-use characteristics (e.g., deapplication-related adhesive characteristics) are tested also. Such tests include removability (dry removability, removability of hot-melt PSAs, removability of water-based PSAs, test for adhesive residue, test for repositionability, test for readherability, test for wet removability, and test for repulpability).

The test for time–temperature-dependent stability of adhesive characteristics must also be carried out. Testing methods for aging include the PSA (common aging tests, special aging tests, accelerated weathering testing, aging test for the release force, aging of PVC face stock, PSP migration test, PSP edge-ooze test, plasticizer resistance test, aging tests for release liner) and the carrier.

Application technology-related tests for finished PSPs (labels, tapes, protective films, etc.) must also be carried out.

End use-related tests for finished PSPs include products for general use, products used in specific chemical environments (e.g., test for water resistance/solubility, water whitening, wet anchorage, humidity resistance, water removability, washing machine resistance, wash-off, and solvent resistance) and products used at specific temperatures (tests of low-temperature adhesion, static shear load at low temperatures, cold-tack and high-temperature adhesion). Products for special application fields (e.g., products for electrical use, products for medical use, special tapes for insulation, and special protective films) must also be tested.

19.7 SUMMARY

A critical look at pressure-sensitives must include the statement that a gap exists between the fundamentals of pressure-sensitives and the technology of self-adhesives. An examination of the large volume of scientific literature concerning pressure-sensitives reveals that in the last two decades, the probe-tack test has become a routine investigation method for PSAs, and unfortunately the majority of theoretical investigations uses it only as a way to collect data without clarifying the fine chemistry- and mechanics-related details of the method.

As is known, there are numerous tack-testing methods used in practice [89], but from the point of view of the theory, only the probe-tack test is carried out relatively simply, under mechanically reproducible and mathematically evaluable conditions. It uses a quantified surface, a quantified bonding and debonding rate (force applied perpendicularly), and a well-characterizable substrate quality. This instrumental method gives a tack curve, where the maximum is accepted as tack. Unfortunately, there are PSAs that deviate from this description.

Another problem is caused by the use of too many tack-testing methods [120]. The preference of the technologist for other tack-testing methods than the probe-tack test is due to their simplicity (RBT), usability for a wide range of industrial products with various carrier materials and coating weights (RBT and LT), and simulation of industrial labeling practice (e.g., blow-touch labeling) [120,19] in such (loop) tack measurement.

There is another reason also for such a preference, namely, that RBT and LT give comparable (univocal) tack values for formulations with various raw materials; for example, for natural rubber-based, block copolymer-based, or viscoelastomer-based formulations. This is a very important practical criterion, and ensures the control of the tack using different testing methods and, thus, the interchangeability of a PSP based on a given formulation with another one based on a different formulation, using the adhesive properties as selection criteria.

The use of probe tack as a characterization method for pressure-sensitives was enforced especially after demonstrating that this method allowed the microscopical examination of debonding, and thus, the phenomena of cavitation and fibrillation were discovered.

In fact, both phenomena were known since the beginning of pressure-sensitive formulation, without giving them much importance. Cavitation has been observed in debonding of plastic films also [121], and the buildup of fibrils is a phenomenon known as *legging* from the formulation of removable adhesives [89]. Such fibrils can be observed via RBT also.

A systematic investigation of such phenomena gives rise to the hypothesis that PSAs debond by fibrillation, that is, that fibrillation is a criterion for pressure sensitivity. On the other hand, in the last years, the study of contact of patterned surfaces, where the adhesive itself possesses fibrils *ab ovo*, confirmed the role of such structures in energy dissipation. Jagota and Bennison [122] pointed out that for sufficiently long and thin fibrils, the elastic strain energy stored in the fibrils was not recoverable at pull-off.

Although the fibrillation theory can be considered as the most important breakthrough in the characterization of the debonding process, there are some discrepancies concerning its actual usability. (They concern even its main parameters, that is, the modulus and the surface geometry.) Admitting the strong influence of asperities on fibrillation results, and the discrepancies of the probe-tack tests for "hard" adhesives *ad absurdum*, it can be assumed that patterning and fibrillation are correlated.

Unfortunately, adopting the definition of pressure-sensitives related to low application pressure and high tack (instantaneous adhesion), a lot of products (e.g., protective films, separation films, masking tapes, etc.) based on PSAs cannot be considered as pressure-sensitives, because they do not possess an instantaneous adhesion by slight pressure and a high tack [84]. (In most cases, such products must be laminated on the substrate by elevated pressure, similarly to their plastomer-based competitors.) However, in a manner different from their plastomer-based competitors [13], such products possess a measurable tack; this is the so-called *application tack* [89]. As is known, some of such products possess a low tack because they have a low coating weight (thickness) of the adhesive. Thus, from the technological point of view, the theoretical definition of tack requires the quantification of the adhesive thickness as a parameter also [89].

Ad absurdum, the carrier is nondeformable and the adhesive must be deformable. The development of nonpaper carrier materials provided webs that do not fulfill this condition: they possess deformability and self-adhesivity. Thus, the role of the carrier as a nondeformable, nonadhesive component is no longer valid.

In practice, the dependence of the peel resistance on the coating weight is complex, and the plot (an S-shaped curve) presenting this dependence includes different domains, depending on the adherend surface, carrier material and adhesive characteristics [89,123]. The reduction of the coating weight decreases the peel resistance and, at the extreme, it can lead to a critical thickness of the adhesive layer that does not provide sufficient adhesion. Thus, the dependence of the peel resistance (P) on the coating weight (C_w) is not monotonous, but obeys different mathematical laws in different coating weight domains, and changes in bond force as well as bond failure are controlled by the coating weight.

There is also a *critical deformability* of the carrier that leads to a critical coating weight. Thus, the nondestructive deformation of the carrier material (strain) can lead to a destructive modification of the adhesive joint, due to the change in the adhesive thickness.

For a technologist, the peel resistance, that is, the resistance to debonding of a PSP, is more important than its tack. Peel resistance is an important parameter *per se*, but in actual applications of PSPs, the time–temperature dependence and environment dependence of the peel has gained a special importance also. In particular, the removability [13,19] and water resistance [82,83] of PSPs play a decisive role in their common use.

Theoretical peel studies have considered constant peel force. However, in practice, in some applications a change in peel force (i.e., in peel resistance) is required during debonding [84]. In such applications, a low delamination peel allows fast debonding of the protective layer from large surfaces (e.g., of protective films). Such a variable peel behavior (*zipp-peel*, *zipper effect*) can be achieved using various modalities. One of them consists in the use of an incompatible adhesive formulation containing components with low and very high glass transition temperatures. Another possibility is based on the use of a soft, elastic carrier based on a plastic film that allows deformation of the carrier material and, thus, energy dissipation. Thus, the so-called zipper effect can be due to the extreme deformation of the carrier. Such *easy-to-delaminate* protective tapes with acrylic adhesive have been developed according to [124]. This adhesive (or carrier) displays the zipper effect; that is, the higher the removal speed, the easier to peel. As a consequence, removing the protective tape will no longer be a time-consuming operation.

Further investigations are required to optimize the critical carrier thickness as a function of the carrier material and its construction; to optimize the critical carrier thickness as a function of the adhesive; to mathemathicize the correlation between the mechanical strength of the carrier and peel resistance; to measure the adhesive and carrier deformations during joint failure caused by carrier deformation; and to elaborate an adequate testing method for the static usability of thin protective films.

Summa summarum, PSPs must be discussed as composites on the macromolecular, formulation, and construction levels, where the buildup of the components decisively influences the final properties of the product. On the macromolecular and formulation scales, the buildup of network structures (physical or chemical) renders the use of common rheological notions and dynamic mechanical analysis (DMA) questionable, and on the manufacturing level, the geometry of the adhesive and of the carrier material limits the use of standard testing methods.

ABBREVIATIONS

1. Compounds

AC	acrylic
APP	atactic polypropylene
BOPP	biaxially oriented polypropylene
CR	chloroprene
EVAc	ethylene–vinyl acetate
HC	hydrocarbon
HDPE	high-density polyethylene
HMA	hot-melt adhesive
HMPSA	hot-melt PSA

H-PVC	hard PVC
LDPE	low-density polyethylene
M	maleinate
NR	natural rubber
NRL	natural rubber latex
OPP	oriented polypropylene
PA	polyamide
PB	polybutylene
PE	polyethylene
PET	poly(ethylene terephthalate)
P1B	polyisobutylene
PSP	pressure-sensitive product
PVAc	poly(vinyl acetate)
PVC	poly(vinyl chloride)
PVP	polyvinylpyrrolidone
RR	rubber/resin
SBC	styrene block copolymer
SBR	styrene–butadiene–rubber
SBS	styrene–butadiene–styrene
SEES	styrene–(ethylene/butene)–styrene
SIS	styrene–isoprene–styrene
TPE	thermoplastic elastomer
VAc	vinyl acetate
VAc/E	vinyl acetate/ethylene

2. Terms

AFERA	Association des Fabricants Européens de Rubans Auto-Adhesifs
ASTM	American Society for the Testing of Materials
BGA	Bundesgesundheitsamt
C_w	coating weight
DIN	German standard
DMA	dynamic mechanical analysis
DSC	differential scanning calorimetry
DWA	debonding wet adhesion
E	modulus
EB	electron beam
EN	European norms
EPSMA	European Pressure-Sensitive Manufacturers Association
FINAT	Federation Internationale des Fabricants Transformateurs, d'Adhesifs et Thermocollants sur Papiers et Autres Supports
FIPAGO	Federation Internationale des Fabricants de Papiers Gommés
FTM	Final Test Method
G'	storage modulus
G''	loss modulus
HM	hot-melt
HMPSA	hot-melt pressure-sensitive adhesive
HS	hot shear
IR	infrared
ISO	International Standards Organization

LT	loop tack
MW	molecular weight
MWD	molecular weight distribution
M_e	entanglements molecular weight
M_n	number average molecular weight
M_w	weight average molecular weight
MP	melting point
mPa.s	milliPascal second
PN	Williams plasticity number
PS	pressure-sensitive
PSA	pressure-sensitive adhesive
PSP	pressure-sensitive product
PSTC	Pressure-Sensitive Tape Council
RB	rolling ball
RBT	rolling ball tack
RCT	rolling cylinder tack
RF	radio-frequency
RH	relative humidity
RT	room temperature
SAF	self-adhesive film
SAFT	shear adhesion failure temperature
SB	solvent-based
SF	solvent-free
SH	shear
SL	solventless
SP	softening point
STS	stainless steel
T	temperature
T_g	glass transition temperature
TLMI	Tag and Label Manufacturers Institute
TMA	thermomechanical analysis
UV	ultraviolet
W	work
W_a	work of adhesion
WB	water-based
WBA	wet bonding adhesion

REFERENCES

1. I. Benedek, Introduction. In: *Developments in Pressure-Sensitive Products*, I. Benedek (Ed.), Taylor & Francis, Boca Raton, FL (2006).
2. J. Andres. *Allgemeine Papier-Rundschau* (16), 444 (1986). In: I. Benedek, *Pressure Sensitive Adhesives and Applications*, 2nd edn, Marcel Dekker, New York (2004).
3. R. Jordan. Haftklebstoffe: Eine Einführung. *Adhäsion*, 1–2, 17 (1987).
4. Anonymous. *Coating Magazine* 11, 46 (1990).
5. Anonymous. *Coating Magazine* 11, 307 (1990).
6. R. Gutte. *Reichold Albert Nachrichten* (4), 26 (1970). In: *Developments In Pressure-Sensitive Products*, 2nd edn, I. Benedek (Ed.), CRC Press, Boca Raton, FL (2005).
7. P. Foreman and P. Mudge. EVA-based waterborne pressure-sensitive adhesives. In: *Technical 12, Technical Seminar Proceedings*, p.203, Itasca, IL (1989).
8. M. Fairley. *Labels and Labelling International Magazine* (5/6) 76 (1997).
9. Anonymous. *Etiketten: Labels Magazine*, 3, 8 (1995).

10. Anonymous. *Der Siebdruck Magazine*, 3, 70 (1986).

11. I. Benedek. Manufacture of PSPs. In: *Development and Manufacture of Pressure-Sensitive Products*, Marcel Dekker, New York (1998).

12. I. Benedek. *Developments in Pressure-Sensitive Products*, Taylor & Francis, Boca Raton, FL (2006).

13. I. Benedek. Build-up and classification of pressure-sensitive products. In: *Development and Manufacture of Pressure-Sensitive Products*, Marcel Dekker, New York, (1998),

14. I. Benedek. European Adhesives and Sealants, (2), 25 (1996).

15. J. H. S. Chang. Water-based pressure-sensitive adhesives. European patent EP 0179628, assigned to Merck & Co. (1984).

16. I. Benedek and L. J. Heymans. Comparison. In: *Pressure-Sensitive Adhesives Technology*, Marcel Dekker, New York (1997).

17. I. Benedek. End-use of pressure-sensitive products. In: *Developments in Pressure-Sensitive Products*, I. Benedek (Ed.), Taylor & Francis, Boca Raton, FL(2006).

18. I. Benedek. Competitors for pressure-sensitive adhesives and products. In: *Handbook of Pressure-Sensitive Adhesives and Products*, Vol. 3, I. Benedek (Ed.), Taylor & Francis, Boca Raton, FL (2009).

19. I. Benedek. Build-up and classification of pressure-sensitive products. In: *Developments in Pressure-Sensitive Products*, I. Benedek (Ed.), Taylor & Francis, Boca Raton, FL (2006).

20. I. Benedek. Construction and classes of pressure-sensitive products. In: *Handbook of Pressure-Sensitive Adhesives and Products*, Vol. 3, I. Benedek (Ed.), Taylor & Francis, Boca Raton, FL (2009).

21. I. Benedek and L. J. Heymans. Definition and construction of pressure-sensitive laminate. *Pressure-Sensitive Adhesives Technology*, Marcel Dekker, New York (1997).

22. I. Benedek. End-use domains and application technology of pressure-sensitive adhesives and products. In: *Handbook of Pressure-Sensitive Adhesives and Products*, Vol. 3, I. Benedek (Ed.), Taylor & Francis, Boca Raton, FL (2009).

23. I. Benedek. Rheology of pressure-sensitive adhesives. In: *Pressure-Sensitive Adhesives and Applications*, Marcel Dekker, New York (2004).

24. I. Benedek and L. J. Heymans. Influence of composite structure on stiffness. In: *Pressure-Sensitive Adhesives Technology*, Marcel Dekker, New York (1997).

25. I. Benedek and L. J. Heymans. Definition and construction of pressure-sensitive laminate. In: *Pressure-Sensitive Adhesives Technology*, Marcel Dekker, New York (1997).

26. I. Benedek. *End-Use of PSPs, Development and Manufacture of Pressure-Sensitive Products*, Marcel Dekker, New York (1998).

27. C. M. Brooke. *Finat News*, (3), 34, (1987). In: I. Benedek, *Development and Manufacture of Pressure Sensitive Products*, Marcel Dekker, New York (1998).

28. H. Brus, C. Weitemeyer and J. Jachmann, *Finat News*, (3), 84 (1987). In: I. Benedek, *Development and Manufacture of Pressure-Sensitive Products*, Marcel Dekker, New York (1998).

29. Anonymous. *Adhesives Age*, (8), 28 (1986) In: *Developments in Pressure-Sensitive Products*, 2nd edn, I. Benedek (Ed.), Taylor & Francis, Boca Raton, FL (2006).

30. I. Benedek and M. M. Feldstein (Eds.). Fundamentals of pressure sensitivity, *Handbook of Pressure-Sensitive Adhesives and Products*, Vol. 1., Taylor & Francis, Boca Raton, FL (2009).

31. O. A. Soboleva, A. V. Semakov, S. V. Antonov and V. G. Kulichikhin. Surface phenomena on a solid-liquid interface, and rheology of pressure-sensitivity. In: *Handbook of Pressure-Sensitive Adhesives and Products*, Vol. 1, I. Benedek and M. M. Feldstein (Eds.), Taylor & Francis, Boca Raton, FL (2009).

32. C. Creton and R. Schach. Diffusion and adhesion. In: *Handbook of Pressure-Sensitive Adhesives and Products*, Vol. 1, I. Benedek and M. M. Feldstein (Eds.), Taylor & Francis, Boca Raton, FL (2009).

33. A. E. Chalikh and A. A. Scherbina. Transition zones in adhesive joints. In: *Handbook of Pressure-Sensitive Adhesives and Products*, Vol. 1, I. Benedek and M. M. Feldstein (Eds.), Taylor & Francis, Boca Raton, FL (2009).

34. C. Derail and G. Marin. Role of viscoelastic behavior in pressure-sensitive adhesives. In: *Handbook of Pressure-Sensitive Adhesives and Products*, Vol. 1, I. Benedek and M. M. Feldstein (Eds.), Taylor & Francis, Boca Raton, FL (2009).

35. E.-P. Chang. Viscoelastic properties and windows of pressure-sensitive adhesives. In: *Handbook of Pressure-Sensitive Adhesives and Products*, Vol. 1, I. Benedek and M. M. Feldstein (Eds.), Taylor & Francis, Boca Raton, FL (2009).

36. C. Creton and K. R. Shull. Probe tack, in: *Handbook of Pressure-Sensitive Adhesives and Products*, Vol. 1, I. Benedek and M. M. Feldstein (Eds.), Taylor & Francis, Boca Raton, FL (2009).

37. M. M. Feldstein. Molecular fundamentals of pressure-sensitive adhesion. In: *Development in Pressure-Sensitive Products*, I. Benedek (Ed.), Taylor & Francis, Boca Raton, FL (2006).

38. M. M. Feldstein. Molecular fundamentals of pressure-sensitive adhesion. In: *Development in Pressure-Sensitive Products*, I. Benedek (Ed.), Taylor & Francis, Boca Raton, FL (2006).

39. M. M. Feldstein and C. Creton. Pressure-sensitive adhesion as a material property and as a process. In: *Pressure-Sensitive Design, Theoretical Aspects*, I. Benedek (Ed.), VSP, Utrecht , the Netherlands (2006).

40. S. V. Kotomin. Durability of viscoelastic adhesive joints. In: *Handbook of Pressure-Sensitive Adhesives and Products*, Vol. 1, I. Benedek and M. M. Feldstein (Eds.), Taylor & Francis, Boca Raton, FL (2009).

41. M. M. Feldstein, M. B. Novikov and C. Creton. Significance of relaxation for adhesion of pressure-sensitive adhesives. In: *Handbook of Pressure-Sensitive Adhesives and Products*, Vol. 1, I. Benedek and M. M. Feldstein (Eds.), Taylor & Francis, Boca Raton, FL (2009).

42. C. A. Dahlquist. *Adhesion: Fundamentals and Practice*, p.143, Elsevier, Amsterdam (1970).

43. I. Benedek. Physical Basis of PSPs. In: *Developments and Manufacture of Pressure-Sensitive Products*, Marcel Dekker, New York (1999).

44. I. Benedek and L. J. Heymans. Rheology of PSA solutions and dispersions. In: *Pressure-Sensitive Adhesives Technology*, Marcel Dekker, New York (1997).

45. I. Benedek and L. J. Heymans. Rheology of the pressure-sensitive laminate. In: *Pressure-Sensitive Adhesives Technology*, Marcel Dekker, New York (1997).

46. I. Benedek and L. J. Heymans. Chemical composition of PSAs. In: *Pressure-Sensitive Adhesives Technology*, Marcel Dekker, New York (1997).

47. I. Benedek. Chemical composition of PSAs. In: *Pressure-Sensitive Adhesives and Applications*, Marcel Dekker, New York (2004).

48. I. Benedek. Chemical basis. In: *Development and Manufacture of Pressure-Sensitive Products*, Marcel Dekker, New York (1999).

49. I. Benedek. Chemical basis of pressure-sensitive products. In: *Development in Pressure-Sensitive Products*, I. Benedek (Ed.), Taylor & Francis, Boca Raton, FL (2006).

50. D. J. St. Clair and J. T. Harlan. Rubber-styrene block copolymers in adhesives. *Adhesives Age* (12), 39 (1975).

51. D. R. Tucker. *Adhäsion* (7), 248 (1971). In: *Development and Manufacture of Pressure-Sensitive Products*, Marcel Dekker, New York (1999).

52. F. M. Rosenbaum. *Adhesives Age* (6), 32 (1972). In: *Development and Manufacture of Pressure-Sensitive Products*, Marcel Dekker, New York (1999).

53. I. Benedek and L. J. Heymans. Introduction. In: *Pressure-Sensitive Adhesives Technology*, Marcel Dekker, New York (1997).

54. I. Benedek. Pressure-sensitive raw materials. In: *Handbook of Pressure-Sensitive Adhesives and Products, Vol. 2: Technology of Pressure-Sensitive Adhesives and Products*, I. Benedek and M. M. Feldstein (Eds.), Taylor & Francis, Boca Raton, FL (2009).

55. I. Benedek. Elastomer-based formulation. In: *Pressure-Sensitive Design, Theoretical Aspects*, Vol. 1, I. Benedek (Ed.), VSP, Utrecht, the Netherlands (2006).

56. Y. Hu and C. W. Paul. Block copolymer-based hot-melt pressure-sensitive adhesives. In: *Handbook of Pressure-Sensitive Adhesives and Products, Vol. 2: Technology of Pressure-Sensitive Adhesives and Products*, I. Benedek and M. M. Feldstein (Eds.), Taylor & Francis, Boca Raton, FL (2009).

57. N. Willenbacher and O. V. Lebedeva. Polyisobutene-based pressure-sensitive adhesives. in: *Handbook of Pressure-Sensitive Adhesives and Products, Vol.2: Technology of Pressure-Sensitive Adhesives and Products*, I. Benedek and M. M. Feldstein (Eds.), Taylor & Francis, Boca Raton, FL (2009).

58. P. B. Foreman. Acrylic adhesives. In: *Handbook of Pressure-Sensitive Adhesives and Products, Vol. 2: Technology of Pressure-Sensitive Adhesives and Products*, I. Benedek and M. M. Feldstein (Eds.), Taylor & Francis, Boca Raton, FL (2009).

59. S. B. Lin, L. D. Durfee, A. A. Knott and G. K. Schalau II. Silicone pressure-sensitive adhesives. In: *Handbook of Pressure-Sensitive Adhesives and Products, Vol. 2: Technology of Pressure-Sensitive Adhesives and Products*, I. Benedek and M. M. Feldstein (Eds.), Taylor & Francis, Boca Raton, FL (2009).

60. M. M. Feldstein, P. Singh and G. W. Cleary. Hydrophylic adhesives. In: *Handbook of Pressure-Sensitive Adhesives and Products, Vol. 2: Technology of Pressure-Sensitive Adhesives and Products*, I. Benedek and M. M. Feldstein (Eds.), Taylor & Francis, Boca Raton, FL (2009).

61. Z. Czech and R. Hinterwaldner. Pressure-sensitive adhesives based on polyurethanes. In: *Handbook of Pressure-Sensitive Adhesives and Products, Vol. 2: Technology of Pressure-Sensitive Adhesives and Products*, I. Benedek and M. M. Feldstein (Eds.), Taylor & Francis, Boca Raton, FL (2009).

62. I. Benedek. Role and methods of formulation. In: *Handbook of Pressure-Sensitive Adhesives and Products, Vol. 2: Technology of Pressure-Sensitive Adhesives and Products*, I. Benedek and M. M. Feldstein (Eds.), Taylor & Francis, Boca Raton, FL (2009).

63. I. Benedek. The role of formulation of pressure-sensitive products. In*: Pressure-Sensitive Formulation*, VSP, Utrecht , the Netherlands (2000).

64. I. Benedek. Formulation principles. In: *Pressure-Sensitive Formulation*, VSP, Utrecht, the Netherlands (2000).

65. I. Benedek. Design and formulation for mixing, in: *Pressure-Sensitive Design, Theoretical Aspects*, I. Benedek (Ed.), VSP, Utrecht, the Netherlands (2006).

66. I. Benedek. Manufacture of pressure-sensitive products. In: *Handbook of Pressure-Sensitive Adhesives and Products, Vol. 2: Technology of Pressure-Sensitive Adhesives and Products*, I. Benedek and M. M. Feldstein (Eds.), Taylor & Francis, Boca Raton, FL (2009).

67. I. Benedek. Manufacture of PSAs. In: *Pressure-Sensitive Adhesives and Applications*, Marcel Dekker, New York (2004).

68. I. Benedek. Manufacture of pressure-sensitive labels. In: *Pressure-Sensitive Adhesives and Applications*, Marcel Dekker, New York (2004).

69. I. Benedek. Formulation basis. In: *Pressure-Sensitive Formulation*, VSP, Utrecht, the Netherlands (2000).

70. D. H. Lim and H. J. Kim. General performance of pressure-sensitive adhesives. In: *Pressure-Sensitive Design, Theoretical Aspects*, I. Benedek (Ed.), VSP, Utrecht , the Netherlands (2006).

71. I. Benedek. Converting properties of PSAs. In: *Pressure-Sensitive Adhesives and Applications*, Marcel Dekker, New York (2004).

72. I. Benedek. The role of the formulation for the adhesive properties. In: *Pressure-Sensitive Formulation*, VSP, Utrecht, the Netherlands (2000).

73. I. Benedek and L. J. Heymans. Adhesion-cohesion balance. In: *Pressure-Sensitive Adhesives Technology*, Marcel Dekker, New York (1997).

74. I. Benedek, Adhesive properties of PSPs. In: *Developments and Manufacture of Pressure-Sensitive Products*, Marcel Dekker New York (1999).

75. I. Benedek. Adhesive properties of pressure-sensitive products. In: *Developments in Pressure-Sensitive Products*, I. Benedek (Ed.), Taylor & Francis, Boca Raton (2006).

76. G. R. Hamed and C. H. Hsieh, Relationship between cohesive strength and the tack of elastomers, *J. Polym. Sci., Polym. Phys.*, 21, 1415–1425 (1983).

77. I. Benedek. The influence of the formulation on the confectionating properties. In: *Pressure-Sensitive Formulation*, VSP, Utrecht, the Netherlands (2000).

78. I. Benedek and L. J. Heymans. Tack. In: *Pressure-Sensitive Adhesives Technology*, Marcel Dekker, New York (1997).

79. I. Benedek and L. J. Heymans. Peel adhesion. In: *Pressure-Sensitive Adhesives Technology*, Marcel Dekker, New York (1997).

80. I. Benedek and L. J. Heymans. Definition and construction of pressure-sensitive laminate. In: *Pressure-Sensitive Adhesives Technology*, Marcel Dekker, New York (1997).

81. I. Benedek and L. J. Heymans. Printability of pressure-sensitive laminate. In: *Pressure-Sensitive Adhesives Technology*, Marcel Dekker, New York (1997).

82. I. Benedek. Manufacture of PSPs. In: *Developments and Manufacture of Pressure-Sensitive Products*, Marcel Dekker, New York (1999).

83. I. Benedek. Chemical composition (Chapter 5). In: *Pressure-Sensitive Adhesives and Applications*, Marcel Dekker, New York, 2004.

84. I. Benedek. End use of PSP. In: *Developments in Pressure-Sensitive Products*, Taylor & Francis, Boca Raton, FL (2006).

85. I. Benedek. End-use domains and application technology of pressure-sensitive adhesives and products. In: *Handbook of Pressure-Sensitive Adhesives and Products*, Vol. 3, I. Benedek and M. M. Feldstein (Eds.), Taylor & Francis, Boca Raton, FL (2009).

86. I. Benedek and L. J. Heymans. End-use properties. In: *Pressure-Sensitive Adhesives Technology*, Marcel Dekker, New York (1997).

87. I. Benedek and L. J. Heymans. Label application technology. In: *Pressure-Sensitive Adhesives Technology*, Marcel Dekker, New York (1997).

88. I. Benedek. Competitors for pressure-sensitive adhesives and products. In: *Handbook of Pressure-Sensitive Adhesives and Products*, Vol. 3, I. Benedek and M. M. Feldstein (Eds.), Taylor & Francis, Boca Raton, FL (2009).

89. I. Benedek. Adhesive performance characteristics. In: *Pressure-Sensitive Adhesives and Applications*, Marcel Dekker, New York (2004).

90. I. Benedek. Conversion properties of PSAs. In: *Pressure-Sensitive Adhesives and Applications*, Marcel Dekker, New York (2004).

91. I. Benedek. Pressure-sensitive formulation in practice. In: *Pressure-Sensitive Design and Formulation, Application*, VSP, Utrecht, the Netherlands (2006).

92. M. M. Feldstein, G. W. Cleary and P. Singh. Pressure-sensitive adhesives of controlled water-absorbing capacity. In: *Pressure-Sensitive Design, Theoretical Aspects*, I. Benedek (Ed.), VSP, Utrecht, the Netherlands (2006).

93. I. Benedek. Construction and classes of pressure-sensitive products. In: *Handbook of Pressure-Sensitive Adhesives and Products*, Vol. 3, I. Benedek and M. M. Feldstein (Eds.), Taylor & Francis, Boca Raton, FL (2009).

94. I. Benedek. Manufacture of pressure-sensitive products. In: *Development in Pressure-Sensitive Products*, I. Benedek (Ed.), Taylor & Francis, Boca Raton, FL (2006).

95. I. Benedek. Formulation principles, off-line/in-line. In: *Pressure-Sensitive Formulation*, VSP, Utrecht, the Netherlands (2000).

96. I. Benedek. Principles of pressure-sensitive design and formulation. In: *Pressure-Sensitive Design, Theoretical Aspects*, I. Benedek (Ed.), VSP, Utrecht, the Netherlands (2006).

97. I. Benedek. Pressure-sensitive raw materials. In: *Handbook of Pressure-Sensitive Adhesives and Products, Vol. 2: Technology of Pressure-Sensitive Adhesives and Products*, I. Benedek and M. M. Feldstein (Eds.), Taylor & Francis, Boca Raton, FL (2009).

98. I. Benedek. Physical basis for the viscoelastic behavior of pressure-sensitive adhesives. In: *Pressure-Sensitive Adhesives and Applications*, Marcel Dekker, New York (2004).

99. I. Benedek. Practical formulation. In: *Pressure-Sensitive Formulation*, VSP, Utrecht, the Netherlands (2000).

100. Z. Czech. Synthesis, properties and application of water-soluble pressure-sensitive adhesives. In: *Pressure-Sensitive Design, Theoretical Aspects*, I. Benedek (Ed.), VSP, Utrecht, the Netherlands (2006).

101. I. Benedek. In-situ synthesis. In: *Pressure-Sensitive Formulation*, VSP, Utrecht, the Netherlands (2000).

102. H.-S. Do and H.-J. Kim. UV-curable pressure-sensitive adhesives. In: *Pressure-Sensitive Design, Theoretical Aspects*, I. Benedek (Ed.), VSP, Utrecht, the Netherlands (2006).

103. I. Benedek. Pressure-sensitive raw materials, in-line synthesis. In: *Handbook of Pressure-Sensitive Adhesives and Products, Vol. 2: Technology of Pressure-Sensitive Adhesives and Products*, I. Benedek and M. M. Feldstein (Eds.), Taylor & Francis, Boca Raton, FL (2009).

104. I. Benedek and L. J. Heymans. Simultaneous manufacture of PSA and PSA laminates. In: *Formulation, Pressure-Sensitive Adhesives Technology*, Marcel Dekker, New York (1997).

105. I. Benedek. *Pressure-Sensitive Formulation*, VSP, Utrecht, the Netherlands (2000).

106. I. Benedek. *Pressure-Sensitive Design, Theoretical Aspects*, VSP, Utrecht, the Netherlands (2006).

107. I. Benedek. *Pressure-Sensitive Design and Formulation, Application*, VSP, Utrecht, the Netherlands (2006).

108. I. Benedek and L. J. Heymans. Formulation. In: *Pressure-Sensitive Adhesives Technology*, Marcel Dekker, New York (1997).

109. I. Benedek. Principles of pressure-sensitive design and formulation. In: *Pressure-Sensitive Design, Theoretical Aspects*, Vol. 1, I. Benedek (Ed.), VSP, Utrecht, the Netherlands (2006).

110. Z. Czech. Removable and repositionable pressure-sensitive materials. In: *Pressure-Sensitive Design, Theoretical Aspects*, I. Benedek (Ed.), VSP, Utrecht, the Netherlands (2006).

111. I. Benedek and L. J. Heymans. Definition and construction of pressure-sensitive laminate. In: *Pressure-Sensitive Adhesives Technology*, Marcel Dekker, New York (1997).

112. I. Benedek. Pressure-sensitive design and formulation in practice. In: *Pressure-Sensitive Design, Theoretical Aspects*, I. Benedek (Ed.), VSP, Utrecht, the Netherlands (2006).

113. Y. J. Park. Hot-melt PSA based on styrenic polymer. In: *Pressure-Sensitive Design, Theoretical Aspects*, I. Benedek (Ed.), VSP, Utrecht, the Netherlands (2006).

114. I. Benedek. Manufacture of the finished product by carrier coating. In: *Developments and Manufacture of Pressure-Sensitive Products*, Marcel Dekker, New York (1999).

115. I. Benedek and L. J. Heymans. Tack, drying of the coating. In: *Pressure-Sensitive Adhesives Technology*, Marcel Dekker, New York (1997).

116. I. Benedek. The influence of the formulation on the manufacturing technology. In: *Pressure-Sensitive Formulation*, VSP, Utrecht, the Netherlands (2006).

117. I. Benedek. Test of pressure-sensitive adhesives and products. In: *Handbook of Pressure-Sensitive Adhesives and Products, Vol. 3: Applications of Pressure-Sensitive Adhesives and Products*, I. Benedek and M. M. Feldstein (Eds.), Taylor & Francis, Boca Raton, FL (2009).

118. Y. Y. Lin and C. Y. Hui. Modeling the failure of an adhesive layer in a peel test. *Proceedings of 25th Annual Meeting of Adhesion Society, and the Second World Congress on Adhesion and Related Phenomena*, p. 230, Orlando, FL (2002).

119. C. Verdier and J. M. Piau. Understanding peeling of PSAs by use visualization. *Proceeding of the 23*rd *Annual Meeting of the Adhesion Society*, p. 36, Myrtle Beach, SC (2000).

120. I. Benedek. Test methods. In: *Pressure-Sensitive Adhesives and Applications*, 2nd edn, Marcel Dekker, New York (2004).

121. A. Sharma, V. Shennoy and J. Sharkar. Pattern, forces and metastable pathways in debonding of elastic films. *Proceedings of the 23rd Annual Meeting of the Adhesion Society*, p. 335, Wilmington, DE (2004).

122. A. Jagota and S. J. Bennison. Mechanics of adhesion through a fibrillar microstructure. *Integr. Comp. Biol.* 42, 1140–1145 (2002).

123. T. Yamaguchi, H. Morita and M. Doi. Modeling on debonding dynamics of pressure-sensitive adhesives. *Proceedings of the 29th Annual Meeting of the Adhesion Society*, p. 297, Jacksonville, FL (2006).

124. A. J. Crosby, K. R. Shull, H. Lakrout and C. Creton. Deformation and failure modes of adhesively bonded elastic layers. J. Appl. Phys. 88, 2956–2966 (2000).

Part 3

Applications of Adhesives

20 Adhesives in the Wood Industry

Manfred Dunky

CONTENTS

20.1   INTRODUCTION

Adhesives play a central role in the production of wood-based panels, such as particleboard (PB), medium- and high-density fiberboard (MDF/HDF), and oriented strandboard (OSB). Despite many attempts to introduce other raw materials, urea, formaldehyde and, to a smaller extent, melamine are still the main raw materials for adhesive resins in this industry; phenol is only used rather seldom. Use of isocyanate has grown in the last two decades, mainly in the OSB production.

The search for cheaper, faster curing/faster development of bond strength, and more complex materials drives the permanent development of adhesives for wood components [1–3].The first two requirements are caused by the strong competition in the wood industry and the necessary efforts to minimize costs at a required level of product quality and performance. The requirement "more complex" stands for new and specialized products and processes, including fulfilling all requirements concerning low emission of formaldehyde from the adhesives and of other volatile components from adhesives and from the wood substance itself, or special requirements concerning application in wet environments.

The quality of bonding and, hence, the properties of the wood-based panels are determined to a great extent by the type and quality of the adhesive. Development in bonded wood products, therefore, is always linked to development in adhesives. The greater part of the adhesives currently used for wood-based panels, however, is produced with just a few raw materials, as mentioned. The resin production and the formulation of the ready-to-use adhesive mixes have become more and more sophisticated and are key factors in meeting the actual requirements in their application.

The quality of bonding and, hence, the properties and performance of the wood-based panels are determined by three main parameters:

- The wood component, especially the wood surface, including the interphase (as the joint region of wood material and adhesive) and the interface at molecular level between the wood surface and the adhesive layer (bondline)
- The applied adhesive
- The working conditions and process parameters

Adhesives can increase the strength and stiffness of the composite by effectively transferring and distributing stresses. All types of wood-based composites can be reduced to the formation of bondlines between the various structural elements (solid wood plies, veneers, strands, particles, or even between fibers) of the composites, including the bonding process, moisture balance during the press process, and moisture content in the interphase. Adhesive bonds in a wood structure are formed by a combination of even or spot-like distribution of the adhesive on the surface, by its wetting and penetration behavior, and finally its solidification, which can be achieved physically (drying) or chemically (chemical hardening). The adhesive must diffuse partly into each of the two surfaces to establish an intimate contact with the porous inner surface of the wood. The area of interface is to be maximized, so that the intermolecular forces of attraction between the adhesive and the wood molecules can reach their maximum.

The most important features of wood that are strongly associated with adhesion and bonding performance are its porous structure, the roughness of its surface, its chemical composition, and its hygroscopic nature. The penetration of the adhesive directly affects the performance of adhesive bonds in wood. Most investigations of adhesive penetration that have been carried out in the past have focused on the two-dimensional analysis of microtome sections prepared from bonded wood samples. The preparation of such microtome sections, however, might affect the bondline. Furthermore, these studies can only speculate about the pathways of liquid adhesive flow in the wood substance during bond formation. New analysis methods such as high-resolution computer tomography (CT) (e.g., submicrometer x-ray CT) with resolutions of a few micrometers (and aiming to achieve values <1 μm) have significantly improved the current knowledge of adhesive penetration.

The ideal case of bonding of wood occurs between two flat surfaces of solid wood, showing (1) the behavior of wood as macroscopic material and (2) properties based on the microscopic scale of wood cells representing the structure of solid wood. The bonding mechanisms are based on the contact between two wooden surfaces, although the individual contact areas of bonding might be only a few square centimeters or only square millimeters in size. As long as the original wood structure still remains at the surface of the strands or the particles used for these boards, the fundamental regularities of bonding of two wood surfaces are still valid. For fiberboards, the individual bonding areas rather represent the cross point between two fibers; but still, the fiber structure offers possibilities for the penetration of the adhesive, and the intramolecular forces at the interface will be similar to solid wood bonding. For all wood-based composites, deformations of the wood and fiber structure can occur in the bondline when heat, moisture, and mechanical pressure are applied.

This chapter is an updated version of the chapter "Wood Adhesives" [1] and concentrates mainly on the new literature of the last 15 years. Older references are preferably to be found in [1].

20.2 TYPES OF WOOD ADHESIVES

In the wood-based panels industry, a great variety of adhesives is currently in use. Condensation resins based on formaldehyde represent, by far, the largest volume within the wood adhesives and are prepared by the reaction of formaldehyde with various chemicals such as urea, melamine, phenol, resorcinol, or combinations thereof. When used at the board plant, these adhesive resins are mainly liquid and consist of linear or branched oligomers and polymers in aqueous solution or dispersion. During hardening and gelling, they convert to three-dimensional cross-linked and, therefore, insoluble and nonmeltable networks. The hardening conditions used can be acidic (for aminoplastic resins), highly alkaline (for phenolic resins), or neutral to slightly alkaline (for resorcinol resins). The so-called *polymeric 4,4'-diphenylmethane diisocyanate (PMDI)*, as the main representative of isocyanates in the wood industry, is another important chemical compound used for various applications in the wood industry, especially for water-resistant bonds. Table 20.1 summarizes the main wood adhesives actually in use and their various applications. Aminoplastic adhesive resins represent by far the largest volume in wood adhesives, especially for the production of PB and MDF/ HDF, and partly also for OSB, plywood, blockboards, and cross-laminated timber (CLT). They are also used in the furniture industry as well as in carpenters' shops.

TABLE 20.1
Fields of Application for Various Wood Adhesives

Adhesive Type	UF	MUF (MUPF)	PF PUF	PMDI	PVAc	Old Natural Adhesives	Natural Adhesives	Inorganic Adhesives
Dry use	x	x[a]						
Use in moist conditions		x	x	x			x	x
Hardboards			x					x[b]
MDF	x	x[a]	x	x			x	
HDF								
Plywood	x	x	x					
Beams	(x)	x	x[c]			x		
Solid wood panels	(x)	x			x			
Veneering	x[d]				x			
Furniture	x				x			

Note: UF, urea–formaldehyde resin; mUF, melamine-fortified UF resin (low content of melamine <8%); MUF, melamine–urea–formaldehyde resin (content of melamine usually >14%); MUPF, melamine–urea–phenol–formaldehyde resin; PF/PUF, phenol and phenol–urea–formaldehyde resin; (P)RF, resorcinol–(phenol–) formaldehyde resin; PMDI, polymeric methylene diisocyanate; PVAc, poly(vinyl acetate) adhesive; Old natural adhesives (historic natural adhesives) such as starch, gluten, casein adhesives; Natural adhesives (adhesives based on natural resources), such as tannin, lignin, carbohydrates; Inorganic adhesives, such as cement and gypsum.
[a]Boards with reduced thickness swelling, for example for laminate flooring.
[b]Special production method.
[c]RF or PRF resins.
[d]Partly powder resins.

20.3   OVERVIEW ON REQUIREMENTS CONCERNING WOOD ADHESIVES

The necessity to achieve shorter press times is omnipresent in the woodworking industry, to keep production costs low. Shorter press times in a given production line can be achieved by, among others:

- Highly reactive adhesive resins possessing rapid gelling and hardening and steep increase in bonding strength, even at a low degree of chemical curing.
- Highly reactive adhesive mixes obtained by the addition of accelerators, special hardeners, cross-linkers, and other additives.
- Preheating of the mat just before entering the press by various means, such as high frequency, microwaves, or steam injection into the mat.
- Optimization of the pressing process, for example, by increasing the effect of the steam shock by (1) higher press temperatures, (2) a bigger difference in the moisture content between the surface and the core layer of the mat before hot pressing, or (3) an additional steam injection step in the hot press.
- Keeping the production process as constant as possible, especially mat weight and mat moisture conditions.

Table 20.2 describes the actual requirements in the production and in the development of wood adhesives.

Table 20.3 summarizes chances and challenges of various adhesives used in the production of wood-based panels. These technical parameters mainly decide, together with the costs for the

TABLE 20.2

Actual Requirements for Wood Adhesives

Production capacity	Reactivity of the adhesives, shorter press times, shorter cycle times
Better hygroscopic behavior of boards	Lower thickness swelling, higher resistance against the influence of humidity and water, better outdoor performance
Cheaper raw materials and alternative products	Target: reduction of production costs, such as minimization of the melamine content in a MUF resin; impeding factors (often temporary) can be the shortage of raw materials for the adhesives
Ecological requirements	Life cycle assessment, energy and raw material balances; recycling also concerns used adhesive resins, since they are one of the major raw materials for wood-based panels
	Waste water and effluent management in adhesive production
	Reduction of gas emission during panel production and from the finished boards

TABLE 20.3

Evaluation of Various Adhesive Types with Respect to Various Parameters

Parameter	UF	MUF	PF	PMDI
Price	Low	Medium to high	Medium	High
Necessary hardening temperature	Low	Medium	High	Low
Press time	Short	Medium	Medium	Medium
Influence of wood species during use	High	Medium	Low	Low
Efficiency (needed amount of adhesive)	Low	Medium	Medium	High
Manipulation of the adhesive in board production	Easy	Easy	Easy	Difficult
Resistance against hydrolysis (boiling water)	No	Medium	High	High
Use in humid conditions	No	Partly yes	Yes	Yes
Formaldehyde emission from the adhesive (emission of formaldehyde from the wood substance itself is not considered)	E1[a]	E1[a]; even lower emission possible[b]	More or less no	No
Tolerance against variation of humidity during process	Medium	Medium	Medium	Good
Suitability for difficult substrates	Poor	Poor	Poor	Good
Release from press platens	Good	Good	Good	Bad
Cold tack	Good	Good	Good	Bad
Toxicity during processing	Good	Good	Good (with restrictions)	Bad
Necessary changes on usual board plants	No	No	Partly	Partly

[a] Emission class for low subsequent formaldehyde emission.

[b] Such as F**** according to JIS A 5908 [4] and A 5905 [5] or North American ULEF specification (Ultra Low Emitting Formaldehyde Resins) [6].

adhesive system per board volume unit (rather than only the price as such for the adhesive), which system should be selected for a certain application. For several applications, various adhesives can often be used in principle and are in competition with each other.

The proper choice of the adhesive depends on the required properties of the wood-based panels and the working conditions during the production, as well as on the costs for the adhesive system; this means not only the net price of the adhesive, but rather the costs of the adhesive system per volume unit of produced boards, the capacity of the line (determined by necessary press time), as well as environmental aspects (formaldehyde emission, volatile organic compounds [VOCs], life cycle balances, recycling) and national requirements.

Gas emission from wood-based panels during their production can be caused by wood-inherent chemicals themselves (such as terpenes or free acids), as well as by volatile compounds and residual monomers from the adhesive. The emission of formaldehyde especially is a matter of concern, and so are possible emissions and discharges of free phenols or other materials. The subsequent formaldehyde emission depends, on the one hand, on the residual, unreacted formaldehyde present mainly in aminoplastic resin-bonded boards (trapped as gas in the wood or dissolved in the moisture present in the panel). On the other hand, the hydrolysis of weakly bonded formaldehyde from N-methylol groups, acetals, and hemiacetals, as well as in more severe cases of hydrolysis (e.g., at high relative humidity) from methylene-ether bridges, increases again the content of emittable formaldehyde after resin hardening [7]. In contrast to phenolic resins, a permanent reservoir of potentially emittable formaldehyde is the consequence of the presence of these weakly bonded structures. This explains the continuous, yet low, release of formaldehyde from urea–formaldehyde (UF)-bonded wood-based panels even over long periods. However, the level of emission depends on the environmental conditions, a fact which may be described by the resin hydrolysis rate, which indicates if this formaldehyde reservoir will or will not lead to unpleasantly high emission. The higher this hydrolysis rate is, the more it can contribute to subsequent formaldehyde emission. For formaldehyde emission after adhesive hardening in board manufacture, stringent emission regulations are in force worldwide, and further reduction of limits is being discussed. However, it is important that not only the boards themselves, but also veneering and carpenters' adhesives, lacquers, varnishes, and other sources of formaldehyde are controlled, since they might also contribute to the formaldehyde steady-state concentration.

20.4 AMINOPLASTIC AND PHENOPLASTIC ADHESIVE RESINS

Aminoplastic and phenoplastic adhesive resins are formed by the reaction of formaldehyde mainly with urea, melamine, and phenol.

| Urea | Melamine | Phenol | Formaldehyde |

The production procedure of formaldehyde condensation resins mainly involves two steps:

- Methylolation: Strongly exothermic reaction, mainly monomeric methylols are formed, in most cases in alkaline conditions
  - UF resin:

- MF resin:

- PF resin:

- Condensation: Formation of methylene and ether linkages in acidic (aminoplastic resins) or alkaline conditions (PF resins). The reaction has to be stopped when the resin is still liquid (pH ⇒ alkaline conditions or by cooling); therefore, the determination of the proper end point of the reaction is essential.

- Methylene bridges:

$$R-CH_2OH \quad + \quad H_2N-R' \xrightarrow[-H_2O]{} R-CH_2-NH-R'$$

- Ether bridges with further reaction to methylene bridges:

$$R-CH_2OH \quad + \quad OHCH_2-R'' \xrightarrow[\substack{-CH_2O \\ -H_2O}]{} R-CH_2-R''$$

The final urea addition, as an important step for the reactivity of aminoplastic resins, enables adjustment of the usually low molar ratio of the final resin and is done in one or several steps, including also if necessary holding steps at a certain temperature and for certain times above room temperature. The addition of urea to a phenolic resin (1) decreases the content of free formaldehyde and the viscosity of the resin by cleavage of hydrogen bonds and by the dilution effect and (2) accelerates the hardening reaction via the possible higher degree of condensation of the PF base resin; however, more or less no cocondensation of this postadded urea with the phenolic resin occurs.

Based on the raw materials used, various types of adhesive resins can be prepared (Table 20.4):

The application in the board production is given mainly in liquid form; only for special applications are powdered (spray-dried) types also used, such as partly for OSB. The resins consist of linear or branched oligomeric and polymeric molecules in an aqueous solution and partly as a dispersion of molecules in these aqueous solutions. They show a duroplastic hardening behavior, which leads to a three-dimensional cross-linking and hence to insoluble and nonmeltable networks.

The molecular characterization of formaldehyde-based resins comprises the determination of:

- The concentration and the molar ratios of the main components; for example, F:U, F:(NH$_2$)$_2$, F:P:NaOH.
- The content of reactive sites and functional groups and their distribution in the resin, including the type of bridges between the aromatic rings of the phenol molecule, as well as branching sites; analysis can be done by infrared (IR) spectroscopy, ^{1}H-nuclear magnetic resonance (NMR) spectroscopy, ^{13}C-NMR, or ^{15}N-NMR.
- The degree of condensation and the molar mass distribution, investigated by gel permeation chromatography (GPC), also called size exclusion chromatography (SEC).
- The monitoring of the condensation process by diluting the resin with water, taking the water tolerance as measure for the degree of condensation or the increase in molar masses (shift of the molar mass distribution). Still an unsolved task is the prediction of board properties based on the analysis results of the resins used. A few correlation equations between chemical structures of various resins with different compositions and various board properties (dry or wet internal bond [IB] strength, subsequent formaldehyde emission) have been described in the literature (see Section 20.4.4), but a general correlation

TABLE 20.4
Types of Industrial Wood Adhesives

UF	Urea–formaldehyde resin
MF	Melamine–formaldehyde resin (only minor role as adhesive resin and always mixed with UF resins)
mUF	Melamine-fortified UF resins (content of melamine up to approx. 8% based on liquid resin)
MUF	Melamine–urea–formaldehyde cocondensation resin
MUPF, PMUF	Melamine–urea–phenol–formaldehyde cocondensation resins, with different proportions of the various raw materials (only minor role as adhesive)

for various resins and various panels does not exist. The definition of the main parameters, however, is valuable, even though the numbers in individual correlation equations might differ.

The molar mass distributions of formaldehyde condensation resins are much broader than for other polymers. The low molar mass monomers comprise free formaldehyde (molar mass = 30) as well as residual and nonreacted species such as urea (molar mass = 60), melamine (molar mass = 126), or phenol (molar mass = 94), respectively. Monomeric methylols are generated, for example, by the reaction of postadded urea with the free formaldehyde or as monomeric methylolphenols. The oligomeric compounds with two to five molecules of urea, melamine, or phenol are linked by methylene or methylene-ether bridges. The more or less polymeric structures with higher molar masses are the resin molecules in the closer sense of the word and mainly responsible to build up the bondline.

Due to the equilibrium reactions in all formaldehyde-based resins, some small amounts of residual monomers can always be present. Aminoplastic resins for particleboard and MDF usually contain less than 0.1% free formaldehyde [2]. Monomer concentrations for PF resins are in the range of <0.3% mass for the free formaldehyde and <0.1% mass for free phenol. The content of monomers should always be minimized by the proper production procedure to minimize environmental and health concerns. On the other hand, for aminoplastic resins, the free formaldehyde is necessary to induce the hardening reaction via the reaction with the so-called *hardener* (mainly ammonium salts); it is an additional cross-linker besides methylol groups. However, it also causes a certain formaldehyde emission during the press cycle and some residual formaldehyde leads to the subsequent formaldehyde emission from the pressed boards. Due to the stringent regulations worldwide concerning subsequent formaldehyde emissions, the molar ratios F:U or F:$(NH_2)_2$ were already decreased significantly in the 1980s. No problems occur with phenolic resins due to the strong C–C bonds in the resin (no hydrolysis and, hence, no further liberation of formaldehyde).

20.4.1   UF Resins

UF resins [1–3,8–12] are based on a series of consecutive reactions of urea and formaldehyde. Using different conditions of reaction and preparation, a large variety of condensed chemical structures are possible. UF resins are thermosetting resins and consist of linear or partly branched oligomers and polymers, always admixed with some amount of monomers, such as urea. The presence of a certain proportion of unreacted urea is the consequence of addition of urea at a later stage of the resin production (*final urea*); this is done to adjust the molar ratio to the required low values to be able to fulfill the requirements concerning formaldehyde emission. The presence of free formaldehyde is necessary to induce the hardening reaction; on the other hand, it causes a certain level of formaldehyde emission during the hot-press resin hardening cycle and from the boards while in service. This fact has changed significantly the composition and formulation of UF resins during the last decades (starting already in the mid-1970s to fulfill the steadily increasing demands for low emission products). After hardening, UF resins consist of insoluble, three-dimensional networks that cannot be melted or thermoformed again. In their application stage, UF resins consist of a mix of water solution and dispersion. Spray-dried powders, in most cases, have to be redissolved and redispersed in water before application.

Even with only the two components, urea and formaldehyde, a broad variety of possible reactions and resin structures can be achieved. The basic characteristics of UF resins can be ascribed at a molecular level to their high reactivity, their waterborne state (no organic solvents in use), but also the reversibility of the aminomethylene bridge formation; this feature explains the low resistance of UF resins to water and moisture attack, especially at higher temperatures. This hydrolysis also causes a certain contribution to the subsequent formaldehyde emission from UF-bonded boards [7]. The reaction of urea and formaldehyde is, basically, a two-step process, with methylolation (hydroxymethylation) as the first step, followed by the acid condensation step. Depending on the molar ratio used in this step, the methylolation reaction can be performed at various pH levels, such as in the strong acidic range, in the slightly acidic range, or, in the traditional way, at alkaline pH levels. Depending on the pH at this stage, the molar ratio must be selected accordingly to assure methylolation before the condensation reaction starts. Depending on this molar ratio (F:U) the addition of up to three (in theory, four) molecules of the bifunctional formaldehyde to one molecule of urea gives the various methylolureas; each methylolation step has its own rate constant k_i, with different values for the forward and the backward reactions.

The condensed structure of the UF resin as such with its molar mass distribution (MWD) is formed in the condensation step, which is always performed in the acidic range; pH, reaction temperature, and duration of this step determine the degree of condensation (DOC) and the MWD. The molar ratio in this step usually is in the range F:U = 1.8–2.5 and is adjusted by a relevant urea addition, if necessary, after the methylolation step. The methylol groups (as formed in the methylolation step), urea, and the free formaldehyde react during this condensation step to form linear and partially branched molecules, usually with a broad and polydisperse MWD characteristic for UF resins (and formaldehyde-based resins in general). Molar ratios lower than approximately 1.8 during this acid condensation step tend to cause resin precipitation. The condensation reaction, with an increase in the molar masses, can be monitored by GPC [13,14]. In a longer acid condensation step, molecules with higher molar masses are formed; the GPC peak, hence, shifts to lower elution volumes. The final UF resin has a low F:U molar ratio obtained by the addition of additional urea, which might also be added in several steps [11,12]. Conditions during this addition of urea can heavily influence the final properties and performance of the produced resins. To achieve the desired solid content, the last step in UF resin production includes the distillation of the resin solution, usually to 65%–68% resin solids content; distillation is performed by vacuum distillation in the reactor itself or in a thin layer evaporator. Manufacturing procedures and various UF recipes are described in the chemical literature and in patents [10,11,15–19].

The type of bonding between the urea molecules depends on the conditions used in the acidic step: Low temperatures and slightly acid pH favor the formation of methylene-ether bridges ($-CH_2-O-CH_2-$), whereas higher temperatures and lower pH rather create the more stable methylene bridges ($-CH_2-$). Ether bridges can be rearranged to methylene bridges by splitting off formaldehyde. One ether bridge needs two formaldehyde molecules and is not even as stable as a methylene bridge; hence, it is highly recommended to follow procedures that minimize the formation of such ether groups in UF resins. Besides the abovementioned ways of resin production, the literature also reports on various special procedures. Some of these yield uron structures in high proportion [13,20–26] or triazinone rings in the resins [21]. The latter are formed by the reaction of ammonia or a primary or secondary amine, with urea in excess of formaldehyde under alkaline conditions. These resins are used, for example, to enhance the wet strength of paper.

The large variety of (1) structural elements, such as methylene bridges, ether bridges, methylols, amide groups, or even cyclic derivatives such as uron rings as well as (2) possible reactions make the study of UF resins difficult. The content of these structural elements in the still uncured resins has an obvious influence on their curing rate and on the structure and the mechanical properties of the final network. The analysis of the MWD of UF resins is restricted by the low solubility of highly condensed structures in most of the usual solvents. Whereas the low molar mass species can be dissolved easily and monitored by GPC (SEC) and even by high-performance liquid chromatography (HPLC), the higher molar mass species are only soluble in dimethylsulfoxide (DMSO). Looking at the performance of an UF resin, it is not enough to concentrate on the low molar masses (which determine mainly the reactivity of the resin and the adhesion to the adherend); clarification of the condensed backbone of the resin is also important, because these structures determine the behavior of the resin when applied to a wood surface with its inherent porosity. The more the MWD tends to higher molar masses, the more the resin will remain at the wood surface, creating the UF bondline between the two adherends.

The most important parameters for the aminoplastic resins (UF, melamine-fortified UF [mUF], melamine–urea–formaldehyde [MUF]) are:

- The type of monomers used.
- The molar ratio of the various monomers in the resin, such as F:U, F:M, or F:$(NH_2)_2$, whereby urea contributes two NH_2 groups, and melamine three NH_2 groups; the molar ratio is sometimes also expressed as F:(U + M); F, U, M, and NH_2 represent moles.
- The purity of the different raw materials, such as the level of residual methanol or formic acid in formaldehyde, biuret in urea, or ammeline and ammelide in melamine.
- The parameters of the reaction procedure used, such as the pH and temperature sequence, the type and amount of alkaline and acidic catalysts, the sequence of addition of the different raw materials, and the duration of the different reaction steps in the production procedure.
- The achieved DOC and the MWD in the final resin.

The following chemical species are present in UF resins:

- Free formaldehyde, which is in steady state with the remaining methylol groups and the postadded urea, depending on the final molar ratio (F:U).
- Monomeric methylol groups, which have been formed mainly by the reaction of the post-added urea with high content of free formaldehyde at the still high molar ratio in the acid condensation step.
- Oligomeric methylol groups, which have not reacted further in the acid condensation reaction or which have been formed by the reaction of several molecules of the postadded urea with formaldehyde.
- Molecules with higher molar masses, which constitute the condensed polymer portion as such.

Because of the necessity of limiting the subsequent formaldehyde emission, the molar ratio F:U has been decreased constantly over the years [2]. UF resins with high and low molar ratios differ in their reactivity, due to the different contents of free formaldehyde and in their degree of cross-linking in the cured network. Decreasing the molar ratio F:U means lowering the degree of branching and cross-linking in the hardened network, causing lower cohesive bond strength. The degree of cross-linking is directly related to the molar ratio of the two components. The decrease in the molar ratio was quite substantial; a standard UF resin for particleboard had a molar ratio F:U = 1.6 at the end of the 1970s. Actual molar ratios F:U and F:$(NH_2)_2$, respectively, of UF and mUF resins are in the range of 1.0–1.1, in case further addition of formaldehyde catchers, such as urea, is necessary as well as modification of UF resins with melamine. Availability of free formaldehyde is essential for the curing reactivity of the resin, because it determines the speed of the reaction with the so-called hardener (ammonium salt) creating the acid that enhances acid-induced further condensation until reaching the cured state. The higher the molar ratio F:U, the higher the content of free formaldehyde in the resin. In a straight UF resin, it is approximately 0.1% at F:U = 1.1% and 1% at F:U = 1.8 [2]. It decreases with the storage time of the resin, due to further reactions of this formaldehyde. Table 20.5 summarizes the influence of the molar ratio F:U on various properties of wood-based panels and their production.

Ferra et al. [13] compared two processes for producing UF resins: (1) an alkaline–acid process involving three steps (procedure with an alkaline methylolation followed by an acid condensation and the addition of a final amount of urea); (2) a strongly acid process involving four steps (a strongly acid condensation, followed by an alkaline methylolation, a second condensation under a moderately acid pH, and finally methylolation and neutralization reaching a slight alkaline pH). The

TABLE 20.5
Influence of the Molar Ratio F:U on Various Properties of UF-Bonded Wood-Based Panels

Decreasing the molar ratio F:U leads to

A decrease of	Formaldehyde emission during the production of wood-based panels
	Subsequent formaldehyde emission
	Mechanical properties
	Degree of hardening
An increase of	Thickness swelling and water absorption
	Necessary press time
	Susceptibility to hydrolysis

two resins differ greatly in chemical structure, composition (molar mass distribution), viscosity, and reactivity.

UF resins prepared with combinations of three reaction pH levels (1.0, 4.8, and 8.0) and four initial molar ratios F:U with stepwise addition of urea to formaldehyde showed an increased proportion of high molar masses in the resins at the low reaction pH [27]. Furthermore, the molar ratio F:U of the first step (methylolation) influenced the optimal pH during the second step (condensation). At acidic pH, the high molar mass products increased as F:U decreased; while, at alkaline pH, little difference was evident between the high or low molar masses at various values of F:U. The formation of a high percentage of uron derivatives under strong acidic conditions also indicated that these resins differed considerably from conventional UF resins. Based on the bond strength and formaldehyde emission data, however, the weak acid catalysts seem to provide the best compromise. A similar attempt [28] investigated the varying polymer structures of UF resins resulting from different values of F:U used in the first step of the resin manufacture by ^{13}C NMR. As the initial F:U decreased from 2.40 to 1.80, the viscosity increased due to a faster condensation reaction, causing a lesser degree of branching and an increased proportion of free urea amide groups in the resin structure.

For the investigation of the effects of the reaction pH conditions and hardener type on the reactivity, chemical structure, and adhesion performance of UF resins, three different reaction pH conditions (such as traditional alkaline–acid at pH 7.5 ⟶ 4.5, weak acid at pH 4.5, and strong acid at pH = 1.0) were used to synthesize UF resins, which were then cured by adding four different types of hardeners (ammonium chloride, ammonium sulfate, ammonium citrate, and zinc nitrate) [19]. The gel time of the UF resins decreased with an increasing amount of the three ammonium hardeners, but increased for zinc nitrate. The strong acid pH condition produced uronic structures in UF resin, while both alkaline–acid and weak acid conditions produced quite similar chemical species in the resins. The maximum adhesion strength, however, was reached with the resin prepared under the strong acid pH condition, whereas the weak acid reaction condition rather provided a balance between increasing resin reactivity and improving adhesion strength of UF resin.

The reaction pH during the UF resin production influences the thermal behavior of such resins, and also the achievable reduction of formaldehyde emission of particleboard bonded with them, as well as the bond strength [29]. The resins were synthesized at three different reaction pH conditions, such as alkaline (pH = 7.5), weak acid (pH = 4.5), and strong acid (pH = 1.0). The activation energy (E_a) increased as the reaction pH decreased from alkaline to strong acid conditions. Gel time was the shortest at the weak acidic preparation pH. The formaldehyde emission of particleboards was the lowest for the UF resins prepared under strong acid, but had the poorest bond strength. These results indicated that the thermal curing behavior is related to chemical species, affecting the formaldehyde emission, while the poor bond strength is believed to be related to the high molecular mobility (i.e., less rigidity) of the resin used.

During storage, molecular rearrangement can take place [30]. Typical UF resins, synthesized at different condensation molar ratios F:U = 1.80, 2.10, and 2.40 but with identical final molar ratio F:U = 1.15, were followed for 50 days at room temperature. The ^{13}C-NMR samples taken during the storage period showed gradual migration of hydroxymethyl groups from the polymeric first-urea components to the monomeric second-urea components, and also an advancing degree of polymerization of resins by forming methylene and methylene-ether groups involving the second urea. The different condensation molar ratios caused different polymer branching structures as the effect of the first step of resin synthesis.

As a part of abating the formaldehyde emission of UF resins, the effects of the F:U molar ratio on the thermal curing behavior (differential scanning calorimetry [DSC]) of UF resins and on the properties of bonded PB were investigated [15,31]. As the F:U molar ratio decreased, the gel time, onset and peak temperatures, and heat of reaction (ΔH) increased, while the activation energy (E_a) and rate constant (k) decreased. The amount of free formaldehyde of the UF resin and the formaldehyde emission from PBs decreased with decreasing F:U molar ratio. The mechanical and hygroscopic properties became slightly worse with decreasing F:U.

The molar mass distribution of UF resins is determined by the DOC and by the addition of urea (and sometimes also other components) after the condensation step; this again shifts the resin mass distribution toward lower average molar masses. For this reason, the molar mass distribution is much broader than for other polymers, reaching from low molar mass monomers (the molecular weight of formaldehyde is 30, for urea it is 60) up to more or less polymerized structures. However, it is not clearly known what the highest molar masses in a UF resin are [14]. When a UF resin is diluted with a surplus of water (usually 100%–300% of the amount of the liquid resin), precipitation of parts of the resin can occur. These parts preferably contain the higher molar mass molecules of the resin, and their relative proportion increases at higher degrees of condensation. Information on correlations between the molar mass distribution (DOC) and mechanical and hygroscopic properties of the boards produced, however, is rather rare and often equivocal [2].

The molar mass distribution (and the DOC) is one of the most important characteristics of the resin, and it determines several properties of the resin. Consequences of highly condensed resin structures (high molar masses) are:

- The viscosity at a given solids content increases [32]
- The flowing ability is reduced [32]
- The wetting behavior on a wood surface becomes worse [32]
- The penetration into the wood surface is reduced [33]
- The distribution of the resin on the wood material (particles, fibers) worsens
- The water dilutability of the resin becomes lower
- The portion of the resin that remains soluble in water decreases

The DOC mainly influences the application behavior (wetting behavior; penetration into a wood surface). At the higher temperatures during the hot-press cycle, the viscosity of the resin drops, before the onset of hardening again leads to an increase in viscosity. With this temporary low viscosity, the adhesive wetting behavior improves significantly, but also its substrate penetration behavior increases. The reactivity of an aminoplastic resin at a certain molar ratio seems to be independent of its viscosity (DOC).

20.4.1.1 Replacement of Formaldehyde in UF Resins

The water tolerance and thickness swelling of PBs bonded with UF resins were markedly improved by introducing small amounts of UF–propanal (UFP) polycondensates into the UF resin [34]. The ^{13}C NMR of the UFP resin showed that urea and propanal can react up to the formation of dimers. However, the water repellency of the alkyl chain of propanal imparted by insertion into the resin limits the proportion of propanal that can be used. According to GPC, UF and UFP resins exist as a steady state between two separate interpenetrating networks, namely, one in solution and the second in a state of physical gelation. The latter one is different from the state of physical gelation observed on aging or advancement of formaldehyde-only-based polycondensation resins. Formaldehyde in a UF resin can also be replaced by furfural [35], hydroxymethylfurfural [36], isobutyraldehyde [37], dimethoxyethanol [38,39], or succinaldehyde [40]; so far, no industrial use of such alternatives is known.

20.4.1.2   UF Resins + PMDI

Wieland et al. [41] showed, by investigating hardened resins, that urethane bridges derived from the reaction of the isocyanate group with the hydroxymethyl group of urea are still formed even at short curing times. Polyureas and biurets obtained from the reaction of isocyanate with water are the predominant cross-linking reactions of PMDI alone and in UF–PMDI resin systems under fastcuring conditions. Residual, unreacted isocyanate groups in the hardened network were consistently observed. Their proportion markedly decreased at higher UF:PMDI ratios. Under these fastcuring conditions, the UF resin appears to self-condense through an unusually high proportion of methylene-ether links rather than methylene bridges alone. Direct NMR tests on thin hardboard bonded under fast pressing conditions with different proportions of UF:PMDI confirmed that cross-linking due to polyurea and biuret formation is predominant in the cross-linking of PMDI when alone or in UF–PMDI resin systems, but still showing residual, unreacted isocyanate groups in the panel. Their proportion is high when the proportion of PMDI in the system is high. The presence or absence of urethanes could not be confirmed directly on the panels, as the relevant peaks were masked by the wood carbohydrate signals of wood cellulose and hemicelluloses. For further literature on UF + PMDI see [42–44].

20.4.2   Improvement of the Hygroscopic Behavior of Boards by Melamine-Fortified UF Resins, Melamine-Urea-Formaldehyde (MUF) Resins, and Phenol-Melamine-Urea-Formaldehyde (MUPF, PMUF) Resins

UF resins are mainly used for interior boards (for use in dry conditions, e.g., in furniture manufacturing); a higher resistance against moisture and water can be achieved by incorporating melamine into the resin (mUF resins, MUF, MUPF, PMUF) [45–54]. The level of melamine addition, and especially the resin manufacturing sequence used in relation to how melamine is incorporated into the resin, can be very different and influences the performance of the resin and the properties of the boards produced herewith. The degree of resistance of these resins against hydrolysis is based on their behavior at molecular level. The methylene bridge linking the nitrogens of amido groups can be hydrolyzed rather easily by water attack in UF resins. This effect is lower in the case of M(U)F resins, mainly due to the much lower water solubility of melamine itself as a consequence of the water repellency characteristic for the triazine ring of melamine. Melamine-containing resins, however, are more expensive due to the significantly higher price of melamine compared with urea. Therefore, the content of melamine in these resins must be as high as strictly necessary, but always as low as possible. The higher the content of melamine, the higher the stability of the hardened resin toward the influence of humidity and water (hydrolysis resistance).

Resins containing melamine can be characterized by calculation of the various molar ratios, such as $F:(NH_2)_2$ or $F:(U+M)$, or of the triple molar ratio $F:U:M$. The different calculation modes are based on different assumptions regarding how many of the six hydrogen atoms in the three NH_2 groups will be replaced by a methylol group. The mass portion of melamine in the resin can be described based on (1) the liquid resin, (2) the resin solids content, or (3) the sum of urea and melamine in the resin. The actual $F:(NH_2)_2$ molar ratios used for wood-based panels are similar to those indicated for the UF resins; the addition of melamine as such improves the chemical stability of the resin, but does not decrease formaldehyde emission at parity of molar ratio.

MUF resins may be formed by real cocondensation, or as two independent networks of UF and MF that only penetrate each other. Mercer and Pizzi [49] assigned two low-intensity resonances in NMR spectra of MUF resins to copolymerized species. Tomita and Hse [55] concluded the presence of cross-linked MUF moieties from NMR signals. The individual species, however, were not characterized, and the resonances overlapped with those of MF and UF resins; this fact limited the precise analyses, particularly where line-broadening occurred after extensive polymerization [55]. Philbrook et al. [56] used ^{15}NMR in the investigation of MUF resins; through the identification of resonances unique to cross-linked species, copolymerization could be proven. This helped to determine the extent of cocondensation in industrial MUF resins, to correlate resin structure with function and performance of the resin and to optimize the manufacturing process [57].

The application of MUF resins is very similar to the UF resins; only the level of hardener addition is usually much higher, to overcome the higher buffer capacity of the MUF resins. Up to approximately 12% melamine (based on liquid resin), the gel time increases; at further increased content of melamine the better cross-linking ability of the melamine reduces gel times again. An MUF resin yields a smaller pH drop after addition of the hardener than a UF resin, due to the buffer capacity of the triazine ring of melamine; however, this also causes a decrease of the hardening rate of the resin up to a certain content of melamine, as mentioned, with the consequence of a longer gel time and longer necessary press times. This is also seen in the shifts of the exothermic DSC peaks of hardening, which are observed in thermal experiments [58].

The deterioration of a bondline and, hence, its durability under influence of moisture or water is determined essentially by:

- Failure of the resin: low hydrolysis resistance and degradation of the hardened resin, causing loss of bond strength. In UF resins, the aminomethylene link is susceptible to hydrolysis and, therefore, it is unstable at higher relative humidity, especially at elevated temperatures [59]. Water also causes degradation of the UF resin; the higher the temperature of the water in which the boards are immersed, the stronger is this effect.

- Failure at the interface between the resin and the wood surface: replacement of physical bonds between resin and reactive wood surface sites by water or other nonresin chemicals. The adhesion of UF resins to cellulose is sensitive to water, not only due to the already mentioned instability to hydrolysis of the methylene bridge and of its partial reversibility, but also because theoretical calculations have shown that, on most cellulose sites, the average adhesion of water to cellulose is stronger than that of UF oligomers [11,60]. Thus, water can displace hardened UF resins from the surface of wood. The opposite effect is valid for PF resins [11,61].

- Mechanical collapse of bondlines due to mechanical forces and stresses: Water causes swelling and, therefore, movement of the structural components of the wood-based panels (cyclic stresses due to swelling and shrinking, including stress rupture).

Hardened UF resins can be hydrolyzed by moisture or water, due to the relative weakness of the bond between the nitrogen of the urea and the carbon of the methylene bridge, especially at higher temperatures. During this reaction, the methylene bridge is emitted as formaldehyde [62]. The amount of liberated formaldehyde can be taken, under certain circumstances, as a measure of the resistance of the resin against hydrolysis [63]. The main parameters influencing the rate and extent of the hydrolysis are temperature, residual pH in the bondline, and degree of hardening of the resin. The acid, which induces hardening of the resin and causes the still low pH in the cured UF bondline, leads to such hydrolysis and, hence, loss of bond strength. Myers [64] pointed out that in the case of such an acid hardening system, the decrease in the durability of adhesive bonds could be initiated by the hydrolysis of the wood cell-wall polymers adjacent to the bondline as well as in the case of UF-bonded products by acid-catalyzed resin degradation. The resistance of UF resins against hydrolysis, hence, might be increased, if the acid residues in the bondline could be removed.

The amount of hardener (acids, acidic substances, latent hardeners), therefore, should always be adjusted to the desired hardening conditions (press temperature, press time) but excess dosage must be avoided. A neutral pH bondline, therefore, should show a distinctly higher hydrolysis resistance. However, bondline neutralization must not take place as long as the hardening reaction is still ongoing; otherwise, this would delay or even prevent curing. This aspect is quite a challenge that, in practice, has never been solved. Yamaguchi et al. [59] found that a complete removal of acidic substances by soaking plywood test specimens in an aqueous sodium bicarbonate solution resulted in considerable increase in water resistance of UF bondlines.

Zhang et al. [65] compared a UF resin with two MUF resins with the melamine added at two different synthesis stages concerning performance and curing behavior. The curing behavior and functional groups of the resins were measured by DSC and Fourier transform infrared (FT-IR) spectroscopy. The melamine addition and the addition stage had significant influences on the characteristics of resins. Early addition of melamine showed the highest bond strength and the lowest formaldehyde emission, with DSC activation energy similar to the UF resin. Late addition of melamine (at lower intermediate molar ratio when melamine is added) increased the activation energy and showed reduced amounts of methylene and methylene-ether bonds in cured resin, meaning that the cross-linking degree of this resin was lower. In similar attempts, low molar ratio MUF resins were synthesized with 2.5% and 5.0% melamine levels, added at the beginning, middle, and end points of the first alkaline step of the typical UF resin synthesis procedure, and were compared with typical MUF resins synthesized with melamine additions carried out at the final alkaline step [66]. Early addition of melamine improved mechanical and hygroscopic properties of PBs, but showed slightly higher formaldehyde emission. MUF resins synthesized with melamine addition carried out at the middle and at the end of the first alkaline step were highly turbid and showed different chemical structures and very short storage stability. Increase in methylene-ether groups at various combinations of reaction conditions showed partial improvement in resin properties, but also an increase in formaldehyde emission. It looks as if methylene-ether-type groups present in significant levels in UF and MUF resins are strong sources for subsequent formaldehyde emission [23,67].

During synthesis of MUF resins, melamine and urea react at different rates with formaldehyde; depending on the extent of preadvancement of the UF part or the MF component, the resin properties vary. In particular, preadvancing the UF base resin component to an appropriate extent was found to be a key to synthesizing various low-level melamine-modified MUF resins [68,69]. Different sequences of addition of melamine in the production of MUF resins can influence chemical structure, curing behavior, and cross-sectional morphology of the resins [70]. Addition of melamine at levels up to 12% has been found to provide lower formaldehyde emission and maintain the physical properties of boards [71]. This, however, is only the case if the molar ratios including melamine are reported as F:(U+M), with all numbers in the equation signifying the number of moles in a certain amount of resin. This implies that, on average, only two of the three NH_2 groups of the melamine will react with formaldehyde. Recalculating the molar ratio as $F:(NH_2)_2$ yields a common connection between the molar ratio and the so-called *perforator value* as a measure of the content of unreacted formaldehyde in a board, independent of the content of melamine. It might be argued that the free formaldehyde can react with all reactive sites of melamine.

Strength improvement induced by addition of acetals such as methylal and ethylal in MUF resins can be mostly ascribed to the increased effectiveness and participation of the melamine in resin cross-linking, due to its now preferentially homogeneous rather than heterogeneous reactions [72,73]. This phenomenon is due to (1) the increased solubility in water afforded by the acetal cosolvents of both the unreacted melamine and the normally very much lower soluble higher molar mass and lower methylolated oligomers fraction, and (2) the effect that acetals (such as methylals) have on the size distribution of the resin colloidal particles on decreasing the average colloidal particle diameter of the resin. This latter effect appears to be due to the disruption of the molecular clustering of the MUF resin colloidal particles.

Hexamine sulfate has been shown to markedly improve the water resistance of hardened MUF resins used as wood adhesives. So-called *iminoamino methylene bases intermediates* are obtained by the decomposition of hexamine sulfate (hexamine stabilized by the presence of strong anions such as SO_4^{2-} and HSO_4^-); the effect of the hexamine sulfate is closely linked to the strong buffering effect on MUF resins, and its role is mainly to induce stability of conditions during resin networking, due to the buffer. Shifting of the polycondensation $\rightleftarrows$ degradation equilibrium to the left appeared to be the determinant factor. This was a consequence of maintaining a higher, constant pH during curing, due to the buffer action; the effect was shown to be induced by very small amounts, between 1% and 5 wt% of this material based on resin solid content. The effects induced by hexamine sulfate are of longer duration than those of other potential buffers. This is due to the heat stability of the hexamine sulfate under the standard hot-curing conditions of the resin. The network formed is then more cross-linked and more resistant against degradation when curing occurs in the correct pH range. The result is a much better performance of the wood board after water attack. This strong effect allowed the use of MUF resins with much lower melamine content [74–76].

20.4.3 REACTIVITY AND HARDENING REACTIONS OF AMINOPLASTIC RESINS

During curing, a three-dimensional network is built up. This leads to an insoluble resin that is no longer thermoformable. The hardening reaction is the continuation of the acid condensation process during resin production. The acid hardening conditions can be adjusted (1) by the addition of a so-called *latent hardener* (usually ammonium salts such as ammonium sulfate or ammonium nitrate) or (2) by direct addition of acids (maleic acid, maleic acid anhydride, formic acid, phosphoric acid, and others) or of acidic substances that dissociate in water (e.g., aluminum sulfate or urea phosphate).

Ammonium sulfate reacts with the free formaldehyde in the resin to generate sulfuric acid, which decreases the pH; this low pH and, hence, the acid conditions enable the condensation reaction to restart; finally, the hardening of the resin takes place. The pH decrease takes place with a rate depending on the relative amounts of available free formaldehyde and hardener, and is greatly accelerated by heat. UF resins show high reactivity, and hence short hot-press times. To increase the capacity of a production line, especially by shortening the panel hot-press times, adhesive resins with a reactivity as high as possible should be used. This involves two parameters: (1) a short gel time and (2) a rapid and instantaneous bond strength development, even at a low degree of chemical curing. Differences in the reactivity can be followed by various methods (see Section 20.8). Higher additions of ammonium salt-based hardeners decrease the gel time; this effect is significant mainly at levels of addition up to approximately 0.5%–1.0% solid hardener based on resin solid. At higher levels, this decrease is still significant but less so [77]. A high surplus of ammonium ions might even increase the gel time slightly again, due to the decrease of the overall $F:(NH_2)_2$ molar ratio including the hardener ammonium ions in the calculation.

The direct addition of an acid (such as acetic acid) promotes the acidic hardening reaction by decreasing the pH to the acidic range and, hence, shortens the gel time. At lower temperatures (60°C and 80°C) this initial lower pH guarantees a quicker hardening compared with the ammonium salts, which still need to generate the acid to start the hardening process. At the lower temperatures, this reaction of the ammonium salt hardener is restricted by the low availability of free formaldehyde. The amount of hydrogen ions supplied by the addition of the acid remains more or less constant even at higher temperatures, since no additional acid is generated during the gel test due to the absence of ammonium ions. This effect explains the distinctly lower activation energy for the addition of an acid compared with the addition of an ammonium salt as hardener, as shown in Figure 20.1, reflecting the temperature dependence of the gel time. In the case of the ammonium salt hardeners, the reaction of the ammonium ions with the free formaldehyde of the resin takes too long to be able to compete at low temperatures with the gel time of the directly added acid. Only at high temperatures (such as 100°C) are the generation of acid and, hence, the decrease in the pH quick enough to enable curing in a short time and to result in even shorter gel times compared with the fixed amounts of the added acetic acid; the acetic acid provides the acidic hydrogen ions directly, without the need for a temperature-controlled reaction with the free formaldehyde. The pH level of a UF resin during hardening at these temperatures using ammonium salt hardeners usually drops down to values of

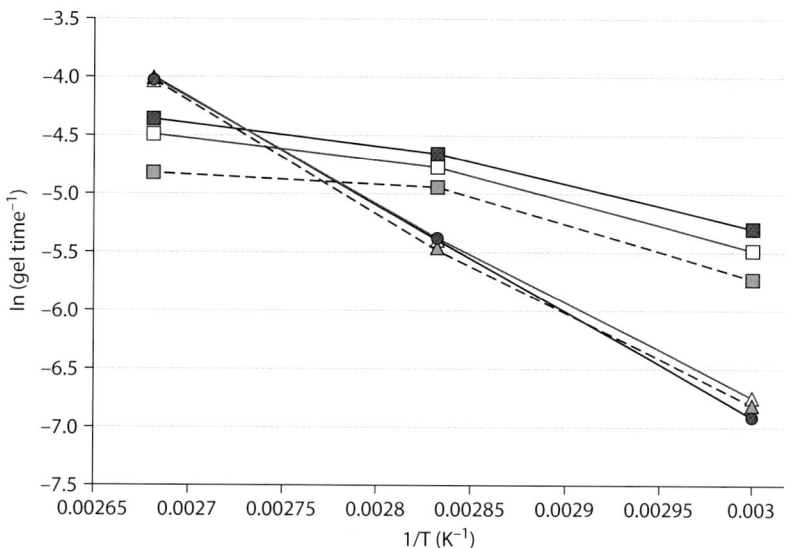

FIGURE 20.1 Arrhenius plots of log inverse gel time (with gel time expressed in seconds) against the reciprocal absolute temperature (1/K) for a UF resin as a measure of the hardening rate with various hardeners at three temperatures. ■ 3.2% acetic acid, □ 2.3% acetic acid, ■ 1.6% acetic acid; △ 2.3% ammonium sulfate, ▲ 1.2% ammonium sulfate; ● 2.3% ammonium chloride. (From Stefke, B. and Dunky, M., *J. Adhes. Sci. Technol.*, 20, 761–785, 2006.)

2–3. The content of free formaldehyde in the liquid resin is reduced considerably with a lower molar ratio F:U; considering this fact also, a strong correlation between the gel time and the content of free formaldehyde in the liquid aminoplastic resin exists. For UF resins with low molar ratio F:U, as used for the production of particleboards or MDF, the content of free formaldehyde is below 0.1%, based on the liquid resin at room temperature. Even though this content increases at higher temperatures, the free formaldehyde remains the rate-determining parameter.

The hardening rate (Φ) of UF resins when creating an Arrhenius plot has been defined and measured in various forms, such as (1) the thermal reaction in the DSC [29,78–80], (2) the inverse gel time for different hardener systems [77,81], (3) an increase of bond strength achieved under certain conditions [82], or (4) based on rheological measurements [81] (see further information in Section 20.8.3). If these tests are performed at different temperatures, the correlation between the hardening rate Φ and the temperature can be described according to the Arrhenius equation as:

$$\Phi = A \times \exp\left(-E_a / RT\right)$$

An apparent activation energy E_a can be calculated from the slope of the linear regression line in the plot of the natural logarithm of the hardening rate versus the reciprocal of absolute temperature. In the case of gel-time measurements, the reciprocal of gel time can be assumed as a suitable measure for the hardening rate (Figure 20.1) [77].

For the three ammonium salt tests, the apparent activation energy calculated from Figure 20.1 is in the range of 70–75 kJ/mol; for the acetic acid as hardener, considerably lower values for this activation energy were found in the range of 29–32 kJ/mol. The low activation energy, as in the case with the acetic acid, means that the temperature dependence of the reaction is small. In the case of acetic acid as hardener, the curing process induced by low pH can start immediately; this process is temperature dependent only regarding the hardening reaction itself, but not regarding the necessary decrease of the pH to induce this curing process. In the opposite case of the addition of ammonium ions, two temperature-controlled reactions are necessary: (1) the formation of an acid from the reaction of the ammonium ions with the free formaldehyde of the resin, and (2) the gelation process itself. In this second case, the temperature dependence can be expected to be considerably stronger. At lower temperatures, the acetic acid is the more effective H^+ ion-donating system with shorter gel times than the ammonium salt-based hardeners. At high temperature (such as 100°C), enough free formaldehyde is available to enable the ammonium salt to react in a sufficient manner to produce a stronger acid, hence resulting in even shorter gel times than observed with acetic acid. It can be argued that the reaction of the ammonium salt with the free formaldehyde at 100°C causes even lower pH values than obtained with the acetic acid.

Activation energies evaluated from monitoring the increase in mechanical strength of the resin adhesive during hardening are significantly lower than those evaluated from gel-time tests or thermal methods such as DSC [82,83]. Obviously, effects such as adhesive wetting, spreading, and penetration into the substrate, along with the curing reaction, play an important role. Additionally, when the bond strength generated is measured immediately after the press time has elapsed, this test is performed on still more or less hot bondlines; the higher the press temperature, the higher the temperature of the samples during the test phase. This causes pronounced temperature-dependent loss in measured bond strength. Tested at the same low temperature, the bond strengths of the samples prepared at the higher press temperatures would be higher; hence, this would give a higher apparent activation energy.

For many applications, such as MDF, the curing behavior of the resin mix must be carefully adjusted to the process parameters; the aim of reducing the necessary press time needs very reactive adhesive systems. On the other hand, precuring can occur in the MDF process, because resin is applied onto the wet fibers immediately after the generation of the fibers in the refiner, and the resin goes through the blowline and dryer, where heat is applied to evaporate most of the water from the refiner [84,85]. An ideal hardener system would not start any reaction before a temperature of 70°C–80°C is reached in the hot press, but should be then very efficient in reducing the pH to start the acid-driven curing of aminoplastic adhesive resins.

20.4.3.1 Catalytic Influence of the Wood Substrate on the Curing of UF Resins

Bonding two wood surfaces includes possible impact of the wood surfaces on the performance of the resin adhesives. Wood extractives can influence the adhesion process in two ways: physically—for example by causing poor wettability and hindering the necessary resin penetration into the wood surface (chemical weak boundary layer)—and chemically. In particular, the acidic nature of the wood extractives can accelerate the hardening process of aminoplastic resins due to a decrease of the pH of the resin adhesive after application onto a wood surface; this fact has to be taken into consideration when optimizing the curing behavior of resin adhesives in the production of wood-based panels. The chemical behavior of wood is mainly determined by its pH and the buffering capacity of aqueous suspensions of wood. Large differences in the measured pH can occur among different wood species due to various influences, such as the season, origin of the sample (core or sap wood, year ring direction, position in length direction of the tree), pH of the soil, age of the tree, storage time after cutting until measurement, and the drying and processing conditions [86]. The pH of aqueous wood suspensions has been reported between 3.0 and nearly 6.0 [86]. The buffering capacity, as a measure of the resistance of the wood substance and of the wood extractives against changes in the pH, determines the rate of possible changes in the pH in the aminoplastic adhesive bondlines during curing. According to Xing et al. [87], no clear relationship was found between the pH and the acidic or alkaline buffering capacity.

The acidity of wood and the acid generating hardener added to the resin adhesive are responsible for the generation of the acidic pH conditions that are necessary to achieve rapid curing of the aminoplastic resin adhesive; this effect might also be influenced by the nature of the wood surface and the wood extractives. To obtain the ultimate bond strength, the press time and the press temperature must be adjusted for the prevailing pH conditions, otherwise precuring, undercuring, or overcuring of the bondline will result in low bond strength.

Usually an ammonium salt is used as the so-called latent hardener. On addition to the aminoplastic resin, immediate reaction of the ammonium ions with the free formaldehyde in the resin generates the relevant acid, which then decreases the pH of the resin and initiates the acidic hardening reaction. If this decrease in the pH is retarded by the buffering capacity of the wood, the cure rate is reduced. The buffering capacity of the aminoplastic resin itself also determines the hardening behavior and depends on the composition and the production procedure of the resin. For further results on the influence of wood extractives and their chemical composition on the hardening behavior of aminoplastic condensation resin adhesives, see [77,88–94], where acceleration as well as retardation of the hardening process are reported.

20.4.4 CORRELATIONS BETWEEN THE COMPOSITION OF AMINOPLASTIC RESINS AND THE PROPERTIES OF THE WOOD-BASED PANELS

Clear correlations between the composition of aminoplastic resins and the properties of the wood-based panels produced by means of these resins as adhesives are still under investigation. It is well known, at least for PB and MDF, how the molar ratio influences the properties of the boards. This, however, is always linked to the type of reaction in resin production. A prediction of bond strength and other board properties based on the results of the analysis of the adhesive resin in its liquid state is still missing. Only in a few cases up till now have equations been derived that correlate (1) the chemical structures in various resins having different molar ratios and different types of preparations with (2) the achievable IB strength of the boards, as well as the formaldehyde emission measured after resin hardening. Such investigations have been done for UF [95,96], MF [97], and MUF resins [98,99]. The basic aim of such experiments is the prediction of the properties of the wood-based panels, hence of the adhesive resin in its hardened state, based on the composition and the properties of the liquid resins used before their hardening. For this purpose, various structural components (such as free urea, methylene bridges, unreacted melamine, or methylols) were determined by means of NMR spectroscopy and the ratios of the amounts of the various structural components were calculated. These ratios are then correlated to various properties of the wood-based panels, such as IB strength or subsequent formaldehyde emission. Although for certain boards some good correlations exist, as has been shown in the literature [95–99], it must be assumed that a general correlation for various resins with different production procedures and various panels will not exist.

20.4.5 ADHESIVE RESIN MIXES

Table 20.6 summarizes some resin mixes for different applications in the production of PB and MDF.

20.5 PHENOLIC RESINS

20.5.1 CHEMISTRY OF PF RESINS

Phenolic resins (phenol–formaldehyde [PF] resins) [2,48,100–104] show high resistance to hydrolysis of the C–C bond between the aromatic nucleus and the methylene bridge and, therefore, can be used for water- and weather-resistant bondlines and boards such as PB, OSB, MDF, or plywood for use in humid conditions or under exterior weather conditions. Another advantage of phenolic resins is the very low formaldehyde emission in service, after hardening, due to the stability of the methylene bridges between the aromatic nuclei. The main disadvantages of phenolic resins are (1) the distinctly longer press times necessary for hardening when compared with UF, but also MUF resins, (2) the dark color of the bondline and of the board surface, and (3) a higher equilibrium moisture content of the boards due to the hygroscopic behavior of the alkali content of the board. For various PF resin mixes for different applications, see [1].

Similarly to the other formaldehyde-based resins, the preparation procedure of a phenolic resin is also a multistage process, characterized by time, sequence, and amounts (in this case in several

TABLE 20.6
UF Resin Mixes for the Production of Particleboard and MDF (Numbers Given as Parts by Weight)

Components/Resin Mixes	PB-CL[a]	PB-FL[a]	PB-CL[b]	PB-FL[b]	MDF[a]
PB–UF resin[a,c]	100	100	—	—	—
MDF–UF resin[d]	—	—	—	—	100
MUF resin[e]	—	—	100	100	—
Water	—	10–20	—	10–20	20–50
Hardener solution[f]	8	2	15	6	2
Urea solution[g]	Up to 5	Up to 10	—	Up to 5	Up to 15

Note: PB, particleboard; CL, core layer; FL, face layer.
[a] For use in dry conditions.
[b] For use in humid conditions.
[c] UF resin with molar F/U = 1.0–1.1.
[d] UF resin with molar F/U approx. 1.0.
[e] MUF resin with molar $F/(NH_2)_2 = 1.0$–1.1.
[f] Ammonium sulfate solution (20%).
[g] Urea solution (40%).

steps) of the additions of phenol, formaldehyde, and alkali as the main raw materials. Two main reactions are given:

- Methylolation: No special preference for ortho- or para-substitution, but it can be achieved with special catalysts [105–108]. The methylolation reaction is strongly exothermic.
- Condensation: Methylene and methylene-ether linkages are formed; the latter do not exist at high alkaline conditions. During this stage, chains are formed, still carrying free methylol groups. The reaction is stopped just by cooling down the resin, thus preventing gelling.

Phenolic resins contain oligomeric and polymeric chains as well as monomeric methylol–phenols, free formaldehyde, and unreacted phenol. The content of both monomers (phenol and formaldehyde) has to be minimized by a proper preparation procedure, as described in the literature [2,3,10,100,101,104]. The production reaction is performed in an aqueous system without addition of organic solvents. The properties of the resins are determined mainly by the F:P molar ratio, the concentrations of phenol and formaldehyde as such in the resin, the type and amount of the catalyst (in most cases NaOH), and the reaction conditions. The reactivity of a PF resin mainly depends on the DOC already achieved. The higher the molar ratio F:P, the higher the reactivity of the resin also and, hence, the higher the hardening rate [109], the degree of branching, and the three-dimensional cross-linking. At lower F:P, linear molecules are preferably formed. The molar ratio F:P of PF resins is usually between 1.8 and 2.5, depending on the type of resin. The higher the molar ratio, the higher its storage stability. Usually, sodium hydroxide (NaOH) is used as a catalyst, in an amount up to 1 mol per mole of phenol (molar ratio NaOH:P), which corresponds to a portion of alkali in the liquid resin of approximately 10% by mass. The pH of a phenolic resin is in the range 10–13. The dominating part of the alkali is free NaOH; only a smaller part is present as sodium phenolate. The alkali is necessary to keep the resin water-soluble via the phenolate ion formation; this enables the achievement of a high DOC at still sufficient low viscosity. Additionally, the alkali content significantly lowers the viscosity of the reaction mixture. Thus, the higher the content of alkali in the resin, the higher its possible DOC still keeping the resin in solution, the greater the reactivity of the resin, the higher its hardening rate, and the shorter the necessary press time. However, maximum NaOH content is limited by various quality standards. Both methylolation- and condensation-stage alkalinities in PF resin production procedures influence chemical structure [110]. The methylolation alkalinity affects the distribution of the structural groups of PF resins, but not to the same extent as does the total condensation alkalinity.

The condensation process of PF resins can be followed by monitoring the increase in viscosity and measuring the molar mass distribution by GPC [111,112]. The lower molar masses are responsible for good wettability; molar masses that are too low can cause overpenetration and hence starved bondlines. Contact angles of phenolic resins on wood increase strongly with increased viscosity of the resin, which increases with the molar masses. The higher molar masses remain at the wood surface and build the bondline, but they will not penetrate and anchor as well in the wood surface. Depending on the porosity of the wood surface, a certain portion with higher molar masses must be present to avoid overpenetration into the wood and causing a starved bondline; this means a certain ratio between low and high molar masses is necessary [112–115]. The penetration behavior of resins into the wood surface is also influenced by various other parameters, such as wood species, amount of resin spread, press temperature, pressure, and hardening time. The temperature at the wood surface and in the bondline and, hence, the viscosity of the resin (which itself also depends on the degree of advancement of the resin at the time of measurement) influences the penetration behavior of the resin. For usual properties of various PF resins see [1].

The content of free monomers (formaldehyde, phenol) depends on the type of the resin and the preparation procedure. Usual values are <0.3% mass for the free formaldehyde and <0.1% mass for free phenol.

The molecular characterization of PF adhesive resins is done in a similar way to that of all the other condensation resins by determining, among others:

- The molar ratios of the main components: F:P, F:P:NaOH, NaOH:P
- The composition of the resins, based on the liquid form of delivery
- The DOC and molar mass distribution, molar mass averages
- The content of reactive sites and functional groups and their distribution in the resin
- The type of bridges between the aromatic rings of the phenol molecule
- The number of branching sites

20.5.2 REACTIVITY AND HARDENING REACTIONS OF PHENOLIC RESINS

PF core-layer resins usually have the highest molar masses and, hence, a high reactivity enabling short press times. They contain higher amounts of alkali than face-layer resins to keep the resin soluble even at these higher degrees of condensation. The higher the DOC during the resin production process (the higher the viscosity), the shorter the gel time [116]. The upper limit of the DOC of the resin during its production process is given by (1) the viscosity of the resin (the resin must be able to be pumped; a certain storage stability as well as a proper distribution of the resin on the particles during blending is required) and (2) the flow behavior of the resin under heat, assuring wetting of the second wood surface without direct resin application and a sufficient penetration into the wood surface. The hardening of a PF resin is the transformation of molecules of different sizes via chain lengthening, branching, and cross-linking to a hardened, three-dimensional network. Alkaline PF resins contain free reactive methylol groups in sufficient numbers, and can continue to react, and finally harden, without any further addition of formaldehyde or catalysts. The methylol groups react to form methylene and methylene-ether bridges. Under high temperatures, methylene-ether bridges can rearrange to methylene bridges. The hardening rate depends on various parameters, such as the molar mass of the liquid PF resin already reached when applied to the wood surfaces, the molecular structure of the resin (mainly the degree of branching), the proportions of various structural elements (especially of the reactive methylol group), as well as possible catalysts and additives. The hardening reaction is initiated by heat only. The lowest temperature for a technically sufficient fast gel rate is approximately 100°C, seen as the temperature in most inner bondlines of plywood or in the core layer of a PB. In some cases, potash (potassium carbonate) in aqueous solution is added to the core-layer resin mix with approximately 3%–5% potash solid based on resin solids content, causing some but not extreme acceleration of the hardening reaction. Since phenolic resins for wood bonding harden only thermally, the postcuring effect during hot-stacking at temperatures as high as possible (limited by discoloration and partial deterioration of the wood substance) helps to improve the strength of PF bonded boards due to the ongoing additional curing reaction during hot-stacking.

The main drawback of phenolic resins is the slower curing rate compared with amino-type resins, causing a reduction in the production efficiency. Phenolic resins, therefore, have lost a substantial part of their market share due to their low reactivity to MUF resins (especially in PB) and to PMDI (especially in OSB). Many attempts have been made to accelerate the curing process of PF resins, such as using different catalysts, additives, or modified resin formulas. An acceleration of the hardening reaction can be achieved by the addition of accelerating esters [117,118], such as propylene carbonate [118]. The mechanism of this acceleration may be linked to the hydrogen carbonate ion of propylene carbonate after hydrolysis [119], or to the formation of hydroxybenzyl alcohols and temporary aromatic carbonyl groups [117]. The higher the addition of esters such as propylene carbonate, the shorter the gel time of the PF resin [117]. Sodium carbonate has been shown to be the most effective catalyst to promote the curing of PF resins by shortening 30% of curing time [120,121].

The curing process can be monitored via DSC, Automatic Bonding Evaluation System (ABES) [122], or differential mechanical analysis (DMA). The chemical changes during hardening can also be followed by solid-state NMR, by monitoring (1) the increase in the amount of methylene bridges based on the amount of aromatic rings [123,124], (2) the portion of 2-, 4-, 6-three-substituted phenols [124], or (3) the ratio between methylol groups and methylene bridges [125,126]. This degree of hardening, however, differs from the degree of hardening as monitored by DSC. The hardening pattern of a resin can also be described by correlations between the degree of chemical hardening and the degree of mechanical hardening (such as measured via ABES or DMA) [124,127,128].

An analysis of PF resins obtained by the addition of 0.5%–5% glycerol triacetate (triacetin) as an accelerator during resin preparation showed the presence of intermediates involved in the acceleration mechanism [129]. This mechanism involved appeared considerably more complex and different from any of the mechanisms presented previously, involving the phenolate ion of the resin apparently giving a carbonyl or carboxyl group attached to the aromatic ring. Either directly or by subsequent rapid rearrangement after the initial attack, these C=O groups were found on sites different from the ortho position, on the para site and, surprisingly, also on the meta sites of the phenolic ring. Also, propylene carbonate was reported to accelerate PF curing [130].

The cure properties of cure-accelerated phenol–urea–formaldehyde (PUF) resins with different catalysts, such as calcium oxide (CaO), sodium carbonate (Na_2CO_3), zinc oxide (ZnO), and magnesium oxide (MgO) were investigated by gel tests, DSC, and liquid ^{13}C NMR. The formation of methylene bridges was considered to be the main reaction during curing. The results indicated that catalysts such as Na_2CO_3, ZnO, and MgO were capable of increasing the curing rate and decreasing the curing temperature of PUF resins. The activation energies (E_a) of cure-accelerated PUF resins other than CaO acceleration were much lower than the control resin. NMR analysis indicated that the catalyst, such as CaO, seemed to retard the polycondensation reaction of phenolic components with urea units, while the Na_2CO_3 appeared to promote the self-condensation reaction of phenolic methylol groups at the para position toward the formation of para–para methylene linkage. Hot-press temperature had a strong influence on the wet shear strength, as did the catalysts. Among the catalysts, MgO had a more significant effect on improving both the curing process and the wet shear strength of PUF resin-bonded plywood. Both ZnO and MgO in PUF resins promoted the self-condensation reaction of para methylol groups and the condensation reaction of ortho methylol groups with para methylol groups [131,132].

20.5.2.1 Catalytic Influence of the Wood Substrate on the Curing of PF Resins

Lignocellulosic substrates show a distinct influence on the hardening behavior of PF resins [133], whereby the activation energy of the hardening process was much lower than for the resin alone and the hardening rate much faster [134]. The reason is a catalytic activation of the PF condensation by carbohydrates such as crystalline and amorphous cellulose and hemicellulose. Covalent bonds between the PF resin and the various wood components such as lignin, however, play only a minor role.

The curing kinetics of resol PF resin and resin–wood interactions in the presence of wood substrates have been studied by DSC and FT-IR spectroscopy [135]. The activation energy of cure of a PF resin generally increases when the PF resin is mixed with wood, mainly due to the decrease of the pH level resulting from the presence of wood. However, wood decreases the curing enthalpy of PF resin through diffusion and the change in the phase of the curing system; this suggests that the curing reactions reached a lower final degree of conversion for the mixtures of PF resin with wood than for the PF resin alone. However, almost no chemical reactions occur between PF resin and wood; only secondary force interactions of hydroxyl groups between PF resin and wood have been detected. These significant secondary forces can also catalyze the self-condensation reactions of PF resin, although their effect is minor to the curing kinetics of PF resin [135]. For further information concerning the effect of the wood substrate on the PF curing, see [136–139].

20.5.3 Modification of Phenolic Resins

20.5.3.1 Postaddition of Urea

The addition of urea to a phenolic resin causes (1) decrease of the content of free formaldehyde, (2) decrease of the viscosity of the adhesive resin, (3) acceleration of the hardening reaction via the possible higher DOC of the PF part of the resin, and (4) reduction in the costs of the resin. Urea can be added during or after the PF condensation step. The distinct decrease of viscosity observed when urea is added after the PF condensation is caused by cleavage of hydrogen bonds and by the dilution effect, rather than by cocondensation of this postadded urea with the phenolic resin moieties. Urea reacts only with the free formaldehyde of the resin to methylols; these, however, do not react further due to the high pH level [140]. Only at high temperatures may some phenol–urea cocondensation occur, as shown for the case of urea and the much more reactive resorcinol [141]. The postadded urea has some negative impact on the properties of the boards, due to dilution of the PF resin. Using such PUF resins, the adhesive solids content should preferably be calculated based only on the PF resin solids content in the PUF resin.

20.5.3.2 Cocondensation between Phenol and Urea

A real cocondensation between phenol and urea can be performed in three ways:

- Reaction of methylolphenols with urea [142–144].
- Acidic reaction of urea–formaldehyde concentrate (UFC) with phenol followed by an alkaline reaction [145].
- Reaction between urea and phenol under alkaline conditions in competition with each other, leading (1) to reaction of methylolureas with phenol and PF oligomers, (2) to reactions of methylolphenols with each other, as well as (3) to reactions of methylolphenols with urea [145–147].

PUF resins were synthesized by reacting a mixture of methylolureas (MMU), phenol, and formaldehyde. The cocondensed PUF resins had no free formaldehyde. Methylolureas, originally from the alkaline reaction between urea and formaldehyde, were well incorporated into the cocondensed resins by reacting with phenol to form cocondensed methylene bridges [148]. Schmidt et al. [149] developed a new three-step synthesis for PUF copolymers. In the first step, two precondensates of phenol with formaldehyde and urea with formaldehyde, respectively, were produced. In the second step, the two precondensates were mixed and condensed using a heterogeneous catalyst in a tube reactor at 90°C. The last step was a vacuum distillation to reach the final copolymer compositions.

PUF resins of high urea content were prepared at different p-methylol OH:P and F:(P + U) molar ratios [150]. NMR indicated that an increase in the F:(P + U) mole ratio decreased the amount of unreacted urea and monosubstituted urea, and promoted the formation of polysubstituted urea. According to DSC, a higher F:(P + U) mole ratio resulted in a lower curing temperature, whereby F:(P + U) showed a strong accelerating effect on the curing reaction. PUF resins of high urea content can also be prepared by adding different forms of urea to the reaction system. Urea added in the form of methylolureas was well incorporated into the cocondensed resins by reacting with phenolic methylols to form cocondensed methylene bridges; these PUF resins had no free formaldehyde. Unreacted urea and monosubstituted urea of PUF resins play a dominant role in the curing behavior and water resistance of resins. The DSC peak temperature, curing time, and curing enthalpy (ΔH) increased; the water resistance decreased when the urea content in PUF resins increased; however, up to 47% replacement by mass still showed ability to produce exterior-grade plywood [151]. Lee and Kim [152] added 10% urea (based on liquid resin) at different stages: (1) at the beginning, (2) at three different middle stages of the condensation reaction, and (3) at the end of PF resin synthesis. No significant cocondensation between the urea and the PF resin components occurred according to ^{13}C NMR. The various urea addition methods resulted in resins that slightly differed in the various tests, due to the urea's temporary holding capacity for formaldehyde. The preferred stage of urea addition was found to be in the later part of PF resin synthesis for convenience, consistency, and slightly better overall performance. For further literature on PUF cocondensation resins, see [153,154].

20.5.3.3 Addition of Isocyanates

PMDI has only been used as a fortifier for phenolic resins in rare cases, mainly due to precuring reactions between the isocyanate and the phenolic resin. Hse et al. [155] found good results with an isocyanate and a PF resin added separately to wood particles. Pizzi and Walton [156] investigated the reactions and their mechanisms of PF resins premixed with nonemulsifiable water-based diisocyanate adhesives for exterior plywood. Pizzi and coworkers [157,158] reported on the industrial applications of such systems (PF + PMDI + partially tannin accelerator; UF + PMDI) and the curing acceleration of the PF resin by the isocyanate.

20.5.3.4 Biobased Cocondensation Components for Phenolic Resins

Green resole phenolic resins for laminating applications use cardanol (CD) and condensed tannin at partial substitutes of up to 40% of the phenol. The resins were synthesized with different molar ratios phenol to cardanol (P:CD) and F:(P+CD). Higher CD content resulted in a proportional increase in the flexibility and fracture toughness of the cured CD–phenol–formaldehyde (CPF) resins. Also, a direct proportionality was found between increasing CD content and decreased cross-linking density of the CPF resins. Tannin was incorporated into the CPF resins, and the fracture toughness and flexibility values of the cured tannin–CPF resins were found to proportionally increase with the tannin content. However, glass transition point T_g, flexural stress, and flexural modulus of the CPF resins decreased with the tannin content [159].

20.5.4 Correlations between the Composition of Phenolic Resins and the Properties of the Wood-Based Panels

Similarly to the investigations described for aminoplastic resins, NMR results for liquid phenolic resins can be correlated to some extent with certain board properties [160]. Again, various structural components have been determined by means of ^{13}C-NMR and the ratios of the amounts of the various structural components are calculated, such as (1) the ratio of methylol groups to methylene bridges, (2) the ratio of free ortho and para sites in relation to all possible reaction sites, (3) methylol groups in relation to all possible reaction sites, or (4) methylene bridges in relation to all possible reaction sites and also ether bridges in relation to all reaction sites. These ratios are then correlated to various properties of the wood-based panels. Also, for phenolic resins, it is not clear if universally valid correlation equations will ever exist.

20.6 ISOCYANATES

20.6.1 ADHESIVES FOR WOOD-BASED PANELS (POLYISOCYANATES)

Adhesives based on isocyanates (especially PMDI) [2,48,161] have been used already for several decades in the wood-based panel industry, but still have low consumption volume compared with systems based on formaldehyde condensation resins. Only for the production of OSB, PMDI is the main adhesive. Additional applications are small volumes of waterproof PB and MDF, but also the production of panels from raw materials that are difficult to bond, such as straw, rice shells, or sugar cane bagasse. During hot pressing, the viscosity of PMDI is lowered, allowing it to flow across and penetrate below the wood surface, locking in the wood subsurface [162]. Table 20.7 summarizes advantages and disadvantages of PMDI.

PMDI is produced in combination with monomeric MDI. The PMDI produced industrially by phosgenation of mainly di-, tri-, and higher amines contains a mixture of the three different MDI isomers, triisocyanates, and partially higher oligomers; the structure and the average molar mass depend on the number of phenyl groups. This distribution influences, to a great extent, the reactivity, but also the usual characteristics such as viscosity, as well as flow and wetting behavior and the penetration into the wood surface. For PMDI, the distribution of the three monomeric isomers has a great influence on the quality, because the reactivity of the various isomers (4,4'-, 2,4'-, and 2,2'-MDI) differs significantly [164]. The greater the portion of the 2,2'- and 2,4'-isomers, the lower the reactivity. This can lead to different bond strengths as well as to residual low reactive isomers in the wood-based panels produced. In the monomeric form (MDI), the functionality is 2 and the NCO content is 33.5%, while PMDI has an average functionality of 2.7 with an NCO content of approximately 30.5%. The residual HCl content is usually below 200 ppm. PMDI is cheaper than pure MDI and has a lower melting point (liquid at room temperature) due to the increased asymmetry. It is less prone to dimerization and, as a consequence, it is more stable during storage than pure MDI.

The excellent application properties of PMDI and the good performance of wood-based panels produced with it are especially based on the excellent wetting behavior of a wood surface when compared with waterborne polycondensation resins (much lower contact angles for PMDI on

TABLE 20.7
Advantages and Disadvantages of PMDI Compared with Other Adhesives, Especially UF Resins

Advantages	Higher storage stability
	Formaldehyde-free bonding, despite the fact that formaldehyde is used in the production of MDI/PMDI
	Higher bonding strength
	Higher tolerance against humidity
	Lower consumption of adhesive
Disadvantages	Higher price, but this is compensated by the low adhesive consumption
	Adhesion to many surfaces, such as metal press platens. This imposes the use of (1) internal or external release agents, (2) special self-releasing types of PMDI [163] or (3) the use of adhesives other than isocyanates on the board surface, such as the combination of PMDI in CL and MUF resin in FL of OSB
	Necessary use of special emulsifiers (EMDI) or special dosing systems
	More stringent worker protection requirements due to the toxicity and the low, but nevertheless existing, vapor pressure of monomeric MDI

various surfaces compared with UF resins). Due to this fact, surfaces with poor wetting behavior such as straw can also be bonded. Additionally, PMDI shows rapid and high penetration into the wood surface, due to the low molar mass of PMDI when compared with polycondensation resins. Marcinko et al. [165] reported that PMDI could penetrate 5–10 times further into wood than PF resins, measured by solid-state ^{13}C-NMR, DSC, fluorescence microscopy, and DMA. PMDI not only penetrates the macroscopic pores of the wood substance, but also penetrates the cell wall, enabling good mechanical anchoring. The good wetting and penetration behavior of PMDI, however, can sometimes cause starved bondlines. Due to the high reactivity of PMDI and its low molar mass, a special interfacial layer between the wood surface and the adhesive appears to be established. Also, Johns [166] showed that isocyanates spread easily on a wood surface; 2%–3% of isocyanate was enough to form a film completely covering the wood strands, which is not possible even with 6% of a phenolic resin. The good mobility of MDI is based on several parameters [166]:

- MDI does not contain water; therefore, it cannot lose its mobility during adsorption on the wood surface.
- Its lower surface tension (ca. 50 mN/m) compared with water (72 mN/m).
- Its low viscosity.

The impossibility of diluting PMDI with water was solved by the introduction of emulsified PMDI, often called *EMDI*, which allows an even distribution of the adhesive during the application of PMDI. EMDI is a product of the reaction of PMDI with polyglycols and manufactured under high pressure and is dispersed in water.

The isocyanate group in PMDI is characterized by high reactivity toward all substances that contain active hydrogens. The main hardening reaction proceeds via a two-step reaction: First, it reacts with water to an amide structure, while at the same time CO_2 is split off. The water necessary to induce the hardening reaction is applied together with the PMDI (spraying together with the PMDI or spraying of an aqueous dispersion of PMDI in water) or is present in wood in sufficient amount. The amide group then reacts further with another isocyanate group to form a urea structure as repeating molecular unit in the molecule:

$$R-N=C=O + H_2O \longrightarrow R-NH_2 + CO_2$$

$$R-NH_2 + O=C=N-R' \longrightarrow R-NH-(C=O)-NH-R'$$

The reaction of an isocyanate group with a hydroxyl group leads to the so-called *urethane structure* as a repeating unit:

$$R-N=C=O + HO-R' \longrightarrow R-NH-(C=O)-O-R'$$

Such a reaction can theoretically also occur between an isocyanate group and an OH group of cellulose or lignin to form a covalent bond. Such chemical bonds are usually more durable than purely physical bonds, especially showing a higher resistance against the influence of humidity. This reaction, however, will only occur in an industrially useful short curing time, when the reaction of the isocyanate groups of the PMDI with water is suppressed; hence, the probability of the formation of such covalent bonds is very low, due to the excess of water in the system. If another isocyanate group reacts with amide hydrogen of an already formed polyurea structure, a branching point is formed (biuret group):

$$R''-N=C=O + R-NH-(C=O)-NH-R' \longrightarrow R-N-(C=O)-NH-R'$$

$$|$$

$$(C=O)-NH-R''$$

During the hardening of PMDI, Frazier et al. [167] found the formation of urethanes, polyureas, biurets, and triurets/polyurets. The proportions of the various compounds depend on the working and hardening conditions. The forming of the network is especially influenced by the ratio between isocyanate and water. The formation of a urethane seems to be possible for low molar mass isocyanates, such as the usual industrial PMDIs, under slightly alkaline conditions. It can also be assumed that the forming of a urethane especially occurs by reaction with lignin. This bond, however, seems not to be stable at higher temperatures (120°C) for longer periods.

Hydrophobic polyols should be able to repel and eliminate water from the wood surface and, therefore, to fortify the reaction of the isocyanate group with the hydroxyl groups of the wood surface. Umemura et al. [168] compared the reaction of isocyanate with water and small amounts of polyols using DMA. The bond strength and the thermal stability increased by adding dipropylene glycol with a molar mass in the 400–1000 range. Usually, no hardeners are added during the production of wood-based panels using PMDI as adhesive. Special additives can give a distinct acceleration of the hardening reaction and, hence, shorter press times or lower press temperatures. This fact is especially interesting for cold-setting systems as well as for the production of OSB. Possible catalysts are tertiary amines (such as triethanolamine, triethylamine, or N,N-dimethylcyclohexylamine) and metal catalysts, based on organic compounds of tin, lead, cobalt, and mercury [169,170].

Jakes et al. [171] assessed the properties of cell walls in PMDI-infiltrated wood by nanoindentation (NI). The elastic modulus and hardness of the cell walls increased, suggesting that PMDI infiltrated the cell walls to form a polyurea interpenetrating network inside the cell walls. Two-dimensional solution-state NMR did not detect covalent bonds between wood polymers and PMDI under the experimental conditions used [172].

20.6.2 Polyurethane (PUR) Adhesives

PUR adhesives [173] are formed by the reaction of various types of isocyanates with polyols, with both components being at least bifunctional. The polar urethane group enables bonding to various surfaces. Depending on the raw materials used, bondlines with either rubber-like or elastic-to-brittle behavior can be achieved. The end groups determine the type of the adhesive, whether it is a reactively or a physically hardening adhesive.

One-component isocyanate adhesive systems consist of chains with isocyanate groups on chain ends or on branching sites. These isocyanate groups can react with the moisture content of the wood surfaces to be bonded, and a hardened system forms from this addition reaction with release of CO_2. Thus, at least one of the two surfaces must contain the amount of water necessary for hardening. Due to the high viscosity of these adhesives, dilution with organic solvents or higher application temperatures is necessary. Additionally, the adhesive may contain various other components, such as flowing agents, fillers, antioxidants, bactericides, and dyes. The bondline reaches the necessary green strength within a few hours and hardens over a few days. During the reaction of the isocyanate group with the moisture content of the wood, CO_2 is formed, which causes some foaming of the bondline and, hence, a certain gap filling ability. The bondlines themselves are more or less resistant against humidity and water.

Two-component PUR systems consist of (1) a polyol or polyamine and (2) an isocyanate that is at least bifunctional. Hardening starts with the mixing of the two components. Due to the low viscosities of these components, they can be used without addition of solvents. The mass ratio between the two components determines the properties of the bondline. Linear polyols and low amounts of isocyanates give flexible bondlines, whereas branched polyols and high amounts of isocyanates result in hard and brittle bondlines. The pot-life of the two-component systems is determined by the reactivity of these components, the temperature, and the addition of catalysts; it can vary between 0.5 and 24 h.

20.6.3 Emulsion Polymer Isocyanates (EPIs)

EPI adhesives [174–177] combine the advantages of thermoplastic dispersion adhesives with duroplastic cross-linking systems. The thermoplastic properties, which mainly have negative impact concerning moisture resistance and creep behavior, are improved by the additional cross-linking. Since the cross-linking component is based on isocyanates, a certain adhesion to metals can occur. EPIs are, in most cases, two-component systems combining an emulsion component and an isocyanate functional cross-linking component; they consist of a base emulsion polymer and a protected polymeric isocyanate on the basis of MDI or hexamethylene diisocyanate (HDI) as cross-linking agent. The emulsion component in the EPI adhesive consists of a mixture of one or several water-based emulsions, poly(vinyl alcohol) (PVA), filler(s), and additives to obtain good dispersion of the filler and to avoid foaming and fungi attack. Emulsions on the basis of polymerized vinyl acetate, ethylene–vinyl acetate, vinyl acetate–acrylate, acrylic–styrene, or styrene–butadiene with post-cross-linking groups give optimal bondline quality. Depending on the application, a great variety of compositions can be used; detailed recipes, however, are usually proprietary. The PVA is a viscosity/rheology modifier; it prevents the sedimentation of filler and plays an important part in the cross-linking reactions with the isocyanates to give a water-resistant bondline. Calcium carbonate is the most frequently used filler in EPI adhesives, but other organic or inorganic materials may also be used. The functions of the filler in the formulations are to increase the solid content, improve the heat resistance, and give gap filling properties in the final bondline. The cross-linking component is an isocyanate with, preferably, two or more NCO groups. To prevent exposure of isocyanate in the production area, it is important to choose a low volatile and preferably polymeric compound. MDI and/or PMDI are the most common isocyanates used for this purpose nowadays, due to their low vapor pressure and high reactivity.

The setting speed, or how fast the initial strength is built up, depends mainly on how fast the water penetrates into the wood, so that the emulsion particles can form an adhesive film. This physical process to gain initial strength mainly depends on the moisture content in the wood; final curing of EPI adhesives to obtain a water-resistant bondline is then a chemical process that is predominantly influenced by temperature.

The cross-linker is mixed with the base emulsion to provide an adhesive with a working life of up to several hours. When the adhesive is applied to the wood, water migrates from the bondline and the cross-linker is free to react with components of the base emulsion, the bound water in the wood, and also with the cellulosic substrate, forming a highly durable bond. The addition of the isocyanate component is in the range of 15%, based on the emulsion polymer. Naturally, the use of heat will accelerate the reaction. The curing characteristics of EPI adhesives are quite complex and include film formation of the emulsion adhesive (which is a physical process) as well as chemical reactions of the highly reactive isocyanate toward water, as well as hydroxy-, amine-, and carboxy groups of the polymers.

EPI is usually free of solvents, but can contain additional additives to adjust proper processing and adhesion properties. The advantages obtained by the use of EPI adhesives are fast setting speed, cold curing ability, light-colored bondline, high flexibility, low creep of the bondline, and does not contain formaldehyde; it gives excellent water resistance in both cold and boiling water and good heat resistance. As the setting and curing behaviors are different from today's primarily used adhesives for laminated beams (phenol–resorcinol–formaldehyde [PRF], MF, MUF, one-component polyurethane [1K-PUR]), the optimal production parameters have to be determined. EPI adhesives were introduced in the early 1970s into the Japanese market for the adhesive bonding of wood-based products [175]. EPI is used in many parts of the world for production of different types of wood-based products, such as solid wood panels of different types, parquet, window frames, furniture parts, plywood, fingerjoints, and load-bearing constructions such as laminated beams and I-beams. EPI adhesives give very good adhesion and are thus well suited for bonding difficult wood species such as hardwood. EPI adhesives can also be used for gluing of wood to metal. An application, for example, is furniture production in Asia, when high relative humidity is present during use.

20.7 INFLUENCE OF THE ADHESIVE ON THE BONDING PROCESS AND THE PROPERTIES OF WOOD PRODUCTS

20.7.1 VISCOSITY, FLOW, AND WETTING BEHAVIOR

The viscosity of a resin mix is determined by the viscosity of the resin (mainly depending on the DOC and the resin solids content) and the composition of the resin mix. If the viscosity or the DOC of the resin is too low, a large portion of the resin may penetrate into the wood, causing a starved bondline. In such a case, a sufficient bondline cannot be formed and, hence, only low or no bond strength will be obtained (see Section 20.7.2). Conversely, at too high a viscosity there may be a lack of proper wetting by the adhesive, especially on the wood surface opposite to that surface where the adhesive has been applied; consequently, no or very low penetration into the wood surface will occur (narrow interphase and less mechanical interlocking of the adhesive into the substrate). Poor bond strength will also be obtained in such a case.

A higher dilution of the resin gives a higher volume to be spread, and thus a better distribution of the resin on the particles or fibers, and better bond strength. The ability of a resin to flow depends on its viscosity and the solids content, as well as the changes in the viscosity at elevated temperatures in the hardening bondline. Low flow causes poor penetration of the resin into the wood surface and low bond strengths. Too strong a flow, on the other hand, leads to overpenetration of the resin into the wood tissue and, hence, to starved bondlines. Flow and hardening act against one another during the hot-press curing process.

Surface tension and wetting behavior of aqueous adhesive resins are similar to water. For UF resins, the wetting behavior strongly depends on their molecular composition [32]. The higher the molar ratio F:U, the lower the surface tension. The proper wetting of the wood surface is a precondition for achieving high adhesion strength between the resin and the wood surface. Mittal [178] discussed the relation between wetting (surface free energy) and bond strength.

20.7.2 PENETRATION INTO WOOD TISSUE

Penetration is the ability of an adhesive to enter into the lumen and into cell walls as a process of fluid movement [179]. Adhesive movement through wood is usually classified as penetration into wood tissue and cell-wall penetration. The first refers to movement of the adhesive through the large voids in the porous structure, whereas cell-wall penetration involves movement of adhesive through the microvoids within the wood cell walls [180]. Penetration of adhesives is given mainly as filling up the lumen by bulk adhesive material. Cell-wall penetration [181–183] is possible in principle, but rather only for substances with low molar masses, such as impregnating resins, and highly controlled by the size of the molecules of the adhesive. Low molar mass portions of the adhesive can easily penetrate; this effect is enhanced by the decrease in viscosity due to the increased temperature in the bondline during hot pressing.

The interphase region of the adhesive bond is defined as the volume containing both wood cells and adhesive. It is created by penetration of the adhesive into the wood surface and partial filling of the lumens. This process is determined by (1) wood-related parameters (such as the lumen diameter and the grain slope on the wood surface), (2) the properties of the resin and the adhesive mix (such as the molar mass distribution, adhesive mix composition, viscosity, the amount of adhesive spread, the hardening time, and the rate of resin curing), and (3) the bonding processing parameters (such as the assembly time, the press temperature and pressure, and the moisture level). The temperature of the wood surface and of the bondline and the viscosity of the resin (which itself also depends on the degree of hardening already reached) influence the penetration behavior of the resin, and hence bond performance.

Adhesive penetration and the formation of the interphase as well as of the interface as the contact zone between the internal surface of the wood and the adhesive strongly determines the bonding effect. Wood composite structural performance is directly influenced by adhesive penetration away from the joint contact area into the porous wood structure. An adequate penetration of the resin into the wood surface enables formation of a sufficiently large bonding interface; this penetration must take place before curing of the resin has occurred. Shallow penetration may result in poor stress transfer and reduced dimensional stability, while overpenetration can yield a starved bondline. Although it is generally known that a certain penetration helps in creating strong bonds, it is not clear whether penetration into the lumens or into the cell walls is more critical. Adhesive bond performance between wood elements, hence, is presumed to be significantly influenced by the degree of penetration of the adhesive into the porous network of interconnected cells in the wood material. Low bond strengths will result from either under- or overpenetration. Underpenetration means that the adhesive is not able to move into the wood substance enough to create a large active bonding surface (interface) within the interphase and, hence, strong interaction between the wood and the adhesive. In contrast, overpenetration occurs if a high portion of the adhesive can penetrate into the wood substance, causing starved joints; as a consequence, insufficient amount of adhesive remains in the bondline to bridge between the wood surfaces and to establish bond strength. To solve these problems, the viscosity of the adhesive, especially in dependence of the temperature in the bondline during the press cycle, has to be adjusted by proper composition and molecular structure (molar mass distribution, DOC) of the adhesive. Furthermore, penetration depth is highly dependent on multiple adhesive and wood characteristics, as well as material processing and bonding conditions. Thus, the complex relationship between adhesive penetration and joint performance is difficult to define quantitatively.

The influence of the DOC of the resin is mainly given during the application and the hardening reaction. As long as full curing is secured, differences in achieved bond strengths should depend less on factors linked to the reactivity of the resin, but should rather be influenced by the distribution of the resin during application and the penetration behavior. Penetration is highly controlled by the size of the molecules of the adhesive; this effect is enhanced by the decrease in viscosity due to the increasing temperature in the bondline during hot pressing. Low molar masses of the adhesive are responsible for good wettability and can easily penetrate. At the same time, this can cause overpenetration and, consequently, starved bondlines. The higher molar masses preferably remain at the wood surface and form the bondline, which is necessary to achieve good bond strengths. Stephens and Kutscha [184] investigated the effect of the resin molecular weight on the bonding behavior of flakeboards. A commercial PF resin was separated into two molar mass fractions by diafiltration, and aspen flakeboards were prepared with the separate fractions as well as with mixtures of the resin fractions. Penetration characteristics of the different fractions were determined microscopically. Results indicate that both low and high molar mass components of the resin are needed to achieve boards with acceptable properties. Hare and Kutscha [185] had already indicated that for a strong bond the adhesive must penetrate deep enough into the wood substrate to be able to reinforce weakened cells and to obtain a large contact surface. Penetration occurs easily into fiber cells that are physically ruptured, for example, during the veneering process.

Bolton et al. [186] detected UF resin penetration in cell walls in PBs by using electron microscopy coupled with energy-dispersive x-ray analysis (EDAX). Gindl and coworkers [181,187] utilized UV microscopy to investigate the diffusion of an MF resin into cell walls; the authors clearly showed resin diffusion into the cell wall, and this finding can be seen as a potential mechanism of mechanical interlocking in wood bonding. Konnerth et al. [182] used scanning thermal microscopy (SThM) to investigate if cell-wall penetration might occur with various adhesives. It was shown that the investigated PRF resin could penetrate the cell wall, but not the PUR adhesive. Gindl [188] showed by using scanning electron microscopy (SEM) that adhesives penetrate predominantly through cut and open tracheids and rays. Bordered pits block the resin flow from one cell to another cell, whereas simple pits are only a minor obstacle. Penetration of resin compounds into the wood cell wall can enhance the durability of adhesive bonds. Gindl et al. [189] investigated the diffusion of a PF resin and of PMDI, using the UV absorbance typical for both resins and following cell-wall mechanical properties by NI. Significant amounts of PF resin diffused into the cell wall, whereas no PMDI could be detected. For both adhesives examined, cell walls at the immediate surface were damaged during machine planing. The diffusion of adhesive into undamaged cell walls did not change their elastic indentation modulus significantly, but clearly increased the hardness of PF-impregnated cell walls. Cell walls damaged due to planing showed a significantly reduced hardness and indentation modulus. To provide good bond stability, it seems to be necessary that the adhesive penetrates deep enough into the wood structure to bond to undamaged cell walls.

The DOC, and hence the molar mass distribution, is one of the most important characteristics of a condensation resin and determines, among other properties, the flowing ability; this is reduced at higher DOC; hence, penetration into the wood tissue will be less. The higher the molar masses (more or less equivalent to the viscosity of the resin at the same solid content), the worse the wettability [32] and the lower the penetration into the wood surface. UF resins with higher degrees of condensation. and hence higher viscosities, show less penetration into wood. A good correlation is given between the viscosity of the resins and the initial rate of penetration into the surface, shown as slower decrease of the droplet volume on the surface with time [33]. Penetration into latewood with narrow lumens is slower than into earlywood with wider lumens. The lower molar masses with stronger polar characteristic are responsible for good wettability; however, a too high portion of low molar masses can cause overpenetration, and hence starved bondlines, with only an insufficient amount of adhesive remaining in the bondline to bridge between the wood surfaces and to establish bond strength. The higher molar masses preferably remain at the wood surface and form the bondline. In the extreme case of underpenetration, the adhesive is not able to move into the wood substance enough to create a large active bonding surface (interface) within the interphase.

The extent of lumen penetration into earlywood, latewood, and rays is preferably determined by examination of the cross-section of a bondline, with several techniques having been successfully used for these purposes:

- Light microscopy: Light visible or fluorescent dyes or pigments for enhancement of the subsequent visualization of the adhesive distribution in the bondline and the adjacent wood material [155,190–194].
- Transmitted and reflected microscopy [180,185].
- (Epi)fluorescence microscopy [180,195–201].
- (Fluorescence) confocal laser scanning microscopy (CLSM) [183,186,192,193,198,201–204]. For penetration into fibers see [205–207].
- SEM [185,188,191–194].
- Transmission electron microscopy (TEM) [194,208,209].
- SEM in combination with an energy-dispersive analyzer for x- rays (SEM/EDAX, TEM-EDXS) [186,194,210].
- X-ray microscopy [211].
- X-ray photoelectron spectroscopy (XPS) [202].
- Neutron radiography [191].
- (X-ray) microtomography [212–221].
- Synchrotron radiation x-ray microtomography (SRXTM) [213].
- Synchrotron-based x-ray fluorescence microscopy [222].
- SThM [182,223].

Hass et al. [213] investigated the penetration of various adhesives [1K-PUR, poly(vinyl acetate) (PVAc), and UF resins] into beech, using SRXTM. Using suitable gray-value segmentation, the distribution of the adhesive in the wood tissue adjacent to the bondline was determined, monitoring maximum penetration depth and the adhesive saturation of the available pore space. As a measure to describe the adhesive distribution inside the bondline, the authors recommended the degree of saturation of the pore space rather than the commonly used maximum penetration depth. With this method, isolated vessels causing high penetration depth were neglected, because they have only a minor influence on the bond strength, and a good correlation between penetration behavior and bond strength was established.

Johnson and Kamke [180] used three PF resins with different degrees of condensation to investigate the penetration behavior into the cell lumens and also the big vessels of hardwood (yellow poplar). The higher the molar mass, the less the penetration into the wood flakes. Adhesive penetration into hardwood is likely to be dominated by flow into vessel elements. In another paper, the same authors [224] investigated the effect of the molar mass distribution of a liquid PF resole on the adhesive flow during steam injection pressing, and hence the mechanical properties of flakeboards. Deeper penetration of the adhesive likely occurred during exposure to the steam injection environment, as the viscosity declined due to an increase of temperature and dilution of the adhesive. Brady and Kamke [195] investigated the effect of hot-pressing parameters on the penetration of PF resins into thin wood flakes using fluorescence microscopy. Resin penetration was influenced rather by the natural variability of the wood material than by temperature, moisture content, time, or pressure, all of these mainly controlling the viscosity of the resin. Pressure influences penetration because it is the driving force for hydrodynamic flow. Moisture content, as well as temperature and time, influence resin penetration by affecting the viscosity of the resin due to the hardening reaction. Penetration was about three times greater in Douglas fir earlywood than in latewood. Cell-wall fractures enhanced penetration by providing additional paths for hydrodynamic flow. Wilson et al. [225] reported that increased portions of higher molar mass fractions in PF resins increased IB strength, whereas low and medium molar mass fractions rather tended to overpenetrate into the wood substrate, causing starved bondlines.

Gavrilovic-Grmusa and coworkers [226–230] investigated the tangential and radial penetration behaviors of three UF resins with different DOCs into various wood species (silver fir, beech, and poplar): UF I with lowest DOC; UF II with medium DOC; UF III with highest DOC. The various DOCs gave different viscosities of the resins themselves, as well as of the resin mixes used for the penetration tests. The measurements considered (1) penetration depth and distribution of the resins under conditions of hot bonding of solid wood blocks, as well as (2) the resultant tensile shear strengths.

The penetration was followed by epifluorescence microscopy on microtome slides; the wood species considered here were coniferous as well as hardwood (deciduous) with high and low density (Figures 20.2 through 20.4). Since penetration mostly depends on the permeability and porosity of the wood surface, as well as on the molecular size of the UF adhesives, the higher the DOC, the less chance for penetration. Evaluation was done (1) for the maximum as well as the average penetration depth as distance from the apparent bondline and (2) for the size of the interphase region and the proportion of filled lumens. Due to the size of the adhesive resin molecules used in these experiments, cell-wall penetration was less likely.

Figure 20.5 shows characteristic epifluorescence microphotographs of radial and tangential penetration for the adhesive mix of UF I into poplar at three different applied pressures [230]. The light-colored sections on both sides of the bondline represent the UF adhesive mix, which had penetrated to a certain extent into the wood material. Depending on the anatomical structure of poplar, the adhesive mix mainly filled the lumens of the vessels, as well as the rays. Only bulk penetration was investigated and evaluated, but not wood cell penetration.

Qin et al. [201] used a modified UF resin to prepare glued laminated timber and measure gross penetration by fluorescence microscopy (FM); the UF passed through 1.5–3.5 earlywood tracheids (with an average penetration depth of 89 μm) or 0.5–4.0 latewood tracheids (with an average penetration depth of 36 μm). In addition, the distribution of cell-wall penetration was observed clearly by CLSM. The adhesive was found to diffuse into the cell walls of surface tissues embedded in the UF. To verify the results from CLSM, the mechanical properties of cell walls with and without adhesive penetration were measured through NI. The reduced elastic modulus of exposed cell walls was roughly equal to that of fully filled cell walls, but significantly greater than that of unfilled ones.

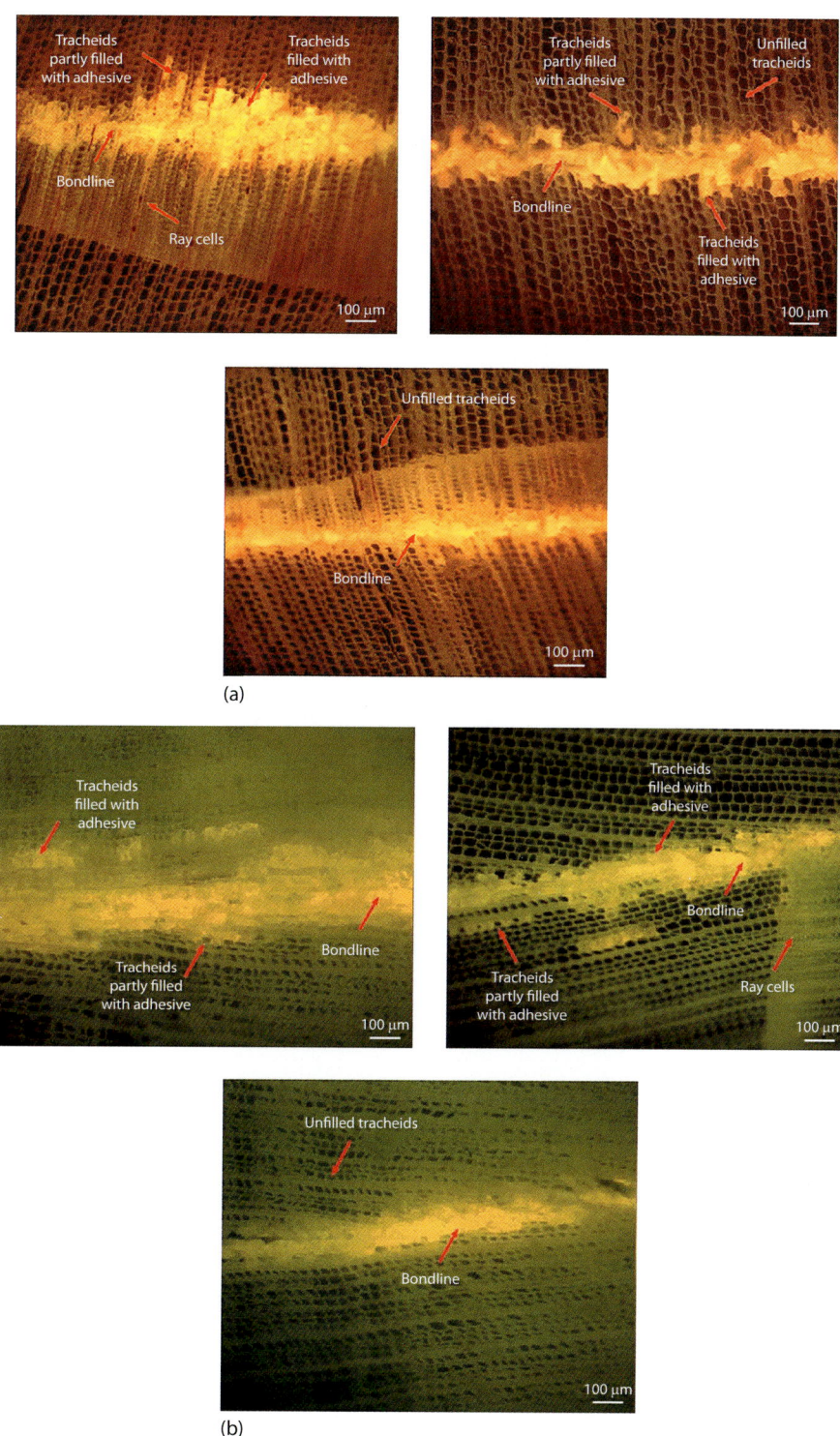

FIGURE 20.2 (a) Photomicrographs of the three resin mixes for radial penetration into fir, using epifluorescence. Left: UF I; middle: UF II; right: UF III. (b) Photomicrographs of the three resin mixes for tangential penetration into fir, using epifluorescence. Left: UF I; middle: UF II; right: UF III. (From Gavrilovic-Grmusa, I. et al., *Eur. J. Wood Prod.,* 70, 655–665, 2012.)

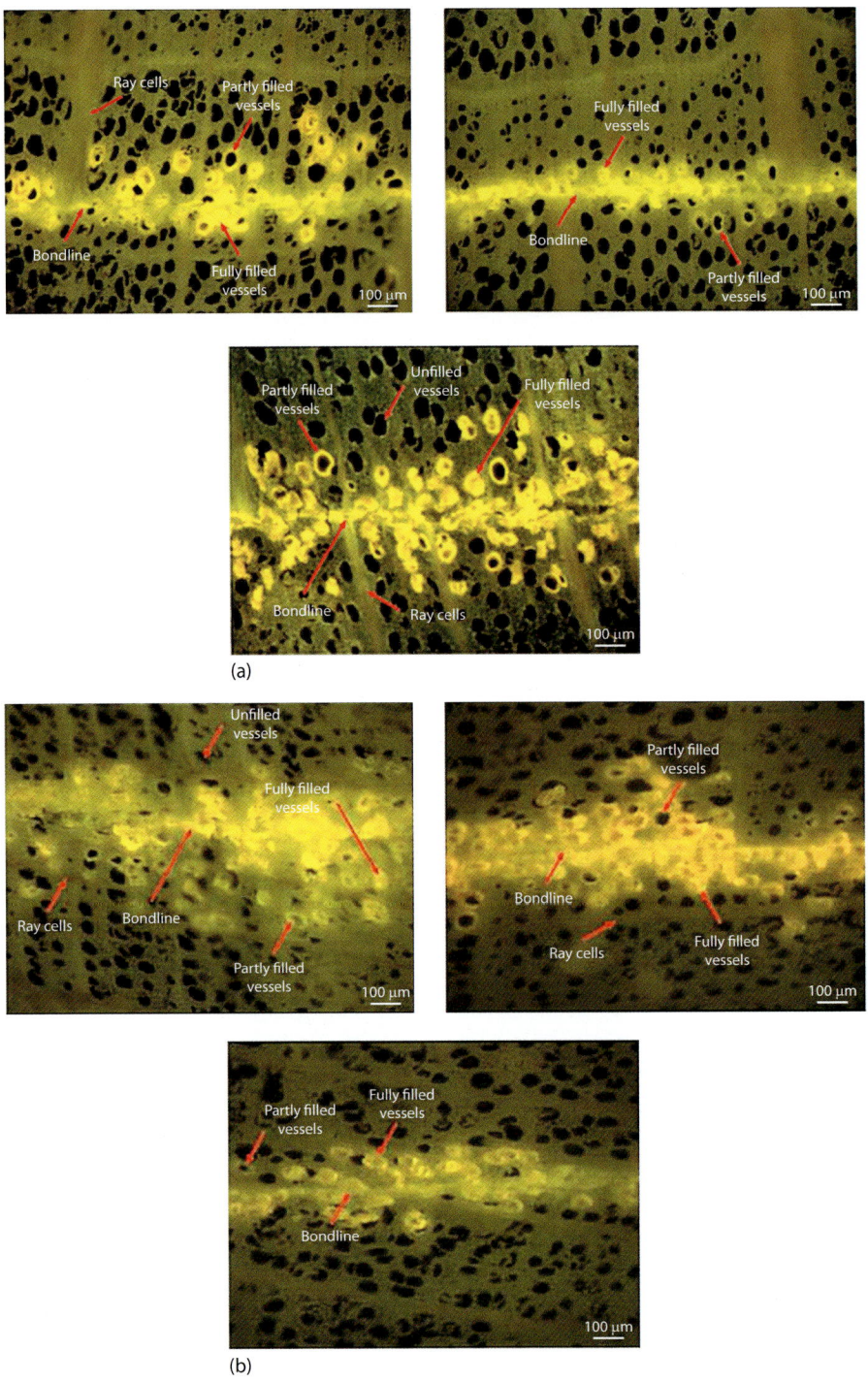

FIGURE 20.3 (a) Photomicrographs of the three resin mixes for radial penetration into beech, using epifluorescence. Left: UF I; middle: UF II; right: UF III. (b) Photomicrographs of the three resin mixes for tangential penetration into beech, using epifluorescence. Left: UF I; middle: UF II; right: UF III. (From Gavrilovic-Grmusa, I. et al., J. Adhes. Sci. Technol., 24, 1753–1768, 2010.)

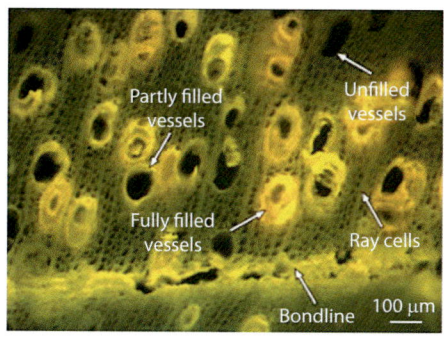
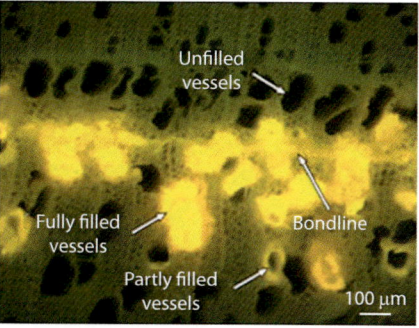
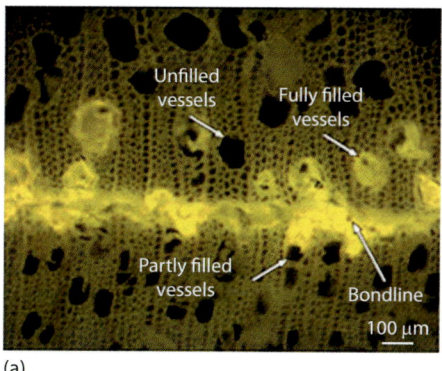

(a)

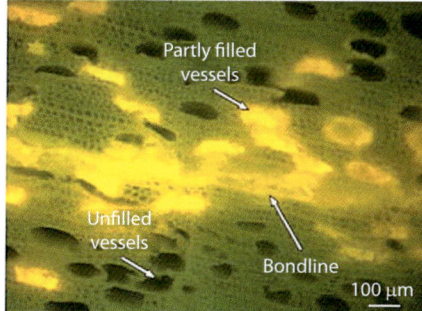
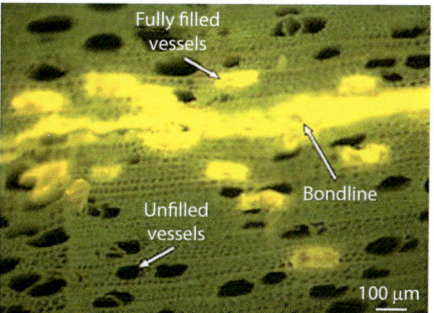
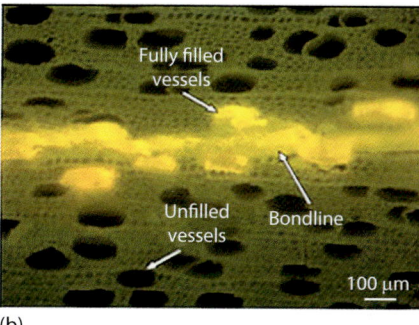

(b)

FIGURE 20.4 (a) Photomicrographs of the three resin mixes for radial penetration into poplar, using epifluorescence. Left: UF I; middle: UF II; right: UF III. (b) Photomicrographs of the three resin mixes for tangential penetration into poplar, using epifluorescence. Left: UF I; middle: UF II; right: UF III. (From Gavrilovic, I. et al., *Holzforschung*, 66, 849–856, 2012.)

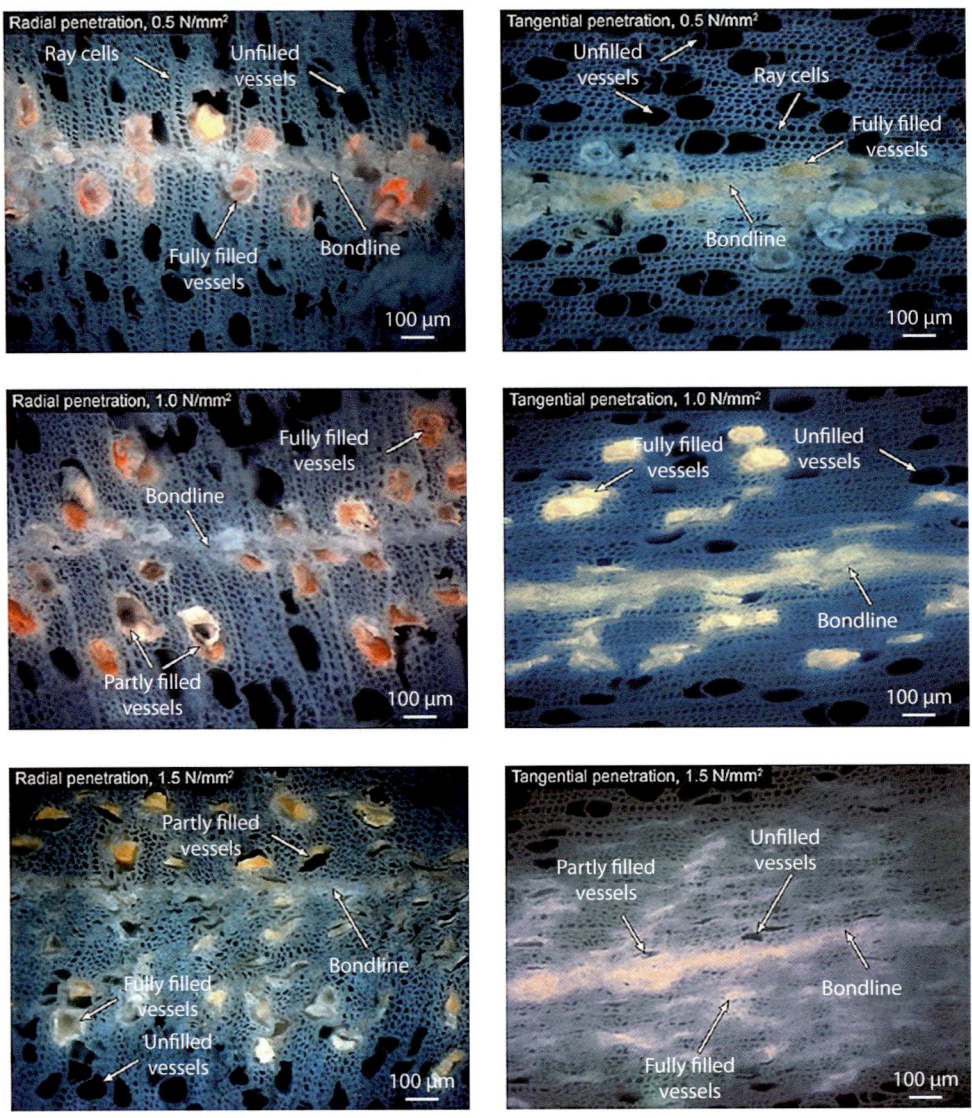

FIGURE 20.5   Epifluorescence microphotographs showing penetration of UF resin I into poplar at three different pressures applied during the press cycle: 0.5 N/mm² (lowest level of pressure; top), 1.0 N/mm² (middle), and 1.5 N/mm² (highest level of pressure; bottom) for radial (left) and tangential penetration (right). (From Gavrilovic-Grmusa, I. et al., *Bioresources*, 11, 2238–2255, 2016.)

To understand the impact of the molar ratio F:U on the penetration characteristics of a UF resin into softwood tissues, Nuryawan et al. [193] tested different F:U resins (1.6, 1.4, 1.2, and 1.0) with different viscosities and two levels of hardener (NH_4Cl). A decrease in F:U that proportionately decreased the resin viscosity resulted in an increase in the average resin penetration and a decrease in the bondline thickness. Higher hardener level provided a greater resin penetration for all values of F:U. These results demonstrated that the F:U mole ratio had an impact on the penetration and bondline thickness of UF resins, owing to differences in the reactivity of resins, with higher F:U mole ratio resins being more reactive. The results indicate that the average resin penetration depth decreases with an increase in F:U and a consequent increase in the viscosity of UF resin adhesive. The higher F:U resin, being more highly branched and also more viscous, resulted in poorer penetration into wood tissues and a thicker bondline. Conversely, lower F:U mole ratio UF resin adhesive, being more linear, resulted in deeper penetration from the bondline and formation of a thinner bondline, which may be attributed to greater abundance of low molar mass chemical species in F:U = 1.0 compared with 1.6. Although an initially higher level of hardener rendered the resin less viscous (simply by the dilution effect) and enhanced resin penetration, in later stages the rate of resin penetration was drastically reduced because of the gelling effect of the resin.

Paris and coworkers [216,219,220] investigated 3-D wood anatomy, bondline, and adhesive distribution using x-ray computed tomography (XCT) as a nondestructive technique on the microscale for multiple bonded specimens prepared from three wood species and for three common wood adhesive types, namely, PF resin, PMDI, and PVAc.

20.7.3 Reactivity

The reactivity of the adhesive mix must be in line with the whole process of production of wood-based panels. The objective of the development of adhesive resins is to achieve the highest possible reactivity, while maintaining within acceptable limits other properties, such as the storage stability of the resin or the pot-life of the ready-to-use adhesive mix and avoiding the risks of precuring in the process.

The reactivity of a formaldehyde-based condensation resin is determined by various parameters, such as type of resin, composition of the resin mix, type and amount of hardeners, additives that might accelerate or retard the hardening process, hardening temperature (press temperature, temperature in the bondline, temperature in the core layer of PB or MDF), and the properties of the wood surface.

20.8 ANALYSIS OF WOOD ADHESIVES BASED ON FORMALDEHYDE CONDENSATION RESINS

In the last 15 years, considerable progress in the characterization of formaldehyde-based condensation resins has been made. The polydisperse nature of the resins, as well as the individual structural elements in the resins, can be determined quantitatively. The curing reaction is usually followed by means of various methods, whereas a simple and quick universal method is still missing. The main topics of analysis are (1) curing reaction and buildup of bond strength, (2) evaluation and monitoring of the DOC and the molar mass distribution, and (3) analysis of the chemical composition of the resins and of their structural components. The characterization of formaldehyde condensation resins comprises basic chemical methods, including elemental analysis as well as various spectroscopic and chromatographic methods. One of the main problems in the analysis is that condensation resins themselves are systems that may still change during their preparation for analysis or during the analysis itself. Furthermore, the polar character and the relatively low solubility of the resins often render their analysis problematic. The basic properties of resins [1] are determined after their production as well as at the customer's site. This includes solid content, viscosity, pH, gel time, and pot-life; for reactivity tests, see 20.8.3. The various components and raw materials of formaldehyde-based resins can be determined using different chemical methods [1]. This also includes (1) the determination of residual monomers, such as free formaldehyde, unreacted urea, or free phenol by chemical analysis or by HPLC; (2) the calculation of the various molar ratios, as well as (3) the determination of low molar mass moieties, such as monomethylolurea (MMU) or dimethylolurea (DMU) [231].

20.8.1 Structural Components

Different spectroscopic methods (Table 20.8) such as IR, ^{1}H-NMR, ^{13}C-NMR, ^{31}P-NMR, or ^{15}N-NMR enable the analysis of the structural compounds and provide insight into the structural

TABLE 20.8
Spectroscopic Analysis Methods for Structural Components

Test Method	Adhesive Type and References
IR	UF [232]; MUF [233]; PF [234]; PUF [235]
NIR	UF [236–239]; MUF [240]
FT-IR	UF [241]; MUF [242]; PF [154]
^{1}H NMR	UF [243]; PUF [149]
^{13}C NMR	UF [17,19,26,243–245]; MUF [246–249]; PF [131,139]; PUF [250]
Solid-state ^{13}C NMR	UF [19]
^{15}N NMR	UF [243]; MUF [56]
Raman spectroscopy	UF [245]; MUF [249], PF [131,132,251]; PUF [149,250]
X-ray diffraction	[252]

Source: Spectroscopic Analysis Methods for Structural Components (for further references see [1] and [2])

composition of resins. These results are the basis for correlations between the molar composition of resins, their structural composition (type of bridges between the monomers or the degree of branching), their preparation procedure, and the properties of the panels produced.

20.8.2 MOLAR MASS DISTRIBUTION (MWD) AND MOLAR MASS AVERAGES

The MWD can be determined by means of GPC (SEC). This method separates the molecules according to their hydrodynamic volume, which is (at the same degree of branching) proportional to their molar mass. A certain drawback in the chromatography of formaldehyde condensation resins is the poor solubility of the resins in some solvents usually used in GPC; hence, the proper choice of the solvent and the mobile phase is important. This choice influences the solubility of the resin, as well as the behavior of the chromatographic columns and the effectiveness of detection. For lower molar mass PF resins, tetrahydrofuran (THF) is still a suitable solvent [111]; for higher molar mass phenolics and for MF resins, dimethylformamide (DMF) can be recommended, sometimes modified by addition of small amounts of ammonium formate or other salts such as LiCl [111]. UF resins are only soluble in DMF (but some undissolved higher molar mass portions can occur) and in dimethylsulfoxide (DMSO). DMSO shortens the lifetime of the chromatographic columns and causes problems (1) with high pressures, because of its higher viscosity in comparison with other organic solvents (it requires higher temperatures in the column oven such as 60°C) and (2) low refractive index increments. The high reactivity of the functional groups of the resins additionally requires the use of the correct solvent and mobile phase, especially concerning sample preparation, to obtain satisfactory reproducibility of the results.

Another problem with GPC of condensation resins is the calibration of the columns. In the oligomeric and polymeric regions of the resins, no compounds with a special and singular molar mass and a well-defined molecular structure are available; therefore, similar or chemically related substances have to be used as calibration standards. However, differences in the hydrodynamic volumes even at the same molar mass cannot be excluded. This fact also induces a certain uncertainty in the calculation of molar mass averages on the basis of the chromatograms obtained.

Molar mass distributions are reported in the literature: UF [18,231], MUF [13,50], PF [111,113,131,140,253,254], PUF [149]; for further references see [1] and [2]. GPC can also follow the buildup of the molecules during the condensation phase, shifting the chromatograms to lower elution volumes [13,111].

20.8.3 Monitoring of Gelling and Hardening

The bond strength of thermosetting adhesives develops during the hardening or curing process, which is usually carried out in a hot press at a defined pressure and temperature, and for a defined period of time [255]. During gelling and hardening of the condensation resins during the hot-press procedure, the chemical advancement of the condensation reaction (buildup of the three-dimensional network) and the development of the mechanical strength of the joint (increase in cohesive bond strength) take place. These two reactions, however, do not proceed at the same rate. Adequate bond strength and long-term performance of the wood-based panels such as PB, MDF, OSB, and plywood is achieved with sufficient adhesive curing during pressing. At the end of the press time—when the press opens or when the board leaves the continuous press—a certain degree of bond strength (i.e., certain mechanical hardening) is necessary to withstand the various internal forces in the board (springback) due to (1) mechanical stresses caused by deformation of the wood substance (reduced by relaxation processes within the board) and (2) the internal steam pressure buildup due to the addition of energy and increase of temperature during the hot-press process. Full chemical curing, however, can be completed afterward outside of the press (e.g., during hot-stacking). If sufficient adhesive bond strength is not developed, the compressed panel will exhibit either a low performance (low IB strength, low shear strength, etc.) or will even delaminate (blister). A stronger adhesive bond can be achieved by prolonging (to a certain limit) the press time, but this increases the cost of heat energy supplied and reduces production capacity. Manufacturers of wood-based panels are thus continually trying to find the optimal press time as a balance between the required performance of the panel and acceptable production costs. To enable shorter press times and, therefore, reduced production costs, an early, quick and instantaneous development of the bond strength is important.

The bond strength between wood substrate and adhesive develops during the hardening or curing of the adhesive, which involves conversion of a liquid adhesive through gelation and vitrification to a fully cured adhesive. Gelation marks the transition from liquid to a rubbery state and retards macroscopic flow. The gel corresponds to the formation of an infinite network in which polymer molecules are cross-linked and form a macroscopic molecule. The viscosity and modulus of a polymer increase dramatically when the gel point is achieved. Vitrification occurs when the glass transition temperature (T_g) rises to the temperature of cure. It marks the transformation from a rubber to a gelled glass or from a liquid to an ungelled glass. The T_g is the critical temperature that separates glassy behavior from rubbery behavior. The strength of adhesion between adherends increases linearly as the amount of interlinking between the two adherends increases. Cross-linking affects many of the physical and mechanical characteristics of thermosetting adhesives. It improves their strength and durability, and also increases resistance against solvents and high temperature.

Hardening and gelling of thermosetting adhesives and resins can usually be followed by monitoring the various reactions and changes ongoing during hardening and gelling (Table 20.9):

- Exo- or endothermic behavior during gelling and hardening (*chemical curing*); suitable test methods are differential thermal analysis (DTA) or DSC.
- Further reactions of the thermosetting resins (chain elongation, branching, and cross-linking) to a more or less three-dimensional network with a theoretically endless high molar mass, generating an insoluble resin that is no longer thermoformable (thermoplastic). This can be followed by means of various spectroscopic methods; changes in the portion of various specific structural elements (such as the remaining methylol groups) in the hardened adhesive can be seen as a measure for the degree of curing. Suitable methods are IR, especially ATR-IR, NIR, ^{13}C NMR, or solid-state NMR.
- Solidification of the adhesive during curing by building up the three-dimensional network, also described by the achievable degree of cross-linking; the usual test methods are DMA, thermomechanical analysis (TMA), torsional braid analysis (TBA), and thermal scanning

TABLE 20.9

Test Methods Used to Follow Curing and Formation of Bond Strength

Test Method	Description	References
Differential thermal analysis (DTA)	Measures the difference in temperature between two cells; these two cells are heated at a certain heating rate. One of the two cells contains the sample under investigation	[236, 258]
Differential scanning calorimetry (DSC)	Uses a similar type of instrument as DTA, but measures directly the heat flows of the exothermic and endothermic reactions occurring. The data obtained that are of interest are: shape of the curve, onset and peak temperatures of an exothermic or an endothermic peak, slope of the up-curve, width of the peak	[21, 29, 79, 94, 153, 259-264]
Differential mechanical analysis (DMA, DMTA)	Uses a small sheet of glass fiber mats as a substrate that is impregnated with the resin. This sample then undergoes periodic oscillations in various modes; at the same time the sample is heated following a special temperature program. The curing of the resin leads to an increase in the strength of the sample that then can be correlated with the increase of the cohesive bond strength	[128, 137, 153, 263, 265-266]
Thermomechanical analysis	Similar to DMA, but follows the adhesive hardening *in situ* on the real wood substrate (rather than on glass fiber). Thin wood strips are used to sandwich a liquid adhesive bondline that is then hardened. The curing of the resin leads to an increase in the strength of the sample that can then be correlated with the increase of the cohesive bond strength, as well as with the IB strength of wood PB using the same adhesive. It has been used both at constant heating rate and in isothermal mode	[128,267–269]
Torsional braid analysis (TBA)	The damping behavior of the torsion of a glass fiber probe impregnated with the resin is characteristic for the increase of stiffness	[145,270]
Spectroscopical monitoring	Following the changes in the proportion of various specific structural elements in the adhesive as measure for the degree of hardening, such as the proportion of methylol groups to the carboxyl groups of urea	[258]
	Investigation of cured UF resins by FT-IR or attenuated total reflectance IR (ATR-IR)	[241]
	In case various methods such as FT-IR or (FT-)NIR used for monitoring, the condensation reactions during the preparation of UF and MUF resins may be used for the monitoring of the ongoing reactions during hardening and gelling	[236–238,242]
Automatic Bonding Evaluation System (ABES)	Consists of a small press and a test machine in a single unit. It enables bonds to be formed under highly controlled conditions; the joints that contain the bonds that are to be measured are pressed against heated blocks for a certain time, cooled within a few seconds, and pulled immediately thereafter in shear mode. Repetition of this procedure at different curing times and temperatures yields the points (a point for each specimen) of a near-isothermal strength development curve	[122,128,271,272]

(Continued)

TABLE 20.9 (CONTINUED)

Test Methods Used to Follow Curing and Formation of Bond Strength

Test Method	Description	References
Thermogravimetry (TG)-DTA	Combination of DTA and TG for monitoring thermal hardening and mass loss due to evaporation of water during high-temperature curing	[241,247,248,273,274]
Rheology	Increase in viscosity during the curing process; application of various temperature programs is possible to simulate the temperature feature of a process, such as for an MDF process: quick heating up (simulating blowline and dryer), cooling down and constant temperature for a certain period (simulating fiber cyclone, forming station, and transport of the mat to the hot press), and finally, quick heating up to (simulating the behavior in the hot press)	[77,275,276]
Composite Testing System (ComTeS)	Pressing of two small fiberboard samples and testing via shear strength; press device is incorporated into an ABES machine	[256]
Integrated Pressing and Testing System (IPATES)	A fiber mat of 100 mm in diameter is pressed and then immediately tested in cross-tensile mode (IB) within the press setup after the preselected press time is reached	[256,272]
Laboratory boards pressed under defined condition	Different board types (particleboard, plywood, simple lap shear test boards) pressed at varying parameters, such as temperature or press time. Test according to standard methods (such as lap shear test) after defined conditioning or special pretreatment	
Dielectric analysis (DEA)	As continuous monitoring method, involves measuring the changes in the dielectric properties of an adhesive during curing, which allows calculation of the degree of cure as a relative measure of the adhesive conversion from the liquid to the solid phase	[277–280]
Nanoindentation (NI)	Determination of hardness and elastic modulus of the hardened adhesive resin, either as bulk (such as filling the cell lumen) or penetrated into the cell wall (representing the strength increase of the cell wall by the penetrated adhesive)	[183,189,281–289]
Kinetic models	Simple empirical model, considering a homogeneous irreversible reaction of a single kind of methylol group and urea with rate constants depending on their degree of substitution. Although such a model can provide a better understanding of the composition and the cure of a UF resin, several issues remain open, such as the influence of reversibility of reactions taking place during the curing process, as well as the possible formation of cyclic groups in the resin	[245]

Source: Test Methods Used to Follow Curing and Formation of Bond Strength (for further literature see [1] and [2])

rheometry [77], as well as various gel test methods at different temperatures and using various hardeners [77] with the moment (or period) of formation of the gelled state as a measure. During curing of the resin, the cohesive bond strength develops step by step (defined as the degree of *mechanical curing*).

- Formation of the bond strength between two adherends; this can be followed by methods such as ABES [122], the Composite Testing System (ComTeS [256]), the Integrated

Pressing and Testing System (IPATES [256]) or also any other test methods where bonds are created under certain conditions (especially in dependence on time) and then tested [257]. The easiest procedure, although time consuming, is to press a series of lab boards (plywood, PB, MDF) for different press times and test the already generated bond strength using the usual test methods, such as for shear strength or IB strength.

At the end of the press cycle in board manufacturing, a certain level of mechanical hardening, and with this, a certain bond strength, are necessary to withstand the internal steam pressure in the pressed board and the springback forces due to the deformation of the wood material during hot pressing. The full chemical curing, however, can be attained outside of the press during hot-stacking. A quicker development of the bond strength already at the same degree of chemical curing will enable shorter press times and will, therefore, increase the production capacity and reduce production costs. Graphs with chemical (such as measured by DSC) and mechanical degrees of curing (e.g., measured by TMA, DMA, or ABES) show different hardening behaviors of various resins [127,290].

With a higher temperature in the bondline or in the mat, a quicker hardening and curing and, hence, development of the cohesive bond strength is reached. Performing tests at different temperatures and recalculating the relevant speed of the reaction (such as reciprocal gel time or cohesive bond strength development) enables the establishment of Arrhenius graphs and calculate (apparent) activation energies (kilojoules per mole). Such evaluations have been done for gel-time measurements [77,81], bond strength development [82], DSC [79,259], TMA [83], and preparation of lap shear joints [291]. A summary of results for activation energies is given in [2].

20.8.4 SIMULATION OF RESIN HARDENING DURING A PRESS CYCLE

The mat temperature developing in a vertical direction from each of the surface layers to the core layer and also horizontally depends on press parameters such as densification speed, venting, and second densification steps, as well as on a variety of mat properties [292]. A variety of analytical models have been developed within the last decades to describe physical processes in wood furnish mats during the hot-pressing cycle. One of the first models was already presented in 1982 by Humphrey [293]. He described the fundamental physical processes of heat and moisture transfer, considering vapor convection, heat conduction and convection, and phase changes. Heat and moisture transfer processes, as well as the rheological response due to the densification process of dry-formed wood-based composites, have been summarized in a series of publications by Bolton, Humphrey and Kavvouras [294–297] and Humphrey and Ren [298]. Further results deal with simulating the physical processes of temperature development in different layers and gas-pressure values during the hot-press cycle of dry-formed wood-fiber mats [299,300]. The numerous equations describing physical principles during hot pressing have been derived from empirical tests published by Haas [301].

Heinemann et al. [292] calculated the degree of chemical cure during the hot-press process, based on the simulated temperature development within various layers of the wood composite mat combined with the kinetic parameters of the exothermic condensation reaction (Figure 20.6). The degree of chemical curing as a partial area of the DSC curve was calculated for two PBs with different initial moisture content in the face layers ($u = 10\%$ and $u = 14\%$) and for two different resin systems (UF and MUPF).

Zombori [302] presented a model for the transient effect during hot pressing of wood-based composites; a numerical model based on fundamental engineering principles was developed and validated to establish a relationship between process parameters and the final properties of wood-based composites. The heat and mass transfer, as one part of the model, predicts the change in density, moisture content, and temperature at designated positions of the cross-section of the mat. Garcia and coworkers [303,304] published experimental and theoretical studies of the effects of flake mat

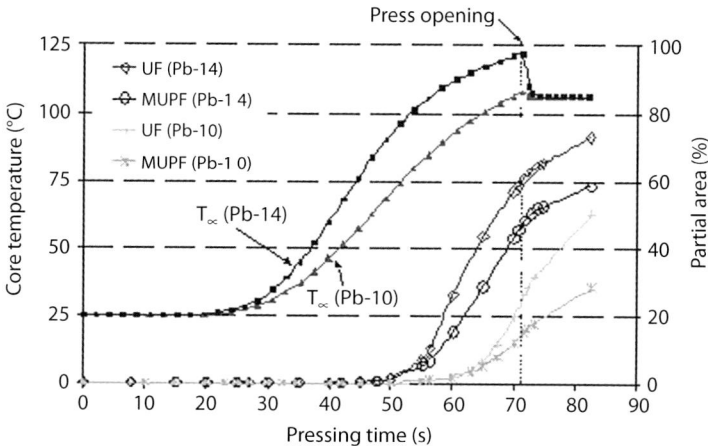

FIGURE 20.6   Degree of cure (partial area of DSC curve) based on simulated temperature development in the core layer of a three-layer particle mat with different initial moisture contents in the face layers ($u = 10\%$ and $u = 14\%$) and for two different adhesive systems (UF and MUPF). (From Heinemann, C. et al., *Proceedings of the Eighth European Panel Products Symposium*, Llandudno, UK, 2004.)

structure on three-dimensional heat and mass transfer during OSB hot pressing. The modeling involves numerically solving a coupled system of differential equations.

A simulation of the physical, and also chemical, aspects of heat and mass transfer and rheological behavior during the hot-pressing process of MDF was prepared by Pereira et al. [305]. In addition to other models, a few detailed features of adhesive curing—for example the exothermic curing reaction and the production of water due to the condensation reaction—have been considered in the model.

The knowledge about the physical principles of heat and mass transfer, as well as the consolidation process during hot pressing, is well understood and described in detail. But, appropriate integration of the curing process of thermosetting adhesive systems within a wood-particle matrix into the model still has to be implemented. The first approach linking simulated physical and measured chemical/physical processes of curing and bonding was introduced by Humphrey [293]. Simulated interparticle bond strength development as a function of temperature and press time during a hot-press cycle is based on bondline strength measurements by ABES. Heinemann et al. [256] simulated the IB strength values using regression equations describing bond strength increasing with press time for a variety of different parameters, such as temperature, mat density, and adhesive content. The tests were performed with IPATES (see Section 20.8.3). It was the first time that the IB strength development, as an industrially relevant property, was predicted for a hot-press cycle. However, the general validity of the predicted IB strength values, as they had been determined with defined fiber material and certain adhesive systems, could not be proven so far. The transferability to other wood particles with certain properties and to different adhesive systems used in the hot-press process is still under investigation. Curing of phenolic thermosetting adhesive systems in the neat state and in the presence of cellulosic material has been determined by using DSC [306], using a variety of different kinetic models. Among these, a model-free isoconversional method has been found to fit the DSC data in the best way. For further literature and description of models for the simulation, see [307–311].

20.9   WOOD AS AN INFLUENTIAL PARAMETER IN WOOD BONDING

The properties of wood-based panels are determined by three main parameters: wood, adhesive, and processing conditions. Proper bonding results require correct balancing of these three parameters. Wood, especially the wood surface and its interface with the bondline, plays a crucial role in the quality of bonding and, therefore, in the properties of the wood-based panels. Low or even no bond strength can be caused by unfavorable properties of the wood surface, such as low wettability. The influence of wood involves several factors. Bonded wood is often described as a chain of several links: the wood (substance), the wood surface, the interphase as the joint region of wood material and adhesive, the interface as the molecular-sized contact zone between wood molecules and adhesive molecules, and the adhesive (as cohesive layer) itself. The strength of an adhesive bond depends on various parameters, such as (1) the strength of the adhesive bondline and its behavior against stresses, (2) the influence of humidity and wood moisture content, (3) the wood properties, which can influence the strength of the bondline and may cause internal stresses, and (4) the mechanical properties of the wood material as the adherend.

20.9.1 WOOD SPECIES AND PARTICLE SIZE AND SHAPE BEFORE PRESSING

In the wood-based panels industry, a great variety of wood species is used as raw material. The choice of the wood species used is often determined just by the availability and the price of the raw material. Furthermore, large amounts of wood residues from the primary wood processing industry (such as sawmill waste) as well as recycled wood from very different sources (pallets, cable drums, demolition of wooden constructions, old furniture, and urban waste) are used. It is more than just a proverb to say that the quality of a wood-based panel is determined at the so-called log yard. The mills generally try to keep the composition of the wood species mix constant, as well as the mix of wood origins and preparation modes for a certain board type. For various board types, different wood mixes (species, shape, and size of the particles) are preferred but difficult to put into industrial practice. Economic reasons (availability of wood, price) also play an important role in the choices made. Many papers deal with special wood species in the production of wood-based panels, but comprehensive information available on this subject is still not yet really satisfactory, and in most cases proprietary information within the industry.

The strength of a bond in a wood panel increases with the wood density for the usual range of approx. 500–800 kg/m³ (depending on the board type). The performance and properties of wood-based panels are strongly influenced by the properties of the wood used. Thus, wood anisotropy, as well as its heterogeneous nature, the variability of its properties, and its hygroscopic behavior, have to be taken into account in all bonding processes. Equally, the orientation of the wood fibers and the fiber angle in bonding solid wood have to be considered. Particles as raw material for PBs show a great variety in wood species, origin, method of preparation, age, and especially size and shape. If wood is comminuted into particles, a mixture of particles of very different sizes and shapes is always obtained.

20.9.2 CHEMICAL COMPOSITION OF WOOD

Wood extractives can influence the bonding process physically and chemically. The chemical composition of a wood surface after processing may be different due to the concentration of polar and nonpolar substances at the surface coming from the wood itself [312]. Extractives soluble in water or steam can migrate during the drying process to the wood surface and can decrease its wettability. In particular, fatty substances and waxes may cover the wood surface. As a consequence of this, a so-called *chemical weak boundary layer* (CWBL) is formed [313]. A chemically induced effect can also occur if the wood extractives have a strong acidic or alkaline behavior. This may cause acceleration or retardation of the hardening process of the adhesives based on polycondensation resins [77,91]. Wood species can show great differences in pH as well as in buffering capacity [86]. Even within the same wood species, additional differences might occur due to seasonal variations, position of origin within the tree log, pH of the soil, age of the tree, time span after cutting, and drying and processing parameters.

20.9.3 WOOD SURFACE

The wood surface is a complex and heterogeneous mixture of polymeric substances such as cellulose, hemicelluloses, and lignin. It is also influenced by factors such as polymer morphology, wood extractives, and processing parameters. During the processing of wood and the generation of new surfaces, damage to the wood surface can occur, which may cause low bond strengths. This often is linked to low percentage wood failure or only a thin fiber layer on one of the two adherends. The reason for this is a mechanical destruction of the uppermost wood layer, causing damaged wood cells, and is usually described as a so-called *mechanical weak boundary layer* (MWBL) [314–317]. A fracture of a bond at the interface between the wood and the adhesive can be caused by a cohesive fracture of such a weak boundary layer or by a real adhesion failure at the interface [318].

A second mode of MWBL can occur directly during the press process, when high temperature, high moisture content, and high mechanical pressure can cause cell-wall buckling or even cell distortion, yielding a zone of lower strength within the interphase [319,320].

20.9.3.1 Contact Angles of Wood Surfaces

Wetting [321,322] by the liquid or liquified adhesive is a precondition for the adhesive bonding of two wood surfaces. Here, wetting includes the formation of a contact angle and the spreading of the liquid on the surface; in most cases, this is linked with partial penetration of the liquid into the porous wood adherend. Sufficient wetting enables the creation of high adhesion forces between the wood surface and the adhesive. However, direct correlations between the contact angle and the bond strength do not exist. Low contact angles ($\theta < 45°$) indicate good wetting behavior. Contact angles greater than 90° lead to incomplete wetting with the risk of low bond strengths. The main parameters influencing the surface tension of the adhesive after application onto the wood surface and, therefore, the possible bond strengths, are:

- Wood species [32,33,323]
- Roughness of the surface [22,324–326]
- Cutting direction (radial/tangential) and fiber angle [32,33]
- Surface preparation [323]
- Earlywood, latewood [32,33,327]
- Direction of spreading of the droplet during measurement of the contact angle (along or lateral to the direction of the fibers) [328]
- Wood moisture content [329]
- Mode of preparation of the surfaces [22,324,330]
- Age of the wood surface [22,324,330–334]
- pH of the wood surface [335,336]
- Type and amount of wood extractives [334]
- Pretreatment of the surface, such as extraction with various solvents [337]; preservative treatments [323]; physical and chemical treatments [338]
- Densification of wood (such as by the viscoelastic thermal compression (VTC) process [339]) [340]
- Type of adhesive: UF resins [21]; PF resins [325]

During the production of wood-based panels, a certain portion of the adhesive penetrates into the wood surface. Overpenetration causes starved bond lines, whereas too low a penetration limits the contact surface between the wood and the adhesive (interface); low penetration is often the consequence of a poor wetting behavior.

20.9.3.2 Modifications of the Wood Surface

Modification of the wood surface [341] can be implemented using various physical, mechanical, and chemical treatments. Chemical treatments are performed in particular to enhance dimensional stability of the panel, but also to improve physical and mechanical properties, or to yield a higher resistance against physical, chemical, and biological attack and the following degradation. Acetylation decreases the number of hydrophilic sites and increases the hydrophobic behavior of the wood substrate [342,343]. The OH groups of the cellulose react with acetic anhydride, forming an ester. The hygroscopic behavior of the wood substrate decreases, and so do the swelling and shrinking of the panel. Use of acetylated fibers for the production of MDF boards showed marked reduction in their thickness swelling [344,345]. Also, seasonal variations of wood quality can influence board quality [346]. Hydrophobic wood extractives and components oxidize or polymerize during storage after harvesting, as can also be seen from their lower extraction ability [347]. Because of this effect, the ability of wood extractives to migrate to a new surface is also reduced.

20.10 PROCESSING CONDITIONS DURING PRODUCTION AS PARAMETERS INFLUENCING WOOD BONDING

20.10.1 Adhesive Consumption and Resin Spread in the Production of Particleboards

Several aspects regarding the proportion of adhesive in the production of PBs must be considered:

- Proportion of adhesive on individual particles
- Proportion of adhesive in particle mixtures and fractions proportion of adhesive in the total particle mix
- Distribution of the adhesive on the surface of the particles
- Proportion of the particle surface area covered with adhesive

The resin loading on wood, as a measure of the consumption of adhesive, is one of the most important parameters. From a technological standpoint, a certain minimum amount of resin is necessary to obtain the desired properties of the boards, resulting in sufficient bonding of the individual particles. Excessive resin loading, on the other hand, increases costs and even causes technological disadvantages, such as high moisture content and problems with high vapor pressure during hot pressing. For economic reasons, the consumption of adhesive should be as low as possible, as the resin contributes considerably to the costs of the boards produced. The resin load, however, is only an overall average of the total mixture of particles, without considering differences in particle size distributions and the shapes of the individual particles. Moreover, the resin load gives no direct indication of the area-specific consumption of the adhesive, which is the amount of resin solids content based on the surface area of the particles. It is well known that fine particles show exceedingly high consumption of adhesive based on the mass of particles, owing to their large surface area [348].

The resin load on wood chips can be described in the following two ways:

- *Mass resin load*: Mass percentage of resin solids based on dry particles
- *Surface specific resin load*: Amount of resin solids content per square unit of surface area

If one of these two terms is known, the other can be calculated, assuming a uniform distribution of the resin on the particle surfaces and estimating the total surface area of the particles. Samples of industrial core-layer and face-layer particles, before and after the blender process, can be fractionated by sieving. In the case of aminoplastic adhesives, each particle fraction, with or without resin, can be investigated for its nitrogen content. By knowing (1) the content of nitrogen, as well as the resin solids content in the resin mix, and (2) the moisture content of the particles in the various fractions, the mass resin factor of each particle size fraction can be calculated. The nitrogen content of the unresinated particles must be considered, a fact that can render this type of analysis difficult when recycling wood material containing a certain nitrogen base load is used as raw material. Resin loading of coarse particles is based on mass, due to the lower specific surface area, much lower than the average resin loading, whereas the fine particles show excessive mass-based resin loading. Such results can also be verified by model calculations as mentioned [348]. An exact screening and classification of the particles before blending can improve the distribution of the resin on the particle surfaces and can help to spare some resin. A lower consumption of resin not only means lower costs of raw materials, but also helps to avoid various technological disadvantages. With the resin, water is also applied to the particles; as long as this amount of water is low enough, especially in the core layer, no problem should occur with high vapor pressure during hot pressing. A high vapor pressure in the board at the end of the press cycle tends to expand the fresh board (springback effect); if venting is not done carefully, blistering of the boards at the end of the continuous press or after the opening of the press can occur. Additionally, the heat transfer by steam shock can be delayed if the vapor pressure difference between the face layer and the core layer is only small. If the moisture content of the resinated core-layer particles is high, the moisture in the resinated face-layer particles must be reduced accordingly. Also, spraying water onto the belt before the forming station and onto the surface of the formed mat might then not be possible due to the problems with high moisture content in the mat and, hence, with high vapor pressure.

The application of resin to the particles is usually done in quickly rotating blenders by spraying the resin mix into the blender or already in special entrance funnels in front of the blender cylinder. Blending OSB strands is done in slowly rotating large blender drums with a diameter of several meters. The liquid adhesive is distributed by several atomizers or nozzles within this blender drum. Application of the adhesive to fibers in MDF production is usually performed in the so-called *blowline* as the connection tube between the refiner and the dryer tube. The advantage of this method is the absence of resin spots on the surface of the board. The disadvantage, however, comes from the fact that the resin passes the dryer and can partially precure. This causes some loss of active resin (approximately 0.5%–2% in absolute figures as resin loading); therefore, resin consumption in blowline blending is higher than in the so-called *mechanical blending* of fibers. The theory of turbulent flow in the blowline operation still is not clearly defined; some equations attempting to describe it have been presented [349]. Chapman [350] developed a design approach for the blowline in MDF production with the target to optimize resin performance. Mass and energy flows in the refiner determine the flow in the blowline, and it is necessary to determine these numbers if the blowline is to be optimized to minimize the required resin consumption. Once the flow in the blowline is established, then the blowline size can be selected. This depends on the length of the blowline that is largely determined by the locations of the refiner and the dryer and the number and location of bends. The other significant factor that determines the resin requirement is the initial size of the droplets sprayed into the blowline. Chapman [350] also proposed a steam-atomized nozzle to provide a uniform droplet size over a wide flow range. Optimization of both the blowline velocities and the location of the resin injection nozzles is necessary to achieve the best resin result. The droplet size can be calculated based on the dimensions of the blowline and the nozzle conditions, as well as the limits that the blowline velocity places on the resin atomization. Comparing the resin-droplet size with the characteristics of a standard fiber allows an assessment of the number of potential resin bonds available to each fiber. Extending this ratio to the number of bonds between the fibers into the MDF fiber network allows a hypothetical model of the failure mode in the IB test to be developed, and this model can be extended to the machinability response of the panel.

The developed blowline blending model identifies an interaction between the blowline steam flow and the resin jet as the reason why blowline blending requires increased resin rates. If the blowline velocity is too low, the resin jet can reach the far wall of the blowline. This prevents the use of effective nozzle atomization of the resin where the blowline diameter is larger than necessary. Resin-droplet size is usually an order of magnitude greater where atomization is achieved by interaction between the blowline steam flow and the resin jet compared with atomization in a nozzle. For the same nozzle conditions, the steam-atomized nozzle produces smaller resin droplets than does a pressure-atomized nozzle. Resin droplets should be small to improve the transmission of the tension forces from the resin bridge to the adjacent structure of the fiber. Interactions between the resin droplets and the fibers occur as they travel along the blowline.

Based on a mass–energy balance in the refiner determining the steam flow in the blowline, this steam flow can be calculated. This allows the blowline to be sized to achieve maximum steam velocities in the resin blending section. The steam flow, together with the blowline configuration, determines the pressure drop profile from the refiner housing to the dryer, as well as the velocity profile in the blowline. Steam velocities between 25 and 100 m/s at resin injection are assumed, with the higher velocities giving the best resin performance. Resin atomization, both in the nozzle and through interaction with the steam flow, determines resin-droplet size, while the relationships between resin-droplet size, fiber size, and steam velocity are important in determining the outcome of the blowline resin blending process.

Pakdel et al. [202] used fluorescence CLSM and XPS to quantify the amount of UF resin in wood fibers after application of the resin, using various fluorescence staining agents for the CLSM investigations. To visualize the UF resin distribution on fibers and within MDF panels, Loxton et al. [351] chemically attached a fluorescent label to the resin, and analyzed digital images of resin-ated fibers, generated via CLSM. Results indicate that this technique can be used to quantify UF resin coverage and distribution, as well as to provide information on resin film thickness on MDF fiber before pressing and as well as in the produced panels. The technique can distinguish between different methods of resination and was employed to determine if these processes could result in different degrees of surface coverage of UF resin on MDF fiber. Resin injected at the end of the blowline gave significantly less resin coverage of fiber than that which was injected at the start of the blowline. UF resin droplets were also relatively bigger and less dispersed when injected at the end of the blowline. Visualization of UF resin also illustrated resin distribution changes on pressing the fiber, particularly in the presence of wax. This result has important implications for future studies targeting optimization of resin deposition, since the droplet size distribution, as applied to the fiber, may not correspond to the droplet size distribution of resin in the panel. He and Riedl [135] developed a technique to show the extent of distribution of the adhesive on the fiber surface. This involves treatment of the resinated fibers with the dye Toluidine Blue O; this quenches the autofluo-rescence of the wood material and only shows the fluorescence of the UF resin in the presence of the dye. The results indicate that this method is a simple and practical way to determine UF resin coverage on wood fibers. With image analysis, the adhesive has been shown to cover only a small percentage (3.5%) of the fiber surface, although the percentage by weight (based on wood fiber) of resin was much higher (13%). The fiber size strongly affects the resin coverage ratio and the chance to collect resin droplets; the resin-droplet size distribution, however, has no significant influence. Further information on the distribution of resin on fibers and in MDF are given in [352–356] and on distribution on OSB strands is given in [357].

20.10.2 Wood Moisture Content

The wood moisture content influences several important processes such as wetting, flow of the adhesive, penetration into the wood surface, and hardening of the adhesive in the production of wood-based panels. In bonding solid wood, a wood moisture content of 6%–14% is usually seen as optimal. Lower wood moisture content can cause quick dry-out of the applied resin, due to strong absorption of water into the wood surface, as well as wetting problems, especially at the surface of the second adherend. High moisture contents can lead to high flow and enhanced penetration into the wood tissue, causing starved bondlines. Additionally, high steam pressure can be generated with the risk of blistering when the press opens or at the end of the continuous press. Also, the hardening of a condensation resin may be retarded or even hindered [358,359].

During the hot-press cycle of the PB or MDF production, quick changes of temperature, moisture content, and steam pressure occur. The gradients of temperature and moisture content significantly determine the hardening rate of the resin, the generation of the cohesive bond strength of the adhesive, and hence, the final board properties. These gradients, together with the mechanical pressure applied to consolidate the mat, generate the density profile and determine the properties and performance of the boards. The higher the moisture content of the glued face-layer particles, the steeper the moisture gradient between the surface and the core of the mat and the quicker the heating up of the mat occurs. In the fiber mat in MDF production, no differences are given in the moisture content of the outer layer and the inner layer; due to the temperature applied to the mat, a vapor pressure gradient nevertheless occurs. Press factors, however, are generally longer due to the lower moisture gradient. Spraying of water can introduce a certain (limited) three-layer effect, with higher moisture content in the outer layer. The moisture content of the particles after blending is the sum of the wood moisture content and the water that is part of the applied resin mix. Therefore, the moisture content of the resinated particles mainly depends on the resin loading. Usual moisture contents of resinated particles are: (1) for UF, 5%–7% in the core layer and 10%–13% in the face layer; (2) for PF, 10%–13% in the core layer and 13%–17% in the face layer. The optimal moisture content of blowline-resinated and dried MDF fibers in the mat before pressing is 9%–11%. The higher the moisture content of the particles, the easier the face layer can be densified when the press cycle starts; this then leads to a lower density in the core layer.

Blistering at the end of the press cycle or at the end of the continuous press occurs if the steam pressure within the fresh, and still hot, board exceeds the *in situ* IB strength of the board. It should be noted that this bond strength at higher temperature is always significantly lower than after cooling of the board. Careful venting, as well as decreasing the moisture content of the resinated particles and reducing the press temperature, will help. Prolonging of the press time will help as long as the venting effect and/or the increase in bond strength (based on full curing of the resin) are stronger than the additional generation of steam.

20.10.3 Press Cycle

During the hot-press cycle, (1) curing of the resin and (2) generation of mechanical strength by bonding the wood substance occur. The main influential parameters are the press temperature and moisture content in the mat, determining the temperature development in the mat. Additional parameters are the wood density, porosity, swelling and shrinking behavior of the wood, surface structure, and wetting behavior. During the press cycle, several processes take place: (1) transport of heat and moisture; (2) densification, increasing internal stresses, followed by relaxation processes; (3) adhesion between the particles or fibers; and (4) increase of the bond strength in the bondline (cohesion). Theoretical models can describe the change of various physical and mechanical parameters during the press cycle, such as temperature or moisture content in various layers. These take into consideration various conditions occurring during the hot-press cycle such as heat transfer, temperature gradients, and moisture content, development of steam pressure and bond strength, and possible postcuring effects [298,360–364]. Heating the mat is performed by the so-called *steam shock effect*. The precondition for this is the high permeability of the particle or fiber mat [362,363,365] for steam and gases. High moisture content of the face layers and spraying of water on the surface layers support this effect. The press temperature influences the press time and, hence, the capacity of the production line. The minimum press time has to guarantee that the bond strength of the still hot board can withstand the internal steam pressure, as well as the elastic springback in board thickness at press opening. Measurement of the mat core temperature and the vapor pressure within the mat are standard methods, yielding important insight into the behavior of the mat [366–368]. The vertical density profile can be measured online at the end of the press or in the laboratory. The formation of the density profile during the press process has been monitored at various fixed positions relative to the board thickness [369–372], or continuously for the whole cross-section during densification of the mat [373].

20.11   SUMMARY

Wood adhesives technology is an advanced science, merging adhesive preparation and formulation with several advanced application technologies to create new and improved wood products and production processes. As in all applications of adhesives, cohesion and adhesion are the main features to be followed, in combination with proper application technology and press conditions, irrespective of the applied temperature.

Though extensive information is available on adhesives and adhesion, basic research on the adhesion behavior of wood, including the evaluation of the formed microstructure of the bondline as the main performance parameter within the bonded products produced, still needs to be promoted. This is to be seen as a fundamental approach, as well as an application to industrial bonding processes in a great variety of modes, from single-layer laminated wood preparation in a carpenter's shop to industrial production of wood-based panels of several thousand cubic meters per day. Aspects such as the surface properties of wood related to bondability, adhesive characteristics concerning wetting and penetration, the formation of the interphase of wood and adhesive, interfacial reactions between the wood surface and the adhesive, chemical curing or physical solidification of the adhesive, the behavior of the adhesive in the formed bondline, the microstructure of the bondline and the wood-based composites, and the properties of the yielded bondline as the basis for the performance of wood-based composites are, among others, the main features of bonding.

REFERENCES

1. M. Dunky. Wood adhesives. In: *Handbook of Adhesive Technology*, 2nd edn, A. Pizzi and K. L. Mittal (Eds.), pp. 887–956, CRC Press, Boca Raton, FL (2003).
2. M. Dunky and P. Niemz. *Wood Based Panels and Resins: Technology and Influence Parameters* (in German). Springer, Heidelberg (2002).
3. M. Dunky and A. Pizzi. Wood adhesives. In: *Adhesion Science and Engineering, Volume 2: Issue Surfaces, Chemistry and Applications*, D. A. Dillard and A. V. Pocius (Eds.), 1039–1103, Elsevier Science B.V., Amsterdam, The Netherlands. (2003).
4. Particleboard, JIS A5908, Japanese Standards Association, Tokyo, Japan (2003).
5. Fiberboards, JIS A5905, Japanese Standards Association, Tokyo, Japan (2003).
6. Airborne Toxic Control Measure (ATCM). Title 17 of the California Code of Regulations, Section 93120.3(d). Air Resources Board (ARB). http://www.arb.ca.gov (2007).
7. M. Dunky. The long term formaldehyde emission from particleboards. *Holz Roh Werkst.* 48, 371–375 (1990).
8. M. Dunky. Urea-formaldehyde glue resins. In: *Polymeric Materials Encyclopedia*, Vol. 11, J. C. Salamone (Ed.), pp. 8502–8511, CRC Press, Boca Raton, FL (1996).
9. M. Dunky. Urea-formaldehyde (UF-) glue resins. *Int. J. Adhes. Adhes.* 18, 95–107 (1998).
10. A. Pizzi. Aminoresin wood adhesves. In: *Wood Adhesives: Chemistry and Technology*, A. Pizzi (Ed.), pp. 59–104, CRC Press, Boca Raton, FL (1983).
11. A. Pizzi. *Advanced Wood Adhesives Technology*, pp. 19–66, CRC Press, Boca Raton, FL (1994).
12. A. Pizzi. Urea-formaldehyde adhesives. In: *Handbook of Adhesive Technology*, 2nd edn, A. Pizzi and K. L. Mittal (Eds.), pp. 635–652, CRC Press, Boca Raton, FL (2003).
13. J. M. M. Ferra, A. Henriques, A. M. Mendes, M. R. N. Costa, L. H. Carvalho and F. D. Magalhaes. Comparison of UF synthesis by alkaline-acid and strongly acid processes. *J. Appl. Polym. Sci.* 123, 1764–1772 (2012).
14. J. Billiani, K. Lederer and M. Dunky. Investigations of the molar mass distribution of urea formaldehyde adhesive resins with GPC coupled with light scattering (in German). *Angew. Makromol. Chem.* 180, 199–208 (1990).
15. B. D. Park, E. C. Kang and J. Y. Park. Effects of formaldehyde to urea mole ratio on thermal curing behavior of urea–formaldehyde resin and properties of particleboard. *J. Appl. Polym. Sci.* 101, 1787–1792 (2006).

16. M. G. Kim. Examination of selected synthesis parameters for typical wood adhesive-type urea–formaldehyde resins by 13C-NMR spectroscopy. II. *J. Appl. Polym. Sci.* 75, 1243–1254 (2000).

17. P. Christianson, T. Pehk and K. Siimer. Hydroxymethylation and polycondensation reactions in urea–formaldehyde resin synthesis. *J. Appl. Polym. Sci.* 100, 1673–1680 (2006).

18. R. N. Kumar, T. L. Han, H. D. Rozman, W. R. W. Daud and M. S. Ibrahim. Studies in the process optimization and characterization of low formaldehyde emission urea-formaldehyde resin by response surface methodology. *J. Appl. Polym. Sci.* 103, 2709–2719 (2007).

19. B. D. Park, Y. S. Kim, A. P. Singh and K. P. Lim. Reactivity, chemical structure, and molecular mobility of urea–formaldehyde adhesives synthesized under different conditions using FTIR and solid-state ^{13}C CP/MAS NMR spectroscopy. *J. Appl. Polym. Sci.* 88, 2677–2687 (2003).

20. C. Soulard, C. Kamoun and A. Pizzi. Uron and uron-urea-formaldehyde resins. *J. Appl. Polym. Sci.* 72, 277–289 (1999).

21. Y. Su, Q. Ran, W. Wu and X. Mao. H_2O_2 catalyzed cure of UF resins with different structures. *Thermochim. Acta* 253, 307–316 (1995).

22. Z. Qin, H. Chen, Q. Gao, S. Zhang and J. Li. Wettability of sanded and aged fast-growing poplar wood surfaces: I. Surface free energy. *BioResources* 10, 1008–1023 (2015).

23. A. Mao, E. B. Hassan and M. G. Kim. Low mole ratio UF and UMF resins entailing uron-type methylene-ether groups and their low formaldehyde emission potentials. *BioResources* 8, 2470–2486 (2013).

24. A. Mao, E. B. Hassan and M. G. Kim, Low mole ratio urea–melamine–formaldehyde resins entailing increased methylene-ether group contents and their formaldehyde emission potentials of wood composite boards. *BioResources* 8, 4659–4675 (2013).

25. W. Gao and J. Li. Influence of uron resins on the performance of UF resins as adhesives for plywood. Maderas. *Ciencia y Tecnol.* 14, 3–12 (2012).

26. N. A. Costa, D. Martins, J. Pereira, J. Martins, J. Ferra, P. Cruz, A. Mendes, F. D. Magalhaes and L. H. Carvalho. ^{13}C NMR study of presence of uron structures in amino adhesives and relation with wood-based panels performance. *J. Appl. Polym. Sci.* 130, 4500–4507 (2013).

27. C. Y. Hse, Z. Y. Xia and B. Tomita. Effects of reaction pH on properties and performance of urea-formaldehyde resins. *Holzforschung* 48, 527–532 (1994).

28. M. G. Kim. Examination of selected synthesis parameters for wood adhesive-type urea-formaldehyde resins by ^{13}C NMR spectroscopy. *J. Appl. Polym. Sci.* 80, 2800–2814 (2001).

29. B.D. Park, E.C. Kang and J.Y. Park. Differential scanning calorimetry of urea–formaldehyde adhesive resins, synthesized under different pH conditions. *J. Appl. Polym. Sci.* 100, 422–427 (2006).

30. M. G. Kim, H. Wan, B. Y. No and W-L. Nieh. Examination of selected synthesis and room-temperature storage parameters for wood adhesive-type urea-formaldehyde resins by ^{13}C-NMR spectroscopy. IV. *J. Appl. Polym. Sci.* 82, 1155–1169 (2001).

31. B. D. Park. Properties of urea-formaldehyde resin adhesives with different formaldehyde to urea mole ratios. *J. Korean Wood Sci. Technol.* 35, 5, 67–75 (2007).

32. M. Scheikl and M. Dunky. Urea formaldehyde resins as liquid phase when wetting wood (in German). *Holzforsch. Holzverwert.* 48, 55–57 (1996).

33. M. Scheikl and M. Dunky. Measurement of dynamic and static contact angles on wood for the determination of its surface tension and the penetration of liquids into the wood surface. *Holzforschung* 52, 89–94 (1998).

34. H. R. Mansouri and A. Pizzi. Urea-formaldehyde-propionaldehyde physical gelation resins for improved swelling in water. *J. Appl. Polym. Sci.* 102, 5131–5136 (2006).

35. J. Zhang, H. Chen, A. Pizzi, Y. Li, Q. Gao and J. Li. Characterization and application of urea-formaldehyde-furfural co-condensed resins as wood adhesives. *BioResources* 9, 6267–6276 (2014).

36. N. Esmaeili, M. J. Zohuriaan-Mehr, S. Mohajeri, K. Kabiri and H. Bouhendi. Hydroxymethyl furfural-modified urea-formaldehyde resin: Synthesis and properties. *Eur. J. Wood Prod.* 75, 71–80 (2017).

37. H. Li, Y. Zhang and X. Zeng. Two-step synthesis and characterization of urea–isobutyraldehyde–formaldehyde resins. *Prog. Org. Coat* 66, 167–172 (2009).

38. A. Despres, A. Pizzi, C. Vu and L. Delmotte. Colourless formaldehyde-free urea resin adhesives for wood panels. *Eur. J. Wood Prod.* 68, 13–20 (2010).

39. M. Properzi, S. Wieland, F. Pichelin, A. Pizzi and A. Despres. Formaldehyde-free dimethoxyethanal-derived resins for wood-based panels. *J. Adhes. Sci. Technol.* 24, 1787–1799 (2010).

40. S. Wang and A. Pizzi. Succinaldehyde induced water resistance improvements of UF wood adhesives. *Holz Roh Werkst.* 55, 9–12 (1997).

41. S. Wieland, A. Pizzi, S. Hill, W. Grigsby and F. Pichelin. The reaction in water of UF resins with isocyanates at short curing times: A ^{13}C NMR investigation. *J. Appl. Polym. Sci.* 100, 1624–1632 (2006).

42. C. Simon, B. George and A. Pizzi. UF/pMDI wood adhesives: Networks blend versus copolymerization. *Holzforschung* 56, 327–334 (2002).

43. H. R. Mansouri, A. Pizzi, and J. M. Leban. Improved water resistance of UF adhesives for plywood by small pMDI additions. *Holz Roh Werkst.*, 64, 218–220 (2006).

44. S. Wieland, A. Pizzi, W. Grigsby, J. Warnes and F. Pichelin. Microcrystallinity and colloidal peculiarities of UF/Isocyanate hybrid resins. *J. Appl. Polym. Sci.* 104, 2633–2636 (2007).

45. A. Pizzi. Melamine-formaldehyde adhesives. In: *Handbook of Adhesive Technology*, 2nd edn, A. Pizzi and K. L. Mittal (Eds.), pp. 653–680, CRC Press, Boca Raton, FL (2003).

46. W. H. Binder and M. Dunky. Melamine–formaldehyde resins. In: *Encyclopedia of Polymer Science and Technology*, 3rd edn, H. F. Mark (Ed.), John Wiley & Sons, Inc., Hoboken, NJ, (10), 369–384 (2004).

47. W. H. Binder, M. Dunky and S. Jahromi. Melamine resins. In: *Kirk-Othmer Encyclopedia of Chemical Technology*, 5th edn, Kirk-Othmer (Ed.), 15, 773–779, John Wiley & Sons, Inc., Hoboken, NJ DOI: 10.1002/0471238961.melabind.a01 (2005).

48. A. Pizzi. Synthetic adhesives for wood panels: Chemistry and technology: A critical review. *Rev. Adhes. Adhes.* 1, 85–126 (2014).

49. T. A. Mercer and A. Pizzi. Consideration on the principles of preparation of melamine-urea-formaldehyde adhesive resins for particleboard. *Holzforsch. Holzverwert.* 46, 51–54 (1994).

50. N. T. Paiva, J. Pereira, J. M. M. Ferra, P. Cruz, L. H. Carvalho and F. D. Magalhaes. Study of influence of synthesis conditions on properties of melamine–urea formaldehyde resins. *Int. Wood Prod. J.* 3, 51–57 (2012).

51. N. T. Paiva, A. Henriques, P. Cruz, J. M. M. Ferra, L. H. Carvalho and F. D. Magalhaes. Production of melamine fortified urea-formaldehyde resins with low formaldehyde emission. *J. Appl. Polym. Sci.* 124, 2311–2317 (2012).

52. Q. N. Sun, C. Y. Hse and T. F. Shupe. Characterization and performance of melamine enhanced urea formaldehyde resin for bonding southern pine particleboard. *J. Appl. Polym. Sci.* 119, 3538–3543 (2011).

53. Q. N. Sun, C. Y. Hse and T. F. Shupe. Effect of different catalysts on urea–formaldehyde resin synthesis. *J. Appl. Polym. Sci.* 131, 406–444 (2014).

54. C. Y. Hse. Development of melamine modified urea formaldehyde resins based on strong acidic pH catalyzed urea formaldehyde polymer. *For. Prod. J.* 59, 5, 19 – 24 (2009).

55. B. Tomita and C. Y. Hse. Analyses of cocondensation of melamine and urea through formaldehyde with ^{13}C nuclear magnetic resonance spectroscopy. l: *Mok. Gakk.* 41, 349–354 (1995). II: *Mok. Gakk.* 41, 490–497 (1995).

56. A. Philbrook, C. J. Blake, N. Dunlop, C. J. Easton, M. A. Keniry and J. S. Simpson. Demonstration of co-polymerization in melamine–urea–formaldehyde reactions using 15N NMR correlation spectroscopy. *Polymer* 46, 2153–2156 (2005).

57. A. Philbrook, S. Earnshaw, C. J. Easton, M. A. Keniry and M. J. Latter. Co-polymerization analysis of thermosetting resins using ^{1}H-^{15}N-^{13}C triple resonance NMR spectroscopy. *J. Appl. Polym. Sci.* 128, 3375–3381 (2013).

58. G. E. Troughton and S. Chow. Effect of fortifier addition on the curing reactions of urea formaldehyde adhesives. *Holzforschung* 29, 214–217 (1975).

59. H. Yamaguchi, M. Higuchi and I. Sakata. Durability of urea formaldehyde resin adhesive. Effect of pH on the adhesive resin layer. *Mok. Gakk.* 35, 801–806 (1989).

60. A. Pizzi. A molecular mechanics approach to the adhesion of urea-formaldehyde resins to cellulose. Part 1. Crystalline cellulose I. *J. Adhes. Sci. Technol.* 4, 573–578 (1990).

61. A. Pizzi. A conformational analysis approach to phenol-formaldehyde resins adhesion to wood cellulose. *J. Adhes. Sci. Technol.* 1, 191–200 (1987).

62. G. E. Myers and J.A. Koutsky. Formaldehyde liberation and cure behavior of urea formaldehyde resins. *Holzforschung* 44, 117–126 (1990).

63. O. Ringena, R. Janzon, G. Pfizenmayer, M. Schulte and R. Lehnen. Estimating the hydrolytic durability of cured wood adhesives by measuring formaldehyde liberation and structural stability. *Holz Roh Werkst.* 64, 321–326 (2006).

64. G. E. Myers. Use of acid scavengers to improve durability of acid-catalyzed adhesive wood bonds. *Forest Prod. J.* 33, 4, 49–57 (1983).

65. J. Zhang, X. Wang, S. Zhang, Q. Gao and J. Li. Effects of melamine addition stage on the performance and curing behavior of melamine-urea-formaldehyde (MUF) resin. *BioResources* 8, 5500–5514 (2013).

66. A. Mao, E. B. Hassan and M. G. Kim. The effects of adding melamine at different resin synthesis points of low mole ratio urea-melamine-formaldehyde (UMF) resins. *BioResources* 8, 5733–5748 (2013).

67. A. Mao, E. B. Hassan and M. G. Kim. Investigation of low mole ratio UF and UMF resins aimed at lowering the formaldehyde emission potential of wood composite boards. *BioResources* 8, 2453–2469 (2013).

68. B. Y. No and M. G. Kim. Syntheses and properties of low-level melamine-modified urea–melamine–formaldehyde resins. *J. Appl. Polym. Sci.* 93, 2559–2569 (2004).

69. B. Y. No and M. G. Kim. Curing of low level melamine-modified urea-formaldehyde particleboard binder resins studied with dynamic mechanical analysis (DMA). *J. Appl. Polym. Sci.* 97, 377–389 (2005).

70. J. Luo, J. Zhang, J. Luo, J. Li and Q. Gao. Effect of melamine allocation proportion on chemical structures and properties of melamine-urea-formaldehyde resins, *BioResoures* 10, 3265–3276 (2015).

71. B. Y. No, M. G. Kim. Evaluation of melamine-modified urea-formaldehyde resins as particleboard binders. *J. Appl. Polym. Sci.* 106, 4148–4156 (2007).

72. A. Pizzi, M. Beaujean, C. Zhao, M. Properzi and Z. Huang. Acetal-induced strength increases and lower resin content of MUF and other polycondensation adhesives. *J. Appl. Polym. Sci.* 84, 2561–2571 (2002).

73. M. Zanetti, A. Pizzi, M. Beaujean, H. Pasch, K. Rode and P. Dalet. Acetals-induced strength increase of melamine-urea-formaldehyde (MUF) polycondensation adhesives. II. Solubility and colloidal state disruption. *J. Appl. Polym. Sci.* 86, 1855–1862 (2002).

74. M. Zanetti and A. Pizzi. Upgrading of MUF polycondensation resins by buffering additives. II. Hexamine sulfate mechanisms and alternate buffers. *J. Appl. Polym. Sci.* 90, 215–226 (2003).

75. C. Kamoun, A. Pizzi and M. Zanetti. Upgrading melamine–urea–formaldehyde polycondensation resins with buffering additives. I. The effect of hexamine sulfate and its limits. *J. Appl. Polym. Sci.* 90, 203–214 (2003).

76. M. Zanetti and A. Pizzi. Dependence on the adhesive formulation of the upgrading of MUF particleboard adhesives and decrease of melamine content by buffer and additives. *Holz Roh Werkst.* 62. 445–451 (2003).

77. B. Stefke and M. Dunky. Catalytic influence of wood on gelling of formaldehyde based glue resins. *J. Adhes. Sci. Technol.* 20, 8, 761–785 (2006).

78. S. Kim, H. J. Kim, H. S. Kim, Y. K. Lee and H. S Yang. Thermal analysis study of viscoelastic properties and activation energy of melamine-modified urea-formaldehyde resins. *J. Adhes. Sci. Technol.* 20, 8, 803–816 (2006).

79. A. Nuryawan, B. D. Park and A. P. Singh. Comparison of thermal curing behavior of liquid and solid urea–formaldehyde resins with different formaldehyde/urea mole ratios. *J. Therm. Anal. Calorim.*, 118, 397–404 (2014).

80. Y. K. Lee and H. J. Kim. Relationship between curing activation energy and free formaldehyde content in urea-formaldehyde resins. *J. Adhes. Sci. Technol.* 27, 598–609 (2013).

81. L. Halasz, O. Vorster, A. Pizzi and J. van Alphen. A rheological study of the gelling of UF polycondensates. *J. Appl. Polym. Sci.* 75, 1296–1302 (2000).

82. P. E. Humphrey and A. J. Bolton. Urea formaldehyde resin bond strength development with reference to wood particleboard manufacture. *Holzforschung* 33, 129–133 (1979).

83. S. Yin, X. Deglise and D. Masson. Thermomechanical analysis of wood/aminoplastic adhesives joints cross-linking: UF, MUF, PMUF. *Holzforschung* 49, 575–580 (1995).

84. C. Xing, B. Riedl, A. Cloutier, J. Deng and S. Y. Zhang. UF resin efficiency of MDF as affected by resin content loss, coverage level and pre-cure. *Holz Roh Werkst.* 64, 221–226 (2006).

85. C. Xing, B. Riedl, A. Cloutier and G. He. The effect of urea-formaldehyde resin pre-cure on the internal bond of medium density fibreboard. *Holz Roh Werkst.* 62, 439–444 (2004).

86. W. Sanderkamp and M. Rothkamm. On the determination of pH-values in timbers and their practical importance (in German). *Holz Roh Werkst.* 17, 433–440 (1959).

87. C. Xing, S. Y. Zhang and J. Deng. Effect of wood acidity and catalyst on UF resin gel time. *Holzforschung* 58, 408–412 (2004).

88. R. O. Albritton and P. H. Short. Effects of extractives from pressure-refined hardwood fiber on the gel time of urea-formaldehyde resin. *For. Prod. J.* 29, 2, 40–41 (1979).

89. G. Han, K. Umemura, S. Kawai and H. Kajita. Improvement mechanism of bondability in UF-bonded reed and wheat straw boards by silane coupling agent and extraction treatments. *J. Wood Sci.* 45, 299–305 (1999).

90. G. Han. Development of high-performance reed and wheat straw composite panels. *Wood Res.* 88, 19–39 (2001).

91. S. Medved and J. Resnik. Influence of the acidity and size of beech particles on the hardening of the urea-formaldehyde adhesive. *Acta Chim. Sloven.* 51, 353–360 (2004).

92. J. R. Slay, P. H. Short and D. C. Wright. Catalytic effects of extractives from pressure-refined hardwood fiber on the gel time of urea-formaldehyde resin. *For. Prod. J.* 30, 3, 22–23 (1980).

93. Z. Gao, X.-M. Wang, H. Wan and Y. Liu. Curing characteristics of urea–formaldehyde resin in the presence of various amounts of wood extracts and catalysts. *J. Appl. Polym. Sci.*, 107, 1555–1562 (2008).

94. C. Xing, J. Deng, S. Y. Zhang, B. Riedl and A. Cloutier. Differential scanning calorimetry characterization of urea–formaldehyde resin curing behavior as affected by less desirable wood material and catalyst content. *J. Appl. Polym. Sci.* 98, 2027–2032 (2005).

95. E. E. Ferg, A. Pizzi and D. C. Levendis. ^{13}C NMR analysis method for urea–formaldehyde resin strength and formaldehyde emission. *J. Appl. Polym. Sci.* 50, 907–915 (1993).

96. E. E. Ferg, A. Pizzi and D. C. Levendis. Correlation of particleboard strength and formaldehyde emission with urea step additions and ^{13}C NMR of UF resins. *Holzforsch. Holzverwert.* 45, 88–92 (1993).

97. T. A. Mercer and A. Pizzi. A ^{13}C-NMR analysis method for MF and MUF resins strength and formaldehyde emission from wood particleboard. II. MF resins. *J. Appl. Polym. Sci.* 61, 1697–1702 (1996).

98. T. A. Mercer and A. Pizzi. A ^{13}C-NMR analysis method for MF and MUF resins strength and formaldehyde emission from wood particleboard. I. MUF resins. *J. Appl. Polym. Sci.* 61, 1687–1695 (1996).

99. L. A. Panamgama and A. Pizzi. A ^{13}C-NMR analysis method for MUF and MF resin strength and formaldehyde emission. *J. Appl. Polym. Sci.* 59, 2055–2068 (1996).

100. A. Pizzi. Phenolic resin adhesives. In: *Handbook of Adhesive Technology*, 2nd edn, A. Pizzi and K. L. Mittal (Eds.), pp. 541–572, CRC Press, Boca Raton, FL (2003).

101. A. Knop and L. A. Pilato. *Phenolic Resins: Chemistry, Applications and Performance*, Springer Science & Business Media Berlin and Heidelberg, Germany (2013).

102. A. Gardziella, L. A. Pilato and A. Knop. *Phenolic Resins: Chemistry, Applications, Standardization, Safety and Ecology*, Springer Science & Business Media Berlin and Heidelberg, Germany (2013).

103. A. Pizzi. Phenolic resin wood adhesives. In: *Wood Adhesives: Chemistry and Technology*, A. Pizzi (Ed.), pp. 105–176, CRC Press, Boca Raton, FL (1983).

104. A. Pizzi (Ed.). *Advanced Wood Adhesives Technology*, pp. 89–148, CRC Press, Boca Raton, FL (1994).

105. G. Astarloa-Aierbe, J. M. Echeverria, M. D. Martin and I. Mondragon. Kinetics of phenolic resol resin formation by HPLC. 2. Barium hydroxide. *Polymer* 39, 3467–3472 (1998).

106. G. Astarloa-Aierbe, J. M. Echeverria and I. Mondragon. Kinetics of phenolic resol resin formation by HPLC. 3. Zinc acetate. *Polymer* 40, 5873–5878 (1999).

107. G. Astarloa-Aierbe, J. M. Echeverria, M. D. Martin, A. M. Etxeberria and I. Mondragon. Influence of the initial formaldehyde to phenol molar ratio (F/P) on the formation of a phenolic resol resin catalyzed with amine. *Polymer* 41, 6797–6802 (2000).

108. G. Astarloa-Aierbe, J. M. Echeverria, A. Vazquez and I. Mondragon. Influence of the amount of catalyst and initial pH on the phenolic resol resin formation. *Polymer* 41, 3311–3315 (2000).

109. S. Tohmura, M. Higuchi and I. Sakata. Acceleration of the cure of phenolic resin adhesives. II. Kinetics of the curing reactions (1). *Mok. Gakk.* 38, 59–66 (1992).

110. P. Luukko, L. Alvila, T. Holopainen, J. Rainio and T. T. Pakkanen. Effect of alkalinity on the structure of phenol–formaldehyde resol resins. *J. Appl. Polym. Sci.* 82, 258–262 (2001).

111. G. Gobec, M. Dunky, T. Zich and K. Lederer. Gel permeation chromatography and calibration on resolic phenol-formaldehyde-condensates (in German). *Angew. Makromol. Chem.* 251, 171–179 (1997).

112. W. L. S. Nieh and T. Sellers, Jr. Performance of flakeboard bonded with three phenol formaldehyde resins of different mole ratios and molecular weights. *For. Prod. J.* 41, 6, 49–53 (1991).

113. S. Ellis. The performance of waferboard bonded with powdered phenol-formaldehyde resins with selected molecular weight distributions. *For. Prod. J.* 43, 2, 66–68 (1993).

114. S. Ellis and P. R. Steiner. Some effects of the chemical and physical characteristics of powdered phenol-formaldehyde resins on their adhesive performance. *For. Prod. J.* 42, 1, 8–14 (1992).

115. B. D. Park, B. Riedl, E. W. Hsu and J. Shields. Effects of weight average molecular mass of phenol-formaldehyde adhesives on medium density fiberboard performance. *Holz Roh Werkst.* 56, 155–161 (1998).

116. R. A. Haupt and T. Sellers, Jr. Characterizations of phenol-formaldehyde resol resins. *Ind. Eng. Chem. Res.* 33, 693–697 (1994).

117. A. Pizzi and A. Stephanou. On the chemistry, behavior, and cure acceleration of phenol-formaldehyde resins under very alkaline conditions. *J. Appl. Polym. Sci.* 49, 2157–2170 (1993).

118. A. Pizzi, R. Garcia and S. Wang. On the networking mechanisms of additives-accelerated phenol–formaldehyde polycondensates. *J. Appl. Polym. Sci.* 66, 255–266 (1997).

119. S. Tohmura and M. Higuchi. Acceleration of the cure of phenolic resin adhesives VI. Core accelerating action of propylene carbonate. *Mok. Gakk.* 41, 1109–1114 (1995).

120. S. Kim, H. S. Kim, H. J. Kim and H. S. Yang. Fast curing PF resin mixed with various resins and accelerators for building composite materials. *Constr. Build. Mater.* 22, 10, 2141–2146 (2008).

121. Y. Chen, D. Fan, T. Qin and F. Chu. Thermal degradation and stability of accelerated-curing phenol-formaldehyde resin. *BioResources* 9, 4063–4075 (2014).

122. P. E. Humphrey. Device for testing adhesive bonds. US Patent 5176028, assigned to the State of Oregon (1990).

123. S. So and A. Rudin. Study of the curing reactions of phenolic resins by high field carbon-13 CP-MAS NMR spectroscopy. *J. Polym. Sci. Polym. Lett. Ed.* 23, 403–407 (1985).

124. R. H. Young. Adhesive cure as determined by dynamic mechanical analysis and its effect on wood composite performance. In: *Proceedings of International Conference Wood Adhesives 1985: Status and Needs*, Madison, WI, pp. 267–276 (1985).

125. R. G. Schmidt and C. E. Frazier. ^{13}C CP/MAS NMR as a direct probe of the wood-phenol formaldehyde adhesive bondline. *Wood Fiber Sci.* 30, 250–258 (1998).

126. R. G. Schmidt and C. E. Frazier. Network characterization of phenol–formaldehyde thermosetting wood adhesive. *Int. J. Adhes. Adhes.* 18, 139–146 (1998).

127. A. W. Christiansen, R. A. Follensbee, R. L. Geimer, J. A. Koutsky and G. E. Myers. Phenol-formaldehyde resin curing and bonding in steam-injection pressing, part II. Differences between rates of chemical and mechanical response to resin cure. *Holzforschung* 47, 1, 72–83 (1993).

128. M. Lecourt, A. Pizzi and P. E. Humphrey. Comparison of TMA and ABES as forecasting systems of wood bonding effectiveness. *Holz Roh Werkst.* 61, 75–76 (2003).

129. H. Lei, A. Pizzi, A. Despres, H. Pasch and G. Du. Ester acceleration mechanisms in phenol–formaldehyde resin adhesives. *J. Appl. Polym. Sci.* 100, 3075–3093 (2006).

130. J. Zhang, A. Pizzi, J. Li and W. Zhang. MALDI-TOF MS analysis of the acceleration of the curing of phenol–formaldehyde (PF) resins induced by propylene carbonate. *Eur. J. Wood Prod.* 73, 135–138 (2015).

131. D. Fan, J. Chang, J. Li, B. Xia and Z. Sang. Cure properties and adhesive performances of cure-accelerated phenol-urea-formaldehyde resins. *Eur. J. Wood Prod.* 69, 2, 213–220 (2011).

132. D. Fan, J. Li and J. Chang. On the structure and cure acceleration of phenol–urea–formaldehyde resins with different catalysts. *Eur. Polym. J.* 45, 2849–2857 (2009).

133. X. Lu and A. Pizzi. Curing conditions effects on the characteristics of thermosetting adhesives-bonded wood joints: Part 1: Substrate influence on TTT and CHT curing diagrams of wood adhesives. *Holz Roh Werkst.* 56, 339–346 (1998).

134. A. Pizzi, B. Mtsweni and W. Parsons. Wood-induced catalytic activation of PF adhesives autopolymerization vs. PF/wood covalent bonding. *J. Appl. Polym. Sci.* 52, 1847–1856 (1994).

135. G. He and B. Riedl. Curing kinetics of phenol formaldehyde resin and wood-resin interactions in the presence of wood substrates. *Wood Sci. Technol.* 38, 69–81 (2004).

136. Y. Lei and Q. Wu. Cure kinetics of aqueous phenol-formaldehyde resins used for oriented strandboard manufacturing: Effect of wood flour. *J. Appl. Polym. Sci.* 102, 3774–3781 (2006).

137. G. He and N. Yan. Effect of wood species and molecular weight of phenolic resins on curing behavior and bonding development. *Holzforschung* 59, 635–640 (2005).

138. G. He and N. Yan. Effect of wood on the curing behavior of commercial phenolic resin systems. *J. Appl. Polym. Sci.* 95, 185–192 (2005).

139. M. P. G. Laborie and C. E. Frazier. ^{13}C CP/MAS NMR study of a wood/phenol–formaldehyde resin bondline. *J. Mater. Sci.* 41, 6001–6005 (2006).

140. M. G. Kim, L. W. Amos and E. E. Barnes. Study of the reaction rates and structures of a phenol-formaldehyde resol resin by carbon-13 NMR and gel permeation chromatography. *Ind. Eng. Chem. Res.* 29, 2032–2037 (1990).

141. E. Scopelitis and A. Pizzi. Urea–resorcinol–formaldehyde adhesives of low resorcinol content. *J. Appl. Polym. Sci.* 48, 2135–2146 (1993).

142. B. Tomita and C. Y. Hse. Cocondensation of urea with methylolphenols in acidic conditions. *J. Polym. Sci. Part A, Polym. Chem.* 30, 1615–1624 (1992).

143. B. Tomita, M. Ohyama and C. Y. Hse. Synthesis of phenol-urea-formaldehyde cocondensed resins from UF-concentrate and phenol. *Holzforschung* 48, 522–526 (1994).

144. B. Tomita and C. Y. Hse. Phenol–urea–formaldehyde (PUF) co-condensed wood adhesives. *Int. J. Adhes. Adhes.* 18, 69–79 (1998).

145. M. Ohyama, B. Tomita and C. Y. Hse. Curing property and plywood adhesive performance of resol-type phenol-urea-formaldehyde cocondensed resins. *Holzforschung* 49, 87–91 (1995).

146. C. Zhao, A. Pizzi, A. Kuhn and S. Garnier. Fast advancement and hardening acceleration of low condensation alkaline phenol-formaldehyde resins by esters and copolymerized urea. II. Esters during resin reaction and effect of guanidine salts. *J. Appl. Polym. Sci.* 77, 249–259 (1999).

147. A. Pizzi, A. Stephanou, I. Antunes and G. de Beer. Alkaline PF resins linear extension by urea condensation with hydroxybenzylalcohol groups. *J. Appl. Polym. Sci.* 50, 2201–2207 (1993).

148. D. Fan, J. Chang, J. Li, A. Mao and L. Zhang. ^{13}C-NMR study on the structure of phenol-urea-formaldehyde resins prepared by methylolureas and phenol. *J Appl. Polym. Sci.* 112, 2195–2202 (2009).

149. K. Schmidt, D. Grunwald and H. Pasch. Preparation of phenol–urea–formaldehyde copolymer adhesives under heterogeneous catalysis. *J. Appl. Polym. Sci.* 102, 2946–2952 (2006).

150. D. Fan, F. Chu, T. Qin and J. Li. Effect of synthesis conditions on the structure and curing characteristics of high-urea content PUF resin. *J. Adhes.* 87, 1191–1203 (2011).

151. D. Fan, J. Li, J. Chang, J. Gou and J. Jiang. Chemical structure and curing behavior of phenol–urea–formaldehyde cocondensed resins of high urea content. *J. Adhes. Sci. Technol.* 23, 1787–1797 (2009).

152. S. M. Lee and M. G. Kim. Effects of urea and curing catalysts added to the strand board core-layer binder phenol–formaldehyde resin. *J. Appl. Polym. Sci.* 105, 1144–1155 (2007).

153. G. He and B. Riedl. Phenol-urea-formaldehyde cocondensed resol resins: Their synthesis, curing kinetics, and network properties. *J. Polym. Sci.: Part B: Polym. Phys.* 41, 1929–1938 (2003).

154. I. Poljansek, U. Sebenik and M. Krajnc. Characterization of phenol–urea–formaldehyde resin by inline FTIR spectroscopy. *J. Appl. Polym. Sci.* 99, 2016–2028 (2006).

155. C. Y. Hse, R. L. Geimer, W. E. Hsu and R. C. Tang. Effect of resin type on properties of steam-press-cured flakeboards. *For. Prod. J.* 45, 1, 57–62 (1995).

156. A. Pizzi and T. Walton. Non-emulsifiable, water-based, mixed diisocyanate adhesive systems for exterior plywood: Part I. Novel reaction mechanisms and their chemical evidence. *Holzforschung* 46, 541–547 (1992).

157. A. Pizzi, J. Valenzuela and C. Westermeyer. Non-emulsifiable, water-based, mixed diisocyanate adhesive systems for exterior plywood. Part II. Theory application and industrial results. *Holzforschung* 47, 69–72 (1993).

158. C. Simon, B. George and A. Pizzi. UF/pMDI wood adhesives: Networks blend versus copolymerization. *J. Appl. Polym. Sci.* 86, 3681–3688 (2002).

159. F. Cardona and M. Thariq bin Hamid Sultan. Characterization of Environmentally sustainable resole phenolic resins synthesized with plant-based bioresources. *BioResources* 11, 965–983 (2016).

160. L. A. Panamgama and A. Pizzi. A ^{13}C-NMR analysis method for phenol–formaldehyde resin strength and formaldehyde emission. *J. Appl. Polym. Sci.* 55, 1007–1015 (1995).

161. C. E. Frazier. Isocyanate wood binders. In: *Handbook of Adhesive Technology*, 2nd edn, A. Pizzi, K. L. Mittal (Eds.), pp. 681–694, CRC Press, Boca Raton, FL (2003).

162. H. Roll. Microtechnological investigations on the behaviour of the adhesive polymeric diphenyl methane-4,4′ diisocyanate on wood surfaces with regard to its distribution in particleboards. PhD thesis, University of Munich, Munich, Germany (1993).

163. M. F. Sonnenschein and B. L. Wendt. Efficacy of polymeric MDI/Polyol mixtures for binding wood boards. *Wood Sci. Technol.* 39, 27–36 (2005).

164. M. N. Schreyer, W. D. Domke and S. Stini. HPLC determination of the isomeric ratios of diphenylmethane diisocyanate in modified isocyanates. *J. Chromatogr. Sci.* 27, 262–266 (1989).

165. J. J. Marcinko, W. H. Newman, C. Phanopoulos and M. A. Sander, The nature of the MDI/wood bond. In: *Proceedings of the 29th Washington State University International Particleboard/Composite Materials Symposium*, Pullman, WA, pp. 175–183 (1995).

166. W. E. Johns. The chemical bonding of wood. In: *Wood Adhesives: Chemistry and Technology*, Vol.2, A. Pizzi (Ed.), pp. 75–96, CRC Press, Boca Raton, FL (1989).

167. C. E. Frazier, R. G. Schmidt and J. Ni. Towards a molecular understanding of the wood-isocyanate adhesive bondline. In: *Proceedings of the Third Pacific Rim Bio-Based Composites Symposium*, Kyoto, Japan, pp. 383–391 (1996).

168. K. Umemura, A. Takahashi and S. Kawai. Durability of isocyanate resin adhesives for wood. II. Effect of the addition of several polyols on the thermal properties. *J. Appl. Polym. Sci.* 74, 1807–1814 (1999).

169. F. W. Abbate and H. Ulrich. Urethanes. I. Organometallic catalysis of the reaction of alcohols with isocyanates. *J. Appl. Polym. Sci.* 13, 1929–1936 (1969).

170. K. C. Frisch, L. P. Rumao and A. Pizzi. Diisocyanates as wood adhesives. In: *Wood Adhesives: Chemistry and Technology*, A. Pizzi (Ed.) pp. 289–318, CRC Press, Boca Raton, FL (1983).

171. J. E. Jakes, D. Y. Yelle, J. F. Beecher, C. R. Frihart and D. S. Stone. Characterizing polymeric methylene diphenyl diisocyanate reactions with wood: 2. Nano-indentation. *Proc. Wood Adhes.*, 366–373 (2009).

172. D. J. Yelle. A solution-state NMR approach to elucidating pMDI-wood bonding mechanisms in loblolly pine. PhD thesis, University of Wisconsin-Madison, Madison, WI (2009).

173. D. G. Lay and P. Cranley. Polyurethane adhesives. In: *Handbook of Adhesive Technology*, 2nd edn, A. Pizzi and K. L. Mittal (Eds.), pp. 695–718, CRC Press, Boca Raton, FL (2003).

174. K. Grostad and R. Bredesen. EPI for glued laminated timber. In: *Materials and Joints in Timber Structure, Recent Developments of Technology*, S. Aicher, H.W. Reinhardt and H. Garrecht (Eds.), RILEM Bookseries 9, pp. 355–364, Springer-Business Media B.V., Dordrecht, The Netherlands (2014).

175. S. Sakurada, H. Miyazaki, T. Hattori, M. Shiraishi and T. Inoue. Adhesive composition consisting of polyvinylalcohol solution or polyvinylacetate latex modified with hydrophobic solution of isocyanate compound. US patent 3931088, assigned to Kuraray and Koyo Sangyo (1976).

176. H. F. Pagel and E. R. Luckman. EPI (Emulsion Polymer/Isocyanate): A new structural adhesive. In: *Proceedings of the Wood Adhesives 1980*, pp.122–132, USDA Forest Service, Forest Products Laboratory, Madison, WI (1980).

177. K. Grostad and A. Pedersen. Emulsion polymer isocyanates as wood adhesive: A review. *J. Adhes. Sci. Technol.* 24, 1357–1381 (2010).

178. K. L. Mittal. The role of the interface in adhesion phenomena. *Polym. Eng. Sci.* 17, 467–473 (1977).

179. A. Marra. *Technology of Wood Bonding Principles in Practice*, Van Nostrand Reinhold, New York (1992).

180. S. E. Johnson and F. A. Kamke. Quantitative analysis of gross adhesive penetration in wood using fluorescence microscopy. *J. Adhes.* 40, 47–61 (1992).

181. W. Gindl, F. Zagar-Yaghubi, and R. Wimmer. Impregnation of softwood cell walls with melamine-formaldehyde resin. *Bioresour. Technol.* 87, 325–330 (2003).

182. J. Konnerth, D. Harper, S. H. Lee, T. Rials and W. Gindl. Adhesive penetration of wood cell walls investigated by scanning thermal microscopy (SThM), *Holzforschung* 62, 91–98 (2008).

183. J. Konnerth and W. Gindl. Mechanical characterization of wood-adhesive interphase cell walls by nanoindentation. *Holzforschung* 60, 429–433 (2006).

184. R. S. Stephens and N. P. Kutscha. Effect of resin molecular weight on bonding flakeboard. *Wood Fiber Sci.* 19, 353–361 (1987).

185. D. A. Hare and N. P. Kutscha. Microscopy of eastern spruce plywood gluelines. *Wood Sci.* 6, 294–304 (1974).

186. A. J. Bolton, J. M. Dinwoodie and D. A. Davies. The validity of the use of SEM/EDAX as a tool for the detection of UF resin penetration into wood cell walls in particleboard. *Wood Sci. Technol.* 22, 345–356 (1988).

187. W. Gindl, E. Dessipri and R. Wimmer. Using UV-microscopy to study diffusion of melamine-urea-formaldehyde resin in cell walls of spruce wood. *Holzforschung* 56, 103–107 (2002).

188. W. Gindl. SEM and UV-microscopic investigation of glue lines in Parallam® PSL. *Holz Roh Werkst.* 59, 211–214 (2001).

189. W. Gindl, T. Schöberl and G. Jeronimidis. The interphase in phenol–formaldehyde and polymeric methylene di-phenyl-di-isocyanate glue lines in wood. *Int. J. Adhes. Adhes.* 24, 279–286 (2004).

190. E. Mahrdt, F. Stoeckel, H.W.G. van Herwijnen, U. Mueller, W. Kantner, J. Moser and W. Gindl-Altmutter. Light microscopic detection of UF adhesive in industrial particle board. *Wood Sci. Technol.* 49 (3), 517–526 (2015).

191. P. Niemz, D. Mannes, E. Lehmann, P. Vontobel, and S. Haase. Investigations of the distribution of adhesive in the area of the bond line using neutron radiography and microscopy (in German). *Holz Roh Werkst.* 62, 424–432 (2004).

192. A. P. Singh, B. Dawson, C. Rickard, J. Bond and A. Singh. Light, confocal and scanning electron microscopy of wood-adhesive interface. *Microsc. Anal.* 22 (3), 5–8 (2008).

193. A. Nuryawan, B. D. Park and A. P. Singh. Penetration of urea–formaldehyde resins with different formaldehyde/urea mole ratios into softwood tissues. *Wood Sci. Technol.* 48, 889–902 (2014).

194. H. Wan and M. G. Kim. Distribution of phenol-formaldehyde resin in impregnated southern pine and effects on stabilization. *Wood Fiber Sci.* 40, 181–189 (2008).

195. E. Brady and F. Kamke. Effects of hot-pressing parameters on resin penetration. *For. Prod. J.* 38, 11–12, 63–68 (1988).

196. H. Edalat, M. Faezipour, V. Thole and F. A. Kamke. A new quantitative method for evaluation of adhesive penetration pattern in particulate wood-based composites: Elemental counting method. *Wood Sci. Technol.* 48, 703–712 (2014).

197. F. Kamke and J. Lee. Adhesive penetration in wood: A review. *Wood Fiber Sci.* 39, 205–220 (2007).

198. M. Sernek, J. Resnik and F. A. Kamke. Penetration of liquid urea-formaldehyde adhesive into beech wood. *Wood Fiber Sci.* 3, 11, 41–48 (1999).

199. J. Zheng, S. C. Fox and C. E. Frazier. Rheological, wood penetration, and fracture performance studies of PF/pMDI hybrid resins. *For. Prod. J.* 54, 10, 74–81 (2004).

200. T. M. Gruver and N. R. Brown. Penetration and performance of isocyanate wood binders on selected wood species. *BioResources* 1, 233–247 (2006).

201. L. Qin, L. Lin and F. Fu. Microstructural and micromechanical characterization of modified urea-formaldehyde resin penetration into wood. *BioResources* 11, 182–194 (2016).

202. H. Pakdel, P. L. Cyr, B. Riedl and J. Deng. Quantification of urea formaldehyde resin in wood fibers using X-ray photoelectron spectroscopy and confocal laser scanning microscopy. *Wood Sci. Technol.* 42, 133–148 (2008).

203. S. Lee, T. F. Shupe, L. H. Groom and C. Y. Hse. Wetting behaviors of phenol- and urea-formaldehyde resins as compatibilizers. *Wood Fiber Sci.* 39, 482–492 (2007).

204. N. Gierlinger, C. Hansmann, T. Roder, H. Sixta, W. Gindl and R. Wimmer. Comparison of UV and confocal Raman microscopy to measure the melamine–formaldehyde resin content within cell walls of impregnated spruce wood. *Holzforschung* 59, 210–213 (2005).

205. P. L. Cyr, B. Riedl and X. M. Wang. Investigation of urea-melamine-formaldehyde (UMF) resin penetration in medium-density fiberboard (MDF) by high resolution confocal laser scanning microscopy. *Holz Roh Werkst.* 66, 129–134 (2008).

206. P. Cyr, B. Riedl, X. Wang and S. Shaler. Urea-melamine-formaldehyde (UMF) resin penetration in medium-density fiberboard (MDF) wood fibers. *J. Adhes. Sci. Technol.* 20, 787–801 (2006).

207. C. Xing, B. Riedl, A. Cloutier and S. M. Shaler. Characterization of urea–formaldehyde resin penetration into medium density fiberboard fiber. *Wood Sci. Technol.* 39, 374–384 (2005).

208. A. P. Singh, E. A. Dunningham and D. V. Plackett. Assessing the performance of a commercial wood stain by transmission electron microscopy. *Holzforschung* 49, 255–258 (1995).

209. A. P. Singh, C. R. Anderson, J. M. Warnes and J. Matsumura. The effect of planing on the microscopic structure of Pinus radiata wood cells in relation to penetration of PVA glue. *Holz Roh Werkst.* 60, 333–341 (2002).

210. A. P. Singh, A. Nuryawan, B. D. Park and K. H. Lee. Urea-formaldehyde resin penetration into Pinus radiata tracheid walls assessed by TEM-EDXS. *Holzforschung* 69, 303–306 (2015).

211. C. J. Buckley, C. Phanopoulos, N. Khaleque, A. Engelen, M. E. J. Holwill and A. G. Michette. Examination of the penetration of polymeric methylene di-phenyl-di-isocyanate (pMDI) into wood structure using chemical-state X-ray microscopy. *Holzforschung* 56, 215–222 (2002).

212. P. Hass, W. Falk, M. Stampanoni, A. Kaestner and P. Niemz. Penetration pathway of adhesives into wood. In: *Proceedings of the International Conference on Wood Adhesives 2009*, Lake Tahoe, CA (2009).

213. P. Hass, F. K. Wittel, M. Mendoza, H. J. Herrmann and P. Niemz. Adhesive penetration in beech wood: Experiments. *Wood Sci. Technol.* 46, 243–256 (2012).

214. P. D. Evans, O. Morrison, T. J. Senden, S. Vollmer, R. J. Roberts, A. Limaye, C. H. Arns, H. Averdunk, A. Lowe and M. A. Knackstedt. Visualization and numerical analysis of adhesive distribution in particleboard using X-ray micro-computed tomography. *Int. J. Adhes. Adhes.* 30, 754–762 (2010).

215. G. Modzel, F. A. Kamke and F. De Carlo. Comparative analysis of a wood: Adhesive bondline. *Wood Sci. Technol.* 45, 147–158 (2011).

216. J. L. Paris, F. A. Kamke, J. A. Nairn, M. Schwarzkopf and L. Muszynski. Wood-adhesive penetration: Non-destructive, 3D visualization and quantification. In: *Proceedings of the International Conference on Wood Adhesives 2013*, Toronto, Canada (2013).

217. J. L. Paris, F. A. Kamke, R. Mbachu and S. K. Gibson. Phenol formaldehyde adhesives formulated for advanced X-ray imaging in wood-composite bondlines. *J. Mater. Sci.* 49, 580–591 (2014).

218. F. A. Kamke, J. A. Nairn, L. Muszynski, J. L. Paris, M. Schwarzkopf and X. Xiao. Methodology for micromechanical analysis of wood adhesive bonds using X-ray computed tomography and numerical modeling. *Wood Fiber Sci.* 46, 15–28 (2014).

219. J. L. Paris and F. A. Kamke, Quantitative wood–adhesive penetration with X-ray computed tomography. *Int. J. Adhes. Adhes.* 61, 71–80 (2015).

220. J. L. Paris, F.A. Kamke and X. Xiao. X-ray computed tomography of wood-adhesive bondlines: Attenuation and phase-contrast effects. *Wood Sci. Technol.* 49, 1185–1208 (2015).

221. P. E. McKinley, D. J. Ching, F. A. Kamke, M. Zauner and X. Xiao. Micro x-ray computed tomography of adhesive bonds in wood. *Wood Fiber Sci.* 48 (2015 Convention, Special Issue), 2–16 (2016).

222. J. E. Jakes, C. G. Hunt, D. J. Yelle, L. Lorenz, K. Hirth, S. C. Gleber, S. Vogt, W. J. Grigsby and C. R. Frihart. Synchrotron-based X-ray fluorescence microscopy in conjunction with nanoindentation to study molecular-scale interactions of phenol-formaldehyde in wood cell walls. *ACS Appl. Mater. Interfaces* 7, 6584–6589 (2015).

223. D. Xu, Y. Zhang, H. Zhou, Y. Meng and S. Wang. Characterization of adhesive penetration in wood bond by means of scanning thermal microscopy (SThM). *Holzforschung* 70, 323–330 (2015).

224. S. E. Johnson and F. A. Kamke. Characteristics of phenol-formaldehyde adhesive bonds in steam injection pressed flakeboard. *Wood Fiber Sci.* 26, 259–269 (1994).

225. J. B. Wilson, G. L. Jay and R. L. Krahmer. Using resin properties to predict bond strength of oak particleboard. *Adhes. Age* 22, 26–30 (1979).

226. I. Gavrilovic-Grmusa, M. Dunky, J. Miljković and M. Điporović-Momčilović. Radial penetration of urea-formaldehyde adhesive resins into beech (*Fagus Moesiaca*). *J. Adhes. Sci. Technol.* 24, 1753–1768 (2010).

227. I. Gavrilovic-Grmusa, J. Miljković and M. Điporović-Momčilović. Influence of the degree of condensation on the radial penetration of urea formaldehyde adhesives into silver fir (*Abies alba, Mill.*) wood tissue. *J. Adhes. Sci. Technol.* 24, 1437–1453 (2010).

228. I. Gavrilovic-Grmusa, M. Dunky, J. Miljković and M. Điporović-Momčilović. Influence of the degree of condensation of urea-formaldehyde adhesives on the tangential penetration into beech and fir and on the shear strength of the adhesive joints. *Eur. J. Wood Prod.* 70, 655–665 (2012).

229. I. Gavrilovic, M. Dunky, J. Miljković and M. Điporović-Momčilović. Influence of the degree of condensation of urea-formaldehyde adhesive on radial and tangential penetration into poplar and the shear strength of adhesive joints. *Holzforschung* 66, 849–856 (2012).

230. I. Gavrilovic-Grmusa, M. Dunky, M. Điporović-Momčilović, M. Popovic and J. Popovic. Influence of pressure on the radial and tangential penetration into poplar wood and on shear strength of adhesive joints. *Bioresources* 11, 2238–2255 (2016).

231. J. M. M. Ferra, A. M. Mendes, M. R. N. Costa, F. D. Magalhaes and L. H. Carvalho. Characterization of urea-formaldehyde resins by GPC/SEC and HPLC techniques: Effect of ageing. *J. Adhes. Sci. Technol.* 24, 1535–1551 (2010).

232. R. Ghafari, K. DoostHosseini, A. Abdulkhani and S. A. Mirshokraie. Replacing formaldehyde by furfural in urea formaldehyde resin: Effect on formaldehyde emission and physical–mechanical properties of particleboards. *Eur. J. Wood Prod.* 74, 609–616 (2016).

233. M. Singh. Preparation, molecular weight determination and structure elucidation of melamine-urea-formaldehyde, melamine-methylurea-formaldehyde and melamine-dimethylurea-formaldehyde polymer resins with IR spectroscopy. *Indian J. Chem.* 43A, 1696–1700 (2004).

234. T. Holopainen, L. Alvila, J. Rainio and T. T. Pakkanen. IR spectroscopy as a quantitative and predictive analysis method of phenol–formaldehyde resol resins. *J. Appl. Polym. Sci.* 69, 2175–2185 (1998).

235. C. L. So, T. L. Eberhardt, E. Hsu, B. K. Via and C. Y. Hse. Infrared spectroscopic monitoring of urea addition to oriented strandboard resins. *J. Appl. Polym. Sci.* 105, 733–738 (2007).

236. E. Minopoulou, E. Dessipri, G. D. Chryssikos, V. Gionis, A. Paipetis and C. Panayiotou. Use of NIR for structural characterization of urea–formaldehyde resins. *Int. J. Adhes. Adhes.* 23, 473–484 (2003).

237. A. Henriques, P. Cruz, J. Martins, J. M. M. Ferra, F. D. Magalhaes and L. H. Carvalho. Determination of formaldehyde/urea molar ratio in amino resins by near-infrared spectroscopy. *J. Appl. Polym. Sci.* 124, 2441–2448 (2012).

238. A. Henriques, J. M. M. Ferra, P. Cruz, J. Martins, F. D. Magalhaes and L. H. Carvalho. Viscosity determination of amino resins during synthesis using near-infrared spectroscopy. *Int. Wood Prod. J.* 3, 64–66 (2012).

239. N. Costa, S. Amaral, R. Alvim, M. Nogueira, M. Schwanninger and J. Rodrigues. Assessment of resin formulations and determination of the formaldehyde to urea molar ratio by near- and mid-infrared spectroscopy and multivariate data analysis. *J. Appl. Polym. Sci.* 128, 498–508 (2013).

240. A. Henriques, P. Cruz, J. M. M. Ferra, J. Martins, F. D. Magalhaes and L. H. Carvalho. Determination of melamine content in amino resins by near-infrared spectroscopy. *Wood Sci Technol.* 47, 939–948 (2013).

241. T. Zorba, E. Papadopoulou, A. Hatjiissaak, K. M. Paraskevopoulos and K. Chrissafis. Urea-formaldehyde resins characterized by thermal analysis and FTIR method. *J. Therm. Anal. Calorim.* 92, 29–33 (2008).

242. A. Kandelbauer, A. Despres, A. Pizzi and I. Taudes. Testing by Fourier transform infrared species variation during melamine–urea–formaldehyde resin preparation. *J. Appl. Polym. Sci.* 106, 2192–2197 (2007).

243. O. Steinhof, E. J. Kibrik, G. Scherr and H. Hasse. Quantitative and qualitative ^{1}H, ^{13}C, and ^{15}N NMR spectroscopic investigation of the urea–formaldehyde resin synthesis. *Magn. Reson. Chem.* 52, 138–162 (2014).

244. K. Siimer, T. Pehk and P. Christjanson. Study of the structural changes in urea-formaldehyde condensates during synthesis. *Macromol. Symp.* 148, 149–156 (1999).

245. L. M. H. Carvalho, M. R. P. F. N. Costa and C. A. V. Costa. A very simple empirical kinetic model of the acid catalyzed cure of urea–formaldehyde resins. *J. Appl. Polym. Sci.* 102, 5977 – 5987 (2006).

246. K. Siimer, P. Christjanson, T. Kaljuvee, T. Pehk, I. Lasn and I. Saks. TG-DTA study of melamine-urea-formaldehyde resins. *J. Therm. Anal. Calorim.* 92, 19–27 (2008).

247. K. Siimer, T. Kaljuvee, T. Pehk and I. Lasn. Thermal behavior of melamine-modified urea–formaldehyde resins. *J. Therm. Anal. Calorim.* 99, 755–762 (2010).

248. K. Siimer, T. Kaljuvee, P. Christjanson and T. Pehk. Changes in curing behavior of aminoresins during storage. *J. Therm. Anal. Calorim.* 80, 123–130 (2005).

249. A. Despres, A. Pizzi, H. Pasch and A. Kandelbauer. Comparative 13C-NMR and matrix-assisted laser desorption/ionization time-of-flight analyses of species variation and structure maintenance during melamine–urea–formaldehyde resin preparation. *J. Appl. Polym. Sci.* 106, 1106–1128 (2007).

250. G. Du, H. Lei, A. Pizzi and H. Pasch. Synthesis-structure-performance relationship of cocondensed phenol-urea-formaldehyde resins by MALDI-TOF and ^{13}C NMR. *J. Appl. Polym. Sci.* 110, 1182–1194 (2008).

251. A. Pizzi, H. Pasch, C. Simon and K. Rode. Structure of resorcinol, phenol, and furan resins by MALDI-TOF mass spectrometry and ^{13}C NMR. *J. Appl. Polym. Sci.* 92, 2665–2674 (2004).

252. B. D. Park and H. W. Jeong. Hydrolytic stability and crystallinity of cured urea–formaldehyde resin adhesives with different formaldehyde/urea mole ratios. *Int. J. Adhes. Adhes.* 31, 524–529 (2011).

253. T. Furuno, Y. Imamura and H. Kajita. The modification of wood by treatment with low molecular weight phenol-formaldehyde resin: A properties enhancement with neutralized phenolic-resin and resin penetration into wood cell walls. *Wood Sci. Technol.* 37, 349–361 (2004).

254. M. G. Kim, W. L. Nieh, T. Sellers, Jr., W. W. Wilson and J. W. Mays. Polymer solution properties of a phenol-formaldehyde resol resin by gel permeation chromatography, intrinsic viscosity, static light scattering, and vapor pressure osmometric methods. *Ind. Eng. Chem. Res.* 31, 973–979 (1992).

255. M. Sernek and M. Dunky. Adhesive bond strength development. In: *Wood-Based Panels. An Introduction for Specialists.*H. Thoemen, M. Irla and M. Sernek (Eds.), Brunel University Press, COST Office Brussels, London, UK (2010).

256. C. Heinemann, A. Fruehwald and P. E. Humphrey. Evaluation of adhesive cure during hot-pressing of wood-based composites. In: *Proceedings of the COST E13 3rd Workshop* 167–174, Vienna, Austria (2002).

257. M. Sernek, A. Kokalj and M. Jost. The development of adhesive bond strength during phenol-formaldehyde resin curing. Wood resources and panel properties. In: *Proceedings COST Action E44–E49, Valencia, Spain*, pp. 89–96, AIDIMA, Furniture, Wood and Packaging Technology Institute, Valencia, Spain (2006).

258. S. Chow and P. R. Steiner. Catalytic, exothermic reactions of urea-formaldehyde resin. *Holzforschung* 29, 4–10 (1975).

259. B. D. Park, B. Riedl, H. J. Bae and Y. S. Kim. Differential scanning calorimetry of phenol-formaldehyde (PF) adhesives. *J. Wood Chem. Technol.* 19, 265–286 (1999).

260. G. He, B. Riedl and A. Aït-Kadi. Model-free kinetics: Curing behavior of phenol formaldehyde resins by differential scanning calorimetry. *J. Appl Polym. Sci.* 87, 433–440 (2003).

261. Y. Lei, Q. Wu and K. Lian. Cure kinetics of aqueous phenol–formaldehyde resins used for oriented strandboard manufacturing: Analytical technique. *J. Appl. Polym. Sci.* 100, 1642–1650 (2006).

262. Y. Lei and Q. Wu. Cure kinetics of aqueous phenol–formaldehyde resins used for oriented strandboard manufacturing: Effect of zinc borate. *J. Appl. Polym. Sci.* 101, 3886–3894 (2006).

263. J. Wang, M. P. G. Laborie and M. P. Wolcott. Correlation of mechanical and chemical cure development for phenol–formaldehyde resin bonded wood joints. *Thermochim. Acta* 513, 20–25 (2011).

264. C. Xing, S. Y. Zhang, J. Deng, B. Riedl and A. Cloutier. Medium-density fiberboard performance as affected by wood fiber acidity, bulk density, and size distribution. *Wood Sci. Technol.* 40, 637–646 (2006).

265. H. Lei and C. E. Frazier. Curing behavior of melamine-urea-formaldehyde (MUF) resin adhesive. *Int. J. Adhes. Adhes.* 62, 40–44 (2015).

266. J. Wang, M. P. G. Laborie and M. P. Wolcott. Application of beam mechanics to sensing the cure development of wood-phenolic joints by dynamic mechanical analysis. *Thermochim. Acta* 465, 18–24 (2007).

267. A. Pizzi. On the correlation of some theoretical and experimental parameters in polycondensation cross-linked networks. *J. Appl. Polym. Sci.* 63, 603–617 (1997).

268. A. Pizzi, R. Garcia and X. Deglise. Thermomechanical analysis of entanglement networks: Correlation of some calculated and experimental parameters. *J. Appl. Polym. Sci.* 67, 1673–1678 (1998).

269. C. Kamoun and A. Pizzi. Particleboard I.B. forecast by TMA bending in MUF adhesives curing. *Holz Roh Werkst.* 58, 288–289 (2000).

270. P. R. Steiner and S. R. Warren. Behavior of urea-formaldehyde wood adhesives during early stages of cure. *For. Prod. J.* 37, 1, 20–22 (1987).

271. N. A. Costa, J. Pereira, J. Ferra, P. Cruz, J. Martins, F. D. Magalhaes, A. Mendes and L. H. Carvalho. Evaluation of bonding performance of amino polymers using ABES. *J. Adhes.* 90, 80–88 (2014).

272. J. M. M. Ferra, M. Ohlmeyer, A. M. Mendes, M. R. N. Costa, L. H. Carvalho and F. D. Magalhaes. Evaluation of urea-formaldehyde adhesives performance by recently developed mechanical tests. *Int. J. Adhes. Adhes.* 31, 127–134 (2011).

273. K. Siimer, T. Kaljuvee and P. Christjanson. Thermal behavior of urea-formaldehyde resins during curing. *J. Therm. Anal. Calorim.* 72, 607–617 (2003).

274. K. Siimer, T. Kaljuvee, P. Christjanson and I. Lasn. Curing of urea-formaldehyde resins on a wood substrate. *J. Therm. Anal. Calorim.* 84, 71–77 (2006).

275. M. Witt. Novel plate rheometer configuration allows monitoring real-time wood adhesive curing behavior. *J. Adhes. Sci. Technol.* 18, 893–904 (2004).

276. M. Schmidt, M. Knorz and B. Wilmes. A novel method for monitoring real-time curing behavior. *Wood Sci. Technol.* 44, 407–420 (2010).

277. S. Wang and P. M. Winistorfer. Monitoring resin cure during particleboard manufacture using a dielectric system. *Wood Fiber Sci.* 35, 532–539 (2003).

278. M. Jost and M. Sernek. Shear strength development of the phenol–formaldehyde adhesive bond during cure. *Wood Sci. Technol.* 43, 153–166 (2009).

279. M. Sernek and F. A. Kamke. Application of dielectric analysis for monitoring the cure process of phenol formaldehyde adhesive. *Int. J. Adhes. Adhes.* 27, 562–567 (2007).

280. C. Pretschuh, U. Müller, G. Wuzella, F. Dorner and R. Eckmann. Dielectric analysis as curing control for aminoplast resins: Correlation with DMA. *Eur. J. Wood Prod.* 70, 749–753 (2012).

281. B. D. Park, C. R. Frihart, Y. Yu and A. P. Singh. Hardness evaluation of cured urea–formaldehyde resins with different formaldehyde/urea mole ratios using nanoindentation method. *Eur. Polym. J.* 49, 3089–3094 (2013).

282. J. Konnerth, A. Jäger, J. Eberhardsteiner, U. Müller and W. Gindl. Elastic properties of adhesive polymers. II. Polymer films and bond lines by means of nanoindentation. *J. Appl. Polym. Sci.* 102, 1234–1239 (2006).

283. J. Konnerth, A. Valla and W. Gindl. Nanoindentation mapping of a wood-adhesive bond. *Appl. Phys. A* 88, 371–375 (2007).

284. J. Konnerth and W. Gindl. Observation of the influence of temperature on the mechanical properties of wood adhesives by nanoindentation. *Holzforschung* 62, 714–717 (2008).

285. F. Stoeckel, J. Konnerth, W. Kantner, J. Moser and W. Gindl. Tensile shear strength of UF- and MUF-bonded veneer related to data of adhesives and cell walls measured by nanoindentation. *Holzforschung* 64, 337–342 (2010).

286. F. Stoeckel, J. Konnerth, W. Kantner, J. Moser and W. Gindl. Mechanical characterisation of adhesives in particle boards by means of nanoindentation. *Eur. J. Wood Prod.* 68, 421–426 (2010).

287. F. Stoeckel, J. Konnerth, J. Moser, W. Kantner and W. Gindl-Altmutter. Micromechanical properties of the interphase in pMDI and UF bond lines. *Wood Sci. Technol.* 46, 611–620 (2012).

288. F. Stoeckel, J. Konnerth and W. Gindl-Altmutter. Mechanical properties of adhesives for bonding wood: A review. *Int. J. Adhes. Adhes.* 45, 32–41 (2013).

289. J. Konnerth, F. Stöckel, U. Müller and W. Gindl. Elastic properties of adhesive polymers. III. Adhesive polymer films under dry and wet conditions characterized by means of nanoindentation. *J. Appl. Polym. Sci.* 118, 1331–1334 (2010).

290. R. L. Geimer and A. W. Christiansen. Critical variables in the rapid cure and bonding of phenolic resins. *For. Prod. J.* 46 (11/12), 67–72 (1996).

291. X. M. Wang, B. Riedl, R. L. Geimer and A. W. Christiansen. Phenol-formaldehyde resin curing and bonding under dynamic conditions. *Wood Sci. Technol.* 30, 423–446 (1996).

292. C. Heinemann, R. Mitter and M. Dunky. Thermokinetic simulation of a hot press cycle in the production of particleboards. In: *Proceedings of the Eighth European Panel Products Symposium*, Llandudno, UK (2004).

293. P. E. Humphrey. Fundamental aspects of wood particleboard manufacture. PhD thesis, University of Wales, UK (1982).

294. A. J. Bolton and P. E. Humphrey. The hot pressing of dry-formed wood-based composites. Part I. A review of literature, identifying the primary physical processes and the nature of their interaction. *Holzforschung* 42, 403–406 (1988).

295. P. E. Humphrey and A. J. Bolton. The hot pressing of dry-formed wood-based composites. Part II. A simulation model for heat and moisture transfer. *Holzforschung* 43, 199–206 (1989).

296. A. J. Bolton, P. E. Humphrey and P. K. Kavvouras. The hot pressing of dry-formed wood-based composites. Part III. Predicted vapour pressure and temperature variation with time compared with experimental data for laboratory boards. *Holzforschung* 43, 265–274 (1989).

297. A. J. Bolton, P. E. Humphrey and P. Kavvouras. The hot pressing of dry-formed wood-based composites. Part IV. Predicted variations of mattress moisture content with time. *Holzforschung* 43, 345–349 (1989).

298. P. E. Humphrey and S. Ren. Bonding kinetics of thermosetting adhesive systems used in wood-based composites: The combined effect of temperature and moisture content. *J. Adhes. Sci. Technol.* 3, 397–413 (1989).

299. H. Thoemen. Modeling the physical processes in natural fiber composites during batch and continuous pressing. PhD thesis, Oregon State University, Corvallis, OR (2000).

300. H. Thoemen and P. E. Humphrey. Modeling the continuous pressing process for wood-based composites. *Wood Fiber Sci.* 35, 456–468 (2003).

301. G. V. Haas Investigations on the hot pressing of wood based panels mats under special consideration of the densification behavior, the permeability, the temperature conductivity, and the sorption rate (in German). PhD thesis, University of Hamburg, Germany (1998).

302. B. G. Zombori. Modeling the transient effects during hot-pressing of wood-based composites. PhD thesis, Virginia Polytechnic Institute and State University, Blacksburg, VA (2001).

303. P. J. Garcia, S. Avramidis and F. Lam. Three-dimensional heat and mass transfer during oriented-strandboard hot-pressing. In: *Proceedings of the 6th Pacific Rim Bio-Based Composite Symposium & Workshop on The Chemical Modification of Cellulosics*, pp. 263–277, Portland, OR (2002).

304. P. J. Garcia. Three-dimensional heat and mass transfer during oriented strandboard hot-pressing. PhD thesis, University of British Columbia, Vancouver, Canada (2002).

305. C. Pereira, L. H. Carvalho and C. A. V. Costa. Simulation of the effect of a cooling zone in the internal conditions of MDF continuous hot-pressing process. In: *Proceedings of the International Symposium on Wood Based Material: Wood Composites and Chemistry*, pp. 117–123, Vienna, Austria, (2002).

306. J. Wang, M. P. G. Laborie and M. P. Wolcott. Comparison of model-fitting kinetics for predicting the cure behavior of commercial phenol–formaldehyde resins. *J. Appl. Polym. Sci.* 105, 1289–1296 (2007).

307. J. N. Lee, F. A. Kamke and L. T. Watson. Simulation of the hot-pressing of a multi-layered wood strand composite. *J. Compos. Mater.* 41, 879–904 (2007).

308. C. A. Lenth and F. A. Kamke. Investigations of flakeboard mat consolidation. Part I. Characterizing the cellular structure. *Wood Fiber Sci.* 28, 153–167 (1996).

309. B. G. Zombori, F. A. Kamke and L. T. Watson. Simulation of the internal conditions during the hot-pressing process. *Wood Fiber Sci.* 35, 2–23 (2003).

310. J. Shu, L. T. Watson, B. G. Zombori and F. A. Kamke. WBCSim: An environment for modeling wood-based composites manufacture. *Eng. Comput.* 21, 259–271 (2006).

311. L. M. H. Carvalho, M. R. N. Costa and C. A. V. Costa: A global model for the hot-pressing of MDF. *Wood Sci. Technol.* 37, 241–258 (2003).

312. M. Sernek. Comparative analysis of inactivated wood surfaces. PhD thesis, Virginia Polytechnic Institute and State University, Blacksburg, VA (2002).

313. J. J. Bikerman. *The Science of Adhesive Joints*, Academic Press, New York (1961).

314. M. Stehr and I. Johansson. Weak boundary layers on wood surfaces. *J. Adhes. Sci. Technol.* 14, 1211–1224 (2000).

315. M. Stehr, J. Seltman and I. Johansson. Laser ablation of machined wood surfaces. 1. Effect on end-grain gluing of pine (*Pinus silvestris L.*) and spruce (*Picea abies Karst.*). *Holzforschung* 53, 93–103 (1999).

316. M. Stehr. Adhesion to machined and laser ablated wood surfaces. PhD thesis, Royal Technical University, Stockholm, Sweden (1999).

317. M. Stehr. Laser ablation of machined wood surfaces. 2. Effect on end-grain gluing of pine (*Pinus silvestris L.*). *Holzforschung* 53, 655–661 (1999).

318. J. J. Bikerman. Causes of poor adhesion: Weak boundary layers. *Ind. Eng. Chem.* 59, 9, 40–44 (1967).

319. M. P. Wolcott, F. A. Kamke and D. A. Dillard. Fundamental aspects of wood deformation pertaining to manufacture of wood-based composites. *Wood Fiber Sci.* 26, 496–511 (1994).

320. M. A. Irle. Cell wall collapse and the pressing of particleboard. In: *Proceedings of the 4th European Panel Products Symposium*, pp. 70–80, Llandudno, UK (2000).

321. C. Piao, J. E. Winandy and T. F. Shupe. From hydrophilicity to hydrophobicity: A critical review: Part I. Wettability and surface behavior. *Wood Fiber Sci.* 42, 490–510 (2010).

322. M. Petric and P. Oven. Determination of wettability of wood and its significance in wood science and technology: A critical review. *Rev Adhes. Adhes.* 3, 121–187 (2015).

323. J. Custodio, J. Broughton and H. Cruz. A review of factors influencing the durability of structural bonded timber joints. *Int. J. Adhes. Adhes.* 29, 173–185 (2009).

324. Z. Qin, Q. Zhang, Q. Gao, S. Zhang and J. Li. Wettability of sanded and aged fast-growing poplar wood surfaces: II. dynamic wetting models. *BioResources* 9, 7176–7188 (2014).

325. C. Y. Hse. Surface tension of phenol-formaldehyde wood adhesives. *Holzforschung* 26, 82–85 (1972).

326. D. E. Packham. Surface energy, surface topography and adhesion. *Int. J. Adhes. Adhes.* 23, 437–448 (2003).

327. T. E. Shupe, C. Y. Hse, E. T. Choong and L. H. Groom. Effect of wood grain and veneer side on loblolly pine veneer wettability. *For. Prod. J.* 48, 6, 95–97 (1998).

328. Q. Shen, J. Nylund and J. B. Rosenholm. Estimation of the surface energy and acid-base properties of wood by means of wetting method. *Holzforschung* 52, 521–529 (1998).

329. J. D. Wellons. Wettability and gluability of Douglas-fir veneer. *For. Prod. J.* 30, 7, 53–55 (1980).

330. Z. Qin, Q. Gao, S. Zhang and J. Li. Surface free energy and dynamic wettability of differently machined poplar woods. *BioResources* 9, 3088–3103 (2014).

331. A. Herczeg. Wettability of wood. *For. Prod. J.* 15, 499–505 (1965).

332. T. Nguyen and W. E. Johns. The effects of aging and extraction on the surface free energy of Douglas fir and redwood. *Wood Sci. Technol.* 13, 29–40 (1979).

333. M. Gindl, A. Reiterer, G. Sinn and S. E. Stanzl-Tschegg. Effects of surface ageing on wettability, surface chemistry, and adhesion of wood. *Holz Roh Werkst.* 62, 273–280 (2004).

334. M. E. P. Wålinder and I. Johansson. Measurement of wood wettability by the Wilhelmy method. Part 1. Contamination of probe liquids by extractives. *Holzforschung* 55, 21–32 (2001).

335. T. Sellers, Jr., J. R. McSween and W. T. Nearn. Gluing of eastern hardwoods: A review. General Technical Report SO-71. U.S. Department of Agriculture, Forest Service, Southern Forest Experiment Station, New Orleans, LA (1988).

336. M. Gindl and S. Tschegg. Significance of the acidity of wood to the surface free energy components of different wood species. *Langmuir* 18, 3209–3212 (2002).

337. D. C. Maldas and D. P. Kamdem. Wettability of extracted southern pine. *For. Prod. J.* 49 (11/12), 91–93 (1999).

338. J. Custodio, J. Broughton, H. Cruz and A. Hutchinson. A review of adhesion promotion techniques for solid timber substrates. *J. Adhes.* 84, 502–529 (2008).

339. F. A. Kamke and H. Sizemore. Viscoelastic thermal compression of wood. US Patent 7404422, assigned to Eagle Analytical Corporation, Inc. (2004).

340. A. Kutnar, L. Rautkari, K. Laine and M. Hughes. Thermodynamic characteristics of surface densified solid Scots pine wood. *Eur. J. Wood Prod.* 70, 727–734 (2012).

341. M. Petric. Surface modification of wood: A critical review. *Rev. Adhes. Adhes.* 1, 216–247 (2013).

342. R. M. Rowell. Acetylation of wood. *For. Prod. J.* 56, 9, 4–12 (2006).

343. R.M. Rowell. Chemical modification of wood: A short review. *Wood Mater. Sci. Eng.* 1, 29–33 (2006).

344. J. Gomez-Bueso, J. Westin, R. Torgilsson, P. O. Olesen and R. Simonson. Composites made from acetylated lignocellulosic fibers of different origin. Part I. Properties of dry-formed fiberboards. *Holz Roh Werkst.* 57, 433–438 (1999).

345. J. Gomez-Bueso, J. Westin, R. Torgilsson, P. O. Olesen and R. Simonson. Composites made from acetylated lignocellulosic fibers of different origin. Part II. The effect of nonwoven fiber mat composition upon molding ability. *Holz Roh Werkst.* 57, 178–184 (2000).

346. P. Hanetho. Seasonable quality problems for particleboards due to use of raw wood. In: *Proceedings of the FESYP Conference, Federation Europeenne du Syndicat des fabricants de Panneaux*, pp. 129–136, Munich, Germany (1987).

347. E. L. Back. Oxidative activation of wood surfaces for glue bonding. *For. Prod. J.* 41, (2), 30–36 (1991).

348. M. Dunky. Particle size distribution and glue resin consumption: How to spare costs. In: *Proceedings of the Second European Panel Products Symposium*, pp. 206–217, Llandudno, UK (1998)

349. D. Robson, M. Riepen, J. Hague, C. Loxton and R. Quinney. Looking into the "black box": The application of flow models to blow line blending of MDF fibres. In: *Proceedings of the First European Panel Products Symposium*, pp. 203–210, Llandudno, UK (1997).

350. K. M. Chapman. Improved resin performance in MDF through development of blowline blending process. *Int. Wood Prod. J.* 3, 9–14 (2012).

351. C. Loxton, A. Thumm, W. J. Grigsby, T. A. Adams and R. M. Ede. Resin distribution in medium density fibreboard. Quantification of UF resin distribution on blow line- and dry blended MDF fiber and panels. *Wood Fiber Sci.* 35, 370–380 (2003).

352. A. Thumm, A. G. McDonald and L. A. Donaldson. Visualisation of UF resin in MDF by cathodoluminescence/scanning electron microscopy. *Holz Roh Werkst.* 59, 215–216 (2001).

353. F. A. Kamke, K. A. Scott and R. E. Smith. Measurement of resin distribution on MDF fiber: A mill trial. *Proc. 6th Pac. Rim Bio-Based Compo. Symp.* 1, 86–93 (2002).

354. W. Grigsby, A. G. McDonald, A. Thumm and C. Loxton. X-ray photoelectron spectroscopy determination of urea formaldehyde resin coverage on MDF fibre. *Holz Roh Werkst.* 62, 358–364 (2004).

355. W. J. Grigsby and A. Thumm. Resin and wax distribution and mobility during medium density fibreboard manufacture. *Eur. J. Wood Prod.* 70, 337–348 (2012).

356. M. Riegler, W. Gindl-Altmutter, M. Hauptmann and U. Mueller. Detection of UF resin on wood particles and in particleboards: Potential of selected methods for practice-oriented offline detection. *Eur. J. Wood Prod.* 70, 829–837 (2012).

357. X. Zhang, L. Muszynski and D. J. Gardner, Spinning disc atomization of wood resin-adhesives: I. Spray characteristics, atomization mechanism, and resin efficiency. *For. Prod. J.* 58, 11, 62–68 (2008).

358. K. Umemura. Curing behavior and bonding performance of wood adhesives under high-pressure steam. *Wood Res.* 84, 130–173 (1997).

359. K. Umemura, S. Kawai, H. Sasaki, R. Hamada and Y. Mizuno. Curing behavior of wood adhesives under high steam pressure. *J. Adhes.* 59, 87–100 (1996).

360. P. E. Humphrey and D. Zavala. A technique to evaluate the bonding reactivity of thermosetting adhesives. *J. Test. Eval.* 17, 323–328 (1989).

361. X. Lu and A. Pizzi. Curing conditions effects on the characteristics of thermosetting adhesives-bonded wood joints. Part 2: Hot postcuring improvement of UF particleboards and its temperature forecasting model. *Holz Roh Werkst.* 56, 393–401 (1998).

362. F. Pichelin, A. Pizzi, A. Fruehwald and P. Triboulot. Exterior OSB preparation technology at high moisture content: Part 1: Transfer mechanisms and pressing parameters. *Holz Roh Werkst.* 59, 256–265 (2001).

363. F. Pichelin, A. Pizzi, A. Fruehwald and P. Triboulot. Exterior OSB preparation technology at high moisture content. Part 2: Transfer mechanisms and pressing parameters. *Holz Roh Werkst.* 60, 9–17 (2002).

364. H. Thoemen and C. Ruf. Measuring and simulation the effects of the pressing schedule on the density profile development in wood-based composites. *Wood Fiber Sci.* 40, 325–338 (2008).

365. H. Thoemen and A. Klueppel. An investigation on the permeability of different wood furnish materials. *Holzforschung* 62, 215–222 (2008).

366. N. Meyer and H. Thoemen. Gas pressure measurements during continuous hot pressing of particleboard. *Holz Roh Werkst.* 65, 49–55 (2007).

367. M. N. Rofii, N. Yamamoto, S. Ueda, Y. Kojima and S. Suzuki. The temperature behaviour inside the mat of wood-based panel during hot pressing under various manufacturing conditions. *J. Wood Sci.* 60, 414–420 (2014).

368. M. N. Rofii, S. Kubota, H. Kobori, Y. Kojima and S. Suzuki. Furnish type and mat density effects on temperature and vapor pressure of wood-based panels during hot pressing. *J. Wood Sci.* 62, 168–173 (2016).

369. S. Wang and P. M. Winistorfer. Fundamentals of vertical density profile formation in wood composites. Part II. Methodology of vertical density formation under dynamic conditions. *Wood Fiber Sci.* 32, 220–238 (2000).

370. P. M. Winistorfer, W. W. Moschler, Jr., S. Wang, E. DePaula and B. L. Bledsoe. Fundamentals of vertical density profile formation in wood composites. Part I. In-situ density measurement of the consolidation process. *Wood Fiber Sci.* 32, 209–219 (2000).

371. S. Wang, P. M. Winistorfer and T. M. Young. Fundamentals of vertical density profile formation in wood composites. Part III. MDF density formation during hot-pressing. *Wood Fiber Sci.* 36, 17–25 (2004).

372. S. Wang and P. M. Winistorfer. Monitoring in-situ density change for in-process measurement and control of hot-pressing. *For. Prod. J.* 52 (7/8), 77–82 (2002).

373. M. Gruchot. In situ investigations of the density profile of wood based panels with MDF as example (in German). PhD thesis, University of Hamburg, Germany (2009).

21 Bioadhesives in Drug Delivery

Inderbir Singh* and Paramjot

CONTENTS

21.1 INTRODUCTION

Bioadhesion is defined as the state in which two materials, at least one of which is of biological nature, are held together for extended periods of time by interfacial forces [1]. For drug delivery purposes, the term *bioadhesion* implies attachment of a drug carrier system to a specific biological location. The biological surface can be epithelial tissue or the mucous coat on the surface of a tissue. If adhesive attachment is to a mucous coat, the phenomenon is referred to as *mucoadhesion* [2]. A mucous coat includes the mucosal linings of the nasal, rectal, esophageal, vaginal, ocular, and oral cavities. The idea of "sticking" dosage forms to the site of application and/or drug absorption has stimulated researchers all over the world [3]. Originally, the advantages of bioadhesive drug delivery systems (DDSs) were seen in their potential (i) to prolong the residence time at the site of drug absorption (e.g., to reduce the dosing frequency for bioadhesive controlled release formulations) and (ii) to intensify contact with the underlying mucosal epithelial barrier (e.g., to enhance the epithelial transport of usually poorly absorbed drugs, such as peptides and proteins). The tight and close contact of the DDS with the absorptive mucosa should generate a steeper concentration gradient, thus increasing the absorption rate [4]. This principle, in particular, supported hopes of increased bioavailability of peptide drugs. Later, it was discovered that some mucoadhesive polymers can also modulate the permeability of epithelial tissues by loosening the tight intercellular junctions, and that some mucoadhesive polymers can also act as inhibitors of proteolytic enzymes [5].

Mucoadhesive DDSs offer a number of advantages including [5,6]:

1. Prolonged residence time of the dosage form at the site of absorption
2. Increased drug bioavailability due to decreased first-pass metabolism
3. Increased absorption due to increased residence time and, hence, increased therapeutic efficacy of the drug

---

* Indicates corresponding author

4. Rapid absorption as a result of the enormous blood supply and blood flow rates at the mucous membranes
5. Improved patient compliance and ease of drug administration

21.2 THEORIES OF BIOADHESION

Mucoadhesion may be regarded as a two-step process, involving the contact and the consolidation stages [1]. In the contact stage, an intimate contact between the polymers of the DDS and the mucous membrane start to develop. The consolidation stage involves interpenetration or entanglement of the polymeric and mucin chains. Moisture plasticizes the system, allowing mucoadhesive molecules to become free and to involve in bond formation with the mucosal surface, predominantly by weak van der Waals forces and hydrogen bonding, although ionic interactions can also occur in some cases.

Five general theories have been adapted to explain the phenomenon of mucoadhesion [1].

1. *Wetting theory*: This theory takes into consideration the surface and interfacial free energies. It involves the ability of a liquid to spread spontaneously onto a surface as a prerequisite for the development of adhesion.
2. *Diffusion theory*: This theory describes the interdiffusion of the polymer and the mucin chains leading to the development of mucoadhesive bonds. This process is driven by concentration gradients and is affected by the available molecular chain lengths, and the mobility and compatibility of the polymers and the mucin.
3. *Adsorption theory*: This theory describes the involvement of hydrogen bonding and van der Waals forces in mucoadhesive interaction.
4. *Electronic theory*: This theory assumes that the electrostatic attractive forces between the glycoprotein of the mucin and the mucoadhesive polymer are responsible for the phenomenon of mucoadhesion. It suggests that electron transfer leading to the formation of electrical double layer occurs at the interface.
5. *Mechanical theory*: This theory assumes that mechanical interlocking and viscoelastic and plastic dissipation of energy are responsible for the development of mucoadhesion.

The complex nature of a mucin molecule permits multiple interaction mechanisms (between the polymer and the mucin) ranging from molecular interactions, which include various types of bonding, to electrostatic, physical entanglement, and hydrophobic interactions.

21.3 MUCOADHESIVE POLYMERS

Mucoadhesive polymers are polymers that exhibit a sufficient mucoadhesive property of strong intermolecular bonding with the mucosal layer. Mucoadhesive DDSs have been explored for the localization of the active agents to a particular location/site. The polymers used in the DDSs play an important role in designing such systems, so as to increase the residence time of the active agent at the desired location. The polymer penetrates into the mucous network or tissue crevices. The easy wetting of the mucosal layer and high molecular weight of the polymer chain facilitate mucoadhesion.

Mucoadhesive polymers are either water-soluble or water-insoluble polymers. Mucoadhesive polymers that adhere to the mucosal surface can be divided into two broad classes [7]:

1. Polymers that become sticky when placed in water and owe their mucoadhesion to stickiness
2. Polymers that adhere through nonspecific, noncovalent interactions that are primarily electrostatic in nature (although hydrogen and hydrophobic bonding may be significant)

21.3.1 Properties of an Ideal Mucoadhesive Polymer

An ideal mucoadhesive polymer should have the following characteristics [8]:

1. It should be nontoxic and nonirritant to the mucous membrane.
2. It should be nonabsorbable through the gastrointestinal tract.
3. It should preferably form a strong bond with the epithelial cell surfaces.
4. It should adhere quickly to most tissues and should possess some site specificity.
5. It should be easily incorporated into the drug and offer no hindrance to the active agent.
6. It must not decompose on storage or during the shelf life of the dosage form.
7. It should not be costly, so that the prepared dosage form remains competitive.

21.3.2 Classification of Mucoadhesive Polymers

21.3.2.1 First-Generation Polymers

These may be called the *traditional* or *nonspecific* polymers. Based on the presence of specific functional groups, these may be further categorized as anionic or cationic polymers.

21.3.2.1.1 Anionic Polymers

These polymers are characterized by the presence of carboxyl and sulfate functional groups that give rise to a net overall negative charge at pH values exceeding the pKa of the polymer. These polymers are most widely used in the development of DDSs because of their high mucoadhesive functionality and low toxicity. Examples of anionic polymers (Table 21.1) include poly(acrylic acid) and its weakly cross-linked derivatives and sodium carboxymethyl cellulose [9].

21.3.2.1.2 Cationic Polymers

These polymers exhibit net positive surface charge because of the presence of cationic groups in the polymeric structure. Because of the presence of a positive charge, these polymers demonstrate strong ionic interaction with negatively charged mucin molecules, thereby providing significant bond strength required to develop mucoadhesive DDSs. Chitosan is the most widely investigated cationic polymer for its mucoadhesive property. Good biocompatibility, high biodegradability, and low toxicological properties make chitosan a favorable candidate for drug delivery applications. Chitosan is produced commercially by the deacetylation of chitin, a naturally occurring polysaccharide, which is the structural element in the exoskeleton of crustaceans (crabs, shrimps, etc.). The presence of reactive primary amino groups renders the polymer useful for pharmaceutical applications [10].

Chitosan has been reported to bind to mucous membrane via ionic interactions between primary amino functional groups and the sialic acid/sulfonic acid groups of mucus. Additionally, hydrogen bonding via the interaction of hydroxyl and amino groups and polymer chain linearity and flexibility are responsible for the interpenetration of polymer and mucus chain, and contribute toward the mucoadhesive interaction of the polymer. Various DDSs, namely, microspheres, nanoparticles, tablets, gels, films, and so on have been investigated while employing chitosan as a mucoadhesive polymer. The major advantage of using chitosan is the ease of tailoring the polymer as per specific requirements for its pharmaceutical application. Various chemical groups may be added, in particular to the C-2 position of chitosan, allowing the formation of novel polymers with added functionality [11]. Other cationic polymers are listed in Table 21.1.

21.3.2.2 Second-Generation Polymers

The major disadvantage of first-generation polymers is their nonspecific attachment to the mucous substrate. This problem has been overcome with the advent of second-generation polymers, which showed more specific (site-specific) binding, more precisely called *cytoadhesives*. Moreover, these

TABLE 21.1

Classification, Categories, and Examples of Different Mucoadhesive Polymers

Classification basis	Category	Examples
Generation	First-generation	Cationic, anionic, nonionic
	Second-generation	Lectins, thiomers, amino acid sequences
Source	Natural	Agarose, chitosan, gelatin, pectin, sodium alginate, various gums (guar, xanthan, gellan, carrageenan)
	Synthetic	*Cellulose Derivatives*
		Carboxymethyl cellulose, sodium carboxymethyl cellulose, thiolated carboxymethyl cellulose, hydroxylethyl cellulose, hydroxylpropyl cellulose, hydroxylpropyl methylcellulose, methyl cellulose, methylhydroxyethyl cellulose
		Poly(acrylic acid)-based polymers
		Carbopol, polycarbophil, poly(acrylic acid), polyacrylates, poly(methylvinylether-co-methacrylic acid), poly(2-hydroxyethyl methacrylate), poly(acrylic acid-co-ethylhexylacrylate), poly(methacrylate), poly(alkylcyanoacrylate), poly(isohexylcyanoacrylate), poly(isobutylcyanoacrylate), copolymer of acrylic acid and poly(ethylene glycol)
Charge	Cationic	Aminodextran, chitosan, trimethylated chitosan, dimethylaminoethyl dextran,
	Anionic	Chitosan–ethylenediaminetetraacetic acid, carbopol, carboxymethyl cellulose, pectin, poly(acrylic acid), polycarbophil, sodium alginate, sodium carboxymethyl cellulose, xanthan gum
	Nonionic	Hydroxyethyl starch, hydroxypropyl cellulose, poly(ethylene oxide), poly(vinyl alcohol), poly(vinyl pyrrolidone), scleroglucan
Solubility	Water-soluble	Carbopol, hydroxylethyl cellulose, hydroxypropyl cellulose, hydroxypropyl methylcellulose, poly(acrylic acid), sodium carboxymethyl cellulose, sodium alginate
	Water insoluble	Chitosan, ethyl cellulose, polycarbophil
Mucoadhesive interaction	Electrostatic interaction	Chitosan
	Covalent bonding	Cyanoacrylate
	Hydrogen bonding	Acrylates [hydroxylated methacrylate, poly(methacrylic acid)], carbopol, polycarbophil, poly(vinyl alcohol)

polymers are less susceptible to mucous turnover rate, providing added advantage for the development of mucoadhesive DDSs. Second-generation polymers have been developed by surface modification of existing polymers; for example, lectin-modified/functionalized polymers, amino acid sequences, and so on. Chemical modification of polymers—for example, introduction of a thiol group or thiolation—is another currently investigated method for enhancing the mucoadhesive property of polymers [12].

21.4 MUCOADHESIVE DRUG DELIVERY THROUGH NASAL ROUTE

The nasal route has recently gained high importance for drug administration, as the nasal epithelium monolayer is highly permeable to various drugs, allowing them to pass through it easily. Secondly, its submucosa layer is richly vascularized, which itself acts as a channel for the transfer of drug molecules. The nasal route also has a large surface area as well as high blood flow, which promotes speedy absorption. Drugs absorbed through the nasal route do not move through the hepatic first-pass metabolism. Nasal administration of drugs, to achieve their systemic administration, provides an interesting alternative to the oral and parenteral routes. The oral route shows unacceptably low bioavailability and the parenteral route has low patient compliance. The nasal route is helpful in avoiding first-pass metabolism, first-pass elimination, gut-wall metabolism, and the destruction of drugs occurring through the gastrointestinal tract. The nasal route is a noninvasive, painless route, whose rate and extent of absorption and also plasma concentration versus time profiles are comparable to intravenous (IV) administration.

The nasal cavity has linings of pseudostratified ciliated columnar epithelium, similar to the rest of the respiratory tract, in which every cell has around 200 cilia. The nasal cavity is helpful in warming, humidifying, and filtering inhaled air before it reaches the lower airway. The cavity is covered with a mucous layer and hairs, so any inhaled particles or microorganisms are trapped by these hairs or a mucous layer. This mucous layer is found all over the respiratory area of the nasal cavity and, hence, can be used to achieve mucoadhesive drug delivery.

Various dosage forms administered for nasal drug delivery (Table 21.2) include solutions, sprays, suspensions, and powders, as well as some semisolid dosage forms. Some examples of mucoadhesive polymers used in nasal delivery are cellulose derivatives such as methyl cellulose, hydroxypropyl methylcellulose, microcrystalline cellulose, ethyl cellulose, and hydropropyl cellulose, polyacrylates such as carbopol 971P, carbopol 974P, starch, chitosan, and so on.

Li et al. [13] prepared polymeric microspheres for nasal vaccination against *Actinobacillus pleuropneumoniae* bacteria by using the polymers mannan and hydroxypropyl methylcellulose phthalate, and found the final formulation to be effective in *in vivo* studies. The results showed that nasal vaccination by the final prepared formulation had higher levels of IgG and IgA serum in the blood against the antigen.

Chand et al. [14] prepared a nasal gel of the drug rizatriptan benzoate using the polymer polysaccharide obtained from *hibiscus rosa*. The drug is mainly effective in the treatment of migraine headaches and nausea, and its nasal delivery showed better and high bioavailability of the drug with enhanced therapeutic effects. *In vitro* studies showed that the final prepared polymeric gel had a higher half-life (95 min) as compared with the control (23 min). The histopathological and cilitoxicity studies showed the final drug formulation to be safe, making it a suitable candidate for drug delivery.

Datta and Bandyopadhyay [15] prepared a nasal gel of diazepam, using a natural mucoadhesive polymer obtained from fenugreek (*Trigonella foenum graecum L.*) in combination with synthetic polymers such as hydroxylpropyl methylcellulose. The drug is mainly used for the treatment of anxiety disorders and confusion. The final prepared gel formulation showed *in vitro* drug release of about 98% in a time interval of 3 h 45 min. The full drug release occurs after approximately 4 h. However, by using penetration enhancers such as sodium taurocholate, 100% of drug release was seen just after 90 min.

Das et al. [16] prepared a nasal hydrogel of penciclovir by using poly-N-vinyl-2-pyrrolidone (PVP) in combination with PEG 600. The drug is mainly effective against various viral disorders, including herpesvirus infections. It was found that the final prepared hydrogel showed higher release rates of PVP as compared with chitosan and carbopol. The release rate was further enhanced by using glycerol. Histopathological studies showed the final formulation as a stable hydrogel, making it promising for nasal drug delivery. The studies showed the final formulated drug formulation to be safe, making it a suitable candidate for drug delivery [16].

Alsarra et al. [17] prepared a nasal hydrogel of acyclovir by using PVP in combination with PEG 600. The drug is mainly effective against various viral disorders, including herpesvirus infections and varicella-zoster virus (VZV) infections. It was found that the final prepared hydrogel showed higher release rates when the mucoadhesive agent used was PVP hydrogel as compared with hydrogels prepared from chitosan and carbopol. The release rate was further enhanced by using glycerol. Histopathological studies showed the final formulation as a stable hydrogel, making it promising for nasal drug delivery. The histopathological and cytotoxicity studies showed the final formulated drug formulation to be safe, making it a suitable candidate for drug delivery.

TABLE 21.2
Recently Used Mucoadhesive Systems through Different Routes

Item Number	Polymer Used	Drug Used	System Used	Uses	Results	Ref.
			Bioadhesive Drug Delivery via Nasal Route			
1	Mannan and hydroxypropyl methylcellulose phthalate	—	Microspheres	Nasal vaccine against *Actinobacillus pleurapneumoniae*	*In vivo* studies revealed higher serum levels of IgG and IgA, making it a promising candidate for nasal vaccination	[13]
2	Polysaccharide obtained from *hibiscus rosa*	Rizatriptan benzoate	Nasal gel	Treatment of migraine headache, nausea	*In vitro* showed increase in plasma concentration of the drug. Histopathological study revealed safety of the prepared formulation	[14]
3	Natural mucoadhesive polymer obtained from fenugreek (*Trigonella foenum graecum L.*) in combination with synthetic polymers such as HPMC	Diazepam	Nasal gel	Treatment of anxiety disorders	The gel showed *in vitro* drug release of about 98% in 3 h 45 min. However, by using sodium taurocholate, 100% of drug release was seen after 90 min	[15]
4	Poly-N-vinyl-2-pyrrolidone (PVP) in combination with PEG 600	Penciclovir	Nasal hydrogel	Antiviral drug used in the treatment of herpesvirus infections	Final prepared hydrogel showed higher drug release rates that were further improved by using glycerol. Histopathological studies showed the formulation to be safe for nasal drug delivery	[16]
5	Poly-N-vinyl-2-pyrrolidone (PVP) in combination with PEG 600	Acyclovir	Nasal hydrogel	Used in the treatment of HSV and VZV infections	Final prepared hydrogel showed higher release rates and were found safe for nasal drug delivery	[17]
6	Polymeric mucilage obtained from *Ficus carica*	Midazolam	Nasal gel	Used in the treatment of generalized tonic–clonic seizures	*The formulation exhibited* improved bioavailability and drug release profiles	[18]
7	Chitosan	Ebastine	Mucoadhesive nanoparticulate system	Used as H1 antagonist for chronic urticaria and allergic rhinitis	*Nanoparticulate system showed improved mucoadhesive power and entrapment, thus making it promising for nasal DDS*	[19]

(Continued)

TABLE 21.2 (CONTINUED)

Recently Used Mucoadhesive Systems through Different Routes

Item Number	Polymer Used	Drug Used	System Used	Uses	Results	Ref.
			Bioadhesive Drug Delivery via Colon Route			
8	Thiolated chitosan/ alginate polysaccharides	5-amino salicylic acid and curcumin	Mucoadhesive microparticulates	Treatment of ulcerative colitis	Mucoadhesive microparticles displayed significant mucoadhesion strength and controlled drug release property	[20]
9	Combination of Carbopol 934P, pectin, polyvinyl pyrrolidone	Prednisolone	Mucoadhesive tablets	Used in the treatment of inflammatory and autoimmune disorders	Mucoadhesive tablets exhibited sustained drug release behavior	[21]
10	Chitosan	Deflazacort	Mucoadhesive microspheres	Treatment of ulcerative colitis and Crohn's disease	Mucoadhesive microspheres displayed significant drug delivery to the target site with better drug stability and sustained release behavior	[22]
11	Eudragit S100-coated chitosan	Mesalamine	Microspheres	Treatment of ulcerative colitis	Microspheres showed no drug release in gastric pH and maximum drug release in colonic environment, making it promising DDS for the treatment of colonic disorders	[23]
12	Chitosan	Ondansetron	Mucoadhesive microspheres	Used in treatment of ulcerative colitis	Mucoadhesive microspheres showed *in vitro* drug release with a burst in first hour. Microspheres showed good drug entrapment efficiency and controlled *in vitro* drug release behavior	[24]
			Bioadhesive Drug Delivery through Vaginal Route			
13	Poly(ethylene glycol) in combination with 1,2-distearoyl-*sn*-glycero-3-phospho ethanolamine	siRNA as targeting agent	Polymeric lipoplexes	Vaccine for prevention of HSV infections	Lipoplexes showed significant adhesion with vaginal mucosa, resulting in prolonged delivery of siRNA	[26]

(Continued)

TABLE 21.2 (CONTINUED)

Recently Used Mucoadhesive Systems through Different Routes

Item Number	Polymer Used	Drug Used	System Used	Uses	Results	Ref.
14	Poly(ethylene glycol), PEGylated lipoplex-entrapped alginate scaffold	siRNA to target vagina	PEGylated lipoplex-entrapped alginate scaffolds	Vaccine for prevention of HSV infections	The formulation showed sixfold increased uptake into the vaginal epithelium after intravaginal administration in mice	[27]
15	Chitosan and carbopol	Curcumin	Mucoadhesive liposomes	Treatment of vaginal infections	Use of bioadhesive polymers significantly increased the therapeutic efficacy of curcumin for the treatment of vaginal infections	[28]
16	Carboxymethyl cellulose, poly(ethyleneoxide), poly(acrylic acid)	—	Nanofibers	Treatment of various vaginal infections	Nanofibers structure size was reported to be less than 100 nm. Significant mucoadhesive ability was induced in formulation with mucoadhesive materials	[29]
17	L-cysteine and cysteamine covalently attached to poly(acrylic acid)	Nystatin	Mucoadhesive tablets and gels	Treatment of vaginal infections	Formulations exhibited good stability and increased residence in vaginal mucosa with controlled release of the drug	[30]
18	Hydroxypropyl methylcellulose K100M or hydroxypropyl methyl cellulose E50	Clotrimazole	Vaginal gel	Treatment of vaginal candidiasis	*In vivo* studies showed that gel formulation remained in vaginal mucosa for about 24 h after application	[31]
Bioadhesive Drug Delivery through Rectal Route						
19	Combination of mucin–gelatin	Ceftriaxone sodium	Mucin–gelatin mucoadhesive microspheres	Treatment of rectal infections	Controlled drug release and increased residence time of formulation for treatment of rectal infections	[32]
20	Chitosan, hydroxypropyl methyl cellulose, Carbopol 934	Diclofenac sodium	Mucoadhesive hydrogels	—	Controlled drug release, increased residence time with low skin irritation of formulation was reported	[33]

(Continued)

TABLE 21.2 (CONTINUED)
Recently Used Mucoadhesive Systems through Different Routes

Item Number	Polymer Used	Drug Used	System Used	Uses	Results	Ref.
21	Methyl cellulose, hydroxyethyl cellulose, hydroxypropyl methylcellulose, polyvinylpyrrolidone, and carbopol	Etolodac	Polymeric gel in the form of suppositories	For treatment of rectal inflammation	Final formulation showed better gel strength, gelation temperature, and bioadhesive ability. Better therapeutic effect without any damage to rectal tissues was reported	[34]
22	Hydroxypropyl methylcellulose	Quinine hydrochloride	Polymeric gel	Treatment of rectal infections	Formulation showed better rectal availability and controlled release of the drug	[35]

Basu and Bandyopadhyay [18] prepared a nasal gel of midazolam from polymeric mucilage obtained from *Ficus carica*. The drug was found to be mainly effective in treatment of seizures and generalized tonic–clonic seizures. *In vivo* studies conducted on rabbits of the final prepared formulation showed better bioavailability of the drug as compared with the gels that were prepared from synthetic mucoadhesive polymers. The final hydrogel formulation showed better drug release profiles compared with synthetic polymeric gels. The studies showed that the final drug formulation gave promising results against different related disorders, making it a suitable candidate for drug delivery [18].

Khom et al. [19] prepared a mucoadhesive nanoparticulate system of Ebastine by using chitosan as polymer. The drug Ebastine is mainly effective as an H1 antagonist against chronic urticaria and allergic rhinitis. It was found that the final prepared nanoparticulate system had a spherical shape and smooth surface, and showed better mucoadhesive capacity of about 78.6% and displayed a high entrapment of the drug in the chitosan polymer, making it promising for nasal drug delivery.

21.5 MUCOADHESIVE DRUG DELIVERY TO COLON

These days, delivery of drugs via a mucoadhesive approach to the colon has gained much popularity. As regards colon-targeted DDSs, different novel systems have been designed, comprising a core that has mucoadhesive properties and carriers that have properties of rapid drug release. Different carbomers, namely, carbopol 971P, polycarbophil AA-1, carbopol 974P, and some other related carbomers were used to prepare pellets of different drugs, with or without organic acids, by the extrusion–spheronization method. To achieve sustained drug release, these pellets need to be enterically coated with double coating systems, using the desired polymeric systems.

Colon drug delivery has recently become popular for targeting proteins and peptides such as insulin. The human colon is rich in lymphoid tissues. The mast cells found in the colonic mucosa can achieve rapid uptake of the antigens, which further results in the production of local antibodies, helping better vaccine delivery. The human colon has a less hostile environment and lesser diversity and intensity of different activities as compared with the stomach and large intestine. Moreover, the human colon also has a longer retention time than other organs, making it suitable for drug delivery.

The simplest and easiest method to target drugs to the colon is by achieving sustained drug release, which can be done by using enteric coatings of the prepared formulations by conventional thick coating materials. Some examples of mucoadhesive systems for drug delivery to the colon are listed in Table 21.2.

Varum et al. [25] prepared pellets of prednisolone by using different carbomers, namely, carbopol 971P, polycarbophil AA-1, and carbopol 974P. The pellets of prednisolone were prepared by the extrusion–spheronization method. The prepared pellets were coated using a double coating system so that the drug release from the pellets could be achieved at pH 7. Such pellets are comprised of two different layers, that is, the inner and outer layers. The inner layer was composed of buffer salts and partially neutralized Eudragit S, while the outer layer of pellets was coated with standard Eudragit S. The pellets showed better mucoadhesive ability and better drug release property for 75 min in the colon-specific region. Moreover, the mucoadhesive property of the prepared double-coated pellets was higher than that of single-coated pellets. This provides a novel platform to target the colon and to obtain better a mucoadhesive property and overcome the unpredictability in the transit of the drug, and to achieve better drug release and bioavailability.

Ahmed and Aljaeid [20] prepared a mucoadhesive nanoparticulate system of a combination of the drugs 5-amino salicylic acid and curcumin, by using thiolated chitosan and alginate polysaccharides. The formulation was prepared to test its efficiency against ulcerative colitis. An ex vivo test on the colon mucosa of rats showed better mucoadhesive ability, and the in vivo results showed superior efficiency of the drug in treatment of colitis of rats. The prepared coated microparticles showed better controlled release, mucoadhesion, and pH-dependent drug delivery, and hence should show promising results when given in the treatment of human inflammatory bowel diseases.

Reddy and Begum [21] prepared mucoadhesive tablets of prednisolone by using a combination of the polymers carbopol 934P, pectin, and polyvinylpyrrolidone. It is mainly used in the treatment of inflammatory and autoimmune disorders. The final prepared mucoadhesive tablets, containing 100 mg of pectin, 80 mg of lactose, and 50 mg of carbopol showed better sustained release of prednisolone for a longer period of time, due to being bilayered.

Gawde et al. [22] prepared mucoadhesive microspheres of deflazacort by using chitosan as the polymer. The formulation was found to be effective in the treatment of ulcerative colitis, Crohn's disease, and various other colon-related disorders. The final prepared microspheres showed better drug delivery to the target site, and reduced drug concentration to nontarget organs and, hence, a reduction in the amount of side effects, better therapeutic effects, and sustained release of the drug, making it promising for various colon-related disorders. The studies showed the final drug formulation to give promising results against different colon-related disorders, making it a suitable candidate for drug delivery.

Badhana et al. [23] prepared Eudragit S100-coated chitosan microspheres to check efficacy against ulcerative colitis. It was found that the final prepared microspheres showed a rough surface in a scanning electron microscope. The entrapment efficiency of the microspheres was found to be 43.72%–82.27%, drug loading was found to be 20.28%–33.26%, and the size of microspheres was found to be 61.22–90.41 μm. However, it was also seen that with the increase in the polymer ratio, the size of the microspheres kept on increasing. The Eudragit S100-coated microspheres showed no release in gastric pH, a negligible release in intestinal pH, and maximum release in the colonic environment, making it promising for the treatment of colon disorders.

Jose et al. [24] prepared mucoadhesive microspheres of ondansetron using chitosan. The final prepared mucoadhesive microspheres showed rapid in vitro drug release in the first hour. The formulation containing a 1:10 core:coat ratio showed better results compared with all other formulations. The formulations demonstrated the best drug entrapment efficiency, particle size, and in vitro drug release rates. The studies showed the final drug formulation to give promising results against different colon-related disorders, making it a suitable candidate for drug delivery.

21.6 MUCOADHESIVE DRUG DELIVERY THROUGH VAGINAL ROUTE

The human vagina is basically a fibromuscular tube of about 10 cm in length that is an important organ of the human reproductive tract. It has a large permeation area, relatively low enzymatic activity, and also rich vascularization. Various drugs can be administered by using the vaginal route of drug delivery, as this route will help the drug to avoid the first-pass metabolism with relatively low drug concentration as compared with the oral route. Drugs in the vagina are mainly delivered through two different routes: intravaginally and transvaginally. Intravaginally, the drug is delivered through the vaginal epithelium, while in the transvaginal case, the drug is delivered through vaginal mucosa to the uterus, from where it moves to systemic circulation. The human vagina has a specific blood flow system. Blood flows either via the portal type of circulation, or by lymphatic and venous channels, allowing the blood to bypass gastrointestinal absorption and liver detoxification. This allows the drug molecules to move from the vagina to the uterus and then into systemic circulation. A number of antifungal drugs such as clotrimazole, miconazole, and tioconazole have been successfully administered to treat vaginal infections.

Vaccination can also be achieved by using the vaginal route of drug administration. For this, mucoadhesion plays a very important role. Different polymers that can be used for mucoadhesion via the vaginal route are sodium alginate, sulfated polysaccharides, hyaluronic acid derivatives, pectin, tragacanth, poly(ethylene glycol) (PEG), polyacrylates, starch, sulfated polysaccharides, chitosan, cellulose derivatives, and so on.

The vaginal mucosal cavity is a safe, feasible, and attractive site for drug delivery. Vaginal delivery has various advantages, such as a drop in the severity of gastrointestinal tract side effects, less tissue damage, lower risk of infections, avoidance of pain, prolonged contact; self-insertion is possible, and removal of the dosage form is also possible if needed. It has also been seen that prolonged adhesion can be achieved with vaginal mucosa as compared with intestinal and rectal mucosa. Mucoadhesive systems for the vaginal route provide a better contact between the dosage form and the vaginal mucosa, which results in a high concentration of drug in the mucosa, further resulting in a high drug flux through the mucosa. However, it should be noted that the efficacy of vaginal mucoadhesive DDSs is highly affected by the nature of the polymer and the drug, and also by the biological environment of the vagina.

Furst et al. [26] prepared freeze-dried polymeric lipoplexes for vaginal delivery of siRNA by using different polymeric conjugated derivatives, namely poly(ethylene glycol) in combination with 1,2-distearoyl-*sn*-glycero-3-phospho ethanolamine-*N*-[methoxy poly(ethylene glycol)-2000] (DPSE-PEG$_{2000}$), 1,2-distearoyl-*sn*-glycero-3-phospho ethanolamine-*N*-[methoxy poly(ethylene glycol)-750] (DPSE-PEG$_{750}$), and ceramide-PEG$_{2000}$. The formulation was mainly prepared for vaccination for the prevention of the herpes simplex virus (HSV) infection. The final prepared topical vaginal siRNA formulation showed sustained release. The prepared formulation was found to be capable of attaching to the vaginal mucosa, which resulted in prolonged delivery of siRNA, and hence was responsible for an increase in the effectiveness of the therapy. A polymeric solid mucoadhesive system was successfully prepared that was loaded with lipoplexes, and had the ability to rehydrate via the vaginal mucosa or fluids to form a new hydrogel, so delivery to the vagina became possible.

Wu et al. [27] prepared PEGylated lipoplex-entrapped alginate scaffolds to target siRNA in the vagina by using poly(ethylene glycol) and PEGylated lipoplex-entrapped alginate polymeric scaffolds. The final prepared formulation showed an entrapment efficiency of about 50%, The formulation showed about a sixfold increase in uptake into the vaginal epithelium after intravaginal administration in mice, as compared with existing transfection systems. The results showed it to be a promising carrier system for the delivery of siRNA/oligonucleotides to the vaginal epithelium.

Berginc et al. [28] prepared mucoadhesive liposomes of curcumin by using a combination of natural and synthetic polymers, namely chitosan and carbopol, in different proportions. The strength or ability of the mucoadhesion of the prepared final mucoadhesive liposomes was determined by using isolated bovine serum mucus. The use of bioadhesive polymers significantly increased the curcumin release in the drug as compared with the control. The results also showed that coating the prepared liposomes with polymer resulted in increased mucoadhesion ability. The result showed the final prepared formulation to be a promising carrier system for the delivery of curcumin in the vaginal cavity.

Brako et al. [29] prepared nanofibers using blends of different polymers, namely, carboxymethyl cellulose, poly(ethylene oxide), poly(acrylic acid), and sodium alginate for the treatment of various vaginal infections. The size of nanofibers was reported to be less than 100 nm. Scanning electron microscopy studies confirmed the uniform cylindrical shapes of the prepared nanofibers. It was seen that the use of carboxymethyl cellulose in nanofibers gives the best mucoadhesion ability.

Hombach et al. [30] prepared mucoadhesive tablets and gels using different polymers in which L-cysteine and cysteamine were covalently attached to the polymer poly(acrylic acid), and two different thiolated polymers were also used. The final prepared tablets of thiolated polymers showed high stability in vaginal fluids at pH 4.2. The time of mucoadhesion of the prepared final tablets on freshly prepared vaginal mucosa increased, because of the formation of disulfide bonds between the two. Nystatin, used in the final formulation, was released more slowly from the thiomer tablets and gels as compared with the control tablets and gels. Hence, these results confirmed the final formulation to be effective in the delivery of nystatin for prolonged and sustained release for the treatment of vaginal infections.

Rencber et al. [31] prepared a vaginal gel of clotrimazole using the polymers hydroxypropyl methylcellulose K100M or hydroxypropyl methylcellulose E50. The formulation was prepared to check the activity of the drug against vaginal candidiasis. The prepared formulation of clotrimazole showed good mucoadhesive ability for the treatment of vaginal candidiasis. Poloxamer 407 and Poloxamer 188 were used to prepare gels. HPMC K100M and E30 were added to ensure the mucoadhesion property of the final formulation. The final prepared formulation contained about 20% of Poloxamer 188, 20% of Poloxamer 407, and 0.5% of hydroxypropyl methylcellulose K100M or hydroxypropyl methylcellulose E50. *In vivo* studies showed that the final prepared gel formulation remained in the vaginal mucosa for about 24 h after application of the gel to the vaginal mucosa. The results confirmed the final formulation to be a better carrier system for the delivery of clotrimazole.

21.7 MUCOADHESIVE DRUG DELIVERY THROUGH RECTAL ROUTE

The human rectum is usually about 15–20 cm long and is highly vascularized, having adequate blood vessels and lymph vessels. For an ideal drug absorption property, the site should have an adequate blood supply, permeability, and surface area. The human rectum has all these ideal characteristics that favor the systemic absorption of drugs. Secondly, rectal delivery is also helpful in avoiding first-pass metabolism and first-pass elimination. Drugs absorbed through rectal veins directly enter the inferior vena cava, which is able to bypass the first-pass metabolism. However, drugs absorbed through the superior rectal vein enters the portal vein, and is subjected to the first-pass metabolism through the liver.

Various advantages of the rectal route compared with the oral route include avoidance of degradation and metabolism caused by the gastrointestinal tract, avoidance of stomach irritation, and avoidance of first-pass metabolism. Different systems for rectal delivery are used to achieve better drug release such as Osmet, Diastat, Hycore, and so on. In the Osmet system, osmotic tablets are used to achieve controlled delivery of drug at a zero-order rate by an osmotic pump mechanism. Various drugs such as theophylline, propranolol, nifedipine, antipyrine, and so on can be given through the Osmet system for rectal delivery. Diastat is a nonsterile rectal gel that contains about 5 mg/ml diazepam, which can be given through the rectal route. This route is acceptable because the drug is used in the treatment of seizures, and during seizures the oral administration of drugs is not possible. The bioavailability of the drug reaches about 90% and the time of action of the drug was found to be 5–15 min, which makes rectal delivery of drug suitable as compared with the oral route. Hycore is another example of controlled release of drugs in the form of inserts.

Various mucoadhesive polymers and their formulations for rectal delivery are shown in Table 21.2. Hydrogels are mainly administered rectally for drug delivery.

Ofokansi et al. [32] prepared mucoadhesive microspheres of ceftriaxone sodium for the treatment of rectal infections using combinations of different polymers, mucin, and gelatin. The results showed that the active drug, that is, ceftriaxone sodium, could be successfully delivered through the rectal route when the drug was embedded in microspheres that were formulated from a single polymer or a combination of polymers, that is, either type A gelatin alone or a mixture with porcine mucin. Hence, from the results, it can be concluded that the prepared formulation can be a better alternative route for the delivery of such types of acid-labile third-generation antibiotics (cephalosporins).

El-Leithy et al. [33] prepared mucoadhesive hydrogels of diclofenac sodium using chitosan, hydroxypropyl methylcellulose, and carbopol 934. *In vitro* drug release of the final prepared mucoadhesive hydrogel formulation showed controlled release of the drug, with a release efficiency of 34.6–39.7% after 6 h of administration. The mucoadhesion time of the final prepared formulation, with 7% w/w HPMC hydrogel, was found to be 330 min, which is enough time to attach the loaded microspheres to the rectal surface. Histopathological studies showed that the formulation had the lowest skin irritation effect. These results confirm that the formulation has promising effects for the rectal delivery of the drug.

Barakat [34] prepared a polymeric gel containing suppositories of etodolac using different polymers such as methyl cellulose, hydroxyethyl cellulose, hydroxypropyl methylcellulose, polyvinylpyrrolidone, and carbopol polymers. The formulation was mainly prepared to check the efficacy of the drug in the treatment of rectal inflammation. The final formulation showed higher gel strength, gelation temperature, and bioadhesive ability compared with other formulations. The final formulation showed better anti-inflammatory effect as compared with the control in rats during *in vivo* studies. Studies also showed that the formulation did not show any damage to the morphology of rectal tissues, making it promising for the treatment of rectal inflammation.

Koffi et al. [35] prepared a quinine hydrochloride drug containing a polymeric gel using hydroxypropyl methylcellulose. The formulation was mainly prepared to check the efficacy of the drug in the treatment of rectal inflammation. With the use of adhesive and thermosensitive polymers, the formulation exhibited significant improvement in rectal bioavailability of the drug. Moreover, good rectal tolerance was reported for the formulation.

21.8 MUCOADHESIVE DRUG DELIVERY THROUGH OCULAR ROUTE

Various processes in the eye such as tear production, the flow of tears, the blinking of the eye, and so on are responsible for the protection of the eye from various harmful agents. All these processes make it very difficult to deliver drugs through an ocular delivery system, since all such processes result in the washing off of the drug from the human eye. With various drugs, when given through conventional dosage forms such as solutions, suspensions, and so on, most of the dosage forms readily wash off from the cornea of the eye. However, if they are given in highly viscous forms such as ointments (semisolid dosage forms), so that these can be retained in the eye for a longer period of time, then these may alter the tear refractive index and may cause blurred vision.

However, by using the bioadhesive/mucoadhesive property, the drug retention capacity can be increased in the eye (Table 21.2). Sensoy et al. [36] prepared microspheres of sulfacetamide sodium that had bioadhesive ability to increase the residence time on the ocular area and also enhance treatment efficacy. The spray-drying method was used to prepare mucoadhesive microspheres using polymers such as polycarbophil, HPMC, and pectin in different ratios.

De la Fuente et al. [37] prepared a bioadhesive DNA nanocarrier system made from chitosan and hyaluronan for ophthalmic delivery. Ocular inserts can also be prepared to achieve better eye delivery of the drug. Ocular inserts are usually applied at the back of the eyelid. However, the tendency of the ocular inserts to move on the surface of the eyelid may cause irritation in the eye. This problem can be solved by using mucoadhesive polymers, which result in reduced movement of the inserts within the eye, thus minimizing other sensations such as irritation and burning caused by the movement of eye inserts.

Baeyens et al. [38] prepared soluble mucoadhesive/bioadhesive ophthalmic drug inserts to treat ophthalmic diseases such as conjunctivitis, corneal ulcers, and keratoconjunctivitis in dogs. The results showed that the prepared mucoadhesive ophthalmic drug inserts demonstrated better results in a single application as compared with eye drops of the drug. The results showed better patient compliance, thus making the formulation promising to achieve higher and prolonged drug release and improved treatment for eye disorders.

Kalam [39] prepared chitosan nanoparticles of the drug dexamethasone using chitosan and hyaluronic acid. The formulation was mainly used to test the efficacy of the product in the treatment of eye irritation, redness, and inflammation. *In vivo* testing of the final prepared nanoparticles on rabbit corneal mucosa showed the drug in sufficiently high concentration in the aqueous humor for 24 h after application. The final prepared hyaluronic acid-coated nanoparticles showed higher bioavailability of the drug as compared with the simple drug solution. About a 1.83–2.14-fold higher area under the curve was observed when using chitosan nanoparticles and hyaluronic acid-coated chitosan nanoparticles, respectively. The increased bioavailability of the drug was supposed to be due to the high mucoadhesive characteristics of both chitosan and hyaluronic acid. The presence of these polymers also speeds up the cellular uptake of drug by receptor-mediated endocytosis, which results in increased bioavailability.

21.9 SUMMARY

In the recent past, numerous new mucoadhesive polymers have been used for the development of mucoadhesive DDSs. Mucoadhesive polymers are responsible for the development of bonds between the delivery system and the mucosal surface. Therefore, researchers are concentrating on developing novel mucoadhesive polymers with enhanced mucoadhesive strength. Natural mucoadhesive polymers and grafted mucoadhesive materials are specifically being examined for their applications in developing mucoadhesive DDSs. A very promising and novel concept is the development of cytoadhesives for more accurate and reliable drug delivery to particular cells. In future, functionalized polymers and cellular adhesives will gain importance for specific targeting and controlled release of therapeutic agents. However, strict toxicological and regulatory issues need to be addressed before these novel mucoadhesive materials can be accepted for commercial application.

REFERENCES

1. R. Shaikh, R. Thakur, R. Singh, M. J. Harland, A. D. Woolfson and R.F. Donnelly. Mucoadhesive drug delivery systems. *J. Pharm. Bioallied Sci.* 3, 89–100 (2011).
2. S. Tyagi, N. Sharma, S. K. Gupta, A. Sharma, A. Bhatnagar, N. Kumar and G.T. Kulkarni. Development and gamma scintigraphical clearance study of novel *Hibiscus rosasinensis* polysaccharide based mucoadhesive nasal gel of rizatriptan benzoate. *Drug Delivery Technology* 30, 100–106 (2015).
3. F. Laffleur. Mucoadhesive therapeutic compositions: A patent review (2011–2014). *Expert Opinion Therapeutic Patents* 26, 377–388 (2016).
4. H. He, X. Cao and L. J. Lee. Design of a novel hydrogel-based intelligent system for controlled drug release. *J. Controlled Release* 95, 391–402 (2004).
5. K. Hansen, G. Kim, K.G. Desai, H. Patel, K.F. Olsen, J. Curtis-Fisk, E. Tocce, S. Jordan and S.P. Schwendeman. Feasibility investigation of cellulose polymers for mucoadhesive nasal drug delivery applications. *Mol. Pharm* 12, 2732–2741 (2015).
6. A. Abruzzo, T. Cerchiara, F. Bigucci, M. C. Gallucci and B. Luppi. Mucoadhesive buccal tablets based on chitosan/gelatin microparticles for delivery of propranolol hydrochloride. *J. Pharm. Sci.* 104, 4365–4372 (2015).

7. K. Netsomboon and A. Bernkop-Schnurch. Mucoadhesive vs. mucopenetrating particulate drug delivery. *Eur. J. Pharm. Biopharm.* 98, 76–89 (2016).

8. M. T. Cook and V. V. Khutoryanskiy. Mucoadhesion and mucosa-mimetic materials: A mini-review. *Int. J. Pharm.* 495, 991–998 (2015).

9. S. Agarwal. Mucoadhesive polymeric platform for drug delivery: A comprehensive review. *Curr. Drug Deliv.* 12, 139–156 (2015).

10. A. Sosnik and M. Menaker Raskin. Polymeric micelles in mucosal drug delivery: Challenges towards clinical translation. *Biotechnol. Adv.* 33, 1380–1392 (2015).

11. M. L. Bruschi, L. M. de Francisco and F. B. Borghi. An overview of recent patents on composition of mucoadhesive drug delivery systems. *Recent Patents Drug Deliv. Formul.* 9, 79–87 (2015).

12. I. Singh and V. Rana. Enhancement of mucoadhesive property of polymers for drug delivery applications: A critical review. *Rev. Adhesion Adhesives* 1, 271–290 (2013).

13. H. S. Li, M. K. Shin, B. Singh, S. Maharjan, T. E. Park, Z. S. Hong, C. S. Cho and Y. J. Choi. Nasal immunization with mannan-decorated mucoadhesive HPMCP microspheres containing ApxIIA toxin induces protective immunity against challenge infection with *Actinobacillus pleuropneumoiae* in mice. *J. Controlled Release* 233, 114–125 (2016).

14. R. Chand, A. A. Naik and H. A. Nair. Thermoreversible biogels for intranasal delivery of rizatriptan benzoate. *Indian J. Pharm. Sci.* 36, 723–725 (2009).

15. R. Datta and A. K. Bandyopadhyay. Development of a new *nasal* drug *delivery* system of *diazepam* with natural mucoadhesive agent from *Trigonella foenum-graecum* L. *J. Sci. Indus. Res.* 64, 973–977 (2005).

16. A. Das, B. K. Gupta and B. Nath Mucoadhesive polymeric hydrogels for nasal delivery of penciclovir. *Appl. Pharma. Sci.* 35, 158–163 (2012).

17. I. A. Alsarra, A. Y. Hamed, G. M. Mahrous, E. L. Maghraby, A. A. Robayan and F. K. Alanazi. Mucoadhesive polymeric hydrogels for nasal delivery of acyclovir. *Drug Dev. Ind. Pharm.* 35, 352–362 (2009).

18. S. Basu and A. K. Bandyopadhyay. Development and characterization of mucoadhesive in situ nasal gel of Midazolam prepared with Ficus carica mucilage. *AAPS Pharm. Sci. Technol.* 11, 1223–1231 (2010).

19. T. C. Khom, K. S. Hemant, A. Raizaday, N. Manne, H. S. Kumar and S. N. Kumar. Development of mucoadhesive nanoparticulate system of ebastine for nasal drug delivery. *Tropical Pharma. Res.* 13, 1013–1019 (2013).

20. T. A. Ahmed and B. M. Aljaeid. Preparation, characterization, and potential application of chitosan, chitosan derivatives, and chitosan metal nanoparticles in pharmaceutical drug delivery. *Drug Devel. Therapy* 10, 483–507 (2016).

21. R. K. Reddy and R. H. Begum. Preparation and in-vitro evaluation of Prednisolone mucoadhesive tablets for colon targeted drug delivery system. *Int. J. Innovative Pharma. Res.* 6, 485–490 (2015).

22. P. Gawde, S. Agarwal and P. Jain. Development of mucoadhesive microsphere for colon delivery. *Int. J. Pharma. Biol. Archives* 3, 440–442 (2012).

23. S. Badhana, N. Garud and A. Garud. Colon specific drug delivery of Mesalamine using Eudragit S100-coated chitosan microspheres for the treatment of ulcerative colitis. *Int. Current Pharma. J.* 2, 42–48 (2013).

24. S. Jose, K. Dhanya, T. A. Cinu and N. A. Aleykutty. Multiparticulate system for colon targeted delivery of ondansetron. *Indian J. Pharm. Sci.* 72, 58–64 (2010).

25. F. J. Varum, F. Veiga, J. S. Sousa and A. W. Basit. Mucoadhesive platforms for targeted delivery to the colon. *Int. J. Pharm.* 420, 11–19 (2011).

26. T. Furst, G. R. Dakwar, E. Zagato, A. Lechanteur, K. Remaut, B. Evrard, K. Braeckmans and G. Piel. Freeze-dried mucoadhesive polymeric system containing pegylated lipoplexes: Towards a vaginal sustained released system for siRNA. *J. Controlled Release* 236, 68–78 (2016).

27. S. Y. Wu, S. I. Chang, M. Burgess and N. A. McMillan. Vaginal delivery of siRNA using a novel PEGylated lipoplex-entrapped alginate scaffold system. *J. Controlled Release* 155, 418–426 (2011).

28. K. Berginc, S. Suljakovic, B. N. Skalko and A. Kristl Mucoadhesive liposomes as new formulation for vaginal delivery of curcumin. *Eur. J. Pharm. Biopharm.* 87, 40–46 (2014).

29. F. Brako, B. R. Abraham, S. Mahalingum, Q. M. Duncan and E Mohan. Making nanofibres of mucoadhesive polymer blends for vaginal therapies. *European Polym. J.* 70, 186–196 (2015).

30. J. Hombach, T. F. Palmberger and A. Bernkop-Schnurch. Development and *in vitro* evaluation of a mucoadhesive vaginal delivery system for nystatin. *J. Pharm. Sci.* 98, 555–564 (2009).

31. S. Rencber, S. Y. Karavana, Z. A. Senyigit, B. Erac, M. H. Limoncu and E. Baloglu. Mucoadhesive in situ gel formulation for vaginal delivery of clotrimazole: Formulation, preparation, and *in vitro/in vivo* evaluation. *Pharm. Dev. Technol.* 12, 1–11 (2016).

32. K. C. Ofokansi, M. U. Adikwu and V. C. Okore. Preparation and evaluation of mucin-gelatin mucoadhesive microspheres for rectal delivery of ceftriaxone sodium. *Drug Dev. Ind. Pharm.* 33, 691–700 (2007).

33. E. S. El-Leithy, D. S. Shaker, M. K. Ghorab and R. S. Abdel-Rashid. Evaluation of mucoadhesive hydrogels loaded with diclofenac sodium-chitosan microspheres for rectal administration. *AAPS Pharm. Sci. Technol.* 11, 1695–1702 (2010).

34. N. S. Barakat. *In vitro* and *in vivo* characteristics of a thermogelling rectal delivery system of etodolac. *AAPS Pharm Sci Technol.* 10, 724–731 (2009).

35. A. A. Koffi, F. Agnely, M. Besnard, J. Kablan Brou, J. L. Grossiord and G. Ponchel. *In vitro* and *in vivo* characteristics of a thermogelling and bioadhesive delivery system intended for rectal administration of quinine in children. *Eur. J. Pharm. Biopharm.* 69, 167–175 (2008).

36. D. Sensoy, E. Cevher, A. Sarici, M. Yilmaz, A. Ozdamar and N. Bergişadi. Bioadhesive sulfacetamide sodium microspheres: Evaluation of their effectiveness in the treatment of bacterial keratitis caused by *Staphylococcus aureus* and *Pseudomonas aeruginosa* in a rabbit model. *Eur. J. Pharm. Biopharm.* 72, 487–495 (2009).

37. M. de la Fuente, B. Seijo and M. J. Alonso. Bioadhesive hyaluronan-chitosan nanoparticles can transport genes across the ocular mucosa and transfect ocular tissue. *Gene Therapy* 15, 668–676 (2008).

38. V. Baeyens, O. Felt-Baeyens, S. Rougier, S. Pheulpin, B. Boisrame and R. Gurny. Clinical evaluation of bioadhesive ophthalmic drug inserts (BODI) for the treatment of external ocular infections in dogs. *J. Controlled Release* 85, 163–168 (2002).

39. M. A. Kalam. The potential application of hyaluronic acid coated chitosan nanoparticles in ocular delivery of dexamethasone *Int. J. Biol. Macromol.* 89, 559–568 (2016).

22 Adhesives in Dentistry

Erdem Özdemir

CONTENTS

22.1 INTRODUCTION

Adhesives in dentistry improved after the 1990s, and adhesive luting resins could be found in dental markets after 2003. Before adhesive technology in dentistry, conventional cements were being used, such as polycarboxylate cement, glass ionomer cement, and zinc phosphate cement. Today, these conventional cements are still being used for crown cementation and under the fillings (especially amalgam). The conventional cements are used in dentistry as insulators between dentin tubules and crowns/filling surfaces.

Before the introduction of adhesive dentistry, the *extension for prevention* rule was in effect to give stability and retention to fillings in dental cavities. Following the principle of this rule, dentists extended the cavity to make the filling more stable. These extended cavity walls are usually non-deformed dentin walls, and this wider cavity makes teeth more fragile. Following the development of adhesive systems in dentistry, the extension for prevention rule was stopped from being used. However, amalgam material, which has a black color, was being used as a filling material in dental cavities within this rule. Patients want dentists to restore their teeth not only anatomically and functionally, but also esthetically. The development of adhesive dentistry also decreased the use of amalgam. Today, the *minimally invasive* rule is used instead of extension for prevention; with the improvements to adhesive systems, tooth-colored restorations are more popular than amalgam due to these new systems.

Adhesive systems were classified as *etch-and-rinse*, *self-etch*, and *glass ionomer* at the end of the 1990s [1], but at the beginning of the year 2000, the self-adhesive system was developed as a new adhesive luting resin, and today, glass ionomer cement is not used as an adhesive resin cement.

22.2 VARIOUS ADHESIVE SYSTEMS USED

22.2.1 ETCH-AND-RINSE SYSTEM

The etch-and-rinse system involves three steps. These are: acid-etching, priming, and bonding [2,3]. Acid-etching, which demineralizes the dentin surface, increases the surface free energy of dentin and also makes the dentin wall microporous. An increased surface area is a prerequisite for successful bonding [3]. The primer is composed of hydrophilic monomers, thus allowing the adhesive agent to bond to the wet dentin surface. The bonding, which is used prior to the resin cement application, is the hydrophobic step (the step that is not consistent with water) of the adhesive application.

TABLE 22.1

Definition of Adhesive Systems According to Application Steps

	Three-Step (Acid-Etching, Priming, Bonding)	Two-Step (Acid-Etching and Priming, Bonding)	One-Step (All-in-One Bottle)(Acid-Etching, Priming, and Bonding)
Etch-and-rinse	+	+	−
Self-etch	−	+	+
Self-adhesive (requires no application step)	−	−	−

Adhesive systems can achieve bonding to dentin in two different ways: by removing or modifying the smear layer. Etch-and-rinse adhesive systems require an etching step where application of the phosphoric acid at a concentration of 32%–37% is required and then rinsed away; the acid application step removes the smear layer. Etch-and-rinse adhesive systems may consist of two or three steps (Table 22.1). This depends on whether or not the priming and the bonding steps are combined.

Adhesion to tooth structure occurs by replacing inorganic tooth material with resin. This exchange process occurs via two mechanisms. One is the removal of calcium phosphate from dentin or enamel, which results in microporosity (as mentioned) [1]. The second is hybridization, which refers to the penetration of the resin luting agent into the microporous dentin surface. This incorporation mechanism between the resin luting agent and dentin walls denotes interlocking; in other words, it produces a hybrid layer (Figure 22.1). This hybridization can be obtained through *resin tags* [4,5]. Following the adhesive resin application procedure (priming and bonding), the resin material spreads into the cavity made by the acid-etching procedure and resin tags occur. These are the parts of hybrid layer that have an important role in adhesive dentistry (Figure 22.2). The complicated three-step adhesive procedure of the etch-and-rinse adhesive system makes it more technically complex than other systems. The three separate steps make adhesive application technically more difficult and sensitive.

As a result of the polymerization of the resin bonding agent in the dentin tubules, macro- and microresin tags are created. Besides the technical complexity of the etch-and-rinse systems, the best resin tags can be formed by this system. However, successful adhesion between the dentin and adhesive resin depends on good interlocking, without detachment or polymerization shrinkage [4].

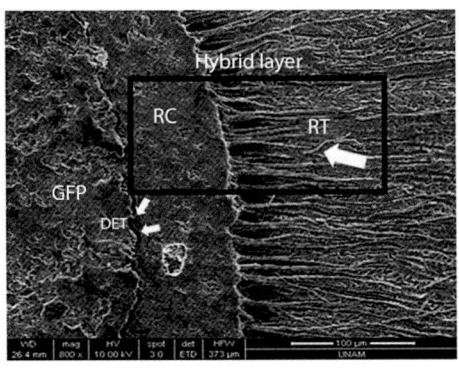

FIGURE 22.1   SEM micrograph of hybrid layer. RC: resin cement; RT: resin tag.

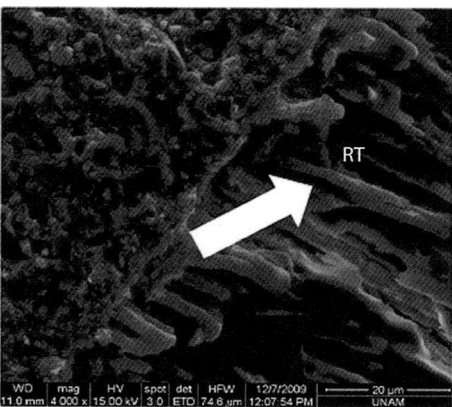

FIGURE 22.2 SEM micrograph of resin tag (RT).

22.2.2 SELF-ETCH ADHESIVE SYSTEM

This approach does not require the acid-etching step and is beneficial not only due to shortening of the application time, but also because of reduction of the risk of mistakes during application and technical sensitivity. The self-etch adhesive system, which only modifies the smear layer, may have one or two steps (Table 22.1) [1,6]. In the two-step system, the etching and priming steps are combined and the bonding step is separate, whereas one-step self-etching adhesives require all the agents (namely, etchant, primer, and bonding agent) combined in a single bottle [7,8].

Depending on the aggressiveness of the etching, the self-etch adhesives can be subdivided into strong and mild. Strong self-etch adhesives, which have a deep demineralization effect, usually have a pH value of 1 or below [1]. In this condition, the dentin collagen is exposed and a large part of the hydroxyapatite is dissolved. The strong self-etch adhesive system is similar to the etch-and-rinse adhesive system, in which the retentive property basically depends on penetration into the dentin tubules (Figure 22.3).

Mild self-etch adhesive systems demineralize the dentin to approximately 1 μm and have a pH value of around 2 [1]. This surface demineralization occurs only partially, keeping the residual hydroxyapatite attached to the collagen. Nevertheless, surface porosity is created to obtain micromechanical interlocking through hybridization. The thickness of the hybrid layer is less than that produced by the strong self-etch or etch-and-rinse approaches (Figure 22.4). Mild self-etch systems

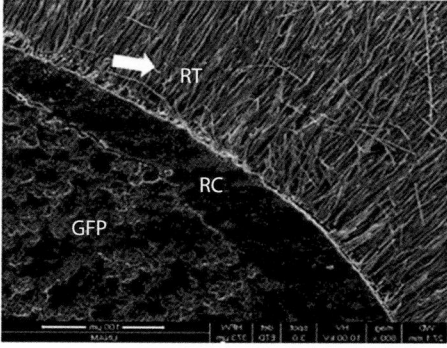

FIGURE 22.3 The view of long and numerous resin tag formations in strong acid-etched surface. RC: resin cement; RT: resin tag; GFP: glass fiber post.

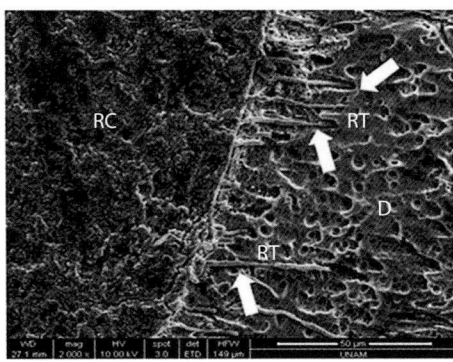

FIGURE 22.4   Hybrid layer of a mild self-etch adhesive resin (Panavia F, Kuraray). RT: resin tag; RC: resin cement; D: dentin.

have been shown to be of minor importance with regard to actual bonding effectiveness [9,10]. Table 22.2 shows mild and strong self-etch resin products according to their pH levels.

The mild self-etching effect is mandatory to (1) deal with the smear layer resulting from cavity preparation, (2) achieve micromechanical interlocking within the etched pits in the enamel, and (3) achieve shallow micromechanical interlocking through hybridization in the dentin [1]. Micromechanical retention is thought to be necessary to resist acute debonding forces. The exposed hydroxyapatite enamel surface and the hydroxyapatite crystals that surround the collagen (in the case of a mild self-etching) are particularly advantageous. They have an intimate chemical interaction with the functional monomers on a molecular level and may help to prevent or retard marginal leakage. The challenge is to have the functional monomers interact with the hydroxyapatite, so that the resulting calcium carboxylate or calcium phosphate bonds are stable in a hydrophilic environment for the long term [1].

Clearfil esthetic cement and Panavia should be treated separately from the other adhesive resin cements. They are mild self-etch adhesive cements and contain the phosphate-based functional monomer 10-methacryloxydecyl dihydrogen phosphate (10-MDP) [4,5]. This molecule can interact with the hydroxyapatite left around the collagen within the hybrid layer, and it has been reported that this interaction is important for long-term bonding stability. The low solubility of the MDP calcium salt in water causes the MDP–hydroxyapatite bonding to be stronger [11,12].

TABLE 22.2
pH Values of Some Self-Etch Adhesive Resins

Adhesive	Classification	Primer pH
AdperPrompt L-Pop (3M ESPE)	Self-etch	0.4
Xeno III (Dentsply)	Self-etch	1.4
I-Bond (Kulzer)	Self-etch	1.6
AdheSE primer (Vivadent)	Self-etch	1.4
OptiBond Solo Plus SE primer (Kerr)	Self-etch	1.5
Clearfil SE Bond primer (Kuraray)	Self-etch	1.9
Unifil Bond primer (GC)	Self-etch	2.2
Panavia ED primer (Kuraray)	Self-etch	2.6

22.2.3 Self-Adhesive Resin Cements

In 2002, a self-adhesive resin luting cement (RelyX Unicem, 3M ESPE, St. Paul, MN) was introduced to eliminate the technical sensitivity of resin luting cements that may use etch-and-rinse and self-etching adhesives. The mechanism of the etch-and-rinse technique involves the removal of the smear layer on the dentin surface and causes the dentin surface to be more penetrable by the resin luting agent. This mechanism dissolves the hydroxyapatite and the smear layer using 37% orthophosphoric acid. On the other hand, the self-etching resin mechanism does not dissolve the hydroxyapatite crystals and just modifies the smear layer. The two mechanisms have been compared many times by different researchers, who have expressed different views. However, logically and clinically, the main criterion for selection of the adhesive material should be based on the sensitivity of the tooth after application and long-term maintenance. Scotti et al. [13] compared a three-step etch-and-rinse adhesive system and a two-step self-etching adhesive system in an *in vivo* study (vital teeth). They compared dentin sealing and postoperative sensitivity on use of the two different adhesive systems. According to their results, on medium-deep cavities, both systems (self-etch and etch-and-rinse) have effective dentin sealing capability and prevent the teeth from postoperative sensitivity. In another study, Kensche et al. [14] compared shear bond strengths of both etch-and-rinse and self-etch adhesive systems and, according to their results, both adhesive systems have acceptable results for long-term clinical usage. Some commercial dental adhesive resin cements are listed in Table 22.3.

Unlike the abovementioned adhesive systems (self-etch/etch-and-rinse), self-adhesive resin luting cements do not need any prior steps to achieve adhesion. With acidic and hydrophilic monomers in their compositions, the enamel and dentin can be demineralized and micromechanical bonding can be achieved [15]. The adhesion mechanism of self-adhesive resin luting cements depends on the micromechanical retention and chemical adhesion to the hydroxyapatite. However, the secondary chemical mechanism should be responsible for high moisture tolerance. In transmission electron microscopy (TEM) studies, less demineralization and hybridization of dentin and no hybrid layer were found when compared with the etch-and-rinse and self-etch resin luting adhesive cements [16,17]. Although less penetration and no resin tags were seen in dentinal tubules with self-adhesive resin luting cements, higher bond strengths were found by Bitter et al. [18] (Figure 22.5).

The easy application of self-adhesive resin luting cements makes them more popular in clinical use. Clinically, adhesive resin cements are usually used for (partial or full) crown cementation, postcementation, or adhesive resin-bonded fixed dentures cementation. In the use of etch-and-rinse or self-etch resin luting cements, more steps are needed, especially for postcementation into root canals, as the application of etching and bonding steps in root canals is more difficult. The bonding of posts into root canals with adhesive cements is also affected by anatomical differences, the orientation of the dentinal tubules at different levels of the root canal, the use of Eugenol-containing sealer ($C_{10}H_{12}O_2$—a type of phenol that is used for analgesic and local anesthetic) at the time of endodontic treatment, the cavity configuration of the root canal, and the hydration degree of the root canal dentin [19,20].

Self-adhesive resin cements do not remove the smear layer and the main adhesion mechanism of a self-adhesive resin cement is attributed to a chemical reaction between phosphate methacrylates and hydroxyapatite [21].

Minimal postoperative sensitivity is expected after using this type of cement. Self-adhesive resin cements, which are produced by 13 different companies, have a confusing chemical composition [22]. Etch-and-rinse resin luting cements use an acid-etch system, as mentioned, and make the dentin surface microporous. On the other hand, self-etch adhesive resin luting systems do not have a separate etching system, but rather etching agents combined with primers, which results in modification of but not removal of the smear layer.

However, self-adhesive resin cements do not require any pretreatments of the dentin. Phosphoric acid groups combined with multifunctional monomers demineralize the dentin; thus, they can

TABLE 22.3

Brand Names, Names of Suppliers, and Chemical Compositions of Current Commercial Adhesive Resins

Material/Manufacturer	Composition
RelyX Unicem/3M ESPE, St. Paul, MN	Powder: glass powder, initiator, silica, substituted pyrimidine, calcium hydroxide, peroxy compound, pigments
	Liquid: methacrylate phosphoric ester, dimethacrylate, acetate, stabilizer, initiators (pH 2.1)
Maxcem Elite/Kerr Corp, Orange, CA	Methacrylate ester, monomers, nonhazardous inert mineral fillers, ytterbium fluoride, activators, stabilizers, colorants (pH 2.5)
Multilink Sprint/Ivoclar Vivadent, Schaan, Liechtenstein	Base and catalyst: pastes of dimethacrylates (24%–26%), inorganic fillers, ytterbium trifluoride, initiators, stabilizers and pigments (5% methacrylate phosphoric acid ester) (pH 2.8)
Rely X U200/3M ESPE, St Paul, MN	Base paste: Methacrylate monomers containing phosphoric acid groups, methacrylate monomers, silanated fillers, initiator components, stabilizers, rheological additives
	Catalyst paste: Methacrylate monomers, alkaline (basic) fillers, silanated fillers, initiator components, stabilizers, pigments, rheological additives
BiFix SE/Voco, Cuxhaven, Germany	Bis-GMA, UDMA, Gly-DMA, phosphate monomers, initiators, stabilizers, glass fillers, aerosol silica (filler = 70 wt%; 61 vol%.; avg = 2 μm)
Clearfil SA Cement/Kuraray, Osaka, Japan	10-MDP, hydrophobic aromatic dimethacrylate, hydrophobic aliphatic dimethacrylate, colloidal silica, barium glass (filler = 66 wt%; 45 vol%.; avg = 25 μm)
RelyX ARC/3M ESPE, St Paul, MN	Paste A: Silane-treated ceramic, Bis-GMA, TEGDMA, photoinitiators, amine, silane-treated silica, functionalized dimethacrylate polymer
	Paste B: Silane-treated ceramic, Bis-GMA, TEGDMA, silane-treated silica benzoyl peroxide, functionalized dimethacrylate polymer
Panavia F/Kuraray, Osaka, Japan	Paste A: 10-MDP, hydrophobic aromatic dimethacrylate, hydrophobic aliphatic dimethacrylate, hydrophilic dimethacrylate, silanated silica, photoinitiator, dibenzoylperoxide
	Paste B: hydrophobic aromatic dimethacrylate, hydrophobic aliphatic dimethacrylate, hydrophilic dimethacrylate, sodium aromatic sulfinate, accelerator, sodium fluoride, silanated barium glass
Variolink II/Ivoclar Vivadent, Schaan, Liechtenstein	Matrix: Bis-GMA, UDMA, TEGDMA
	Inorganic fillers: Barium glass, ytterbium fluoride, Ba-Al fluorosilicate glass, spheroid mixed oxide, catalysts, stabilizers, pigments
Calibra/DentsplyDeTrey GmbH, Konstanz, Germany	Base: Ba-B fluoroaluminosilicate glass, bisphenol A diglycidyl methacrylate, polymerizable dimethacrylate resin, hydrophobic amorphous fumed silica, titanium dioxide, dl-camphoroquinone
	Catalyst: Ba-B fluoroaluminosilicate glass, bisphenol A diglycidyl methacrylate, polymerizable dimethacrylate resin, hydrophobic amorphous fumed silica, titanium dioxide, benzoyl peroxide

penetrate into the dentin tubules. Following polymerization of self-adhesive resin luting cements, extensive cross-links of cement monomers are obtained and high molecular weight polymers are created. Additionally, after neutralization of the acidic system, a glass ionomer system is applied, resulting in a pH increase from 1 to 6 through reactions between phosphoric acid groups and alkaline filler [23]. Phosphoric acid groups also react with the tooth apatite. During this neutralization process, water is formed and exposed in the dentinal tubules; normally, water causes the bonding of resin luting cement to dentin to be difficult for classical resin cement types. However, self-adhesive

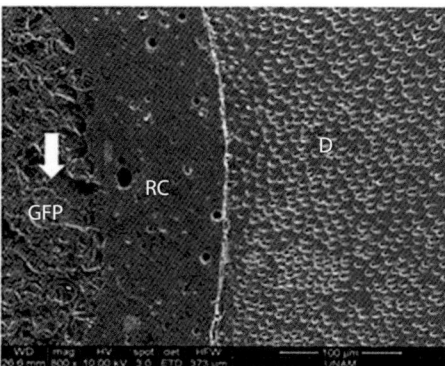

FIGURE 22.5 View of no resin tag but successful bonding between resin luting cement and dentin. RC: resin cement; RT: resin tag; D: dentin; GFP: glass fiber post.

resin luting cements' properties of hydrophilicity and moisture tolerance provide improved adaptation to the tooth structure. Subsequently, water is expected to be reused via reaction with acidic functional groups and during the cement's reaction with ion-releasing basic filler particles [24]. Such a reaction finally results in a shift to a hydrophobic matrix. The adhesion obtained is claimed to rely on micromechanical retention and chemical interaction between monomer acidic groups and hydroxyapatite.

A number of studies have been conducted on adhesion to dentin, predominantly using RelyX Unicem, but also with other self-adhesive resin cements [8,25,26,27]. RelyX Unicem presents both a new methacrylate monomer formulation and technique to initiate polymerization in an acidic environment. These methacrylate monomers contain phosphoric acid esters that simultaneously demineralize and infiltrate into both the smear layer and the underlying dentin, providing micromechanical bonding [28]. The potential for chemical interaction between RelyX Unicem and hydroxyapatite was investigated, and a strong chemical interaction with calcium derived from hydroxyapatite was confirmed. Hence, it was proven that this adhesion strategy relies not only on micromechanical retention (resin tag free), but also on chemical interactions between monomer acidic groups and hydroxyapatite (Figure 22.5) [28].

Based on optical micrographs of cement–dentin interfaces stained with Masson's trichrome, Monticelli et al. [8] found no signs of demineralization at the cement–dentin interface of RelyX Unicem, G-Cem, and Bis-Cem (Figure 22.6). From these images, one can conclude that these cements were capable of demineralizing/dissolving the smear layer completely. Similarly, De Munck et al. [26] revealed that although the pH of RelyX Unicem is very low (<2 during the first minute), almost no demineralization of the dentin surface was noticed. According to evaluation by field-emission scanning electron microscopy (FE-SEM), RelyX Unicem only interacted on the surface of the underlying dentin. Neither a hybrid layer nor resin tags were observed [26] (Figure 22.7). The authors reported that almost no demineralized dentin with relatively high viscosity was obtained.

22.3 COMPARISON OF ADHESİVE SYSTEMS USED IN DENTISTRY

In the last decade, these newly introduced adhesive systems (self-adhesives) have gained popularity because of their easy application process, as mentioned. Since these cements do not necessitate pretreatment of the dentin for adhesion, penetration into and interaction with the underlying dentin are not necessary. Their bond strength to dentin was evaluated and compared with other adhesive systems in many studies. Some of them have reported that self-adhesive cements promote adequate bond strength to dentin [25,26–30], whereas others have reported low bond strength of these materials to dentin when they were compared with conventional

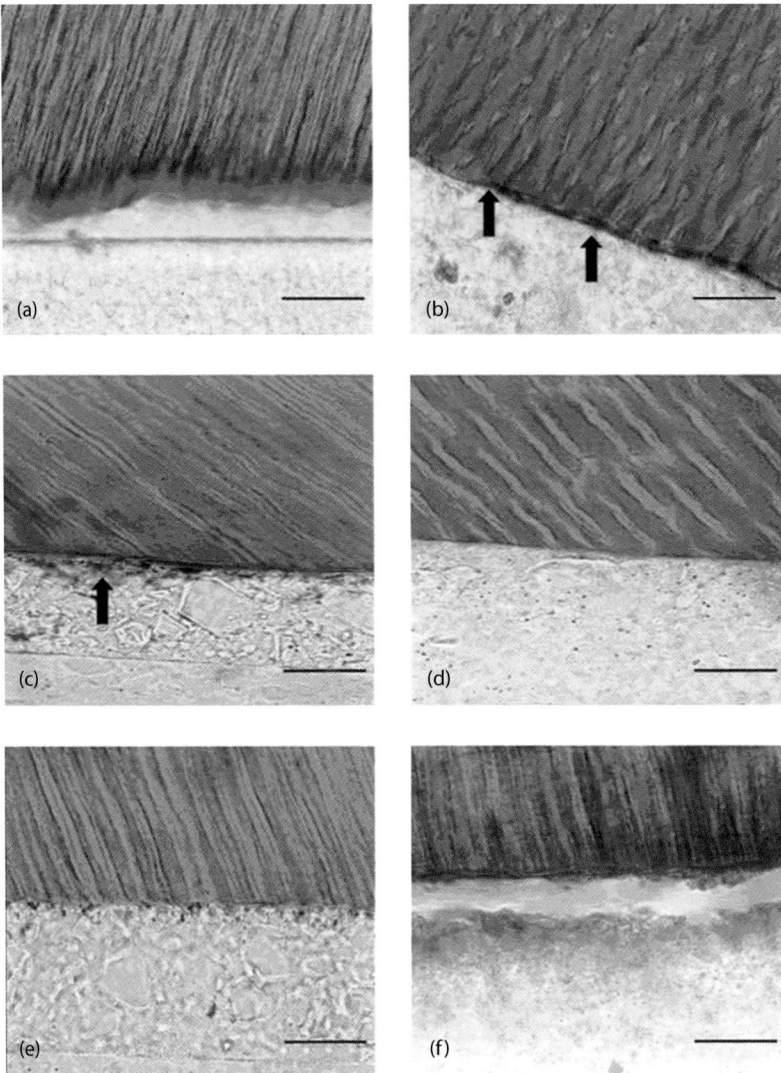

FIGURE 22.6 Representative optical micrographs of cement/dentin interfaces stained with Masson's trichrome: mineralized dentin (green), resin cement (clear with filler particles), exposed protein (red). (a) A distinct red zone of exposed protein was identified in the sections recovered from specimens etched with phosphoric acid (Calibra). (b, c) Representative collagen partially reacted with resin cement is detectable at the interface between dentin and self-etching primer (b; Panavia F 2.0) or Multilink Sprint (c) (arrows). (d–f) No signs of demineralization and/or exposure (red stain) are detectable at the cement–dentin interface of Rely X Unicem (d), G-Cem (e), and Bis-Cem (f). Magnification, 100×; bar = 10 μm. (From Monticelli, F. et al., *J. Dental Res.*, 87, 974–979, 2008.)

resin cements [31,32]. Hikita et al. [25] showed that etch-and-rinse, self-etch, and self-adhesive resin cements were equally effective in bonding when a correct adhesive procedure was carried out. The authors advised avoidance of dentin etching before using RelyX Unicem [25]. Prieto et al. [33] found similar microtensile bond strength values for RelyX Unicem and RelyX ARC. Similarly, in two different studies, no statistically significant difference was observed between RelyX Unicem and Panavia F [26,27]. Microtensile bond strength values for RelyX Unicem and Panavia F are shown in Table 22.4 [26]. Importantly, the researchers advised that RelyX Unicem should always be applied with some pressure to ensure that the relatively high-viscosity

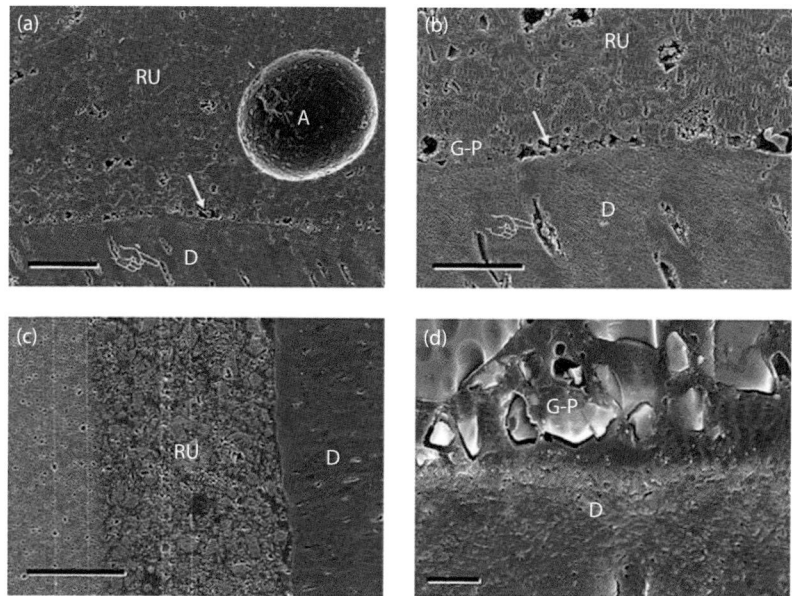

FIGURE 22.7 FE-SEM micrographs of RelyX Unicem bonded to bur-cut dentin. (a) RelyX Unicem (RU) applied to a flat surface. At dentin (D), tubules were not obturated by resin tags (hand). Within the RelyX Unicem cement and especially at the interface with dentin, many porosities can be detected (arrow). A: Air bubble resulting from mixing. (b) Higher magnification of (a), clearly showing porosities at the interface (arrow). Within the cement, glass particles (G-P) as well as voids may be noted. (c) FE-SEM overview of RelyX Unicem (RU)–dentin (D) interface (inlay cemented under pressure). The cement is approximately 80 μm thick, packed with glass particles ranging in size from 1 to 5 μm. No voids can be observed at the interface with the dentin. (d) Higher magnification of (c), though for RelyX Unicem dentin, no distinct morphological manifestation of interaction with unaffected dentin can be observed. (From De Munck, J. et al., *Dental Mater*, 20, 963–971, 2004.)

cement intimately adapts to the cavity wall [26]. Moreover, Walter et al. [29] showed enhanced microtensile bond strength for RelyX Unicem, followed by Panavia F and FujiCEM. In this study, the specimens were tested after storage in distilled water for 24 h. However, it is known that with time, bonding to restorative materials as well to tooth structures may undergo hydrolytic degradation, resulting in bond failure [34]. However, Abo-Hamar et al. [35] showed that thermocycling did not influence the bond strength of RelyX Unicem to dentin. On the other hand, Holderegger et al. [36] showed that RelyX Unicem exhibited the lowest shear bond strength to

TABLE 22.4
Microtensile Bond Strength Results of Different Adhesive Systems

	RelyX Unicem (Self-Adhesive)	Panavia F (Self-Etch)
Mean	15.9	17.5
SD	3.9	5.9
N	11	10
Ptf	1	0

Source: De Munck, J. et al., *Dental Mater.*, 20, 963–971, 2004.

Note: SD: standard deviation; N: total number of specimens prepared; Ptf: pretesting failures ($p < 0.05$, Scheffe multiple comparisons).

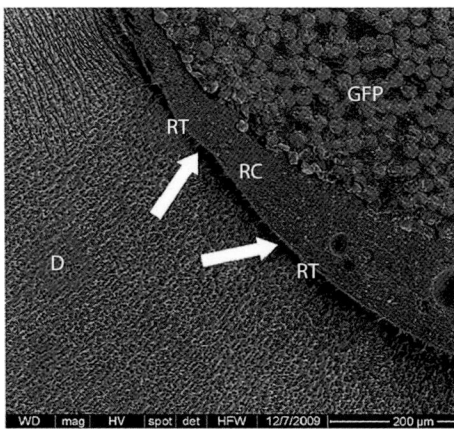

FIGURE 22.8   The view of the detachment between resin cement and dentin because of polymerization shrinkage. D: dentin; RT: resin tag; RC: resin cement; GFP: glass fiber post.

dentin compared with generally used conventional resin cements such as RelyX ARC, Multilink, and Panavia 21. However, they stated that the bonding results of this self-adhesive resin cement were more reliable and least influenced by operator skill and the aging process. The authors explained the lower influence of aging on RelyX Unicem compared with other resin cements as due to the absence of a priming solution and no distinct hybridization of the dentin. The relatively low bond strength values for RelyX Unicem in the study mentioned [36] were explained by the cement being tested in the self-curing setting mode. It is known that dual-curing cements achieve an optimal degree of conversion with additional photoactivation [4,26]. In another study, Özdemir et al. [4] cemented glass fiber posts with different adhesive resin cements, dynamically loaded the samples, and compared microleakage results. According to their study, Panavia F had lower microleakage than other self-adhesive (RelyX Unicem) and etch-and-rinse (RelyX ARC) resin cements. Sailer et al. [31] compared the shear bond strengths of three conventional resin cements (Panavia 21, Variolink, and RelyX Unicem). They reported that the shear bond strength of RelyX Unicem (5.5 ± 2.0 MPa) was lower than that of Panavia 21 (16.3 ± 3.5 MPa) and not significantly different from Variolink (3.9 ± 2.4 MPa).

Besides the technical difficulties in bonding, the *C factor* is another important microleakage issue. High polymerization shrinkage stress of adhesive luting resins can occur in the intraradicular system during polymerization of the luting agent. This is defined as the C factor [4]. This polymerization stress causes detachment of resin luting agents from the dentin walls and results in microleakage. Figure 22.8 shows the detachment between resin luting cement and dentin. This SEM view also depicts unsuccessful bonding to dentin.

22.4 SUMMARY

Adhesive science has brought renewed concepts to dental treatments. Fillings of cavities can be made smaller than previously and cementations (posts, crowns, bridges, etc.) are more reliable and sensible. The chemical connection between adhesive systems and dentin is stronger than in conventional cementation systems. In cases where color is very important, dentists can easily select the color of adhesive resin cement according to the patient's teeth. However, the polymerization shrinkage of adhesive systems, sophisticated application steps (technical sensitivity), and the price of such systems compared with conventional cementation systems are still serious concerns.

REFERENCES

1. V. B. Meerbeek, J. De Munck, Y. Yoshida, S. Inoue, M. Vargas, P. Vijay, K. Van Landuyt, P. Lambrechts and G. Vanherle. Buonocore memorial lecture. Adhesion to enamel and dentin: Current status and future challenges. *Oper. Dentistry* 28, 215–235 (2003).
2. S. N. White, J. A. Sorensen, S. K. Kang and A. A. Caputo. Microleakage of new crown and fixed partial denture luting agents. *J. Prosthet. Dentistry* 67, 156–161 (1992).
3. S. Pavan, P. H. Dos Santos, S. Berger and A. K. Bedran-Russo. The effect of dentin pretreatment on the microtensile bond strength of self-adhesive resin cements. *J. Prosthet. Dentistry* 104, 258–264 (2010).
4. E. Özdemir, S. Erkut, K. Gülsahi, W. S. Lin and H. Oruçoğlu. Influence of dynamic loading and different adhesive systems on the microleakage in root canals. *J. Adhesion Sci. Technol.* 26, 2517–2530 (2012).
5. O. Acar and E. Ozdemir. Bonding strength of self-adhesive resin cements to human dentin: A critical review. *Rev. Adhesion Adhesives* 1, 346–364 (2013).
6. L. Breschi, A. Mazzoni, A. Ruggeri, M. Cadenaro, R. Di Lenarda and E. De Stefano Dorigo. Dental adhesion review: Aging and stability of the bonded interface. Dental Mater. 24, 90–101 (2008).
7. F. Shafiei and M. Memarpour. Effect of chlorhexidine application on long-term shear bond strength of resin cements to dentin. *J. Prosthodont. Res.* 54, 153–158 (2010).
8. F. Monticelli, R. Osorio, C. Mazzitelli, M. Ferrari and M. Toledano. Limited decalcification/diffusion of self-adhesive cements into dentin. *J. Dental Res.* 87, 974–979 (2008).
9. S. Inoue, M. A. Vargas, Y. Abe, Y. Yoshida, P. Lambrechts, G. Vanherle, H. Sano and B. Van Meerbeek. Micro-tensile bond strength of eleven contemporary modern adhesives to dentin. *J. Adhesive Dentistry* 3, 237–245 (2001).
10. J. De Munck, B. Van Meerbeek, S. Inoue, M. Vargas, Y. Yoshida, S. Armstrong, P. Lambrechts and G. Vanherle. Micro-tensile bond strength of one- and two-step self-etch adhesives to bur-cut enamel and dentin. *Am. J. Dentistry* 16, 414–420 (2003).
11. F. Zicari, E. Couthino, J. De Munck, A. Poitevin, R. Scotti, I. Naert and B. Van Meerbeek. Bonding effectiveness and sealing ability of fiber-post bonding. *Dental Mater.* 24, 967–977 (2008).
12. Y. Yoshida, K. Nagakane, R. Fukuda, Y. Nakayama, M. Okazaki, H. Shintani, S. Inoue, Y. Tagawa, K. Suzuki, J. De Munck and B. Van Meerbeek. Comparative study on adhesive performance of functional monomers. *J. Dental Res.* 83, 454–458 (2004).
13. N. Scotti, E. Bergantin, R. Giovannini, L. Delbosco, L. Breschi, G. Migliaretti, D. Pasqualini and E. Berutti. Influence of multi-step etch-and-rinse versus self-etch adhesive systems on the post-operative sensitivity in medium-depth carious lesions: An in vivo study. *Am. J. Dentistry* 28, 214–218 (2015).
14. A. Kensche, F. Dähne, C. Wagenschwanz, G. Richter, G. Viergutz and C. Hannig. Shear bond strength of different types of adhesive systems to dentin and enamel of deciduous teeth in vitro. *Clin. Oral Investig.* 20, 831–840 (2015).
15. B. Dejak and A. Mlotkowski. Finite element analysis of strength and adhesion of castposts compared to glass fiber-reinforced composite resin posts in anterior teeth. *J. Prosthet. Dentistry* 105, 115–126 (2011).
16. P. Stona, G. A. Borges, M. A. Montes, L. H. Junior, J. B. Weber and A. M. Spohr. Effect of polyacrylic acid on the interface and bond strength of self-adhesive resin cements to dentin. *J. Adhesive Dentistry* 15, 221–227 (2013).
17. K. Bitter, J. Perdigão, M. Exner, K. Neumann, A. Kielbassa and G. Sterzenbach. Reliability of fiber post bonding to root canal dentin after simulated clinical function in vitro. *Oper. Dentistry* 37, 397–405 (2012).
18. K. Bitter, H. K. Meyer-Lueckel Priehn, J. P. Kanjuparambil, K. Neumann and A. M. Kielbassa. Effects of luting agent and thermocycling on bond strengths to root canal dentine. *Int. Endodontics J.* 39, 809–818 (2006).
19. S. Bouillaguet, S. Troesch, J. C. Wataha, I. Krejci, J. M. Meyer and D. H. Pashley. Microtensile bond strength between adhesive cements and root canal dentin. *Dental Mater.* 19, 199–205 (2003).
20. M. Vano, A. H. Cury, C. Goracci, N. Chieffi, M. Gabriele, F. R. Tay and M. Ferrari. The effect of immediate versus delayed cementation on the retention of different types of fiber post in canals obturated using an Eugenol sealer. *J. Endodontics* 32, 882–885 (2006).
21. K. Al-Assaf, M. Chakmakchi, G. Palaghias, A. Karanika-Kouma and G. Eliades. Interfacial characteristics of adhesive luting resins and composites with dentine. *Dental Mater* 23, 829–839 (2007).
22. J. L. Ferracane, J. W. Stansbury and F. J. Burke. Self-adhesive resin cements: Chemistry, properties and clinical considerations. *J. Oral Rehabil.* 38, 295–314 (2011).

23. G. Ibarra, G. H. Johnson, W. Geurtsen and M. A. Vargas. Microleakage of porcelain veneer restorations bonded to enamel and dentin with a new self-adhesive resin-based dental cement. *Dental Mater.* 23, 218–225 (2007).

24. K. Nagakane, Y. Yoshida, I. Hirata, R. Fukuda, Y. Nakayama, K. Shirai, T. Ogawa, K. Suzuki, B. Van Meerbeek and M. Okazaki. Analysis of chemical interaction of 4-MET with hydroxyapatite using XPS. *Dental Mater. J.* 25, 645–649 (2006).

25. K. Hikita, B. Van Meerbeek, J. De Munck, T. Ikeda, K. Van Landuyt, T. Maida, P. Lambrechts and M. Peumans. Bonding effectiveness of adhesive luting agents to enamel and dentin. *Dental Mater.* 23, 71–80 (2007).

26. J. De Munck, M. Vargas, K. Van Landuyt, K. Hikita, P. Lambrechts and B. Van Meerbeek. Bonding of an auto-adhesive luting material to enamel and dentin. *Dental Mater.* 20, 963–971 (2004).

27. J. Johnson, J. Burgess and M. Blatz. Bond of new resin cements to enamel, dentin, and alumina. *J. Dental Res.* 83, 474–479 (2004).

28. H. U. Gerth, T. Dammaschke, H. Zuchner, and E. Schafer. Chemical analysis and bonding reaction of RelyX Unicem and Bifix composites-A comparative study. *Dental Mater.* 22, 934–941 (2006).

29. R. Walter, P. A. Miguez and P. N. Pereira. Microtensile bond strength of luting materials to coronal and root dentin. *J. Esthet. Restor. Dentistry* 17, 165–171 (2005).

30. J. P. Matinlinna and K. L. Mittal (Eds.). *Adhesion Aspects in Dentistry*, CRC Press, Boca Raton, FL (2009).

31. I. Sailer, A. E. Oendra, B. Stawarczyk and C. H. Hammerle. The effects of desensitizing resin, resin sealing, and provisional cement on the bond strength of dentin luted with self-adhesive and conventional resin cements. *J. Prosthet. Dentistry* 107, 252–260 (2012).

32. R. R. Vaz, V. D. Hipolito, P. H. D'Alpino and M. F. Goes. Bond strength and interfacial micromorphology of etch-and-rinse and self-adhesive resin cements to dentin. *J. Prosthodont.* 21, 101–111 (2012).

33. L. T. Prieto, M. M. C. Humel, E. J. Souza-Junior and L. A. M. S. Paulillo. Influence of selective acid etching on microtensile bond strength of a self-adhesive resin cement to enamel and dentin. *Braz. J. Oral Sci.* 9, 455–458 (2010).

34. F. R. Tay, M. Hashimoto, D. H. Pashley, M. C. Peters, S. C. Lai, C. K. Yiu and C. Cheong. Aging affects two modes of nanoleakage expression in bonded dentin. *J. Dental Res.* 82, 537–541 (2003).

35. S. E. Abo-Hamar, K. A. Hiller, H. Jung, M. Federlin, K. H. Friedl and G. Schmalz. Bond strength of a new universal self-adhesive resin luting cement to dentin and enamel. *Clin. Oral Investig.* 9, 161–167 (2005).

36. C. Holderegger, I. Sailer, C. Schuhmacher, R. Schlapfer, C. Hammerle and J. Fischer. Shear bond strength of resin cements to human dentin. *Dental Mater.* 24, 944–950 (2008).

23 New Adhesive Technologies in the Footwear Industry

Elena Orgilés-Calpena*, Francisca Arán-Aís,
Ana M. Torró-Palau, and César Orgilés-Barceló

CONTENTS

23.1 THE FOOTWEAR INDUSTRY

The manufacturing of shoes can be divided into seven main operating processes [1]: modeling, cutting, stitching, lasting, assembling, finishing, and packaging. The shoe is designed in the modeling stage, which is determined by fashion trends. Materials, colors, molds, and size scales to be produced are well defined from the beginning. The cutting step is followed by the stitching process, where the upper parts are joined together using sewing machines. The next step is lasting and assembling, where the upper-to-sole bonding process is carried out. In this sense, the joining of the upper to the sole is the most exigent process in footwear manufacturing, as it demands high bond strength. The upper includes the vamp, which covers the front, the toe, and the quarters. The sole is the underside of the shoe, comprising the insole, the midsole, and the heel (Figure 23.1).

This chapter mainly deals with upper-sole bonding. In this regard, the adhesion between the upper, the adhesive, and the sole surfaces must be properly optimized to produce adequate joints. Adhesion in shoe bonding can be improved by surface modification of the upper and/or sole materials, by modifying the adhesive formulation, or both.

The proper choice of adhesive is fundamental to ensure the bond strength required by both shoe manufacturers and standards. Nowadays, contact adhesives based on polychloroprene (PCP) and polyurethane (PU) are the most commonly used ones in the footwear industry. Formerly, PCP adhesives were more extensively used in upper-sole bonding, but nowadays PU adhesives are preferred due to their performance and versatility with new fashion materials. Depending on the type of shoe, different bonding performances are required. While, on the one hand, sports or safety shoes are extremely exigent in terms of bond strength, on the other hand, requirements for fashion or casual footwear are not so stringent.

---

* Indicates corresponding author

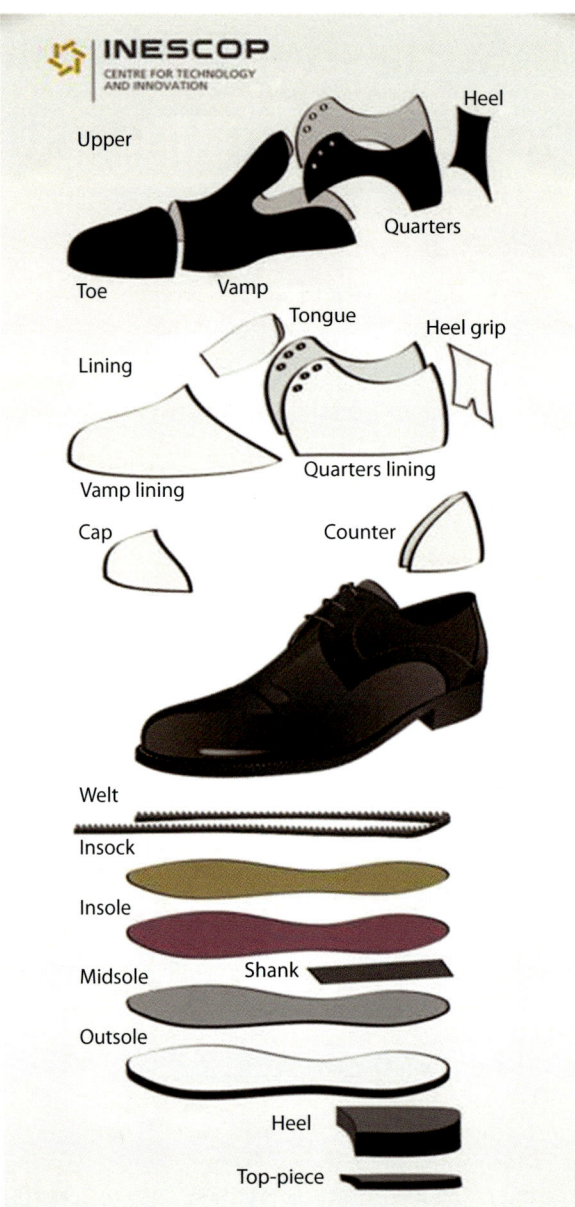

FIGURE 23.1 Main parts of a shoe.

23.1.1 Substrates and Adhesives

A wide range of materials are used in the footwear industry. Even every season new fashion materials appear, which makes it particularly difficult to standardize the adhesive bond formation, and adhesive formulations must be continuously adapted to new materials' features. Among upper materials, chrome- and vegetable-tanned leathers are the most common materials used in shoe manufacturing. The porous nature of leather facilitates adhesive bonding, but its weak grain layer must be removed by roughening. Moreover, good adhesion in thin or grassy leathers can be obtained by applying a primer. Furthermore, several synthetic upper materials such as canvas, textiles, nylons, and so forth, are used with prior roughening and/or solvent wiping followed by the application of two consecutive adhesive layers to produce adequate joints.

Outsoles can be made of leather, natural (crepe rubber) or synthetic polymers, cork, and so on. The synthetic polymer materials used for the manufacture of footwear outsoles are based on PU, thermoplastic polyurethane (TPU), thermoplastic rubber (TR), styrene–butadiene rubber (SBR), styrene–butadiene–styrene rubber (SBS), ethylene–vinyl acetate (EVA), polyamide (PA), poly(vinyl chloride) (PVC), polystyrene (PS), or acrylonitrile–butadiene–styrene (ABS). However, PS and ABS materials are mainly used for heels. The selection of these materials will depend on their mechanical properties, price, and design suitability, thereby determining the strength, quality, and comfort desired for the final product.

In regard to adhesives for the footwear industry, PCP adhesives show good results with leather, textiles, and other materials, such as vulcanized rubbers. However, the emergence of new fashion and plastic materials for the footwear industry in the 1970s containing large quantities of plasticizers made it necessary to search for alternative adhesives. For this reason, PU adhesives were introduced, which are currently the most common adhesives used for upper-sole joints due to their high versatility and performance [2].

PCP and PU adhesives are mainly used for highly demanding joints such as the upper-sole bond. Furthermore, other adhesives based on synthetic and natural polymers, such as styrene–isoprene–styrene (SIS), SBS, SBR latex, hot-melts (polyamide, EVA-based), and so on, are used in the different footwear operations. The most suitable adhesive is selected according to the technical requirements of each operation.

In the case of PU adhesives, after being applied and dried, they must be activated via temperature (adhesive softening) to acquire tack to allow the joint of both adhesive layers to be formed. Then, the substrates are bonded by pressing the materials and the adhesive joint is cooled down in seconds, although 24 h are required for curing the adhesive. After a few minutes, the crystallization of the PU adhesive occurs and the curing process is almost complete. When an isocyanate-based cross-linker is added to a PU adhesive (a two-component [2K] adhesive), the curing process is irreversible, which means that the adhesive is permanently cross-linked and the adhesive joint durability increases greatly.

23.1.2 Surface Treatments

Certainly, the joining of the upper to the sole is the most demanding operation in the footwear industry. As for any adhesive joint, the adhesion between the upper, the adhesive, and the sole surfaces must be properly optimized to produce adequate joints. In this sense, most of the upper and sole materials cannot be directly joined using current adhesives, due to their intrinsic low surface free energy or the presence of contaminants, for example. Therefore, a surface preparation is required. In this sense, there are four types of treatments usually employed in the footwear industry: physical, chemical (halogenation), primer application, and solvent wiping.

Considering that most of the upper materials are porous and the adhesives used in shoe bonding are applied in liquid form, acceptable penetration of the adhesive into the upper is expected after roughening and, consequently, mechanical adhesion is produced. However, rubber soling materials are nonporous and have, in general, a relatively low surface free energy; therefore, both mechanical (e.g., roughening) and chemical (e.g., halogenation) surface preparations are necessary.

The following objectives must be reached through the application of a surface treatment:

- Removing contaminants and preventing migration to the surface of substances that may cause generation of weak boundary layers on the substrates
- Increasing the degree of molecular contact between the adhesive and the substrate in the bonding process
- Generating a specific surface topography on the substrate
- Protecting the substrate surface prior to joining
- Preventing a poor joint and reducing the risk of material surface delamination

Surface treatment with solvents favors sole–adhesive interaction and can be applied by wiping, spraying, or immersion, or in an ultrasonic bath. In some cases, or when they have been stored for a long period, two consecutive applications of solvent are necessary for very contaminated soles. After treatment, residual solvent must be removed by evaporation in the open air for at least 30 min.

Roughening of footwear materials removes surface contaminants (oils, mold-release agents, anti-adhesion moieties, etc.) and creates roughness. After roughening, residues must be removed by solvent wiping or pressurized air. The adhesive must be applied on freshly roughened surfaces to avoid further contamination or oxidation of the surface. The removal of weak boundary layers and the creation of surface heterogeneities improve the strength of upper-sole joints [3].

Finally, as for chemical surface treatments, halogenation and cyclization (fumaric acid, sulfuric acid) are the most common treatments for sole materials. However, chlorination is the most effective in improving the adhesion of rubbers, because it is cheap and easy to apply. Solutions of tri-chloroisocyanuric acid (TCI) in different organic solvents are commonly employed in the footwear industry to improve the adhesion of rubber soles to PU adhesive. Moreover, the application of TCI solution on a previously roughened rubber enhances the effects due to chlorination [4,5].

However, the growing awareness of the importance to reduce the emissions of volatile organic compounds (VOCs) has forced the footwear industry to devote a major part of its research efforts to this matter. Due to health and environmental issues in the footwear industry, solvent-free surface treatments are a feasible alternative, as they are neither flammable nor toxic. In this sense, to avoid the use of organic solvents in the chlorinating solutions, several water-based chlorinating treatments have been investigated [6,7], such as chlorine bleach, inorganic chlorine compounds, organic chlorine donors or salts, and so on. The surface modifications produced by treatment of SBS rubber with the aqueous chlorinating agents are comparable to those produced by the current solvent-based chlorinating treatment (3 wt%. TCI/methyl ethyl ketone [MEK]) [4].

23.1.3 Adhesives Requirements

In the footwear industry, more specifically in the upper-sole bonding process, the proper choice of adhesive is important to ensure the bond strength required by the standards. In this sense, two important mechanical properties are controlled: the peel strength and the creep strength. These properties are standardized by European standards (EN) such as EN 1392 [8] and EN 17708 [9], and the technical requirements for the different types of footwear regarding upper-sole bondability are defined in EN 15307 [10].

Performance requirements for footwear materials to assess the suitability for different end uses (sports, casual, town, fashion, infants', indoor, etc.) are established in *Technical Report ISO/TR 20879* [11]. This report also establishes the test methods to be used to evaluate compliance with requirements.

T-peel tests are the most commonly used to analyze adhesive properties in shoe bonding. A minimum of 24 h are required for curing the adhesive, 72 h being the optimum time. Two rectangular pieces of substrate are peeled off at a rate of 100 mm/min and the strength should be expressed in kilonewtons per meter, although newtons per millimeter are also used. The strength of the upper-sole bond must meet minimum specifications according to reference adhesion values depending on the type of shoe, as shown in Table 23.1.

To ensure the durability of the upper-sole joint, it is necessary to determine the peel strength after aging under extreme conditions of temperature and humidity according to EN 15062 [12]. In these cases, the upper-sole bond should retain at least 80% of its initial strength.

Furthermore, it is important to consider not only the T-peel strength value, but also the way in which the separation occurs; that is, the appearance of the separated surfaces. The type of failure can provide useful information about the performance of the adhesive joint or which parameter should be improved (e.g., material cohesion, heat activation, pressure, surface treatment, adhesive viscosity, adhesive initial strength, etc.). The loci of failure of the joints (Figure 23.2) are generally expressed as follows:

TABLE 23.1
Minimum Upper-Sole Bonding Requirements in the Footwear Industry According to Standard EN 15307.

Type of Shoe	Initial Peel Strength*	Final Peel Strength**
Class A: Low stress on sole-upper bond in use (infants, indoor, and fashion footwear)	≥1 N/mm	≥2.5 N/mm
Class B: Medium stress on sole-upper bond in use (men's town, women's town, cold weather, and casual footwear)		≥3.0 N/mm or ≥2.5 N/mm if material failure occurs
Class C: High stress on sole-upper bond in use (children and general sports footwear)		≥4.0 N/mm or ≥3.0 N/mm if material failure occurs
Class D: Very high stress on sole-upper bond in use (mountain footwear)		≥5.0 N/mm or ≥3.5 N/mm if material failure occurs

* Initial peel strength 2 min after assembling at 23°C.

** Final peel strength after 4-day storage in standard atmosphere (23°C and 50%RH) according to ISO 554.

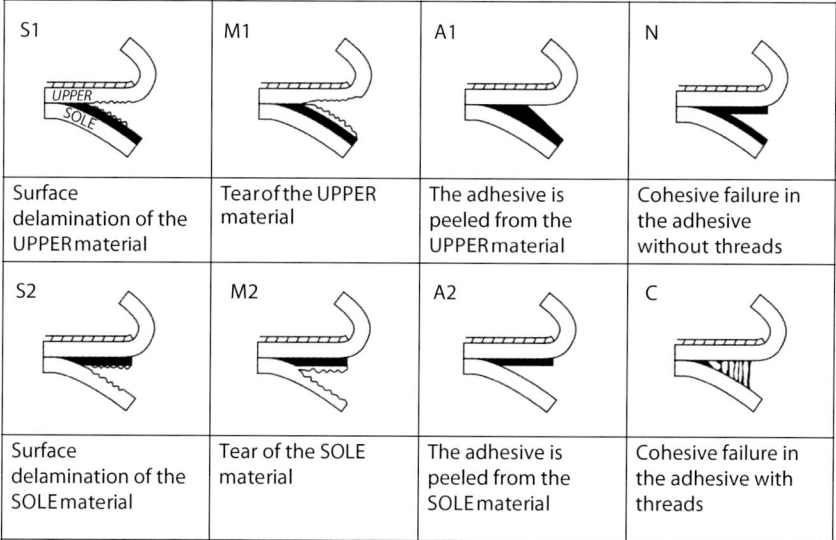

FIGURE 23.2 Types of failures after footwear upper-sole bond separation.

A: Adhesion failure to either substrate.

C: Cohesive failure in the adhesive with threads. This failure is related to immediate adhesion.

N: Cohesive failure in the adhesive without threads. This failure is related to final adhesion (>24 h).

S: Surface failure of either substrate. Delamination of upper or sole material.

M: Tearing through either substrate or deep failure in one the substrates.

23.1.4 FUTURE TRENDS

Future trends in the footwear industry are based on three concepts: sustainability, functionalization, and automatic application of adhesives by robots. These aspects could involve the production of high-added-value and marketable footwear materials to satisfy the needs and expectations of the final consumers.

Sustainable processes and products are a great challenge for the footwear industry. In this sense, water-based and hot-melt adhesives are environmentally friendly, with several advantages compared with solvent-based ones. The current trend is toward the elimination or substitution of organic solvents, both in adhesives and surface treatments. Among the future trends in solvent-free surface treatments for footwear are ultraviolet and/or ozone irradiation [13,14], corona discharge [15], or plasma [16] for modification of rubbers to obtain good final adhesion results. A moderate increase in the peel strength of the joints is produced after these safe rubber treatments, depending on treatment time. The modifications produced by these sustainable processes are comparable to those produced by the solvent-based halogenation treatment; however, their high price compared with halogenation causes their low usage.

Furthermore, the design of polymers from renewable resources is a topic of increasing interest, more specifically with regard to the development of biodegradable materials to reduce the dependence on petroleum and its negative impact on the environment. In this sense, the use of polyols derived from vegetable oils [17–20] for the synthesis of PU adhesives represents a sustainable alternative for the footwear industry. Biobased PU adhesives were obtained from a soy-based polyol [21] and met the requirements established by the standards for footwear regarding upper-sole bondability. Furthermore, the use of carbon dioxide (CO_2) as a feedstock for the chemical industry is an interesting alternative to oil, because it is inexpensive and abundant in the atmosphere. PU reactive hot-melt adhesives were synthesized starting from polyols derived from CO_2 [22,23] with mixtures of polyadipate of 1,4-butanediol and 4,4′-diphenylmethane diisocyanate (MDI), meeting the technical requirements for footwear [24]. This is a new generation of future materials that bring about economic, environmental, and technical advantages [25]. Moreover, nonisocyanate polyurethanes (NIPUs) based on cyclic carbonate oligomers are currently under investigation [26,27] to pursue the trend toward renewable alternatives to petroleum.

In addition, different bonding systems have been found in nature that allow living organisms to adhere firmly to different surfaces, even in highly adverse conditions where other traditional adhesives would present problems. This is the case for geckos, some arachnids, crustaceans, or mollusks, as well as microorganisms. For this purpose, proteins for bioinspired adhesive formulations were studied, such as *Bacillus subtilis*. This protein has been produced in an overexpressing strain, with the aim to assess its involvement in biofilm formation and adhesion mechanisms. The bioadhesion assay showed that the overexpressing strain presented enhanced adhesion to abiotic surfaces. The results confirmed the relationship between the expression of the selected protein and an increase in biofilm adhesion to abiotic surfaces, which led to consideration of the use of this protein as a base polymer in the formulation of bioinspired adhesives [28].

Moreover, the functionalization of adhesives using nanoparticles to provide improved properties and new functions is an emerging technological innovation in the footwear industry. The incorporation of certain additives into the adhesive formulation can also be used to reduce or prevent its degradation. For example, several conditions such as heat, humidity, and long periods of storage make water-based adhesives excellent candidates for microorganism proliferation. Therefore, to adequately protect the waterborne adhesive, silver nanoparticles (AgNPs) may be added [29,30].

Furthermore, to achieve flame retardance, modify thermal conductivity or electrical conductivity, or improve mechanical or hydrophobic properties, a wide range of nanoparticles can be added to adhesive formulations. However, the incorporation of nanoadditives implies a great difficulty, because of their tendency to form aggregates, which prevents efficient dispersion. Therefore, it is necessary to insert functional groups on the surface of the nanoparticles to improve their interaction with the adhesive polymer, as in the case of chemical functionalization of carbon nanofibers (CNFs) with concentrated acid prior to their incorporation into PU adhesives [31]. Moreover, the addition of carbon nanotubes (CNTs) to PU dispersions contributed to the improvement of their rheological, viscoelastic, thermal, and mechanical properties and even imparted some electrical-conductive character to adhesives [32,33].

Finally, the automation of the bonding process through robotics will be one of the big challenges for the footwear industry in the near future. Both hot-melt and water-based adhesives can be applied by robots with several advantages, such as the control of the quantity of adhesive applied, the high quality of the upper-sole adhesive bond, and the increase in the productivity, efficiency, and competitiveness of the industry.

23.2 SUSTAINABLE ADHESIVES FOR THE FOOTWEAR INDUSTRY

23.2.1 WATER-BASED ADHESIVES

Water-based adhesives are dispersions or emulsions of a natural or synthetic polymer in water. They represent an environmentally sustainable alternative to solvent-based adhesives and imply minor changes in the current existing technology in the footwear industry; that is why they are widely used today. They are composed of a base polymer, and some additives are incorporated to achieve suitable properties for a successful application and bonding, such as antioxidants, metal oxides, thickeners [34,35], fillers, plasticizers, and so forth. Generally, their curing mechanism occurs by evaporation of water contained in their formulation; therefore, the environmental conditions may affect the drying time of the adhesive. Nevertheless, these adhesives have a higher price due to a higher solids content than solvent-based ones, although the amount of adhesive applied should be lower. As a result, suitable and similar drying times are obtained, which results in optimum yield and efficiency.

The application procedure for water-based adhesives in the footwear industry is very similar to that of solvent-based adhesives, with the advantage that they can be applied not only by brush but also by gun or roller. Their advantages are as follows:

- Absence of organic solvents in their formulation, which avoids risks of flammability when stored, transported, or applied.
- Improvement of working conditions, as they do not involve environmental hazards. It is not necessary to use exhaust cabins for the application of these adhesives.
- No generation of hazardous waste, thus avoiding its management.
- Higher solids content than solvent-based adhesives, which means better performance.
- Formation of strong and durable bonds that meet the requirements of any type of footwear: children's, men's, women's, leisure, mountain, and so on.

The most important water-based adhesives used in the footwear industry are:

Latex adhesives: Natural or synthetic rubber aqueous dispersions characterized by their easy handling due to their adequate viscosity, tack, and open time. The application of these adhesives by a gun allows adequate dosage in terms of the optimal quantity and uniform application of the adhesive. These adhesives are used in the following steps: stitching of uppers, covering, and folding materials. These joints are soft and simple, and form part of auxiliary manufacturing stages in footwear production.

Furthermore, water-based PCP adhesives may be used in the covering and folding steps, and acrylic and vinylic adhesives may be used for insole attachment.

Water-based PUs: These adhesives are based on PU dispersions. Their structure consists of linear thermoplastic PU chains dispersed in water (Figure 23.3) due to the presence of ionic groups in their structure (i.e., PU ionomer) that act as internal emulsifiers [36]. Waterborne PU adhesives generally have a pH value between 6 and 9, higher solids content (35%–50%), and lower viscosity (about 100 MPa·s) than solvent-based adhesives. They are contact adhesives, so the adhesive must be applied to both substrates, and the water evaporates at room temperature in about 30 min or inside a hot oven (50°C–60°C) in a few minutes. Then, the film adhesive is activated at 80°C by infrared radiation.

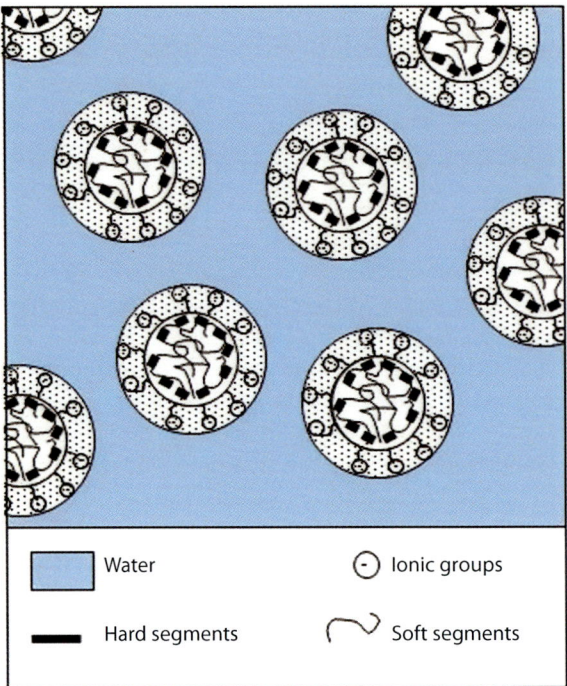

FIGURE 23.3 Ionomeric structure of a PU dispersion. The particles form a stable dispersion in water owing to the external ionic groups, inside of which are soft and hard segments of PU.

Water-based PU adhesives are the ones most used in upper-sole bonding, due to their environmental benefits and because their performance is similar to that of their solvent-based analogues. In the case of PCP adhesives, they are limited by their poor heat resistance and bond strength.

The total global production of PU dispersions was estimated at 290,600 t in 2015 [37], which accounted for an overall increase of about 9.2% since 2012. It is expected to continue to grow at an average rate of 5.6% per annum over the next 5 years, driven by the ever-increasing environmental concerns leading to tightening legislation in the key markets for PU dispersions, including Western Europe, China, and the United States. The second most important driver for growth is that PU dispersions demonstrate excellent performance properties, leading to heightened awareness and a preference among customers for using these dispersions in many applications, specifically in footwear. Nevertheless, a lack of environmental legislation in some emerging and developing economies, and higher prices of PU dispersions compared with alternatives such as acrylic dispersions, are still creating major hurdles to a faster development of the PU dispersions market.

PU dispersions find their widest usage in industrial coatings, occupying a ~45% share of total production; the largest application in this respect is notably in the wood coatings sector, followed by automotive coatings. Leather and textile applications also use substantial quantities of PU dispersions, with Asia showing the highest growth rate. The global production of water-based PU adhesives is forecast to increase in the short term, driven by the footwear industry [37].

23.2.2 Hot-Melt Adhesives

Hot-melt adhesives, popularly known as hot glues, are designed in such a way that they require heating through appropriate application equipment, showing an important advantage of fast processing over other adhesives. Unlike water-based and solvent-based adhesives, these adhesives do not need drying after application. Hot-melt adhesives begin bonding immediately after they are applied, because they cool down to their solidification point immediately. Such a fast solidification feature of hot-melt adhesives is ideal for industrial applications that require a high-speed manufacturing line, bonding versatility, large gap filling, fast green strength, and minimal shrinkage.

Hot-melts are composed of three main components: a high molecular weight polymer that provides the adhesive's main structure and mechanical properties, a resin/tackifier that provides wetting and adhesion properties, and a wax that controls viscosity [38]. The polymer does not provide the optimum required characteristics on its own, and thus it is necessary to incorporate additives to improve the adhesive formulation, such as antioxidants and plasticizers.

The global hot-melt adhesives market can be classified into ethylene acrylic acid adhesives, polyolefins, PUs, amorphous poly-alpha-olefins, EVAs [39,40], styrenic block copolymers, polyamides, metallocenes, polyester adhesives, and others. They are mostly applied by the packaging industry. Other applications of hot-melt adhesives include building and construction, footwear and leather goods, woodworking, tiles and floor adhesive products, bookbinding, clear case bonding, soap and bubble gum wrapper coating, stock and tape manufacturing, bottle labeling, automotive headlights, and so on, among others.

North America has the largest market for hot-melt adhesives, followed by Europe and the Asia-Pacific region. China is the largest consumer of hot-melt adhesive products, owing to a greater demand from the manufacturing industry. North America is expected to maintain its dominance in the foreseeable future. Europe is expected to show average growth. However, the Asia-Pacific region is expected to witness the highest growth in the near future. Increasing numbers of manufacturing units in developing regions such as India are expected to boost the Asian market. The global market for hot-melt adhesives is expected to reach 2,379.9 kt by 2020, growing at an estimated compound annual growth rate (CAGR) of 5.1% from 2014 to 2020. The hot-melt adhesives market is expected to be worth US$7 billion by 2018 [41].

Given that hot-melts do not require drying time, they boast the distinct advantage of very fast processing times. Furthermore, they are considered a best available technology (BAT) for the footwear industry [42]. Hot-melt adhesives are 100% solid at room temperature but become fluid at higher temperatures. They require heating through appropriate application equipment; then, the temperature is reduced and the adhesive solidifies, developing maximum strength. Due to their advantages over solvent/water-based adhesives, they allow quick production, reducing the time taken in the assembly steps. There are important reasons for employing hot-melt adhesives in the footwear industry, such as:

- Ease of application, handling, storage, and transport
- No drying time
- Formation of strong, permanent, and durable bonds within a few seconds after application
- No environmental hazard and minimal wastage, as they are 100% solid
- Absence of highly volatile or flammable ingredients
- Excellent adhesion to a large variety of materials
- Cost-effectiveness

There are two main types of hot-melt adhesives suitable for the footwear industry: nonreactive and reactive hot-melt adhesives. Nonreactive hot-melts are physically curing adhesives (Figure 23.4) that are heated to melting point prior to their application. Because they cure by physical means in one step (the material hardens by passing from the molten to the cold, solid state), they are characterized by high early strength. For the footwear industry, specifically nonreactive hot-melt adhesives based on polyamides are used in folding and lasting machines; polyester hot-melts are employed in lasting machines and EVA hot-melts are used for attaching the heel and insole. Furthermore, stiffeners and toe puffs are used as reinforcing elements of the footwear structure and are made of EVA, polyamide, and thermoplastic PU [43].

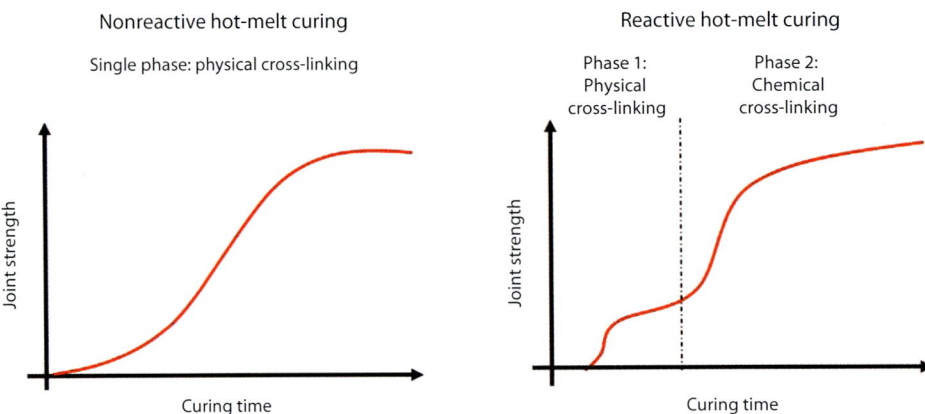

FIGURE 23.4   Joint strength as a function of curing time for nonreactive and reactive hot-melt adhesives.

Reactive hot-melt adhesives are mainly based on PU as a combination of a conventional hot-melt with PU chemistry. Their curing mechanism consists of two stages [44,45]: In the first phase, called *physical cross-linking* (Figure 23.4), a green strength is developed as a result of the cooling process (as also happens with a conventional hot-melt), together with the crystallization of the soft segments that can be found in the polyol component. In the second phase, or *chemical cross-linking*, the isocyanate groups start to react with ambient moisture and/or the moisture of the substrates. As a result, an increase in the molecular weight is produced and a fully reacted polymer, where urethane and urea groups alternate, is obtained after some time. This allows stronger and lasting adhesive bonds that cannot be melted again, as happens with conventional hot-melts.

Although conventional hot-melt adhesives are employed as footwear components, the use of hot-melt PU as an adhesive for the upper-sole bonding stage nowadays poses a great challenge for the footwear industry, in that this is one of the most important and critical adhesive bonds with respect to the final product's technical requirements. In this sense, PU hot-melt adhesives provide some key advantages compared with adhesives currently used in footwear manufacturing, which are solvent and water-based. One of these advantages is their fast processing, because they do not require drying time. Moreover, they are applied to one side and are one-component (1K) adhesives, unlike solvent- and water-based adhesives. Furthermore, their versatility, durability, and good mechanical properties make PU hot-melts good candidates for the footwear industry. The benefits offered by PU hot-melts versus conventional hot-melts are due to their cross-linked structure, which implies adequate strength, excellent resistance, and also excellent adhesion to a wide range of materials.

The following variables should be considered when PU hot-melts are applied to footwear materials:

- Quantity of adhesive: A larger quantity of adhesive retains more heat, which may affect the curing speed [46], so the optimization of the quantity of adhesive must be considered.
- Application temperature: The melt flow of the adhesive depends on the temperature, so the optimum application temperature must be considered.
- Relative humidity: The bond strength of PU hot-melts is strongly dependent on moisture, so the curing speed will be different in different humidity environments.
- The wide versatility of upper and soling materials employed in the footwear industry is another important factor to consider, as the behavior of hot-melt adhesives may be different depending on the nature of the material.
- On the application of PU hot-melts, monomeric isocyanate vapors are given off, which requires the use of personal protective equipment and vapor exhaust units; however, recently new PU hot-melt adhesives called *microemission hot-melts* have been developed

[47]. These contain less than 0.1% isocyanate monomer and, therefore, reduce emissions by up to 90%. With a significant improvement of the occupational safety and environmental protection, they show a reduction up to 90% of hazardous, isocyanate-containing vapors and are labeled according to EU Directive 2008/58/EG and EU Regulation 1272/2008.

Due to the following advantages provided by hot-melt adhesives—freedom from solvents, instantaneous adhesion, high structural strength, and excellent resistance to aging—they represent a viable alternative for the upper-sole bonding stage in footwear, both from the environmental and the process points of view.

23.3 FUNCTIONALIZATION OF ADHESIVES FOR THE FOOTWEAR INDUSTRY

The use of nanotechnology enables the production of polymeric materials with improved performance and new features, contributing to the development of new applications in traditional and emerging sectors, such as footwear. Specifically, the incorporation of nanofillers into polymers constitutes a new generation of nanostructured materials. In this sense, CNTs have become a very attractive material for use as reinforcement at the nanoscale [31,32]. The possibility of incorporating nanotubes into adhesives offers a new range of innovative solutions, due to their unique structure and properties: high aspect ratio, high Young's modulus, excellent mechanical strength, and good thermal and electrical conductivity. Moreover, CNTs are capable of producing a multitude of changes in polymer properties just by being added in very small quantities. In this sense, the addition of CNTs to waterborne PU adhesives influences their rheological, thermal, viscoelastic, and adhesive properties [31]. The formation of an enhanced structure with CNT–PU interactions and a noticeable increase in the electrical conductivity of the PU matrix were demonstrated [32], especially above the electrical percolation threshold, where CNTs must be approximated to form the electrical-conductive network [31]. Furthermore, the addition of CNTs during PU synthesis by *in situ* polymerization produces adhesives with improved properties, different from those achieved through external addition. In addition, the existence of interactions between hydrogen bonds in PU adhesives with CNTs prepared by *in situ* polymerization was confirmed [48].

Controlled release of additives and adhesives is a key functionality for the footwear industry that can be provided by microencapsulation technology. A timely and targeted release improves the effectiveness of additives and adhesives, broadens their application range, and ensures optimal dosage, thereby improving cost-effectiveness for the adhesive manufacturer. Reactive, sensitive, or volatile additives (catalysts, prepolymers, solvents, antioxidants, preservatives, adhesion promoters, biocides, flame retardants, dyes, etc.), as well as the total formulation of the adhesive, can be turned into stable ones through microencapsulation [49].

To obtain innovative, intelligent, and multifunctional adhesives, the design of microcapsules should be undertaken so as to be able to fulfill certain criteria for further adhesive applications. Microencapsulated adhesives are often conveniently classified based on the mode of activation, the extent of component microencapsulation, the adhesive chemistry, or the suitability for various surfaces. They involve solvent-based and waterborne systems or reactive and curable resin systems. Either the total adhesive formulation or a single component can be encapsulated. Currently, encapsulated adhesives or those formulations containing encapsulating ingredients are used in such applications as threaded or retaining elements, labels, adhesive tapes, packaging, industrial processing, electronics, and so on.

In the footwear industry, 2K PU adhesives are used to ensure the durability of the upper-sole bond through the addition of a cross-linking agent, which improves hydrolysis and resistance to organic solvents and plasticizers, and prevents migration of oils and fats. The useful lifetime (*pot-life*) of a 2K PU adhesive as a function of time after the addition of the isocyanate cross-linker was studied [50]. At least during the first 6 h from mixing the cross-linker with the PU adhesive, the cross-linker remained stable, ensuring that the properties of the resulting film did not vary during this period.

However, 1 day after the realization of the mixture, some NCO groups were consumed and the viscosity and adhesive properties were not suitable. In this sense, for better handling, safety, and durability of isocyanate cross-linkers, recent works deal with the microencapsulated isocyanates as adhesion promoters for latent waterborne PU adhesives [50–53].

The use of an encapsulated adhesion promoter is an effective approach to produce latent adhesives. The term *latent adhesive* is used for an adhesive that is initially applied to a substrate and is then dried to evaporate all the water from the adhesive, generally to prevent the adhesion promoter moieties from becoming reactive. In this manner, the surface of such a prebonded substrate is nontacky, meaning that the substrate can be easily handled and stored. At any time, the two prebonded parts may be assembled by simultaneous pressing and heat activation, allowing the adhesion promoter to become reactive. The use of delayed-action encapsulated catalysts in reactive curing systems is certainly not a new concept. A specific example is the encapsulation of a high-toxicity catalyst used in PU resin curing, such as dibutyltin dilaurate (DBTL). This catalyst is frequently encapsulated using different shell materials such as waxes or other low molecular weight polymers. The microencapsulated catalyst is released by an applied pressure or temperature, accelerating PU gelation [49].

Microbiological degradation of waterborne adhesives can cause deterioration of the shelf life and product quality of the adhesive, which may manifest itself by a loss of viscosity, gas formation, undesirable odors, discoloration, coagulation of the emulsion, loss of adhesion, and so forth [30,54,55]. To provide antimicrobial properties to footwear materials and reduce or prevent their degradation, the use of silver-based silica nanocomposites ($Ag@SiO_2$) has recently been studied [29] to inhibit microorganism growth in waterborne PU adhesives. For this purpose, colloidal silver nanoparticles were prepared and embedded in a SiO_2-based matrix to decrease the silver aggregation tendency. The protection of AgNPs with silica was also intended to improve the stability of the adhesive toward metallic particles. With regard to adhesion properties, the addition of silica nanospheres or the $Ag@SiO_2$ nanocomposites maintained or even increased the bond strength values compared with unmodified adhesives. Last, but not least, $Ag@SiO_2$ enhances safer handling of the silver nanoparticles due to their immobilization within the silica matrix [29].

Typical biocides are based on combining formaldehyde-releasing agents with other toxic preservatives, and they are increasingly avoided by many industries, since they are organic products, harmful to humans and the environment, that are considered carcinogenic, cytotoxic, irritant, and/or allergenic [30]. For this reason, new alternative biocides that are less damaging, more sustainable, and biocompatible have been investigated. Another aspect to be taken into account is that the cytotoxicity of silver nanoparticles (AgNPs) may be reduced if they are embedded in a polymer matrix, similarly to the PU film formed after the drying process of waterborne adhesives [30]. AgNPs were synthesized from $AgNO_3$ using a natural polymer-like gelatin as a reducing and stabilizing agent. The AgNPs exhibited a strong antimicrobial activity at very low concentrations.

23.4 ROBOTIC APPLICATION OF ADHESIVES IN THE FOOTWEAR INDUSTRY

Hot-melt adhesives require heating to be applied using specific equipment; therefore, the upper-sole bonding stage in the footwear manufacturing process can be automated via industrial robotics (Figure 23.5). The advantage of using robots for adhesive application, apart from the inherent advantages of process automation, is that the amount of adhesive applied is controlled accurately, resulting in adequate quality of the upper-sole adhesive bonding. Furthermore, the automation of adhesive application means an increase in productivity, efficiency, and competitiveness.

Taking into account the complex geometry of the sole, the sizes, and the huge quantity of different models in the footwear industry, it is necessary to use intelligent systems that identify the geometry and generate the robot path in an automatic way. Before applying the adhesive automatically, the sole must be digitized to define the bonding area. In this regard, INESCOP (the Spanish Footwear Technology Institute) has developed a complete system that digitizes the 3-D sole geometry online

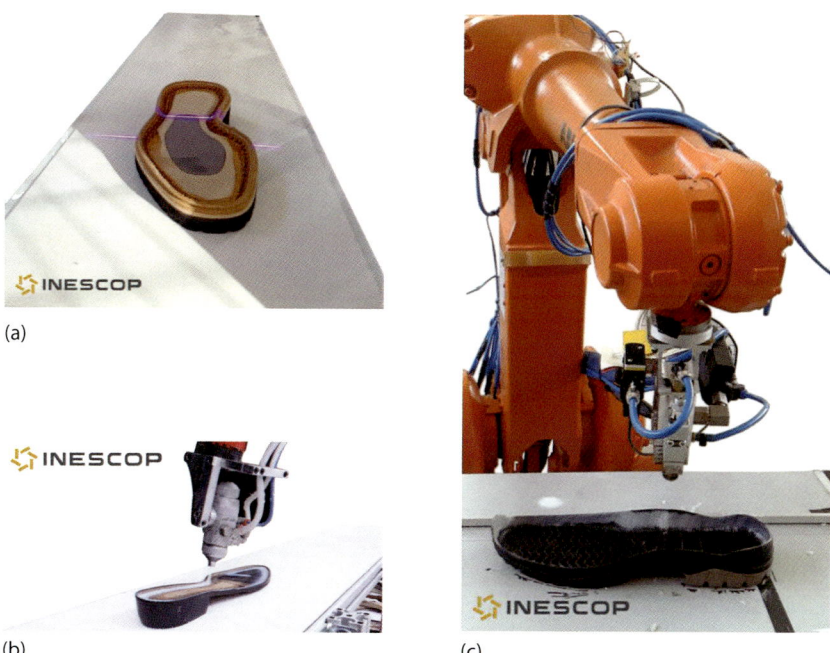

FIGURE 23.5   (a) Digitizing system for sole geometry; (b) hot-melt adhesive applied by a robot; (c) water-based adhesive applied by a robot.

and generates the robot's path in real time, which allows any kind of sole to be processed, such as flat, medium, or high-heeled soles. In addition to hot-melt adhesives, water-based adhesives can also be applied by robots.

The robotic application of adhesives fits into the so-called *Industry 4.0*, the aim of which is the implementation of *smart factories* with greater ability to adapt themselves to different needs and production processes, as well as a more efficient allocation of resources, thus opening the way to a fourth industrial revolution. It is a new milestone in industrial development that will certainly involve major social changes in the coming years, making intensive use of the Internet and advanced technologies, such as artificial intelligence, advanced robotics, 3-D printing, virtual augmented reality, and so on, where adhesives will also play a key role in the future footwear industry.

23.5 SUMMARY

A proper choice of adhesive is crucial to ensure the bond strength required in footwear manufacturing; however, the adhesive joint performance depends on the nature of the different substrates used as footwear materials. For this reason, it is important to consider the joint design, surface treatment, and adhesives properties that depend on their formulation. Furthermore, due to fashion trends and the large variety of materials used, the adhesives industry has developed a wide range of new products based on consumer requirements.

Solvent-based PU and PCP adhesives are the most commonly used ones today. However, due to environmental and occupational risks, one of the objectives of the footwear industry in the near future is the use of more sustainable adhesives, such as water-based and hot-melt adhesives. Furthermore, the automation of the bonding process through robotics is one of the big challenges for the footwear industry, as well as the functionalization of adhesives using nanoparticles to provide improved properties.

ACKNOWLEDGMENTS

The authors thank Ms. Esperanza Almodóvar from the staff of INESCOP for her assistance in the translation. Also, thanks to the Valencian Institute of Business Competitiveness (IVACE) of Generalitat Valenciana, Spain, and the European Union, via the European Regional Development Fund (FEDER) through its R&D Programme for Technology Institutes. Project file number IMDEEA/2017/2-CO2PUSHOE.

REFERENCES

1. R. M. M. Paiva, E. A. S. Marques, L. F. M. da Silva, C. A. C. António and F. Arán-Ais. Adhesives in the footwear industry. *Proc. IMechE Part L: J Mater.: Design and Appl.* 230, 357–374 (2016).
2. G. Oertel. *Polyurethane Handbook: Chemistry, Raw Materials, Processing, Applications, Properties*, 2nd edn, Hanser, Munich (1993).
3. M. D. Romero-Sánchez, M. M. Pastor-Blas and J. M. Martín-Martínez. Improved peel strength in vulcanized SBR rubber roughened before chlorination with tri-chloroisocyanuric acid. *J. Adhesion* 78, 15–38 (2002).
4. M. D. Romero-Sánchez, M. M. Pastor-Blas, T. P. Ferrándiz-Gómez and J. M. Martín-Martínez. Durability of the halogenation in synthetic rubber. *Int. J. Adhesion Adhesives* 21, 101–106 (2001).
5. R. Torregrosa-Coque, S. Álvarez-García and J. M. Martín-Martínez. Effect of temperature on the extent of migration of low molecular weight moieties to rubber surface. *Int. J. Adhesion Adhesives* 31, 20–28 (2011).
6. C. M. Cepeda-Jiménez, M. Mercedes Pastor-Blas, J. M. Martín-Martínez and P. Gottschalk. Treatment of thermoplastic rubber with bleach as an alternative halogenation treatment in the footwear industry. *J. Adhesion* 79, 207–237 (2003).
7. M. V. Navarro Bañón, M. Mercedes Pastor-Blas and J. M. Martín-Martínez. Water-based chlorination treatment of SBS rubber soles to improve their adhesion to waterborne polyurethane adhesives in the footwear industry. *J. Adhesion Sci. Technol.* 19, 947–974 (2005).
8. EN 1392:2006. Adhesives for leather and footwear materials. Solvent-based and dispersion adhesives. Test methods for measuring the bond strength under specified conditions.
9. EN ISOe 17708:2003. Footwear. Test methods for whole shoe. Upper sole adhesion.
10. ISO EN 15307:2014. Adhesives for leather and footwear materials. Sole-upper bonds. Minimum strength requirements.
11. Technical Report ISO/TR 20879:2007. Footwear. Performance requirements for components for footwear: Uppers, 1st edn.
12. ISO EN 15062:2006. Adhesives for leather and footwear materials. Solvent-based and dispersion adhesives. Testing ageing of bonds under specified conditions.
13. M. D. Romero-Sánchez, M. M. Pastor-Blas, J. M. Martín-Martínez and M. J. Walzak. Addition of ozone in the UV radiation treatment of a synthetic styrene-butadiene-styrene (SBS) rubber. *Int. J. Adhesion Adhesives* 25, 358–370 (2005).
14. M. A. Moyano and J. M. Martín-Martínez. Surface treatment with UV-ozone to improve adhesion of vulcanized rubber formulated with an excess of processing oil. *Int. J. Adhesion Adhesives* 55, 106–113 (2014).
15. M. D. Romero-Sánchez, M. M. Pastor-Blas and J. M. Martín-Martínez. Treatment of a styrene-butadiene-styrene rubber with corona discharge to improve the adhesion to polyurethane adhesive. *Int. J. Adhesion Adhesives* 23, 49–57 (2003).
16. J. Tyczkowski, I. Krawczyk-Kłys, S. Kuberski and P. Makowski. Chemical nature of adhesion: Plasma modified styrene–butadiene elastomer and polyurethane adhesive joints. *Eur. Polym. J.* 46, 767–773 (2010).
17. Z. S. Petrović, Y. Xu and W. Zhang. Segmented polyurethanes from vegetable oil-based polyols. *Polym. Preprints* 48(2), 852–853 (2007).
18. Z. S. Petrović, Y. Xu, J. Milic, G. Glenn and A. Klamczynsky. Biodegradation of thermoplastic polyurethanes from vegetables oils. *J. Polym. Environ.* 18, 94–97 (2010).
19. Z. S. Petrović. Polyurethanes from vegetable oils. *Polym. Rev.* 48, 109–155 (2008).
20. M. A. Corcuera, L. Rueda, A. Saralegui, M. D. Martín, B. Fernández-D'Arlas, I. Mondragon and A. Eceiza. Effect of diisocyanate structure on properties and microstructure of polyurethanes based on polyols derived from renewable resources. *J. Appl. Polym. Sci* 122, 3677–3685 (2011).

21. E. Orgilés-Calpena, F Arán-Aís, A. M. Torró–Palau and C. Orgilés-Barceló. Biodegradable polyurethane adhesives based on polyols derived from renewable resources. *Proc. IMechE Part L: J. Mater.: Design and Appl.* 228, 125–136 (2014).

22. S. Waddington. CO_2 based polyols give enhanced properties in reactive hot-melts. In: Proc. FEICA Conference (Association of the European Adhesive & Sealant Industry), Berlín (2014).

23. G. Verlag. CO_2 as a polyol intermediate: The dream becomes a reality. *Polyurethane Magazine* 10, 4, 236–239 (2013).

24. E. Orgilés-Calpena, F. Arán-Aís, A. M. Torró–Palau and C. Orgilés-Barceló. Novel polyurethane reactive hot-melt adhesives based on polycarbonate polyols derived from CO_2 for the footwear industry. *Int. J. Adhesion Adhesives* 70, 218–224 (2016).

25. A. Cherian. Novel CO_2-based polycarbonate polyols for high performance polyurethane hot-melt adhesives. In: Proc. Polyurethanes Technical Conference 2014, Dallas, TX, pp. 223–238 (2014).

26. O. Figovsky, L. Shapovalov, A. Leykin, O. Birukova and R. Potashnikova. Advances in the field of non-isocyanate polyurethanes based on cyclic carbonates. *Chem. Chem. Technol.* 79, 79–87 (2013).

27. M. Blain, L. Jean-Gérard, R. Auvergne, D. Benazet, S. Caillol and B. Andrioletti. Rational investigations in the ring opening of cyclic carbonates by amines. *Green Chem.* 16, 4286–4291 (2014).

28. N. Cuesta, M. J. Escoto, F. Arán and C. Orgilés. Proteins for bioinspired adhesive formulations. *J. Adhesion* 88, 294–307 (2012).

29. M. A. Pérez-Limiñana, F. Arán-Aís and C. Orgilés-Barceló. Novel waterborne polyurethane adhesives based on Ag@SiO2 nanocomposites. *J. Adhesion* 90, 437–456 (2014).

30. M. A. Pérez-Limiñana, F. Arán-Aís and C. Orgilés-Barceló. Waterborne polyurethane adhesives based on gelatine-stabilized AgNPs with improved antimicrobial properties. *J. Adhesion* 90, 860–876 (2014).

31. E. Orgilés-Calpena, F. Arán-Aís, A. M. Torró-Palau and C. Orgilés-Barceló. Chemical functionalization and dispersion of carbon nanofibers in waterborne polyurethane adhesives. *J. Adhesion* 89, 174–191 (2013).

32. E. Orgilés-Calpena, F. Arán-Aís, A. M. Torró-Palau and C. Orgilés-Barceló. Effect of amount of carbon nanotubes in polyurethane dispersions. *Macromol. Symp.* 321–322, 135–139 (2012).

33. E. Orgilés-Calpena, F. Arán-Aís, A. M. Torró-Palau and C. Orgilés-Barceló. Conductive polyurethane adhesives based on carbon nanotubes. *J. Nanostruct. Polym. Nanocomposites* 9, 36–39 (2013).

34. E. Orgilés-Calpena, F. Arán-Aís, A. M. Torró-Palau, C. Orgilés-Barceló and J. M. Martín-Martínez. Addition of different amounts of a urethane-based thickener to waterborne polyurethane adhesives. *Int. J. Adhesion Adhesives* 29, 309–318 (2009).

35. E. Orgilés-Calpena, F. Arán-Aís, A. M. Torró-Palau, C. Orgilés-Barceló and J. M. Martín-Martínez. Influence of the chemical structure of urethane-based thickeners on the properties of waterborne polyurethane. *J. Adhesion* 85, 665–689 (2009).

36. M. A. Pérez-Limiñana, F. Arán-Aís, A. M. Torró-Palau, C. Orgilés-Barceló and J. M. Martín-Martínez. Characterization of waterborne polyurethane adhesives containing different amounts of ionic groups. *Int. J. Adhesion Adhesives* 25, 507–517 (2005).

37. Global overview of the polyurethane dispersions (PUD) market. Report of IAL Consultants, London, UK (2015).

38. J. P. Kalish, S. Ramalingam, H. Bao, D. Hall, O. Wamuo, S. L. Hsu, C. W. Paul, A. Eodice and L. Yew-Guan. An analysis of the role of wax in hot-melt adhesives. *Int. J. Adhesion Adhesives* 60, 63–68 (2015).

39. P. Young-Jun, J. Hyo-Sook, K. Hyun-Joong and L. Young-Kyu. Adhesion and rheological properties of EVA-based hot-melt adhesives. *Int. J. Adhesion Adhesives* 26, 571–576 (2006).

40. P. Young-Jun and K. Hyun-Joong. Hot-melt adhesive properties of EVA/aromatic hydrocarbon resin blend. *Int. J. Adhesion Adhesives*, 23, 383–392 (2003).

41. Hot Melt Adhesives Market. Report of Markets and Markets, Dallas, TX (2016).

42. Life ShoeBAT project web. Promotion of the best available techniques in the European footwear and tanning sectors. http://www.life-shoebat.eu (2014).

43. M. A. Martínez, J. Santamaría, J. Ferrer and A. Zapatero. Thermoplastic polyurethane (TPU) from renewable sources applied in footwear. In: Proceedings of Green Polymer Chemistry International Conference (2012).

44. Y. Cui, D. Chen, X. Wang and X. Tang. Crystalline structure in isocyanate reactive hot-melt adhesives. *Int. J. Adhesion Adhesives* 22, 317–322 (2002).

45. J. Comyn, F. Brady, R. A. Dust, M. Graham and A. Haward. Mechanism of moisture-cure of isocyanate reactive hot-melt adhesives. *Int. J. Adhesion Adhesion* 18, 51–60 (1998).

46. E. M. Petrie. Reactive hot-melt adhesives for better structural bonding. *Metal Finishing* 39–43 (May 2008).

47. S. Hampe and P. Lienkamp. Emission-reduced PUR hot-melts from Henkel are increasingly establishing themselves in the market. In DRUPA 2008 Press Release, Henkel AG & Co. KGaA, Corporate Communications, Düsseldorf, Germany (2008).

48. E. Orgilés-Calpena, F. Arán-Aís, A. M. Torró-Palau and C. Orgilés-Barceló. Studies of thermoplastic polyurethane adhesives with carbon nanotubes obtained by two different routes and their interaction mechanisms. *Appl. Polym. Composites* 1, 9–25 (2013).

49. F. Arán-Ais, M. A. Pérez-Liminana, M. M. Sánchez-Navarro and C. Orgilés-Barceló. Developments in microencapsulation technology to improve adhesive formulations. *J. Adhesion* 88, 391–405 (2012).

50. M. A. Pérez-Liminana, F. Arán-Aís, A. M. Torró-Palau and C. Orgilés-Barceló. Determination of the life or pot life of a two-component polyurethane adhesive in aqueous dispersion. *Rev. Plást. Mod.* 614, 134–137 (2007).

51. H. Yang, S. K. Mendon and J. W. Rawlins. Nanoencapsulation of blocked isocyanates through aqueous emulsion polymerization. *Express Polym. Lett.* 2, 349–356 (2008).

52. O. R. Ganster, J. Buechner and J. F. Dormish. Under the microscope: Analyzing the development and properties of one-component PU-dispersion adhesives. *Adhesives Age* 45, 36–40 (August 2002).

53. M. M. Sanchez-Navarro, M. A. Pérez-Liminana, F. Arán-Aís and C. Orgilés-Barceló. Microencapsulation of isocyanates for use in adhesives. Trends in Adhesion and Adhesives III, Universidad Pontificia Comillas, Madrid, pp. 271–278 (2010).

54. H. K. Engeldinger, Storage of adhesives. In: *Handbook of Adhesion Technology*, L. F. M da Silva, A. Öchsner and R. D. Adams (Eds.), pp. 921–940, Springer-Verlag, Berlin, Heidelberg (2011).

55. G. T. Howard. Biodegradation of polyurethane: A review. *Int. Biodeterior. Biodegrad.* 49, 245–252 (2002).

24 Adhesives in the Automotive Industry

Klaus Dilger* and Michael Frauenhofer

CONTENTS

24.1 INTRODUCTION

The use of adhesives in automobile structures started more than 50 years ago as sealants for various purposes. Until now, this has been the main target for the application of adhesives and sealants in modern cars. Additionally, adhesives nowadays are used for different structural needs such as the stiffening of the body and a better crash performance [1,2].

Classically, the applications of adhesives in automobiles can be divided according to use into:

- Body in white (BIW), including hand-on parts
- Painted body
- Power train

---

* Indicates corresponding author

TABLE 24.1
Advantages of the Use of Adhesives with Regard to Challenges Faced in Terms of Joining

Challenges with Regard to Joining	Advantages of Adhesive Use
Lightweight design requires joining of components of different materials, e.g., metals, plastics, painted substrates.	Adhesives have the ability to adhere to different substrates and, therefore, have the ability to bond different materials.
Joining of materials that do not withstand the high temperature during the cataphoretic paint process.	Use of cold-curing adhesives enables a transfer of joining processes from the body shop to the assembly line.
Stiffening of the outer skin of body parts without affecting the class A surface, while also avoiding read-through of the fixation.	Adhesives that expand during curing provide antifluttering of skin components while preventing read-though.
Improvement of crash performance.	Use of high-strength adhesives, specially toughened for crash resistance, allows improved crash properties.
Joining of components with different coefficients of thermal expansion.	Elastic adhesives have the ability to compensate for tolerances or gaps resulting from different thermal expansions of parts.
Joining of components with different electrical potential.	Adhesives can be used to create an electrically isolating layer.
Prevention of parts from corrosion.	Use of adhesives for sealing prevents water from penetrating into gaps or protects of trimmed edges.

Table 24.1 shows an overview of the key challenges during joining and the advantages provided by the application of adhesives. These applications, as well as additional ones, are explained in further detail in this chapter.

24.2   THE AUTOMOTIVE PROCESS CHAIN

In a classic car production, there are four different production steps.
These are:

- Press shop
- Body shop
- Paint shop
- Trim assembly

24.2.1   PRESS SHOP

In the press shop, the supplied sheets, which are mainly made out of steel and aluminum, are formed by a deep-draw procedure. In large presses, the sheets are normally cold formed in large tools with a weight of often more than 40 tn. To obtain a better deforming behavior, avoid cracks, and achieve a proper surface, different lubricants are used. These can be oil or solid layers used as dry lubricants. In the recent past, a new hot-stamping process was introduced to produce steel parts with a very high yield strength. In this process, the deformation step is combined with defined fast cooling to harden the parts in the same production step. The sheets have to be heated up to a temperature of about 850°C to render the hardening possible. To avoid scaling and corrosion, the steel surface has to be covered by an intermetallic layer of aluminum and silicon. A trend in the body shop is the production of fiber-reinforced plastic parts with glass or carbon fibers. Here, typically, thermoset materials such as epoxy or unsaturated polyester are processed. Nowadays, the use of thermoplastic materials (mainly polyamide) has started where parts are produced in a pressing process. To facilitate the demolding of the parts from the mold, release agents are used. After demolding, the release agent remains as a residue on the surface [3].

24.2.2 Body Shop

After having formed the parts, almost all joining activities take place in the body shop. For the welding process, it is essential that the process is carried out on the blank sheet before the paint layer is applied in the paint shop. For mechanical joining techniques such as clinching or riveting, the process is similar. These could be used on painted sheets as well, but this leads to damage of the lacquer layer, and thus to amplified corrosion, which would require a greater effort to protect the surface. Adhesive bonding is used in the body shop for both metals and plastics. Additionally, this allows the possibility of transferring the joining procedure to the assembly line, because adhesive bonding can be used to join painted parts as well [4].

In the body shop, the metal sheets are oily or carry a layer of a dry lubricant, and the plastics normally have residues of release agent on their surface. It has to be taken into account that in the paint shop, a process will be conducted subsequently at 180°C for about 30 min. Therefore, only those materials can be used that can withstand these high-temperature loads. On the other hand, the heat can be used to cure the adhesives and to harden the metals (bake hardening, deposition) [5].

24.2.3 Paint Shop

Before the lacquer is applied, different pretreatments such as cleaning and phosphating are executed. The paint is applied in different layers, starting with the application of the electrolytic cataphoretic paint and the curing of this layer for about 30 min at a temperature of 180°C. Afterward, more paint layers are applied and cured at temperatures from 110°C to 150°C. The resulting layer structure is shown in Figure 24.1.

24.3 CONSTRAINTS FOR ADHESIVE BONDING

24.3.1 Adhesive Bonding in BIW

An automobile structure has to fulfill many different demands. Besides the structural integrity and/or the deformation behavior under different load conditions, fatigue or crash corrosion, for instance, has to be avoided, and a high stiffness of the body is necessary to improve driving performance. These properties have to be achieved under the constraints of high noise damping and a lightweight design (Figure 24.2).

The required properties, shown in Figure 24.2, are significantly affected by the joining technology used, the joining processes, and the defined process chain. The material of the component obviously has the main impact on the choice of joining technology. But it also has to be taken into account that in the case of joining different materials, the relative thermal expansion is compensated, (contact) corrosion is avoided, and the structure can be repaired after an accident. Additionally, it

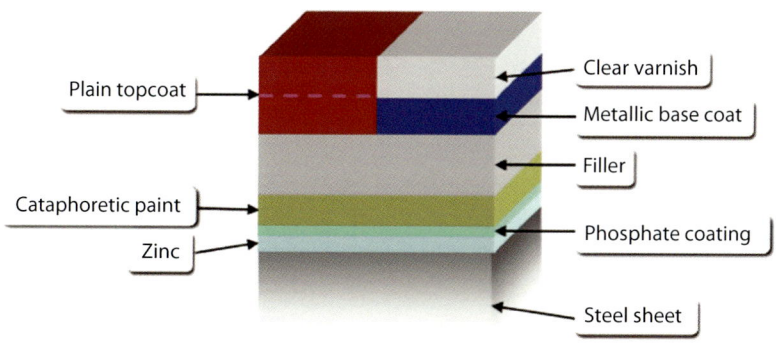

FIGURE 24.1 Paint layers on a painted car body. (From Keßel, A., and Dilger, K., *J. Oberflächentechnik* 46, 9, 56–59, 2006.)

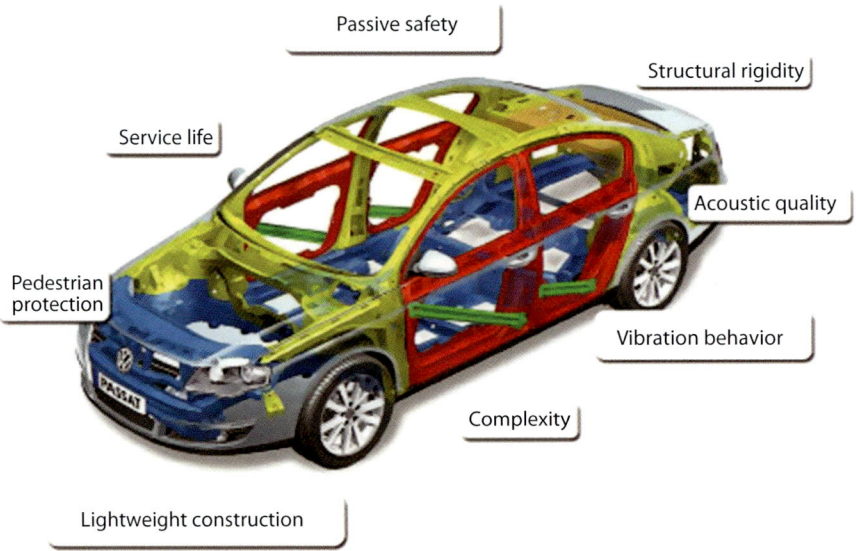

FIGURE 24.2 Demands on a modern car structure (Volkswagen AG).

has to be considered that the properties of the joined parts do not change and are durable as far as possible under the severe ambient conditions. In Figure 24.3, the different demands on a joining process of an automobile body are described [7].

In the body structure and in hang-on parts, adhesives are used in the hem flange area to guarantee the geometry of the parts during production, to prevent the flanges from corrosion and to enhance the stiffness for good crash behavior. Besides the adhesive in the flange, a so-called *cosmetic sealer* is used to prevent water from penetrating into the flange. For structural purposes, the parts are connected in flanges by hybrid joining, which means that spot welding, laser welding, riveting, or clinching are combined with the use of an adhesive. A further application of adhesives

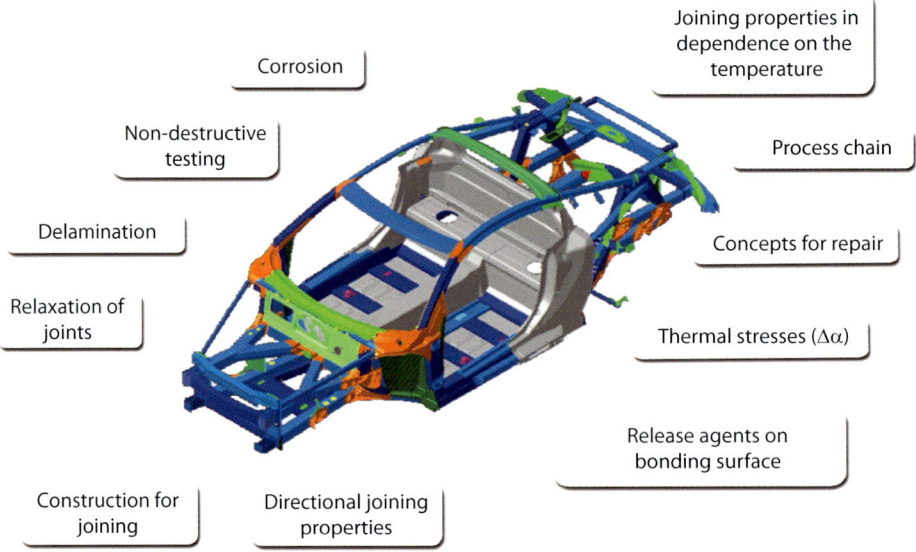

FIGURE 24.3 Demands on the joining processes for a car body (Audi AG).

Spot welding and bonding
Metal active-gas welding
Laser beam welding and plasma soldering
Antiflutter adhesive

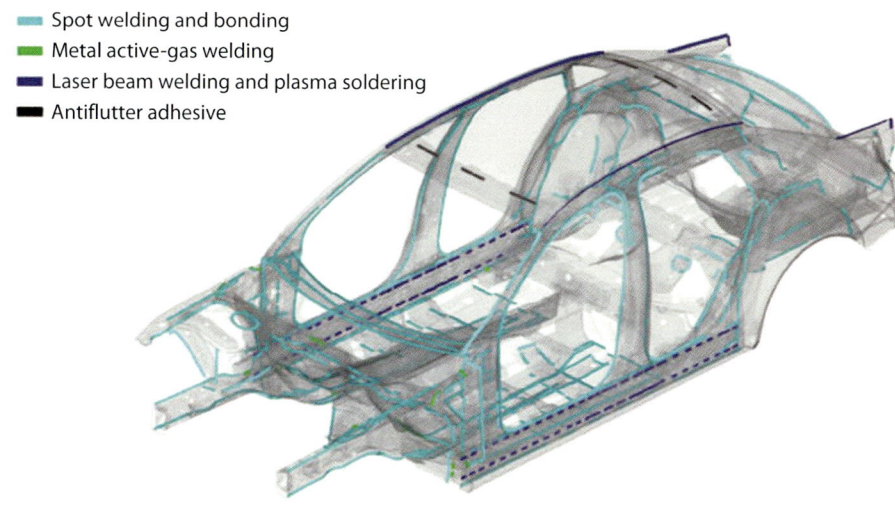

FIGURE 24.4 Adhesives in the Audi A4 B8 (Audi AG).

is their use as antiflutter adhesive to stiffen the outer skin without the formation of sink marks and deformations that have a negative effect on the so-called *class A surface* of the part. Near the tank nozzle, sealants are used to avoid the penetration of fuel into the body. To demonstrate the use of adhesives in a modern car, the different adhesives in the body of an Audi A4 are shown in Figure 24.4. Here, structural adhesive bonds—combined with spot welding in a hybrid joint—are shown in light blue and antiflutter adhesive bonds are shown in black (Figure 24.4).

24.3.2 ADHESIVE BONDING IN THE PAINTED BODY

After the geometrical definition of the body in the body shop by welding and/or mechanical joining in combination with adhesive bonding, the body is transferred to the paint shop, where some sealing applications are carried out and the painting process takes place. Here, the first coating is applied by an electro-cataphoretic (EC) coating process. This coating is cured at a temperature of about 180°C for 30 min. Afterward, different paint layers are applied and cured at lower temperatures (e.g., 110°C). From the paint shop, the painted body goes to the trim assembly where the final assembly takes place. Typical bonding operations here are the bonding of the windshield and the nonmovable windows and the different plastic parts. A major difference from the application in the body shop is that no additional fixation by means of mechanical joining can be used, and the adhesives have to be cold-cured because no process in an oven will follow as the parts used are sensitive to elevated temperatures. With an expanding use of plastic parts in the body and the assembly of preassembled, painted parts, some joining operations move to the assembly line, because they cannot withstand the oven process [8,9].

24.3.3 ADHESIVE BONDING IN THE POWER TRAIN

In the power train, adhesives are used to join gears and shafts, fix screws, and seal parts of the engine or the gearbox. The application in the power train is completely different from the other applications in an automobile, mainly because of the higher temperatures and the contact with different fluids such as gear oil.

24.3.4 THE TREND: MULTIMATERIAL DESIGN

The need to reduce the weight of a car leads to the use of alternative materials that have to be joined with a similar material or a different material in a so-called *hybrid construction*. Here, traditional

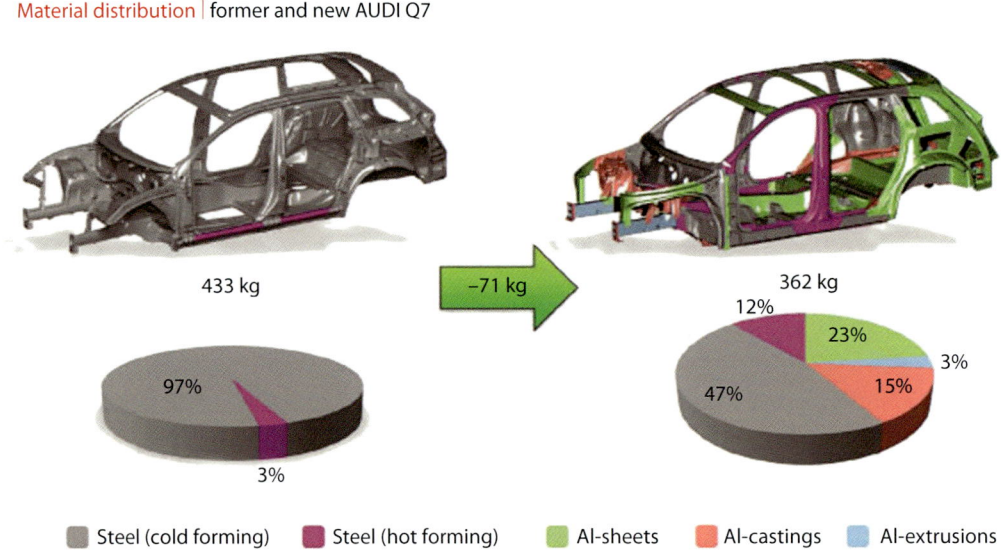

FIGURE 24.5 Material mix in a modern car structure (Audi AG). (From Alber, U., *Joining in Car Body Engineering. Module I: Adhesive Bonding and Hybrid Bonding*, Proceedings of Automotive Circle Conference at Bad Nauheim, Germany, pp. 175–186, 2015.)

joining and welding processes are often not suitable for different reasons. One is the difference between the thermal expansion coefficients of the materials, another is the electrical potential of the materials. Adhesives are suitable for bonding almost any material and reduce the mentioned problems, so that the use of adhesives in automobiles has increased significantly in the last 10 years [3]. In Figure 24.5, the material mix in a modern car structure is shown.

Here, adhesives with a high modulus and a high strength are used in a thin adhesive layer to stiffen the construction and to improve the crash behavior; elastic adhesives with a thick adhesive layer are used to dampen the noise, to compensate for tolerances, and for different thermal expansions [11,12].

24.4 USE OF ADHESIVES IN AN AUTOMOBILE

24.4.1 Adhesives in Body in White (BIW)

24.4.1.1 Materials

Typical materials that are used in a car body are shown in Figure 24.5. In the body shop, parts formed from these materials are to be joined. Due to the forming processes and/or to protect the materials from corrosion during production, different auxiliary substances are used. Steel has an oily surface in the body shop, because the oil is needed to reduce friction in the deep-draw operation and to prevent the parts from corrosion. Light metals such as aluminum or magnesium alloys have different lubricant layers on their surfaces if they are formed in a deep-draw process; if they are produced by a casting procedure, the surface is covered with an oxide layer that contains a high amount of release agents. Plastic parts that are used in the body shop have to have a high thermal resistance, because of the thermal process that follows in the paint shop. Hence, thermoplastics with a high melting temperature such as polyamides and thermoset plastics—for example, epoxy or polyesters—are used. The need to demold the parts leads to the use of release agents that may remain on the surface [13].

24.4.1.2 Surfaces and Surface Pretreatment

No surface pretreatment is needed for oily steel blanks. Modern body shop adhesives have good adhesion to oily surfaces of blanks and zinc-coated steel sheets. Aluminum and magnesium alloys

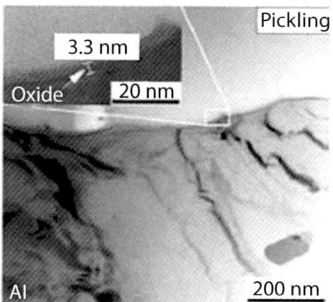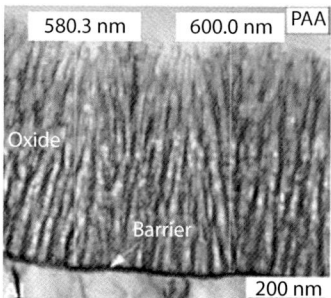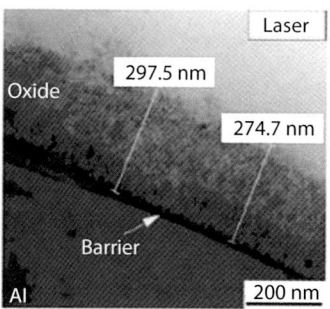

FIGURE 24.6 Visible increase of the oxide layer thickness resulting from pickling, phosphoric acid anodization (PAA) and laser pretreatments. (From Rechner, R., Laseroberflächenvorbehandlung von Aluminium zur Optimierung der Oxidschichteigenschaften für das strukturelle Kleben. Dissertation, Technische Universität Dresden, Dresden, Germany, 2011.)

have to be passivated by an anodizing process, where acid solutions of titanium and zirconium complexes are used to form stable oxide layers. The passivation can be applied either to the sheets or the pressed parts. A new surface treatment to produce a corrosion-resistant local surface layer is laser pretreatment [14]. Here, by the use of disk or fiber lasers, a corrosion-resistant oxide layer can be achieved (Figure 24.6).

Additionally, an adhesion-friendly surface morphology results from the laser pretreatment. The resulting surface structure is shown in Figure 24.7.

The increased application of die-cast aluminum parts in the car structure requires removing the release agents from the surface. Due to the use of different release agents, different surface conditions can result [3,7,16] (Figure 24.8).

Due to the molding process and the geometry of the parts, the allocation of the release agent on the surface varies, which results in a locally different thickness of the agent's layer. The thickness of the layer can be determined by EDX (Energy Dispersive X-Ray) mapping (Figure 24.9).

For fiber-reinforced plastic parts, a sandblasting process is used to remove the release agents. A new process producing less dust, and therefore fitting better into the production process chain, is

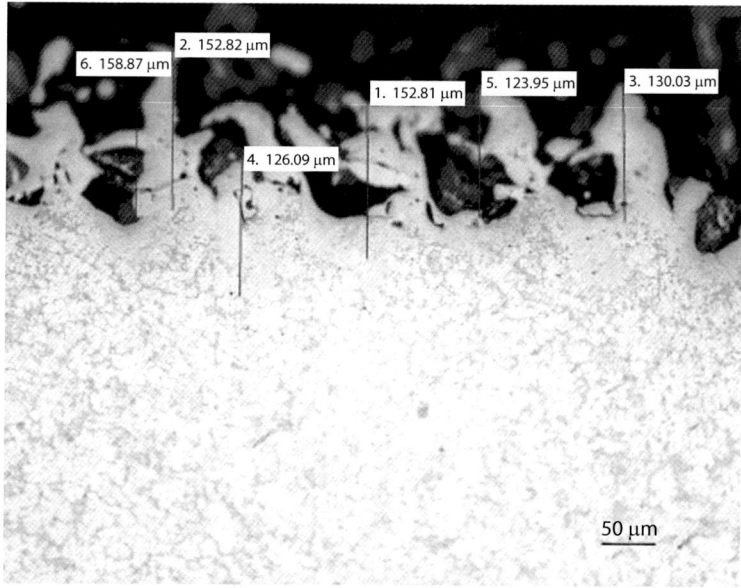

FIGURE 24.7 Surface structure resulting from the laser pretreatment.

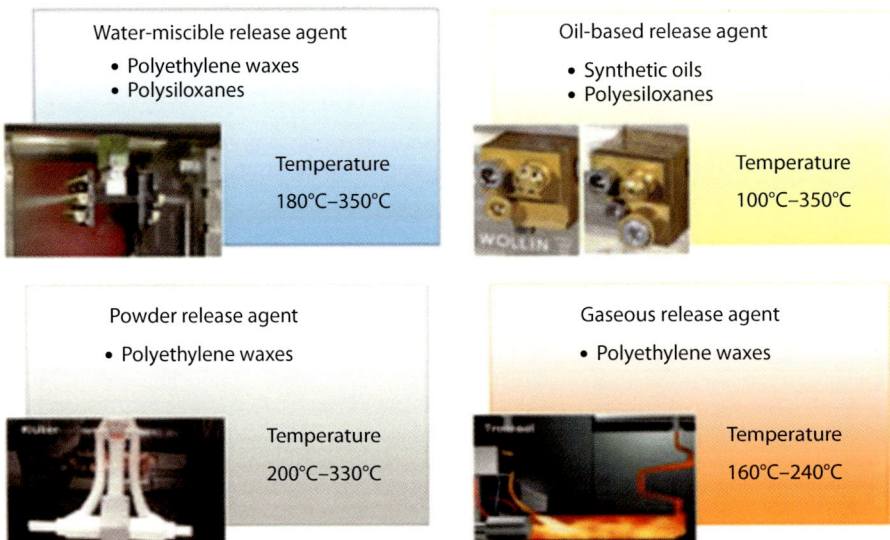

FIGURE 24.8 Surface conditions resulting from the use of different release agents (figure at lower left from Klueber Lubrication, Munich, Germany; figure at lower right from Trennsol, Hanover, Germany).

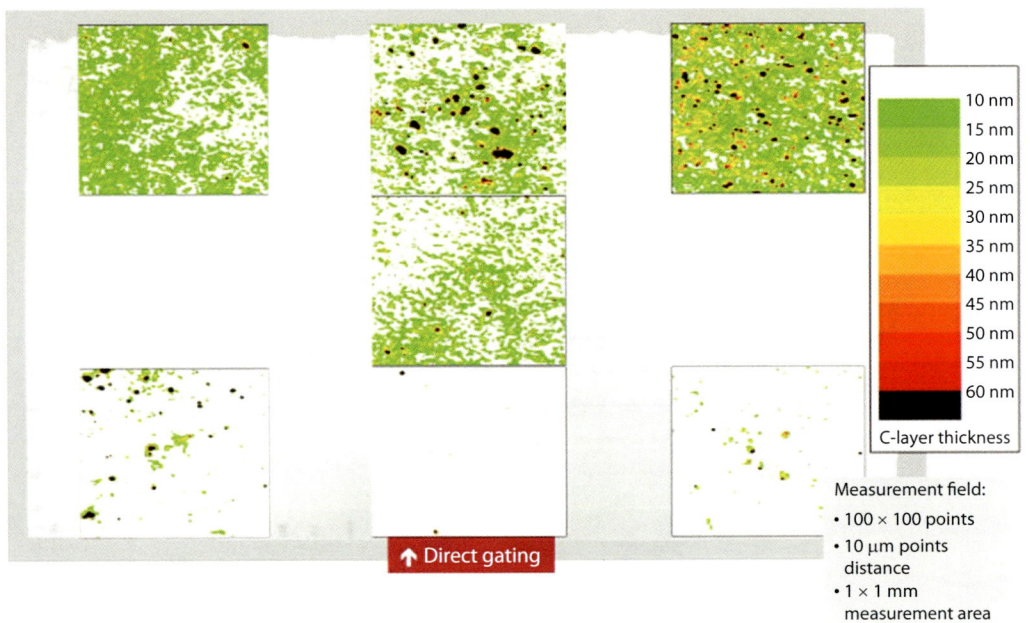

FIGURE 24.9 Visualization of release agent on the surface of a part produced by die-casting. EDX-mapping shows the layer thickness.

vacuum suction blasting, as shown in Figure 24.10. Due to the pressure difference of p (underneath the hood) and p_u (the surrounding pressure), particles are removed in this process.

A technique that is currently investigated is the removal of the release agent by laser treatment. Depending on the laser parameters, there is the possibility to remove either the release agent only or the release agent plus matrix material from the surface. The latter makes it possible to bond directly to the fibers (Figure 24.11).

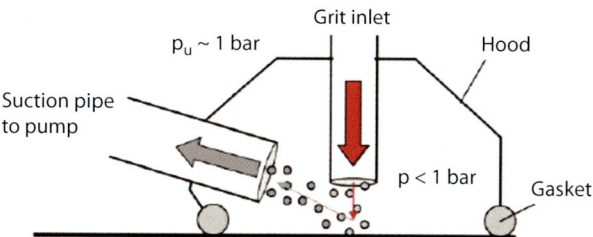

FIGURE 24.10   Vacuum suction blasting of a carbon fiber-reinforced plastic part. (From Kreling, S. et al., *Adhesion, Adhesives & Sealants*, 10, 2, 34–39, 2013.)

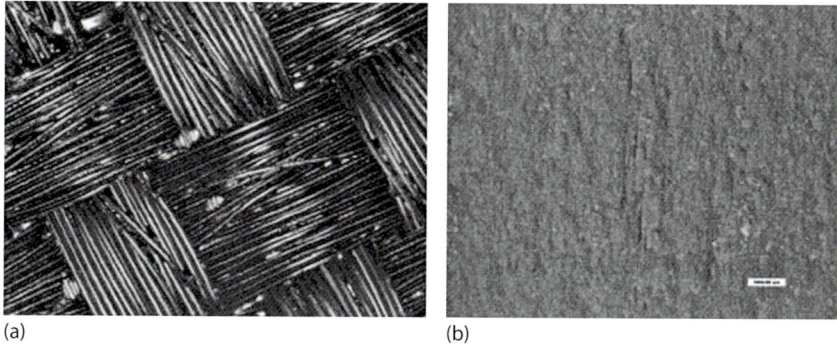

(a)   (b)

FIGURE 24.11   Different surface structures of a CFRP component (a) surface structure without treatment; (b) surface structure after pretreatment).

In the near future, hot-stamped press-hardened parts will be used more and more, due to their high lightweight potential and the relative low cost of the material. For the press-hardening procedure, it is necessary to heat up the parts to a temperature of about 850°C to achieve high strength and high yield stress. Hence, at 850°C, in the presence of oxygen, the steel parts tend to form a thick oxide layer that is not suitable for the following processes and the durability of the parts during use. Therefore, an intermetallic layer of Al and Si is applied in advance to prevent the parts from corrosion. Thus, the adhesive is applied to this layer and the Al–Si layer becomes a structural component of the system. Additionally, it has to be taken into account that the properties of this layer are highly dependent on the production process and the part geometry [18]. A typical area of the cross-section of a press-hardened part is shown in Figure 24.12.

Due to its high strength, a stiff adhesive layer with stress concentrations at the ends of the overlap will lead to a crack initiation in the surface layer that already propagates at a low energy rate when load is applied to the component. This will result in a bad crash behavior. Toughened adhesives show a cohesive failure and improved crash properties.

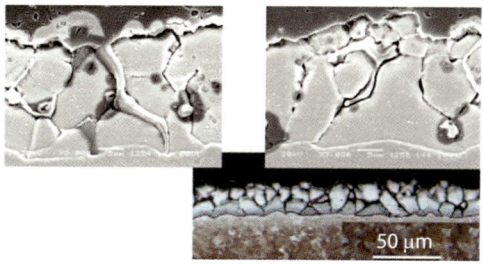

FIGURE 24.12   The figure shows cross-sections of the surface layers of a press-hardened part, including enlargements of the top layers.

24.5 APPLICATIONS AND PROPERTIES

24.5.1 Body in White (BIW)

24.5.1.1 Structural Bonded Flanges

In the BIW, adhesives are used in flanges combined with spot welding, self-piercing riveting, or clinching to improve the stiffness of the body and/or the crash behavior [19]. The static stiffness can be increased up to 30%, or even more, depending on the car design and construction. Another advantage of the bonding of structural flanges is the improvement of the fatigue behavior of the structure [20]. The behavior of a purely spot-welded flange in comparison with a hybrid-bonded flange is shown in Figure 24.13.

Whereas, to just improve the stiffness, conventional more or less brittle epoxy adhesives can be applied, the requirement for an improved crash behavior makes the use of special toughened crash-resistant epoxy adhesives necessary [22,23]. These epoxy adhesives consist of a strong, highly cross-linked epoxy matrix with a microscale rubber phase to enhance crack propagation energy, due to the so-called *crack pinning* where the cracks are trapped in the soft phases and a large crack is divided into different small cracks. In Figure 24.14, the morphology of an epoxy adhesive with dispersed rubber phase is shown [7].

By using these adhesives in a body structure or a closure, the absorbed energy in a crash scenario can be raised, and the structure will be intact because the adhesive layers can be deformed with the plates and do not break brittlely. In Figure 24.15, two doors are shown after a crash, where one door is bonded with a "normal" structural adhesive (to stiffen the body) and the other one is bonded with a crash-resistant adhesive.

24.5.1.2 Hem Flange Bonding

The adhesive bonding of a hem flange has different purposes. These are, among others:

1. Sealing of the gap to prevent water from penetrating
2. Protection of the cutting edges
3. Fixing of the relative position between outer and inner panels
4. Stiffness of the part
5. Improved crash behavior of the part

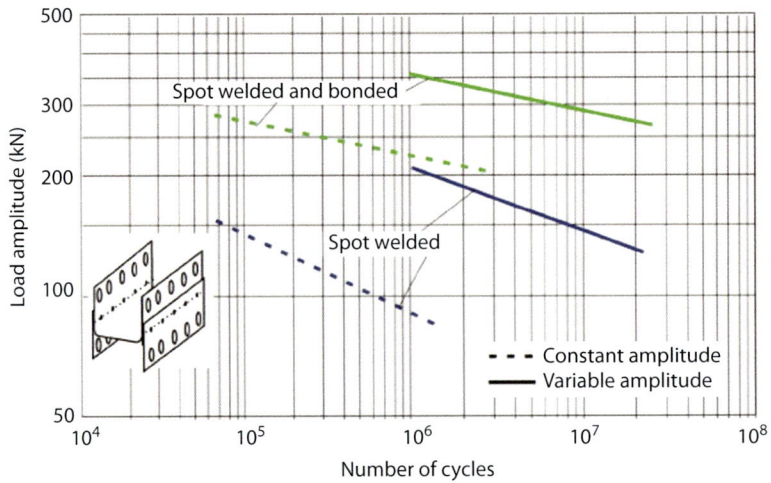

FIGURE 24.13 Improvement in the fatigue behavior of a component through the use of adhesives. (From Grubisic, V., *DVM-Bericht*, 120, 47–61, 1994.)

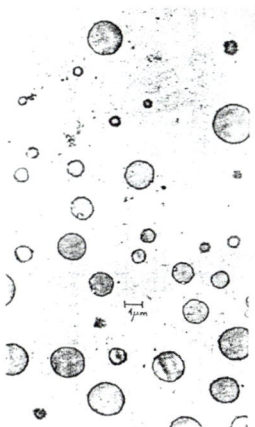

FIGURE 24.14   Modified epoxy adhesive: The dispersed rubber phase (round particles) improves the crash resistance. [From Mühlhaupt, R., *Chimia* 44, 43–52, 1990; Mulhaupt, R. et al., Phenol-terminated poly-urethane or polyurea (urethane) with epoxy resin, US Patent 5278257, assigned to Ciba-Geigy Corporation (January 11, 1994); Mulhaupt, R., and Rufenacht, W., Composition of butadiene/polar comonomer copolymer, aromatic reactive end group-containing prepolymer and epoxy resin, US Patent 5073601 A, assigned to Ciba-Geigy Corporation (July 25, 1989).]

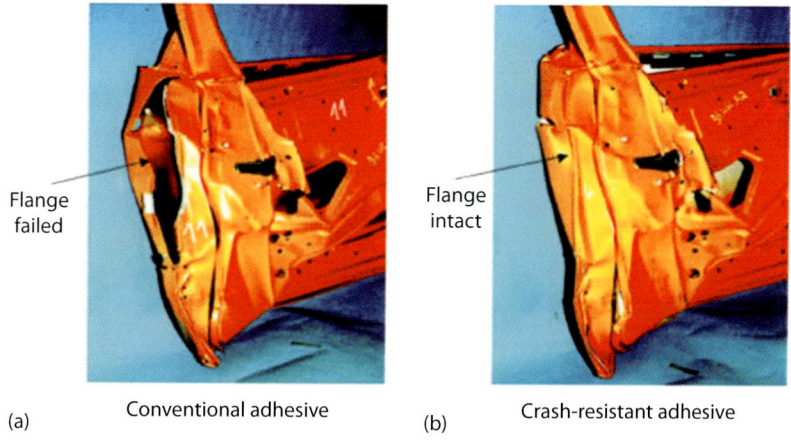

(a) Conventional adhesive   (b) Crash-resistant adhesive

FIGURE 24.15   Doors of a car bonded with different adhesives (Mercedes Benz): (a) flange bonded with a conventional adhesive failed; (b) flange bonded with crash-resistant adhesive is intact.

The stiffness and the crash behavior are similar to the bonding of structural flanges. As a hem flange without an adhesive has no strength in the plane of the plate, the adhesive is used to achieve this fixation. In practice, the required strength is well described by the stiffness and the crash behavior. To fasten the components' geometry in production, it is necessary to gel or precure the adhesive during or directly after the hemming procedure. This avoids any movement of the parts that may occur due to handling operations. Corrosion protection is achieved by both sealing the gap and protecting the cutting edges. In this context, different problems can appear. One is that the cutting edge of the outer part is not covered by the adhesive; another is that the best corrosion resistance is achieved by a complete filling of the gap. This, however, leads to contamination of the hemming tools by the adhesives, which will cause cost-intensive rework. Therefore, the flange, in general, is not filled completely. A typical filling is 130%, which means that one side of the hem flange is filled completely, the other one only to 30%. To prevent water from penetrating into the unfilled gap, another adhesive, the so-called cosmetic sealer, is applied. By the application of a sealer bead, the

FIGURE 24.16   Hem flange bonding and sealing (From Dilger, K., *Adhesive Bonding: Science, Technology and Applications*, CRC Press, Boca Raton, FL, pp. 357–385, 2005.)

free cutting edge can also be protected. The configuration of hem flange bonded with adhesive and cosmetic sealer is shown in Figure 24.16.

24.5.1.3   Antiflutter Bonding

The relatively thin outer skin of a car body has to be stiffened by an inner structure. Because tolerances have to be compensated for and the optical impression of the class A surface has to be guaranteed, adhesive bonding is the only way to join these parts. To avoid optical irregularities on the surface, so-called *read-through*, the shrinkage of the adhesives used is of importance. To compensate for the deformation of the outer skin that can result from shrinkage, adhesives that expand during curing can be used. It is important to have a more or less constant thickness of the adhesive layer, because different volumes of adhesive will lead to variations in the shrinkage, which can be seen on the painted surface. To achieve good stiffening and to avoid read-through, the stiffness of the adhesive (Young's modulus, shear modulus) has to be optimized. With a coupled process and structure simulation by means of FE (Finite Element) analysis, a prediction of the distortion and an optimization of the geometry can be achieved [27]. The simulation can also be used to choose a proper adhesive (Figure 24.17).

24.5.2   Carbon Fiber-Reinforced Polymer-Dominated Body Shop

In modern lightweight designs, carbon fiber-reinforced polymer (CFRP) parts are bonded with other parts of the same type or with metal parts in hybrid constructions. The CFRP surfaces are not covered by an oil layer, so different kinds of adhesives can be used. Besides epoxy adhesives, acrylics and polyurethanes are applied. The use of very stiff adhesives can lead to delamination of the first layer of the CFRP part. The stress concentration at the end of the overlap can lead to crack initiation in the CFRP that propagates near the surface of the CFRP part. This failure is very brittle and has only a low energy absorption that is correlated with a bad crash behavior. Thus, the crash performance can be enhanced by the use of an adhesive with a lower Young´s modulus. To achieve a sufficient stiffness, structural adhesives with a midmodulus between 50 and 500 MPa are

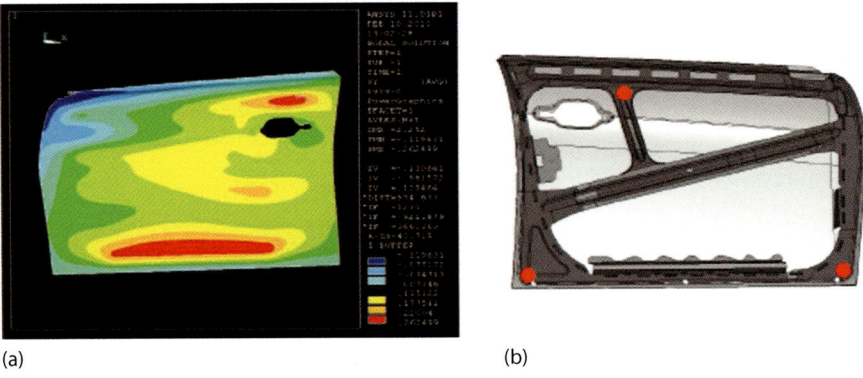

(a)   (b)

FIGURE 24.17   FE-analysis of a car door to determine the distortion after curing of the adhesive: (a) results of an FE-analysis; (b) door structure.

$\alpha_{CFRP} = 0 * 1/K$

$\alpha_{Al} = 23 * 10^{-6}\ 1/K$

FIGURE 24.18 The life module (top module, consisting mostly of CFRP materials) and the drive module (bottom module, consisting mostly of metal materials) of a BMW i3 with their different coefficients of thermal expansion. (BMW Group).

used. Commonly, they are based on polyurethane. A typical application of adhesives in CFRP bodies is the bonding of studs to the CFRP surface. The classical stud-welding process that leads to a high immediate strength on metals cannot be used, so a fast bonding process has to be introduced. Therefore, studs with a transparent foot are applied to allow fast curing of an acrylic adhesive by ultraviolet (UV) radiation. An alternative method is the curing of an on-the-stud preapplied epoxy by fast inductive heating. The thermal expansion coefficient α_K of CFRP is 0; steel has an α_K of $10 \cdot 10^{-6}$ 1/K and aluminum $23 \cdot 10^{-6}$ 1/K. This leads to a thermal expansion mismatch that can occur during the paint curing process in the paint oven or during use. The different thermal expansions can lead to a distortion of the parts and/or to high stresses in the adhesive layer. To reduce stress, thick layers of an adhesive with a low modulus are applied. In Figure 24.18, the life and the drive modules of the BMW i3 are shown as an example for the need to join parts with different thermal extensions [28,29,30,31]. The so-called *life module* of the vehicle carries the passengers, whereas the *drive module* provides the vehicle with motion.

24.5.3 PAINT SHOP

There is no use of structural adhesives in the paint shop. Only sealing is carried out. Here, poly(vinyl chloride) plastisols are applied, where prejelling is often used to increase the viscosity with the target to guarantee a washout stability of the adhesive during the paint processes. Foams are used as so-called *pillar fillers* to seal closed profiles.

24.5.4 ASSEMBLY LINE

In the trim assembly, inner parts are generally made out of plastics such as acrylonitrile–butadiene–styrene copolymer (ABS) and polycarbonate (PC) or combinations of these, and are bonded to one another or to the structure. Here, butyl rubbers, hot-melts, and adhesive tapes are applied. If a higher heat resistance is needed, reactive hot-melts are applied. For optical reasons, the surfaces of the interior plastic parts are coated with a decorative fabric or leather. This lamination is made by water-based reactive polyurethane adhesives that build a solid adhesive layer after the spray application of the emulsion. Within a time period of some hours, the adhesive layer can be activated by heat and the parts are joined while the adhesive is tacky due to heat activation.

On the assembly line, the adhesive is applied on the painted parts. Hence, for (semi)structural applications, such as the adhesive bonding of the windshield and the rear window, moisture-curing polyurethane adhesives are used that have good adhesion to the paint. In Figure 24.19, typical semi-structural bonding in the cabin of a truck is shown [3,7].

The adhesives used here show a very elastic behavior and have a Young´s modulus of about 5 MPa. They are applied in a triangular-shaped bead with a height of more than 10 mm and pressed to the final geometry in the joining process. The thickness of the remaining adhesive layer is from

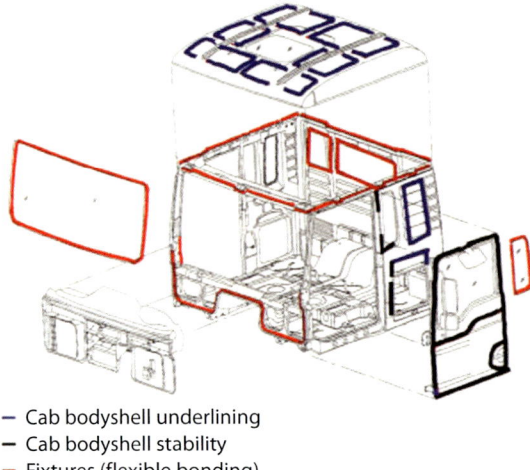

— Cab bodyshell underlining
— Cab bodyshell stability
— Fixtures (flexible bonding)

FIGURE 24.19   Areas of adhesive use for semistructural bonding in a truck cabin.

a few millimeters to centimeters. Such a thick, hyperelastic adhesive layer is used to compensate for the joining tolerances between the window and the structure, thus sealing the gap between the window and the body and stiffening the body while forming a shear plane. For this purpose, the adhesive has to have a certain shear modulus to achieve stiffening on the one hand, but on the other hand it should not be so stiff that a torsion of the car body leads to a cracking of the glass. The influence of the modulus of the adhesive on the stiffness of the body is shown in Figure 24.20.

24.5.5 POWER TRAIN

In the power train, gear shaft connections and the fixation of bearings are made by an anaerobic acrylic adhesive. The same class of adhesive is used as a threadlocker. To seal the flanges of, for example, the gearbox, anaerobic acrylics as well as silicones are used. As the application of silicones in the car body is strictly forbidden, because of the possible contamination of the skin and the "antiadhesive" effect of this contamination on the paint adhesion, the use of silicones is tolerated

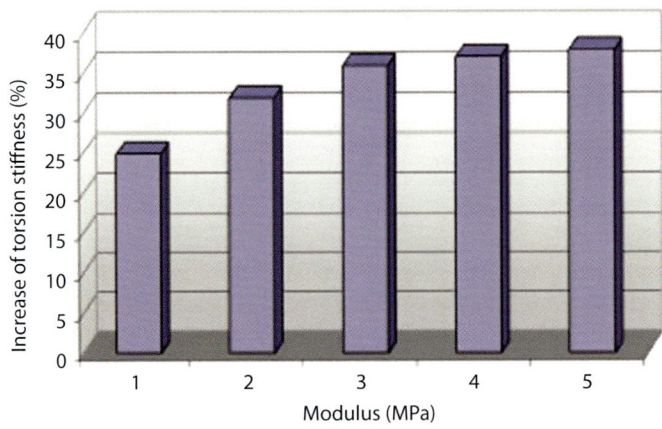

FIGURE 24.20   Relation between torsion stiffness of a car body and the modulus of the window adhesive (From Hirthammer, M., Proceedings of Glass Processing Days, the 5th International Conference on Architectural and Automotive Glass, Now and in the Future, pp. 363–365, 1997.)

only in the assembly of the power train components. To accelerate hardening, UV curing silicones are often applied [35].

24.6 SUMMARY

The use of adhesives in an automobile is a common joining technique, with certain advantages compared with other joining technologies. In particular, good corrosion protection, increase in body stiffness, and good crash behavior inspire the use of adhesives. Additionally, modern multimaterial design is not possible without adhesive bonding.

REFERENCES

1. H. Schenkel. Adhesive bonding in car body. DVS-Report 218, 163–172 (2001).
2. J. Daniels. Design implications of adhesive bonding in car body construction. *Intl. J. Adhesion Adhesives*, 4, 5–8 (1984).
3. K. Dilger. Kleben im Leichtbau. In: *Handbuch Leichtb: Methoden, Werkstoffe, Fertigung*, F. Henning and E. Moeller (Eds.), pp. 899–939, Carl Hanser Verlag, Munich (2011).
4. L. Papadakis, V. Vassiliou, M. Menicou, M. Schiel and K. Dilger. Adhesive bonding on painted car bodies in automotive production lines: Alternatives and cost analysis. In: *Proceedings of the World Congress on Engineering 2012* Vol III, pp. 1382–1387 (2012).
5. M. Greiveldinger, M. E. R. Shanahan, D. Jacquet and D. Verchere. Oil-covered substrates: A model study of the evolution in the interphase during cure of an epoxy adhesive. *J. of Adhesion*, 73, 179–195 (2000).
6. A. Keßel and K. Dilger. Einfluss der Lacktrocknung auf die Hafteigenschaften "Kleben auf Lack." *J. Oberflächentechnik* 46, 9, 56–59 (2006).
7. K. Dilger. Automobiles. In: *Adhesive Bonding: Science, Technology and Applications*, R. D. Adams (Ed.), pp. 357–385, CRC Press, Boca Raton, FL (2005).
8. T. Koll, U. Eggers and M. Höfemann. Leaner manufacturing with precoated high steels. In: *Proceedings of SAE-P*, Vol P-369, pp. 171–175 (2001)
9. L. Papadakis, V. Vassiliou, M. Menicou, M. Schiel and K. Dilger. Adhesive bonding of attachments in automotive final assembly. In: *IAENG Trans. Eng. Technologies*, Special Volume of the World Congress on Engineering 2012, G-C. Yang, S-I Ao and L. Gelman (Eds.), pp. 739–752, Springer, Dordrecht (2013).
10. U. Alber. Innovative joining technologies in the new AUDI Q7. In: *Joining in Car Body Engineering. Module I: Adhesive Bonding and Hybrid Bonding*, Proceedings of Automotive Circle Conference at Bad Nauheim, Germany, pp. 175–186 (2015).
11. E. Wiese, D. Schlingmann and R. Eichleiter. Hybrid joining for multimaterial designs. In: *Joining in Car Body Engineering. Module I: Adhesive Bonding and Hybrid Bonding*, Proceedings of Automotive Circle Conference at Bad Nauheim, Germany, pp. 235–252 (2015).
12. M. Frauenhofer and J. Schaefer. Adhesive joining as an enabler for CFRP mixed material structures: The MMS platform example. In: *Joining in Car Body Engineering. Module I: Adhesive Bonding and Hybrid Bonding*, Proceedings of Automotive Circle Conference at Bad Nauheim, Germany, pp. 33–46 (2015).
13. M. Pfestorf and P. Müller. Application of aluminium in automotive structure and hang on parts. In: *Proceedings of Materials Week 2001*, International Congress on Adv. Materials, their Processes and Applications, pp. 1–8 (2001).
14. S. Böhm and K. Dilger. Pulsed Nd: YAG lasers for the treatment of adhesively bonded parts. In: *Proceedings of ICMAT International Conference for Mechanical and Automotive Technologies*, pp. 113–119 (2005).
15. R. Rechner. Laseroberflächenvorbehandlung von Aluminium zur Optimierung der Oxidschichteigenschaften für das strukturelle Kleben. Dissertation, Technische Universität Dresden, Dresden, Germany (2011).
16. H. Gehmecker. Chemical pretreatment of multi-metal and all-aluminium car bodies. In: *Proceedings of Aluminium 2000*, 4th World Congress on Aluminium, pp. 342–352 (2000).
17. S. Kreling, D. Blass, F. Fischer and K. Dilger. Bonding pretreatment of CFRP: Clean and reliable, using low-pressure blasting. *Adhesion, Adhesives & Sealants*, 10, 2, 34–39 (2013).

18. A. Wieczorek. Laser pretreatment of the AlSi-coating of hot formed steels to improve its bonding strength. In: *Joining in Car Body Engineering. Module I: Adhesive Bonding and Hybrid Bonding*, Proceedings of Automotive Circle Conference at Bad Nauheim, Germany, pp. 29–38 (2016).

19. T. A. Barnes and I. R. Pashby. Joining techniques for aluminium spaceframes used in automobiles, Part II: Adhesive bonding and mechanical fasteners. *J. Mater. Process. Technol.*, 99, 72–79 (2000).

20. P. C. Wang, S. K. Chisholm, G. Banas and F. V. Lawrence. The role of failure mode, resistance spot weld and adhesive on the fatigue behavior of welded-bonded aluminum. *Welding J.* 74, 2, 41s–47s (1995).

21. V. Grubisic, H. Beenken, K. Biswas, C. P. Bork, E. J. Drewes, B. Engl, J. Hünecke, A. Rupp, G. Schmid and S. Singh. Verbesserung der Bauteilfestigkeit für höherfeste Feinblechwerkstoffe und optimierte Fügeverfahren, Fügen im Leichtbau. *DVM-Bericht* 120, 47–61 (1994).

22. M. Schiel, S. Kreling, C. Unger, F. Fischer and K. Dilger. Behavior of adhesively bonded coated steel for automotive applications under impact loads. *Intl. J. Adhesion Adhesives* 56, 32–40 (2014).

23. A. J. Kinloch, G. A. Kodokian and M. B. Jamarani. Impact properties of epoxy polymers. *J. Mater. Sci.* 22, 4111–4120 (1987).

24. R. Mühlhaupt. Flexibility or toughness? The design of thermoset toughening agents. *Chimia* 44, 43–52 (1990).

25. R. Mulhaupt, J. H. Powell, C. S. Adderley and W. Rufenacht. Phenol-terminated polyurethane or poly-urea (urethane) with epoxy resin. US Patent 5278257, assigned to Ciba-Geigy Corporation (January 11, 1994).

26. R. Mulhaupt and W. Rufenacht, Composition of butadiene/polar comonomer copolymer, aromatic reactive end group-containing prepolymer and epoxy resin. US Patent 5073601 A, assigned to Ciba-Geigy Corporation (July 25, 1989).

27. M. Schiel, F. Fischer, L. Papadakis and K. Dilger. Determination of mechanical properties of adhesive joints on painted substrates and numerical analysis. In: Proceedings of 36th Annual Meeting of The Adhesion Society in Daytona Beach, FL (2013).

28. S. Gramsch-Kempkes. Automated stud gluing with closed-loop process control. In: *Joining in Car Body Engineering. Module I: Adhesive Bonding and Hybrid Bonding*. Proceedings of Automotive Circle Conference at Bad Nauheim, Germany, pp. 41–50 (2016).

29. J. Sczepanski. BMW 7 Series: Challenge lightweight design carbon core. In: *Joining in Car Body Engineering. Module I: Adhesive Bonding and Hybrid Bonding*, Proceedings of Automotive Circle Conference at Bad Nauheim, Germany, pp 1–8. , (2016).

30. S. Mailaender. Semi-structural adhesives as enabler for CFRP multimaterial design: challenges and opportunities. In: *Joining in Car Body Engineering. Module I: Adhesive Bonding and Hybrid Bonding*. Proceedings of Automotive Circle Conference at Bad Nauheim, Germany, pp. 51–66 (2016).

31. A. Lutz and S. Schmatloch. High modulus structural adhesives for lightweight trim shop applications. In: *Joining in Car Body Engineering. Module I: Adhesive Bonding and Hybrid Bonding*. Proceedings of Automotive Circle Conference at Bad Nauheim, Germany, pp. 67–76 (2016).

32. A. Luger. BMWi: innovative joining technology for lightweight car bodies. In: *Joining in Car Body Engineering. Module I: Adhesive Bonding and Hybrid Bonding*, Proceedings of Automotive Circle Conference at Bad Nauheim, Germany, pp. 61–71, (2014).

33. M. Hirthammer. Advanced adhesives for direct glazing applications. Proceedings of Glass Processing Days, the 5th International Conference on Architectural and Automotive Glass, Now and in the Future, pp. 363–365 (1997).

34. C. Baer. Scheibenklebstofftypen: Einfluss auf die Fertigung. In: Proceedings of Conference Kleben im Automobilbau, pp. 71–86 (1997).

35. G. Habenicht. *Kleben. Grundlagen, Technologien, Anwendung*, 4th edn. Springer-Verlag, Heidelberg (2002).

Index

irst published in 2025 by OH
HEADLINE PUBLISHING GROUP LIMITED

1

Disclaimer:

tation Data is available from the British Library

ISBN 978-1-03542-262-3

ed and written by: Katie Meegan
torial: Saneaah Muhammad
d typeset in Avenir by: Stephen Cary
ect manager: Russell Porter
oduction: Arlene Lestrade
inted and bound in China

E PUBLISHING GROUP LIMITED
Hachette UK Company
Victoria Embankment, London EC4Y 0DZ

in the EEA is Hachette Ireland, 8 Castlecourt Centre,
XTP3, Ireland (email: info@hbgi.ie)

ine.co.uk   www.hachette.co.uk

V

THE LITTLE GUIDE TO

VALENTINO

STYLE TO LIVE BY
Unofficial and Unauthorized

CONTENTS

INTRODUCTION

There are few designers more synonymous with elegance than Valentino – so much so that he is almost exclusively known by his first name.

From an early age, Valentino was besotted with the elegance of silver screen actresses, opera singers and the women around him. Upon making his way to Paris to study fashion (rare for an Italian to do at the time), Valentino learnt the ins and outs of the fashion trade and, at the age of just 28, he returned to Rome where, with the help of his lifelong business partner Giancarlo Giammetti, he set up his very first atelier.

Valentino's collections, known for their intricate craftsmanship, refined silhouettes and timeless appeal, have graced the most prestigious fashion runways for decades. His signature colour, "Valentino red", became a symbol of femininity and passion, and his designs have dressed royalty, film stars and political figures alike. A master of haute couture, Valentino's approach to design was both artistic and meticulous, believing in fashion as an expression of beauty, romance and individuality.

In 2008, Valentino shocked the fashion world by announcing his retirement after 45 years of designing, marking the end of an era. His final haute couture show, in Paris, was a grand celebration of his legacy, featuring some of his most iconic designs. Yet, his departure from the industry didn't mark the end of his influence. The House of Valentino continued to thrive, with a new creative vision brought in by his successors, Pierpaolo Piccioli and Maria Grazia Chiuri.

In 2016, Piccioli took sole creative control and successfully blended Valentino's heritage with fresh, contemporary design. Then, in April 2024, it was announced that visionary designer Alessandro Michele would be taking over creative control, having achieved great success at Gucci.

Whether you're an aspiring designer or simply an admirer of the artistry behind luxury fashion, this little book is a tribute to Valentino's unmatched creativity, enduring influence and commitment to excellence, offering an intimate look at the man who turned his passion for design into a global empire.

CHAPTER
ONE

VOGHERA TO VOGUE: VALENTINO'S HUMBLE BEGINNINGS

BORN IN A TINY VILLAGE IN NORTHERN ITALY, VALENTINO GARAVANI WAS DESTINED FOR GREATNESS FROM AN EARLY AGE.

FROM PURSUING HIS DREAMS IN PARIS AND OPENING HIS FIRST ATELIER AT THE AGE OF 28, VALENTINO BECAME THE LEGEND HE IS TODAY THROUGH SHEER TALENT AND HARD WORK.

On May 11, 1923, Valentino Clemente Ludovico Garavani was born in the small town of Voghera in the Lombardy region of Italy.

As a young child, Valentino fell in love with the sirens of the silver screen – Lana Turner, Judy Garland and Hedy Lamarr.

Since my young, young age I like beautiful things. For me it's very important.

Valentino

huffpost.com, November 19, 2014

Even as a young boy, my passion was to design, and I have been very lucky to do what I love all my life. Few gifts are greater than that.

Valentino

On his lifelong pursuit of beauty, system-magazine.com, Issue 5, 2015

There are few designers more synonymous with glamour than Valentino Garavani.

Karen Homer

The Little Book of Valentino, Karen Homer, 2022

I'm Italian. I love pastas and rice.
I love fish. I eat lots of chocolate.

Valentino

Always very proud of his Italian heritage, huffpost.com,
November 19, 2014

My cousins used to dress very, very well, and every time they wore an evening gown, I was there staring at them.

Valentino

Recalling his childhood and the impact of the women around him, *The Little Book of Valentino*, Karen Homer, 2022

Inspired by fashion from an early age, Valentino moved to Milan in 1949 to study fashion illustration at the Santa Maria Institute.

As Paris was the centre of haute couture, it was Valentino's goal to move to fashion's French epicentre. Still a teenager, he started studies at the École des Beaux-Arts and the École de Chambre Syndicale de la Couture Parisienne.

I was 15 when I arrived in Paris, and by 18, I was already learning to draw at the Chambre Syndicale de la Couture.

Valentino

Remembering his start in the fashion industry, system-magazine.com, Issue 5, 2015

My salary was nothing.
My parents used to send me money.
I was completely broke!

Valentino

His hardships during his early years as a designer are at odds with the luxury his brand is synonymous with, huffpost.com, November 19, 2014

Compared to this, the rest of us are making rags.

Karl Lagerfeld

To Valentino at the latter's retirement show in 2007. Lagerfeld was a fellow winner of the prestigious International Wool Secretariat fashion award. Winning this award helped Valentino secure important internships to set his career in haute couture in motion, time.com, March 18, 2009

I learned from the best: Balenciaga, Jacques Fath and Guy Laroche.

Valentino

On his early inspirations, system-magazine.com, Issue 5, 2015

I am especially grateful that I have been able to keep my own style over the decades, in spite of the many changes that have taken place in the world of fashion and in its business.

Valentino

Reflecting on how his style has both changed and remained constant since the early days of the brand

Upon winning the 1950 International Wool Secretariat fashion award, Valentino secured several coveted apprenticeships with Jacques Fath, Balenciaga and the society couturier Jean Dessès.

In 1957, he joined Guy Laroche's new label, where he learned the ropes of starting a fashion house.

I dealt with everything and learned more and more. I handled all sorts of things – drawings, dressing the models for the runway and getting into taxis to go and pick up dresses.

Valentino

To curator, fashion historian and author Pamela Golbin about the early days working for Guy Laroche, *The Little Book of Valentino*, Karen Homer, 2022

Balenciaga taught me to focus on simplicity, and that has stayed with me.

Valentino

Paying homage to the designers that he learnt from, system-magazine.com, Issue 5, 2015

I know what women want. They want to be beautiful.

Valentino

His (arguably) most iconic quote

My debut was in Rome. It was overwhelming, but I thrived on the excitement.

Valentino

Recalling his opening-night nerves for his first solo collection, system-magazine.com, Issue 5, 2015

Valentino's time in Paris was formative. Growing in experience and confidence, he finally felt ready to take the next step: opening his own atelier on Rome's famous Via Condotti.

His first ready-to-wear collection launched in 1959, received positive reviews and introduced what would become his signature colour: Valentino red.

There is something about drawing by hand that is irreplaceable. The computer can't capture the soul.

Valentino

On the classic methods of designing fashion,
system-magazine.com, Issue 5, 2015

My first year, people were
interested in me because I was new.
Then the press started to come.

Valentino

In the 50s and early 60s, Rome was an amazing place – so many incredible artists, like Cy Twombly, were drawn to it.

Valentino

Reminisces on the time and place of his fashion debut, system-magazine.com, Issue 5, 2015

[My first collection was] full of ideas, but had no personality.

Valentino

Admitted to Eugenia Sheppard on his first collection, "Ibis" that he felt that there was room for improvement, *The Little Book of Valentino*, Karen Homer, 2022

In the following year, the trajectory of Valentino's life changed forever when he met architecture student Giancarlo Giammetti.

While the pair were together romantically for several years, they remained business partners for life.

"

He asked if he could see the fashion house, and he said, 'What a boring life I have, to be an architect, to go home to study. I would love to come here and do something for the fashion house.' He came, and he took care of everything that was not part of the creation. Little by little, we arrived at something quite solid.

Valentino

Recalls the early days of his relationship with Giancarlo Giammetti, instyle.com, November 15, 2022

Giancarlo always helped me stay calm and focused. Without him, it would have been much harder.

Valentino

On his relationship with his business partner, system-magazine.com, Issue 5, 2015

Giancarlo and I understand each other, but his character is the opposite of mine.

Valentino

On he and Giancarlo being opposites that attract, *The Little Book of Valentino*, Karen Homer, 2022

Giancarlo Giammetti gave me the possibility to work very calmly and without shocks. I was locked in my studio, drawing.

Valentino

On how he and Giancarlo worked together to become one of the many great duos of fashion history, system-magazine.com, Issue 5, 2015

"

When Valentino and I started the
business we were two young kids.
There was no fashion organization,
there was nothing to imitate.
So we had to create a lot.

"

Giancarlo Giammetti

Reminisces on the early days of working with Valentino,
archive.nytimes.com, November 21, 2007

"

Valentino and I have been lucky in many ways. We were very lucky because we didn't become super famous – especially as a company – in one shot. Our life has been more one of steps, one after the other, and when that happens, I think you learn to have different tastes and to be better as a person.

"

Giancarlo Giammetti

On the "Valentino boom" of the 1960s, interviewmagazine.com, November 4, 2013

Despite nearly facing financial ruin just five years after the first opening of Valentino, the designer and his partner managed to turn their fortunes around, with the particular help of a very unlikely ally – the film industry.

CHAPTER
TWO

VIVA VALENTINO: A TRANSATLANTIC LOVE AFFAIR

DESPITE HIGH PRAISE FROM FASHION CRITICS AND BUYERS OF THE TIME, THE BEGINNINGS OF THE VALENTINO ATELIER WERE ANYTHING BUT SMOOTH.

THAT WAS, UNTIL IT CAUGHT THE ATTENTION OF SOME VERY CONSPICUOUS CHARACTERS.

While Paris had dominated haute couture, Rome was the perfect place to be a burgeoning designer in the 1960s.

Thanks to the success of *La Dolce Vita* in 1960, Rome was fast becoming the chic home of European cinema.

I think he would be very, very happy because I was always a very big dreamer. I was always dreaming of beauty. My mother used to say, 'You think of stupid things all the time.'

Valentino

When asked if his younger self would be proud of where he is today, the-talks.com, 2024

The jet set started flying to Europe and Valentino was very hooked into that. He had a big impact on making Italian fashion utterly first class.

Caroline Rennolds Milbank

Fashion historian and author, on the lucky timing of Valentino's rise to fame among the jet set, time.com, August 7, 2000

Rome in the 1950s was magical.
There was Fellini and all these
incredible creatives.

Valentino

On the magic of the Rome where his atelier began,
system-magazine.com, Issue 5, 2015

I like things simple.

Valentino

On his patented attention to detail, brainyquote.com, 2024

I have my favourite fashion decade, yes, yes, yes: the 60s. It was a sort of little revolution; the clothes were amazing but not too exaggerated.

Valentino

Always had a soft spot for the decade in which he made his debut, harpersbazaar.com, November 19, 2014

The first celebrity to brace the upcoming designer was the actress Elizabeth Taylor.

Attending a film opening in Rome, Taylor was the first high-profile celebrity to visit the atelier and buy a dress.

Valentino's reputation soon began to spread, particularly among American clientele.

My mouth was open in front of this beauty. She was unbelievable… [She said] 'I would love to wear an evening gown from you. But I have to tell you, I already have a gown from Dior. If the dress is beautiful, it suits me and I like it, I'm going to wear your dress.' So I tried my best. And she did wear it.

Valentino

Recalling his first meeting with Elizabeth Taylor, valentinogaravanimuseum.com, 2024

I am somebody with taste, of course, and when I don't like things in the collection I don't include them. I just want the essence of beauty.

Valentino

On his commitment to taste, the-talks.com, 2024

No one understands glamour
like Valentino.

Sophia Loren

Actress and key figure of the resurgence of Italian cinema in the
1960s, harpersbazaar.com, November 19, 2014

I try to love my clothes as much as possible.

Valentino

system-magazine.com, Issue 5, 2015

Fashion can change so quickly, but true elegance lasts forever.

Valentino

officielibiza.com, 2023

The fashion press soon took notice of Valentino's rising star, making him the first Italian designer to have a design on the cover of French *Vogue*, something that was unheard of at the time.

I never wanted to be just another designer; I wanted to make a mark on history.

Valentino

On his legacy, harpersbazaar.com, November 19, 2014

As a creator, beauty is the most important. Since I was a child I loved the way a dress looks, I admired a great face, a lovely body. I enjoy the beauty in a woman, in a man, in a child, in a painting. Beautiful things are important and make life important.

Valentino

On the importance of beauty in his work and life, the-talks.com, 2024

What Valentino does is make outrageously pretty dresses without making them seem silly.

Harold Koda

Costume historian, time.com, August 7, 2000

I did a very important collection
in 1968 and I realized that I was
quite good. All the magazines and
everybody came to visit me.
I became quite well known for my
glamour and my femininity and
women started to love my clothes.

Valentino

On the collection that started it all, the-talks.com, 2024

In 1968, I had a huge success, and my name went around the world.

Valentino

On his global success, system-magazine.com, Issue 5, 2015

A turning point came in 1967, when First Lady Jackie Kennedy commissioned Valentino to create a collection of twelve white dresses.

This proved the inspiration for Valentino's Spring/Summer 1968 collection: the Collezione Bianca, or White Collection.

These accomplished and intricate designs, all in white, were a revelation.

Elegance is the balance between proportion, emotion and surprise.

Valentino

On his secret formula for timeless design,
grandmagazine.argusmedia.sk, December 30, 2017

For me, white is a colour. It's the lightness of summer, purity.

Valentino

On his seminal 1968 White Collection, glamobserver.com, June 22, 2023

His crisp whites, his lacy whites, his soft and creamy whites, all shown together white on white.

Valentino

On the infamous White Collection, *The Little Book of Valentino*, Karen Homer, 2022

The 60s White Collection was when everything exploded for me.

Valentino

Acknowledging the role that the White Collection played in his global expansion, system-magazine.com, Issue 5, 2015

I wasn't interested in fame; I was interested in making beautiful clothes.

Valentino

On shying away from the spotlight, brainyquote.com, 2024

His place in haute couture confirmed, Valentino returned to Paris, opening a boutique on 42 Avenue Montaigne.

A decade after leaving the city of couture, the Italian master had returned to claim his French throne.

Haute couture is about more than fabric – it's about creating a feeling.

Valentino

harpersbazaar.com, November 19, 2014

I wanted my fashion to be timeless, not just for a season.

Valentino

On the importance of timeless fashion, system-magazine.com, Issue 5, 2015

No one has been able to sustain that level of chic and elegance like Valentino.

Anna Wintour

Vogue editor-in-chief, harpersbazaar.com, November 19, 2014

CHAPTER THREE

A GLOBAL ICON

FROM ROME TO NEW YORK AND PARIS TO L.A., VALENTINO HAS DRESSED THE WORLD'S MOST POWERFUL WOMEN.

WITH THE COUTURE HOUSE'S COMMITMENT TO ELEGANCE AND SUBSTANCE, IT'S EASY TO SEE WHY THE STARS FLOCK TO THE FASHION HOUSE.

My designs are meant to make women feel elegant and confident.

Valentino

His commitment to making women feel great about themselves has never wavered, system-magazine.com, Issue 5, 2015

Valentino's first encounter with Elizabeth Taylor marked the beginning of a lifelong friendship between the actress and the designer.

As well as Sophia Loren, Taylor was one of the first proponents of Valentino; however, it was former First Lady Jackie Kennedy who broke Valentino into the mainstream by wearing the brand almost exclusively in the year following JFK's death.

She inspired me, supported me during difficult times. With her style, she was able to get people talking about me by getting them to talk about her.

Valentino

On his decades-long friendship with Jackie Kennedy, glamobserver.com, June 22, 2023

I was always a very big dreamer…
I was not nervous at all because
I loved what I did. I said to myself,
'Maybe there will be people who
don't like it, but who cares? I love it,
I created it.'

Valentino

On his 1960s collections, racked.com, November 19, 2014

I prefer timeless beauty over trends that fade away.

Valentino

system-magazine.com, Issue 5, 2015

"

Tell them, Valentino, that I don't like your clothes – I love them!

"

Gloria Guinness

Style icon of the 1960s, time.com, March 10, 1967

Success came after much hard work, but it was never about the fame. It was about the clothes.

Valentino

On his work ethic, lofficielibiza.com, December 5, 2023

Valentino and Jackie Kennedy first met when he offered to create a mourning wardrobe for her after the assassination of her husband, John F. Kennedy. He designed a collection of black and white outfits for her, a palette that the pair returned to again and again.

For several years, Kennedy almost exclusively wore Valentino, and he even designed Kennedy's second wedding dress when she wed Greek shipping magnate Aristotle Onassis.

Diana Vreeland, we called her the Chinese Empress because she loved China and the colour red. She was the most important lady in the world at the time, and taught me a way to see and appreciate pictures. I learned a lot from her.

Valentino

Talking about another influential figure in securing his reputation among the New York elite – *Vogue* editor-in-chief Diana Vreeland, valentinogaravanimuseum.com, 2024

Mrs Vreeland famously adored us –
until she died, she was still calling
us 'the boys.'

Giancarlo Giammetti

interviewmagazine.com, November 4, 2013

She was encouraging without imposing ideas. Just her presence was an inspiration. She did not need words.

Valentino

On legendary *Vogue* editor-in-chief Diana Vreeland, valentinogaravanimuseum.com, October 2, 2024

Even at birth, genius always stands out. I see genius in you. Good luck!

Diana Vreeland

In a note that she once famously wrote to Valentino, racked.com, November 19, 2014

She has the most beautiful arms in the world.

Valentino

Praising one of the most influential women that he dressed, Michelle Obama, huffpost.com, November 19, 2014

Throughout the decades, Valentino has been responsible for some of the most iconic red carpet looks for many A-list stars, from Anne Hathaway to Liza Minnelli, Cate Blanchett, Susan Sarandon and, more recently, Florence Welch and Zendaya.

The thing is: I have to love my collection; I have to create my own personal things for the season. If I like it, then movie stars and the ladies around me are also very fond of it.

Valentino

On designing for the stars, the-talks.com, 2024

Jennifer Lawrence, Keira Knightley…
[these are] many of my friends that
I love and adore. Gisele [Bündchen]
for me is the top of the top.

Valentino

On some of the A-listers he's designed for, huffpost.com,
November 19, 2014

When we first met, I was understandably intimidated, but you and Valentino put me at ease because you're such great fun.

Anne Hathaway

In conversation with Giancarlo Giammetti, recalling the first time she met the designer, interviewmagazine.com, November 4, 2013

I get a lot of inspiration from
these sorts of things [museum and
theatre], but never from a woman.
I do the dress for a woman, but
I don't take inspiration from them.

Valentino

On taking inspiration from his art, the-talks.com, 2024

Given his speciality in beautiful gowns, it's hardly surprising that Valentino has had a hand in some of the most show-stopping wedding dresses of the 20th and 21st centuries.

A woman with taste. She must know what she wants, because it's very frustrating for me if somebody says, 'Listen, I'll let you do what you want; I'm there like a piece of glass and you have to do something nice for me.' I think it's better if a woman comes and she discusses with you and she has personality… This is the kind of woman I love.

Valentino

On the perfect Valentino woman, the-talks.com, 2024

"

I felt like Cinderella.

"

Princess Madeleine of Sweden

On her Valentino wedding gown, royalcentral.co.uk,
October 19, 2018

A dress that defies adjectives.

Gwyneth Paltrow

On her 2018 Valentino wedding dress, theknot.com, July 1, 2022

I never pay so much attention
to things that I don't like.

Valentino

Only acknowledges beauty in his life, theguardian.com,
November 27, 2012

Valentino makes you feel like you're the centre of the universe when you wear his creations.

Rihanna

Singer and beauty mogul, wwd.com, May 1, 2023

I want people to remember me for what I contributed to fashion, not for anything else.

Valentino

On his legacy, harpersbazaar.com, November 19, 2014

Many models over the years have also inspired Valentino, particularly Gisele Bündchen and Naomi Campbell.

When Valentino retired in 2008, it was Naomi Campbell who closed the show.

So where the world was once about Jackie Kennedy, today we're at Kim Kardashian. It's there and you have to accept it.

Giancarlo Giammetti

On the changing face of celebrity, interviewmagazine.com, November 4, 2013

With a new generation of celebrities flocking to Valentino, and with vintage pieces still in high demand, it is safe to say that the couture house will be dressing the red carpet for many years to come.

CHAPTER
FOUR

VALENTINO RED: A SIGNATURE OF ELEGANCE

CHANEL HAS BLACK,
YVES SAINT LAURENT HAS
CERULEAN BLUE,
HERMÈS HAS ORANGE.

BUT VALENTINO?
VALENTINO OWNS RED.

In his first ready-to-wear collection, presented in 1959, Valentino included a bright red "Fiesta" dress.

In 2005, Jennifer Aniston revived the original fiesta dress at the premier of *Along Came Polly*.

From 1959 to today, every single Valentino collection has featured a red dress.

Enthralled, I saw a woman with grey hair in one of the boxes, very beautiful, dressed in red velvet. Among all the colours worn by the other women, she looked unique, isolated in her splendour… I told myself that if I were ever going to become a designer, I would do lots of red.

Valentino

Recalls the first time he was struck by a woman in red at a theatre, glamobserver.com, June 22, 2023

Red has guts... it's deep, strong, dramatic.

Valentino

savoirflair.com, June 19, 2024

He did red once, and now you have red in every collection. Most of our statements came to be because we are romantic; we don't like to throw away things we like or that bring good luck.

Giancarlo Giammetti

Speaking to *Vogue* in 1985 about Valentino's "lucky" red, glamobserver.com, June 22, 2023

Red was there, everywhere around her. I just brought a bit more of it to her life.

Valentino

On legendary *Vogue* editor Diana Vreeland, who was also a fan of red, valentinogaravanimuseum.com, 2024

Valentino would say that when a woman puts on red, it is like making a statement – bold, fearless.

Giancarlo Giammetti

lofficielibiza.com, December 5, 2023

Valentino red is so unique that it has its own Pantone colour: a mix of 100% magenta, 100% yellow, and 10% black.

Red is a colour that takes me back to my childhood. It has such vitality and allure that I don't just like seeing it on clothes, but on houses, in flowers, on objects, in details. It's my good-luck charm.

Valentino

Valentino Rosso, Charlie Porter, 2022

I think that a woman dressed in red is always magnificent.

Valentino

valentinogaravanimuseum.com, 2024

There is a shade of red for every woman.

Audrey Hepburn

Valentino Rosso, Charlie Porter, 2022

As well as gracing the catwalk, Valentino red has also been a staple of red carpet looks over the years.

Some of the A-listers to have donned Valentino red on the carpet include Jennifer Aniston, Claudia Schiffer, Penélope Cruz, Anne Hathaway, Emma Watson and Zendaya.

I love my beauty. It's not my fault.

Valentino

Only wear clothes that make you feel alive.

Valentino

Expresses that red is the colour of life, azquotes.com, 2024

Red is passion, red is power.
When Valentino dresses me in red,
I feel invincible.

Lady Gaga

instyle.com, November 15, 2022

I am not so enchanted when I see lots of people dressed in black on the street.

Valentino

Not a fan of muted colour palettes, instead leaning toward more vibrant colours, huffpost.com, November 19, 2014

"

All colour has always been there. It is just for humans to find it, connect with it, name it, claim it. Time is irrelevant, colour transcends time.

"

Charlie Porter

Author of *Valentino Rosso*, Charlie Porter, 2022

It is a strong colour but at the same time it is a non-colour, it is neutral: like black, brown, blue, white.

Valentino

On the House's signature colour, esquire.com/it, February 2018

Red was the first colour to have a big impact on me.

Valentino

system-magazine.com, Issue 5, 2015

Valentino's retirement show in 2008, marking the close of his 45-year fashion career, ended with an unforgettable parade of 30 models in identical red gowns.

The crowd rose for a standing ovation – a crowd that included Uma Thurman, Miuccia Prada and Claudia Schiffer.

It is not pale green, it is not a pastel shade. It gives a lot of energy, a lot of polish.

Valentino

On the House's signature colour, esquire.com/it, February 2018

Valentino's red – there's magic in it. It's the kind of red that makes you believe in fairy tales.

Anne Hathaway

lofficielibiza.com, December 5, 2023

A woman dressed in red never makes a mistake: it is a colour that suits, it looks good on everyone.

Valentino

esquire.com/it, February 2018

Red is life, passion, love, it is the remedy against sadness.

Valentino

esquire.com/it, February 2018

"

I love the signature red of Valentino, but I do like it not just as a symbol of power and glamour, but as something personal and romantic. Giving red new perception… it's a good thing.

"

Pierpaolo Piccioli

Valentino Rosso, Charlie Porter, 2022

Each successor of Valentino Garavani has presented red in every Valentino collection.

There has yet to be a creative director to break this tradition and step away from the colour that was the favourite of the legend himself.

I think that a woman dressed
in red, especially in the evening, is
wonderful. She is, in the crowd, the
perfect image of a heroine.

Valentino

esquire.com/it, February 2018

Red is embedded in the DNA of the Valentino house. It is more than a tradition – it's a symbol of love and passion.

Pierpaolo Piccioli

Valentino creative director 2009–2024, villa88.com,
October 24, 2022

[The runway is] a dream. And this is what I did create, for all my life. This is what I drank all my life. And what I can tell you, that I became quite a bit perfectionist. For my work, for the houses, for everything, you just want to see beauty, beauty, beauty, beauty, beauty. I am in this way.

Valentino

On the manifestation of his designs on the catwalk, theguardian.com, November 27, 2012

CHAPTER
FIVE

THE LAST EMPEROR

VALENTINO GARAVANI'S
LEGACY IS BUILT ON STYLE,
SUBSTANCE AND AN ENDURING
COMMITMENT TO PROVIDING
ELEGANT GLAMOUR TO RED
CARPETS AND CATWALKS ALL
OVER THE GLOBE.

THANKS TO HIS RELENTLESS
PASSION FOR HAUTE COUTURE,
VALENTINO NOT ONLY JOINED THE
RANKS OF FASHION DESIGNERS
THAT SHAPED THE MODERN
WARDROBE, BUT ROSE TO BECOME
THE COURTIER SINGULARLY
KNOWN AS THE "EMPEROR".

The Valentino atelier experienced significant evolution in the 1970s.

During this decade, the brand became a favourite among celebrities and high society, establishing a strong presence in the fashion world, particularly with the fashionable set of high society New York.

The designer was often papped in the company of the likes of Diana Vreeland and Andy Warhol.

I am happy that thousands of students, young designers and fashion people will be able to see and study my work.

Valentino

On his legacy

I love many of them, but if I flash back, I have a special, special feeling for Gisele. For me she is now the most beautiful woman in the world; first of all, she is a real super-star model. I also love Claudia a lot.

Valentino

On his favourite supermodels to work with, the-talks.com, 2024

He never accepted failure – even when we were in the process of giving up our fashion house and our collections, which were moments that were maybe not the best for us. But Valentino made me convinced that everything was working well. He can be very convincing. I believe everything he says – always.

Giancarlo Giammetti

interviewmagazine.com, November 4, 2013

I dream about dresses sometimes. Then I turn the light on at my night table and I draw.

Valentino

the-talks.com, 2024

Valentino's clothes have a soul
that speaks to you.

Claudia Schiffer

system-magazine.com, Issue 5, 2015

In the 1980s, Valentino expanded globally and introduced perfumes.

The brand continued to innovate, embracing the dramatic silhouettes of the era.

Joan Collins in particular was a fan of Valentino, consistently wearing the designer on the hit TV show *Dynasty*, which served to further raise the profile of the designer among the general public.

I didn't like the 80s at all; it was a vulgar moment of fashion.

Valentino

On 80s fashion, despite his continued success throughout this decade

The dresses were out of proportion. The hair was terrible. I hated the shoes.

Valentino

Continues to slate the fashion of the 1980s, huffpost.com,
November 19, 2014

An evening dress that reveals a woman's ankles while walking is the most disgusting thing I have ever seen.

Valentino

On dresses that revealed the ankles, theguardian.com, June 5, 2010

Valentino's secret is knowing what women want and making them feel like the most beautiful versions of themselves.

Giancarlo Giammetti

system-magazine.com, Issue 5, 2015

Fashion has changed every decade very strongly. I like the 60s quite a bit, I like the 70s very much. But I hated the 80s… Ladies that were crazy to buy clothes but the looks were not beautiful: big shoulders, short dresses, hair like mountains… I never liked it. But I enjoyed the 90s quite a bit.

Valentino

the-talks.com, 2024

The 1990s marked a period of strategic growth as Valentino embraced collaborations and ventured into new markets, launching successful diffusion lines that helped attract a new generation of clientele.

I am like a freight train. Working on the details, twisting them and playing with them over the years, but always staying on the same track.

Valentino

newsweek.com, June 21, 2008

I didn't pay much attention to the documentary they were filming about me. I was too busy working.

Valentino

On the documentary, *Valentino: The Last Emperor*, in which the production crew followed him as he got ready for his final show, system-magazine.com, Issue 5, 2015

For me, the greatest compliment when you look at this dress from 1965, is telling me that it would be perfect for tomorrow night.

Valentino

On the timelessness of his designs, newsweek.com, June 21, 2008

> "
>
> Mr Valentino is a true visionary whose boundless creativity, innovative designs and dedication to craftsmanship have revolutionized the fashion industry.
>
> "

Caroline Rush

Chief Executive of the British Fashion Council. Valentino was honoured with the Lifetime Achievement Award in 2023, graziamagazine.com, 2024

Just doing a little cameo, being in this film with Meryl Streep was a great honour.

Valentino

On his iconic cameo in *The Devil Wears Prada*, marieclaire.co.uk, September 23, 2024

While Valentino first announced his retirement in 2002, it was not until 2007 that he felt fully ready to hand over the reins to the business that he and his partner, Giancarlo Giammetti, had built.

Valentino delivered his last women's ready-to-wear show in Paris on October 4, 2007, and his last haute couture show at the Musée Rodin on January 23, 2008.

When I stopped designing,
I continued drawing, just for myself.

Valentino

On continuing to create after his retirement from fashion,
system-magazine.com, Issue 5, 2015

I love the 2000s because everyone started to love haute couture.

Valentino

On the resurgence of haute couture in the 2000s, harpersbazaar.com, November 19, 2014

He established his stylistic
vocabulary early on, and over
49 years he refined it.

Pamela Golbin

Fashion historian and curator, newsweek.com, June 21, 2008

I have to tell you I am so happy that I stopped at the last moment. Because I thought that I did everything. After 45 years I did everything, I did all my collection that I wanted to do.

Valentino

On knowing when to retire, theguardian.com, November 27, 2012

I think when you are at the top,
as Valentino has been for 47 years,
that's the time to go out.

Giancarlo

On Valentino's retirement, archive.nytimes.com,
November 21, 2007

Throughout his prestigious career, Valentino has been awarded several honours, including the Chevalier de la Légion d'honneur, the Medal of the City of Paris, the Couture Council Award for Artistry of Fashion, the Golden Plate Award of the American Academy of Achievement and the "Outstanding Achievement Award" at the Fashion Awards.

Valentino continues to mentor and inspire the next generation of designers.

Valentino made me feel like art whenever I wore his gowns.

Isabella Rossellini

system-magazine.com, Issue 5, 2015

I love him as a stylist, as a friend, as a magician of beauty, of femininity, of joy.

Marie-Hélène de Rothschild

French socialite, lofficielibiza.com, December 5, 2023

Valentino's clothes transform you.
There's a magic to them.

Kate Moss

Supermodel, wmagazine.com, October 31, 2019

CHAPTER SIX

A NEW GENERATION: BIG SHOES TO FILL

WHEN VALENTINO ANNOUNCED HIS INTENTION TO RETIRE FROM FASHION IN 2002, IT THREW THE FASHION WORLD INTO TURMOIL.

WHO COULD POSSIBLY REPLACE THE EMPEROR OF HAUTE COUTURE?

TO DATE, FOUR CREATIVE DIRECTORS HAVE TRIED, TO VARYING DEGREES OF SUCCESS.

In 2008, when Valentino Garavani officially retired, he appointed Alessandra Facchinetti as his replacement.

While her designs showcased softer silhouettes and innovative fabrics, she faced challenges in aligning her vision with the brand's established identity and so her tenure at Valentino lasted only a year.

"

I wanted to move away from the cliché of pure luxury to focus on a modern and elegant campaign and a dynamic woman who gets around. It's a starting point for us and offers the world of Valentino at 360 degrees.

"

Alessandra Facchinetti

On her first collection for Valentino, Autumn/Winter 2008–9, vogue.co.uk, May 30, 2008

There is an existing archive with thousands of dresses where [a designer] can draw and take inspiration from to create a Valentino product that is relevant today. It is a shame that [Facchinetti] didn't feel this need.

Valentino

On his reason for unexpectedly dismissing Alessandra Facchinetti from Valentino, just 24 hours after presenting her second collection with the brand, guardian.com, October 6, 2008

Following Facchinetti's unceremonious firing, the House saw the arrival of the dynamic duo, Pierpaolo Piccioli and Maria Grazia Chiuri.

Initially serving as co-creative directors, they revitalized the unsteady label, celebrating the bold colours that Valentino is known for.

I don't think that couture has to be modern in the way you think, in terms of form or shape. To me, modernity means embracing the world that is around us. Couture works if, at the end, it is worn by humans who are of today.

Pierpaolo Piccioli

Creative director from 2008–24, wmagazine.com, December 6, 2022

I was so shy. I even found it so difficult to talk. I had to teach myself English, every night with books at home.

Maria Grazia Chiuri

On her first year as creative co-director of Valentino, vogue.com, May 27, 2015

At the beginning, we really had to clarify our vision. We wanted to give an idea of the beauty of our time, and to find the best language to talk about it. The first year was [about] looking to balance that with the heritage, which was kind of heavy.

Pierpaolo Piccioli

On the early days as creative co-director at Valentino, vogue.com, May 27, 2015

Couture is pure as a process, as an approach, so when you generate attention with couture, it can have a bigger impact than with other means of expression.

Pierpaolo Piccioli

wmagazine.com, December 6, 2022

66

His reign in the House of Valentino
has been a lesson in grace. He feels
no need to erase the past but builds
on its foundation and by doing
so makes the future of the House
stronger, braver. His inclusionary
vision is evidenced by the way he
honours the craftspeople in his
atelier and shares a meal with them
after the couture events.

99

Frances McDormand

The Oscar-winning actress on Pierpaolo Piccioli, time.com, 2024

Piccioli and Chiuri also brought an avant-garde flair to the House of Valentino, possibly best signified by the "Rockstud" collection.

We are so lucky to have Pierpaolo.
He is seriously enamoured of
Valentino and what it can bring to
today's world.

Giancarlo Giammetti

wmagazine.com, December 6, 2022

Valentino today is about now.

Pierpaolo Piccioli

Valentino Rosso, Charlie Porter, 2022

It is plain to see that Chiui and Piccioli had done their research on classic couture shapes, however abbreviated they might be here. But that will be scant consolation to mournful clients of the *ancien régime*.

Vogue

On Chiuri and Piccioli's Autumn/Winter 2010 collection, *The Little Book of Valentino*, Karen Homer, 2022

Fashion is a dream, and in this moment we need dreams.

Pierpaolo Piccioli

vogue.com, October 3, 2011

In 2016, Maria Grazia Chiuri departed to lead Dior, leaving Piccioli as the sole creative director.

The pair had worked together for nearly a quarter of a century, over 17 years together in Valentino, with eight of them as co-directors and several years in accessories at Fendi before that.

Everything achieved in these years would have been impossible without Maria Grazia Chiuri and Pierpaolo Piccioli's talent, determination and vision that together have contributed into making Valentino one of the most successful fashion companies.

Stefano Sassi

Valentino CEO, on Chiuri's departure, harpersbazaar.com,
July 7, 2016

I really wanted to have people in that show who exemplified big societal changes… It was very classical, very couture, with ruffles and bows, very much in the Valentino vocabulary, but actually, it was a big 'fuck you' to a traditional kind of beauty, to all the conservatism, the reactionaries of the moment.

Pierpaolo Piccioli

On his Spring/Summer 2023 collection, wmagazine.com, December 6, 2022

Now the sole head, Piccioli emphasized inclusivity and diversity, showcasing a range of body types and ethnicities on the runway.

Under his leadership, Valentino continued to thrive, becoming synonymous with both elegance and modernity and dressing some of the hottest stars for the red carpet and runway.

"

In short, he aced it, not only meeting expectations but surpassing them too.

"

Vogue

On Piccioli's first solo Spring/Summer 2017 collection,
The Little Book of Valentino, Karen Homer, 2022

Being yourself is the coolest thing you can do.

Pierpaolo Piccioli

Valentino Rosso, Charlie Porter, 2022

If you have a good relationship with clothes, you can really glorify life.

Alessandro Michele

gq.com, September 30, 2024

In another shock move,
it was announced on April 2,
2024 that Piccioli was being
replaced by former
head of design at Gucci,
Alessandro Michele.

Known for his avant garde
and ultra-modern approach,
Michele will certainly shake
things up at Valentino.

He made clothes for friends
and acquaintances and people
belonging to his sentimental world.
I didn't think he was working, I think
he was simply living.

Alessandro Michele

On the legacy of Valentino, gq.com, September 30, 2024

I strongly believe that with his unique creativity and sensibility, he will continue the elevation of the brand's everlasting heritage and unique Italian Maison de Couture identity.

Rachid Mohamed Rachid

Chairman of Valentino, valentino.com, April 2, 2024

I wanted to tell the new generation
that it's possible to be weirdly chic,
in an unruly way.

Alessandro Michele

On his first Valentino collection, gq.com, September 30, 2024

We sort of pigeonholed him as something classic, but he was revolutionary, a gay man in fashion, one of the first who displayed his life with great elegance and great savoir-faire.

Alessandro Michele

On the legacy of Valentino, gq.com, September 30, 2024

Valentino spends his retirement between homes in Rome, Gstaad, New York and Paris.

He is never without his four pugs: Margot, Maude, Maggie and Monty.

I always draw every day, because I absolutely love to draw. With a good pen, I can do it very, very fast, in five seconds.

Valentino

On how he spends his retirement, system-magazine.com, Issue 5, 2015

I have always accepted with joy
all the names and the titles
that they have given me: The King,
The Emperor, The Icon. I am
Valentino. I live in my own world.
My life didn't change, it's always
been the same.

Valentino

the-talks.com, 2024

With the House of Valentino firmly in a new era under Alessandro Michele, there is no doubt that Valentino will continue to dress the most influential people in the world.

A timeless icon, Valentino Garavani's designs are a testament to elegance, artistic vision and the enduring magic of haute couture.

I hope people will say,
'Mr Valentino, he did something
for fashion, no?'

Valentino

newsweek.com, June 21, 2008